PRINCIPLES of
General, Organic, & Biological Chemistry

Third Edition

Janice Gorzynski Smith

University of Hawai'i at Mānoa

McGraw Hill

PRINCIPLES OF GENERAL, ORGANIC, & BIOLOGICAL CHEMISTRY, THIRD EDITION

Published by McGraw Hill LLC, 1325 Avenue of the Americas, New York, NY 10019. Copyright ©2023 by McGraw Hill LLC. All rights reserved. Printed in the United States of America. Previous editions ©2015 and 2012. No part of this publication may be reproduced or distributed in any form or by any means, or stored in a database or retrieval system, without the prior written consent of McGraw Hill LLC, including, but not limited to, in any network or other electronic storage or transmission, or broadcast for distance learning.

Some ancillaries, including electronic and print components, may not be available to customers outside the United States.

This book is printed on acid-free paper.

1 2 3 4 5 6 7 8 9 LWI 26 25 24 23 22

ISBN 978-1-266-51712-9 (bound edition)
MHID 1-266-51712-X (bound edition)
ISBN 978-1-264-45182-1 (loose-leaf edition)
MHID 1-264-45182-2 (loose-leaf edition)

Associate Portfolio Manager: *Hannah Downing*
Senior Product Developer: *Mary Hurley*
Marketing Manager: *Cassie Cloutier*
Content Project Managers: *Laura Bies, Rachael Hillebrand*
Buyer: *Laura Fuller*
Designer: *David W. Hash*
Content Licensing Specialists: *Beth Cray*
Cover Image: *David Merron Photography/Getty Images*
Compositor: *Aptara®, Inc.*

All credits appearing on page or at the end of the book are considered to be an extension of the copyright page.

Library of Congress Cataloging-in-Publication Data

Names: Smith, Janice Gorzynski, author.
Title: Principles of general, organic, & biological chemistry / Janice
  Gorzynski Smith, University of Hawai'i at Mānoa.
Other titles: Principles of general, organic, and biological chemistry
Description: Third edition. | New York, NY : McGraw Hill Education, [2023] |
  Includes index.
Identifiers: LCCN 2021030648 (print) | LCCN 2021030649 (ebook) | ISBN
  9781266517129 (hardcover) | ISBN 9781264451821 (spiral bound) | ISBN
  9781264452385 (pdf)
Subjects: LCSH: Chemistry–Textbooks. | Chemistry, Inorganic–Textbooks. |
  Biochemistry–Textbooks.
Classification: LCC QD31.3 .S634 2023 (print) | LCC QD31.3 (ebook) | DDC
  540–dc23
LC record available at https://lccn.loc.gov/2021030648
LC ebook record available at https://lccn.loc.gov/2021030649

The Internet addresses listed in the text were accurate at the time of publication. The inclusion of a website does not indicate an endorsement by the authors or McGraw Hill LLC, and McGraw Hill LLC does not guarantee the accuracy of the information presented at these sites.

About the Author

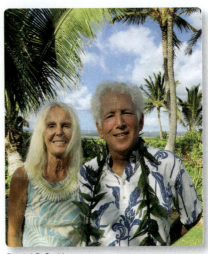

Daniel C. Smith

Janice Gorzynski Smith was born in Schenectady, New York. She received an A.B. degree *summa cum laude* in chemistry at Cornell University and a Ph.D. in Organic Chemistry from Harvard University under the direction of Nobel Laureate E. J. Corey. During her tenure with the Corey group she completed the total synthesis of the plant growth hormone gibberellic acid.

Following her postdoctoral work, Jan joined the faculty of Mount Holyoke College, where she was employed for 21 years. During this time, she was active in teaching organic chemistry lecture and lab courses, conducting a research program in organic synthesis, and serving as department chair. Her organic chemistry class was named one of Mount Holyoke's "Don't-miss courses" in a survey by *Boston* magazine. After spending two sabbaticals amidst the natural beauty and diversity of Hawai'i in the 1990s, Jan and her family moved there permanently in 2000. She has been a faculty member at the University of Hawai'i at Mānoa, where she has taught a one-semester organic and biological chemistry course for nursing students, as well as the two-semester organic chemistry lecture and lab courses. In 2003, she received the Chancellor's Citation for Meritorious Teaching.

Jan resides in Hawai'i with her husband Dan, an emergency medicine physician, pictured with her at an event on Oahu. She has four children and nine grandchildren. When not teaching, writing, or enjoying her family, Jan bikes, hikes, snorkels, and scuba dives in sunny Hawai'i, and time permitting, enjoys travel and Hawaiian quilting.

*T*o my family

Contents in Brief

Contents

Buttchi 3 Sha Life/Shutterstock

Terry Vine/Blend Images LLC

Jill Braaten

6　Gases　195

Daniel C. Smith

7　Solutions　225

*Science Photo Library/Alamy
Stock Photo*

8　Acids and Bases　260

Mark Dierker/McGraw Hill

ClarkandCompany/Getty Images

cmcderm1/Getty Images

Hin255/iStock/Getty Images

John A. Rizzo/Getty Images

12 Organic Compounds That Contain Oxygen, Halogen, or Sulfur 402

Daniel C. Smith

13 Carboxylic Acids, Esters, Amines, and Amides 447

MaraZe/Shutterstock

Fcafotodigital/iStock/Getty Images

Daniel C. Smith

17   Nucleic Acids and Protein Synthesis   595

Rafe Swan/Cultura Creative/Alamy

18   Energy and Metabolism   630

Daniel C. Smith

Appendices

Preface

This textbook is written for students who have an interest in nursing, nutrition, environmental science, food science, and a wide variety of other health-related professions. The content of this book is designed for an introductory chemistry course with no chemistry prerequisite and is suitable for either a one- or two-semester course. This text relates the principal concepts of general, organic, and biological chemistry to the world around us, and in this way illustrates how chemistry explains many aspects of daily life.

The learning style of today's students relies heavily on visual imagery. In this text, new concepts are introduced one at a time, keeping the basic themes in focus, and breaking down complex problems into manageable chunks of information. Relevant applications are provided for all basic chemical concepts, and molecular art illustrates and explains common everyday phenomena. Students learn step-by-step problem solving throughout the chapter within sample problems and *How To* boxes. Students are given enough detail to understand basic concepts so that they may acquire a new appreciation of the human body and the larger world around them.

New to This Edition

General

Problem Solving Sample Problems are now paired with Practice Problems to allow students to apply what they have just learned. The Practice Problems are followed by More Practice lists that point students to end-of-chapter problems that are similar in concept. Concept Check problems have replaced other in-chapter problems to give students an immediate check on the topic that has just been presented. The answers to all Practice Problems, all Concept Check problems, and odd-numbered end-of-chapter problems are given at the end of each chapter.

Chapter Review Chapter Review, which replaces Chapter Highlights at the end of each chapter, consists of Key Terms that are defined in the Glossary, Key Concepts, Key Equations, Key Reactions, and Key Skills. The Key Concepts and Key Skills sections use art and chemical structures to more clearly explain the key features detailed within the chapter. Key Skills, which presents the steps needed to solve important topics within the chapter, should be especially valuable for students learning stepwise processes.

Self-Test Each chapter contains a Self-Test, which consists of short-answer questions that test an understanding of definitions, equations, and other material encountered within the chapter. Answers to each question are provided at the end of the chapter.

Study Tips Brief Study Tips have been added to the margins in Chapters 1, 3, 6, 10, and 14 to help students develop general methods for solving recurrent types of problems, such as those that require a specific equation.

Photos Three-fourths of the chapter-opening photos have been replaced with photos emphasizing relevant material within the chapter. More marginal photos of applications have also been added on topics including non-contact thermometers (Chapter 1), radioactive seeds for cancer treatment (Chapter 9), leghemoglobin in plant-based burgers (Chapter 16), and many others.

Problems Over 150 new problems have been added.

Other New Coverage

Some of the new material added within specific chapters is listed below.

- Coverage on using a scientific calculator with scientific notation and logarithms has been expanded in Sections 1.6B and 8.5B. Tables with art that indicates what buttons should be pressed and what calculator displays will be shown are given.
- Chapter 3 opens by presenting a new current topic, the effect of sunscreens like oxybenzone on the bleaching of coral reefs. Recent research on oxybenzone is also discussed in Section 11.10.
- A new section on determining types of reactions—combination, decomposition, single replacement, and double replacement—has been added to Chapter 5. The chapter has been reorganized to place oxidation and reduction reactions immediately following this section, so that all of these different reaction types are in proximity.
- The discussion of dialysis and osmosis in Section 7.8 has been edited to emphasize the distinction between these related concepts. Three new problems on this subject have been added within the chapter.
- New material on using PET scans to visualize the brain in Alzheimer's patients has been added to Section 9.5.
- Material on methane, a greenhouse gas, has been expanded in Section 10.8.
- New material on the human milk oligosaccharides in breast milk has been added in Section 14.6D.
- Figure 15.1 now presents data on saturated fats, unsaturated oils, and trans fats in bar graph form for easier visualization of lipid content.
- The material on enzymes has been expanded into two sections (Sections 16.9 and 16.10), which include classes of enzymes, naming enzymes, and factors that affect enzyme activity.
- A section on the Human Genome Project (Section 17.10B) has been added.
- Section 17.11 on viruses has been expanded with material on coronaviruses and mRNA vaccines.

The Construction of a Learning System

Writing a textbook and its supporting learning tools is a multifaceted endeavor. McGraw Hill's development process is an ongoing, market-oriented approach to building accurate and innovative learning systems. It is dedicated to continual large scale and incremental improvement, driven by multiple customer feedback loops and checkpoints. This is initiated during the early planning stages of new products and intensifies during the development and production stages, and then begins again upon publication, in anticipation of the next version of each print and digital product. This process is designed to provide a broad, comprehensive spectrum of feedback for refinement and innovation of learning tools for both student and instructor. The development process includes market research, content reviews, faculty and student focus groups, course- and product-specific symposia, accuracy checks, and art reviews.

The Learning System Used in *Principles of General, Organic, & Biological Chemistry,* Third Edition

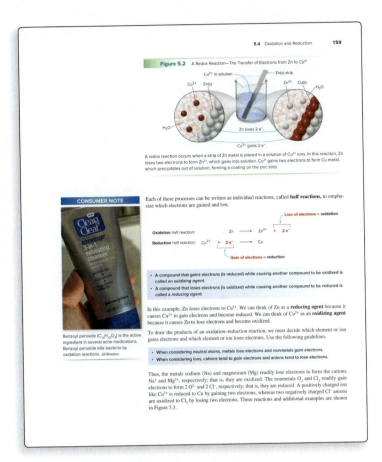

Writing Style

A succinct writing style weaves together key points of general, organic, and biological chemistry, along with attention-grabbing applications to consumer, environmental, and health-related fields. Concepts and topics are broken into small chunks of information that are more easily learned.

Chapter Goals, Tied to End-of-Chapter Review

Chapter Goals at the beginning of each chapter identify what students will learn, and are tied to the end-of-chapter Key Concepts and Key Skills, which serve as bulleted summaries of the most important concepts for study.

Macro-to-Micro Illustrations

Visualizing molecular-level representations of macroscopic phenomena is critical to the understanding of any chemistry course. Many illustrations in this text include photos or drawings of everyday objects, paired with their molecular representation, to help students visualize and understand the chemistry behind ordinary things. Many illustrations of the human body include magnifications for specific anatomic regions, as well as representations at the microscopic level, for today's visual learners.

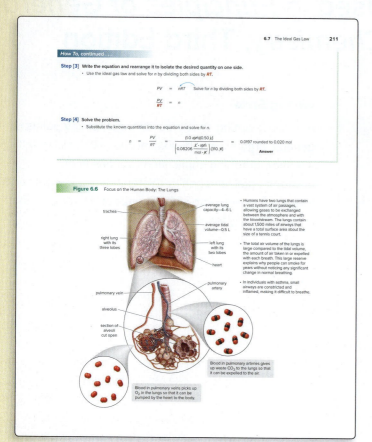

Applications

Relevant, interesting applications of chemistry to everyday life are included for all basic chemical concepts. These are interspersed in margin-placed Health Notes, Consumer Notes, and Environmental Notes, as well as sections entitled "Focus on Health & Medicine," "Focus on the Environment," and "Focus on the Human Body."

How To's

Key processes are taught to students in a straightforward and easy-to-understand manner by using examples and multiple, detailed steps to solving problems.

Problem Solving

Stepwise sample problems lead students through the thought process tied to successful problem solving by employing *Analysis* and *Solution* steps. Sample Problems are categorized sequentially by topic to match chapter organization, and are paired with practice problems to allow students to apply what they have just learned. Students can immediately verify their answers to the follow-up problems in the answers at the end of each chapter.

ALEKS®

Create More Lightbulb Moments.

Every student has different needs and enters your course with varied levels of preparation. ALEKS® pinpoints what students already know, what they don't and, most importantly, what they're ready to learn next. Optimize your class engagement by aligning your course objectives to ALEKS® topics and layer on our textbook as an additional resource for students.

ALEKS® Creates a Personalized and Dynamic Learning Path

ALEKS® creates an optimized path with an ongoing cycle of learning and assessment, celebrating students' small wins along the way with positive real-time feedback. Rooted in research and analytics, ALEKS® improves student outcomes by fostering better preparation, increased motivation and knowledge retention.

*visit **bit.ly/whatmakesALEKSunique** to learn more about the science behind the most powerful adaptive learning tool in education!

Preparation & Retention

The more prepared your students are, the more effective your instruction is. Because ALEKS® understands the prerequisite skills necessary for mastery, students are better prepared when a topic is presented to them. ALEKS® provides personalized practice and guides students to what they need to learn next to achieve mastery. ALEKS® improves knowledge and student retention through periodic knowledge checks and personalized learning paths. This cycle of learning and assessment ensures that students remember topics they have learned, are better prepared for exams, and are ready to learn new content as they continue into their next course.

Flexible Implementation: Your Class Your Way!

ALEKS® enables you to structure your course regardless of your instruction style and format. From a traditional classroom, to various co-requisite models, to an online prep course before the start of the term, ALEKS® can supplement your instruction or play a lead role in delivering the content.

*visit **bit.ly/ALEKScasestudies** to see how your peers are delivering better outcomes across various course models!

Outcomes & Efficacy

Our commitment to improve student outcomes services a wide variety of implementation models and best practices, from lecture-based to labs and co-reqs to summer prep courses. Our case studies illustrate our commitment to help you reach your course goals, and our research demonstrates our drive to support all students, regardless of their math background and preparation level.

*visit **bit.ly/outcomesandefficacy** to review empirical data from ALEKS® users around the country

Turn Data Into Actionable Insights

ALEKS® Reports are designed to inform your instruction and create more meaningful interactions with your students when they need it the most. ALEKS® Insights alert you when students might be at risk of falling behind so that you can take immediate action. Insights summarize students exhibiting at least one of four negative behaviors that may require intervention including Failed Topics, Decreased Learning, Unusual Learning, and Procrastination & Cramming.

DIGITAL EDGE50 AWARDS ‹2019›

Winner of 2019 Digital Edge 50 Award for Data Analytics!

bit.ly/ALEKS_MHE

Learning Resources for Instructors and Students

ALEKS (Assessment and LEarning in Knowledge Spaces) is a web-based system for individualized assessment and learning available 24/7 over the Internet. ALEKS uses artificial intelligence to accurately determine a student's knowledge and then guides her to the material that she is most ready to learn. ALEKS offers immediate feedback and access to ALEKSPedia—an interactive text that contains concise entries on chemistry topics. ALEKS is also a full-featured course management system with rich reporting features that allow instructors to monitor individual and class performance, set student goals, assign/grade online quizzes, and more. ALEKS allows instructors to spend more time on concepts while ALEKS teaches students practical problem-solving skills. And with ALEKS 360, your student also has access to this text's eBook. Learn more at **www.aleks.com/highered/science**

Instructors have access to the following instructor resources:

- **Presentation Tools Table:** Instructors have access to fully editable **accessible PowerPoint lecture outlines,** which appear as ready-made presentations that combine art and lecture notes for each chapter of the text. For instructors who prefer to create their lectures from scratch, all illustrations, photos, and tables are pre-inserted by chapter into blank **PowerPoint slides** and are also available as **downloadable jpeg files.**
- **Instructor's Solutions Manual:** This supplement contains complete, worked out solutions for all the end-of-chapter problems in the text.
- **Computerized Test Bank:** Over 1,800 test questions that accompany *Principles of General, Organic, & Biological Chemistry* are available for creating exams or quizzes.
- **Videos and Animations:** More than 90 videos and animations available through the eBook and the instructor resource center, supplement the textbook material in much the same way as instructor demonstrations. However, they are only a few mouse-clicks away, any time, day or night. Because many students are visual learners, the animations add another dimension of learning; they bring a greater degree of reality to the written word.

McGraw Hill Virtual Labs is a must-see, outcomes-based lab simulation. It assesses a student's knowledge and adaptively corrects deficiencies, allowing the student to learn faster and retain more knowledge with greater success. First, a student's knowledge is adaptively leveled on core learning outcomes: Questioning reveals knowledge deficiencies that are corrected by the delivery of content that is conditional on a student's response. Then, a simulated lab experience requires the student to think and act like a scientist: recording, interpreting, and analyzing data using simulated equipment found in labs and clinics. The student is allowed to make mistakes—a powerful part of the learning experience! A virtual coach provides subtle hints when needed, asks questions about the student's choices, and allows the student to reflect on and correct those mistakes. Whether your need is to overcome the logistical challenges of a traditional lab, provide better lab prep, improve student performance, or make your online experience one that rivals the real world, McGraw Hill Virtual Labs accomplishes it all.

McGraw Hill Create™

With **McGraw Hill Create,** you can easily rearrange chapters, combine material from other content sources, and quickly upload content you have written, like your course syllabus or teaching notes. Find the content you need in Create by searching through thousands of leading McGraw Hill textbooks. Arrange your book to fit your teaching style. Create even allows you to personalize your book's appearance by selecting the cover and adding your name, school, and course information. Order a Create book and you'll receive a complimentary print review copy in three to five business days or a complimentary electronic review copy (eComp) via email in minutes. Go to www.mcgrawhillcreate.com today and register to experience how McGraw Hill Create empowers you to teach *your* students *your* way. www.mcgrawhillcreate.com

Student Solutions Manual

The *Student Solutions Manual* contains the solutions to all in-chapter problems as well as the solutions to all odd-numbered end-of-chapter problems.

Acknowledgments

When I first began textbook writing 20 years ago, I had no idea how many people I would have to rely upon to see a project from manuscript preparation to published text. Special thanks for this edition go to Senior Product Developer Mary Hurley, Senior Core Content Project Manager Laura Bies, and freelance Developmental Editor John Murdzek, who handled all the day-to-day steps needed to publish an accurate and engaging text. Thanks are also due to Executive Portfolio Manager Michelle Hentz and Associate Portfolio Manager Hannah Downing for spearheading the revision of *Principles,* and the art, production, marketing, and sales teams for their support and contributions.

I would also like to acknowledge the following individuals for their masterful authoring of the ancillaries to accompany the third edition: Lauren McMills of Ohio University for the Solutions Manuals; Andrea Leonard of the University of Louisiana, Lafayette for the Accessible PowerPoint Lecture Outlines; and Cari Gigliotti of Sinclair Community College for the Test Bank.

Finally, I thank my family for their support and patience during the long process of publishing a textbook. My husband Dan, an emergency medicine physician, took several photos that appear in the text, and served as a consultant for many medical applications.

Matter and Measurement

1

Determining the weight and length of a newborn are common measurements performed by healthcare professionals.
Daniel C. Smith

CHAPTER OUTLINE

CHAPTER GOALS

In this chapter you will learn how to:

1. Describe the three states of matter
2. Classify matter as a pure substance, mixture, element, or compound
3. Report measurements using the metric units of length, mass, and volume
4. Use significant figures
5. Use scientific notation for very large and very small numbers
6. Use conversion factors to convert one unit to another
7. Convert temperature from one scale to another
8. Define density and specific gravity and use density to calculate the mass or volume of a substance

Why Study ...

Matter and Measurement?

Everything you touch, feel, or taste is composed of chemicals—that is, **matter**—so an understanding of its composition and properties is crucial to our appreciation of the world around us. Some matter—lakes, trees, sand, and soil—is naturally occurring, whereas other examples of matter—aspirin, nylon fabric, plastic syringes, and vaccines—are made by humans. To understand the properties of matter, as well as how one form of matter is converted to another, we must also learn about measurements. Following a recipe, pumping gasoline, and figuring out drug dosages involve manipulating numbers. Thus, Chapter 1 begins our study of chemistry by examining the key concepts of matter and measurement.

1.1 Chemistry—The Science of Everyday Experience

What activities might occupy the day of a typical student? You may have done some or all of the following tasks: eaten some meals, drunk coffee or cola, showered with soap, checked email on a computer, ridden a bike or car to a part-time job, taken an aspirin to relieve a headache, and spent some of the evening having snacks and refreshments with friends. Perhaps, without your awareness, your life was touched by chemistry in each of these activities. What, then, is this discipline we call **chemistry?**

- *Chemistry* is the study of matter—its composition, properties, and transformations.

What is **matter?**

- *Matter* is anything that has mass and takes up volume.

In other words, **chemistry studies anything that we touch, feel, see, smell, or taste,** from simple substances like water or salt, to complex substances like proteins and carbohydrates that combine to form the human body. Some matter—cotton, sand, an apple, and the cardiac drug digoxin—is **naturally occurring,** meaning it is isolated from natural sources. Other substances—nylon, Styrofoam, the plastic used in soft drink bottles, and the pain reliever ibuprofen—are **synthetic,** meaning they are produced by chemists in the laboratory (Figure 1.1).

Figure 1.1

Naturally Occurring and Synthetic Materials

a. Naturally occurring materials

b. Synthetic materials

Matter occurs in nature or is synthesized in the lab. (a) Sand and apples are two examples of natural materials. Cotton fabric is woven from cotton fiber, obtained from the cotton plant. The drug digoxin (trade name Lanoxin), widely prescribed for decades for patients with congestive heart failure, is extracted from the leaves of the woolly foxglove plant. (b) Nylon was the first synthetic fiber made in the laboratory. It quickly replaced the natural fiber silk in parachutes and ladies' stockings. Styrofoam and PET (polyethylene terephthalate), the plastic used for soft drink bottles, are strong yet lightweight synthetic materials used for food storage. Over-the-counter pain relievers like ibuprofen are synthetic. The starting materials for all of these useful products are obtained from petroleum.
(a)–(b): Jill Braaten/McGraw Hill

a.

b.

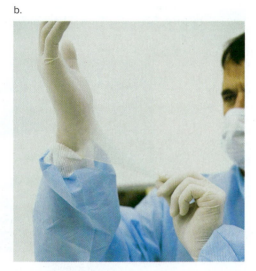

(a) Latex, the sticky liquid that oozes from a rubber tree when it is cut, is too soft for most applications.
(b) Vulcanization converts latex to the stronger, elastic rubber used in tires and other products.
(a): Suphatthra China/Shutterstock; (b): Roy McMahon/Fuse/Getty Images

Sometimes a chemist studies what a substance is made of, whereas at other times, the focus may be how to convert one material into a new material with unique and useful properties. As an example, naturally occurring rubber exists as the sticky liquid latex, which is too soft for most applications. The laboratory process of vulcanization converts it to the stronger, more elastic material used in tires and other products (Figure 1.2).

Chemistry is truly the science of everyday experience. Soaps and detergents, newspapers and DVDs, condoms and oral contraceptives, Tylenol and penicillin—all of these items are products of chemistry. Without a doubt, advances in chemistry have transformed life in modern times.

1.2 States of Matter

Matter exists in three common states—solid, liquid, and gas.

- A *solid* has a definite volume, and maintains its shape regardless of the container in which it is placed. The particles of a solid lie close together, and are arranged in a regular three-dimensional array.
- A *liquid* has a definite volume, but takes on the shape of the container it occupies. The particles of a liquid are close together, but they can randomly move around, sliding past one another.
- A *gas* has no definite shape or volume. The particles of a gas move randomly and are separated by a distance much larger than their size. The particles of a gas expand to fill the volume and assume the shape of whatever container they are put in.

For example, water exists in its solid state as ice or snow, liquid state as liquid water, and gaseous state as steam or water vapor. Blow-up circles like those in Figure 1.3 will be used commonly in this text to indicate the composition and state of the particles that compose a substance. In this molecular art, different types of particles are shown in color-coded spheres, and the distance between the spheres signals its state—solid, liquid, or gas.

Matter is characterized by its **physical properties** and **chemical properties.**

- *Physical properties* are those that can be observed or measured without changing the composition of the material.

Figure 1.3   The Three States of Water—Solid, Liquid, and Gas

a. Solid water

b. Liquid water

c. Gaseous water

- The particles of a solid are close together and highly organized. (Photo: snow-capped Mauna Kea on the Big Island of Hawaii)

- The particles of a liquid are close together but more disorganized than the solid. (Photo: Akaka Falls on the Big Island of Hawaii)

- The particles of a gas are far apart and disorganized. (Photo: steam formed by a lava flow on the Big Island of Hawaii)

Each red sphere joined to two gray spheres represents a single water particle. In proceeding from left to right, from solid to liquid to gas, the molecular art shows that the level of organization of the water particles decreases. Color-coding and the identity of the spheres within the particles will be addressed in Chapter 2. *(a): Alvis Upitis/Getty Images; (b): Daniel C. Smith; (c): Source:T.J. Takahash/USGS*

Common physical properties include melting point (mp), boiling point (bp), solubility, color, and odor. **A *physical change* alters a substance without changing its composition.** The most common physical changes are **changes in state**—that is, the **conversion of matter from one state to another.** Melting an ice cube to form liquid water, and boiling liquid water to form steam are two examples of physical changes. Water is the substance at the beginning and end of both physical changes. More details about physical changes are discussed in Chapter 4.

solid water   liquid water   water vapor

- *Chemical properties* are those that determine how a substance can be converted to another substance.

A *chemical change,* or a *chemical reaction,* **converts one material to another.** The conversion of hydrogen and oxygen to water is a chemical reaction because the composition of the material is different at the beginning and end of the process. Chemical reactions are discussed in Chapter 5.

Concept Check 1.1	Characterize each process as a physical change or a chemical change: (a) making ice cubes; (b) burning natural gas; (c) silver jewelry tarnishing; (d) a pile of snow melting; (e) fermenting grapes to produce wine.
Concept Check 1.2	Does the molecular art represent a chemical change or a physical change? Explain your choice.

1.3   Classification of Matter

All matter can be classified as either a **pure substance** or a **mixture.**

- A *pure substance* is composed of a single component and has a constant composition, regardless of the sample size and the origin of the sample.

A pure substance, such as water or table sugar, can be characterized by its physical properties, because these properties do not change from sample to sample. **A pure substance cannot be broken down to other pure substances by any physical change.**

- A *mixture* is composed of more than one substance. The composition of a mixture can vary depending on the sample.

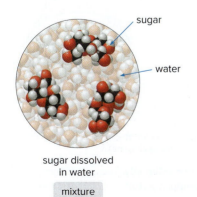

sugar

water

sugar dissolved
in water

mixture

The physical properties of a mixture may also vary from one sample to another. **A mixture can be separated into its components by physical changes.** Dissolving table sugar in water forms a mixture, whose sweetness depends on the amount of sugar added. If the water is allowed to evaporate from the mixture, pure table sugar and pure water are obtained.

Mixtures can be formed from solids, liquids, and gases, as shown in Figure 1.4. The compressed air breathed by a scuba diver consists mainly of the gases oxygen and nitrogen. A saline solution used in an IV bag contains solid sodium chloride (table salt) dissolved in water.

Figure 1.4   Two Examples of Mixtures

a. Two gases

oxygen   nitrogen

b. A solid and a liquid

water   sodium

chloride
(from chlorine)

(a): Daniel C. Smith; (b): Janis Christie/Digital Vision/Alamy Stock Photo

A pure substance is classified as either an **element** or a **compound.**

An alphabetical list of elements is located in Appendix A. The elements are commonly organized into a periodic table, shown in Appendix B, and discussed in much greater detail in Chapter 2.

• An *element* is a pure substance that cannot be broken down into simpler substances by a chemical reaction.
• A *compound* is a pure substance formed by chemically combining (joining together) two or more elements.

Nitrogen gas, aluminum foil, and copper wire are all elements. Water is a compound because it is composed of the elements hydrogen and oxygen. Table salt, sodium chloride, is also a compound because it is formed from the elements sodium and chlorine (Figure 1.5). Although

Figure 1.5   Elements and Compounds

a. Aluminum foil

b. Nitrogen gas

c. Water

d. Table salt

aluminum

nitrogen

hydrogen   oxygen

chloride
(from chlorine)   sodium

• Aluminum foil and nitrogen gas are elements. **The molecular art used for an element shows spheres of *one* color only.** Thus, aluminum is a solid shown with gray spheres, whereas nitrogen is a gas shown with blue spheres. Water and table salt are compounds. Color-coding of the spheres used in the molecular art indicates that water is composed of two elements—hydrogen shown as gray spheres and oxygen shown in red. Likewise, the gray (sodium) and green (chlorine) spheres illustrate that sodium chloride is formed from two elements as well.
(a): Daniel C. Smith; (b): Keith Eng, 2008; (c): Jill Braaten/McGraw Hill; (d): Daniel C. Smith

Figure 1.6

Classification of Matter

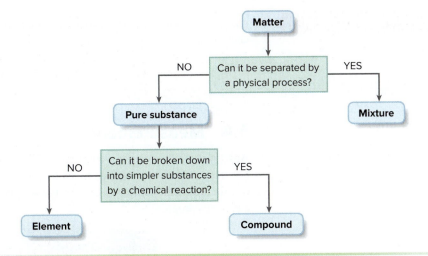

only 118 elements are currently known, over 50 million compounds occur naturally or have been synthesized in the laboratory. We will learn much more about elements and compounds in Chapters 2 and 3.

Concept Check 1.3	Use the list of elements in Appendix A to classify each item as an element or a compound: (a) the gas inside a helium balloon; (b) table sugar; (c) the rust on an iron nail; (d) aspirin.

Figure 1.6 summarizes the categories into which matter is classified.

Sample Problem 1.1	**Using Molecular Art for an Element and a Compound**

Classify each example of molecular art as an element or a compound:

a.   b.

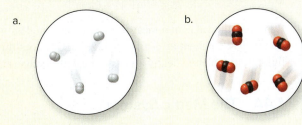

Analysis

In molecular art, an element is composed of spheres of the same color, whereas a compound is composed of spheres of different colors.

Solution

Representation (a) is an element because each particle contains only gray spheres. Representation (b) is a compound because each particle contains both red and black spheres.

Practice Problem 1.1	Classify each example of molecular art as a pure substance or a mixture:

a.   b.

| Practice Problem 1.2 | Classify each item as a pure substance or a mixture: (a) blood; (b) ocean water; (c) a piece of wood; (d) a chunk of ice. |

More Practice: Try Problems 1.1, 1.2, 1.15, 1.16.

1.4   Measurement

Any time you check your weight on a scale, measure the ingredients of a recipe, or figure out how far it is from one location to another, you are measuring a quantity. Measurements are routine for healthcare professionals who use weight, blood pressure, pulse, and temperature to chart a patient's progress.

Length:
21.0 in.
or
53.3 cm

number   **unit**

Mass
(Weight):
8 lb 8 oz
or
3.9 kg

number   **unit**

Roy Hsu/Photographer's Choice RF/Getty Images

In 1960, the **International System of Units** was formally adopted as the uniform system of units for the sciences. **SI units,** as they are called, are based on the metric system, but the system recommends the use of some metric units over others. SI stands for the French words, *Système Internationale.*

- Every measurement is composed of a *number* and a *unit.*

Reporting the value of a measurement is meaningless without its unit. For example, if you were told to give a patient an aspirin dosage of 325, does this mean 325 ounces, pounds, grams, milligrams, or tablets? Clearly there is a huge difference among these quantities.

1.4A   The Metric System

In the United States, most measurements are made with the **English system,** using units like miles (mi), gallons (gal), pounds (lb), and so forth. A disadvantage of this system is that the units are not systematically related to each other and require memorization. For example, 1 lb = 16 oz, 1 gal = 4 qt, and 1 mi = 5,280 ft.

Scientists, health professionals, and people in most other countries use the **metric system,** with units like meter (m) for length, gram (g) for mass, and liter (L) for volume. The metric system is slowly gaining popularity in the United States. The weight of packaged foods is often given in both ounces and grams. Distances on many road signs are shown in miles and kilometers. Most measurements in this text will be reported using the metric system, but learning to convert English units to metric units is also a necessary skill that will be illustrated in Section 1.7.

The important features of the metric system are the following:

- Each type of measurement has a base unit—the meter (m) for length; the gram (g) for mass; the liter (L) for volume; the second (s) for time.
- All other units are related to the base unit by powers of 10.
- The prefix of the unit name indicates if the unit is larger or smaller than the base unit.

Table 1.1 Metric Units

Quantity	Base Unit	Symbol
Length	Meter	m
Mass	Gram	g
Volume	Liter	L
Time	Second	s

The base units of the metric system are summarized in Table 1.1, and the most common prefixes used to convert the base units to smaller or larger units are summarized in Table 1.2. **The same prefixes are used for all types of measurement.** For example, the prefix *kilo-* means 1,000 times as large. Thus,

$$1 \text{ kilometer} = \textbf{1,000} \text{ meters} \quad \text{or} \quad 1 \text{ km} = 1,000 \text{ m}$$
$$1 \text{ kilogram} = \textbf{1,000} \text{ grams} \quad \text{or} \quad 1 \text{ kg} = 1,000 \text{ g}$$
$$1 \text{ kiloliter} = \textbf{1,000} \text{ liters} \quad \text{or} \quad 1 \text{ kL} = 1,000 \text{ L}$$

The prefix *milli-* means one thousandth as large (1/1,000 or 0.001). Thus,

$$1 \text{ millimeter} = \textbf{0.001} \text{ meters} \quad \text{or} \quad 1 \text{ mm} = 0.001 \text{ m}$$
$$1 \text{ milligram} = \textbf{0.001} \text{ grams} \quad \text{or} \quad 1 \text{ mg} = 0.001 \text{ g}$$
$$1 \text{ milliliter} = \textbf{0.001} \text{ liters} \quad \text{or} \quad 1 \text{ mL} = 0.001 \text{ L}$$

Table 1.2 Common Prefixes Used for Metric Units

Prefix	Symbol	Meaning	Numerical Value[a]	Scientific Notation[b]
Giga-	G	Billion	1,000,000,000.	10^9
Mega-	M	Million	1,000,000.	10^6
Kilo-	k	Thousand	1,000.	10^3
Deci-	d	Tenth	0.1	10^{-1}
Centi-	c	Hundredth	0.01	10^{-2}
Milli-	m	Thousandth	0.001	10^{-3}
Micro-	μ[c]	Millionth	0.000 001	10^{-6}
Nano-	n	Billionth	0.000 000 001	10^{-9}

[a]Numbers that contain five or more digits to the right of the decimal point are written with a small space separating each group of three digits.
[b]How to express numbers in scientific notation is explained in Section 1.6.
[c]The symbol μ is the lower case Greek letter mu. The prefix *micro-* is sometimes abbreviated as **mc.**

The metric symbols are all lower case except for the unit **liter** (L) and the prefixes **mega-** (M) and **giga-** (G).

Adam Gault/Science Photo Library RF/Science Source

Concept Check 1.4

What term is used for each of the following units: (a) a million liters; (b) a thousandth of a second; (c) a hundredth of a gram; (d) a tenth of a liter?

1.4B Measuring Length

The base unit of length in the metric system is the *meter* (m). A meter, 39.37 inches in the English system, is slightly longer than a yard (36 inches). Common units derived from a meter are the decimeter (dm), centimeter (cm), and millimeter (mm).

1 meter = 10 decimeters = 100 centimeters = 1,000 millimeters

Note how these values are related to those in Table 1.2. Because a centimeter is one *hundredth* of a meter (0.01 m), there are *100* centimeters in a meter.

1.4C   Measuring Mass

Although the terms mass and weight are often used interchangeably, they really have different meanings.

- *Mass* is a measure of the amount of matter in an object.
- *Weight* is the force that matter feels due to gravity.

The mass of an object is independent of its location. The weight of an object changes slightly with its location on the earth, and drastically when the object is moved from the earth to the moon, where the gravitational pull is only one-sixth that of the earth. Although we often speak of *weighing* an object, we are really *measuring its mass.*

The base unit of mass in the metric system is the *gram* (g), a small quantity compared to the English pound (1 lb = 453.6 g). Two common units derived from a gram are the kilogram (kg) and milligram (mg).

$$1{,}000 \text{ g} = 1 \text{ kg}$$
$$1 \text{ g} = 1{,}000 \text{ mg}$$

Concept Check 1.5

What term is used for each quantity? (For example, 0.001 g = one milligram.)

a. 0.000 001 g   c. 0.000 000 001 s

b. 1,000,000,000 m   d. 0.01 g

Note the difference between the units **cm** and **cm³**. The centimeter (cm) is a unit of length. A cubic centimeter (cm³ or cc) is a unit of volume.

$$1 \text{ mL} = 1 \text{ cm}^3 = 1 \text{ cc}$$

1.4D   Measuring Volume

The base unit of volume in the metric system is the *liter* (L), which is slightly larger than the English quart (1 L = 1.057 qt). One liter is defined as the volume of a cube 10 cm on an edge.

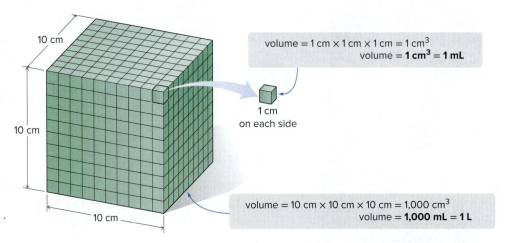

Three common units derived from a liter used in medicine and laboratory research are the deciliter (dL), milliliter (mL), and microliter (μL). **One milliliter is the same as one cubic centimeter (cm³), which is abbreviated as cc.**

$$1 \text{ L} = 10 \text{ dL}$$
$$1 \text{ L} = 1{,}000 \text{ mL}$$
$$1 \text{ L} = 1{,}000{,}000 \text{ μL}$$

Concept Check 1.6

If an IV bag contains 1,000 mL of solution, how many liters does it contain?

Table 1.3 summarizes common metric units of length, mass, and volume. Table 1.4 lists English units of measurement, as well as their metric equivalents.

Table 1.3 **Summary of the Common Metric Units of Length, Mass, and Volume**

Length	Mass	Volume
1 km = 1,000 m	1 kg = 1,000 g	1 L = 10 dL
1 m = 100 cm	1 g = 1,000 mg	1 L = 1,000 mL
1 m = 1,000 mm	1 mg = 1,000 µg	1 L = 1,000,000 µL
1 cm = 10 mm		1 dL = 100 mL
		1 mL = 1 cm³ = 1 cc

Common metric abbreviations and conversion factors are also listed in Appendix C.

Table 1.4 **English Units and Their Metric Equivalents**

Quantity	English Unit	Metric–English Relationship
Length	1 ft = 12 in.	2.54 cm = 1 in.
	1 yd = 3 ft	1 m = 39.37 in.
	1 mi = 5,280. ft	1 km = 0.6214 mi
Mass	1 lb = 16.00 oz	1 kg = 2.205 lb
	1 ton = 2,000 lb	453.6 g = 1 lb
		28.35 g = 1 oz
Volume	1 qt = 4 cups	946.4 mL = 1 qt
	1 qt = 2 pt	1 L = 1.057 qt
	1 qt = 32 fl oz	29.57 mL = 1 fl oz
	1 gal = 4 qt	

Common abbreviations for English units: inch (in.), foot (ft), yard (yd), mile (mi), pound (lb), ounce (oz), gallon (gal), quart (qt), pint (pt), and fluid ounce (fl oz).

Concept Check 1.7

Using the prefixes in Table 1.2, determine which quantity in each pair is larger.

a. 3 mL or 3 cL   c. 5 km or 5 cm

b. 1 ng or 1 µg   d. 2 mL or 2 µL

1.5 Significant Figures

Numbers used in chemistry are either **exact** or **inexact.**

- An *exact* number results from counting objects or is part of a definition.

Our bodies have 10 fingers, 10 toes, and two kidneys. A meter is composed of 100 centimeters. These numbers are exact because there is no uncertainty associated with them.

- An *inexact* number results from a measurement or observation and contains some uncertainty.

A container of 71 macadamia nuts weighs 125 g. The number of nuts (71) is exact, whereas the mass of the nuts (125 g) is inexact. *Zachary D.-K. Smith*

Whenever we measure a quantity there is a degree of uncertainty associated with the result. The last number (farthest to the right) is an estimate, and it depends on the type of measuring device we use to obtain it. For example, the length of a fish caught on a recent outing could be reported as 53 cm or 53.5 cm depending on the tape measure used.

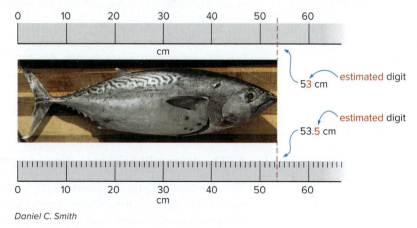

Daniel C. Smith

- *Significant figures* are all the digits in a measured number including *one* estimated digit.

Thus, the length 53 cm has two significant figures, and the length 53.5 cm has three significant figures.

1.5A Determining the Number of Significant Figures

How many significant figures are contained in a number?

- All nonzero digits are always significant.

65.2 g	**three** significant figures
1,265 m	**four** significant figures
25 μL	**two** significant figures
255.345 g	**six** significant figures

Whether a zero counts as a significant figure depends on its location in the number.

Rules to Determine When a Zero Is a Significant Figure

Rule [1] A zero *counts* as a significant figure when it occurs:

- Between two nonzero digits

  29.05 g—four significant figures
  1.0087 mL—five significant figures

- At the end of a number with a decimal point

  25.70 cm—four significant figures
  3.7500 g—five significant figures
  620. lb—three significant figures

Rule [2] A zero does *not* count as a significant figure when it occurs:

- At the beginning of a number

  0.0245 mg—three significant figures
  0.008 mL—one significant figure

- At the end of a number that does *not* have a decimal point

  2,570 m—three significant figures
  1,245,500 m—five significant figures

In reading a number with a decimal point from left to right, **all digits starting with the first nonzero number are significant figures.** The number 0.003 450 120 has seven significant figures, shown in red.

Sample Problem 1.2 — **Determining the Number of Significant Figures**

How many significant figures does each number contain?

  a. 34.08   b. 0.0054   c. 260.00   d. 260

Analysis

All nonzero digits are significant. A zero is significant only if it occurs between two nonzero digits, or at the end of a number with a decimal point.

Solution

Significant figures are shown in red.

  a. 34.08 (four)   b. 0.0054 (two)   c. 260.00 (five)   d. 260 (two)

Practice Problem 1.3

How many significant figures does each number contain?

  a. 23.45   b. 230   c. 0.202   d. 0.003 60   e. 10,040   f. 1,004.00

Practice Problem 1.4

Indicate whether each zero in the following numbers is significant.

  a. 0.003 04   b. 26,045   c. 1,000,034   d. 0.304 00

More Practice: Try Problems 1.5b, 1.6b, 1.23, 1.24.

1.5B Using Significant Figures in Multiplication and Division

We often must perform calculations with numbers that contain a different number of significant figures. The number of significant figures in the answer of a problem depends on the type of mathematical calculation—multiplication (and division) or addition (and subtraction).

- In multiplication and division, the answer has the same number of significant figures as the original number with the *fewest* significant figures.

Let's say you drove a car 351.2 miles in 5.5 hours, and you wanted to calculate how many miles per hour you traveled. Entering these numbers on a calculator would give the following result:

four significant figures

$$\text{Miles per hour} = \frac{351.2 \text{ miles}}{5.5 \text{ hours}} = 63.854\,545 \text{ miles per hour}$$

two significant figures

The answer must contain only **two** significant figures.

The answer to this problem can have only *two* significant figures, because one of the original numbers (5.5 hours) has only *two* significant figures. To write the answer in proper form, we must **round off the number** to give an answer with only two significant figures. Two rules are used in rounding off numbers.

- If the first number that must be dropped is 4 or fewer, *drop it and all remaining numbers.*
- If the first number that must be dropped is 5 or greater, *round the number up* by adding one to the last digit that will be retained.

In this problem:

These digits must be retained.

first digit to be dropped

63.**8**54 545

These digits must be dropped.

- Because the first digit to be dropped is **8** (5 or greater), add 1 to the digit to its left.
- The answer 63.854 545 rounded to **two** digits is **64 miles per hour.**

Table 1.5 gives other examples of rounding off numbers.

When a calculator is used in calculations, sometimes the display shows *more* digits than are significant and sometimes it shows *fewer* digits. For example, in multiplying 23.2 by 1.1, the calculator displays the answer as 25.52. Because the quantity 1.1 has only *two* significant figures, the answer must contain only *two* significant figures and rounded to 26.

23.2	×	1.1	=	25.52	round off →	26
three significant figures		two significant figures		calculator display		two significant figures

Answer

In contrast, dividing 25.0 by 0.50 displays the answer as 50, a quantity with only one significant figure. Because 0.50 has *two* significant figures, the answer must contain *two* significant figures, and this is achieved by adding a decimal point (50.).

25.0	×	0.50	=	50	add decimal →	50.
three significant figures		two significant figures		calculator display		two significant figures

Answer

Table 1.5 Rounding Off Numbers

Original Number	Rounded to	Rounded Number
61.2537	Two places	61
61.2537	Three places	61.3
61.2537	Four places	61.25
61.2537	Five places	61.254

The first number to be dropped is indicated in red in each original number. When this number is 4 or fewer, it and all other digits to its right are dropped. When this number is 5 or greater, 1 is added to the digit to its left.

A scientific calculator is useful in simple mathematical calculations, as well as more complicated functions, such as converting a number to scientific notation (Section 1.6B) or taking the logarithm of a number (Section 8.5B).
S.Narongrit/Shutterstock

Sample Problem 1.3   **Rounding Off Numbers**

Round off each number to three significant figures.

    a. 1.2735   b. 0.002 536 22   c. 3,836.9

Analysis

If the answer is to have *three* significant figures, look at the *fourth* number from the left.

- If this number is 4 or fewer, drop it and all remaining numbers to the right.
- If the fourth number from the left is 5 or greater, round the number up by adding one to the third nonzero digit.

Solution

a. 1.27   b. 0.002 54   c. 3,840 (Omit the decimal point after the 0. The number 3,840. has four significant figures.)

Practice Problem 1.5

Round off each number in Sample Problem 1.3 to two significant figures.

More Practice: Try Problems 1.25, 1.26.

Sample Problem 1.4

Determining Significant Figures in Multiplication and Division

Carry out each calculation and give the answer using the proper number of significant figures.

a. 3.81 × 0.046   b. 120.085 ÷ 106

Analysis

Because these calculations involve multiplication and division, the answer must have the same number of significant figures as the original number with the fewest number of significant figures.

Solution

a. 3.81 × 0.046 = 0.1753   • Because 0.046 has only two significant figures, round the answer to give it two significant figures.

0.1753   Because this number is 5 (5 or greater), round the 7 to its left up by one.

Answer: 0.18

b. 120.085 ÷ 106 = 1.132 877 36   • Because 106 has three significant figures, round the answer to give it three significant figures.

1.132 877 36   Because this number is 2 (4 or fewer), drop it and all numbers to its right.

Answer: 1.13

Practice Problem 1.6

Carry out each calculation and give the answer using the proper number of significant figures.

a. 10.70 × 3.5   b. 0.206 ÷ 25,993   c. 1,300 ÷ 41.2   d. 120.5 × 26

More Practice: Try Problems 1.27a, c; 1.28a, c.

1.5C   Using Significant Figures in Addition and Subtraction

In determining significant figures in addition and subtraction, the decimal place of the last significant digit determines the number of significant figures in the answer.

> • In addition and subtraction, the answer has the same number of decimal places as the original number with the *fewest* decimal places.

Suppose a baby weighed 3.6 kg at birth and 10.11 kg on his first birthday. To figure out how much weight the baby gained in his first year of life, we subtract these two numbers and report the answer using the proper number of significant figures.

weight at one year = 10.11 kg   10.11 kg ← **two** digits after the decimal point

weight at birth = 3.6 kg   − 3.6 kg ← **one** digit after the decimal point

weight gain = 6.**5**1 kg

↑ last significant digit

> • The answer can have only **one** digit after the decimal point.
> • Round 6.51 to 6.5.
> • The baby gained 6.5 kg during his first year of life.

Because 3.6 kg has only one significant figure after the decimal point, the answer can have only one significant figure after the decimal point as well.

| Sample Problem 1.5 | **Determining Significant Figures in Addition and Subtraction** |

While on a diet, a woman lost 3.52 lb the first week, 2.2 lb the second week, and 0.59 lb the third week. How much weight did she lose in all?

Analysis

Add up the amount of weight loss each week to get the total weight loss. When adding, the answer has the same number of decimal places as the original number with the fewest decimal places.

Solution

3.52 lb
2.2 lb   **one** digit after the decimal point
0.59 lb

6.31 lb - - - - → 6.3 lb
     round off

last significant digit

- Because 2.2 lb has only **one** digit after the decimal point, the answer can have only **one** digit after the decimal point.
- Round 6.31 to 6.3.
- Total weight loss: 6.3 lb.

| Practice Problem 1.7 | Carry out each calculation and give the answer using the proper number of significant figures. |

a. 27.8 cm + 0.246 cm

b. 102.66 mL + 0.857 mL + 24.0 mL

c. 54.6 mg − 25 mg

d. 2.35 s − 0.266 s

More Practice: Try Problems 1.27b, d; 1.28b, d.

1.6 Scientific Notation

Healthcare professionals and scientists must often deal with very large and very small numbers. For example, the blood platelet count of a healthy adult might be 250,000 platelets per mL. At the other extreme, the level of the female sex hormone estriol during pregnancy might be 0.000 000 250 g per mL of blood plasma. Estriol is secreted by the placenta and its concentration is used as a measure of the health of the fetus.

1.6A Writing Numbers in Scientific Notation

To write numbers that contain many leading zeros (at the beginning) or trailing zeros (at the end), scientists use **scientific notation.**

Hospital laboratory technicians determine thousands of laboratory results each day. *Stockbyte/Getty Images*

- In scientific notation, a number is written as $y \times 10^x$.
- The term y, called the coefficient, is a number between 1 and 10.
- The value x is an exponent, which can be any positive or negative whole number.

First, let's recall what powers of 10 with *positive* exponents, such as 10^2 or 10^5, mean. These correspond to numbers greater than one, and the positive exponent tells how many zeros are to be written after the number one. Thus, $10^2 = 100$, a number with **two** zeros after the number one.

The product has **two** zeros.

$$10^2 = 10 \times 10 = 100$$

The exponent **2** means "multiply **two** 10s."

The product has **five** zeros.

$$10^5 = 10 \times 10 \times 10 \times 10 \times 10 = 100,000$$

The exponent **5** means "multiply **five** 10s."

Powers of 10 that contain *negative* exponents, such as 10^{-3}, correspond to numbers less than one. In this case the exponent tells how many places (*not* zeros) are located to the right of the decimal point.

The answer has **three** places to the *right* of the decimal point, including the number one.

$$10^{-3} \quad = \quad \frac{1}{10 \times 10 \times 10} \quad = \quad 0.001$$

The exponent **–3** means "divide by **three** 10s."

To write a number in scientific notation, we follow a stepwise procedure.

How To   Convert a Standard Number to Scientific Notation

Example   Write each number in scientific notation: (a) 2,500; (b) 0.036.

Step [1]   Move the decimal point to give a number between 1 and 10.

a. 2500.

Move the decimal point three places to the left to give the number 2.5.

b. 0.036

Move the decimal point two places to the right to give the number 3.6.

Step [2]   Multiply the result by 10^x, where x is the number of places the decimal point was moved.
- If the decimal point is moved to the **left**, x is **positive**.
- If the decimal point is moved to the **right**, x is **negative**.

a. Because the decimal point was moved three places to the **left**, the exponent is +3, and the coefficient is multiplied by 10^3.

    Answer: 2,500 = 2.5 × 10^3

b. Because the decimal point was moved two places to the **right**, the exponent is –2, and the coefficient is multiplied by 10^{-2}.

    Answer: 0.036 = 3.6 × 10^{-2}

Notice that the number of significant figures in the coefficient in scientific notation must equal the number of significant figures in the original number. Thus, the coefficients for both 2,500 and 0.036 need two significant figures and no more. Figure 1.7 shows two more examples of numbers written in standard form and scientific notation.

$$2,500 = 2.5 \times 10^3 \quad \textit{not} \quad 2.50 \times 10^3 \text{ (three significant figures)}$$
$$\textit{not} \quad 2.500 \times 10^3 \text{ (four significant figures)}$$

two significant figures

Figure 1.7

Numbers in Standard Form and Scientific Notation

The human body contains 2×10^{13} red blood cells.

A single red blood cell has a diameter of 6×10^{-6} m.

Cre8tive Studios/Alamy Stock Photos

Very large and very small numbers are more conveniently written in scientific notation.

Quantity	Number	Scientific Notation
Number of red blood cells	20,000,000,000,000	2×10^{13}
Diameter of a red blood cell	0.000 006 m	6×10^{-6} m

| Sample Problem 1.6 | **Writing a Number in Scientific Notation** |

Write the recommended daily dietary intake of each nutrient in scientific notation: (a) sodium, 2,400 mg; (b) vitamin B_{12}, 0.000 006 g.

Analysis

Move the decimal point to give a number between 1 and 10. Multiply the number by 10^x, where x is the number of places the decimal point was moved. The exponent x is **(+)** when the decimal point moves to the **left** and **(−)** when it moves to the **right.**

Solution

a.

$2400. = 2.4 \times 10^3$

the number of places the decimal point was moved to the **left**

Move the decimal point **three** places to the **left**.

- Write the coefficient as 2.4 (**two** significant figures), because 2,400 contains **two** significant figures.

b.

$0.000\ 006 = 6 \times 10^{-6}$

the number of places the decimal point was moved to the **right**

Move the decimal point **six** places to the **right**.

- Write the coefficient as 6 (**one** significant figure), because 0.000 006 contains **one** significant figure.

| Practice Problem 1.8 | Lab results for a routine check-up showed an individual's iron level in the blood to be 0.000 098 g per deciliter, placing it in the normal range. Convert this number to scientific notation. |

| Practice Problem 1.9 | Write each number in scientific notation.
a. 93,200 b. 0.000 725 c. 6,780,000 d. 0.000 030 |

More Practice: Try Problems 1.29, 1.30, 1.35, 1.36.

To convert a number in scientific notation to a standard number, reverse the procedure, as shown in Sample Problem 1.7. It is often necessary to add leading or trailing zeros to write the number.

- **When the exponent x is positive, move the decimal point x places to the *right*.**

2.800×10^2 $2.800 \dashrightarrow 280.0$

Move the decimal point to the *right* **two** places.

- **When the exponent x is negative, move the decimal point x places to the *left*.**

2.80×10^{-2} $002.80 \dashrightarrow 0.0280$

Move the decimal point to the *left* **two** places.

| Sample Problem 1.7 | **Converting a Number in Scientific Notation to a Standard Number** |

The average diameter of a human hair is about 1.2×10^{-4} m. Convert this value to a standard number.

Analysis

The exponent in 10^x tells how many places to move the decimal point in the coefficient to generate a standard number. The decimal point goes to the *right* when x is positive and to the *left* when x is negative.

Solution

1.2×10^{-4} $000\ 01.2 \dashrightarrow 0.000\ 12$ m

Move the decimal point to the *left* **four** places.   **Answer**

The answer, 0.000 12, has two significant figures, just like 1.2×10^{-4}.

Practice Problem 1.10

Convert each number to its standard form.

    a. 6.5×10^3 b. 3.26×10^{-5} c. 3.780×10^{-2} d. 1.04×10^8

More Practice: Try Problems 1.31, 1.32.

1.6B Using a Scientific Calculator for Numbers in Scientific Notation

Several examples involving the use of a scientific calculator are found in Appendix D, Useful Mathematical Concepts.

A scientific calculator can be used to convert a standard number to a number in scientific notation. Described in this section are the steps that can be followed in calculations with *some* calculators. Consult your manual if these steps do *not* produce the stated result.

To convert a number from its standard form to a number in scientific notation, follow the steps in columns [1]–[3].

	① Number to enter	② Buttons to press	③ Display	Meaning
A number ≥ 1:	1,200	2nd SCI	1.2^{03}	1.2×10^3
A **decimal** < 1:	0.052	2nd SCI	5.2^{-02}	5.2×10^{-2}

To enter a number in scientific notation, follow the steps in columns [1]–[3].

	① Number to enter	② Buttons to press	③ Display	Meaning
A number with a **positive** exponent, as in 1.5×10^8:	1.5	EE 8	1.5^{08}	150,000,000
A number with a **negative** exponent, as in 3.5×10^{-4}:	3.5	EE 4 Change sign (+ to –)	3.5^{-04}	0.000 35

Concept Check 1.8

If you used the calculator described in this section, write each of the following calculator displays as a number in standard form and using scientific notation: (a) 5.43^{05}; (b) 4.43^{-06}; (c) 7.4^{04}.

1.7 Problem Solving Using Conversion Factors

Often a measurement is recorded in one unit, and then it must be converted to another unit. For example, a patient may weigh 130 lb, but we may need to know her weight in kilograms to calculate a drug dosage. The recommended daily dietary intake of potassium is 3,500 mg, but we may need to know how many grams this corresponds to.

1.7A Conversion Factors

To convert one unit to another we use one or more **conversion factors.**

original quantity   ×   **conversion factor**   =   desired quantity

These two quantities are equivalent.
Only the units are different.

• A *conversion factor* is a term that converts a quantity in one unit to a quantity in another unit.

Refer to Tables 1.3 and 1.4 for metric and English units needed in problem solving. Common metric and English units are also listed in Appendix C.

A conversion factor is formed by taking an equality, such as 2.205 lb = 1 kg, and writing it as a fraction. We can always write a conversion factor in two different ways.

$$\frac{2.205 \text{ lb}}{1 \text{ kg}} \quad \text{or} \quad \frac{1 \text{ kg}}{2.205 \text{ lb}} \quad \begin{array}{l} \text{numerator} \\[1em] \text{denominator} \end{array}$$

conversion factors for pounds and kilograms

With pounds and kilograms, either of these values can be written above the division line of the fraction (the **numerator**) or below the division line (the **denominator**). The way the conversion factor is written will depend on the problem.

Concept Check 1.9	Write two conversion factors for each pair of units: (a) miles and kilometers; (b) meters and millimeters.

1.7B Solving a Problem Using One Conversion Factor

When using conversion factors to solve a problem, a unit that appears in the numerator in one term and the denominator in another term will *cancel*. **The goal in setting up a problem is to make sure *all unwanted units cancel.***

Let's say we want to convert 130 lb to kilograms.

$$\underset{\text{original quantity}}{130 \text{ lb}} \quad \times \quad \boxed{\textbf{conversion factor}} \quad = \quad \underset{\text{desired quantity}}{? \text{ kg}}$$

Two possible conversion factors: $\dfrac{2.205 \text{ lb}}{1 \text{ kg}}$ or $\dfrac{1 \text{ kg}}{2.205 \text{ lb}}$

How many grams of aspirin are contained in a 325-mg tablet?
Mark Dierker/McGraw Hill

To solve this problem we must use a conversion factor that satisfies two criteria.

- The conversion factor must relate the two quantities in question—pounds and kilograms.
- The conversion factor must cancel out the unwanted unit—pounds.

This means choosing the conversion factor with the unwanted unit—pounds—*in the denominator* to cancel out pounds in the original quantity. This leaves kilograms as the only remaining unit, and the problem is solved.

$$130 \text{ lb} \quad \times \quad \overset{\textbf{conversion factor}}{\frac{1 \text{ kg}}{2.205 \text{ lb}}} \quad = \quad \boxed{59 \text{ kg}} \quad \textbf{answer in kilograms}$$

Pounds (lb) must be the denominator to cancel the unwanted unit (lb) in the original quantity.

We must use the correct number of significant figures in reporting an answer to each problem. In this case, the value 1 kg is *defined* as 2.205 lb; in other words, 1 kg contains the exact number "1" with *no* uncertainty, so it does not limit the number of digits in the answer. Because 130 lb has two significant figures, the answer is rounded to two significant figures (59 kg).

As problems with units get more complicated, keep in mind the following general steps that are useful for solving any problem using conversion factors.

How To Solve a Problem Using Conversion Factors

Example   How many grams of aspirin are contained in a 325-mg tablet?

Step [1]   Identify the original quantity and the desired quantity, including units.

- In this problem the original quantity is reported in milligrams and the desired quantity is in grams.

325 mg   ? g

original quantity   desired quantity

Step [2]   Write out the conversion factor(s) needed to solve the problem.

- We need a conversion factor that relates milligrams and grams (Table 1.3). Because the unwanted unit is in milligrams, **choose the conversion factor that contains milligrams in the denominator so that the** *units cancel.*

Two possible conversion factors: $\dfrac{1000\ mg}{1\ g}$ or $\dfrac{1\ g}{1000\ mg}$ ⟵ Choose this factor to cancel the unwanted unit, mg.

- Sometimes one conversion factor is all that is needed in a problem. At other times (Section 1.7C) more than one conversion factor is needed.
- If the desired answer has a single unit (grams in this case), **the conversion factor must contain the desired unit in the numerator and the unwanted unit in the denominator.**

Step [3]   Set up and solve the problem.

- Multiply the original quantity by the conversion factor to obtain the desired quantity.

conversion factor

$325\ \cancel{mg}\quad \times\quad \dfrac{1\ g}{1000\ \cancel{mg}}\quad =\quad 0.325\ g$ of aspirin

original quantity   desired quantity

The number of mg (unwanted unit) cancels.

Step [4]   Write the answer using the correct number of significant figures and check it by estimation.

- Use the number of significant figures in each inexact (measured) number to determine the number of significant figures in the answer. In this case the answer is limited to three significant figures by the original quantity (325 mg).
- Estimate the answer using a variety of methods. In this case we knew our answer had to be less than one, because it is obtained by dividing 325 by a number larger than itself.

Sample Problem 1.8   **Solving a Problem Using Conversion Factors**

As we will learn in Chapter 11, PET, the plastic used in water bottles, can be recycled into bags, containers, fleece jackets, and carpeting. *Angela Hampton Picture Library/Alamy Stock Photo*

If 1.65×10^{9} lb of PET (polyethylene terephthalate), the plastic used in water and soft drink bottles, is recycled in a year, how many kilograms of PET does this represent?

Analysis and Solution

[1] Identify the original quantity and the desired quantity.

1.65×10^{9} lb   ? kg

original quantity   desired quantity

[2] Write out the conversion factors.

- We need a conversion factor that relates pounds and kilograms (Table 1.4). Choose the conversion factor that places pounds in the denominator, so that the units cancel.

$\dfrac{2.205\ lb}{1\ kg}$ or $\dfrac{1\ kg}{2.205\ lb}$ ⟵ Choose this factor to cancel the unwanted unit, lb.

Whenever a calculation requires conversion of a quantity from one unit to another, follow the three-step *How To:*

1. Identify the units in the original quantity and the desired quantity.
2. Write out the conversion factor(s).
3. Arrange the conversion factors so that unwanted units cancel, then solve.

[3] Solve the problem.

- Multiply the original quantity by the conversion factor to obtain the desired quantity. Set up the problem so that the unwanted unit, pounds, cancels.

conversion factor

$$1.65 \times 10^9 \text{ lb} \quad \times \quad \frac{1 \text{ kg}}{2.205 \text{ lb}} \quad = \quad 7.48 \times 10^8 \text{ kg of PET}$$

The number of lb (unwanted unit) cancels.

Practice Problem 1.11

Carry out each of the following conversions.
  a. 25 L to dL   b. 40.0 oz to g   c. 32 in. to cm   d. 10 cm to mm

---

Practice Problem 1.12

If the mass of the baby in the chapter-opening photograph is 6 lb 12 oz, how many kilograms does the baby weigh?

Practice Problem 1.13

(a) What is the volume of liquid contained in the given 0.5-mL syringe? (b) Convert this value to microliters.

More Practice: Try Problems 1.10; 1.37a, c; 1.38b, c; 1.39; 1.40; 1.41a; 1.42a, b; 1.44; 1.57.

1.7C   Solving a Problem Using Two or More Conversion Factors

Some problems require the use of more than one conversion factor to obtain the desired units in the answer. The same stepwise procedure is followed no matter how many conversion factors are needed. Keep in mind:

- **Always arrange the factors so that the denominator in one term cancels the numerator in the preceding term.**

Sample Problem 1.9 illustrates how to solve a problem with two conversion factors.

Sample Problem 1.9

Solving a Problem Using More Than One Conversion Factor

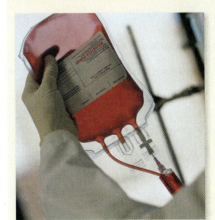

How many liters does this pint of blood contain? *Keith Brofsky/Photodisc/Getty Images*

An individual donated 1.0 pt of blood at the local blood bank. How many liters of blood does this correspond to?

Analysis and Solution

[1] Identify the original quantity and the desired quantity.

$$\underset{\text{original quantity}}{1.0 \text{ pt}} \qquad \underset{\text{desired quantity}}{? \text{ L}}$$

[2] Write out the conversion factors.

- We have no conversion factor that relates pints to liters directly. We do, however, know conversions for pints to quarts, and quarts to liters.

pint–quart conversion

$$\frac{2 \text{ pt}}{1 \text{ qt}} \quad \text{or} \quad \boxed{\frac{1 \text{ qt}}{2 \text{ pt}}}$$

quart–liter conversion

$$\frac{1.057 \text{ qt}}{1 \text{ L}} \quad \text{or} \quad \boxed{\frac{1 \text{ L}}{1.057 \text{ qt}}}$$

Choose the conversion factors with the unwanted units—pt and qt—in the denominator.

[3] Solve the problem.

- To set up the problem so that unwanted units cancel, arrange each term so that the units in the numerator of one term cancel the units of the denominator of the adjacent term. In this problem we need to cancel both pints and quarts to get liters.
- The single desired unit, liters, must be in the **numerator** of one term.

$$1.0 \text{ pt} \times \frac{1 \text{ qt}}{2 \text{ pt}} \times \frac{1 \text{ L}}{1.057 \text{ qt}} = 0.47 \text{ L}$$

Liters do not cancel.

Pints cancel.   Quarts cancel.

[4] Check.

- Because there are two pints in a quart and a quart is about the same size as a liter, one pint should be about half a liter. The answer, 0.47, is just about 0.5.
- Write the answer with two significant figures because one term, 1.0 pt, has two significant figures.

Practice Problem 1.14

Carry out each of the following conversions.

  a. 6,250 ft to km   b. 3 cups to L   c. 4.5 ft to cm

More Practice: Try Problems 1.9; 1.14; 1.37b, d; 1.38a, d; 1.41b; 1.43.

1.8   FOCUS ON HEALTH & MEDICINE
Problem Solving Using Clinical Conversion Factors

Sometimes conversion factors don't have to be looked up in a table; they are stated in the problem. If a drug is sold as a 250-mg tablet, this fact becomes a conversion factor relating milligrams to tablets.

The active ingredient in Children's Tylenol is acetaminophen.
Jill Braaten/McGraw Hill

$$\frac{250 \text{ mg}}{1 \text{ tablet}} \quad \text{or} \quad \frac{1 \text{ tablet}}{250 \text{ mg}}$$

mg–tablet conversion factors

Alternatively, a drug could be sold as a liquid solution with a specific concentration. For example, Children's Tylenol contains 80 mg of the active ingredient acetaminophen in 2.5 mL. This fact becomes a conversion factor relating milligrams to milliliters.

$$\frac{80 \text{ mg}}{2.5 \text{ mL}} \quad \text{or} \quad \frac{2.5 \text{ mL}}{80 \text{ mg}}$$

mg of acetaminophen–mL conversion factors

Sample Problems 1.10 and 1.11 illustrate how these conversion factors are used in determining drug dosages.

Sample Problem 1.10

Using Clinical Conversion Factors to Solve a Problem

A patient is prescribed 1.25 g of amoxicillin, which is available in 250-mg tablets. How many tablets are needed?

Analysis and Solution

[1] Identify the original quantity and the desired quantity.

- We must convert the number of grams of amoxicillin needed to the number of tablets that must be administered.

  1.25 g   ? tablets
  original quantity   desired quantity

[2] Write out the conversion factors.

- We have no conversion factor that relates grams to tablets directly. We do know, however, how to relate grams to milligrams, and milligrams to tablets.

g–mg conversion factors

$$\frac{1 \text{ g}}{1000 \text{ mg}} \quad \text{or} \quad \frac{1000 \text{ mg}}{1 \text{ g}}$$

mg–tablet conversion factors

$$\frac{250 \text{ mg}}{1 \text{ tablet}} \quad \text{or} \quad \frac{1 \text{ tablet}}{250 \text{ mg}}$$

Choose the conversion factors with the unwanted units—g and mg—in the denominator.

[3] Solve the problem.

- Arrange each term so that the **units in the numerator of one term** *cancel* **the units in the denominator of the adjacent term.** In this problem we need to cancel both grams and milligrams to get tablets.
- The single desired unit, tablets, must be located in the **numerator** of one term.

Tablets do not cancel.

$$1.25 \text{ g} \quad \times \quad \frac{1000 \text{ mg}}{1 \text{ g}} \quad \times \quad \frac{1 \text{ tablet}}{250 \text{ mg}} \quad = \quad 5 \text{ tablets}$$

Grams cancel.   Milligrams cancel.

[4] Check.

- The answer of 5 tablets of amoxicillin (not 0.5 or 50) is reasonable. Because the dose in a single tablet (250 mg) is a fraction of a gram, and the required dose is more than a gram, the answer must be greater than one.

Practice Problem 1.15	A patient is prescribed 0.100 mg of a drug that is available in 25-µg tablets. How many tablets are needed?

More Practice: Try Problems 1.13, 1.59.

Sample Problem 1.11	**Using a Clinical Conversion Factor to Solve a Problem**

A dose of 240 mg of acetaminophen is prescribed for a 20-kg child. How many mL of Children's Tylenol (80. mg of acetaminophen per 2.5 mL) are needed?

Analysis and Solution

[1] Identify the original quantity and the desired quantity.

- We must convert the number of milligrams of acetaminophen needed to the number of mL that must be administered.

$$\underset{\text{original quantity}}{240 \text{ mg}} \qquad \underset{\text{desired quantity}}{? \text{ mL}}$$

[2] Write out the conversion factors.

mg of acetaminophen–mL conversion factors

$$\frac{80. \text{ mg}}{2.5 \text{ mL}} \quad \text{or} \quad \frac{2.5 \text{ mL}}{80. \text{ mg}}$$

Choose the conversion factor to cancel mg.

[3] Solve the problem.

- Arrange the terms so that the **units in the numerator of one term** *cancel* **the units of the denominator of the adjacent term.** In this problem we need to cancel milligrams to obtain milliliters.
- In this problem we are given a fact we don't need to use—the child weighs 20 kg. We can ignore this quantity in carrying out the calculation.

$$240 \text{ mg} \quad \times \quad \frac{2.5 \text{ mL}}{80. \text{ mg}} \quad = \quad 7.5 \text{ mL of Children's Tylenol}$$

Milligrams cancel.

[4] Check.

- The answer of 7.5 mL (not 0.75 or 75) is reasonable. Because the required dose is larger than the dose in 2.5 mL, the answer must be larger than 2.5 mL.

Practice Problem 1.16

(a) How many milliliters are contained in the dose of Children's Tylenol shown in the adjacent photo (1 teaspoon = 5 mL)? (b) If Children's Tylenol contains 80. mg of acetaminophen per 2.5 mL, how much acetaminophen (in mg) is contained in the dose?

Practice Problem 1.17

How many milliliters of Children's Motrin (100 mg of ibuprofen per 5 mL) are needed to give a child a dose of 160 mg?

More Practice: Try Problem 1.60.

Mark Dierker/McGraw Hill

1.9 Temperature

Temperature is a measure of how hot or cold an object is. Three temperature scales are used: **Fahrenheit** (most common in the United States), **Celsius** (most commonly used by scientists and countries other than the United States), and **Kelvin** (Figure 1.8).

The Fahrenheit and Celsius scales are both divided into **degrees.** On the Fahrenheit scale, water freezes at 32 °F and boils at 212 °F. On the Celsius scale, water freezes at 0 °C and boils at 100 °C. To convert temperature values from one scale to another, we use two equations, where T_C is the Celsius temperature and T_F is the Fahrenheit temperature.

To convert from Celsius to Fahrenheit:

$$T_F = \frac{9}{5}(T_C) + 32$$

To convert from Fahrenheit to Celsius:

$$T_C = \frac{5}{9}(T_F - 32)$$

Figure 1.8 Fahrenheit, Celsius, and Kelvin Temperature Scales Compared

Because the freezing point and boiling point of water span 180° on the Fahrenheit scale, but only 100° on the Celsius scale, a Fahrenheit degree and a Celsius degree differ in size. The Kelvin scale is divided into kelvins (K), not degrees. Because the freezing point and boiling point of water span 100 kelvins, one kelvin is the same size as one Celsius degree.

Whenever a calculation requires a specific equation:

① Write down the equation.
② Substitute the known quantities.
③ Solve the equation for the unknown quantity. **If there is more than one unknown, you are using the *wrong* equation.**

The Kelvin scale is divided into **kelvins** (K), not degrees. The only difference between the Kelvin scale and the Celsius scale is the zero point. A temperature of –273 °C corresponds to 0 K. The zero point on the Kelvin scale is called **absolute zero,** the lowest temperature possible. To convert temperature values from Celsius to Kelvin, or vice versa, use two equations.

To convert from Celsius to Kelvin:

$$T_K = T_C + 273.15$$

To convert from Kelvin to Celsius:

$$T_C = T_K - 273.15$$

Sample Problem 1.12

Converting Temperature from One Scale to Another

An infant had a temperature of 104 °F. Convert this temperature to both °C and K.

Analysis

First convert the Fahrenheit temperature to degrees Celsius using the equation $T_C = 5/9(T_F - 32)$. Then convert the Celsius temperature to kelvins by adding 273.15.

Solution

[1] Convert T_F to T_C:

$$T_C = \frac{5}{9}(T_F - 32)$$

$$= \frac{5}{9}(104 - 32) = 40. \, °C$$

[2] Convert T_C to T_K:

$$T_K = T_C + 273.15$$

$$= 40. + 273.15 = 313 \text{ K}$$

Practice Problem 1.18

When the human body is exposed to extreme cold, hypothermia can result and the body's temperature can drop to 28.5 °C. Convert this temperature to T_F and T_K.

Practice Problem 1.19

Convert each temperature to the requested temperature scale.
  a. 20. °C to T_F b. 150 °F to T_C c. 75 °C to T_K

More Practice: Try Problems 1.5c, 1.41c, 1.42c, 1.45.

1.10   Density and Specific Gravity

Two additional quantities used to characterize substances are **density** and **specific gravity.**

1.10A   Density

Density **is a physical property that relates the mass of a substance to its volume.** Density is reported in grams per milliliter (g/mL) or grams per cubic centimeter (g/cc).

$$\text{density} = \frac{\text{mass (g)}}{\text{volume (mL or cc)}}$$

The density of a substance depends on temperature. For most substances, the solid state is more dense than the liquid state, and as the temperature increases, the density decreases. This phenomenon occurs because the volume of a sample of a substance generally increases with temperature but the mass is always constant.

Water is an exception to this generalization. Solid water, ice, is *less* dense than liquid water, and from 0 °C to 4 °C, the density of water *increases*. Above 4 °C, water behaves like other liquids and its density decreases. Thus, water's maximum density of 1.00 g/mL occurs at 4 °C. Some representative densities are reported in Table 1.6.

Table 1.6 Representative Densities at 25 °C

Substance	Density [g/(mL or cc)]
Oxygen (0 °C)	0.001 43
Gasoline	0.66
Ice (0 °C)	0.92
Water (4 °C)	1.00
Urine	1.003–1.030
Blood plasma	1.03
Table sugar	1.59
Bone	1.80

The density (not the mass) of a substance determines whether it floats or sinks in a liquid.

- **A less dense substance floats on a more dense liquid.**

Ice floats on water because it is less dense. When petroleum leaks from an oil tanker or gasoline is spilled when fueling a boat, it floats on water because it is less dense. In contrast, a cannonball or torpedo sinks because it is more dense than water.

Although a can of a diet soft drink floats in water because it is less dense, a can of a regular soft drink that contains sugar is more dense than water so it sinks.
Jill Braaten/McGraw Hill

Sample Problem 1.13   Using Density to Solve a Problem

The density of liquid **A** is twice the density of liquid **B.** (a) If you have an equal mass of **A** and **B,** which graduated cylinder ([1] or [2]) corresponds to **A** and which corresponds to **B?** (b) How do the masses of the liquids in graduated cylinders [2] and [3] compare?

Analysis

Density is the number of grams per milliliter (g/mL) or grams per cubic centimeter (g/cc) of a substance.

Solution

a. If the density of **A** is twice the density of **B,** you need twice the volume of **B** to have the same mass as a sample of **A.** Thus, graduated cylinder [1] represents **A** (gold liquid) and graduated cylinder [2] represents **B** (green liquid).

b. Because graduated cylinders [2] and [3] have equal volumes of **A** and **B** but **A** is twice as dense as **B,** the mass of [3] (**A**) must be twice the mass of [2] (**B**).

Practice Problem 1.20

How does the mass of liquid **A** in cylinder [1] compare with the mass of liquid **B** in cylinder [2] in each case (greater than, less than, or equal to)? (a) The densities of **A** and **B** are the same. (b) The density of **A** is twice the density of **B.** (c) The density of **B** is twice the density of **A.**

More Practice: Try Problem 1.50.

Knowing the density of a liquid allows us to convert the volume of a substance to its mass, or the mass of a substance to its volume.

To convert volume (mL) to mass (g):   **To convert mass (g) to volume (mL):**

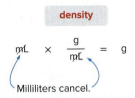

For example, one laboratory synthesis of aspirin uses the liquid acetic acid, which has a density of 1.05 g/mL. If we need 5.0 g for a synthesis, we could use density to convert this mass to a volume that could then be easily measured out using a syringe or pipette.

$$5.0 \text{ g acetic acid} \quad \times \quad \frac{1 \text{ mL}}{1.05 \text{ g}} \quad = \quad 4.8 \text{ mL of acetic acid}$$

Grams cancel.

Sample Problem 1.14	**Using Density and Volume to Determine Mass**

Calculate the mass in grams of 15.0 mL of a saline solution that has a density of 1.05 g/mL.

Analysis

Use density (g/mL) to interconvert the mass and volume of a liquid.

Solution

density

$$15.0 \text{ mL} \quad \times \quad \frac{1.05 \text{ g}}{1 \text{ mL}} \quad = \quad 15.8 \text{ g of saline solution}$$

Milliliters cancel.

The answer, 15.8 g, is rounded to three significant figures to match the number of significant figures in both factors in the problem.

Practice Problem 1.21	Calculate the mass in grams of 10.0 mL of diethyl ether, an anesthetic that has a density of 0.713 g/mL.

Practice Problem 1.22	If a household recycles 10.5 kg of plastic bottles composed of PET in a month, and the density of PET is 1.38 g/mL, what volume in liters does this represent?

More Practice: Try Problems 1.48, 1.49.

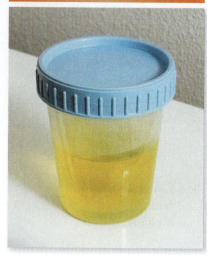

The specific gravity of a urine sample is measured to check if a patient has an imbalance in metabolism. *Nancy R. Cohen/Photodisc/Getty Images*

1.10B   Specific Gravity

Specific gravity is a quantity that compares the density of a substance with the density of water at 4 °C.

$$\text{specific gravity} = \frac{\text{density of a substance (g/mL)}}{\text{density of water (g/mL)}}$$

Unlike most other quantities, specific gravity is a quantity without units, because the units in the numerator (g/mL) cancel the units in the denominator (g/mL). Because the density of water is 1.00 g/mL at 4 °C, **the specific gravity of a substance equals its density, but it contains no units.** For example, if the density of a liquid is 1.5 g/mL at 4 °C, its specific gravity is 1.5.

The specific gravity of urine samples is often measured in a hospital lab. Normal urine has a density in the range of 1.003–1.030 g/mL (Table 1.6), so it has a specific gravity in the range of 1.003–1.030. Consistently high or low values can indicate an imbalance in metabolism. For example, the specific gravity of urine samples from patients with poorly controlled diabetes is abnormally high, because a large amount of glucose is excreted in the urine.

> **Concept Check 1.10**
>
> (a) If the density of a liquid is 0.80 g/mL, what is its specific gravity? (b) If the specific gravity of a substance is 2.3, what is its density?

CHAPTER REVIEW

KEY TERMS

Celsius scale (1.9)

Chemical properties (1.2)

Chemistry (1.1)

Compound (1.3)

Conversion factor (1.7)

Cubic centimeter (1.4)

Density (1.10)

Element (1.3)

English system of measurement (1.4)

Exact number (1.5)

Fahrenheit scale (1.9)

Gas (1.2)

Gram (1.4)

Inexact number (1.5)

Kelvin scale (1.9)

Liquid (1.2)

Liter (1.4)

Mass (1.4)

Matter (1.1)

Meter (1.4)

Metric system (1.4)

Mixture (1.3)

Physical properties (1.2)

Pure substance (1.3)

Scientific notation (1.6)

SI units (1.4)

Significant figures (1.5)

Solid (1.2)

Specific gravity (1.10)

States of matter (1.2)

Temperature (1.9)

Weight (1.4)

KEY CONCEPTS

❶   Three states of matter (1.2)

① Solid	**② Liquid**	**③ Gas**
• The **solid state** is composed of highly organized particles that lie close together.	• The **liquid state** is composed of particles that lie close together but are less organized than the solid state.	• The **gas state** is composed of highly disorganized particles that lie far apart.

2 **Classification of matter (1.3)**

Matter can be classified as a **pure substance** or a **mixture.**

1 Pure substance—Element	**2** Pure substance—Compound	**3** Mixture
• An **element** is a pure substance that cannot be broken down into simpler substances by a chemical reaction.	• A **compound** is a pure substance formed from two or more elements.	• A **mixture** is composed of more than one substance.

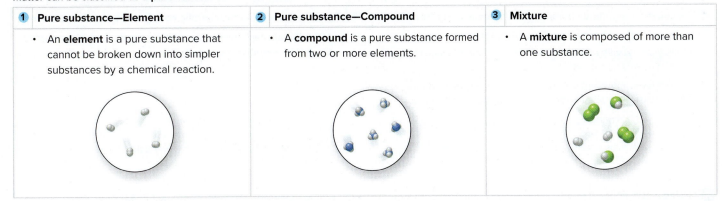

3 **Significant figures (1.5)**

Significant figures are all of the digits in a measured number, including one estimated digit.

1 All nonzero digits are significant.		**2** A zero is significant when it occurs between two nonzero digits, or at the end of a number *with* a decimal point.		**3** A zero is *not* significant at the beginning of a number or at the end of a number *without* a decimal point.	
62.34	62,443	3.02	5,280.	0.05	200
four significant figures	**five** significant figures	**three** significant figures	**four** significant figures	**one** significant figure	**one** significant figure

KEY EQUATIONS

1 Converting temperature from one scale to another (1.9)	**2** Density and specific gravity (1.10)
To convert from Celsius to Fahrenheit: To convert from Fahrenheit to Celsius: $$T_F = \frac{9}{5}(T_C) + 32 \qquad T_C = \frac{5}{9}(T_F - 32)$$ To convert from Celsius to Kelvin: To convert from Kelvin to Celsius: $$T_K = T_C + 273.15 \qquad T_C = T_K - 273.15$$	$$\text{density} = \frac{\text{mass (g)}}{\text{volume (mL or cc)}}$$ $$\text{specific gravity} = \frac{\text{density of a substance (g/mL)}}{\text{density of water (g/mL)}}$$

KEY SKILLS

1 **Rounding off a number (1.5B)**

1 If a number is to have *two* significant figures, look at the *third* significant number from the left. If the third number is 4 or fewer, drop it and all remaining numbers to the right.	**2** If the *third* significant number from the left is 5 or greater, round the number up by adding one to the second digit.

2 **Determining significant figures in a calculation**

1 In multiplication or division, the answer has the same number of significant figures as the original number with the *fewest* significant figures. (1.5B)	**2** In addition and subtraction, the answer has the same number of decimal places as the original number with the *fewest* decimal places. (1.5C)

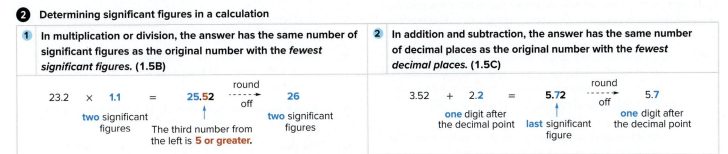

❸ Using scientific notation (1.6)

Scientific notation is a method of writing a number as $y \times 10^x$, where y is a number between 1 and 10, and x is a positive or negative exponent.

❶ **To convert a standard number to a number in scientific notation:**	❷ **To convert a number in scientific notation to a standard number:**
• Move the decimal point to give a number between 1 and 10. • Multiply the result by 10^x, where x is the number of places the decimal point was moved. • When the decimal point is moved to the **left,** x is **positive.** When the decimal point is moved to the **right,** x is **negative.**	• Move the decimal point to the **right** for **positive** exponents, and to the **left** for **negative** exponents.

$$0.000\ 032 \quad = \quad 3.2 \times 10^{-5}$$

Move the decimal point **five** places to the **right.**

$$1.400 \times 10^2 \qquad 1.400 \quad \cdots \cdots \rightarrow \quad 140.0$$

Move the decimal point to the **right two** places.

❹ Using a conversion factor to solve an equation (1.7, 1.8); example: determining how many liters are contained in 4.0 qt of milk

❶ Identify the original quantity and the desired quantity.	❷ Write out the conversion factor that places the unwanted unit in the denominator.	❸ Solve the problem so that the unwanted unit cancels.
4.0 qt ? L original quantity **desired** quantity	$\dfrac{1\ L}{1.057\ qt}$ ← **desired** unit / ← unwanted unit **conversion factor**	$4.0\ \cancel{qt} \quad \times \quad \dfrac{1\ L}{1.057\ \cancel{qt}} \quad = \quad 3.8\ L$ Quarts cancel.

❺ Converting temperature from one scale to another (1.9); example: converting 92 °F to T_C and T_K

❶ Convert the Fahrenheit temperature to degrees Celsius using the equation $T_C = 5/9(T_F - 32)$.	❷ Convert the Celsius temperature to kelvins by adding 273.15.
$T_C = \dfrac{5}{9}(T_F - 32) = \dfrac{5}{9}(92 - 32) = 33\ °C$	$T_K = T_C + 273.15 = 33 + 273.15 = 306\ K$

❻ Using density to solve a problem (1.10A); example: determining the mass of 20.0 mL of an antibiotic solution (density = 1.21 g/mL)

❶ Identify the original quantity and the desired quantity.	❷ Use the density as a conversion factor to place the unwanted unit (mL) in the denominator.	❸ Solve the problem so that the unwanted unit (mL) cancels.
20.0 mL ? g original quantity **desired** quantity	$\dfrac{1.21\ g}{1\ mL}$ ← **desired** unit / ← unwanted unit **conversion factor**	$20.0\ \cancel{mL} \quad \times \quad \dfrac{1.21\ g}{1\ \cancel{mL}} \quad = \quad 24.2\ g$ solution mL cancel.

CHAPTER 1 SELF-TEST

The Self-Test consists of short-answer questions that test an understanding of definitions, equations, and other material encountered within the chapter. Answers to each question are provided at the end of the chapter. For any questions you are unable to answer correctly, revisit the appropriate section(s) of the text; then proceed to working the end-of-chapter problems to further reinforce your understanding of the chapter material.

1. Which term from "Key Terms" answers each question?
   a. Which state of matter consists of close, randomly moving particles that takes on the shape of the container it occupies?
   b. What physical property relates the mass of an object to its volume?
   c. What type of pure substance cannot be broken down into simpler substances by a chemical reaction?
   d. On what temperature scale does water freeze at 0° and boil at 100°?

2. How many significant figures does each number contain: (a) 54,000; (b) 0.067 00; (c) 1,109.0?

3. What term is used for each unit: (a) a thousand seconds; (b) a hundredth of a meter?

4. Which number represents 10,400 written in scientific notation: (a) 1.04×10^{-4}; (b) 1.04×10^4; (c) 1.0400×10^4; (d) 10.4×10^3?

5. Round off each number to two significant figures: (a) 2.7983; (b) 0.010 023 9; (c) 42,980.0

6. The density of liquid **A** is half the density of liquid **B**. What volume of liquid **B** has the same mass as 50 mL of liquid **A**?

7. What is the product of the following calculation, written with the proper number of significant figures; $4.29 \times 0.023 =$: (a) 0.098 67; (b) 0.0987; (c) 0.099?

8. To determine how many liters are contained in 1.0 gal of milk, fill in quantities (a) and (b) needed to solve the following equation.

$$1.0 \text{ gal} \times \frac{4 \text{ qt}}{\text{(a)}} \times \frac{1 \text{ L}}{\text{(b)}} = 3.8 \text{ L}$$

UNDERSTANDING KEY CONCEPTS

1.1 Classify each example of molecular art as a pure element, a pure compound, or a mixture.

a.   c.

b.   d.

1.2 (a) Which representation(s) in Problem 1.1 illustrate a mixture of two elements? (b) Which representation(s) in Problem 1.1 illustrate a mixture of a compound and an element?

1.3 When a chunk of dry ice (solid carbon dioxide) is placed out in the air, the solid gradually disappears and a gas is formed above the solid. Does the molecular art drawn below indicate that a chemical or physical change has occurred? Explain your choice.

solid   gas

1.4 The inexpensive preparation of nitrogen-containing fertilizers begins with mixing together two elements, hydrogen and nitrogen, at high temperature and pressure in the presence of a metal. Does the molecular art depicted below indicate that a chemical or physical change occurs under these conditions? Explain your choice.

metal
heat

1.5

a. What is the temperature on the given Fahrenheit thermometer?
b. How many significant figures does your answer contain?
c. Convert this temperature into T_C.

1.6 (a) What is the length of the given crayon in centimeters? (b) How many significant figures does this value contain? (c) Convert this value to meters, and write the answer in scientific notation.

1.7 The given beaker contains 100 mL of water. Draw an illustration for what would be observed in each circumstance. (a) Hexane (50 mL, density = 0.65 g/mL) is added. (b) Dichloromethane (50 mL, density = 1.33 g/mL) is added.

1.8 (a) What can be said about the density of the liquid in beaker **A**, if the object in the beaker has a density of 2.0 g/cc? (b) What can be said about the density of the liquid in beaker **B**, if the object has a density of 0.90 g/cc?

A   B

1.9 A blood vessel is 0.40 μm in diameter. (a) Convert this quantity to meters and write the answer in scientific notation. (b) Convert this quantity to inches and write the answer in scientific notation.

1.10 On admission to the hospital, a patient weighed 60.6 kg and was 67 in. tall. (a) What is the weight of the patient in pounds? (b) What is the height of the patient in centimeters?

1.11 What is the height in centimeters of a child who is 50. in. tall?

1.12 If the human body has 5.0 qt of blood that contains 2×10^{13} red blood cells, how many red blood cells are present in each μL of blood?

1.13 A woman was told to take a dose of 1.5 g of calcium daily. How many 500-mg tablets should she take?

1.14 The recommended daily calcium intake for a woman over 50 years of age is 1,200 mg. If one cup of milk has 306 mg of calcium, how many cups of milk provide this amount of calcium? (b) How many milliliters of milk does this correspond to? (1 cup = 8 fl oz)

ADDITIONAL PROBLEMS

Matter

1.15 Label each component in the molecular art as an element or a compound.

1.16 (a) Which representation(s) illustrate a mixture of two elements? (b) Which representation(s) illustrate a mixture of a compound and an element?

      A   B   C   D

1.17 Classify each process as a chemical or physical change.
- a. dissolving calcium chloride in water
- b. burning gasoline to power a car
- c. heating wax so that it melts

1.18 Classify each process as a chemical or physical change.
- a. the condensation of water on the outside of a cold glass
- b. mixing a teaspoon of instant coffee with hot water
- c. baking a cake

Measurement

1.19 Which quantity in each pair is larger?
- a. 5 mL or 5 dL
- b. 10 mg or 10 μg
- c. 5 cm or 5 mm
- d. 10 Ms or 10 ms

1.20 Which quantity in each pair is larger?
- a. 10 km or 10 m
- b. 10 L or 10 mL
- c. 10 g or 10 μg
- d. 10 cm or 10 mm

1.21 Label each quantity as an exact or inexact number.
- a. A recipe requires 10 cloves of garlic and two tablespoons of oil.
- b. A child fell and had a 4-cm laceration that required 12 stitches.

1.22 Rank the quantities in each group from smallest to largest.
- a. 100 μL, 100 dL, and 100 mL
- b. 10 g, 100 mg, and 0.1 kg
- c. 1 km, 100 m, and 1,000 cm

Significant Figures

1.23 How many significant figures does each number contain?
- a. 16.00
- b. 160
- c. 0.001 60
- d. 1,600,000
- e. 0.1600
- f. 1.060×10^{10}

1.24 How many significant figures does each number contain?
- a. 160.
- b. 160.0
- c. 0.000 16
- d. 1,600.
- e. 1.060
- f. 1.600×10^{-10}

1.25 Round each number to three significant figures.
- a. 25,401
- b. 1,248,486
- c. 0.001 265 982
- d. 0.123 456

1.26 Round each number in Problem 1.25 to four significant figures.

1.27 Carry out each calculation and report the answer using the proper number of significant figures.
- a. 53.6×0.41
- b. $25.825 - 3.86$
- c. $65.2 \div 12$
- d. $41.0 + 9.135$

1.28 Carry out each calculation and report the answer using the proper number of significant figures.
- a. $49,682 \times 0.80$
- b. $66.815 + 2.82$
- c. $1,000 \div 2.34$
- d. $21 - 0.88$

Scientific Notation

1.29 Write each quantity in scientific notation.
- a. 1,234 g
- b. 0.000 016 2 m
- c. 5,244,000 L
- d. 0.005 62 g

1.30 Write each quantity in scientific notation.
- a. 0.001 25 m
- b. 8,100,000,000 lb
- c. 54,235.6 m
- d. 0.000 001 899 L

1.31 Convert each number to its standard form.
- a. 3.4×10^8
- b. 5.822×10^{-5}
- c. 3×10^2
- d. 6.86×10^{-8}

1.32 Convert each number to its standard form.
- a. 4.02×10^{10}
- b. 2.46×10^{-3}
- c. 6.86×10^9
- d. 1.00×10^{-7}

1.33 Which number in each pair is larger?
- a. 4.44×10^3 or 4.8×10^2
- b. 5.6×10^{-6} or 5.6×10^{-5}
- c. 1.3×10^8 or 52,300,000
- d. 9.8×10^{-4} or 0.000 089

1.34 Rank the numbers in each group from smallest to largest.
- a. 5.06×10^6, 7×10^4, and 2.5×10^8
- b. 6.3×10^{-2}, 2.5×10^{-4}, and 8.6×10^{-6}

1.35 Write the recommended daily intake of each nutrient in scientific notation.
- a. 0.000 400 g of folate
- b. 0.002 g of copper
- c. 0.000 080 g of vitamin K
- d. 3,400 mg of chloride

1.36 A picosecond is one trillionth of a second (0.000 000 000 001 s). (a) Write this number in scientific notation. (b) How many picoseconds are there in one second? Write this answer in scientific notation.

Problem Solving Using Conversion Factors

1.37 Carry out each of the following conversions.
- a. 300 g to mg
- b. 2 L to μL
- c. 5.0 cm to m
- d. 2 ft to m

1.38 Carry out each of the following conversions.

a. 25 μL to mL

b. 35 kg to g

c. 300 mL to qt

d. 3 cups to L

1.39

a. What is the volume of liquid contained in the given 3-mL syringe?

b. Convert this value to liters and write the answer in scientific notation.

1.40 What is the mass in kilograms of an individual whose weight in pounds is shown on the given scale?

1.41 Carry out each of the following conversions.

a. What is the mass in kilograms of an individual who weighs 234 lb?

b. A patient required 3.0 pt of blood during surgery. How many liters does this correspond to?

c. A patient had a body temperature of 37.7 °C. What is his body temperature in T_F?

1.42 Carry out each of the following conversions.

a. What is the mass in pounds of an individual who weighs 53.2 kg?

b. What is the height in inches of a child who is 90. cm tall?

c. A patient had a body temperature of 103.5 °F. What is his body temperature in T_C?

1.43 The concentration of mercury, a toxic pollutant, in some yellowfin tuna is 354 mg in 1.0×10^6 g of fish. The Food and Drug Administration recommends that women who are pregnant should eat no more than 6.0 oz of yellowfin tuna per week. How much mercury (in milligrams) would an individual ingest in one serving of fish?

Nishihama/Shutterstock

1.44 Honeydew melons are high in potassium, a nutrient needed for proper fluid balance [875 mg in one serving (one-quarter of a medium melon)]. If an individual eats one serving each day for a week, how many grams of potassium are obtained?

Brent Hofacker/Shutterstock

Temperature

1.45 Carry out each of the following temperature conversions.

a. An over-the-counter pain reliever melts at 53 °C. Convert this temperature to T_F and T_K.

b. A cake is baked at 350. °F. Convert this temperature to T_C and T_K.

1.46 Which temperature in each pair is higher?

a. −10 °C or 10 °F

b. −50 °C or −50 °F

Density and Specific Gravity

1.47 If a urine sample has a mass of 122 g and a volume of 121 mL, what is its density in g/mL?

1.48 The density of sucrose, table sugar, is 1.56 g/cc. What volume (in cubic centimeters) does 20.0 g of sucrose occupy?

1.49 Isooctane is a high-octane component of gasoline. If the density of isooctane is 0.692 g/mL, what is the mass of 220 mL of gasoline?

1.50 A graduated cylinder contains three liquids: water (density = 1.00 g/mL), corn syrup (density = 1.37 g/mL), and corn oil (density = 0.93 g/mL). (a) Identify the liquid that corresponds to **A, B,** and **C** in the graduated cylinder. (b) How many grams of each liquid are present?

1.51 Which is the upper layer when each of the following liquids is added to water?

a. olive oil (density = 0.92 g/mL)

b. chloroform (density = 1.49 g/mL)

1.52 (a) What is the specific gravity of mercury, the liquid used in thermometers, if it has a density of 13.6 g/mL? (b) What is the density of ethanol if it has a specific gravity of 0.789?

Applications

1.53 A lab test showed an individual's cholesterol level to be 186 mg/dL. (a) Convert this quantity to g/dL. (b) Convert this quantity to mg/L.

1.54 Hemoglobin is a protein that transports oxygen from the lungs to the rest of the body. Lab results indicated a patient had a hemoglobin concentration in the blood of 15.5 g/dL, which is in the normal range. (a) Convert the number of grams to milligrams and write the answer in scientific notation. (b) Convert the number of grams to micrograms and write the answer in scientific notation.

1.55 Liposuction is a cosmetic procedure used to remove excess fat from the abdomen, thigh, or buttocks of a patient. (a) If 2.0 L of fat (density = 0.94 g/mL) is removed, what is the mass (in kg) of this fat? (b) How many pounds of fat have been removed?

1.56 A single 1-oz serving of tortilla chips contains 250 mg of sodium. If an individual ate the entire 13-oz bag, how many grams of sodium would he ingest? If the recommended daily intake of sodium is 2.4 g, does this provide more or less than the recommended daily value, and by how much?

1.57 If a person weighs 232 lb and has 25% body fat, how many kilograms of body fat does this individual possess?

1.58 The average nicotine content of a Camel cigarette is 1.93 mg. (a) Convert this quantity to both grams and micrograms. (b) Nicotine patches, which are used to help quit smoking, release nicotine into the body by absorption through the skin. The patches come with different amounts of nicotine. A smoker begins with the amount of nicotine that matches his typical daily intake. The maximum amount of nicotine in one brand of patch supplies a smoker with 21 mg of nicotine per day. If an individual smoked one pack of 20 Camel cigarettes each day, would a smoker get more or less nicotine per day using this patch?

1.59 A patient is prescribed 2.0 g of a medication to be taken four times a day. If the medicine is available in 500.-mg tablets, how many tablets are needed in a 24-hour period?

1.60 Maalox is the trade name for an antacid and antigas medication used for relief of heartburn, bloating, and acid indigestion. Each 5-mL portion of Maalox contains 400 mg of aluminum hydroxide, 400 mg of magnesium hydroxide, and 40 mg of simethicone. If the recommended dose is two teaspoons four times a day, how many grams of each substance would an individual take in a 24-hour period? (1 teaspoon = 5 mL)

CHALLENGE PROBLEMS

1.61 A soccer player weighed 70.7 kg before a match, drank 1.8 L of liquid (density 1.05 g/mL) during the match, and weighed 69.3 kg after the match. How many pounds of sweat did the soccer player lose?

1.62 A patient receives an intravenous (IV) solution that flows at the rate of 150 mL per hour. (a) How much fluid does the patient receive in 20. min? (b) How long does it take for the patient to receive 90. mL of fluid? (c) If the IV bag holds 600. mL of fluid, how many minutes does it take to empty the bag? (d) If the solution contains 90. mg of glucose per mL, how long will it take to give the patient 2.0 g of glucose?

BEYOND THE CLASSROOM

1.63 A urinometer is a device for measuring the specific gravity of a urine sample. Values above or below the normal range (1.003–1.030) may result from a variety of conditions. Research how each of the following might affect the specific gravity of urine: (a) a high fever; (b) taking a diuretic (a drug that increases urine output); (c) diabetes; (d) dehydration after sustained exercise.

1.64 A typical soda contains 39 g of carbohydrates in a 12-oz can, whereas a diet soda with aspartame contains 180 mg of artificial sweetener. Estimate how many cans of regular or diet soft drinks your household consumes in a 30-day month. Determine how many grams of carbohydrates or milligrams of artificial sweetener this corresponds to. Compare your results with others in the class.

CHAPTER 1 ANSWERS

Concept Check Answers

1.1 a,d: physical change;   b,c,e: chemical change

1.2 chemical change

1.3 a. element;   b,c,d: compound

1.4 a. megaliter   b. millisecond   c. centigram   d. deciliter

1.5 a. one microgram   c. one nanosecond
b. one gigameter   d. one centigram

1.6 1 L

1.7 a. 3 cL   b. 1 μg   c. 5 km   d. 2 mL

1.8 a. 543,000, 5.43×10^5
b. 0.000 004 43, 4.43×10^{-6}
c. 74,000, 7.4×10^4

1.9 a. 0.6214 mi/1 km   and   1 km/0.6214 mi
b. 1000 mm/1 m   and   1 m/1000 mm

1.10 a. 0.80   b. 2.3 g/mL

Practice Problem Answers

1.1   a. pure substance   b. mixture

1.2   a,b,c: mixture   d. pure substance

1.3   a. 4   b. 2   c. 3   d. 3   e. 4   f. 6

1.4   Significant zeros are shown in **red**.
a. 0.003 **0**4   b. 26,**0**45   c. 1,**000,0**34   d. 0.3**04 00**

1.5   a. 1.3   b. 0.0025   c. 3,800

1.6   a. 37   b. 0.000 007 93   c. 32   d. 3,100

1.7   a. 28.0 cm   b. 127.5 mL   c. 30. mg   d. 2.08 s

1.8 9.8×10^{-5} g/dL

1.9   a. 9.32×10^{4} c. 6.78×10^{6}
b. 7.25×10^{-4} d. 3.0×10^{-5}

1.10 a. 6,500   b. 0.000 032 6   c. 0.037 80   d. 104,000,000

1.11 a. 250 dL   b. 1,130 g   c. 81 cm   d. 100 mm

1.12 3.1 kg

1.13 a. 0.46 mL   b. 460 μL

1.14 a. 1.90 km   b. 0.7 L   c. 140 cm

1.15 four tablets

1.16 a. 13 mL   b. 420 mg

1.17 8 mL

1.18 83.3 °F or 302 K

1.19 a. 68 °F   b. 66 °C   c. 348 K

1.20 a,b,c: mass **B** > mass **A**

1.21 7.13 g

1.22 7.61 L

Self-Test Answers

1.   a. liquid   b. density   c. element   d. Celsius

2.   a. 2   b. 4   c. 5

3.   a. kilosecond   b. centimeter

4.   b.

5.   a. 2.8   b. 0.010   c. 43,000

6.   25 mL

7.   c.

8.   a. 1 gal   b. 1.057 qt

Odd-Numbered Problem Answers

1.1   a. pure element   c,d: mixture
b. pure compound

1.3   physical change

1.5   a. 76.5 °F   b. three   c. 24.7 °C

1.7   a. Hexane (50 mL) is on top of water.
b. Water (100 mL) is on top of dichloromethane.

1.9   a. 4.0×10^{-7} m   b. 1.6×10^{-5} in.

1.11 130 cm

1.13 three

1.15 element: **A, D**
compound: **B, C**

1.17 a,c: physical   b. chemical

1.19 a. 5 dL   b. 10 mg   c. 5 cm   d. 10 Ms

1.21 a. 10 cloves: exact number; 2 tablespoons: inexact number
b. 4 cm: inexact number; 12 stitches: exact number

1.23 a. 4   b. 2   c. 3   d. 2   e. 4   f. 4

1.25 a. 25,400   b. 1,250,000   c. 0.001 27   d. 0.123

1.27 a. 22   b. 21.97   c. 5.4   d. 50.1

1.29 a. 1.234×10^{3} g   c. 5.244×10^{6} L
b. 1.62×10^{-5} m   d. 5.62×10^{-3} g

1.31 a. 340,000,000   c. 300
b. 0.000 058 22   d. 0.000 000 068 6

1.33 a. 4.44×10^{3} b. 5.6×10^{-5} c. 1.3×10^{8} d. 9.8×10^{-4}

1.35 a. 4.00×10^{-4} g   c. 8.0×10^{-5} g
b. 2×10^{-3} g   d. 3.4×10^{3} mg

1.37 a. 300,000 mg   b. 2,000,000 μL   c. 0.050 m   d. 0.6 m

1.39 a. 1.4 mL   b. 1.4×10^{-3} L

1.41 a. 106 kg   b. 1.4 L   c. 99.9 °F

1.43 0.060 mg

1.45 a. 127 °F or 326 K   b. 177 °C or 450 K

1.47 1.01 g/mL

1.49 152 g rounded to 150 g

1.51 a. olive oil   b. water

1.53 a. 0.186 g/dL   b. 1,860 mg/L

1.55 a. 1.9 kg   b. 4.2 lb

1.57 26 kg

1.59 16 tablets

1.61 7.3 lb

Design Credits: DNA molecule structure background for Chapter Opener: Watchara/Shutterstock

2 Atoms and the Periodic Table

Most food and beverage cans are made from the element aluminum. Aluminum cans can be recycled an unlimited number of times.
Daniel C. Smith

CHAPTER OUTLINE

CHAPTER GOALS

In this chapter you will learn how to:

1. Identify an element by its symbol and classify it as a metal, nonmetal, or metalloid
2. Describe the basic parts of an atom
3. Distinguish isotopes
4. Describe the basic features of the periodic table
5. Understand the electronic structure of an atom
6. Write an electronic configuration for an element in periods 1–3
7. Relate the location of an element in the periodic table to its number of valence electrons
8. Draw an electron-dot symbol for an atom
9. Use the periodic table to predict the relative size and ionization energy of atoms

Examine the ingredients listed on a box of crackers. They may include flour, added vitamins, sugar for sweetness, a natural or synthetic coloring agent, baking soda, salt for flavor, and BHT as a preservative. No matter how simple or complex each of these substances is, it is composed of the basic building block, the **atom.** The word *atom* comes from the Greek word *atomos* meaning "unable to cut." In Chapter 2, we examine the structure and properties of atoms, the building blocks that comprise all forms of matter.

Elements are named for people, places, and things. For example, *carbon* (C) comes from the Latin word *carbo,* meaning *coal* or *charcoal; neptunium* (Np) was named for the planet Neptune; *einsteinium* (Es) was named for scientist Albert Einstein; and *californium* (Cf) was named for the state of California.

2.1 Elements

You were first introduced to elements in Section 1.3.

> • An *element* is a pure substance that cannot be broken down into simpler substances by a chemical reaction.

Of the 118 elements currently known, 90 are naturally occurring and the remaining 28 have been prepared by scientists in the laboratory. Some elements, like oxygen in the air we breathe and aluminum in a soft drink can, are familiar to you, while others, like samarium and seaborgium, are probably not. An alphabetical list of all elements appears in Appendix A.

Each element is identified by a one- or two-letter symbol. The element carbon is symbolized by the single letter **C,** whereas the element chlorine is symbolized by **Cl.** When two letters are used in the element symbol, the first is upper case and the second is lower case. Thus, **Co** refers to the element cobalt, but CO is carbon monoxide, which is composed of the elements carbon (C) and oxygen (O). Table 2.1 lists common elements and their symbols.

While most element symbols are derived from the first one or two letters of the element name, 11 elements have symbols derived from Latin or German origins. Table 2.2 lists these elements and their symbols.

ENVIRONMENTAL NOTE

Carbon monoxide (CO), formed in small amounts during the combustion of fossil fuels like gasoline, is a toxic component of the smoggy air in many large cities. Carbon monoxide contains the elements carbon and oxygen. We will learn about carbon monoxide in Section 10.10.
Trekandshoot/iStock/Getty Images

Concept Check 2.1

Give the symbol for each element.

a. calcium, a nutrient needed for strong teeth and bones

b. radon, a radioactive gas produced in the soil

c. nitrogen, the main component of the earth's atmosphere

d. gold, a precious metal used in coins and jewelry

Table 2.1 Common Elements and Their Symbols

Element	Symbol	Element	Symbol
Bromine	Br	Magnesium	Mg
Calcium	Ca	Manganese	Mn
Carbon	C	Molybdenum	Mo
Chlorine	Cl	Nitrogen	N
Chromium	Cr	Oxygen	O
Cobalt	Co	Phosphorus	P
Copper	Cu	Potassium	K
Fluorine	F	Sodium	Na
Hydrogen	H	Sulfur	S
Iodine	I	Zinc	Zn
Lead	Pb		

Table 2.2 Unusual Element Symbols

Element	Symbol
Antimony	Sb (stibium)
Copper	Cu (cuprum)
Gold	Au (aurum)
Iron	Fe (ferrum)
Lead	Pb (plumbum)
Mercury	Hg (hydrargyrum)
Potassium	K (kalium)
Silver	Ag (argentum)
Sodium	Na (natrium)
Tin	Sn (stannum)
Tungsten	W (wolfram)

Concept Check 2.2	Give the name corresponding to each element symbol: (a) Ne; (b) S; (c) I; (d) Si; (e) B; (f) Hg.

2.1A Elements and the Periodic Table

A periodic table appears in Appendix B for easy reference.

Long ago it was realized that groups of elements have similar properties, and that these elements could be arranged in a schematic way called the **periodic table** (Figure 2.1). The position of an element in the periodic table tells us much about its chemical properties.

Figure 2.1   The Periodic Table of the Elements

metal	metalloid	nonmetal

- **Metals** are shiny substances that conduct heat and electricity. Metals are ductile, meaning they can be drawn into wires, and malleable, meaning they can be hammered into shapes.
- **Metalloids** have properties intermediate between metals and nonmetals.
- **Nonmetals** are poor conductors of heat and electricity.

The elements in the periodic table are divided into three categories—**metals, nonmetals,** and **metalloids.** The solid line that begins with boron (B) and angles in steps down to tellurium (Te) marks the three regions corresponding to these groups. All metals are located to the *left* of the line. All nonmetals except hydrogen are located to the *right*. Metalloids are located along the steps.

- *Metals* are shiny materials that are good conductors of heat and electricity. All metals are solids at room temperature except for mercury, which is a liquid.
- *Nonmetals* do not have a shiny appearance, and they are generally poor conductors of heat and electricity. Nonmetals like sulfur and carbon are solids at room temperature; bromine is a liquid; and nitrogen, oxygen, and nine other elements are gases.
- *Metalloids* have properties intermediate between metals and nonmetals. Only six elements are categorized as metalloids: boron (B), silicon (Si), germanium (Ge), arsenic (As), antimony (Sb), and tellurium (Te).

Concept Check 2.3	Locate each element in the periodic table and classify it as a metal, nonmetal, or metalloid.

| a. titanium | c. krypton | e. arsenic | g. selenium |
| b. chlorine | d. palladium | f. cesium | h. osmium |

2.1B FOCUS ON THE HUMAN BODY
The Elements of Life

Because living organisms selectively take up elements from their surroundings, the abundance of elements in the human body is very different from the distribution of elements in the earth's crust. **Four nonmetals—oxygen, carbon, hydrogen, and nitrogen—comprise 96% of the mass of the human body, and are called the** *building-block elements* (Figure 2.2). Hydrogen and oxygen are the elements that form water, the most prevalent substance in the body. Carbon, hydrogen, and oxygen are found in the four main types of biological molecules—proteins, carbohydrates, lipids, and nucleic acids. Proteins and nucleic acids contain the element nitrogen as well. These biological molecules are discussed in Chapters 14–17.

Figure 2.2   The Elements of Life

Building-Block Elements

Oxygen (O)
Carbon (C)
Hydrogen (H)
Nitrogen (N)

These four elements compose almost 96% of the mass of the human body. Muscle tissue contains all four building-block elements.

Trace Elements

Arsenic (As)	Fluorine (F)	Nickel (Ni)
Boron (B)	Iodine (I)	Selenium (Se)
Chromium (Cr)	Iron (Fe)	Silicon (Si)
Cobalt (Co)	Manganese (Mn)	Zinc (Zn)
Copper (Cu)	Molybdenum (Mo)	

Each trace element is present in less than 0.1% by mass. A small quantity (15 mg or less) of each element is needed in the daily diet.

Major Minerals

Potassium (K), sodium (Na), and chlorine (Cl) are present in body fluids.

Magnesium (Mg) and sulfur (S) are present in the proteins found in muscle.

Calcium (Ca) and phosphorus (P) are present in teeth and bones.

Each major mineral is present in 0.1–2% by mass. At least 100 mg of each mineral is needed in the daily diet.

Seven other elements, called the **major minerals** or **macronutrients,** are also present in the body in much smaller amounts (0.1–2% by mass). Sodium, potassium, and chlorine are present in body fluids. Magnesium and sulfur occur in proteins, and calcium and phosphorus are present in teeth and bones. Phosphorus is also contained in all nucleic acids, such as the DNA that transfers genetic information from one generation to another. At least 100 mg of each macronutrient is needed in the daily diet.

Many breakfast cereals are fortified with iron to provide the consumer with this essential micronutrient.

Many other elements occur in very small amounts in the body, but are essential to good health. These **trace elements** or **micronutrients** are required in the daily diet in small quantities, usually less than 15 mg. Each trace element has a specialized function that is important for proper cellular function. For example, iron is needed for hemoglobin, the protein that carries oxygen in red blood cells, and myoglobin, the protein that stores oxygen in muscle. Zinc is needed for the proper functioning of many enzymes in the liver and kidneys, and iodine is needed for proper thyroid function. Although most of the trace elements are metals, nonmetals like fluorine and selenium are micronutrients as well.

Concept Check 2.4

Classify each micronutrient in Figure 2.2 as a metal, nonmetal, or metalloid.

2.1C   Compounds

In Section 1.3 we learned that a *compound* **is a pure substance formed by chemically combining two or more elements together.** Element symbols are used to write chemical formulas for compounds.

- A *chemical formula* uses element symbols to show the identity of the elements forming a compound and subscripts to show the ratio of atoms (the building blocks of matter) contained in the compound.

For example, table salt is formed from sodium (Na) and chlorine (Cl) in a ratio of 1:1, so its formula is $NaCl$. Water, on the other hand, is formed from two hydrogens for each oxygen, so its formula is H_2O. The subscript "1" is understood when no subscript is written. Other examples of chemical formulas are shown below.

H_2O — 2 H's for each O
water

CO_2 — 2 O's for each C
carbon dioxide (dry ice)

C_3H_8 — 3 C's — 8 H's
propane

As we learned in Section 1.2, molecular art will often be used to illustrate the composition and state of elements and compounds. Color-coded spheres, shown in Figure 2.3, are used to identify the common elements that form compounds.

Figure 2.3   Common Element Colors Used in Molecular Art

C   H   O   N   F   Cl   Br   I   S   P

For example, a red sphere is used for the element oxygen and gray is used for the element hydrogen, so H_2O is represented as a red sphere joined to two gray spheres. Sometimes the spheres will be connected by "sticks" to generate a **ball-and-stick** representation for a compound. At other times, the spheres will be drawn close together to form a **space-filling** representation. No matter how the spheres are depicted, H_2O always consists of one red sphere for the oxygen and two gray spheres for the two hydrogens.

No matter how it is depicted, water is a compound that contains two hydrogens for each oxygen. *Pixtal/SuperStock*

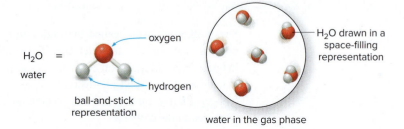

H_2O =
water

oxygen

hydrogen

ball-and-stick representation

H_2O drawn in a space-filling representation

water in the gas phase

Identifying Elements in Molecular Art

Identify the elements used in each example of molecular art.

a.

b.

Analysis

Use Figure 2.3 to determine the identity of the color-coded spheres.

Solution

a. The blue spheres in this space-filling representation correspond to the element nitrogen and the red spheres correspond to the element oxygen. Thus, one "particle" contains two nitrogens, one contains two oxygens, and two "particles" contain one oxygen and one nitrogen.

b. This ball-and-stick representation contains the elements of carbon (black), nitrogen (blue), and hydrogen (gray).

Identify the elements used in each example of molecular art.

a.   b.   c.

Identify the elements in each chemical formula, and give the number of atoms of each element.

a. NaCN (sodium cyanide)   c. C_2H_6 (ethane)   e. CO (carbon monoxide)

b. H_2S (hydrogen sulfide)   d. SnF_2 (stannous fluoride)   f. $C_3H_8O_3$ (glycerol)

Halothane is an inhaled general anesthetic, commonly used since the 1950s. Identify the elements in the ball-and-stick representation of halothane.

halothane

More Practice: Try Problems 2.1, 2.2, 2.16.

2.2   Structure of the Atom

All matter is composed of the same basic building blocks called *atoms*. An atom is much too small to be seen even by the most powerful light microscopes. The period at the end of this sentence holds about 1×10^8 atoms, and a human cheek cell contains about 1×10^{16} atoms. An atom is composed of three subatomic particles.

- A proton, symbolized by p, has a positive (+) charge.
- An electron, symbolized by e⁻, has a negative (−) charge.
- A neutron, symbolized by n, has no charge.

Protons and neutrons have approximately the same, exceedingly small, mass as shown in Table 2.3. The mass of an electron is much less, 1/1,836 the mass of a proton. These subatomic particles are not evenly distributed in the volume of an atom. There are two main components of an atom.

- The *nucleus* is a dense core that contains the protons and neutrons. Most of the mass of an atom resides in the nucleus.
- The *electron cloud* is composed of electrons that move rapidly in the almost empty space surrounding the nucleus. The electron cloud comprises most of the volume of an atom.

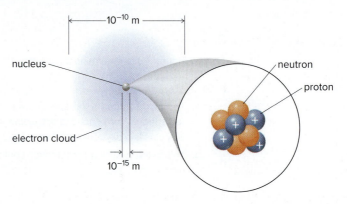

main components of an atom

While the diameter of an atom is about 10^{-10} m, the diameter of a nucleus is only about 10^{-15} m. For a macroscopic analogy, if the nucleus were the size of a baseball, an atom would be the size of Yankee Stadium!

Table 2.3 **Summary: The Properties of the Three Subatomic Particles**

Subatomic Particle	Charge	Mass (g)	Mass (amu)
Proton	+1	1.6726×10^{-24}	1
Neutron	0	1.6749×10^{-24}	1
Electron	−1	9.1094×10^{-28}	Negligible

The charged particles of an atom can either attract or repel each other.

- Opposite charges attract whereas like charges repel each other.

Positive charges repel.

Thus, two electrons or two protons repel each other, whereas a proton and an electron attract each other.

Negative charges repel.

Because the mass of an individual atom is so small (on the order of 10^{-24} g), chemists use a standard mass unit, the **atomic mass unit,** which defines the mass of individual atoms relative to a standard mass.

Opposite charges attract.

- One atomic mass unit (amu) equals one-twelfth the mass of a carbon atom that has six protons and six neutrons; 1 amu = 1.661×10^{-24} g.

Using this scale, one proton has a mass of 1.0073 amu, a value typically rounded to 1 amu. One neutron has a mass of 1.0087 amu, a value also typically rounded to 1 amu. The mass of an electron is so small that it is ignored.

The element lithium is found in many consumer products, from long-lasting lithium batteries to the prescription medication lithium carbonate, used to treat individuals with bipolar disorder.
(top): Jill Braaten; (bottom): Charles D. Winters/Timeframe Photography/McGraw Hill

Every atom of a given type of element always has the *same* **number of protons in the nucleus, a value called the** *atomic number,* symbolized by **Z.** Conversely, two *different* elements have *different* atomic numbers.

- The *atomic number* (**Z**) = the number of protons in the nucleus of an atom.

Thus, the element hydrogen has one proton in its nucleus, so its atomic number is one. Lithium has three protons in its nucleus, so its atomic number is three. **The periodic table is arranged in order of increasing atomic number beginning at the upper left-hand corner.** The atomic number appears just above the element symbol for each entry in the table.

Because a neutral atom has no overall charge:

- **Z** = the number of protons in the nucleus = the number of electrons.

Thus, the atomic number tells us *both* the number of protons in the nucleus and the number of electrons in the electron cloud of a neutral atom.

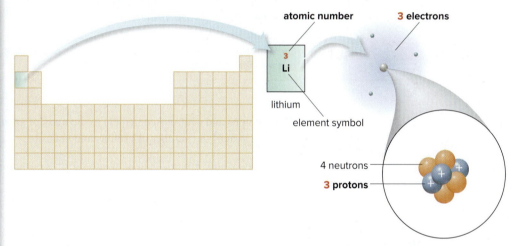

Sample Problem 2.2	Determining the Atomic Number and the Identity of an Element from Molecular Art

For the given atom: (a) determine the number of protons, neutrons, and electrons in the neutral atom; (b) give the atomic number; (c) identify the element.

Analysis

- The **number of protons** = the number of positively charged particles = the **atomic number.**
- The **number of neutrons** = the number of uncharged particles.
- The **number of protons** = the **number of electrons** in a neutral atom.
- **The atomic number determines the identity of an element.**

Solution

a. The element contains two protons and two neutrons. Because a neutral atom has the same number of protons and electrons, the element has two electrons.

b. The two protons give the element an atomic number of two.

c. The element with two protons in the nucleus is **helium.**

Practice Problem 2.4

An element has nine protons and 10 neutrons in the neutral atom. (a) How many electrons are present in the neutral atom? (b) What is the atomic number of this element? (c) Identify the element.

More Practice: Try Problems 2.3a, b, d; 2.4a, b, d.

Sample Problem 2.3

Using the Atomic Number to Determine the Number of Protons and Electrons

Identify the element that has an atomic number of 19, and give the number of protons and electrons in the neutral atom.

Analysis

The atomic number is unique to an element and tells the number of protons in the nucleus and the number of electrons in the electron cloud of a neutral atom.

Solution

According to the periodic table, the element potassium has atomic number 19. A neutral potassium atom has 19 protons and 19 electrons.

Practice Problem 2.5

Identify the element with each atomic number, and give the number of protons and electrons in the neutral atom: (a) 2; (b) 11; (c) 20; (d) 47; (e) 78.

Both protons and neutrons contribute to the mass of an atom. The **mass number,** symbolized by A, is the sum of the number of protons and neutrons.

- Mass number (A) = the number of protons (Z) + the number of neutrons.

For example, a fluorine atom with nine protons and 10 neutrons in the nucleus has a mass number of 19.

Sample Problem 2.4

Determining the Atomic Number, Mass Number, and the Identity of an Element

For the given atom: (a) determine the number of protons and neutrons; (b) give the atomic number and the mass number; (c) identify the element.

Analysis

- The number of protons = the number of positively charged particles = the atomic number. **The atomic number determines the identity of an element.**
- The number of neutrons = the number of uncharged particles.
- **The mass number = the number of protons + the number of neutrons.**

Solution

   a. The element contains six protons and seven neutrons.

   b. The six protons give the element an atomic number of six. The mass number = the number of protons + the number of neutrons = 6 + 7 = 13.

   c. The element with six protons in the nucleus is **carbon.**

More Practice: Try Problems 2.3, 2.4.

| Practice Problem 2.6 | For the given atom: (a) determine the number of protons and neutrons; (b) give the atomic number and the mass number; (c) identify the element. |

| Sample Problem 2.5 | **Using the Atomic Number and Mass Number to Determine the Number of Protons, Neutrons, and Electrons** |

How many protons, neutrons, and electrons are contained in an atom of argon, which has an atomic number of 18 and a mass number of 40?

Analysis

- In a neutral atom, the **atomic number (Z) = the number of protons = the number of electrons.**
- The **mass number (A) = the number of protons + the number of neutrons.**

Solution

The atomic number of 18 means that argon has 18 protons and 18 electrons. To find the number of neutrons, subtract the atomic number (Z) from the mass number (A).

$$
\begin{aligned}
\text{number of neutrons} &= \text{mass number} - \text{atomic number} \\
&= \quad\quad 40 \quad\quad - \quad\quad 18 \\
&= \quad\quad 22 \text{ neutrons}
\end{aligned}
$$

| Practice Problem 2.7 | How many protons, neutrons, and electrons are contained in each atom with the given atomic number and mass number? |

    a. $Z = 17, A = 35$ b. $Z = 14, A = 28$ c. $Z = 92, A = 238$

| Practice Problem 2.8 | What is the mass number of an atom that contains |

    a. 42 protons, 42 electrons, and 53 neutrons?   b. 24 protons, 24 electrons, and 28 neutrons?

More Practice: Try Problems 2.21, 2.22.

2.3 Isotopes

Two atoms of the same element always have the same number of protons, but the number of neutrons can vary.

- *Isotopes* are atoms of the same element having a different number of neutrons.

2.3A Isotopes, Atomic Number, and Mass Number

Most elements in nature exist as a mixture of isotopes. For example, all atoms of the element chlorine contain 17 protons in the nucleus, but some of these atoms have 18 neutrons in the nucleus and some have 20 neutrons. Thus, chlorine has two isotopes with different mass numbers, 35 and 37. These isotopes are often referred to as chlorine-35 (or Cl-35) and chlorine-37 (or Cl-37).

An **isotope symbol** is also written using the element symbol with the atomic number as a subscript and the mass number as a superscript, both to the left.

Two isotopes of the element chlorine

mass number $^{35}_{17}\text{Cl}$ $^{37}_{17}\text{Cl}$
atomic number

chlorine-**35**   chlorine-**37**

Because all chlorine atoms have the same atomic number (17), the subscript is sometimes omitted. Thus, chlorine-35 and chlorine-37 can be written as ^{35}Cl and ^{37}Cl, respectively.

The element hydrogen has three isotopes. Most hydrogen atoms have one proton and no neutrons, giving them a mass number of one. About 1% of hydrogen atoms have one proton and one neutron, giving them a mass number of two. This isotope is called **deuterium,** and it is often symbolized as **D.** An even smaller number of hydrogen atoms contain one proton and two neutrons, giving them a mass number of three. This isotope is called **tritium,** often symbolized as **T.**

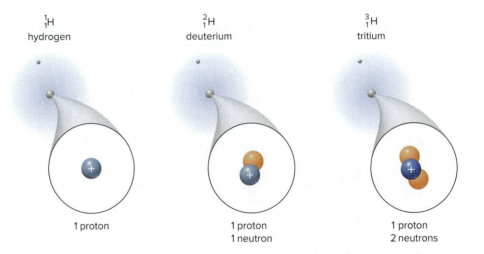

$^{1}_{1}\text{H}$
hydrogen

$^{2}_{1}\text{H}$
deuterium

$^{3}_{1}\text{H}$
tritium

1 proton

1 proton
1 neutron

1 proton
2 neutrons

Sample Problem 2.6	**Using an Isotope Symbol to Determine the Numbers of Protons, Neutrons, and Electrons**

For each atom give the following information: [1] the atomic number; [2] the mass number; [3] the number of protons; [4] the number of neutrons; [5] the number of electrons.

    a. $^{118}_{50}\text{Sn}$ b. ^{195}Pt

Analysis

- The **superscript gives the mass number** and the **subscript gives the atomic number** for each element. When the subscript is omitted, use the identity of the element to determine the atomic number.
- The **atomic number = the number of protons = the number of electrons.**
- The **mass number = the number of protons + the number of neutrons.**

Solution

	Atomic Number	Mass Number	Number of Protons	Number of Neutrons	Number of Electrons
a. $^{118}_{50}\text{Sn}$	50	118	50	118 − 50 = 68	50
b. ^{195}Pt	78	195	78	195 − 78 = 117	78

Practice Problem 2.9	For each atom give the following information: [1] the atomic number; [2] the mass number; [3] the number of protons; [4] the number of neutrons; [5] the number of electrons.

    a. $^{13}_{6}\text{C}$ b. ^{121}Sb

Eloxatin is a platinum-containing drug used to treat cancer patients (Sample Problem 2.6, Practice Problem 2.10).
Dr P. Marazzi/Science Source

Practice Problem 2.10

Write an isotope symbol for the isotope of platinum that contains 116 neutrons.

More Practice: Try Problems 2.27, 2.28.

Sample Problem 2.7 — Determining the Number of Neutrons in an Isotope

Determine the number of neutrons in each isotope: (a) carbon-14; (b) ^{81}Br.

Analysis

- **The identity of the element tells us the atomic number.**
- **The number of neutrons = mass number (A) − atomic number (Z).**

Solution

a. Carbon's atomic number (Z) is 6. Carbon-14 has a mass number (A) of 14.

$$\text{number of neutrons} = A - Z$$
$$= 14 - 6 = 8 \text{ neutrons}$$

b. Bromine's atomic number is 35 and the mass number of the given isotope is 81.

$$\text{number of neutrons} = A - Z$$
$$= 81 - 35 = 46 \text{ neutrons}$$

Practice Problem 2.11

Magnesium has three isotopes that contain 12, 13, and 14 neutrons. For each isotope give the following information: (a) the number of protons; (b) the number of electrons; (c) the atomic number; (d) the mass number. Write the isotope symbol of each isotope.

More Practice: Try Problems 2.5e, 2.6d, 2.29–2.32.

Sample Problem 2.8 — Comparing Elements and Isotopes

Answer the following questions about the three elements whose nuclei are drawn below. (a) Which representation shows an isotope of [1]? (b) Which representation shows a different element from the other two? (c) Which element has the lowest atomic number?

[1]   [2]   [3]

Analysis

- **Two isotopes have the same number of protons but a different number of neutrons.**
- Two different elements have different numbers of protons.
- The number of protons in the nucleus gives the atomic number.

Solution

a. Representation [1] has eight protons and eight neutrons in the nucleus. Representation [3] is an isotope because it has the same number of protons (eight), but it contains a different number of neutrons (nine).
b. Representation [2] is a different element because it contains only seven protons and representations [1] and [3] each have eight protons.
c. Because the number of protons determines the atomic number, representation [2] has the lowest atomic number (seven) because it has only seven protons in its nucleus.

Practice Problem 2.12

Use the representations ([1]–[3]) in Sample Problem 2.8 to answer the following questions.

a. Which species has the highest mass number?
b. How many electrons are present in the neutral atom of each element?
c. Give the isotope symbol for each species.

More Practice: Try Problems 2.33, 2.34, 2.57.

Although gasoline sold in the United States no longer contains lead, leaded gasoline is still used extensively in Asia, Africa, and Latin America. Gasoline exhaust containing lead pollutes the air and soil, and individuals exposed to high lead levels can suffer from circulatory, digestive, and nervous disorders. Lead (Pb) is a metal with atomic number 82 and atomic weight 207.2, as shown in the periodic table. *Lampshade/Shutterstock*

2.3B Atomic Weight

Some elements like fluorine occur naturally as a single isotope. More commonly, an element is a mixture of isotopes, and it is useful to know the average mass, called the **atomic weight** (or **atomic mass**), of the atoms in a sample.

- The *atomic weight* is the weighted average of the mass of the naturally occurring isotopes of a particular element reported in atomic mass units.

The alphabetical list of elements in Appendix A gives the atomic weight for each element. The atomic weight is also given under the element symbol in the periodic table in Appendix B.

lead

Concept Check 2.5

The element cobalt is a micronutrient present in vitamin B_{12}. (a) What is the atomic number and atomic weight of cobalt? (b) How many protons and electrons does a neutral cobalt atom contain?

2.3C FOCUS ON HEALTH & MEDICINE
Isotopes in Medicine

Generally the chemical properties of isotopes are identical. Sometimes, however, one isotope of an element is radioactive—that is, it emits particles or energy as some form of radiation. Radioactive isotopes have both diagnostic and therapeutic uses in medicine.

As an example, iodine-131 is used in at least two different ways for thyroid disease. Iodine is a micronutrient needed by the body to synthesize the thyroid hormone thyroxine, which contains four iodine atoms. To evaluate the thyroid gland, a patient can be given sodium iodide (NaI) that contains radioactive iodine-131. Iodine-131 is taken up in the thyroid gland and as it emits radiation, it produces an image in a thyroid scan, which is then used to determine the condition of the thyroid gland, as shown in Figure 2.4.

Higher doses of iodine-131 can also be used to treat thyroid disease. Because the radioactive isotope is taken up by the thyroid gland, the radiation it emits can kill overactive or cancerous cells in the thyroid.

Other applications of radioactive isotopes in medicine are discussed in Chapter 9.

Figure 2.4   Iodine-131 in Medicine

a.

larynx

thyroid gland

trachea

b.

c.

The thyroid gland is a butterfly-shaped gland in the neck, shown in (a). Uptake of radioactive iodine-131 can reveal the presence of a healthy thyroid as in (b), or an unsymmetrical thyroid gland with dense areas of iodine uptake as in (c), which may be indicative of cancer or other thyroid disease.
(b) BSIP SA/Alamy; (c) Southern Illinois University/Science Source

Iodine-123 and iodine-131 are radioactive isotopes used for the diagnosis or treatment of thyroid disease. Complete the following table for both isotopes.

	Atomic Number	Mass Number	Number of Protons	Number of Neutrons	Isotope Symbol
Iodine-123					
Iodine-131					

2.4 The Periodic Table

Every beginning chemistry text has a periodic table in a prominent location—because it is a valuable list of all known elements organized so that groups of elements with similar characteristics are arranged together. The periodic table evolved over many years, and it resulted from the careful observations and experiments of many brilliant scientists in the nineteenth century. Most prominent was Russian chemist Dmitri Mendeleev, whose arrangement in 1869 of the 60 known elements into groups having similar properties became the precursor of the modern periodic table (Appendix B and Figures 2.1 and 2.5).

2.4A Basic Features of the Periodic Table

The periodic table is arranged into seven horizontal rows and 18 vertical columns. The particular row and column tell us much about the properties of an element.

- A row in the periodic table is called a *period*. Elements in the same row are similar in size.
- A column in the periodic table is called a *group*. Elements in the same group have similar electronic and chemical properties.

The rows in the periodic table are numbered 1–7. The number of elements in each row varies. The first period has just two elements, hydrogen and helium. The second and third rows have eight elements each, and the fourth and fifth rows have 18 elements. Also note that two sets of fifteen elements appear at the bottom of the periodic table. The **lanthanides** begin with the element lanthanum (La, $Z = 57$), immediately following the element barium (Ba, $Z = 56$). The **actinides** begin with the element actinium (Ac, $Z = 89$), immediately following the element radium (Ra, $Z = 88$).

Each column in the periodic table is assigned a **group number.** Groups are numbered in two ways. In one system, the 18 columns of the periodic table are assigned the numbers 1–18, beginning with the column farthest to the left. An older but still widely used system numbers the groups 1–8, followed by the letter A or B.

- The *main group elements* consist of the two columns on the far left and the six columns on the far right of the table. These groups are numbered 1A–8A.
- The *transition metal elements* are contained in the 10 short columns in the middle of the table, numbered 1B–8B.
- The *inner transition metal elements* consist of the lanthanides and actinides, and they are not assigned group numbers.

The periodic table in Figure 2.5 has both systems of numbering groups. For example, the element carbon (C) is located in the second row (period 2) of the periodic table. Its group number is 4A (or 14). One group and one period are highlighted in the periodic table in Figure 2.6.

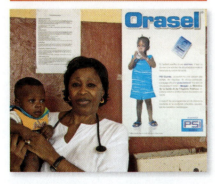

Figure 2.5   Basic Features of the Periodic Table

Period

Group number

Period	1A 1	2A 2	3B 3	4B 4	5B 5	6B 6	7B 7	8B 8	8B 9	8B 10	1B 11	2B 12	3A 13	4A 14	5A 15	6A 16	7A 17	8A 18	
1	H																	He	1
2	Li	Be											B	C	N	O	F	Ne	2
3	Na	Mg											Al	Si	P	S	Cl	Ar	3
4	K	Ca	Sc	Ti	V	Cr	Mn	Fe	Co	Ni	Cu	Zn	Ga	Ge	As	Se	Br	Kr	4
5	Rb	Sr	Y	Zr	Nb	Mo	Tc	Ru	Rh	Pd	Ag	Cd	In	Sn	Sb	Te	I	Xe	5
6	Cs	Ba	Lantha-nides	Hf	Ta	W	Re	Os	Ir	Pt	Au	Hg	Tl	Pb	Bi	Po	At	Rn	6
7	Fr	Ra	Acti-nides	Rf	Db	Sg	Bh	Hs	Mt	Ds	Rg	Cn	Nh	Fl	Mc	Lv	Ts	Og	7

Lanthanides	6	La	Ce	Pr	Nd	Pm	Sm	Eu	Gd	Tb	Dy	Ho	Er	Tm	Yb	Lu	6
Actinides	7	Ac	Th	Pa	U	Np	Pu	Am	Cm	Bk	Cf	Es	Fm	Md	No	Lr	7

Main group elements

Transition metal elements

Inner transition metal elements

- Each element of the periodic table is part of a horizontal row and a vertical column.
- The periodic table consists of seven rows, labeled periods 1–7, and 18 columns that are assigned a group number. Two different numbering systems are indicated.
- Elements are divided into three categories: **main group elements** (groups 1A–8A, shown in light blue), **transition metals** (groups 1B–8B, shown in tan), and **inner transition metals** (shown in light green).
- Atomic numbers and atomic weights for the elements are shown in Figure 2.1.

Figure 2.6   A Group and a Period in the Periodic Table

Group 4A or 14

Period 3

- The elements C, Si, Ge, Sn, Pb, and Fl are part of group 4A (or 14) in the periodic table.
- The elements Na, Mg, Al, Si, P, S, Cl, and Ar are part of the third row, period 3, of the periodic table.

Concept Check 2.7   (a) How many elements are in the third period of the periodic table? (b) How many elements are in group 2A (or 2)?

Sample Problem 2.9   **Determining the Group Number and Period of an Element**

Give the period and group number for each element: (a) aluminum; (b) mercury.

Analysis

Use the element symbol to locate an element in the periodic table. **Count down the rows of elements to determine the period. The group number is located at the top of each column.**

Solution

   a. Aluminum (Al) is located in the third row (period 3), and has group number 3A (or 13).

   b. Mercury (Hg) is located in the sixth row (period 6), and has group number 2B (or 12).

Practice Problem 2.13

Give the period and group number for each element: (a) oxygen; (b) calcium; (c) phosphorus; (d) platinum; (e) iodine.

More Practice: Try Problems 2.5b; 2.6b; 2.23b, c; 2.24[2], [3]; 2.58.

2.4B   Characteristics of Groups 1A, 2A, 7A, and 8A

Four columns of main group elements illustrate an important fact about the periodic table.

• Elements that comprise a particular group have similar chemical properties.

Alkali Metals (Group 1A) and Alkaline Earth Elements (Group 2A)

The alkali metals and the alkaline earth elements are located on the far left side of the periodic table.

The **alkali metals,** located in group 1A (group 1), include lithium (Li), sodium (Na), potassium (K), rubidium (Rb), cesium (Cs), and francium (Fr). Alkali metals share the following characteristics:

• They are soft and shiny and have low melting points.

• They are good conductors of heat and electricity.

• They react readily with water to form basic solutions.

Although hydrogen is also located in group 1A, it is *not* an alkali metal.

The **alkaline earth elements,** located in group 2A (group 2), include beryllium (Be), magnesium (Mg), calcium (Ca), strontium (Sr), barium (Ba), and radium (Ra). Alkaline earth metals are also shiny solids but less reactive than the alkali metals.

None of the metals in groups 1A or 2A exist in nature as pure elements; rather, they are always combined with other elements to form compounds. Examples of compounds from group 1A elements include sodium chloride **(NaCl),** table salt, and potassium iodide **(KI),** an essential nutrient added to make iodized salt. Examples of compounds from group 2A elements include magnesium sulfate **($MgSO_4$),** an anticonvulsant used to prevent seizures in pregnant women; and barium sulfate **($BaSO_4$),** which is used to improve the quality of X-ray images of the gastro-intestinal tract.

Chlorpheniramine is an antihistamine that contains the halogen chlorine. Antihistamines are drugs used to treat the symptoms of the common cold and allergies. We will learn more about antihistamines in Chapter 13.
Jill Braaten

Halogens (Group 7A) and Noble Gases (Group 8A)

The halogens and noble gases are located on the far right side of the periodic table.

Radon detectors are used to measure high levels of radon, a radioactive noble gas linked to an increased incidence of lung cancer. *Keith Eng, 2008*

The **halogens,** located in group 7A (group 17), include fluorine (F), chlorine (Cl), bromine (Br), iodine (I), and the rare radioactive element astatine (At). In their elemental form, halogens contain two atoms joined together—F_2, Cl_2, Br_2, and I_2. Fluorine and chlorine are gases at room temperature, bromine is a liquid, and iodine is a solid. Halogens are very reactive and combine with many other elements to form compounds. In Chapter 12, we will learn about carbon compounds that contain halogen atoms.

The **noble gases,** located in group 8A (group 18), include helium (He), neon (Ne), argon (Ar), krypton (Kr), xenon (Xe), and radon (Rn). Unlike other elements, the noble gases are especially stable as atoms, and so they rarely combine with other elements to form compounds.

The noble gas **radon** has received attention in recent years. Radon is a radioactive gas, and generally its concentration in the air is low and therefore its presence harmless. In some types of soil, however, radon levels can be high and radon detectors are recommended for the basement of homes to monitor radon levels. High radon levels are linked to an increased risk of lung cancer.

Concept Check 2.8

Identify the element fitting each description.

a. an alkali metal in period 4

b. a second-row element in group 7A

c. a noble gas in the third period

d. a main group element in period 5 and group 2A

e. a transition metal in group 12, period 4

f. a transition metal in group 5, period 5

Concept Check 2.9 Identify each highlighted element in the periodic table and give its [1] element name and symbol; [2] group number; [3] period; [4] classification (main group element, transition metal, or inner transition metal).

2.5   Electronic Structure

Why do elements in a group of the periodic table have similar chemical properties? **The chemical properties of an element are determined by the number of *electrons* in an atom.** To understand the properties of an element, therefore, we must learn more about the electrons that surround the nucleus.

The modern description of the electronic structure of an atom is based on the following principle.

- Electrons do not move freely in space; rather, they occupy specific energy levels.

The electrons that surround a nucleus are confined to regions called the **principal energy levels** or **shells.**

- The shells are numbered, n = 1, 2, 3, 4, and so forth, beginning closest to the nucleus.
- Electrons closer to the nucleus are held more tightly and are *lower* in energy.
- Electrons farther from the nucleus are held less tightly and are *higher* in energy.

The number of electrons that can occupy a given shell is determined by the value of n. **The farther an energy level is from the nucleus, the larger its volume becomes, and the more electrons it can hold.** Thus, the first energy level can hold only two electrons, the second holds eight, the third 18, and so forth. The maximum number of electrons is given by the formula $2n^2$, where n = **the shell number.**

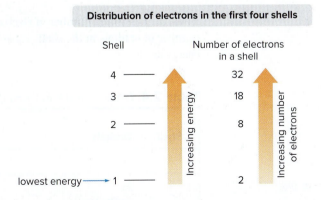

Distribution of electrons in the first four shells

Shell		Number of electrons in a shell	
4 ——		32	
3 ——	Increasing energy	18	Increasing number of electrons
2 ——		8	
lowest energy ⟶ 1 ——		2	

The energy levels consist of a set of **orbitals**, identified by the letters *s, p, d,* and *f.*

- An *orbital* is a region of space where the probability of finding an electron is high. Each orbital can hold *two* electrons.

A particular energy level contains a specific number of a given type of orbital. There can be **one s orbital, three p orbitals, five d orbitals, and seven f orbitals.** The energy of orbitals shows the following trend:

| *s* orbital | *p* orbital | *d* orbital | *f* orbital |
| lowest energy | | | highest energy |

Increasing energy

The first energy level of electrons around a nucleus (n = 1) has only one s orbital. This orbital is called the **1s** orbital because it is the *s* orbital in the first shell. Because each orbital can hold two electrons and the first shell has only one orbital, the **first energy level can hold two electrons.**

shell number   1s   =   the *s* orbital in the first shell
(principal energy level)

The second energy level (n = 2) has two types of orbitals—one s and three p orbitals. These orbitals are called the **2s** and **2p** orbitals because they are located in the second energy level. Because each orbital can hold two electrons and there are four orbitals, the **second energy level can hold eight electrons.**

one 2s orbital   three 2p orbitals

| 2s | 2p | 2p | 2p | = | **4 orbitals** in the second shell |

$$4 \text{ orbitals} \quad \times \quad \frac{2 \text{ electrons}}{1 \text{ orbital}} \quad =$$

Each orbital holds **two** electrons.

8 electrons in the second shell

The third energy level (n = 3) has three types of orbitals—one s, three p, and five d orbitals. These orbitals are called the **3s, 3p,** and **3d** orbitals because they are located in the third energy level. Because each orbital can hold two electrons and the third shell has a total of nine orbitals, the **third energy level can hold 18 electrons.**

one 3s orbital   three 3p orbitals   five 3d orbitals

| 3s | 3p | 3p | 3p | 3d | 3d | 3d | 3d | 3d | = | **9 orbitals** in the third shell |

Each orbital holds **two** electrons.

$$9 \text{ orbitals} \quad \times \quad \frac{2 \text{ electrons}}{1 \text{ orbital}} \quad =$$

18 electrons in the third shell

Thus, the maximum number of electrons that can occupy an energy level is determined by the number of orbitals in the shell. Table 2.4 summarizes the orbitals and electrons in the first three energy levels.

Table 2.4 Orbitals and Electrons Contained in the Principal Energy Levels (n = 1–3)

Shell	Orbitals	Number of Electrons in the Orbitals	Maximum Number of Electrons
1	1s	2	2
2	2s	2	8
	2p 2p 2p	3 × 2 = 6	
3	3s	2	18
	3p 3p 3p	3 × 2 = 6	
	3d 3d 3d 3d 3d	5 × 2 = 10	

Each type of orbital has a particular shape.

> • An *s* orbital has a **sphere** of electron density. It is *lower* in energy than other orbitals in the same shell because electrons are kept closer to the positively charged nucleus.
> • A *p* orbital has a **dumbbell** shape. A *p* orbital is *higher* in energy than an *s* orbital in the same shell because its electron density is farther from the nucleus.

s orbital   p orbital

nucleus   nucleus

lower in energy   higher in energy

All *s* orbitals are spherical, but the orbital gets larger in size as the shell number increases. Thus, both a 1*s* orbital and a 2*s* orbital are spherical, but the 2*s* orbital is larger. The three *p* orbitals in a shell are perpendicular to each other along the *x*, *y*, and *z* axes.

1s   2s   2p$_x$ 2p$_y$ 2p$_z$ all three 2p orbitals drawn on the same set of axes

---

Concept Check 2.10

What is the maximum number of electrons possible for each energy level or orbital?

a. a 2*p* orbital   b. the second energy level   c. a 3*s* orbital   d. the third shell

---

2.6 Electronic Configurations

We can now examine the **electronic configuration** of an individual atom—that is, how the electrons are arranged in an atom's orbitals. **The lowest-energy arrangement of electrons is called the *ground state*.** Two rules are followed.

Rules to Determine the Ground-State Electronic Configuration of an Atom

Rule [1] Electrons are placed in the lowest-energy orbitals beginning with the 1s orbital.

- In comparing similar types of orbitals from one shell to another (e.g., 1s and 2s), an orbital closer to the nucleus is lower in energy. Thus, the energy of a 1s orbital is lower than a 2s orbital.
- Within a shell, orbital energies increase in the following order: s, p, d, f.
- These guidelines result in the following order of energies in the first three periods: 1s, 2s, 2p, 3s, 3p (Figure 2.7).

Rule [2] Each orbital holds a maximum of two electrons.

Figure 2.7

Order of Orbital Filling

Increasing energy

3p — — —
3s —
2p — — —
2s —
1s —

To illustrate how these rules are used, we can write the electronic configuration for several elements. **The electronic configuration shows what orbitals contain electrons and uses a superscript with each orbital to show how many electrons it contains.**

2.6A First-Row Elements (Period 1)

The first row of the periodic table contains only two elements—hydrogen and helium. Because the number of protons in the nucleus equals the number of electrons in a neutral atom, the **atomic number tells us how many electrons must be placed in orbitals.**

A blimp is an airship that contains helium, making it lighter than air. A blimp maneuvers with rudders and propellers and can hover for extended periods, thus making it a useful aircraft for aerial photography at sporting events.
Luismmolina/E+/Getty Images

Hydrogen (H, $Z = 1$) has one electron. In the ground state, this electron is added to the lowest-energy orbital, the $1s$ orbital. The electronic configuration of hydrogen is written as $1s^1$, meaning that the $1s$ orbital contains one electron.

Helium (He, $Z = 2$) has two electrons. In the ground state, both electrons are added to the $1s$ orbital. The electron configuration is written as $1s^2$, meaning the $1s$ orbital has two electrons. Helium has a filled first energy level of electrons.

2.6B   Second-Row Elements (Period 2)

To write electronic configurations for the second-row elements, we must now use the four orbitals in the second energy level—the $2s$ orbital and the three $2p$ orbitals. Because electrons are always added to the lowest-energy orbitals first, all second-row elements have the $1s$ orbital filled with electrons, and then the remaining electrons are added to the orbitals in the second shell. Because the $2s$ orbital is lower in energy than the $2p$ orbitals, it is completely filled before adding electrons to the $2p$ orbitals.

Lithium (Li, $Z = 3$) has three electrons. In the ground state, two electrons are added to the $1s$ orbital and the remaining electron is added to the $2s$ orbital.

Carbon (C, $Z = 6$) has six electrons. In the ground state, two electrons are added to both the $1s$ and $2s$ orbitals. The two remaining electrons are added to the $2p$ orbitals, so the electronic configuration is written as $1s^2 2s^2 2p^2$.

Neon (Ne, $Z = 10$) has 10 electrons. In the ground state, two electrons are added to the $1s$, $2s$, and each of the three $2p$ orbitals, giving a total of six electrons in the $2p$ orbitals.

The electronic configurations of all the first- and second-row elements are listed in Table 2.5. The total number of electrons used for the electronic configuration of a neutral atom is always equal to the atomic number.

Concept Check 2.11

What element(s) in the first and second period fit each description?

a. The element has one electron in the second energy level.

b. There are two electrons in the $2s$ orbital.

c. The electronic configuration is $1s^2 2s^2 2p^5$.

d. The element contains six electrons in the second energy level.

Table 2.5 Electronic Configurations of the First- and Second-Row Elements

Atomic Number	Element	Electronic Configuration	Total Number of Electrons
1	H	$1s^1$	1
2	He	$1s^2$	2
3	Li	$1s^2 2s^1$	3
4	Be	$1s^2 2s^2$	4
5	B	$1s^2 2s^2 2p^1$	5
6	C	$1s^2 2s^2 2p^2$	6
7	N	$1s^2 2s^2 2p^3$	7
8	O	$1s^2 2s^2 2p^4$	8
9	F	$1s^2 2s^2 2p^5$	9
10	Ne	$1s^2 2s^2 2p^6$	10

2.6C Other Elements

Electronic configurations can be written in a similar fashion for other elements in the periodic table. Sample Problems 2.10 and 2.11 illustrate two examples.

ENVIRONMENTAL NOTE

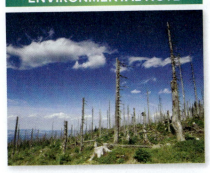

Coal that is high in **sulfur** content burns to form sulfur oxides, which in turn react with water to form sulfurous and sulfuric acids. Rain that contains these acids has destroyed acres of forests worldwide. Sulfur is a third-row element in the periodic table (Sample Problem 2.10).
Marek Mnich/E+/Getty Images

Sample Problem 2.10 | **Writing the Electronic Configuration of an Element**

Give the ground-state electronic configuration of the element sulfur.

Analysis

- Use the atomic number to determine the number of electrons.
- Place electrons two at a time into the lowest-energy orbitals, following the order of orbital filling in Figure 2.7.

Solution

The atomic number of sulfur is 16, so 16 electrons must be placed in orbitals. Twelve electrons are added in pairs to the 1s, 2s, three 2p, and 3s orbitals. The remaining four electrons are then added to the three 3p orbitals.

S
sulfur
16 electrons

Answer:
The electronic configuration is
$1s^2 2s^2 2p^6 3s^2 3p^4$.

Practice Problem 2.14

Give the ground-state electronic configuration for each element: (a) aluminum; (b) chlorine.

More Practice: Try Problems 2.8a, 2.37, 2.38.

Sample Problem 2.11 | **Identifying an Element from an Electronic Configuration**

What element has the ground-state electronic configuration $1s^2 2s^2 2p^6 3s^2$?

Analysis

- Count the number of electrons in the electronic configuration.
- **Because the number of electrons equals the atomic number in a neutral atom,** identify the element from the atomic number in the periodic table.

Solution

The element has a total of 12 electrons (2 + 2 + 6 + 2). The element with an atomic number of 12 is magnesium.

Practice Problem 2.15

What element has each ground-state electronic configuration?

   a. $1s^2 2s^2 2p^6 3s^1$ b. $1s^2 2s^2 2p^6 3s^2 3p^2$

More Practice: Try Problems 2.7a, 2.39 and 2.40 (only identify the element).

2.7   Valence Electrons

The chemical properties of an element depend on the most loosely held electrons—that is, those electrons in the outermost shell, called the **valence shell. The period number tells the number of the valence shell.**

-   The electrons in the outermost shell are called the *valence electrons.*

2.7A   Relating Valence Electrons to Group Number

To identify the electrons in the valence shell, always **look for the shell with the *highest* number.** Thus, beryllium has two valence electrons that occupy the $2s$ orbital. Chlorine has seven valence electrons because it has a total of seven electrons in the third shell, two in the $3s$ orbital and five in the $3p$ orbitals.

Be (beryllium):   Cl (chlorine):

$1s^2 2s^2$ **2** valence electrons $1s^2 2s^2 2p^6 3s^2 3p^5$ **7** valence electrons

valence shell   valence shell

If we examine the electronic configuration of a group in the periodic table, two facts become apparent.

-   Elements in the same group have the same number of valence electrons and similar electronic configurations.
-   The group number (using the 1A–8A system) equals the number of valence electrons for main group elements (except helium).

Thus, the periodic table is organized into groups of elements with similar valence electronic configurations in the *same column*. The valence electronic configurations of the main group elements in the first three rows of the periodic table are given in Table 2.6. As an example, the alkali metals in group **1A** all have **one** valence electron that occupies an s orbital.

-   The chemical properties of a group are similar because these elements contain the same electronic configuration of valence electrons.

Table 2.6   Valence Electronic Configurations for the Main Group Elements in Periods 1–3

Group Number	1A	2A	3A	4A	5A	6A	7A	8A[a]
Period **1**	H $1s^1$							He $1s^2$
Period **2**	Li $2s^1$	Be $2s^2$	B $2s^2 2p^1$	C $2s^2 2p^2$	N $2s^2 2p^3$	O $2s^2 2p^4$	F $2s^2 2p^5$	Ne $2s^2 2p^6$
Period **3**	Na $3s^1$	Mg $3s^2$	Al $3s^2 3p^1$	Si $3s^2 3p^2$	P $3s^2 3p^3$	S $3s^2 3p^4$	Cl $3s^2 3p^5$	Ar $3s^2 3p^6$
General configuration	ns^1	ns^2	$ns^2 np^1$	$ns^2 np^2$	$ns^2 np^3$	$ns^2 np^4$	$ns^2 np^5$	$ns^2 np^6$

[a]The general electronic configuration in group 8A applies to all of the noble gases except helium. Because helium is a first-row element, it has only two electrons, and these occupy the only available orbital in the first shell, the $1s$ orbital.

Take particular note of the electronic configuration of the noble gases in group 8A. **All of these elements have a completely filled outer shell of valence electrons.** Helium has a filled first shell ($1s^2$ configuration). The remaining elements have a completely filled valence shell of s and p orbitals (s^2p^6). This electronic arrangement is especially stable, and as a result, these elements exist in nature as single atoms. We will learn about the consequences of having a completely filled valence shell in Chapter 3.

Sample Problem 2.12

Identifying the Number of Valence Electrons

Identify the total number of electrons, the number of valence electrons, and the name of the element with each electronic configuration.

a. $1s^2 2s^2$ b. $1s^2 2s^2 2p^6 3s^2 3p^2$

Analysis

To obtain the total number of electrons, add up the superscripts. This gives the atomic number and identifies the element. **To determine the number of valence electrons, add up the number of electrons in the shell with the highest number.**

Solution

a.   valence shell

$1s^2\mathbf{2s^2}$

2 valence electrons

Total number of electrons =
2 + 2 = **4**

Answer: Beryllium (Be), 4 total electrons
and 2 valence electrons

b.   valence shell

$1s^2 2s^2 2p^6 \mathbf{3s^2 3p^2}$

4 valence electrons

Total number of electrons =
2 + 2 + 6 + 2 + 2 = **14**

Answer: Silicon (Si), 14 total electrons
and 4 valence electrons

Practice Problem 2.16

Identify the total number of electrons, the number of valence electrons, and the name of the element with each electronic configuration.

a. $1s^2 2s^1$ b. $1s^2 2s^2 2p^6 3s^2$ c. $1s^2 2s^2 2p^6 3s^2 3p^3$

More Practice: Try Problems 2.5d, 2.39, 2.40.

Sample Problem 2.13

Using the Group Number to Determine the Number of Valence Electrons

Determine the number of valence electrons of each element: (a) nitrogen; (b) potassium.

Analysis

The group number of a main group element = the number of valence electrons.

Solution

a. Nitrogen is located in group 5A, so it has five valence electrons.

b. Potassium is located in group 1A, so it has one valence electron.

Practice Problem 2.17

Determine the number of valence electrons of each element: (a) fluorine; (b) krypton; (c) magnesium; (d) germanium.

More Practice: Try Problems 2.41[3], 2.42[3], 2.45, 2.46.

2.7B Electron-Dot Symbols

The number of valence electrons around an atom is often represented by an **electron-dot symbol.** Representative examples are shown.

	H	C	O	Cl
Number of valence electrons:	**1**	**4**	**6**	**7**
Electron-dot symbol:	H·	·Ċ·	·Ö·	·Ċl:

- Each dot represents one valence electron.
- The dots are placed on the four sides of an element symbol.
- For one to four valence electrons, single dots are used. With more than four electrons, the dots are paired.

The location of the dots around the symbol—side, top, or bottom—does *not* matter. Each of the following representations for the five valence electrons of nitrogen is equivalent.

·N̈·   ·N̈·   :N̈·   ·N̈:

Sample Problem 2.14

Writing an Electron-Dot Symbol

Write an electron-dot symbol for each element: (a) sodium; (b) phosphorus.

Analysis

Write the symbol for each element and use the group number to determine the number of valence electrons for a main group element. Represent each valence electron with a dot.

Solution

a. The symbol for sodium is Na. Na is in group 1A and has one valence electron. Electron-dot symbol:

Na·

b. The symbol for phosphorus is P. P is in group 5A and has five valence electrons. Electron-dot symbol:

·P̈·

Practice Problem 2.18

Give the electron-dot symbol for each element: (a) bromine; (b) lithium; (c) aluminum; (d) sulfur; (e) neon.

More Practice: Try Problems 2.8c, 2.47, 2.48.

2.8 Periodic Trends

Many properties of atoms exhibit **periodic trends;** that is, they change in a regular way across a row or down a column of the periodic table. Two properties that illustrate this phenomenon are **atomic size** and **ionization energy.**

2.8A Atomic Size

The size of an atom is measured by its atomic radius—that is, the distance from the nucleus to the outer edge of the valence shell. Two periodic trends characterize the size of atoms.

- The size of atoms *increases* down a column of the periodic table, as the valence electrons are farther from the nucleus.

For group 7A:

- The size of atoms *decreases* across a row of the periodic table as the number of protons in the nucleus increases. An increasing number of protons pulls the electrons closer to the nucleus, so the atom gets smaller.

For period 2:

Sample Problem 2.15	**Using Periodic Trends to Determine the Relative Size of Atoms**

(a) Which of the labeled atoms has the larger atomic radius? (b) Identify the elements.

Analysis

Atomic radius (size) decreases across a row of the periodic table as the number of protons in the nucleus increases.

Solution

  a. Element [1] is farther to the left in the third row, so it has the larger atomic radius.

  b. Element [1] is silicon (period 3, group 4A) and element [2] is phosphorus (period 3, group 5A).

Practice Problem 2.19	Which element in each pair has the larger atomic radius?

  a. nitrogen or arsenic   b. silicon or sulfur   c. fluorine or phosphorus

More Practice: Try Problems 2.9; 2.11a; 2.12a; 2.49; 2.51; 2.53; 2.55c, d.

2.8B Ionization Energy

Because a negatively charged electron is attracted to a positively charged nucleus, **energy is required to remove an electron from a neutral atom.** The more tightly the electron is held, the greater the energy required to remove it. Removing an electron from a neutral atom forms a **cation.**

$$Na \quad + \quad energy \quad \longrightarrow \quad Na^+ \quad + \quad e^-$$

neutral atom   **ionization energy**   **cation**

- The *ionization energy* is the energy needed to remove an electron from a neutral atom.
- A *cation* is positively charged, and has fewer electrons than the neutral atom.

Two periodic trends characterize ionization energy.

- Ionization energies *decrease* down a column of the periodic table as the valence electrons get farther from the positively charged nucleus.

For group 1A:

- Ionization energies generally *increase* across a row of the periodic table as the number of protons in the nucleus increases.

For period 2:

Concept Check 2.12

Which element in each pair has the higher ionization energy?

a. silicon or sodium   b. carbon or silicon   c. sulfur or fluorine

---

Sample Problem 2.16

Determining Periodic Trends

(a) Which of the indicated atoms has the larger atomic radius? (b) Which of the indicated atoms has the higher ionization energy?

Analysis

- Atomic size *decreases* across a row and *increases* down a column of the periodic table.
- Ionization energy *increases* across a row and *decreases* down a column of the periodic table.

Solution

a. Element [2] is farther to the left and farther down the column, so it has the larger atomic radius.

b. Element [1] is farther to the right and closer to the top of the column (lower period number), so it has the higher ionization energy.

Practice Problem 2.20

(a) Which of the indicated atoms has the smaller atomic radius? (b) Which of the indicated atoms has the lower ionization energy? (c) Identify the elements.

More Practice: Try Problems 2.10; 2.11; 2.12; 2.50; 2.52; 2.54; 2.55e, f.

CHAPTER REVIEW

KEY TERMS

Actinide (2.4)

Alkali metal (2.4)

Alkaline earth element (2.4)

Atom (2.2)

Atomic mass unit (2.2)

Atomic number (2.2)

Atomic weight (Atomic mass, 2.3)

Building-block element (2.1)

Cation (2.8)

Chemical formula (2.1)

Compound (2.1)

Deuterium (2.3)

Electron (2.2)

Electron cloud (2.2)

Electron-dot symbol (2.7)

Electronic configuration (2.6)

Element (2.1)

Ground state (2.6)

Group (2.4)

Group number (2.4)

Halogen (2.4)

Inner transition metal
    element (2.4)

Ionization energy (2.8)

Isotope (2.3)

Lanthanide (2.4)

Main group element (2.4)

Major mineral (Macronutrient, 2.1)

Mass number (2.2)

Metal (2.1)

Metalloid (2.1)

Neutron (2.2)

Noble gas (2.4)

Nonmetal (2.1)

Nucleus (2.2)

Orbital (2.5)

Period (2.4)

Periodic table (2.1)

p Orbital (2.5)

Proton (2.2)

Shell (Principal energy level, 2.5)

s Orbital (2.5)

Trace element (Micronutrient, 2.1)

Transition metal element (2.4)

Tritium (2.3)

Valence electron (2.7)

KEY CONCEPTS

① **The periodic table**

① **Classification as metals, nonmetals, and metalloids (2.1)**	**②** **Classification as main group elements, transition metals, and inner transition metals (2.4)**	**③** **Information obtained from an entry in the periodic table (2.2–2.4)**

- A **period** is a row in the periodic table.
- A **group** is a column. Each column is assigned a **group number:** 1–18 or 1–8 followed by the letters A or B.

- Inner transition metals consist of the **lanthanides** and **actinides,** and they are *not* assigned group numbers.

- The **atomic number** = the number of positively charged particles = the number of **protons**.
- The number of **neutrons** = the number of uncharged particles.
- The **mass number** = the number of **protons** + the number of **neutrons**.
- The number of **protons** = the number of **electrons** in a neutral atom.
- The **atomic weight** is the weighted average of the mass of all naturally occurring isotopes.

② **Relating an electronic configuration to the number of valence electrons and element identity (2.6); example: $1s^2 2s^2 2p^6 3s^2 3p^3$**

① **Counting electrons**	**②** **Identifying an element**
$1s^2 2s^2 2p^6 3s^2 3p^3$ Total number of electrons = sum of the **superscripts** $= 2 + 2 + 6 + 2 + 3 = \mathbf{15}$ Number of valence electrons = number of electrons in the shell with the **highest number** $= 2 + 3 = \mathbf{5}$ • The **valence electrons** occupy the shell with the highest number.	• The total number of electrons = the number of protons = 15. • The number of protons identifies the atomic number of the element = 15. **P** **phosphorus** atomic number 15

③ **Periodic trends**

① **Size of an atom (2.8A)**	**②** **Ionization energy of an atom (2.8B)**
• The **size** of an atom *decreases* across a row and *increases* down a column.	• The **ionization energy**—the energy needed to remove an electron from an atom—*increases* across a row and *decreases* down a column.

KEY SKILLS

1 Determining the number of protons, neutrons, and electrons of an atom (2.2); example: an atom of sodium that has a mass of 23

1 Number of protons	**2** Number of neutrons	**3** Number of electrons
Na $Z = 11 \longrightarrow$ **11 protons** sodium • The identity of an atom determines the **atomic number (Z).** • Z = the number of protons	number of neutrons $= A - Z$ $\qquad = 23 - 11 = $ **12 neutrons** • The number of **neutrons** = the mass number (**A**) – the number of protons (**Z**).	**Na** **11 protons** $\longrightarrow$ **11 electrons** sodium • The number of protons = the number of **electrons** in a neutral atom.

2 Using an isotope symbol to determine the number of protons, neutrons, and electrons (2.3); example: bromine-81

1 Atomic number, number of protons, and number of electrons	**2** Mass number	**3** Number of neutrons
 ${}^{81}_{35}\text{Br}$ atomic number = number of **protons** = number of **electrons** = **35** • The **atomic number (Z)** is the subscript in an isotope symbol. • Z = the number of protons • The number of protons = the number of **electrons** in a neutral atom.	${}^{81}_{35}\text{Br}$ mass number = number of **protons** + number of **neutrons** = **81** • The **mass number (A)** is the superscript in an isotope symbol. • A = the number of protons + the number of neutrons.	number of **neutrons** $= A - Z$ $\qquad = 81 - 35 = $ **46** • The number of **neutrons** = the mass number (**A**) – the number of protons (**Z**).

3 Writing an electronic configuration (2.6); example: chlorine

1 Use the atomic number to determine the number of electrons.	**2** Use Figure 2.7 to determine the order of orbitals filled with electrons.
Cl $Z = 17$ chlorine = number of **electrons** • The identity of an atom determines the number of **electrons.** • Chlorine (atomic number = 17) has 17 electrons to be placed in orbitals.	Electronic configuration of Cl: $1s^2 2s^2 2p^6 3s^2 3p^5$ **17 electrons** • From Figure 2.7, the order of orbital filling is **1s, 2s, 2p (three orbitals), 3s,** and **3p (three orbitals).** • Give 1s–3s orbitals two electrons each, and give five electrons to the 3p orbitals.

4 Determining the number of valence electrons and writing an electron-dot symbol using the group number (2.7); example: Ca

1 Use the group number of a main group element to determine the number of valence electrons.	**2** Represent each valence electron with a dot.
Ca group **2A** **two** valence electrons • The **group number** of a main group element = the number of valence electrons. • The element symbol for calcium is Ca. Ca is in group **2**A and has **two** valence electrons.	•Ca• **two** single dots for **two** valence electrons • From one to four electrons, *single* dots are used. With more than four electrons the dots are *paired*.

CHAPTER 2 SELF-TEST

The Self-Test consists of short-answer questions that test an understanding of definitions, equations, and other material encountered within the chapter. Answers to each question are provided at the end of the chapter. For any questions you are unable to answer correctly, revisit the appropriate section(s) of the text; then proceed to working the end-of-chapter problems to further reinforce your understanding of the chapter material.

1. Which term from "Key Terms" answers each question?
   a. In what part of an atom does most of the mass reside?
   b. What term refers to the number of protons in an atom?
   c. What term is used for elements located in groups 1A–8A of the periodic table?
   d. What is a row in the periodic table called?
   e. Two isotopes differ in the number of what subatomic particle?

2. How many protons, neutrons, and electrons are contained in an atom of sodium, which has an atomic number of 11 and a mass number of 23?

3. Give the following information for the isotope silver-109: (a) atomic number; (b) mass number; (c) number of protons; (d) number of neutrons; (e) number of electrons.

4. Give the period and group number of each atom: (a) lead; (b) calcium.

5. Which of the following are proper electron-dot symbols for the element oxygen?

a. $\ddot{\text{O}}:$ b. $\cdot\ddot{\text{O}}\cdot$ c. $:\ddot{\text{O}}:$ d. $:\ddot{\text{O}}\cdot$

6. Label each symbol or group of symbols as an element or compound: (a) Hs; (b) NO; (c) No; (d) Cf; (e) HF.

7. (a) Which element has the larger atomic radius: phosphorus or arsenic? (b) Which element has the higher ionization energy: magnesium or sulfur?

8. Give the ground-state electronic configuration for the element argon.

UNDERSTANDING KEY CONCEPTS

2.1 Identify the elements used in each example of molecular art.

a.   b.

2.2 Write a chemical formula for each example of molecular art.

a.   b.   c.

2.3 Give the following information about the atom shown: (a) the number of protons and neutrons in the nucleus; (b) the atomic number; (c) the mass number; (d) the number of electrons in the neutral atom; and (e) the element symbol.

2.4 Give the following information about the atom shown: (a) the number of protons and neutrons in the nucleus; (b) the atomic number; (c) the mass number; (d) the number of electrons in the neutral atom; and (e) the element symbol.

2.5 Selenium is a micronutrient necessary for certain enzymes that block unwanted oxidation reactions. Selenium is also needed for proper functioning of the thyroid gland. Answer the following questions about the element selenium.
   a. What is its element symbol?
   b. To what group number and period does selenium belong?
   c. Is selenium a main group element, transition metal, or inner transition metal?
   d. How many valence electrons does selenium contain?
   e. Draw an isotope symbol for a selenium atom that contains 46 neutrons in the nucleus.

2.6 Answer the following questions about the element silicon, a micronutrient needed for healthy bones, nails, skin, and hair.
   a. What is its element symbol?
   b. To what group number and period does silicon belong?
   c. Is silicon a main group element, transition metal, or inner transition metal?
   d. Draw an isotope symbol for a silicon atom that contains 14 neutrons in the nucleus.

2.7 (a) What element has the ground-state electronic configuration $1s^2 2s^2 2p^6 3s^2 3p^1$? (b) How many valence electrons does this element contain? (c) Give the group number and the period number for the element.

2.8 (a) Write the ground-state electronic configuration for the element silicon. (b) How many valence electrons does silicon contain? (c) Give an electron-dot symbol for silicon.

2.9 Which element in each pair is larger?
a. bromine or iodine
b. carbon or nitrogen

2.10 Which element in each pair has its valence electrons farther from the nucleus?
a. sodium or magnesium
b. neon or krypton

2.11 Consider the following three spheres, which represent atoms in the same group in the periodic table. (a) Which atom has the highest atomic number? (b) Which atom has the largest ionization energy? (c) Which atom gives up an electron most easily?

 [1]   [2]   [3]

2.12 Consider the following three spheres, which represent atoms in the same period in the periodic table. (a) Which atom has the highest atomic number? (b) Which atom has the largest ionization energy? (c) Which atom gives up an electron most easily?

 [1]   [2]   [3]

ADDITIONAL PROBLEMS

Elements

2.13 Give the name of the elements in each group of three element symbols.
a. Au, At, Ag
b. N, Na, Ni
c. S, Si, Sn
d. Ca, Cr, Cl

2.14 What element(s) are designated by each symbol or group of symbols?
a. CU and Cu
b. Os and OS
c. Ni and NI
d. BIN, BiN, and BIn

2.15 Does each chemical formula represent an element or a compound?
a. H_2 b. H_2O_2 c. S_8 d. Na_2CO_3 e. C_{60}

2.16 Identify the elements in each chemical formula and tell how many atoms of each are present.
a. $K_2Cr_2O_7$
b. $C_5H_8NNaO_4$ (MSG, flavor enhancer)
c. $C_{10}H_{16}N_2O_3S$ (vitamin B_7)

2.17 Identify the element that fits each description.
a. an alkali metal in period 6
b. a transition metal in period 5, group 8
c. a main group element in period 3, group 7A
d. a halogen in period 2

2.18 Identify the element that fits each description.
a. an alkaline earth element in period 3
b. a noble gas in period 6
c. a transition metal in period 4, group 11
d. a transition metal in period 6, group 10

2.19 Give all of the terms that apply to each element:
[1] metal; [2] nonmetal; [3] metalloid; [4] alkali metal; [5] alkaline earth element; [6] halogen; [7] noble gas; [8] main group element; [9] transition metal; [10] inner transition metal.
a. sodium
b. silver
c. xenon
d. platinum

2.20 Give all of the terms that apply to each element:
[1] metal; [2] nonmetal; [3] metalloid; [4] alkali metal; [5] alkaline earth element; [6] halogen; [7] noble gas; [8] main group element; [9] transition metal; [10] inner transition metal.
a. bromine
b. calcium
c. cesium
d. gold

Atomic Structure

2.21 Complete the following table for neutral elements.

Element Symbol	Atomic Number	Mass Number	Number of Protons	Number of Neutrons	Number of Electrons
a. C		12			
b.		31			15
c.				35	30
d. Mg		24			

2.22 For the given atomic number (Z) and mass number (A):
[1] identify the element; [2] give the element symbol; [3] give the number of protons, neutrons, and electrons.
a. $Z = 10, A = 20$
b. $Z = 13, A = 27$
c. $Z = 38, A = 88$
d. $Z = 55, A = 133$

Periodic Table

2.23 Label each region on the periodic table.
a. noble gases
b. period 3
c. group 4A
d. alkaline earth elements
e. transition metals
f. group 10

2.24 Identify each highlighted element in the periodic table and give its [1] element name and symbol; [2] group number; [3] period; [4] classification (i.e., main group element, transition metal, or inner transition metal).

2.25 Classify each element in the fourth row of the periodic table as a metal, nonmetal, or metalloid.

2.26 (a) Are the elements calcium, barium, and radium in the same group or same period in the periodic table? (b) Are the elements rhodium, cadmium, and tin in the same group or period?

Isotopes and Atomic Weight

2.27 Complete the following table for the two most widely occurring isotopes of the element nickel (Ni). The five-cent nickel coin minted in the United States contains the metals nickel and copper in a 1:3 ratio.

	^{58}Ni	^{60}Ni
a. Atomic number		
b. Mass number		
c. Number of protons		
d. Number of neutrons		

Ken Cavanagh/McGraw Hill

2.28 Complete the following table for the two most widely occurring isotopes of the element mercury (Hg). Compact fluorescent bulbs each contain about 4 mg of the toxic metal mercury, so used bulbs should not be discarded in the household trash.

	^{202}Hg	^{200}Hg
a. Atomic number		
b. Mass number		
c. Number of protons		
d. Number of neutrons		

Kelly Redinger/Design Pics/
Getty Images

2.29 How many protons, neutrons, and electrons are contained in each element?
a. $^{27}_{13}$Al   b. $^{35}_{17}$Cl   c. $^{34}_{16}$S

2.30 Give the number of protons, neutrons, and electrons in each element: (a) silver-115; (b) Au-197; (c) Rn-222; (d) osmium-192.

2.31 Write the element symbol that fits each description, using a superscript for the mass number and a subscript for the atomic number.
a. an element that contains 53 protons and 74 neutrons
b. an element with 35 electrons and a mass number of 79

2.32 Write the element symbol that fits each description. Use a superscript for the mass number and a subscript for the atomic number.
a. an element that contains 10 protons and 12 neutrons
b. an element with atomic number 24 and mass number 52

2.33 Can the neutral atoms of two different elements have the same number of electrons? Explain.

2.34 Can the neutral atoms of two different elements have the same number of neutrons? Explain.

Electronic Configuration

2.35 What element(s) in the first three periods fit each description?
a. The element contains five electrons in the $3p$ orbitals.
b. There is one valence electron.
c. The element contains four electrons in the second energy level.

2.36 What element(s) in the first three periods fit each description?
a. The element contains three electrons in the $3p$ orbitals.
b. There are two valence electrons.
c. The element contains five electrons in the second shell.

2.37 Write out the electronic configuration for each element: (a) B; (b) Mg.

2.38 Write out the electronic configuration for each element: (a) nitrogen; (b) argon.

2.39 Give the total number of electrons, the number of valence electrons, and the identity of the element with each electronic configuration.
a. $1s^2 2s^2 2p^6 3s^2 3p^4$ b. $1s^2 2s^2 2p^4$

2.40 Give the total number of electrons, the number of valence electrons, and the identity of the element with each electronic configuration.
a. $1s^2 2s^2 2p^6 3s^2 3p^5$ b. $1s^2 2s^2 2p^3$

2.41 For each element, give the following information: [1] total number of electrons; [2] group number; [3] number of valence electrons; [4] period.
a. carbon   b. calcium   c. krypton

2.42 For each element, give the following information: [1] total number of electrons; [2] group number; [3] number of valence electrons; [4] period.
a. oxygen   b. sodium   c. phosphorus

2.43 How many valence electrons does an element in each group contain: (a) 2A; (b) 4A; (c) 7A?

2.44 In what shell do the valence electrons reside for an element in period: (a) 2; (b) 3; (c) 4?

2.45 Give the number of valence electrons in each element.
a. sulfur   b. chlorine   c. barium

2.46 Give the number of valence electrons in each element.
a. neon   b. rubidium   c. aluminum

2.47 Write an electron-dot symbol for each element: (a) beryllium; (b) iodine; (c) magnesium; (d) argon.

2.48 Write an electron-dot symbol for each element: (a) K; (b) B; (c) F; (d) Ca.

Periodic Trends

2.49 Which element in each pair has its valence electrons farther from the nucleus?

   a. carbon or fluorine   b. argon or bromine

2.50 For each pair of elements in Problem 2.49, label the element from which it is easier to remove an electron.

2.51 Rank the atoms in each group in order of increasing size.

   a. boron, carbon, neon

   b. calcium, magnesium, beryllium

   c. silicon, sulfur, magnesium

   d. krypton, neon, xenon

2.52 Arrange the elements in each group in order of increasing ionization energy.

   a. phosphorus, silicon, sulfur

   b. magnesium, calcium, beryllium

   c. carbon, fluorine, beryllium

   d. neon, krypton, argon

2.53 Rank the following elements in order of increasing size: sulfur, silicon, oxygen, magnesium, and fluorine.

2.54 Rank the following elements in order of increasing ionization energy: nitrogen, fluorine, magnesium, sodium, and phosphorus.

Applications

2.55 Answer the following questions about macronutrients [1], [2], and [3] highlighted in the periodic table.

a. What elements correspond to macronutrients [1]–[3]?

b. Is each element classified as a metal, nonmetal, or metalloid?

c. Which element has the smallest atomic radius?

d. Which element has the largest atomic radius?

e. Which element has the largest ionization energy?

f. Which element has the smallest ionization energy?

g. How many valence electrons does each element possess?

2.56 Enflurane is an inhaled general anesthetic.

a. What is the chemical formula for enflurane?

b. Which element in enflurane has valence electrons that are farthest from the nucleus?

2.57 Carbon-11 is an unnatural isotope used in positron emission tomography (PET) scans. PET scans are used to monitor brain activity and diagnose dementia. How does carbon-11 compare to carbon-12 in terms of the number of protons, neutrons, and electrons? Write the element symbol of carbon-11 using superscripts and subscripts.

2.58 The element titanium is strong and has a low density, making it a "light" metal that is used for surgical implants. Answer the following questions about the element titanium.

a. What is its element symbol?

b. What group number and period are assigned to titanium?

c. What is its atomic number?

d. Is titanium classified as a main group element, transition metal, or inner transition metal?

CHALLENGE PROBLEMS

2.59 If one gallon of gasoline contains about 2.0 g of lead, how much lead (in kg) is emitted into the atmosphere from a car that travels 10,000. miles a year and averages 25 miles per gallon of gasoline?

2.60 Sesame seeds, sunflower seeds, and peanuts are good dietary sources of the trace element copper. Copper is needed for the synthesis of neurotransmitters, compounds that transmit nerve signals from one nerve cell to another. Copper is also needed for the synthesis of collagen, a protein found in bone, tendons, teeth, and blood vessels.

a. Give the element symbol, group number, and period number for copper.

b. Classify copper as a main group element, transition metal, or inner transition metal.

c. If a 60.-kg individual contains 60. mg of copper in his body, how many grams of copper are present in each gram of body mass? Write the number in scientific notation.

BEYOND THE CLASSROOM

2.61 Pick one of the trace elements in Figure 2.2 and research why it is needed in the body. From what dietary sources do we obtain the nutrient and how much do we need? What is the recommended daily intake of the trace element, and what symptoms result from its deficiency?

2.62 Soft drink cans, composed almost entirely of the element aluminum, are a staple of most households in the United States. Research where the aluminum comes from and in what form it is found in the earth. Why is aluminum used instead of other substances like copper, tin, or stainless steel? A soda can contains about 15 g of aluminum. Estimate how many aluminum cans are used in your household or class each month, and how many pounds of aluminum this amounts to. If your state or town has a recycling program, see if you can find out how many pounds of aluminum cans are recycled each month.

CHAPTER 2 ANSWERS

Concept Check Answers

2.1   a. Ca   b. Rn   c. N   d. Au

2.2   a. neon   b. sulfur   c. iodine   d. silicon   e. boron   f. mercury

2.3   a,d,f,h: metals   b,c,g: nonmetals   e. metalloid

2.4   As, B, Si: metalloids
Cr, Co, Cu, Fe, Mn, Mo, Ni, Zn: metals
F, I, Se: nonmetals

2.5   a. atomic number 27 and atomic weight 58.93 amu
b. 27 protons and 27 electrons

2.6   Iodine-123: $Z = 53$; $A = 123$; 53 protons; 70 neutrons; $^{123}_{53}I$
Iodine-131: $Z = 53$; $A = 131$; 53 protons; 78 neutrons; $^{131}_{53}I$

2.7   a. 8   b. 6

2.8   a. K   b. F   c. Ar   d. Sr   e. Zn   f. Nb

2.9   a. titanium, Ti, group 4B (or 4), period 4, transition metal
b. phosphorous, P, group 5A (or 15), period 3, main group element
c. dysprosium, Dy, no group number, period 6, inner transition metal

2.10   a. 2   b. 8   c. 2   d. 18

2.11   a. Li
b. Be, B, C, N, O, F, Ne
c. F
d. O

2.12   a. silicon   b. carbon   c. fluorine

Practice Problem Answers

2.1   a. 4 hydrogens, 1 carbon
b. 3 hydrogens, 1 nitrogen
c. 6 hydrogens, 2 carbons, 1 oxygen

2.2   a. 1 sodium, 1 carbon, 1 nitrogen   d. 1 tin, 2 fluorines
b. 2 hydrogens, 1 sulfur   e. 1 carbon, 1 oxygen
c. 2 carbons, 6 hydrogens   f. 3 carbons, 8 hydrogens, 3 oxygens

2.3   2 C, 1 H, 3 F, 1 Br, 1 Cl

2.4   a. 9   b. 9   c. fluorine

2.5

Atomic Number		Element	Protons	Electrons
a.	2	Helium	2	2
b.	11	Sodium	11	11
c.	20	Calcium	20	20
d.	47	Silver	47	47
e.	78	Platinum	78	78

2.6   a. 4 protons and 5 neutrons
b. atomic number 4, mass number 9
c. beryllium

2.7

	Protons	Neutrons	Electrons
a.	17	18	17
b.	14	14	14
c.	92	146	92

2.8   a. 95   b. 52

2.9

	Atomic Number	Mass Number	Protons	Neutrons	Electrons
a.	6	13	6	7	6
b.	51	121	51	70	51

2.10 $^{194}_{78}Pt$

2.11

	Protons	Electrons	Atomic Number	Mass Number
$^{24}_{12}Mg$	12	12	12	24
$^{25}_{12}Mg$	12	12	12	25
$^{26}_{12}Mg$	12	12	12	26

2.12   a. [3]
b. [1] 8; [2] 7; [3] 8
c. [1] $^{16}_{8}O$; [2] $^{16}_{7}N$; [3] $^{17}_{8}O$

2.13

Element	Period	Group
a. Oxygen	2	6A (or 16)
b. Calcium	4	2A (or 2)
c. Phosphorus	3	5A (or 15)
d. Platinum	6	8B (or 10)
e. Iodine	5	7A (or 17)

2.14   a. $1s^22s^22p^63s^23p^1$ b. $1s^22s^22p^63s^23p^5$

2.15   a. sodium   b. silicon

2.16   a. 3 electrons, 1 valence electron, lithium
b. 12 electrons, 2 valence electrons, magnesium
c. 15 electrons, 5 valence electrons, phosphorus

2.17   a. 7   b. 8   c. 2   d. 4

2.18   a. :B̈r: b. L̇i   c. Ȧl·   d. ·S̈:   e. :N̈e:

2.19   a. arsenic   b. silicon   c. phosphorus

2.20   a. [1]   b. [2]   c. [1] fluorine; [2] magnesium

Self-Test Answers

1.   a. nucleus   c. main group elements   e. neutrons
b. atomic number   d. period

2.   11 protons, 11 electrons, 12 neutrons

3.   a. 47   b. 109   c. 47   d. 62   e. 47

4.   a. Pb: period 6, group number 4A (or 14)
b. Ca: period 4, group number 2A (or 2)

5.   b, d

6.   element: a, c, d / compound: b, e

7.   a. arsenic   b. sulfur

8. $1s^22s^22p^63s^23p^6$

Odd-Numbered Problem Answers

2.1   a. carbon (black), oxygen (red)
b. carbon (black), hydrogen (gray), chlorine (green)

2.3 a. 5 protons and 6 neutrons   d. 5
b. 5   e. B
c. 11

2.5 a. Se   d. 6
b. group number 6A (16) and period 4   e. $^{80}_{34}$Se
c. main group element

2.7 a. aluminum   b. 3   c. group number 3A (13) and period 3

2.9 a. iodine   b. carbon

2.11 a. [3]   b. [1]   c. [3]

2.13 a. gold, astatine, silver
b. nitrogen, sodium, nickel
c. sulfur, silicon, tin
d. calcium, chromium, chlorine

2.15 a,c,e: element   b,d: compound

2.17 a. cesium   b. ruthenium   c. chlorine   d. fluorine

2.19 a. sodium: metal, alkali metal, main group element
b. silver: metal, transition metal
c. xenon: nonmetal, noble gas, main group element
d. platinum: metal, transition metal

2.21

	Element Symbol	Atomic Number	Mass Number	Number of Protons	Number of Neutrons	Number of Electrons
a.	C	6	12	6	6	6
b.	P	15	31	15	16	15
c.	Zn	30	65	30	35	30
d.	Mg	12	24	12	12	12

2.23

2.25 K, Ca, Sc, Ti, V, Cr, Mn, Fe, Co, Ni, Cu, Zn, Ga: metals
Ge, As: metalloids
Se, Br, Kr: nonmetals

2.27 For ^{58}Ni: a. 28   b. 58   c. 28   d. 30
For ^{60}Ni: a. 28   b. 60   c. 28   d. 32

2.29

	Protons	Neutrons	Electrons
a.	13	14	13
b.	17	18	17
c.	16	18	16

2.31 a. $^{127}_{53}$I   b. $^{79}_{35}$Br

2.33 No

2.35 a. chlorine   b. hydrogen, lithium, sodium   c. carbon

2.37 a. $1s^2 2s^2 2p^1$ b. $1s^2 2s^2 2p^6 3s^2$

2.39 a. 16 electrons, 6 valence electrons, sulfur
b. 8 electrons, 6 valence electrons, oxygen

2.41

	Total Electrons	Group Number	Valence Electrons	Period
a. Carbon	6	4A	4	2
b. Calcium	20	2A	2	4
c. Krypton	36	8A	8	4

2.43 a. 2   b. 4   c. 7

2.45 a. 6   b. 7   c. 2

2.47 a. Be·   b. ·Ï: c. ·Mg·   d. :Är:

2.49 a. carbon   b. bromine

2.51 a. neon < carbon < boron
b. beryllium < magnesium < calcium
c. sulfur < silicon < magnesium
d. neon < krypton < xenon

2.53 fluorine, oxygen, sulfur, silicon, magnesium

2.55 [1] sodium, metal, 1 valence electron
[2] potassium, metal, largest radius, lowest ionization energy, 1 valence electron
[3] chlorine, nonmetal, smallest radius, highest ionization energy, 7 valence electrons

2.57 carbon-11: 6 protons, 6 electrons, 5 neutrons, $^{11}_{6}$C
carbon-12: 6 protons, 6 electrons, 6 neutrons

2.59 0.80 kg

3 Ionic and Covalent Compounds

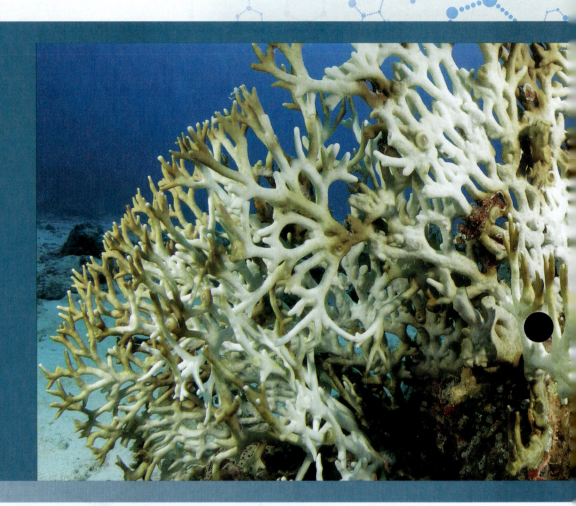

Coral bleaching, a phenomenon in which corals expel symbiotic algae from their tissues, occurs in the presence of minute amounts of some sunscreen components that contain covalent bonds.
Buttchi 3 Sha Life/Shutterstock

CHAPTER OUTLINE

CHAPTER GOALS

In this chapter you will learn how to:

1. Describe the basic features of ionic and covalent bonds
2. Use the periodic table to determine the charge of an ion using the group number
3. Describe the octet rule
4. Write the formula for an ionic compound
5. Name ionic compounds
6. Describe the properties of ionic compounds
7. Recognize the structures of common polyatomic ions and name compounds that contain them
8. Draw Lewis structures for covalent compounds
9. Name covalent compounds that contain two types of elements
10. Predict the shape around an atom in a molecule
11. Use electronegativity to determine whether a bond is polar or nonpolar
12. Determine whether a molecule is polar or nonpolar

Why Study ...

Ionic and Covalent Compounds?

Although much of the discussion in Chapter 2 focused on atoms, individual atoms are rarely encountered in nature. Instead, atoms are far more commonly joined together to form compounds. There are two types of chemical compounds, **ionic** and **covalent**. **Ionic compounds** are composed of positively and negatively charged ions held together by strong **electrostatic forces**—the electrical attraction between oppositely charged ions. Examples of ionic compounds include the sodium chloride (NaCl) in table salt and the calcium carbonate ($CaCO_3$) in snail shells. **Covalent compounds** are composed of individual molecules, discrete groups of atoms that share electrons. Covalent compounds include water (H_2O) and methane (CH_4), the main component of natural gas. Chapter 3 focuses on the structure and properties of ionic and covalent compounds.

You were first introduced to elements and compounds in Section 1.3.

3.1　Introduction to Bonding

It is rare in nature to encounter individual atoms. Instead, anywhere from two to hundreds or thousands of atoms tend to join together. The oxygen we breathe, for instance, consists of two oxygen atoms joined together, whereas the hemoglobin that transports it to our tissues consists of thousands of carbon, hydrogen, oxygen, nitrogen, and sulfur atoms joined together. We say **two atoms are** *bonded* **together.**

* *Bonding* is the joining of two atoms in a stable arrangement.

Only the noble gases in group 8A of the periodic table are particularly stable as individual atoms; that is, the **noble gases do** *not* **readily react to form bonds,** because the electronic configuration of the noble gases is especially stable to begin with. As a result, one overriding principle explains the process of bonding.

* In bonding, elements gain, lose, or share electrons to attain the electronic configuration of the noble gas closest to them in the periodic table.

Bonding involves only the valence electrons of an atom. There are two different kinds of bonding: **ionic** and **covalent.**

Transfer an electron.

ionic bond

Share two electrons.

covalent bond

* *Ionic bonds* result from the transfer of electrons from one element to another.
* *Covalent bonds* result from the sharing of electrons between two atoms.

The position of an element in the periodic table determines the type of bonds it makes. **Ionic bonds form between a metal on the left side of the periodic table and a nonmetal on the right side.** As shown in Figure 3.1, when the metal sodium (Na) bonds to the nonmetal chlorine (Cl_2), the ionic compound sodium chloride (NaCl) forms. Ionic compounds are composed of *ions*—**charged species in which the number of protons and electrons in an atom is** *not* **equal.**

Covalent bonds are formed when two nonmetals combine, or when a metalloid bonds to a nonmetal. **A** *molecule* **is a compound or element containing two or more atoms joined together with covalent bonds.** For example, when two hydrogen atoms bond they form the molecule H_2, and two electrons are shared.

hydrogen atoms

H· 　 ·H 　　 H:H 　　Two electrons are shared in a covalent bond.

hydrogen molecule

Sodium metal   Chlorine gas   Sodium chloride crystals

Na

Cl$_2$

Na$^+$
(sodium ion)

Cl$^-$
(chloride ion)

Sodium metal and chlorine gas are both elements. Sodium chloride is an ionic compound composed of sodium ions and chloride ions. *(left & middle): Stephen Frisch/McGraw Hill; (right): Vladislav Gajic/Shutterstock*

Many consumer products used to whiten teeth contain hydrogen peroxide (Sample Problem 3.1).
Jill Braaten/McGraw Hill

Sample Problem 3.1   **Predicting Whether a Bond Is Ionic or Covalent**

Predict whether the bonds in the following compounds are ionic or covalent: (a) NaI (sodium iodide); (b) H$_2$O$_2$ (hydrogen peroxide).

Analysis

The position of the elements in the periodic table determines the type of bonds they form. When a **metal** and **nonmetal** combine, the bond is **ionic.** When **two nonmetals** combine, or a **metalloid** bonds to a **nonmetal**, the bond is **covalent.**

Solution

a. Because Na is a metal on the left side and I is a nonmetal on the right side of the periodic table, the bonds in NaI are ionic.

b. Because H$_2$O$_2$ contains only the nonmetals hydrogen and oxygen, the bonds must be covalent.

Practice Problem 3.1

Predict whether the bonds in the following species are ionic or covalent.

a. CO   b. CaF$_2$ c. MgO   d. Cl$_2$ e. HF   f. NaF

More Practice: Try Problems 3.1, 3.2, 3.43, 3.44.

Sample Problem 3.2   **Using Molecular Art to Identify a Compound and an Element**

(a) Classify each example of molecular art as a compound or an element. (b) Which of the species are molecules?

A   B

Analysis

- In molecular art, an element is composed of spheres of the *same* color, whereas a compound is composed of spheres of *different* colors. The common element colors are shown in Figure 2.3 and Appendix C.
- **A molecule is a compound or element that contains covalent bonds.**

Solution

a. **A** is an element composed of two fluorines, whereas **B** is a compound composed of hydrogen and fluorine.

b. Both **A** and **B** are molecules because they are composed solely of covalent bonds between nonmetals.

Practice Problem 3.2

(a) Classify each example of molecular art as a compound or element. (b) Which of the species are molecules?

A   B   C   D   E   F

3.2   Ions

Ionic compounds consist of oppositely charged **ions** that have a strong attraction for each other.

3.2A   Cations and Anions

There are two types of ions called **cations** and **anions**.

- *Cations* are positively charged ions. A cation has fewer electrons than protons.
- *Anions* are negatively charged ions. An anion has more electrons than protons.

The charge on an ion depends on the position of an element in the periodic table. In forming an ion, an atom of a main group element loses or gains electrons to obtain the electronic configuration of the noble gas closest to it in the periodic table. This gives the ion an especially stable electronic arrangement in which the **electrons completely fill the shell farthest from the nucleus.**

For example, sodium (group 1A) has an atomic number of 11, giving it 11 protons and 11 electrons in the neutral atom. This gives sodium one *more* electron than neon, the noble gas closest to it in the periodic table. In losing one electron, sodium forms a cation with a +1 charge, which still has 11 protons, but now has only 10 electrons in its electron cloud.

What does this mean in terms of valence electrons? A neutral sodium atom, with an electronic configuration of $1s^2 2s^2 2p^6 3s^1$, has a single valence electron. Loss of this valence electron forms a **sodium cation,** symbolized as **Na⁺,** which has the especially stable electronic configuration of the noble gas neon, $1s^2 2s^2 2p^6$. The sodium cation now has **eight electrons** that fill the $2s$ and three $2p$ orbitals.

Sodium is an example of a metal.

> • Metals form *cations*. By losing one, two, or three electrons, an atom forms a cation with a completely filled outer shell of electrons.

Some metals—notably tin and lead—can lose *four* electrons to form cations.

A neutral chlorine atom (group 7A), on the other hand, has 17 protons and 17 electrons. This gives it one *fewer* electron than argon, the noble gas closest to it in the periodic table. By gaining one electron, chlorine forms an anion with a −1 charge because it still has 17 protons, but now has 18 electrons in its electron cloud.

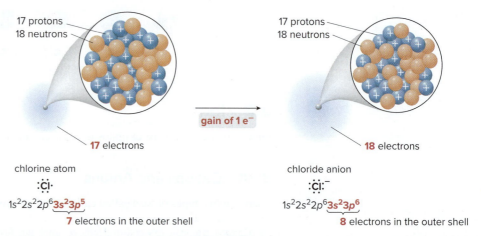

17 protons
18 neutrons

gain of 1 e⁻

17 protons
18 neutrons

17 electrons

18 electrons

chlorine atom

:C̈l·

$1s^2 2s^2 2p^6 \underline{3s^2 3p^5}$

7 electrons in the outer shell

chloride anion

:C̈l:⁻

$1s^2 2s^2 2p^6 \underline{3s^2 3p^6}$

8 electrons in the outer shell

In terms of valence electrons, a neutral chlorine atom, with an electronic configuration of $1s^2 2s^2 2p^6 3s^2 3p^5$, has seven valence electrons. Gain of one electron forms a **chloride anion,** symbolized as **Cl⁻,** which has the especially stable electronic configuration of the noble gas argon, $1s^2 2s^2 2p^6 3s^2 3p^6$. The chloride anion now has **eight valence electrons** that fill the $3s$ and three $3p$ orbitals.

Chlorine is an example of a nonmetal.

> • Nonmetals form *anions*. By gaining one, two, or sometimes three electrons, an atom forms an anion with a completely filled outer shell of electrons.

Ions are written with the element symbol followed by a superscript to indicate the charge. The number "1" is omitted in ions that have a +1 or −1 charge, as in Na⁺ or Cl⁻. When the charge is "2" or greater, it is written as 2+ or 2−, as in Mg²⁺ or O²⁻.

Each of these ions formed from a main group element has **eight valence electrons.** This illustrates the **octet rule.**

> • A main group element is especially stable when it possesses an *octet* of electrons in its outer shell.

Sample Problem 3.3	Writing an Ion Symbol by Counting Protons and Electrons

Identify the ion shown in the molecular art and give its ion symbol.

2 electrons

Analysis

- Count the **number of protons** to determine the **atomic number** and the **identity of the element.**
- **Compare the number of protons with the number of electrons to determine the charge.** When the number of protons is *greater* than the number of electrons, the ion has a net **positive** charge. When the number of electrons is *greater* than the number of protons, the ion has a net **negative** charge.

Solution

The element with three protons in the nucleus is lithium. Because there are three protons in the nucleus and only two electrons in the electron cloud, the ion has a net +1 charge. **Answer:** Li⁺.

Practice Problem 3.3

Give the ion symbol for the ion shown in the molecular art.

10 electrons

More Practice: Try Problem 3.3.

Sample Problem 3.4

Writing an Ion Symbol

Write the ion symbol for an atom with nine protons and 10 electrons.

Analysis

Because the number of protons equals the atomic number (Section 2.2), this quantity identifies the element. The charge is determined by comparing the number of protons and electrons. If the number of electrons is *greater* than the number of protons, the charge is **negative (an anion).** If the number of protons is *greater* than the number of electrons, the charge is **positive (a cation).**

Solution

An element with nine protons has an atomic number of nine, identifying it as fluorine (F). Because there is one more electron than proton (10 vs. 9), the charge is −1.   **Answer:** F⁻.

Practice Problem 3.4

Write the ion symbol for an atom with the given number of protons and electrons.

a. 19 protons and 18 electrons   c. 35 protons and 36 electrons
b. seven protons and 10 electrons   d. 23 protons and 21 electrons

More Practice: Try Problems 3.4, 3.17.

Sample Problem 3.5

Using an Ion Symbol to Count Protons and Electrons

How many protons and electrons are present in each ion: (a) Ca^{2+}; (b) O^{2-}?

Analysis

Use the identity of the element to determine the number of protons. The charge tells how many more or fewer electrons there are compared to the number of protons. A **positive charge** means *more* protons than electrons, whereas a **negative charge** means *more* electrons than protons.

Solution

a. Ca^{2+}: The element calcium (Ca) has an atomic number of 20, so it has 20 protons. Because the charge is +2, there are two more protons than electrons, giving the ion 18 electrons.

b. O^{2-}: The element oxygen (O) has an atomic number of eight, so it has eight protons. Because the charge is −2, there are two more electrons than protons, giving the ion 10 electrons.

Practice Problem 3.5

How many protons and electrons are present in each ion?

a. Ni^{2+} b. Se^{2-} c. Fe^{3+}

More Practice: Try Problem 3.18.

3.2B   Relating Group Number to Ionic Charge for Main Group Elements

Because elements with similar electronic configurations are grouped together in the periodic table, **elements in the same group form ions of similar charge.** The group number of a main group element can be used to determine the charge on an ion derived from that element.

- Metals form cations. For metals in groups 1A, 2A, and 3A, the group number = the charge on the cation.

Group **1**A elements (Li, Na, K, Rb, and Cs) have **one** valence electron. Loss of this electron forms a cation with a **+1** charge. Group **2**A elements (Be, Mg, Ca, Sr, and Ba) have **two** valence electrons. Loss of both electrons forms a cation with a **+2** charge. Group **3**A elements (Al, Ga, In, and Tl) form cations, too, but only aluminum is commonly found in ionic compounds. It has **three** valence electrons, so loss of three electrons from aluminum forms a cation with a **+3** charge.

- Nonmetals form anions. For nonmetals in groups 5A, 6A, and 7A, the anion charge = 8 − (the group number).

Group **5**A elements have **five** valence electrons. A gain of **three** electrons forms an anion with a **−3** charge (anion charge = 8 − 5). Group **6**A elements have **six** valence electrons. A gain of **two** electrons forms an anion with a **−2** charge (anion charge = 8 − 6). Group **7**A elements have **seven** valence electrons. A gain of one electron forms an anion with a **−1** charge (anion charge = 8 − 7).

The periodic table in Figure 3.2 gives the common ions formed by the main group elements.

Sample Problem 3.6

Determining Ion Charge from the Group Number

Use the group number to determine the charge on an ion derived from each element: (a) barium; (b) sulfur.

Analysis

Locate the element in the periodic table.

- A metal in groups 1A, 2A, or 3A forms a cation equal in charge to the **group number.**
- A nonmetal in groups 5A, 6A, and 7A forms an anion whose charge equals **8 − (the group number).**

Solution

a. Barium (Ba) is located in group 2A, so it forms a cation with a +2 charge; Ba^{2+}.

b. Sulfur (S) is located in group 6A, so it forms an anion with a negative charge of 8 − 6 = 2; S^{2-}.

Practice Problem 3.6

Use the group number to determine the charge on an ion derived from each element.

a. magnesium   b. iodine   c. selenium   d. rubidium

More Practice: Try Problem 3.23.

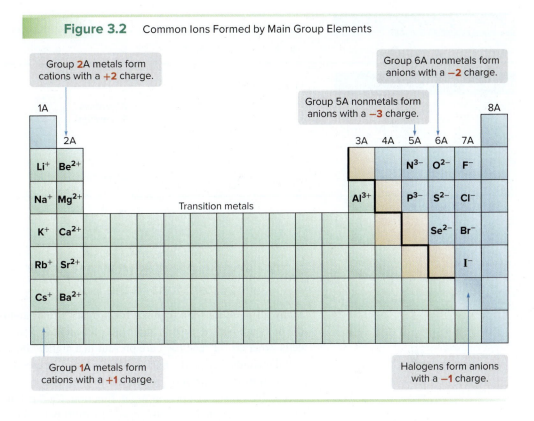

Figure 3.2   Common Ions Formed by Main Group Elements

3.2C   Metals with Variable Charge

The transition metals form cations like other metals, but the magnitude of the charge on the cation is harder to predict. Some transition metals, and a few main group metals as well, form more than one type of cation. For example, iron forms two different cations, Fe^{2+} and Fe^{3+}. Figure 3.3 illustrates the common cations formed from transition metals, as well as some main group elements that form more than one cation.

Figure 3.3   Common Cations Derived from Transition Metals and Group 4A Metals

How many electrons and protons are contained in each cation?

a. Au^+ b. Au^{3+} c. Sn^{2+} d. Sn^{4+}

Identify each ion.

a.   27 protons
32 neutrons
24 electrons

b.   24 protons
28 neutrons
22 electrons

3.2D   FOCUS ON THE HUMAN BODY
Important Ions in the Body

Many different ions are required for proper cellular and organ function. The major cations in the body are Na^+, K^+, Ca^{2+}, and Mg^{2+}. K^+ and Mg^{2+} are present in high concentrations inside cells, whereas Na^+ and Ca^{2+} are present in a higher concentration outside of cells, in the extracellular fluids. Na^+ is the major cation present in blood and extracellular bodily fluids and its concentration is carefully regulated to maintain blood volume and blood pressure within acceptable ranges that permit organ function. Ca^{2+} is found mainly in solid body parts such as teeth and bones, but it is also needed for proper nerve conduction and muscle contraction, as is Mg^{2+}.

In addition to these four cations, Fe^{2+} and Cl^- are also important ions. Fe^{2+} is essential for oxygen transport by red blood cells. Cl^- is present in red blood cells, gastric juices, and other body fluids. Along with Na^+, it plays a major role in regulating the fluid balance in the body.

Although Na^+ is an essential mineral needed in the daily diet, the average American consumes three to five times the recommended daily allowance (RDA) of 2,400 mg. Excess sodium intake is linked to high blood pressure and heart disease. Dietary Na^+ comes from salt, NaCl, added during cooking or at the table. Na^+ is also added during the preparation of processed foods and canned products. For example, one 3.5-oz serving of fresh asparagus has only 1 mg of Na^+, but the same serving size of canned asparagus contains 236 mg of Na^+. Potato chips, snack foods, ketchup, processed meats, and many cheeses are particularly high in Na^+. Table 3.1 lists the Na^+ content of some common foods.

All of these foods are high in sodium.
Jill Braaten/McGraw Hill

Table 3.1 Na^+ Content in Common Foods

Foods High in Na^+		Foods Low in Na^+	
Food	Na^+ (mg)	Food	Na^+ (mg)
Potato chips (30)	276	Banana (1)	1
Hot dog (1)	504	Orange juice (1 cup)	2
Ham, smoked (3 oz)	908	Oatmeal, cooked (1 cup)	2
Chicken soup, canned (1 cup)	1,106	Cereal, shredded wheat (3.5 oz)	3
Tomato sauce, canned (1 cup)	1,402	Raisins, dried (3.5 oz)	27
Parmesan cheese (1 cup)	1,861	Salmon (3 oz)	55

Concept Check 3.3

Mn²⁺ is an essential nutrient needed for blood clotting and the formation of the protein collagen. (a) How many protons and electrons are found in a neutral manganese atom? (b) How many protons and electrons are found in the cation Mn^{2+}?

3.3   Ionic Compounds

When a metal (on the left side of the periodic table) transfers one or more electrons to a nonmetal (on the right side), **ionic bonds** are formed.

- Ionic compounds are composed of cations and anions.

The ions in an ionic compound are arranged to maximize the attractive force between the oppositely charged species. For example, sodium chloride, NaCl, is composed of sodium cations (Na^+) and chloride anions (Cl^-), packed together in a regular arrangement in a crystal lattice. Each Na^+ cation is surrounded by six Cl^- anions, and each Cl^- anion is surrounded by six Na^+ cations.

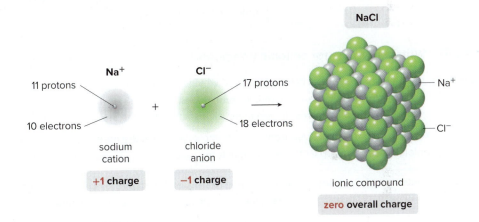

- The sum of the charges in an ionic compound must always be zero overall.

The formula for an ionic compound shows the ratio of ions that combine to give zero charge. Because the sodium cation has a +1 charge and the chloride anion has a −1 charge, there must be one Na^+ cation for each Cl^- anion; thus, the formula is **NaCl.**

When cations and anions having charges of different magnitude combine, the number of cations per anion is not equal. Consider an ionic compound formed from calcium (Ca) and fluorine (F). Because calcium is located in group 2A, it loses two valence electrons to form Ca^{2+}. Because fluorine is located in group 7A, it gains one electron to form F^- like other halogens. When Ca^{2+} combines with the fluorine anion F^-, **there must be two F^- anions for each Ca^{2+} cation to have an overall charge of zero.**

In writing a formula for an ionic compound, we use subscripts when the number of ions needed to achieve zero charge is greater than one. Because two F^- anions are needed for each calcium cation, the formula is **CaF_2.**

Figure 3.4 illustrates additional examples of how cations and anions combine to give zero overall charge.

Figure 3.4

Examples of Ionic Compounds
with Zero Overall Charge

NaCl	Li$_2$O	BaI$_2$	Al$_2$O$_3$
Na$^+$ Cl$^-$	Li$^+$ O^{2-}	Ba^{2+} I$^-$	Al^{3+} O^{2-}
+1 −1	+2 −2	+2 −2	+6 −6

The ratio of oppositely charged ions that combine to form an ionic compound depends on the charge of the ions.

- NaCl: One Na$^+$ cation (+1 charge) combines with one Cl$^-$ anion (−1 charge).
- Li$_2$O: Two Li$^+$ cations (+2 charge total) combine with one O^{2-} anion (−2 charge).
- BaI$_2$: One Ba^{2+} cation (+2 charge) combines with two I$^-$ anions (−2 charge total).
- Al$_2$O$_3$: Two Al^{3+} cations (+6 charge total) combine with three O^{2-} anions (−6 charge total).

3.3A   Formulas for Ionic Compounds

Writing a formula for an ionic compound from two elements is a useful skill that can be practiced by following a series of steps.

How To   Write a Formula for an Ionic Compound

Step [1]   Identify which element is the cation and which is the anion.

- **Metals form *cations* and nonmetals form *anions*.**
- **Use the group number** of a main group element to determine the charge.

An ionic compound derived from calcium and oxygen has the metal calcium as the cation and the nonmetal oxygen as the anion. Calcium (group 2A) loses two electrons to form Ca^{2+}. Oxygen (group 6A) gains two electrons to form O^{2-}.

Step [2]   Determine how many of each ion type are needed for an overall charge of zero.

- When the cation and anion have the *same* charge only *one* of each is needed.

The charges are equal in magnitude,
+2 and −2.

$$Ca^{2+} + O^{2-} \longrightarrow \boxed{CaO}$$

One of each ion is needed
to balance charge.

- When the cation and anion have *different* charges, as is the case with the Ca^{2+} cation and Cl$^-$ anion, use the ion charges to determine the number of ions of each needed. **The charges on the ions tell us how many of the *oppositely* charged ions are needed to balance charge.**
- Write a **subscript for the cation** that is equal in magnitude to the **charge on the anion**. Write a **subscript for the anion** that is equal in magnitude to the **charge on the cation.**

The charges are not equal in
magnitude, +2 and −1.

$$Ca^{2+} \qquad Cl^{1-}$$ The "1" is written for emphasis.

$$Ca_1Cl_2 = \boxed{CaCl_2}$$ 2 Cl$^-$ anions for each Ca^{2+}

Step [3]   To write the formula, place the cation first and then the anion, and omit charges.

- Use subscripts to show the number of each ion needed to have zero overall charge. When no subscript is written it is assumed to be "1."

As shown in step [2], the formula for the ionic compound formed from one calcium cation (Ca^{2+}) and one oxygen anion (O^{2-}) is **CaO**. The formula for the ionic compound formed from one calcium cation (Ca^{2+}) and two chlorine anions (Cl$^-$) is **CaCl$_2$.**

The tarnish on sterling silver is composed of an ionic compound formed from silver and sulfur (Sample Problem 3.7).
Alejandro Rivera/iStock/Getty Images

Sample Problem 3.7   Writing the Formula for an Ionic Compound

When sterling silver tarnishes it forms an ionic compound derived from silver and sulfur. Write the formula for this ionic compound.

Analysis and Solution

[1] **Identify the cation and the anion and use the periodic table to determine their charges.**

- Silver is a metal, so it forms the cation, and from Figure 3.3, the charge on silver is +1 (Ag^+).
- Sulfur (group 6A) is a main group element with a −2 charge (S^{2-}).

[2] **Balance charges.**

- Because the charges are unequal, use their magnitudes to determine the relative number of each ion to give an overall charge of zero.
- **The charges on the ions tell us how many of the *oppositely* charged ions are needed to balance charge.**

The "1" is written for emphasis. Ag_1^{1+} S^{2-}

Ag_2S_1 **2** Ag^+ cations for each S^{2-}

[3] **Write the formula with the cation first and then the anion, and use subscripts to show the number of each ion needed.**

- To balance charge, two Ag^+ cations are needed for each S^{2-} anion, so the formula is **Ag_2S.**

Practice Problem 3.7

Write the formula for the ionic compound formed from each pair of elements.

a. sodium and bromine

b. barium and oxygen

c. magnesium and iodine

d. lithium and oxygen

More Practice: Try Problems 3.5, 3.6, 3.29.

Potassium is a critical cation for normal heart and skeletal muscle function and nerve impulse conduction. Drinking electrolyte replacement beverages like Gatorade or Powerade can replenish K^+ lost in sweat. *Corbis/VCG/Getty Images*

3.3B   FOCUS ON HEALTH & MEDICINE
Ionic Compounds in Consumer Products

Simple ionic compounds are added to food or consumer products to prevent disease or maintain good health. For example, **potassium iodide** (KI) is an essential nutrient added to table salt. Iodine is needed to synthesize thyroid hormones. A deficiency of iodine in the diet can lead to insufficient thyroid hormone production. In an attempt to compensate, the thyroid gland may become enlarged, producing a swollen thyroid referred to as a goiter. **Sodium fluoride** (NaF) is added to toothpaste to strengthen tooth enamel and help prevent tooth decay.

K^+ potassium iodide   sodium fluoride Na^+
I^- in table salt   in toothpaste F^-

Jill Braaten/McGraw Hill

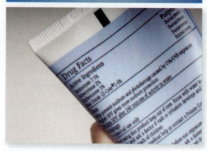
Potassium chloride (KCl), sold under trade names such as K–Dur, Klor–Con, and Micro–K, is an ionic compound used for patients whose potassium levels are low. Potassium chloride can be given as tablets, an oral suspension, or intravenously. Adequate potassium levels are needed for proper fluid balance and organ function. Although potassium is readily obtained from many different food sources (e.g., potatoes, beans, melon, bananas, and spinach), levels can become low when too much potassium is lost in sweat and urine or through the use of certain medications.

Concept Check 3.4

Unlike some sunscreen components like oxybenzone that cause coral bleaching when washed into the ocean, zinc oxide, an ionic compound formed from zinc and oxygen, is a sunblock component considered reef-safe. What is the ionic formula for zinc oxide?

3.4   Naming Ionic Compounds

Now that we have learned how to write the formulas of some simple ionic compounds, we must learn how to name them. Assigning an unambiguous name to each compound is called chemical **nomenclature.**

3.4A   Naming Cations

- Cations of main group metals are given the name of the element from which they are formed.

Na^+	K^+	Ca^{2+}	Mg^{2+}
sodium	potassium	calcium	magnesium

It is common to add the word "ion" after the name of the metal cation to distinguish it from the neutral metal itself. For example, when the concentration of sodium in a blood sample is determined, what is really measured is the concentration of sodium *ions* (Na^+).

When a metal is able to form two different cations, a method is needed to distinguish these cations. Two systems are used, the systematic method and the common method. The systematic method (Method [1]) will largely be followed in this text. Because many ions are still identified by older names, however, the common method (Method [2]) is also given.

- Method [1]: Follow the name of the cation by a Roman numeral in parentheses to indicate its charge.
- Method [2]: Use the suffix *-ous* for the cation with the lesser charge, and the suffix *-ic* for the cation with the higher charge. These suffixes are often added to the Latin names of the elements.

For example, the element iron (Fe) forms two cations, Fe^{2+} and Fe^{3+}, which are named in the following way:

	Systematic Name	Common Name
Fe^{2+}	iron(II)	ferr**ous**
Fe^{3+}	iron(III)	ferr**ic**

Table 3.2 lists the systematic and common names for several cations.

Table 3.2 Systematic and Common Names for Some Metal Ions

Element	Ion Symbol	Systematic Name	Common Name
Copper	Cu^+	Copper(I)	Cuprous
	Cu^{2+}	Copper(II)	Cupric
Chromium	Cr^{2+}	Chromium(II)	Chromous
	Cr^{3+}	Chromium(III)	Chromic
Iron	Fe^{2+}	Iron(II)	Ferrous
	Fe^{3+}	Iron(III)	Ferric
Mercury	Hg_2^{2+}	Mercury(I)[a]	Mercurous
	Hg^{2+}	Mercury(II)	Mercuric
Tin	Sn^{2+}	Tin(II)	Stannous
	Sn^{4+}	Tin(IV)	Stannic

[a]Mercury(I) exists as Hg_2^{2+}, containing two atoms of mercury, each with a +1 charge.

3.4B Naming Anions

- Anions are named by replacing the ending of the element name by the suffix *-ide*. For example:

Cl $---\rightarrow$ Cl^- [Change *-ine* to *-ide*.]

chlor*ine*   chlor*ide*

O $---\rightarrow$ O^{2-} [Change *-ygen* to *-ide*.]

ox*ygen*   ox*ide*

Table 3.3 lists the names of common anions derived from nonmetal elements.

Table 3.3 Names of Common Anions

Element	Ion Symbol	Name
Bromine	Br^-	Bromide
Chlorine	Cl^-	Chloride
Fluorine	F^-	Fluoride
Iodine	I^-	Iodide
Nitrogen	N^{3-}	Nitride
Oxygen	O^{2-}	Oxide
Phosphorus	P^{3-}	Phosphide
Sulfur	S^{2-}	Sulfide

Concept Check 3.5

Give the name of each ion.

a. S^{2-} b. Cu^+ c. Cs^+ d. Al^{3+} e. Sn^{4+}

Concept Check 3.6

Give the symbol for each ion.

a. stannous   b. iodide   c. manganese ion   d. lead(II)

3.4C Naming Ionic Compounds with Cations from Main Group Metals

To name an ionic compound with a main group metal cation whose charge never varies, **name the cation and then the anion.** Do *not* specify the charge on the cation. Do *not* specify how many ions of each type are needed to balance charge.

Na^+ F^- $---\rightarrow$ NaF

sodium   fluoride   sodium fluoride

Mg^{2+} Cl^- $---\rightarrow$ $MgCl_2$

magnesium   chloride   magnesium chloride

Sample Problem 3.8	**Naming an Ionic Compound That Contains a Main Group Metal**

Name each ionic compound: (a) Na_2S; (b) $AlBr_3$.

Analysis
Name the cation and then the anion.

Solution

a. Na_2S: The cation is sodium and the anion is sulfide (derived from sulfur); thus, the name is sodium sulfide.

b. $AlBr_3$: The cation is aluminum and the anion is bromide (derived from bromine); thus, the name is aluminum bromide.

Practice Problem 3.8	Name each ionic compound.

a. NaF   b. MgO   c. $SrBr_2$ d. Li_2O e. TiO_2

More Practice: Try Problems 3.33a, b, d, f; 3.34a, b, f.

3.4D Naming Ionic Compounds Containing Metals with Variable Charge

To name an ionic compound that contains a metal with variable charge, **we must specify the charge on the cation.** The formula of the ionic compound—that is, how many cations there are per anion—allows us to determine the charge on the cation.

How To Name an Ionic Compound That Contains a Metal with Variable Charge

Example Give the name for $CuCl_2$.

Step [1] Determine the charge on the cation.

• Because there are two Cl^- anions, each of which has a −1 charge, the copper cation must have a +2 charge to make the overall charge zero.

$CuCl_2$ **2** Cl^- anions   − − −→   The total negative charge is **−2**.

Cu must have a **+2** charge to balance the **−2** charge of the anions.

Cu^{2+}

Step [2] Name the cation and anion.

• Name the cation using its element name followed by a Roman numeral to indicate its charge. In the common system, use the suffix *-ous* or *-ic* to indicate charge.

• Name the anion by changing the ending of the element name to the suffix *-ide*.

Cu^{2+} − − −→   copper(II)   or   cupric

Cl^- − − −→   chloride

Step [3] Write the name of the cation first, then the anion.

• **Answer:** Copper(II) chloride or cupric chloride.

Sample Problem 3.9 illustrates the difference in naming ionic compounds derived from metals that have fixed or variable charge.

Some toothpastes contain the ionic compounds SnF_2 as a source of fluoride and Al_2O_3 as an abrasive. *Daniel Pangbourne/Digital Vision/Getty Images*

Sample Problem 3.9   **Naming an Ionic Compound**

SnF_2 and Al_2O_3 are both ingredients in commercial toothpastes. SnF_2 contains fluoride, which strengthens tooth enamel. Al_2O_3 is an abrasive that helps to scrub the teeth clean when they are brushed. Give names for (a) SnF_2; (b) Al_2O_3.

Analysis

First determine if the cation has a fixed or variable charge.

- To name an ionic compound that contains a cation that always has the same charge, name the cation and then the anion (using the suffix *-ide*).
- When the metal has a variable charge, use the overall anion charge to determine the charge on the cation. Then name the cation (using a Roman numeral or the suffix *-ous* or *-ic*), followed by the anion.

Solution

a. SnF_2: Sn cations have variable charge so the **overall anion charge determines the cation charge.**

SnF_2 **2** F⁻ anions   - - - →   The total negative charge is **−2**.

Sn must have a **+2** charge to balance the **−2** charge of the anions.

tin(II) or stannous   fluoride

Answer: tin(II) fluoride or stannous fluoride

b. Al_2O_3: Al has a **fixed charge** of +3. To name the compound, name the cation as the element (aluminum), and the anion by changing the ending of the element name to the suffix *-ide* (oxygen → oxide).

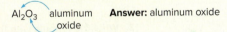

Al_2O_3 aluminum   **Answer:** aluminum oxide
               oxide

Practice Problem 3.9

Name each ionic compound.

a. $CrCl_3$ b. PbS   c. SnF_4 d. PbO_2 e. $FeBr_2$

More Practice: Try Problems 3.33c, e; 3.34c, d, e.

3.4E   Writing a Formula from the Name of an Ionic Compound

Writing a formula from the name of an ionic compound is also a useful skill.

How To Derive a Formula from the Name of an Ionic Compound

Example Write the formula for tin(IV) oxide.

Step [1] Identify the cation and the anion and determine their charges.

- The name of the cation appears *first*, followed by the anion.
- For metals with variable charge, the **Roman numeral gives the charge on the cation.**

In this example, tin is the cation. The Roman numeral tells us that its charge is +4, making the cation Sn^{4+}. Oxide is the name of the oxygen anion, O^{2-} (Table 3.3).

—Continued

How To, continued . . .

Step [2] Balance charges.

- Use the charge on the cation to determine the number of ions of the anion needed to balance charge.

$$Sn^{4+} \quad O^{2-} \quad \text{Two } \mathbf{-2} \text{ anions are needed}$$
$$\text{for each } \mathbf{+4} \text{ cation.}$$
$$\text{cation} \quad \text{anion}$$

Step [3] Write the formula with the cation first, and use subscripts to show the number of each ion needed to have zero overall charge.

Answer: SnO_2

Sample Problem 3.10 — Deriving a Formula from the Name of an Ionic Compound

Write the formula for sodium oxide.

Analysis and Solution

[1] Identify the cation and the anion and determine their charges.

- Because the name of the cation appears first, sodium is the cation and oxide is the anion.
- Sodium has a fixed charge of +1 (Na^+) and oxide has a charge of −2 (O^{2-}).

[2] Balance charges.

- Use the charge on the anion to determine the number of ions of the cation needed to balance charge.

$$\text{Two } \mathbf{+1} \text{ cations are needed} \quad Na^+ \quad O^{2-}$$
$$\text{for each } \mathbf{-2} \text{ anion.} \quad \text{cation} \quad \text{anion}$$

[3] Write the formula with the cation first, and use subscripts to show the number of each ion needed.

- To balance charge, two Na^+ cations are needed for each oxide anion (O^{2-}), so the formula of sodium oxide is $\mathbf{Na_2O}$.

Practice Problem 3.10

Write the formula for each ionic compound.

a. calcium bromide
b. copper(I) iodide
c. ferric bromide
d. magnesium sulfide
e. chromium(II) chloride
f. sodium oxide

3.5 Physical Properties of Ionic Compounds

Ionic compounds are crystalline solids held together by extremely strong interactions of the oppositely charged ions. How is this reflected in the melting point and boiling point of an ionic compound?

When a compound melts to form a liquid, energy is needed to overcome some of the attractive forces of the ordered solid, to form the less ordered liquid phase. Because an ionic compound is held together by very strong electrostatic interactions, it takes a great deal of energy to separate the ions from each other. As a result, **ionic compounds have very high melting points.** For example, the melting point of NaCl is 801 °C.

A great deal of energy is needed to overcome the attractive forces present in the liquid phase, too, to form ions that are far apart and very disorganized in the gas phase, so **ionic compounds have extremely high boiling points.** The boiling point of liquid NaCl is 1,413 °C.

Figure 3.5   Dissolving NaCl in Water

When NaCl dissolves in water, each Na⁺ ion and each Cl⁻ ion are surrounded by water molecules. The interactions of these ions with water molecules provide the energy needed to break apart the ions of the crystal lattice. *Jill Braaten/McGraw Hill*

A great many ionic compounds are soluble in water. When an ionic compound dissolves in water, the ions are separated, and each anion and cation is surrounded by water molecules, as shown in Figure 3.5. The interaction of the water solvent with the ions provides the energy needed to overcome the strong ion–ion attractions of the crystalline lattice.

When an ionic compound dissolves in water, the resulting aqueous solution conducts an electric current. This distinguishes ionic compounds from other compounds that dissolve in water but do not form ions and therefore do not conduct electricity.

An **aqueous solution** contains a substance dissolved in liquid water. Solutions are discussed in greater detail in Chapter 7.

Concept Check 3.7   List four physical properties of ionic compounds.

3.6   Polyatomic Ions

Sometimes ions are composed of more than one element. The ion bears a charge because the total number of electrons it contains is different from the total number of protons in the nuclei of all of the atoms.

- A *polyatomic ion* is a cation or anion that contains more than one atom.

The atoms in the polyatomic ion are held together by covalent bonds, but because the ion bears a charge, it bonds to other ions by ionic bonding. For example, calcium sulfate, $CaSO_4$, is composed of a calcium cation, Ca^{2+}, and the polyatomic anion sulfate, SO_4^{2-}. $CaSO_4$ is used to make plaster casts for broken bones.

H_3O^+
hydronium ion

NH_4^+
ammonium ion

SO_4^{2-}
sulfate

SO_3^{2-}
sulfite

HSO_4^-
hydrogen sulfate
(bisulfate)

We will encounter only two polyatomic cations: **H_3O^+, the hydronium ion,** which will play a key role in the acid–base chemistry discussed in Chapter 8, and **NH_4^+, the ammonium ion.**

In contrast, there are several common polyatomic anions, most of which contain a nonmetal like carbon, sulfur, or phosphorus, usually bonded to one or more oxygen atoms. Common examples include **carbonate (CO_3^{2-}), sulfate (SO_4^{2-}),** and **phosphate (PO_4^{3-}).** Table 3.4 lists the most common polyatomic anions.

Table 3.4 Names of Common Polyatomic Anions

Nonmetal	Formula	Name
Carbon	CO_3^{2-}	**Carbonate**
	HCO_3^-	Hydrogen carbonate or bicarbonate
	$CH_3CO_2^-$	Acetate
	CN^-	Cyanide
Nitrogen	NO_3^-	**Nitrate**
	NO_2^-	Nitrite
Oxygen	OH^-	**Hydroxide**
Phosphorus	PO_4^{3-}	**Phosphate**
	HPO_4^{2-}	Hydrogen phosphate
	$H_2PO_4^-$	Dihydrogen phosphate
Sulfur	SO_4^{2-}	**Sulfate**
	HSO_4^-	Hydrogen sulfate or bisulfate
	SO_3^{2-}	Sulfite
	HSO_3^-	Hydrogen sulfite or bisulfite

The names of most polyatomic anions end in the suffix *-ate*. Exceptions to this generalization include hydroxide (OH^-) and cyanide (CN^-). Two other aspects of nomenclature are worthy of note.

- The suffix *-ite* is used for an anion that has one fewer oxygen atom than a similar anion named with the *-ate* ending. Thus, SO_4^{2-} is sulf*ate*, but SO_3^{2-} is sulf*ite*.
- When two anions differ in the presence of a hydrogen, the word *hydrogen* or the prefix *bi-* is added to the name of the anion. Thus, SO_4^{2-} is sulfate, but HSO_4^- is *hydrogen* sulfate or *bi*sulfate.

3.6A   Writing Formulas for Ionic Compounds with Polyatomic Ions

Writing the formula for an ionic compound with a polyatomic ion is no different than writing a formula for an ion with a single charged atom, so we follow the procedure outlined in Section 3.3A. When the cation and anion have the *same* charge, only *one* of each ion is needed for an overall charge of zero.

The charges are equal in magnitude,
+1 and **−1.**

Na^+ + NO_2^- ⟶ $NaNO_2$
sodium   nitrite   sodium nitrite

One of each ion is needed
to balance charge.

In a compound formed from ions of unequal charge, such as magnesium (Mg^{2+}) and hydroxide (OH^-), **the charges on the ions tell us how many of the *oppositely* charged ions are needed to balance the charge.**

The charges are not equal in magnitude, **+2** and **−1**.

$$Mg^{2+} \quad + \quad OH^- \quad \longrightarrow \quad Mg(OH)_2$$

Two OH⁻ anions are needed to balance charge.

Use parentheses around all atoms of the ion.

Use a subscript outside the parentheses.

Parentheses are used around the polyatomic ion, and a subscript indicates how many of each are needed to balance charge. The formula is written as $Mg(OH)_2$ *not* MgO_2H_2.

Concept Check 3.8

Write the formula for the compound formed when K^+ combines with each anion.

a. OH^- b. NO_2^- c. SO_4^{2-} d. HSO_3^- e. PO_4^{3-}

Concept Check 3.9

Write the formula of the ionic compound formed from each pair of cations and anions.

a. sodium and bicarbonate   c. ammonium and sulfate   e. calcium and bisulfate

b. potassium and nitrate   d. magnesium and phosphate   f. barium and hydroxide

---

3.6B Naming Ionic Compounds with Polyatomic Ions

Naming ionic compounds derived from polyatomic anions follows the same procedures outlined in Sections 3.4C and 3.4D. There is no easy trick for remembering the names and structures of the anions listed in Table 3.4. The names of the anions in boldface type are especially common and should be committed to memory.

Barium sulfate (Sample Problem 3.11) is used to visualize the digestive system during an X-ray procedure.
Samunella/Shutterstock

Sample Problem 3.11   Naming an Ionic Compound That Contains a Polyatomic Ion

Name each ionic compound: (a) $NaHCO_3$, the active ingredient in baking soda; (b) $BaSO_4$, a compound used in X-ray imaging.

Analysis

First determine if the cation has a fixed or variable charge.

- To name an ionic compound that contains a cation that always has the same charge, name the cation and then the anion.
- When the metal has a variable charge, use the overall anion charge to determine the charge on the cation. Then name the cation (using a Roman numeral or the suffix *-ous* or *-ic*), followed by the anion.

Solution

a. $NaHCO_3$: Sodium cations have a fixed charge of +1. The anion HCO_3^- is called bicarbonate or hydrogen carbonate.

   Answer: sodium bicarbonate or sodium hydrogen carbonate

b. $BaSO_4$: Barium cations have a fixed charge of +2. The anion SO_4^{2-} is called sulfate.

   Answer: barium sulfate

Practice Problem 3.11

Name each compound.

a. Na_2CO_3 c. $Mg(NO_3)_2$ e. $Fe(HSO_3)_3$

b. $Ca(OH)_2$ d. $Mn(CH_3CO_2)_2$ f. $Mg_3(PO_4)_2$

More Practice: Try Problems 3.9, 3.37, 3.38.

The shells of oysters and other mollusks are composed largely of calcium carbonate, $CaCO_3$. *Joy Atkinson/Getty Images*

3.6C   FOCUS ON HEALTH & MEDICINE
Useful Ionic Compounds

Ionic compounds are the active ingredients in several over-the-counter drugs. Examples include **calcium carbonate ($CaCO_3$),** the antacid in Tums; **magnesium hydroxide [$Mg(OH)_2$],** one of the active components in the antacids Maalox and milk of magnesia; and **iron(II) sulfate ($FeSO_4$),** an iron supplement used to treat anemia.

| $CaCO_3$ | $Mg(OH)_2$ | $FeSO_4$ |

(left & middle): Jill Braaten/McGraw Hill; (right): Jill Braaten/McGraw Hill Education

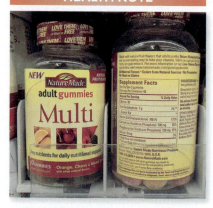

Some calcium supplements contain calcium phosphate (Concept Check 3.10), which provides the consumer with calcium and phosphorus, both needed for healthy teeth and bones.
Mark Dierker/McGraw Hill

Some ionic compounds are given as intravenous drugs. Bicarbonate (HCO_3^-) is an important polyatomic anion that controls the acid–base balance in the blood. When the blood becomes too acidic, sodium bicarbonate ($NaHCO_3$) is administered intravenously to decrease the acidity. Magnesium sulfate ($MgSO_4$), an over-the-counter laxative, is also given intravenously to prevent seizures caused by extremely high blood pressure associated with some pregnancies.

> **Concept Check 3.10**
>
> Write the formula for the dietary supplement formed from calcium and phosphate.

3.7   Covalent Bonding

In Section 3.1 we learned that **covalent bonds result from the *sharing* of electrons between two atoms.** For example, when two hydrogen atoms with one electron each (H·) combine, they form the hydrogen molecule, H_2, with a covalent bond that contains two electrons. We use a **solid line between two element symbols to represent a two-electron bond.**

- A covalent bond is a two-electron bond in which the bonding atoms share the electrons.

Hydrogen is called a **diatomic molecule** because it contains just two atoms. In addition to hydrogen, six other elements exist as diatomic molecules: nitrogen (N_2), oxygen (O_2), fluorine (F_2), chlorine (Cl_2), bromine (Br_2), and iodine (I_2).

Hydrogen fluoride, HF, is a diatomic molecule formed between two different atoms, hydrogen and fluorine. Hydrogen has one valence electron and fluorine has seven. H and F each donate one electron to form a single two-electron bond.

H· + ·F̈: ⟶ H:F̈: or H—F̈:◄— **three lone pairs**

two electrons ⤸   ⤷ **eight electrons**
around H   around F

The resulting molecule gives both H and F a filled valence shell: H is surrounded by two electrons, giving it the noble gas configuration of helium, and F is surrounded by eight electrons, giving it the noble gas configuration of neon. The F atom shares two electrons in one covalent bond, and it also contains three pairs of electrons that it does not share with hydrogen. These unshared electron pairs are called **nonbonded electron pairs** or **lone pairs.**

Nonbonded electron pair = lone pair.

- In covalent bonding, atoms share electrons to attain the electronic configuration of the noble gas closest to them in the periodic table.
- As a result, hydrogen shares two electrons.
- Other main group elements are especially stable when they possess an *octet* of electrons in their outer shell.

Concept Check 3.11   Use electron-dot symbols to show how a hydrogen atom and a chlorine atom form the diatomic molecule HCl.

Methane (CH_4) and water (H_2O) are two examples of covalent molecules in which each main group element is surrounded by eight electrons. Methane, a gas released by the decomposition of organic material in flooded rice fields, contains four covalent carbon–hydrogen bonds, each having two electrons. The oxygen atom in H_2O is also surrounded by an octet because it has two bonds and two lone pairs.

(top): Daniel C. Smith; (bottom): Tim McCabe/USDA Natural Resources Conservation Service

These electron-dot structures for molecules are called Lewis structures. Lewis structures show the location of all valence electrons in a molecule, both the shared electrons in bonds, and the nonbonded electron pairs.

How many covalent bonds will a particular atom typically form? In the first row, hydrogen forms one covalent bond with its one valence electron. Other main group elements generally have no more than eight electrons around them. For neutral molecules, two consequences result.

- Atoms with one, two, or three valence electrons generally form one, two, or three bonds, respectively.
- Atoms with four or more valence electrons form enough bonds to give an octet. Thus, for atoms with four or more valence electrons:

$$\text{Predicted number of bonds} \ = \ 8 \ - \ \text{number of valence electrons}$$

These guidelines are used in Figure 3.6 to summarize the usual number of covalent bonds formed by some common atoms. Except for hydrogen, **the number of bonds plus the number of lone pairs equals four** for common atoms.

Figure 3.6   Bonding Patterns for Common Main Group Elements

Sample Problem 3.12   **Predicting the Number of Covalent Bonds**

Without referring to Figure 3.6, how many covalent bonds are predicted for each atom: (a) B; (b) N?

Analysis

- Atoms with one, two, or three valence electrons form one, two, or three bonds, respectively.
- Atoms with four or more valence electrons form enough bonds to give an octet.

Solution

a. B has three valence electrons. Thus, it is expected to form three bonds.

b. N has five valence electrons. Since it contains more than four valence electrons, it is expected to form $8 - 5 = 3$ bonds.

Practice Problem 3.12   How many covalent bonds are predicted for each atom: (a) F; (b) Si; (c) Br; (d) O; (e) P; (f) S?

3.8 Lewis Structures

A **molecular formula** shows the number and identity of all of the atoms in a compound, but it does not tell us what atoms are bonded to each other. A **Lewis structure,** in contrast, shows the connectivity between the atoms, as well as where all the bonding and nonbonding valence electrons reside.

3.8A Drawing Lewis Structures

There are three general rules for drawing Lewis structures.

> 1. Draw only the *valence electrons*.
> 2. Give every main group element (except hydrogen) an *octet* of electrons.
> 3. Give each hydrogen two electrons.

Sample Problem 3.13 illustrates how to draw a Lewis structure in a molecule that contains only single bonds. In any Lewis structure, **always place hydrogens and halogens on the periphery, because these atoms form only one bond.**

The covalent molecule CH_3Cl (Sample Problem 3.13) is one of many gases released into the air from an erupting volcano. *G.E. Ulrich/HVO/U.S. Geological Survey*

Sample Problem 3.13   Drawing a Lewis Structure

Draw a Lewis structure for chloromethane, CH_3Cl, a compound produced by giant kelp and a component of volcanic emissions.

Analysis

To draw a Lewis structure:

- **Arrange the atoms,** placing hydrogens and halogens on the periphery.
- **Count the valence electrons** from all atoms.
- **Add the bonds, and use the remaining electrons to fill octets with lone pairs.**

Solution

[1] Arrange the atoms.

H
H   C   Cl
H

- Place C in the center and 3 H's and 1 Cl on the periphery.
- In this arrangement, C is surrounded by four atoms, its usual number.

[2] Count the electrons.

- Use the group number to determine the number of valence electrons.

Element	Group Number	Valence Electrons		Number of Atoms		Total Electrons
C	**4**A (14)	**4**	×	1	=	4 e⁻
H	**1**A (1)	**1**	×	3	=	3 e⁻
Cl	**7**A (17)	**7**	×	1	=	7 e⁻
				Total number of valence electrons =		**14 e⁻**

[3] Add the bonds and lone pairs.

First add four single bonds, three C—H bonds and one C—Cl bond. This uses eight valence electrons, and gives carbon an octet (four two-electron bonds) and each hydrogen two electrons. Next, give Cl an octet by adding three lone pairs. This uses all 14 valence electrons, and gives a valid Lewis structure for CH_3Cl.

Practice Problem 3.13

Draw a Lewis structure for each covalent molecule.

a. HBr   b. CH_3F c. H_2O_2 d. N_2H_4 e. C_2H_6 f. CH_2Cl_2

More Practice: Try Problems 3.11; 3.12; 3.45a, b; 3.46a, b.

3.8B Multiple Bonds

Sometimes it is not possible to give every main group element (except hydrogen) an octet of electrons by placing only single bonds in a molecule. In this case, the Lewis structures must contain one or more *multiple* bonds.

For example, in drawing a Lewis structure for N_2, each N has five valence electrons, so there are 10 electrons to place. If there is only one N—N bond, adding lone pairs gives one or both N's fewer than eight electrons.

For N_2:
Each N has five valence electrons.
total 10 e⁻

Add one bond.

N—N ------→ Add lone pairs. N—N or N—N

N has only 4 e⁻.

Each N has only 6 e⁻.

N has 8 e⁻.

In this case, we must convert a lone pair to a bonding pair of electrons to form a multiple bond. Because we have four fewer electrons than needed, **we must convert *two* lone pairs to *two* bonding pairs of electrons and form a *triple* bond.**

For example:

N—N ------→ Move 2 e⁻ to form a double bond. N=N ------→ Move 2 e⁻ to form a triple bond. N≡N

triple bond

Each N has only 6 e⁻.

One N has only 6 e⁻.

Each N now has 8 e⁻.

- A triple bond contains six electrons in three two-electron bonds.

Sample Problem 3.14 illustrates another example of a Lewis structure that contains a double bond.

- A double bond contains four electrons in two two-electron bonds.

- After placing all electrons in bonds and lone pairs, use a lone pair to form a multiple bond if an atom does not have an octet.

Sample Problem 3.14

Drawing a Lewis Structure for a Compound with a Multiple Bond

Draw a Lewis structure for ethylene, a compound of molecular formula C_2H_4 in which each carbon is bonded to two hydrogens.

Analysis and Solution

Follow steps [1]–[3] to draw a Lewis structure.

[1] Arrange the atoms.

H C C H
  H H

- Each C gets 2 H's.

[2] Count the electrons using the group number.

Element	Group Number	Valence Electrons		Number of Atoms		Total Electrons
C	4A (14)	4	×	2	=	8 e⁻
H	1A (1)	1	×	4	=	4 e⁻
				Total number of valence electrons =		**12 e⁻**

[3] Add the bonds and lone pairs.

Add bonds first...   ...then lone pairs.

$$H-C-C-H \quad ---\rightarrow \quad H-C-\overset{..}{C}-H$$

no octet

After placing five bonds between the atoms and adding the two remaining electrons as a lone pair, one C still has no octet.

[4] To give both C's an octet, change *one* lone pair into *one* bonding pair of electrons between the two C atoms, forming a *double* bond.

double bond

Move a lone pair.

$$H-C-\overset{..}{C}-H \quad ----------\rightarrow \quad H-C=C-H$$

ethylene

- Each C now has four bonds.
- Each C is now surrounded by eight electrons.

This uses all 12 electrons, each C has an octet, and each H has two electrons. The Lewis structure is valid. **Ethylene contains a carbon–carbon double bond.**

Practice Problem 3.14

Draw a valid Lewis structure for each compound, using the given arrangement of atoms.

a. HCN   H C N   b. CH_2O H C O

    hydrogen cyanide   formaldehyde   H

Practice Problem 3.15

Draw a Lewis structure for formic acid with the given arrangement of atoms.

O

H C O H

More Practice: Try Problems 3.45c, 3.46c, 3.69b.

Formic acid (CH_2O_2, Practice Problem 3.15) is responsible for the sting of some types of ants. *Ted Nelson/Dembinsky Photo Associates*

3.9 Naming Covalent Compounds

Although some covalent compounds are always referred to by their common names—H_2O (water) and NH_3 (ammonia)—these names tell us nothing about the atoms that the molecule contains. Other covalent compounds with two elements are named to indicate the identity and number of elements they contain.

How To Name a Covalent Molecule with Two Elements

Example Name each covalent molecule: (a) NO_2; (b) N_2O_4.

Step [1] Name the first nonmetal by its element name and the second using the suffix *-ide.*

- In both compounds the first nonmetal is nitrogen.
- To name the second element, change the name oxygen to **ox*ide.***

Step [2] Add prefixes to show the number of atoms of each element.

- Use a prefix from Table 3.5 for each element.
- The prefix *mono-* is omitted when only one atom of the first element is present, but retained for the second element.
- When the prefix and element name would place two vowels next to each other, omit the first vowel. For example, mono- + oxide = monoxide (*not* monooxide).
  a. NO_2 contains one N atom, so the prefix *mono-* is understood. Because NO_2 contains two O atoms, use the prefix *di- → di*oxide. Thus, NO_2 is **nitrogen dioxide.**
  b. N_2O_4 contains two N atoms, so use the prefix *di- → di*nitrogen. Because N_2O_4 contains four O atoms, use the prefix *tetra-* and omit the *a → tetr*oxide (*not* tetraoxide). Thus, N_2O_4 is **dinitrogen tetroxide.**

Table 3.5 Common Prefixes in Nomenclature

Number of Atoms	Prefix
1	Mono
2	Di
3	Tri
4	Tetra
5	Penta
6	Hexa
7	Hepta
8	Octa
9	Nona
10	Deca

Concept Check 3.12

Name each compound: (a) CS_2; (b) SO_2; (c) PCl_5; (d) BF_3.

To write a formula from a name, **write the element symbols in the order of the elements in the name.** Then **use the prefixes to determine the subscripts** of the formula, as shown in Sample Problem 3.15.

Sample Problem 3.15   Writing the Formula from the Name of a Covalent Compound

Give the formula for each compound: (a) silicon tetrafluoride; (b) diphosphorus pentoxide.

Analysis

- Determine the symbols for the elements in the order given in the name.
- Use the prefixes to write the subscripts.

Solution

a. silicon   **tetra**fluoride

        Si   **4** F atoms

Answer: SiF$_4$

b. **di**phosphorus **pent**oxide

        2 P atoms   **5** O atoms

Answer: P$_2$O$_5$

Practice Problem 3.16

Give the formula for each compound: (a) silicon dioxide; (b) phosphorus trichloride; (c) sulfur trioxide; (d) dinitrogen trioxide.

More Practice: Try Problems 3.49–3.52.

3.10 Molecular Shape

What is the bond angle?

H—Ö—H

We can now use Lewis structures to determine the shape around a particular atom in a molecule. Consider the H_2O molecule. The Lewis structure tells us only which atoms are connected to each other, but it implies nothing about the geometry. Is H_2O a bent or linear molecule?

To determine the shape around a given atom, we must first determine how many groups surround the atom. **A group is either an atom or a lone pair of electrons.** Then we use the **valence shell electron pair repulsion (VSEPR) theory** to determine the shape. VSEPR is based on the fact that electron pairs repel each other; thus:

- The most stable arrangement keeps these groups as far away from each other as possible.

In general, an atom has three possible arrangements of the groups that surround it.

linear   trigonal planar   tetrahedral

- An atom surrounded by two groups is linear and has a bond angle of 180°.
- An atom surrounded by three groups is trigonal planar and has bond angles of 120°.
- An atom surrounded by four groups is tetrahedral and has bond angles of 109.5°.

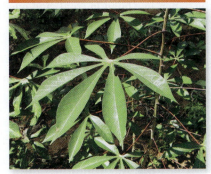

Cassava is a widely grown root crop, first introduced to Africa by Portuguese traders from Brazil in the sixteenth century. The root must be boiled or roasted to remove the compound linamarin before ingestion. Eating the root without processing affords high levels of HCN, a cellular poison with a characteristic almond odor. *Daniel C. Smith*

3.10A   Two Groups Around an Atom

Any atom surrounded by only two groups is linear and has a bond angle of 180°. Two examples illustrating this geometry are **CO_2** (carbon dioxide) and **HCN** (hydrogen cyanide).

The Lewis structure for CO_2 contains a central carbon atom surrounded by two oxygen atoms. To give every atom an octet and the usual number of bonds requires two carbon–oxygen double bonds. Because the carbon atom is surrounded by **two groups,** the molecule is **linear** and the O—C—O bond angle is **180°.**

Carbon dioxide illustrates another important feature of VSEPR theory: ***ignore multiple bonds in predicting geometry.* Count only atoms and lone pairs.**

Similarly, the Lewis structure for HCN contains a central carbon atom surrounded by one hydrogen and one nitrogen. To give carbon and nitrogen an octet and the usual number of bonds requires a carbon–nitrogen triple bond. The carbon atom is surrounded by **two groups,** making the molecule **linear** and the H—C—N bond angle **180°.**

3.10B   Three Groups Around an Atom

Any atom surrounded by three groups is trigonal planar and has bond angles of 120°. Formaldehyde (**$H_2C=O$**) illustrates this geometry.

The carbon atom in $H_2C=O$ is surrounded by three atoms (two H's and one O) and no lone pairs—that is, **three groups.** To keep the three groups as far from each other as possible, they are arranged in a **trigonal planar** fashion, with bond angles of **120°.**

3.10C   Four Groups Around an Atom

Any atom surrounded by four groups is tetrahedral and has bond angles of (approximately) 109.5°. For example, the simple organic compound methane, **CH_4,** has a central carbon atom with four bonds to hydrogen, each pointing to the corners of a tetrahedron.

Over time, some adhesives and insulation made from formaldehyde can decompose back to formaldehyde, a reactive and potentially hazardous substance. Spider plants act as natural air purifiers by removing formaldehyde (H_2CO) from the air. *Daniel C. Smith*

tetrahedral carbon

How can we represent the three-dimensional geometry of a tetrahedron on a two-dimensional piece of paper? **Place two of the bonds in the plane of the paper, one bond in front, and one bond behind,** using the following conventions:

- A solid line is used for bonds in the plane.
- A wedge is used for a bond in front of the plane.
- A dashed wedge is used for a bond behind the plane.

Up to now, each of the groups around the central atom has been another atom. **A group can also be a lone pair of electrons.** NH_3 and H_2O represent two examples of molecules with atoms surrounded by four groups, some of which are lone pairs.

The Lewis structure for ammonia, **NH_3,** has an N atom surrounded by three hydrogen atoms and one lone pair of electrons—**four groups.** To keep four groups as far apart as possible, the three H atoms and the one lone pair around N point to the corners of a tetrahedron. The H—N—H bond angle of 107° is close to the tetrahedral bond angle of 109.5°. This shape is referred to as **trigonal pyramidal,** because one of the groups around the N is a nonbonded electron pair, not another atom.

To determine the geometry around any atom:

1. Draw a valid Lewis structure.
2. Count groups around a given atom.

The Lewis structure for water, **H_2O,** has an O atom surrounded by two hydrogen atoms and two lone pairs of electrons—**four groups.** In **H_2O,** the two H atoms and the two lone pairs around O point to the corners of a tetrahedron. The H—O—H bond angle of 105° is close to the tetrahedral bond angle of 109.5°. Water has a *bent* shape, because two of the groups around oxygen are lone pairs of electrons.

Common molecular shapes are summarized in Table 3.6.

Table 3.6 Common Molecular Shapes Around Atoms

Total Number of Groups	Number of Atoms	Number of Lone Pairs	Shape Around an Atom (A)	Approximate Bond Angle (°)	Example
2	2	0	linear	180	CO_2, $HC\equiv N$
3	3	0	trigonal planar	120	$H_2C=O$
4	4	0	tetrahedral	109.5	CH_4
4	3	1	trigonal pyramidal	~109.5[a]	NH_3
4	2	2	bent	~109.5[a]	H_2O

[a]The symbol "~" means approximately.

Sample Problem 3.16	Determining the Shape Around an Atom

Using the given Lewis structure, determine the shape around the second-row elements in each compound.

a. H—C≡C—H

acetylene

b.
$$\left[\begin{array}{c} H \\ | \\ H—N—H \\ | \\ H \end{array} \right]^{+}$$

ammonium ion

Analysis

To predict the shape around an atom, we **count groups** around the atom in the Lewis structure and use the information in Table 3.6.

Solution

a. Each C in H—C≡C—H is surrounded by two atoms (one C and one H) and no lone pairs—that is, **two groups.** An atom surrounded by two groups is **linear** with a 180° bond angle.

180°

H—C≡C—H

180°

b. The N atom in NH_4^+ is surrounded by four H atoms—that is, **four groups.** An atom surrounded by four groups is **tetrahedral,** with 109.5° bond angles.

109.5°

tetrahedral

Practice Problem 3.17

What is the shape around the indicated atom in each molecule?

a. $H_2\ddot{S}:$ b. CH_2Cl_2 c. $:NCl_3$ d.

More Practice: Try Problems 3.53–3.60, 3.67b, 3.68b, 3.69d, 3.70b.

3.11   Electronegativity and Bond Polarity

When two atoms share electrons in a covalent bond, are the electrons in the bond attracted to both nuclei to the same extent? That depends on the **electronegativity** of the atoms in the bond.

- *Electronegativity* is a measure of an atom's attraction for electrons in a bond. Electronegativity tells us how much a particular atom "*wants*" electrons.

The electronegativity of an atom is assigned a value from 0 to 4; **the *higher* the value, the *more electronegative* an atom is, and the *more it is attracted* to the electrons in a bond.** The electronegativity values for main group elements are shown in Figure 3.7. The noble gases are not assigned values, because they do not typically form bonds.

Figure 3.7   Electronegativity Values for Main Group Elements

Increasing electronegativity →

1A					3A	4A	5A	6A	7A	8A
H 2.2	2A									
Li 1.0	Be 1.6				B 2.0	C 2.6	N 3.0	O 3.4	F 4.0	
Na 0.9	Mg 1.3				Al 1.6	Si 1.9	P 2.2	S 2.6	Cl 3.2	
K 0.8	Ca 1.0				Ga 1.8	Ge 2.0	As 2.0	Se 2.6	Br 3.0	
Rb 0.8	Sr 1.0				In 1.8	Sn 2.0	Sb 2.0	Te 2.1	I 2.7	

Increasing electronegativity ↑

Electronegativity values exhibit periodic trends.

- Electronegativity *increases* across a row of the periodic table as the nuclear charge increases (excluding the noble gases).
- Electronegativity *decreases* down a column of the periodic table as the atomic radius increases, pushing the valence electrons farther from the nucleus.

Concept Check 3.13

Using the trends in the periodic table, rank the following atoms in order of increasing electronegativity.

a. Li, Na, H   b. O, C, Be   c. Cl, I, F   d. B, O, N

Electronegativity values are used as a guideline to indicate whether the electrons in a bond are *equally* shared or *unequally* shared between two atoms. For example, whenever two *identical* atoms are bonded together, each atom attracts the electrons in the bond to the same extent. The electrons are equally shared, and the bond is said to be **nonpolar.** Thus, a **carbon–carbon bond is nonpolar,** as is the fluorine–fluorine bond in F_2. The same is true whenever two different atoms having *similar* electronegativities are bonded together. **C—H bonds are considered to be nonpolar,** because the electronegativity difference between C (2.6) and H (2.2) is small.

nonpolar bond   nonpolar bond

The small electronegativity difference
between C and H is ignored.

In contrast, bonding between atoms of *different* electronegativity results in the *unequal* sharing of electrons. For example, in a C—O bond, the electrons are pulled away from C (2.6) toward the element of higher electronegativity, O (3.4). **The bond is *polar,* or *polar covalent.*** The bond is said to have a **dipole**—that is, **a partial separation of charge.**

A C–O bond is a *polar* bond.

The direction of polarity in a bond is often indicated by an arrow, with the head of the arrow pointing toward the more electronegative element. The tail of the arrow, with a perpendicular line drawn through it, is drawn at the less electronegative element. Alternatively, the lower case Greek letter **delta (δ)** with a positive or negative charge is used, resulting in the symbols $\delta+$ and $\delta-$ to indicate this unequal sharing of electron density.

- The symbol $\delta+$ is given to the *less* electronegative atom.
- The symbol $\delta-$ is given to the *more* electronegative atom.

Students often wonder how large an electronegativity difference must be to consider a bond polar. That's hard to say. **Usually, a polar bond will be one in which the electronegativity difference between two atoms is 0.5 units or greater.**

As the electronegativity difference between the two atoms in a bond increases, the shared electrons are pulled more and more toward the more electronegative element. When the electronegativity difference is larger than 2.0 units, the electrons are essentially transferred from the less electronegative element to the more electronegative element and the bond is considered ionic. Table 3.7 summarizes the relationship between the electronegativity difference of the atoms in a bond and the type of bond formed.

Table 3.7 Electronegativity Difference and Bond Type

Electronegativity Difference	Bond Type	Electron Sharing
Less than 0.5 units	Nonpolar	Electrons are equally shared.
0.5–2.0 units	Polar covalent	Electrons are unequally shared; they are pulled toward the more electronegative element.
Greater than 2.0 units	Ionic	Electrons are transferred from the less electronegative element to the more electronegative element.

Sample Problem 3.17	**Classifying a Bond as Nonpolar, Polar Covalent, or Ionic**

Use electronegativity values to classify each bond as nonpolar, polar covalent, or ionic: (a) Cl_2; (b) HCl; (c) NaCl.

Analysis

Calculate the electronegativity difference between the two atoms and use the following rules: less than 0.5 (nonpolar); 0.5–2.0 (polar covalent); and greater than 2.0 (ionic).

Solution

	Electronegativity Difference	**Bond Type**
a. Cl_2	3.2 (Cl) − 3.2 (Cl) = 0	Nonpolar
b. HCl	3.2 (Cl) − 2.2 (H) = 1.0	Polar covalent
c. NaCl	3.2 (Cl) − 0.9 (Na) = 2.3	Ionic

Practice Problem 3.18	Use electronegativity values to classify the bond(s) in each compound as nonpolar, polar covalent, or ionic.

    a. HF   b. MgO   c. F_2 d. ClF   e. H_2O

More Practice: Try Problems 3.15b, 3.16b, 3.63, 3.64, 3.67c, 3.68c, 3.69c, 3.70c.

3.12   Polarity of Molecules

Thus far, we have been concerned with the polarity of a single bond. Is an entire covalent molecule polar or nonpolar? That depends on two factors: the **polarity of the individual bonds** and the **overall shape.** When a molecule contains zero or one polar bond, the following can be said:

- A molecule with *no* polar bonds is a nonpolar molecule.
- A molecule with *one* polar bond is a polar molecule.

Thus, CH_4 is a nonpolar molecule because all of the C—H bonds are nonpolar. In contrast, CH_3Cl contains only one polar bond, so it is a polar molecule. The dipole is in the same direction as the dipole of the only polar bond.

CH_4
no polar bonds
nonpolar molecule

CH_3Cl
one polar bond
polar molecule

With covalent compounds that have more than one polar bond, the shape of the molecule determines the overall polarity.

- If the individual bond dipoles do *not* cancel, the molecule is polar.
- If the individual bond dipoles *cancel,* the molecule is nonpolar.

To determine the polarity of a molecule that has two or more polar bonds:

1. Identify all polar bonds based on electronegativity differences.
2. Determine the shape around individual atoms by counting groups.
3. Decide if individual dipoles cancel or reinforce.

Figure 3.8 illustrates several examples of polar and nonpolar molecules that contain polar bonds. The net dipole is the sum of all the bond dipoles in a molecule.

Figure 3.8 Examples of Polar and Nonpolar Molecules

one polar bond

three polar bonds
All dipoles cancel.
NO net dipole

net dipole

three polar bonds
All dipoles reinforce.

net dipole

two polar bonds
Two dipoles reinforce.

polar molecule   **nonpolar** molecule   **polar** molecule   **polar** molecule

| Sample Problem 3.18 | Determining the Polarity of a Molecule |

Determine whether each molecule is polar or nonpolar: (a) H_2O; (b) CO_2.

Analysis

To determine the overall polarity of a molecule:

- Identify the **polar bonds.**
- Determine the **shape** around individual atoms.
- Decide if the individual bond dipoles **cancel** or **reinforce.**

Solution

a. **H_2O:** Each O—H bond is polar because the electronegativity difference between O (3.4) and H (2.2) is 1.2. Because the O atom of H_2O has two atoms and two lone pairs around it, H_2O is a **bent** molecule around the O atom. The two dipoles reinforce (both point *up*), so **H_2O has a net dipole;** that is, **H_2O is a polar molecule.**

net dipole

The dipoles reinforce.

b. **CO_2:** Each C—O bond is polar because the electronegativity difference between O (3.4) and C (2.6) is 0.8. The Lewis structure of CO_2 (Section 3.10A) shows that the C atom is surrounded by two groups (two O atoms), making it **linear.** In this case, the two dipoles are equal and opposite in direction so they cancel. Thus, CO_2 is a **nonpolar molecule** with **no net dipole.**

NO net dipole

| Practice Problem 3.19 |

Label the polar bonds in each molecule, and then decide if the molecule is polar or nonpolar.

a. HCl   b. HCN   c. CCl_4

More Practice: Try Problems 3.15c, 3.16c, 3.69e, 3.70d.

HEALTH NOTE

The ethanol in wine is formed by the fermentation of carbohydrates in grapes. The chronic and excessive consumption of alcoholic beverages has become a major health and social crisis in the United States. *John A. Rizzo/Getty Images*

The principles learned in Sections 3.10–3.12 apply to all molecules regardless of size. Count groups around each atom individually to determine its shape. Look for electronegativity differences between the two atoms in a bond to determine polarity.

For example, ethanol, the "alcohol" in alcoholic beverages and the world's most widely abused drug, has molecular formula C_2H_6O. The Lewis structure of ethanol shows that each carbon atom is surrounded by four atoms (four groups), making each carbon tetrahedral. The oxygen atom is surrounded by two atoms and two lone pairs, giving it a bent shape. Ethanol contains two polar bonds, the C—O and the O—H bonds, because of the large electronegativity difference between the atoms in each bond (C and O or O and H).

four atoms around each C

two atoms and **two lone pairs** around O

tetrahedral

bent

ethanol
C_2H_6O

Ethanol contains **two polar bonds**, drawn in red.

Concept Check 3.14	Ethanol is metabolized in the body to acetaldehyde, a toxic compound that produces some of the ill effects of ingesting too much ethanol. Determine the geometry around each carbon in acetaldehyde. Label each bond as polar or nonpolar.

acetaldehyde

CHAPTER REVIEW

KEY TERMS

Ammonium ion (3.6)

Anion (3.2)

Bonding (3.1)

Carbonate (3.6)

Cation (3.2)

Covalent bond (3.7)

Diatomic molecule (3.7)

Dipole (3.11)

Double bond (3.8)

Electronegativity (3.11)

Hydronium ion (3.6)

Hydroxide (3.6)

Ion (3.1)

Ionic bond (3.1)

Lewis structure (3.7)

Lone pair (3.7)

Molecular formula (3.8)

Molecule (3.1)

Nomenclature (3.4)

Nonbonded electron pair (3.7)

Nonpolar bond (3.11)

Octet rule (3.2)

Phosphate (3.6)

Polar bond (3.11)

Polyatomic ion (3.6)

Sulfate (3.6)

Triple bond (3.8)

Valence shell electron pair repulsion (VSEPR) theory (3.10)

KEY CONCEPTS

1 **Ionic bonding (3.1, 3.3)**

General rule: Elements gain, lose, or share electrons to attain the electronic configuration of the noble gas closest to them in the periodic table.

1 **General features**	**2** **Ionic compounds**
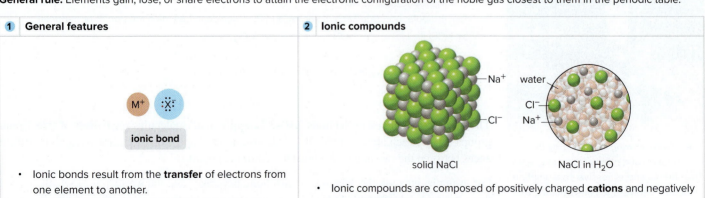M^+ :$\ddot{X}$:⁻ ionic bond	Na⁺ water Cl⁻ Na⁺ solid NaCl NaCl in H_2O
• Ionic bonds result from the **transfer** of electrons from one element to another. • Ionic bonds form between a **metal** and a **nonmetal.** • Oppositely charged ions are held together by **electrostatic attraction.**	• Ionic compounds are composed of positively charged **cations** and negatively charge **anions.** • They are **crystalline solids** with **high melting points** and **high boiling points.** • Most ionic compounds are **soluble in water,** and the resulting **aqueous solution** conducts an electric current.

❷ Covalent bonding (3.1, 3.7)

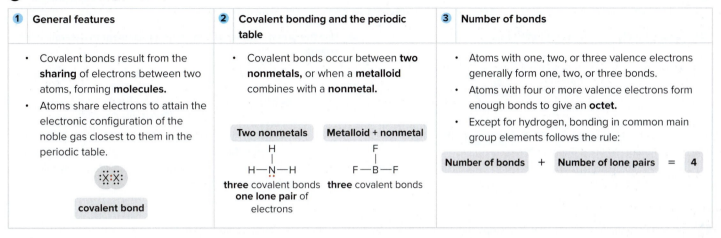

① **General features**	② **Covalent bonding and the periodic table**	③ **Number of bonds**
• Covalent bonds result from the **sharing** of electrons between two atoms, forming **molecules.** • Atoms share electrons to attain the electronic configuration of the noble gas closest to them in the periodic table. covalent bond	• Covalent bonds occur between **two nonmetals,** or when a **metalloid** combines with a **nonmetal.** **Two nonmetals** **Metalloid + nonmetal** H—N—H F—B—F **three** covalent bonds **three** covalent bonds **one lone pair** of electrons	• Atoms with one, two, or three valence electrons generally form one, two, or three bonds. • Atoms with four or more valence electrons form enough bonds to give an **octet.** • Except for hydrogen, bonding in common main group elements follows the rule: **Number of bonds** + **Number of lone pairs** = **4**

❸ Electronegativity and bond polarity (3.11)

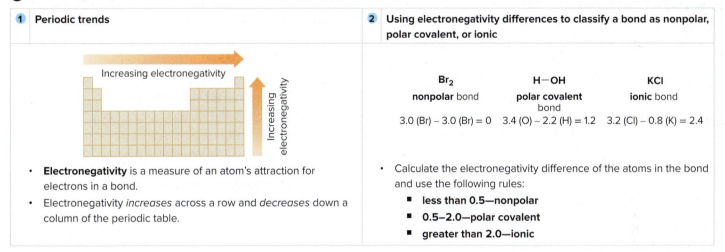

① **Periodic trends**	② **Using electronegativity differences to classify a bond as nonpolar, polar covalent, or ionic**

Increasing electronegativity
 	
Increasing electronegativity
 	
• **Electronegativity** is a measure of an atom's attraction for electrons in a bond.
• Electronegativity *increases* across a row and *decreases* down a column of the periodic table.
	Br_2 H—OH KCl
	nonpolar bond **polar covalent** bond **ionic** bond
	3.0 (Br) − 3.0 (Br) = 0 3.4 (O) − 2.2 (H) = 1.2 3.2 (Cl) − 0.8 (K) = 2.4
	
	• Calculate the electronegativity difference of the atoms in the bond and use the following rules:
	  ▪ **less than 0.5—nonpolar**
	  ▪ **0.5–2.0—polar covalent**
	  ▪ **greater than 2.0—ionic**

KEY SKILLS

❶ Relating the number of protons and electrons to an ion symbol (3.2)

① **Writing an ion symbol from the number of protons and electrons**	② **Using an ion symbol to determine the number of protons and electrons**
11 protons → **Na** 12 neutrons Z = **11** **11** protons (+ charges) **10** electrons (− charges) **Na⁺** one more proton than electron **10** electrons • The number of protons (11) = the atomic number (11), which identifies the element (sodium). • Determine the charge by comparing the number of protons (**+ charges**) with the number of electrons (**− charges**). • If the number of protons is *greater than* the number of electrons, the charge is **positive.** • If the number of electrons is *greater than* the number of protons, the charge is **negative.**	S²⁻ — Two more electrons than protons, so **18** electrons — Atomic number of S = **16,** so **16** protons • The identity of the element (S) determines the atomic number, which equals the number of protons (16). • A **positive** charge means more protons than electrons. • A **negative** charge means more electrons (18) than protons (16).

2 Using the group number to determine ion charge (3.2B)

Octet rule: Main group elements gain or lose one, two, or three electrons to form ions with eight outer shell electrons.

1 Locate the element in the periodic table.	**2 Use the group number of a main group element to determine the charge.**
Ca Se calcium selenium group **2A** group **6A** • A metal in groups 1A, 2A, or 3A forms a **cation** equal in charge to the group number. • A nonmetal in groups 5A, 6A, and 7A forms an **anion** whose charge equals 8 − (the group number).	Ca^{2+} Se^{2-} group **2A** group **6A** cation with **+2** charge $8 - 6 = 2$ anion with **−2** charge • Calcium (Ca) is located in group **2A**, so it forms a cation with a **+2** charge, Ca^{2+}. • Selenium (Se) is located in group **6A**, so it forms an anion with a charge of $8 - 6 = 2$; Se^{2-}.

3 Determining the formula of an ionic compound; example: writing the formula for the ionic compound derived from magnesium and chlorine (3.3A, 3.6A)

General rule: Cations and anions always form ionic compounds that have **zero overall charge**. Write the formula with the cation *first*.

1 Identify the cation and the anion.	**2 When the cation and anion have different charges, use the ion charges to determine the number of ions needed.**
Mg - - -→ Mg^{2+} Cl - - -→ Cl^- magnesium group **2A** chlorine group **7A** metal cation with **+2** charge nonmetal $8 - 7 = 1$ **cation** **anion** anion with **−1** charge • **Metals form cations** and **nonmetals form anions.** • Use the **group number** of a main group element to determine the charge.	Mg^{2+} Cl^{1-} Mg_1Cl_2 = $MgCl_2$ • The charges on the ions tell how many of the *oppositely* charged ions are needed to balance charge. • Write a **subscript for the cation** that is equal in magnitude to the **charge on the anion.** • Write a **subscript for the anion** that is equal in magnitude to the **charge on the cation.**

4 Naming an ionic compound (3.4, 3.6B)

Ionic compounds are always named with the name of the cation *first*.

1 Naming an ionic compound with a cation that has a fixed charge	**2 Naming an ionic compound with a cation that has a variable charge**
$Ca_3(PO_4)_2$ **calcium phosphate** • The cation has the same name as its neutral element. • The name of the anion usually ends in the suffix **-ide** if it is derived from a single atom or **-ate** (or **-ite**) if it is polyatomic.	SnF_2 ← Two F^- anions mean a Sn^{2+} cation is present. **tin(II) fluoride** **stannous fluoride** • Use the **overall anion charge** to determine the charge on the cation. • Name the cation using a **Roman numeral** or the suffix **-ous** (for the ion with the lesser charge) or **-ic** (for the ion with the larger charge).

5 Drawing a valid Lewis structure (3.8); example: CH_3CHO having the atoms arranged as drawn

1 Arrange the atoms with H's on the periphery.	**2 Count valence electrons.**	**3 Add single bonds.**	**4 Complete octets with multiple bonds and lone pairs.**
H H H C C O H	2 C's × 4 e^- = 8 4 H's × 1 e^- = 4 1 O × 6 e^- = 6 ────────── total e^- = 18	H H | | H−C−C−O | H 12 e^- used.	H H | | H−C−C=Ö | ·· H Add one double bond and two lone pairs to complete O and C octets.

6 Naming covalent compounds with two elements (3.9)

1 Deriving a formula from a name; example: sulfur dioxide	**2** Naming a compound from its formula; example: SiF_4
  • Write the symbols for the elements in the order given in the name. • Use the prefixes to write the subscripts.	• Name the first nonmetal by its element name and the second using the suffix -ide. • Add prefixes to show the number of atoms of each element.

7 Predicting geometry from a Lewis structure (3.10)

1 Count groups on each atom.	**2** Use the following rules:		

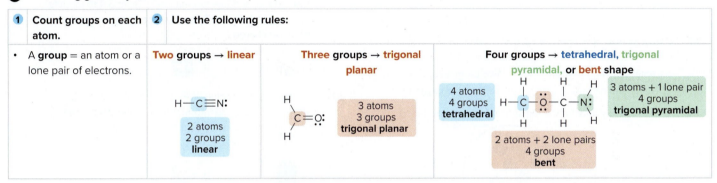

8 Determining if a molecule is polar or nonpolar from a valid Lewis structure (3.12); example: CH_3OH

1 Label polar bonds based on electronegativity differences.	**2** Determine the geometry by counting groups.	**3** If there is more than one polar bond, check if bond dipoles cancel or reinforce.

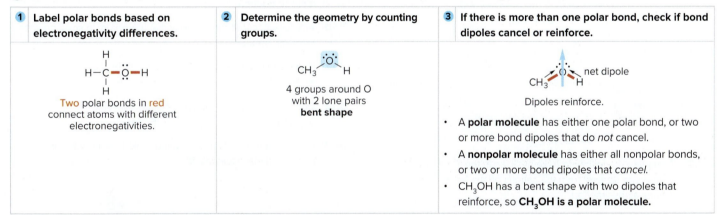

Two polar bonds in red connect atoms with different electronegativities.	4 groups around O with 2 lone pairs **bent shape**	Dipoles reinforce. • A **polar molecule** has either one polar bond, or two or more bond dipoles that do *not* cancel. • A **nonpolar molecule** has either all nonpolar bonds, or two or more bond dipoles that *cancel*. • CH_3OH has a bent shape with two dipoles that reinforce, so **CH_3OH is a polar molecule.**

CHAPTER 3 SELF-TEST

The Self-Test consists of short-answer questions that test an understanding of definitions, equations, and other material encountered within the chapter. Answers to each question are provided at the end of the chapter. For any questions you are unable to answer correctly, revisit the appropriate section(s) of the text; then proceed to working the end-of-chapter problems to further reinforce your understanding of the chapter material.

1. Which term from "Key Terms" answers each question?
   a. What term describes an ion that has more protons than electrons?
   b. What type of bond results from the sharing of four electrons between two atoms?
   c. What is the name for the polyatomic ion with the formula PO_4^{3-}?
   d. What type of bond results from the unequal sharing of electrons between atoms of different electronegativities?
   e. What is another name for a nonbonded electron pair?

2. Label the bonds in each compound as ionic or covalent: (a) $CaCl_2$; (b) CCl_4; (c) HF.

3. How many protons and electrons are present in each ion: (a) Mg^{2+}; (b) Cl^-?

4. Name each ionic compound: (a) KCl; (b) $CaBr_2$; (c) $Fe_2(SO_4)_3$; (d) $FeSO_4$.

5. Write the formula for the ionic compound formed from calcium and bicarbonate.

6. How many nonbonded electron pairs are present in the Lewis structure for CH_2F_2?

7. Give the formula for each compound: (a) selenium dioxide; (b) nitrogen trifluoride.

8. Determine the shape around each indicated atom.

(a)   (b)   (c)

9. Use electronegativity values to classify each bond as nonpolar, polar covalent, or ionic: (a) KBr; (b) Cl_2; (c) H_2O.

UNDERSTANDING KEY CONCEPTS

3.1 Which formulas represent ionic compounds and which represent covalent compounds?
   a. CO_2 b. H_2SO_4 c. KF   d. CH_5N

3.2 Which pairs of elements are likely to form ionic bonds and which pairs are likely to form covalent bonds?
   a. potassium and oxygen   c. two bromine atoms
   b. sulfur and carbon   d. carbon and oxygen

3.3 Give the ion symbol for the ion shown in the molecular art.

2 electrons

3.4 Identify each atom or ion.

a.   56 protons / 81 neutrons / 54 electrons

b.   15 protons / 16 neutrons / 18 electrons

3.5 Complete the following table by filling in the formula of the ionic compound derived from the cations on the left and each of the anions across the top.

	Br^-	OH^-	HCO_3^-	SO_3^{2-}	PO_4^{3-}
Na^+					
Co^{2+}					
Al^{3+}					

3.6 Complete the following table by filling in the formula of the ionic compound derived from the cations on the left and each of the anions across the top.

	I^-	CN^-	NO_3^-	SO_4^{2-}	HPO_4^{2-}
K^+					
Mg^{2+}					
Cr^{3+}					

3.7 Use the element colors in Figure 2.3 (also shown in Appendix C) and the information in Table 3.4 to identify each polyatomic ion. Give the name and proper formula, including charge.

a.   b.

3.8 Use the element colors in Figure 2.3 (also shown in Appendix C) and the information in Table 3.4 to identify each polyatomic ion. Give the name and proper formula, including charge.

a.   b.

3.9 $CaSO_3$ is used to preserve cider and fruit juices. Name this ionic compound.

3.10 Ammonium carbonate is the active ingredient in smelling salts. Write its formula.

3.11 Draw a valid Lewis structure for each molecule.
   a. HI   b. CH_2F_2

3.12 Draw a valid Lewis structure for each molecule.
   a. CH_3Br b. PH_3

3.13 Identify elements **A, B,** and **C** and rank them in order of increasing electronegativity.

3.14 Identify elements **D, E,** and **F** and rank them in order of increasing electronegativity.

3.15 (a) Translate each ball-and-stick model to a Lewis structure and include all nonbonded electron pairs. (b) Label all polar bonds. (c) Is the molecule polar or nonpolar?

vinyl chloride   dichloroacetylene

A   **B**

3.16 (a) Translate each ball-and-stick model to a Lewis structure and include all nonbonded electron pairs. (b) Label all polar bonds. (c) Is the molecule polar or nonpolar?

1,1-dichloroethylene   tetrafluoroethylene

A   **B**

ADDITIONAL PROBLEMS

Ions

3.17 Write the ion symbol for an atom with the given number of protons and electrons.

a. four protons and two electrons

b. 22 protons and 20 electrons

c. 16 protons and 18 electrons

d. 13 protons and 10 electrons

3.18 How many protons and electrons are present in each ion?

a. K^+ b. Mn^{2+} c. Fe^{2+} d. Cs^+

3.19 What species fits each description?

a. a period 2 element that forms a +2 cation

b. an ion from group 7A with 18 electrons

c. a cation from group 1A with 36 electrons

3.20 What species fits each description?

a. a period 3 element that forms an ion with a −1 charge

b. an ion from group 2A with 36 electrons

c. an ion from group 6A with 18 electrons

3.21 Give the ion symbol for each ion.

a. sodium ion   c. gold(III)

b. manganese ion   d. stannic

3.22 Give the ion symbol for each ion.

a. barium ion   c. oxide

b. iron(II)   d. ferrous

3.23 Label each of the following regions in the periodic table.

a. a group that forms cations with a +2 charge

b. a group that forms anions with a −2 charge

c. a group that forms cations with a +1 charge

d. a group that forms anions with a −1 charge

3.24 For each of the general electron-dot formulas for elements, give the following information: [1] the number of valence electrons; [2] the group number of the element; [3] how many electrons

must be gained or lost to achieve a completely filled outer shell of electrons; [4] the charge on the resulting ion.

a. X·   b. ·Z̈·

3.25 Use the element colors in Figure 2.3 (also shown in Appendix C) and the information in Table 3.4 to identify each of the following anions. Give the name and proper formula, including the charge.

a.   b.   c.

3.26 Use the element colors in Figure 2.3 (also shown in Appendix C) and the information in Table 3.4 to identify each of the following anions. Give the name and proper formula, including the charge.

a.   b.   c.

Ionic Compounds

3.27 What is the charge on the cation M in each of the following ionic compounds?

a. MCl_2 b. MO c. M_2O_3 d. M_3N

3.28 What is the charge on the cation M in each of the following ionic compounds?

a. M_2S b. M_3P_2 c. MBr_3 d. MO_2

3.29 Write the formula for the ionic compound formed from each pair of elements.

a. calcium and sulfur   c. lithium and iodine

b. aluminum and bromine   d. nickel and chlorine

3.30 Write the formula for the ionic compound formed from each cation and anion.

a. lithium and nitrite

b. calcium and acetate

c. sodium and bisulfite

3.31 Write the formula for the ionic compound formed from the barium cation (Ba^{2+}) and each anion: (a) CN^-; (b) PO_4^{3-}; (c) HPO_4^{2-}; (d) $H_2PO_4^-$.

3.32 Write the formula for the ionic compound formed from the sulfite anion (SO_3^{2-}) and each cation: (a) K^+; (b) Ba^{2+}; (c) Al^{3+}; (d) Zn^{2+}.

Naming Ionic Compounds

3.33 Name each ionic compound.

    a. Na_2O c. PbS_2 e. $CoBr_2$

    b. BaS d. $AgCl$ f. $RbBr$

3.34 Name each ionic compound.

    a. KF c. Cu_2S e. $AuBr_3$

    b. $ZnCl_2$ d. SnO f. Li_2S

3.35 Write formulas to illustrate the difference between each pair of compounds.

    a. sodium sulfide and sodium sulfate

    b. magnesium oxide and magnesium hydroxide

    c. magnesium sulfate and magnesium bisulfate

3.36 Write formulas to illustrate the difference between each pair of compounds.

    a. lithium sulfite and lithium sulfide

    b. sodium carbonate and sodium hydrogen carbonate

    c. calcium phosphate and calcium dihydrogen phosphate

3.37 Name each ionic compound.

    a. NH_4Cl c. $Cu(NO_3)_2$

    b. $PbSO_4$ d. $Ca(HCO_3)_2$

3.38 Name each ionic compound.

    a. $(NH_4)_2SO_4$ c. $Cr(CH_3CO_2)_3$

    b. NaH_2PO_4 d. $Sn(HPO_4)_2$

3.39 Write a formula from each name.

    a. magnesium carbonate   c. copper(II) hydroxide

    b. nickel sulfate   d. gold(III) nitrate

3.40 Write a formula from each name.

    a. copper(I) sulfite   c. lead(IV) carbonate

    b. aluminum nitrate   d. zinc hydrogen phosphate

Properties of Ionic Compounds

3.41 Label each statement as "true" or "false." Correct any false statement to make it true.

    a. Ionic compounds have high melting points.

    b. Ionic compounds can be solid, liquid, or gas at room temperature.

    c. Most ionic compounds are insoluble in water.

    d. An ionic solid like sodium chloride consists of discrete pairs of sodium cations and chloride anions.

3.42 Label each statement as "true" or "false." Correct any false statement to make it true.

    a. Ionic compounds have high boiling points.

    b. The ions in a crystal lattice are arranged randomly and the overall charge is zero.

    c. When an ionic compound dissolves in water, the solution conducts electricity.

    d. In an ionic crystal, ions having like charges are arranged close to each other.

Covalent Bonding and Lewis Structures

3.43 For each pair of compounds, classify the bonding as ionic or covalent and explain your choice.

    a. LiCl and HCl   b. KBr and HBr

3.44 For each pair of compounds, classify the bonding as ionic or covalent and explain your choice.

    a. BeH_2 and $BeCl_2$ b. Na_3N and NH_3

3.45 Draw a valid Lewis structure for each molecule: (a) H_2Se; (b) C_2Cl_6; (c) C_2H_2 (Assume each C is bonded to one H).

3.46 Draw a valid Lewis structure for each molecule: (a) HCl; (b) SiF_4; (c) CCl_2O (Assume C is bonded to both Cl's and the O atom).

3.47 Convert the 3-D model of alanine, one of the 20 amino acids found in proteins, to a Lewis structure. Include all nonbonded electron pairs.

alanine

3.48 Convert the 3-D model of the general anesthetic methoxyflurane into a Lewis structure and include all nonbonded electron pairs on atoms that contain them.

methoxyflurane

Naming Covalent Compounds

3.49 Name each covalent compound.

    a. PBr_3 b. SO_3 c. NCl_3 d. P_2S_5

3.50 Name each covalent compound.

    a. SF_6 b. CBr_4 c. N_2O d. P_4O_{10}

3.51 Write a formula that corresponds to each name.

    a. selenium dioxide   c. dinitrogen pentoxide

    b. carbon tetrachloride

3.52 Write a formula that corresponds to each name.

    a. silicon tetrafluoride   c. phosphorus triiodide

    b. nitrogen monoxide

Molecular Shape

3.53 Determine the shape around each indicated atom.

    a. H—C—Ö—H

    b. :NF_3

3.54 Determine the shape around each indicated atom.

a.

b. (structure: H—C≡C—C—Cl with H atoms)

3.55 Add lone pairs to the N and O atoms to give octets and then determine the shape around each indicated atom.

a.

b. (structure: H—C—C—OH with O double bond and H)

3.56 Add lone pairs to the N and O atoms to give octets and then determine the shape around each indicated atom.

a. (structure: H—N—N—H with H atoms)

b. (structure: H—C—O—O—H with H atoms)

3.57 Predict the bond angles around the indicated atoms in each compound.

a.

b. (structure: H—C≡C—C=O with H)

3.58 Predict the bond angles around the indicated atoms in each compound.

a. (structure: H—C—C—Cl with H atoms)

b. (structure: H—C≡C—C—N—H with H atoms)

3.59 The tartness of rhubarb is primarily due to malic acid.

rhubarb

malic acid

Mikeledray/Shutterstock

a. How many lone pairs are present in a Lewis structure of malic acid?

b. Determine the molecular shape around atoms **A–C.**

3.60 Match each compound with one of the molecular shapes shown below (**C** or **D**): (a) BBr_3; (b) NBr_3; (c) PBr_3.

C   D

Electronegativity and Polarity

3.61 Rank the atoms in each group in order of increasing electronegativity.

a. Se, O, S
b. P, Na, Cl
c. Cl, S, F
d. O, P, N

3.62 Rank the atoms in each group in order of increasing electronegativity.

a. Si, P, S
b. Be, Mg, Ca
c. Se, Cl, Br
d. Li, Be, Na

3.63 Label the bond formed between carbon and each of the following elements as nonpolar, polar, or ionic.

a. carbon
b. oxygen
c. lithium
d. chlorine
e. hydrogen

3.64 Label the bond formed between fluorine and each of the following elements as nonpolar, polar, or ionic.

a. hydrogen
b. fluorine
c. carbon
d. lithium
e. sulfur

Applications

3.65 Zinc is an essential nutrient needed by many enzymes to maintain proper cellular function. Zinc is obtained in many dietary sources, including oysters, beans, nuts, whole grains, and sunflower seeds. (a) How many protons and electrons are found in a neutral zinc atom? (b) How many electrons and protons are found in the Zn^{2+} cation?

3.66

Liquid dietary supplements with fewer grams of carbohydrates are made for individuals who have diabetes. Write the formula for each of the following ionic compounds, which are ingredients in such a nutritional drink: (a) magnesium phosphate; (b) cupric sulfate; (c) potassium chloride; (d) calcium carbonate.

Helen Sessions/Alamy Stock Photo

3.67 Isobutyl cyanoacrylate is used in medical glues to close wounds, thus avoiding the need for sutures.

isobutyl cyanoacrylate

a. Convert the ball-and-stick model to a Lewis structure with all bonds and lone pairs drawn in.

b. Determine the shape around each carbon atom.

c. Label all polar bonds.

3.68 3-Methyl-2-hexenoic acid is one of the compounds that give human sweat its characteristic odor.

3-methyl-2-hexenoic acid

a. Convert the ball-and-stick model to a Lewis structure with all bonds and lone pairs drawn in.

b. Determine the shape around each carbon atom.

c. Label all polar bonds.

CHALLENGE PROBLEMS

3.69 Answer the following questions about the molecule OCS.

a. How many valence electrons does OCS contain?

b. Draw a valid Lewis structure. Assume that OCS contains a carbon–oxygen double bond, as well as a carbon–sulfur double bond.

c. Label all polar bonds.

d. What is the shape around the C atom?

e. Is OCS a polar molecule? Explain.

3.70 Glycine is a building block used to make proteins, such as those in heart muscle.

glycine

a. Add lone pairs where needed to give the second-row elements octets, and then count the total number of valence electrons in glycine.

b. Determine the shape around the four indicated atoms.

c. Label all of the polar bonds.

d. Is glycine a polar or nonpolar molecule? Explain.

BEYOND THE CLASSROOM

3.71 Energy bars contain protein, carbohydrates, and added nutrients for athletes and others looking for an on-the-go meal. Look at an ingredients label for some common energy bars, concentrating on the vitamins and minerals added in small amounts, itemized at the end of the list. Examples might include calcium phosphate, ferrous fumarate, copper gluconate, riboflavin, and folic acid. Try to determine the chemical formulas of these ingredients and why they are added. Is there any significant difference among popular energy bars? Can you determine which compounds are ionic and which are covalent?

3.72 Although mercury is safely used in dental fillings and thermometers, some forms of mercury, such as methylmercury and mercuric chloride, are highly toxic, and breathing mercury vapor is also a health hazard. In some waters, mercury-contaminated bluefin tuna, albacore, and swordfish have high levels of methylmercury and are considered unsafe to eat. Research the uses of mercury and how it enters the environment. Discuss the different forms of mercury encountered in the environment and why some are much more harmful than others. What laws have been enacted in the United States and other countries to limit an individual's exposure to toxic mercury compounds?

CHAPTER 3 ANSWERS

Concept Check Answers

3.1
a. 79 protons, 78 electrons
b. 79 protons, 76 electrons
c. 50 protons, 48 electrons
d. 50 protons, 46 electrons

3.2 a. Co^{3+} b. Cr^{2+}

3.3
a. 25 protons, 25 electrons
b. 25 protons, 23 electrons

3.4 ZnO

3.5
a. sulfide   c. cesium   e. tin(IV), stannic
b. copper(I), cuprous   d. aluminum

3.6 a. Sn^{2+} b. I^- c. Mn^{2+} d. Pb^{2+}

3.7 Ionic compounds have high melting points and high boiling points. They usually dissolve in water, and their solutions conduct electricity.

3.8 a. KOH   b. KNO_2 c. K_2SO_4 d. $KHSO_3$ e. K_3PO_4

3.9
a. $NaHCO_3$ c. $(NH_4)_2SO_4$ e. $Ca(HSO_4)_2$
b. KNO_3 d. $Mg_3(PO_4)_2$ f. $Ba(OH)_2$

3.10 $Ca_3(PO_4)_2$

3.11 $H\!:\!\ddot{\underset{..}{C}l}\!:$

3.12
a. carbon disulfide   c. phosphorus pentachloride
b. sulfur dioxide   d. boron trifluoride

3.13
a. Na < Li < H   c. I < Cl < F
b. Be < C < O   d. B < N < O

3.14 C bonded to 3 H's and C, tetrahedral
C doubly bonded to O, trigonal planar
C=O polar bond
C–C, C–H nonpolar bonds

Practice Problem Answers

3.1 a,d,e: covalent   b,c,f: ionic

3.2 a. compound: **A, C, D, E**   element: **B, F**
b. All are molecules.

3.3 N^{3-}

3.4 a. K^+ b. N^{3-} c. Br^- d. V^{2+}

3.5 a. 28 protons, 26 electrons   c. 26 protons, 23 electrons
b. 34 protons, 36 electrons

3.6 a. +2   b. −1   c. −2   d. +1

3.7 a. NaBr   b. BaO   c. MgI_2 d. Li_2O

3.8 a. sodium fluoride   d. lithium oxide
   b. magnesium oxide   e. titanium oxide
   c. strontium bromide

3.9 a. chromium(III) chloride, chromic chloride
   b. lead(II) sulfide
   c. tin(IV) fluoride, stannic fluoride
   d. lead(IV) oxide
   e. iron(II) bromide, ferrous bromide

3.10 a. $CaBr_2$ c. $FeBr_3$ e. $CrCl_2$
    b. CuI d. MgS f. Na_2O

3.11 a. sodium carbonate   d. manganese acetate
    b. calcium hydroxide   e. iron(III) hydrogen sulfite, ferric bisulfite
    c. magnesium nitrate   f. magnesium phosphate

3.12 a. 1   b. 4   c. 1   d. 2   e. 3   f. 2

3.13 a. H—B̈r̈:   c. H—Ö—Ö—H   e. H—C—C—H (with H's)

   b. H—C—F̈: (with H's)   d. H—N̈—N̈—H (with H's)   f. H—C—C̈l: (with H's)

3.14 a. H—C≡N̈:   b. H—C=Ö: (with H)

3.15 H—C—Ö—H (with =O)

3.16 a. SiO_2 b. PCl_3 c. SO_3 d. N_2O_3

3.17 a. bent   c. trigonal pyramidal
    b. tetrahedral   d. trigonal planar

3.18 a. polar   c. nonpolar   e. polar
    b. ionic   d. polar

3.19 a. δ+ δ− H—Cl polar   c. nonpolar structure

   b. δ+ δ− H—C≡N polar

Self-Test Answers

1. a. cation   c. phosphate   e. lone pair
    b. double bond   d. polar

2. a. ionic   b,c: covalent

3. a. 12 protons, 10 electrons
    b. 17 protons, 18 electrons

4. a. potassium chloride
    b. calcium bromide
    c. iron(III) sulfate, ferric sulfate
    d. iron(II) sulfate, ferrous sulfate

5. $Ca(HCO_3)_2$

6. six

7. a. SeO_2 b. NF_3

8. a. tetrahedral   b. bent   c. trigonal planar

9. a. ionic   b. nonpolar   c. polar covalent

Odd-Numbered Problem Answers

3.1 a,b,d: covalent   c. ionic

3.3 Be^{2+}

3.5

	Br^-	OH^-	HCO_3^-	SO_3^{2-}	PO_4^{3-}
Na^+	NaBr	NaOH	$NaHCO_3$	Na_2SO_3	Na_3PO_4
Co^{2+}	$CoBr_2$	$Co(OH)_2$	$Co(HCO_3)_2$	$CoSO_3$	$Co_3(PO_4)_2$
Al^{3+}	$AlBr_3$	$Al(OH)_3$	$Al(HCO_3)_3$	$Al_2(SO_3)_3$	$AlPO_4$

3.7 a. OH^-, hydroxide   b. NH_4^+, ammonium

3.9 calcium sulfite

3.11 a. H—Ï: b. H—C—F̈: (with F's)

3.13 **B** (sulfur) < **C** (chlorine) < **A** (fluorine)

3.15 **A** a. C=C structure   **B** a. :C̈l—C≡C—C̈l:
    b. one polar bond   b. two polar bonds
    c. polar molecule   c. nonpolar molecule

3.17 a. Be^{2+} b. Ti^{2+} c. S^{2-} d. Al^{3+}

3.19 a. Be   b. Cl^- c. Rb^+

3.21 a. Na^+ b. Mn^{2+} c. Au^{3+} d. Sn^{4+}

3.23

3.25 a. SO_4^{2-} b. NO_2^- c. S^{2-}
    sulfate   nitrite   sulfide

3.27 a. +2   b. +2   c. +3   d. +1

3.29 a. CaS   b. $AlBr_3$ c. LiI d. $NiCl_2$

3.31 a. $Ba(CN)_2$ c. $BaHPO_4$
    b. $Ba_3(PO_4)_2$ d. $Ba(H_2PO_4)_2$

3.33 a. sodium oxide   d. silver chloride
     b. barium sulfide   e. cobalt bromide
     c. lead(IV) sulfide   f. rubidium bromide

3.35 a. Na_2S Na_2SO_4
     b. MgO $Mg(OH)_2$
     c. $MgSO_4$ $Mg(HSO_4)_2$

3.37 a. ammonium chloride
     b. lead(II) sulfate
     c. copper(II) nitrate, cupric nitrate
     d. calcium bicarbonate, calcium hydrogen carbonate

3.39 a. $MgCO_3$ c. $Cu(OH)_2$
     b. $NiSO_4$ d. $Au(NO_3)_3$

3.41 a. True.
     b. False—ionic compounds are solids at room temperature.
     c. False—most ionic compounds are soluble in water.
     d. False—ionic solids exist as crystalline lattices with the ions arranged to maximize the electrostatic interactions of anions and cations.

3.43 a. LiCl: ionic; the metal Li donates electrons to chlorine.
       HCl: covalent; H and Cl share electrons because both are nonmetals and the electronegativity difference is not large enough for electron transfer to occur.
     b. KBr: ionic; the metal K donates electrons to bromine.
       HBr: covalent; H and Br share electrons because the electronegativity difference is not large enough for electron transfer to occur.

3.45 a. $H-\ddot{Se}-H$ b. structure   c. $H-C{\equiv}C-H$

3.47 structure

3.49 a. phosphorus tribromide   c. nitrogen trichloride
     b. sulfur trioxide   d. diphosphorus pentasulfide

3.51 a. SeO_2 b. CCl_4 c. N_2O_5

3.53 a. C tetrahedral, O bent   b. N trigonal pyramidal

3.55 a. N trigonal pyramidal
       O bent
     b. C (with single bonds) tetrahedral
       C (with double bond) trigonal planar

3.57 a. structure   b. structure (120° 180° 120°)
     both 109.5°

3.59 a. 10   b. **A** trigonal planar
            B bent
            C tetrahedral

3.61 a. $Se < S < O$ c. $S < Cl < F$
     b. $Na < P < Cl$ d. $P < N < O$

3.63 a,e: nonpolar   b,c,d: polar

3.65 a. 30 protons, 30 electrons
     b. 30 protons, 28 electrons

3.67 a. structure
     b. C's with x — trigonal planar
       C's with * — tetrahedral
       C with arrow — linear
     c. All C—N and C—O bonds are polar.

3.69 a. 16 valence electrons [1 O atom (6 valence electrons) + 1 C atom (4 valence electrons) + 1 S atom (6 valence electrons)]
     b,c: $\ddot{O}{=}C{=}\ddot{S}$
     d. linear
     e. Yes

Energy and Matter

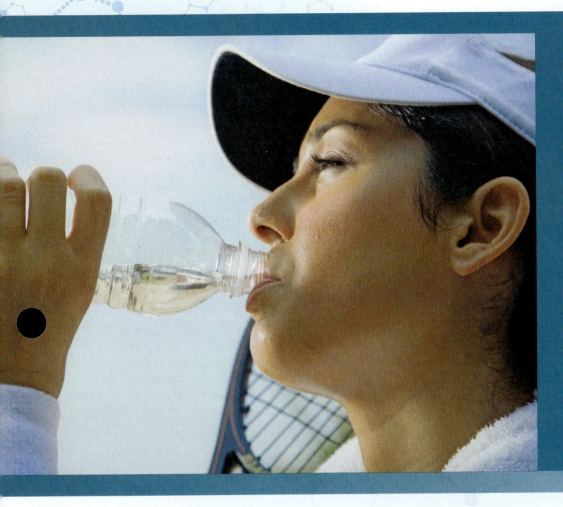

When sweat evaporates, it cools the skin by absorbing heat from the body. *Terry Vine/Blend Images LLC*

CHAPTER OUTLINE

CHAPTER GOALS

In this chapter you will learn how to:

1. Define energy and become familiar with the units of energy

2. Identify the characteristics of the three states of matter

3. Determine what types of intermolecular forces a compound possesses

4. Relate the strength of intermolecular forces to a compound's boiling point and melting point

5. Define specific heat and use specific heat to determine the amount of heat gained or lost by a substance

6. Describe the energy changes that accompany changes of state

7. Interpret the changes depicted in heating and cooling curves

The water at the top of a waterfall has potential energy because of its position. This potential energy becomes kinetic energy as the water falls. *Daniel C. Smith*

hy does water (H_2O) have a much higher boiling point than methane (CH_4), the main component of natural gas? Why does chloroethane, a local anesthetic, numb an injury when it is sprayed on a wound? To answer these questions, we turn our attention in Chapter 4 to energy and how energy changes are related to the forces of attraction that exist between molecules.

4.1 Energy

Energy **is the capacity to do work.** Whenever you throw a ball, ride a bike, or read a newspaper, you use energy to do work. There are two types of energy.

- Potential energy is stored energy.
- Kinetic energy is the energy of motion.

A ball at the top of a hill or the water in a reservoir behind a dam are examples of potential energy. When the ball rolls down the hill or the water flows over the dam, the stored potential energy is converted to the kinetic energy of motion. Although energy can be converted from one form to another, one rule, the **law of conservation of energy,** governs the process.

- The total energy in a system does not change. Energy cannot be created or destroyed.

Throwing a ball into the air illustrates the interplay of potential and kinetic energy (Figure 4.1).

Figure 4.1   Potential and Kinetic Energy

All of the ball's kinetic energy is converted to potential energy.

Kinetic energy is converted to potential energy as the ball rises.

Potential energy is converted to kinetic energy as the ball falls.

Kinetic energy is transferred from the thrower's hand to the baseball.

Some of the kinetic energy of the ball is transferred to the ground.

When a ball is thrown into the air it possesses kinetic energy, which is converted to potential energy as it reaches its maximum height. When the ball descends, the potential energy is converted back to kinetic energy, until the ball rests motionless on the ground.

4.1A   The Units of Energy

Energy can be measured using two different units, **calories (cal)** and **joules (J).** A **calorie** is the amount of energy needed to raise the temperature of 1 g of water 1 °C. Joules and calories are related in the following way.

$$1 \text{ cal} = 4.184 \text{ J}$$

The joule, named after the nineteenth-century English physicist James Prescott Joule, is pronounced *jewel*.

Because both the calorie and the joule are small units of measurement, more often energies in reactions are reported with kilocalories (kcal) and kilojoules (kJ). Recall from Table 1.2 that the prefix *kilo* means 1,000.

$$1 \text{ kcal} = 1,000 \text{ cal}$$
$$1 \text{ kJ} = 1,000 \text{ J}$$
$$1 \text{ kcal} = 4.184 \text{ kJ}$$

To convert a quantity from one unit of measurement to another, set up conversion factors and use the method first shown in Section 1.7B and illustrated in Sample Problem 4.1.

Sample Problem 4.1

Converting Kilojoules to Kilocalories

A reaction releases 421 kJ of energy. How many kilocalories does this correspond to?

Analysis and Solution

[1] Identify the original quantity and the desired quantity.

421 kJ	? kcal
original quantity	desired quantity

[2] Write out the conversion factors.

- Choose the conversion factor that places the unwanted unit, kilojoules, in the denominator so that the units cancel.

kJ–kcal conversion factors

$$\frac{4.184 \text{ kJ}}{1 \text{ kcal}} \quad \text{or} \quad \boxed{\frac{1 \text{ kcal}}{4.184 \text{ kJ}}} \quad \text{Choose this conversion factor to cancel kJ.}$$

[3] Set up and solve the problem.

- Multiply the original quantity by the conversion factor to obtain the desired quantity.

kJ–kcal conversion factor

$$421 \text{ k\cancel{J}} \times \frac{1 \text{ kcal}}{4.184 \text{ k\cancel{J}}} = 100.6 \text{ kcal, rounded to 101 kcal}$$

Kilojoules cancel.   **Answer**

Practice Problem 4.1

Carry out each of the following conversions.

a. 42 J to cal   b. 55.6 kcal to cal   c. 326 kcal to kJ   d. 25.6 kcal to J

Practice Problem 4.2

Combustion of 1 g of gasoline releases 11.5 kcal of energy. How many kilojoules of energy is released? How many joules does this correspond to?

More Practice: Try Problems 4.7, 4.13–4.15.

4.1B FOCUS ON THE HUMAN BODY
Energy and Nutrition

When we eat food, the protein, carbohydrates, and fat (lipid) in the food are metabolized to form small molecules that in turn are used to prepare new molecules that cells need for maintenance and growth. This process also generates the energy needed for the organs to function, allowing the heart to beat, the lungs to breathe, and the brain to think.

WHEATS

Nutrition Facts
Serving Size 1 cup (55 g)
Servings Per Container about 13

Amount Per Serving

Calories 200   Calories from Fat 10

	% Daily Value*
Total Fat 1g	**2%**
Saturated Fat 0g	**0%**
Trans Fat 0g	
Cholesterol 0mg	**0%**
Sodium 0mg	**0%**
Potassium 180mg	**5%**
Total Carbohydrate 42g	**14%**
Dietary Fiber 5g	**20%**
Sugars 12g	
Protein 5g	

Vitamin A 0%	•	Vitamin C 0%
Calcium 0%	•	Iron 90%
Thiamine 25%	•	Riboflavin 25%
Niacin 25%	•	Vitamin B6 25%

Jill Braaten/McGraw Hill

Table 4.1 Caloric Value for Three Classes of Compounds

	Cal/g	cal/g
Protein	4	4,000
Carbohydrate	4	4,000
Fat	9	9,000

One nutritional Calorie (1 Cal) = 1,000 cal = 1 kcal.

CONSUMER NOTE

Knowing the number of grams of protein, carbohydrates, and fat allows you to estimate how many Calories a food product contains. A quarter-pound burger with cheese with 29 g of protein, 40 g of carbohydrates, and 26 g of fat contains 510 Calories. *Foodcollection/StockFood*

The amount of stored energy in food is measured using nutritional Calories (upper case C), where 1 Cal = 1,000 cal. Because 1,000 cal = 1 kcal, the following relationships exist.

Nutritional Calorie

1 Cal   =   1 kcal

1 Cal   =   1,000 cal

Upon metabolism, proteins, carbohydrates, and fat each release a predictable amount of energy, the **caloric value** of the substance. For example, one gram of protein or one gram of carbohydrate typically releases about 4 Cal/g, whereas fat releases 9 Cal/g (Table 4.1). If we know the amount of each of these substances contained in a food product, we can make a first approximation of the number of Calories it contains by using caloric values as conversion factors, as illustrated in Sample Problem 4.2.

When an individual eats more Calories than are needed for normal bodily maintenance, the body stores the excess as fat. The average body fat content for men and women is about 20% and 25%, respectively. This stored fat can fill the body's energy needs for two or three months. Frequent ingestion of a large excess of Calories results in a great deal of stored fat, causing an individual to be overweight.

Sample Problem 4.2   **Calculating Nutritional Calories**

If a baked potato contains 3 g of protein, a trace of fat, and 23 g of carbohydrates, estimate its number of Calories. Round off your answer to the tens place.

Analysis

Use the caloric value (Cal/g) of each class of molecule to form a conversion factor to convert the number of grams to Calories and add up the results.

Solution

[1] Identify the original quantity and the desired quantity.

3 g protein
23 g carbohydrates   ? Cal
original quantities   desired quantity

[2] Write out the conversion factors.

• Write out conversion factors that relate the number of grams to the number of Calories for each substance. Each conversion factor must place the unwanted unit, grams, in the denominator so that the units cancel.

Cal–g conversion factor for **protein**   Cal–g conversion factor for **carbohydrates**

$$\frac{4\ \text{Cal}}{1\ \text{g protein}}$$ $$\frac{4\ \text{Cal}}{1\ \text{g carbohydrate}}$$

[3] Set up and solve the problem.

• Multiply the original quantity by the conversion factor for both protein and carbohydrates and add up the results to obtain the desired quantity.

Calories due to **protein**   Calories due to **carbohydrate**

Total Calories = $3\ \text{g} \times \dfrac{4\ \text{Cal}}{1\ \text{g protein}}$ + $23\ \text{g carbohydrate} \times \dfrac{4\ \text{Cal}}{1\ \text{g carbohydrate}}$

Grams cancel.   Grams cancel.

=   12 Cal   +   92 Cal

Total Calories = 104 Cal, rounded to 1.0×10^2 Cal

Answer

(Practice Problem 4.4)
The Photo Works/Alamy Stock Photo

Practice Problem 4.3

How many Calories are contained in one tablespoon of olive oil, which has 14 g of fat? Round off your answer to the tens place.

Practice Problem 4.4

Use the number of grams of protein, carbohydrates, and fat in the accompanying container of yogurt to estimate the number of Calories in a serving. Round off your answer to the tens place.

More Practice: Try Problems 4.1, 4.2, 4.8, 4.16, 4.17, 4.19, 4.20.

4.2 The Three States of Matter

As we first learned in Section 1.2, matter exists in three common states—**gas, liquid,** and **solid.**

- A gas consists of particles that are far apart and move rapidly and independently from each other.
- A liquid consists of particles that are much closer together but are still somewhat disorganized since they can move about. The particles in a liquid are close enough that they exert a force of attraction on each other.
- A solid consists of particles—atoms, molecules, or ions—that are close to each other and are often highly organized. The particles in a solid have little freedom of motion and are held together by attractive forces.

As shown in Figure 4.2, air is composed largely of N_2 and O_2 molecules, along with small amounts of argon (Ar), carbon dioxide (CO_2), and water molecules that move about rapidly. Liquid water is composed of H_2O molecules that have no particular organization. Sand is a solid composed of SiO_2, which contains a network of covalent silicon–oxygen bonds.

Figure 4.2

The Three States of Matter—Solid, Liquid, and Gas

solid SiO_2 liquid H_2O N_2 and O_2 gases

Most sand is composed of silicon dioxide (SiO_2), which forms a three-dimensional network of covalent bonds. Liquid water is composed of H_2O molecules, which can move past each other but are held close together by a force of attraction (Section 4.3). Air contains primarily N_2 and O_2 molecules that move rapidly with no force of attraction for each other. *Daniel C. Smith*

Whether a substance exists as a gas, liquid, or solid depends on the balance between the kinetic energy of its particles and the strength of the interactions between the particles. In a gas, the kinetic energy of motion is high and the particles are far apart from each other. As a result, the attractive forces between the molecules are negligible and gas molecules move freely. In a liquid, attractive forces hold the molecules much more closely together, so the distance between molecules and the kinetic energy is much less than the gas. In a solid, the attractive forces between molecules are even stronger, so the distance between individual particles is small and there is little freedom of motion. The properties of gases, liquids, and solids are summarized in Table 4.2.

Concept Check 4.1	How do gaseous, liquid, and solid methanol (CH_4O) compare in each of the following features: (a) density; (b) the space between the molecules; (c) the attractive force between the molecules?
Concept Check 4.2	Why is a gas much more easily compressed into a smaller volume than a liquid or solid?

Table 4.2 Properties of Gases, Liquids, and Solids

Property	Gas	Liquid	Solid
Shape and volume	Expands to fill its container	A fixed volume that takes the shape of the container it occupies	A definite shape and volume
Arrangement of particles	Randomly arranged, disorganized, and far apart	Randomly arranged but close	Fixed arrangement of very close particles
Density	Low (< 0.01 g/mL)	High (~1 g/mL)[a]	High (1–10 g/mL)
Particle movement	Very fast	Moderate	Slow
Interaction between particles	None	Strong	Very strong

[a]The symbol "~" means approximately.

4.3 Intermolecular Forces

To understand many of the properties of solids, liquids, and gases, we must learn about the forces of attraction that exist between particles—atoms, molecules, or ions.

Ionic compounds are composed of extensive arrays of oppositely charged ions that are held together by strong electrostatic interactions. **These ionic interactions are much stronger than the forces between covalent molecules,** so it takes a great deal of energy to separate ions from each other (Section 3.5).

In covalent compounds, the nature and strength of the attraction between individual molecules depend on the identity of the atoms.

- Intermolecular forces are the attractive forces that exist *between* molecules.

There are three different types of intermolecular forces in covalent molecules, presented in order of *increasing strength:*

- London dispersion forces
- Dipole–dipole interactions
- Hydrogen bonding

Thus, a compound that exhibits hydrogen bonding has stronger intermolecular forces than a compound of similar size that has dipole–dipole interactions. Likewise, a compound that has dipole–dipole interactions has stronger intermolecular forces than a compound of similar size that has only London dispersion forces. The strength of the intermolecular forces determines whether a compound has a high or low melting point and boiling point, and thus if the compound is a solid, liquid, or gas at a given temperature.

4.3A London Dispersion Forces

London dispersion forces can also be called **van der Waals forces.**

- London dispersion forces are very weak interactions due to the momentary changes in electron density in a molecule.

For example, although a nonpolar methane molecule (CH_4) has no net dipole, at any one instant its electron density may not be completely symmetrical. If more electron density is present in one region of the molecule, less electron density must be present some place else, and this creates a *temporary* dipole. A temporary dipole in one CH_4 molecule induces a temporary dipole in another CH_4 molecule, with the partial positive and negative charges arranged close to each other. **The weak interaction between these temporary dipoles constitutes London dispersion forces.**

London dispersion force between two CH_4 molecules

More electron density in one region creates a partial negative charge (δ−).

Less electron density in one region creates a partial positive charge (δ+).

Although any single interaction is weak, a large number of London dispersion forces creates a strong force. For example, geckos stick to walls and ceilings by London dispersion forces between the surfaces and the 500,000 tiny hairs on each foot. *(top): Matthew D.H. Smith; (bottom): Andrew Syred/Science Source*

All covalent compounds exhibit London dispersion forces. These intermolecular forces are the only intermolecular forces present in nonpolar compounds. The strength of these forces is related to the size of the molecule.

- The *larger* the molecule, the *larger* the attractive force between two molecules, and the *stronger* the intermolecular forces.

Atoms like the noble gases helium and argon exhibit London dispersion forces, too. The attractive forces between argon atoms are stronger than the attractive forces between helium atoms because argon atoms are considerably larger in size.

Concept Check 4.3

Which species in each pair exhibits the stronger London dispersion forces?

a.

$$H-\overset{\overset{\displaystyle H}{|}}{\underset{\underset{\displaystyle H}{|}}{C}}-H$$

or

$$H-\overset{\overset{\displaystyle H}{|}}{\underset{\underset{\displaystyle H}{|}}{C}}-\overset{\overset{\displaystyle H}{|}}{\underset{\underset{\displaystyle H}{|}}{C}}-\overset{\overset{\displaystyle H}{|}}{\underset{\underset{\displaystyle H}{|}}{C}}-H$$

b. He atoms or Ne atoms

4.3B   Dipole–Dipole Interactions

How to determine whether a molecule is polar is shown in Section 3.12.

- Dipole–dipole interactions are the attractive forces between the permanent dipoles of two polar molecules.

For example, the carbon–oxygen bond in formaldehyde, $H_2C{=}O$, is polar because oxygen is more electronegative than carbon. This polar bond gives formaldehyde a permanent dipole, making it a polar molecule. The dipoles in adjacent formaldehyde molecules can align so that the partial positive and partial negative charges are close to each other. These attractive forces due to permanent dipoles are much stronger than London dispersion forces.

formaldehyde

Concept Check 4.4

Draw the individual dipoles of two H—Cl molecules and show how the dipoles are aligned in a dipole–dipole interaction.

4.3C   Hydrogen Bonding

- Hydrogen bonding occurs when a hydrogen atom bonded to O, N, or F, is electrostatically attracted to an O, N, or F atom in another molecule.

hydrogen bond between two H_2O molecules

covalent O—H bond

Hydrogen bonding can occur only when molecules contain a hydrogen atom bonded to a very electronegative atom—that is, oxygen, nitrogen, or fluorine. For example, two H_2O molecules can hydrogen bond to each other: a hydrogen atom is covalently bonded to oxygen in one water molecule, and hydrogen bonded to an oxygen atom in another water molecule. **Hydrogen bonds are the strongest of the three types of intermolecular forces.** Table 4.3 summarizes the three types of intermolecular forces.

Table 4.3 Summary of the Types of Intermolecular Forces

Type of Force	Relative Strength	Exhibited by	Example
London dispersion	Weak	All molecules	CH_4, H_2CO, H_2O
Dipole–dipole	Moderate	Molecules with a net dipole	H_2CO, H_2O
Hydrogen bonding	Strong	Molecules with an O—H, N—H, or H—F bond	H_2O

Hydrogen bonding is important in many biological molecules, including proteins and DNA. DNA, which is contained in the chromosomes of the nucleus of a cell, is responsible for the storage of all genetic information. DNA is composed of two long strands of atoms that are held together by hydrogen bonding as shown in Figure 4.3. A detailed discussion of DNA appears in Chapter 17.

Figure 4.3 Hydrogen Bonding and DNA

DNA double helix

Hydrogen bonding interactions are shown as dashed red lines.

DNA is composed of two long strands of atoms that wind around each other in an arrangement called a double helix. The two strands are held together by an extensive network of hydrogen bonds. In each hydrogen bond, a hydrogen atom of an N—H bond on one chain is intermolecularly hydrogen bonded to an oxygen or nitrogen atom on an adjacent chain. Five hydrogen bonds are indicated.

Sample Problem 4.3

Determining the Types of Intermolecular Forces in a Compound

What types of intermolecular forces are present in each compound: (a) HCl; (b) C_2H_6 (ethane); (c) NH_3?

Analysis

- **London dispersion forces** are present in all covalent compounds.
- **Dipole–dipole interactions** are present only in polar compounds with a permanent dipole.
- **Hydrogen bonding** occurs only in compounds that contain an O—H, N—H, or H—F bond.

Solution

a.

$\delta+$ H—Cl $\delta-$ polar bond

- HCl has London forces like all covalent compounds.
- HCl has a polar bond, so it exhibits dipole–dipole interactions.
- HCl has no H atom on an O, N, or F, so it has no intermolecular hydrogen bonding.

b.

H—C—C—H (with H atoms)

nonpolar molecule

- C_2H_6 is a nonpolar molecule because it has only nonpolar C—C and C—H bonds. Thus, it exhibits only London forces.

c.

$\delta-$ N, H H H $\delta+$ net dipole

- NH_3 has London forces like all covalent compounds.
- NH_3 has a net dipole from its three polar bonds (Section 3.12), so it exhibits dipole–dipole interactions.
- NH_3 has a H atom bonded to N, so it exhibits intermolecular hydrogen bonding.

Practice Problem 4.5

What types of intermolecular forces are present in each molecule?

a. Cl_2 b. HCN c. HF d. CH_3Cl e. H_2

Practice Problem 4.6

Which of the compounds in each pair has stronger intermolecular forces?

a. CO_2 or H_2O b. CO_2 or HBr c. HBr or H_2O d. CH_4 or C_2H_6

More Practice: Try Problems 4.9, 4.10a, 4.23–4.25, 4.26a, 4.27, 4.28, 4.33a, 4.34a.

4.4 Boiling Point and Melting Point

The **boiling point (bp)** of a compound is the temperature at which a liquid is converted to the gas phase, whereas the **melting point (mp)** is the temperature at which a solid is converted to the liquid phase. The strength of the intermolecular forces determines the boiling point and melting point of compounds.

> • The *stronger* the intermolecular forces, the *higher* the boiling point and melting point.

In boiling, energy must be supplied to overcome the attractive forces of the liquid state and separate the molecules to the gas phase. Similarly, in melting, energy must be supplied to overcome the highly ordered solid state and convert it to the less ordered liquid phase. A stronger force of attraction between molecules means that more energy must be supplied to overcome those intermolecular forces, increasing the boiling point and melting point.

In comparing compounds of similar size, the following trend is observed:

Compounds with London dispersion forces only	Compounds with dipole–dipole interactions	Compounds that can hydrogen bond

Increasing strength of intermolecular forces
Increasing boiling point
Increasing melting point

Methane (CH_4) and water (H_2O) are both small molecules with hydrogen atoms bonded to a second-row element, so you might expect them to have similar melting points and boiling points. Methane, however, is a nonpolar molecule that exhibits only London dispersion forces, whereas water is a polar molecule that can form intermolecular hydrogen bonds. As a result, the melting point and boiling point of water are *much higher* than those of methane. In fact, the hydrogen bonds in water are so strong that it is a liquid at room temperature, whereas methane is a gas.

methane

London forces only
bp = −162 °C
mp = −183 °C

water

hydrogen bonding
bp = 100 °C ◄── **stronger forces**
mp = 0 °C ◄── **higher bp and mp**

Methane, the main component of natural gas, is a gas at room temperature because the CH_4 molecules have weak forces of attraction for each other.
Ingram Publishing/SuperStock

In comparing two compounds with similar types of intermolecular forces, the higher-molecular-weight compound generally has more surface area and therefore a larger force of attraction, giving it the higher boiling point and melting point. Thus, propane (C_3H_8) and butane (C_4H_{10}) have only nonpolar bonds and London forces, but butane is larger and therefore has the higher boiling point and melting point.

propane

bp = −42 °C
mp = −190 °C

butane

bp = −0.5 °C
mp = −138 °C

larger molecule
stronger forces
higher bp and mp

Sample Problem 4.4

Predicting the Relative Boiling Point and Melting Point of Two Compounds

(a) Which compound, **A** or **B**, has the higher boiling point? (b) Which compound, **C** or **D**, has the higher melting point?

NH_3

ammonia

A

CH_4

methane

B

H—C—O—H (with H above and H below)

methanol

C

H—C—Cl (with H above and H below)

chloromethane

D

Analysis

Determine the types of intermolecular forces in each compound. The compound with the *stronger* forces has the *higher* boiling point or melting point.

Solution

a. NH_3 (**A**) has an N—H bond, so it exhibits intermolecular hydrogen bonding. CH_4 (**B**) has only London forces because it has only nonpolar C—H bonds. NH_3 has *stronger* forces and the *higher* boiling point.

b. Methanol (**C**) has an O—H bond, so it can intermolecularly hydrogen bond. Chloromethane (**D**) has a polar C—Cl bond, so it has dipole–dipole interactions, but it cannot hydrogen bond. **C** has *stronger* forces, so **C** has the *higher* melting point.

Practice Problem 4.7

Which compound in each pair has the higher boiling point? Which compound in each pair has the higher melting point?

a. CH_4 or C_2H_6 b. C_2H_6 or CH_3OH c. HBr or HCl   d. C_2H_6 or CH_3Br

---

Practice Problem 4.8

Explain why CO_2 is a gas at room temperature but H_2O is a liquid.

More Practice: Try Problems 4.10b; 4.21; 4.22; 4.26b; 4.29–4.32; 4.33b, c; 4.34b, c.

4.5   Specific Heat

In addition to boiling point and melting point, there are many other physical properties that characterize a substance. For example, **specific heat (SH)** is a physical property that reflects the ability of a substance to absorb heat. Specific heat relates energy, mass, and temperature change (ΔT).

- **The specific heat is the amount of heat energy (in calories or joules) needed to raise the temperature of 1 g of a substance by 1 °C.**

$$\text{specific heat} = \frac{\text{heat}}{\text{mass} \times \Delta T} = \frac{\text{cal (or J)}}{\text{g} \cdot °C}$$

Table 4.4 **Specific Heats of Some Substances**

Substance	cal/(g · °C)	J/(g · °C)
Aluminum	0.214	0.897
Carbon (graphite)	0.169	0.709
Copper	0.0920	0.385
Ethanol	0.583	2.44
Gold	0.0308	0.129
Iron	0.107	0.449
Mercury	0.0335	0.140
2-Propanol	0.612	2.56
Rock	0.200	0.837
Sand	0.200	0.837
Silver	0.0560	0.234
Water(l)	1.00	4.18
Water(g)	0.447	1.87
Water(s)	0.486	2.03

The specific heat of water is 1.00 cal/(g · °C), meaning that 1.00 cal of heat must be added to increase the temperature of 1.00 g of water by 1.00 °C. The amount of heat that must be added for a particular temperature increase depends on the amount of sample present. To increase the temperature of 2.00 g of water by 1 °C requires 2.00 cal of heat. Table 4.4 lists the specific heat values for a variety of substances.

- The *larger* the specific heat of a substance, the *less* its temperature will change when it absorbs a particular amount of heat energy.

The specific heat of graphite is about twice that of copper. Adding 1.00 cal of heat will raise the temperature of 1.00 g of graphite by 5.9 °C, but raise the temperature of 1.00 g of copper by 11.1 °C. Because metals like copper have low specific heats, they absorb and transfer heat readily. We use copper, iron, or aluminum cookware because the metals readily transfer heat from the stove to the food in the pan.

The specific heat of water is very high compared to other liquids. As a result, **water absorbs a large amount of heat with only a small change in temperature.** Water has a high specific heat because of its strong intermolecular forces. Because water molecules are held together with an extensive array of hydrogen bonds, it takes a great deal of heat energy to break these bonds and increase the disorder of the water molecules.

Moreover, because the **amount of heat absorbed for a given temperature increase equals the amount of heat released upon cooling,** water releases a great deal of energy when its temperature drops even a few degrees. This explains why the climate in a coastal city is often more moderate than that of a city located 100 miles inland. A large body of water acts as a reservoir to absorb or release heat as temperature increases or decreases, so it moderates the climate of the land nearby.

Sample Problem 4.5 — Using Specific Heat When a Substance Absorbs or Releases Heat

Consider the elements aluminum, copper, gold, and iron in Table 4.4. (a) If 10 kcal of heat is added to 10 g of each element, which element will have the highest temperature? (b) Which element would require the largest amount of heat to raise the temperature of a 5-g sample by 5 °C?

Analysis

- The *larger* the specific heat, the *less* the temperature of a substance will change when it absorbs heat energy.
- The *larger* the specific heat, the *more* heat that must be added to increase the temperature of a substance a given number of degrees.

Solution

The specific heats of the metals in Table 4.4 increase in the following order: gold < copper < iron < aluminum. In part (a), gold has the *lowest* specific heat, so its temperature will be the *highest* if the same amount of heat is added to the same mass of all four elements. In part (b), aluminum has the *largest* specific heat, so it will require the *largest* amount of heat to raise its temperature the same number of degrees as the same mass of the other elements.

Practice Problem 4.9

Consider two beakers that contain equal volumes of two liquids **A** and **B**.

a. If the specific heat of **A** is greater than the specific heat of **B**, which liquid has the higher temperature when the same amount of heat is added to both liquids?

b. If the specific heat of **A** is greater than the specific heat of **B**, which substance requires the larger amount of heat to raise its temperature by 10 °C?

Practice Problem 4.10

The human body is composed of about 70% water. How does this help the body to maintain a steady internal temperature?

More Practice: Try Problems 4.11, 4.12.

We can use the specific heat as a conversion factor to calculate how much heat is absorbed or lost from a substance as long as its mass and change in temperature are known, using the equation:

Sample Problem 4.6 shows how to use specific heat to calculate the amount of heat absorbed by a substance when the mass and temperature change are known.

Sample Problem 4.6   Using Specific Heat in Calculations

How many calories are needed to heat a pot of 1,600 g of water from 25 °C to 100. °C?

Analysis

Use specific heat as a conversion factor to calculate the amount of heat absorbed given the known mass and temperature change.

Solution

[1] Identify the known quantities and the desired quantity.

$$mass = 1{,}600 \text{ g}$$
$$T_1 = 25 \text{ °C}$$
$$T_2 = 100. \text{ °C} \qquad \text{? calories}$$

$$\underbrace{\phantom{mass = 1{,}600 \text{ g}}}_{\text{known quantities}} \qquad \underbrace{}_{\text{desired quantity}}$$

- Subtract the initial temperature (T_1) from the final temperature (T_2) to determine the temperature change: $T_2 - T_1 = \Delta T = 100. - 25 = 75 \text{ °C}$.
- The specific heat of water is 1.00 cal/(g · °C).

[2] Write the equation.

- The specific heat is a conversion factor that relates the heat absorbed to the temperature change (ΔT) and mass.

$$\textbf{heat} \quad = \quad \textbf{mass} \quad \times \quad \Delta\textbf{T} \quad \times \quad \textbf{specific heat}$$

$$\text{cal} \quad = \quad \text{g} \quad \times \quad \text{°C} \quad \times \quad \frac{\text{cal}}{\text{g} \cdot \text{°C}}$$

[3] Solve the equation.

- Substitute the known quantities into the equation and solve for heat in calories.

$$\text{cal} \quad = \quad 1600 \text{ g} \quad \times \quad 75 \text{ °C} \quad \times \quad \frac{1.00 \text{ cal}}{1 \text{ g} \cdot 1 \text{ °C}} \quad = \quad 1.2 \times 10^5 \text{ cal}$$

Answer

Practice Problem 4.11

How much energy is required to heat 28.0 g of iron from 19 °C to 150. °C? Report your answer in calories and joules.

Practice Problem 4.12

How much energy is released when 200. g of water is cooled from 55 °C to 12 °C? Report your answer in calories and kilocalories.

More Practice: Try Problems 4.35–4.40.

4.6 Energy and Phase Changes

In Section 4.4 we learned how the strength of intermolecular forces in a liquid and solid affect a compound's boiling point and melting point. Let's now look in more detail at the energy changes that occur during phase changes.

- When energy is absorbed, a process is said to be *endothermic*.
- When energy is released, a process is said to be *exothermic*.

4.6A Converting a Solid to a Liquid

Converting a solid to a liquid is called *melting*. Melting is a phase change because the highly organized water molecules in the solid phase become more disorganized in the liquid phase, but the chemical bonds do not change. Each water molecule is composed of two O—H bonds in both the solid and the liquid phases.

Melting is an *endothermic* process. Energy must be absorbed to overcome some of the attractive intermolecular forces that hold the organized solid molecules together to form the more random liquid phase. The amount of energy needed to melt 1 g of a substance is called its **heat of fusion.**

ice

liquid water

(left & right): Jill Braaten/McGraw Hill

Freezing is the opposite of melting; that is, *freezing* **converts a liquid to a solid.** Freezing is an *exothermic* process because energy is released as the faster-moving liquid molecules form an organized solid in which particles have little freedom of motion. For a given mass of a particular substance, the **amount of energy released in freezing is the same as the amount of energy absorbed during melting.**

Heats of fusion are reported in calories per gram (cal/g). A heat of fusion can be used as a conversion factor to determine how much energy is absorbed when a particular amount of a substance melts, as shown in Sample Problem 4.7.

Sample Problem 4.7	Using the Heat of Fusion to Determine Energy Changes

How much energy in calories is absorbed when 50.0 g of ice cubes melt? The heat of fusion of H_2O is 79.7 cal/g.

Analysis

Use the heat of fusion as a conversion factor to determine the amount of energy absorbed in melting.

Solution

[1] **Identify the original quantity and the desired quantity.**

$$50.0 \text{ g} \qquad\qquad ? \text{ calories}$$

original quantity   desired quantity

[2] Write out the conversion factors.

- Use the heat of fusion as a conversion factor to convert grams to calories.

g–cal
conversion factors

$$\frac{1\,g}{79.7\,cal} \quad \text{or} \quad \boxed{\frac{79.7\,cal}{1\,g}} \quad \leftarrow \quad \text{Choose this conversion factor to cancel the unwanted unit, g.}$$

[3] Solve the problem.

g–cal
conversion factor

$$50.0\,\cancel{g} \times \frac{79.7\,cal}{1\,\cancel{g}} = \quad 3,985 \text{ cal rounded to 3,990 cal}$$

Grams cancel.

Answer

When an ice cube is added to a liquid at room temperature, the ice cube melts. The energy needed for melting is "pulled" from the warmer liquid molecules and the liquid cools down. *Suzie Ross/McGraw Hill*

Practice Problem 4.13

Use the heat of fusion of water from Sample Problem 4.7 to answer each question.

a. How much energy in calories is released when 50.0 g of water freezes?
b. How much energy in calories is absorbed when 35.0 g of water melts?
c. How much energy in kilocalories is absorbed when 35.0 g of water melts?

More Practice: Try Problems 4.43a, b; 4.45, 4.52.

4.6B   Converting a Liquid to a Gas

Converting a liquid to a gas is called *vaporization.* Vaporization is an *endothermic* process. Energy must be absorbed to overcome the attractive intermolecular forces of the liquid phase to form gas molecules. The amount of energy needed to vaporize 1 g of a substance is called its **heat of vaporization.**

endothermic

vaporization

condensation

exothermic

H_2O

liquid water

steam

(left & right): Jill Braaten/McGraw Hill

Condensation is the opposite of vaporization; that is, **condensation converts a gas to a liquid.** Condensation is an *exothermic* process because energy is released as the faster-moving gas molecules form the more organized liquid phase. For a given mass of a particular substance, the **amount of energy released in condensation equals the amount of energy absorbed during vaporization.**

Chloroethane (CH_3CH_2Cl), commonly called ethyl chloride, is a local anesthetic. When chloroethane is sprayed on a wound it quickly evaporates, causing a cooling sensation that numbs the site of an injury. *Gebauer Company*

Heats of vaporization are reported in calories per gram (cal/g). A high heat of vaporization means that a substance absorbs a great deal of energy as it is converted from a liquid to a gas. **Water has a high heat of vaporization.** As a result, the evaporation of sweat from the skin is a very effective cooling mechanism for the body. The **heat of vaporization can be used as a conversion factor** to determine how much energy is absorbed when a particular amount of a substance vaporizes, as shown in Sample Problem 4.8.

Sample Problem 4.8   **Using the Heat of Vaporization to Determine Energy Changes**

How much heat in kilocalories is absorbed when 22.0 g of 2-propanol, rubbing alcohol, evaporates after being rubbed on the skin? The heat of vaporization of 2-propanol is 159 cal/g.

Analysis

Use the heat of vaporization to convert grams to an energy unit, calories. Calories must also be converted to kilocalories using a cal–kcal conversion factor.

Solution

CONSUMER NOTE

2-Propanol (Sample Problem 4.8) is the major component of rubbing alcohol.
Jill Braaten/McGraw Hill

[1] Identify the original quantity and the desired quantity.

22.0 g	? kilocalories
original quantity	desired quantity

[2] Write out the conversion factors.

- We have no conversion factor that directly relates grams and kilocalories. We do know, however, how to relate grams to calories using the heat of vaporization, and calories to kilocalories.

  g–cal conversion factors   cal–kcal conversion factors

$$\frac{1\ g}{159\ cal} \quad \text{or} \quad \frac{159\ cal}{1\ g} \qquad\qquad \frac{1000\ cal}{1\ kcal} \quad \text{or} \quad \frac{1\ kcal}{1000\ cal}$$

Choose the conversion factors with the unwanted units—g and cal—in the denominator.

[3] Solve the problem.

g–cal	cal–kcal
conversion factor	**conversion factor**

$$22.0\ \cancel{g} \quad \times \quad \frac{159\ \cancel{cal}}{1\ \cancel{g}} \quad \times \quad \frac{1\ kcal}{1000\ \cancel{cal}} \quad = \quad 3.50\ kcal$$

Answer

Grams cancel.   Calories cancel.

Practice Problem 4.14   Answer the following questions about water, which has a heat of vaporization of 540 cal/g.

a. How much energy in calories is absorbed when 42 g of water is vaporized?

b. How much energy in calories is released when 42 g of water is condensed?

More Practice: Try Problems 4.43c, d; 4.45; 4.46.

CONSUMER NOTE

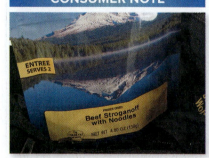

Freeze-drying removes water from foods by the process of sublimation. These products can be stored almost indefinitely, because bacteria cannot grow in them without water. *Jill Braaten/McGraw Hill*

4.6C   Converting a Solid to a Gas

Occasionally a solid phase forms a gas phase without passing through the liquid state. This process is called **sublimation.** The reverse process, conversion of a gas directly to a solid, is called **deposition.** Carbon dioxide is called *dry ice* because solid carbon dioxide (CO_2) sublimes to gaseous CO_2 without forming liquid CO_2.

endothermic

sublimation

deposition

exothermic

$CO_2(s)$ $CO_2(g)$

(left & right): Jill Braaten/McGraw Hill

Carbon dioxide is a good example of a solid that undergoes this process at atmospheric pressure. At reduced pressure other substances sublime. For example, freeze-dried foods are prepared by subliming water from a food product at low pressure.

Concept Check 4.5	Label each process as endothermic or exothermic and explain your reasoning: (a) sublimation; (b) deposition.

Sample Problem 4.9 illustrates how molecular art can be used to depict and identify phase changes.

Sample Problem 4.9	**Determining Energy Changes During Changes of State**

What phase change is shown in the accompanying molecular art? Is the process endothermic or exothermic?

A   B

Analysis

Identify the phase by the distance between the spheres and their level of organization. A solid has closely packed spheres that are well organized; a liquid has closely packed but randomly arranged spheres; a gas has randomly arranged spheres that are far apart. Then, classify the transformation as melting, freezing, vaporization, condensation, sublimation, or deposition, depending on the phases depicted.

Solution

A represents a solid and **B** represents a liquid, so the molecular art represents melting. Melting is an endothermic process because energy must be absorbed to convert the more ordered solid state to the less ordered liquid state.

Practice Problem 4.15	What phase change is shown in the accompanying molecular art? Is the process endothermic or exothermic?

More Practice: Try Problems 4.3, 4.4.

4.7 Heating and Cooling Curves

The changes of state described in Section 4.6 can be illustrated on a single graph called a **heating curve** when heat is added and a **cooling curve** when heat is removed.

4.7A Heating Curves

A **heating curve** shows how the temperature of a substance (plotted on the vertical axis) changes as heat is *added*. A general heating curve is shown in Figure 4.4.

Figure 4.4 Heating Curve

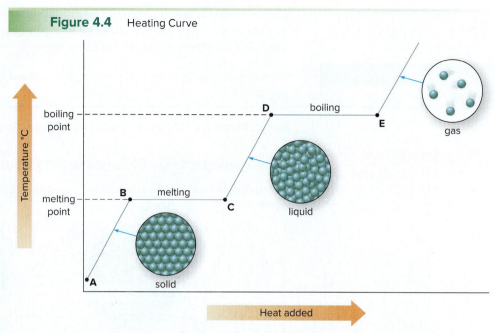

A heating curve shows how the temperature of a substance changes as heat is added. The plateau **B → C** occurs at the melting point, whereas the plateau **D → E** occurs at the boiling point.

A solid is present at point **A.** As the solid is heated it increases in temperature until its melting point is reached at **B.** More heat causes the solid to melt to a liquid, without increasing its temperature (the plateau from **B → C**). Added heat increases the temperature of the liquid until its boiling point is reached at **D.** More heat causes the liquid to boil to form a gas, without increasing its temperature (the plateau from **D → E**). Additional heat then increases the temperature of the gas. Each diagonal line corresponds to the presence of a single phase—solid, liquid, or gas—whereas horizontal lines correspond to phase changes—solid to liquid or liquid to gas.

Concept Check 4.6

Answer the following questions about the graph.

a. What is the melting point of the substance?

b. What is the boiling point of the substance?

c. What phase(s) are present at plateau **B → C?**

d. What phase(s) are present along the diagonal **C → D?**

Concept Check 4.7

If the substance shown in the heating curve in Figure 4.4 has a melting point of 50 °C and a boiling point of 75 °C, what state or states of matter are present at each temperature?

a. 85 °C　　　b. 50 °C　　　c. 65 °C　　　d. 10 °C　　　e. 75 °C

4.7B Cooling Curves

A **cooling curve** illustrates how the temperature of a substance (plotted on the vertical axis) changes as heat is *removed.* A cooling curve for water is shown in Figure 4.5.

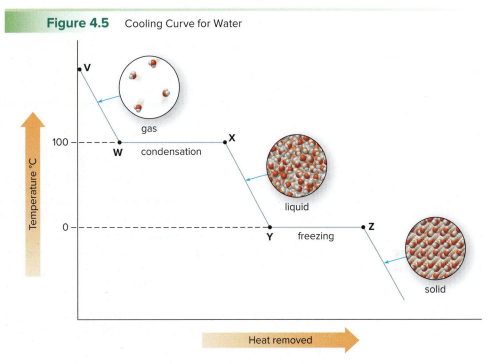

Figure 4.5   Cooling Curve for Water

The cooling curve shows how the temperature of water changes as heat is removed. The plateau **W** ⟶ **X** occurs at the boiling point, whereas the plateau **Y** ⟶ **Z** occurs at the freezing point.

Gaseous water is present at point **V.** As the gas is cooled it decreases in temperature until its boiling point is reached at **W.** Condensation at 100 °C forms liquid water, represented by the plateau from **W** → **X.** Further cooling of the liquid water takes place until its freezing point (melting point) is reached at **Y.** Freezing water forms ice at 0 °C, represented by the plateau from **Y** → **Z.** Cooling the ice further decreases its temperature below its freezing point.

Concept Check 4.8	If Figure 4.5 represented a substance with a melting point of 40 °C and a boiling point of 85 °C, what state or states of matter would be present at each temperature?
	a. 75 °C b. 50 °C c. 65 °C d. 10 °C e. 85 °C

4.7C   Combining Energy Calculations

We have now performed two different types of energy calculations. In Section 4.5 we used specific heat to calculate the amount of energy needed to raise the temperature of a substance in a single phase—such as heating a liquid from a lower to a higher temperature. In Section 4.6 we learned how to calculate energy changes during changes of state using heats of fusion or vaporization. Sometimes these calculations must be combined together to determine the total energy change when both a temperature change and a change of state occur.

Sample Problem 4.10 illustrates how to calculate the total energy needed to convert a given mass of liquid water to steam at its boiling point.

Sample Problem 4.10	Calculating the Total Energy Change of a Multistep Process

How much energy is required to heat 25.0 g of water from 25 °C to a gas at its boiling point of 100. °C? The specific heat of water is 1.00 cal/(g · °C), and the heat of vaporization of water is 540 cal/g.

Analysis

- Use **specific heat** to calculate how much energy is required to heat the given mass of water to its boiling point.
- **Use the heat of vaporization** to calculate how much energy is required to convert liquid water to a gas.

Solution

[1] Identify the original quantities and the desired quantity.

$$\text{mass} = 25.0 \text{ g}$$
$$T_1 = 25 \text{ °C}$$
$$T_2 = 100. \text{ °C} \qquad \text{? calories}$$

known quantities   desired quantity

- Subtract the initial temperature (T_1) from the final temperature (T_2) to determine the temperature change: $T_2 - T_1 = \Delta T = 100. - 25 = 75$ °C.

[2] Write out the conversion factors.

- Conversion factors are needed for both the specific heat and the heat of vaporization.

specific heat
conversion factors

$$\frac{1.00 \text{ cal}}{1 \text{ g} \cdot 1 \text{ °C}} \quad \text{or} \quad \frac{1 \text{ g} \cdot 1 \text{ °C}}{1.00 \text{ cal}}$$

heat of vaporization
conversion factors

$$\frac{540 \text{ cal}}{1 \text{ g}} \quad \text{or} \quad \frac{1 \text{ g}}{540 \text{ cal}}$$

Choose the conversion factors that place the unwanted units—(g · °C) and g—in the denominator.

[3] Solve the problem.

- **Calculate the heat needed to change the temperature** of water 75 °C using specific heat.

$$\text{heat} = \text{mass} \times \Delta T \times \text{specific heat}$$

$$\text{cal} = 25.0 \text{ g} \times 75.0 \text{ °C} \times \frac{1.00 \text{ cal}}{1 \text{ g} \cdot 1 \text{ °C}} = 1{,}875 \text{ cal rounded to } 1{,}900 \text{ cal}$$

- **Calculate the heat needed for the phase change** (liquid water ⟶ gaseous water) using the heat of vaporization.

g–cal
conversion factor

$$\text{cal} = 25.0 \text{ g} \times \frac{540 \text{ cal}}{1 \text{ g}} = 13{,}500 \text{ cal rounded to } 14{,}000 \text{ cal}$$

- **Add the two values together** to obtain the total energy required.

$$\text{Total energy} = 1{,}900 \text{ cal} + 14{,}000 \text{ cal} = 15{,}900 \text{ cal rounded to } 16{,}000 \text{ cal}$$

Answer

Practice Problem 4.16	How much energy (in calories) is released when 50.0 g of water is cooled from 25 °C to solid ice at 0.0 °C? The specific heat of water is 1.00 cal/(g · °C), and the heat of fusion of water is 79.7 cal/g.

Practice Problem 4.17	How much energy (in calories) is required to melt 25.0 g of ice to water at 0.0 °C, heat the liquid water to 100. °C, and vaporize the water to steam at 100. °C? The specific heat of water is 1.00 cal/(g · °C), the heat of fusion of water is 79.7 cal/g, and the heat of vaporization of water is 540 cal/g.

More Practice: Try Problems 4.49, 4.50.

CHAPTER REVIEW

KEY TERMS

Boiling point (bp, 4.4)
Calorie (4.1)
Condensation (4.6)
Cooling curve (4.7)
Deposition (4.6)
Dipole–dipole interactions (4.3)
Endothermic (4.6)

Energy (4.1)
Exothermic (4.6)
Freezing (4.6)
Heating curve (4.7)
Heat of fusion (4.6)
Heat of vaporization (4.6)
Hydrogen bonding (4.3)

Intermolecular forces (4.3)
Joule (4.1)
Kinetic energy (4.1)
Law of conservation of
   energy (4.1)
London dispersion forces (4.3)
Melting (4.6)

Melting point (mp, 4.4)
Potential energy (4.1)
Specific heat (SH, 4.5)
Sublimation (4.6)
Vaporization (4.6)

KEY CONCEPTS

1 General features of energy and matter (4.1, 4.2)

1 Energy (4.1)	**2 The three phases of matter (4.2)**		

• Energy—the capacity to do work
  ○ **Potential energy**—stored energy
  ○ **Kinetic energy**—energy of motion
• Units of energy
  ○ **Calorie (cal)**—the amount of energy needed to raise the temperature of 1 g of water 1 °C.
  ○ **Joule (J)**; 1 cal = 4.184 J
	Gas
	• Randomly arranged and disorganized particles that are far apart and move very fast
	• Low density
	• No intermolecular interaction
		Liquid
		• Randomly arranged, close particles with a definite volume but no definite shape
		• High density
		• Strong intermolecular interactions
			Solid
			• Fixed arrangement of very close, slow-moving particles with a definite shape and volume
			• High density
			• Very strong intermolecular interactions

2 Intermolecular forces and physical properties (4.3, 4.4)

1 Types of intermolecular forces in order of increasing strength (4.3)	**2 Relationship to melting point and boiling point (4.4)**
• **London dispersion forces** are due to momentary changes in the electron density in a molecule. • **Dipole–dipole interactions** are due to permanent dipoles. • **Hydrogen bonding** results when a H atom bonded to an O, N, or F is attracted to an O, N, or F atom in another molecule.	 C_3H_8 — London forces C_4H_{10} — London forces, larger surface area than C_3H_8 $C_2H_6O_2$ — O–H bonds, hydrogen bonding, strongest forces **→ Increasing boiling point and melting point** • The *stronger* the intermolecular forces, the *higher* the melting point and boiling point. • In compounds with similar types of intermolecular forces, the *larger* the surface area, the *higher* the melting point and boiling point.

3 Energy changes associated with solids, liquids, and gases

1 Specific heat (4.5)	**2 Phase changes (4.6)**
$$\text{specific heat} = \frac{\text{heat}}{\text{mass} \times \Delta T} = \frac{\text{cal (or J)}}{g \cdot °C}$$ • **Specific heat is the amount of heat energy needed to raise the temperature of 1 g of a substance 1 °C.** • The *larger* the specific heat, the *less* the temperature will change when it absorbs heat energy. • The *larger* the specific heat, the *more* heat that must be added to increase the temperature of a substance a given number of degrees.	 • Converting a more organized phase to a less organized phase *requires* energy—**endothermic** process. **Melting, vaporization,** and **sublimation** are endothermic. • Converting a less organized phase to a more organized phase *releases* energy—**exothermic** process. **Condensation, freezing,** and **deposition** are exothermic. • **Heat of fusion** = energy needed to melt 1 g of a substance • **Heat of vaporization** = energy needed to vaporize 1 g of a substance

KEY SKILLS

1 Using specific heat in calculations (4.5); example: determining how many calories are needed to heat 160 g of water from 50. °C to 100. °C

1 Identify the original quantities and the desired quantity.	**2** Write the equation using specific heat as a conversion factor.	**3** Solve the problem.
mass = 160 g T_1 = 50. °C T_2 = 100. °C ? calories original quantities **desired** quantity • Determine the **temperature change**: $T_2 - T_1 = \Delta T = 100. - 50. = 50.$ °C. • The specific heat of water is 1.00 cal/(g • °C).	heat = mass × ΔT × **specific heat** cal = g × °C × $\dfrac{cal}{g \cdot °C}$ **conversion factor** • The **specific heat** is a conversion factor that relates the heat absorbed to the temperature change and mass.	cal = 160 g × 50 °C × $\dfrac{1.00\ cal}{1\ g \cdot 1\ °C}$ **conversion factor** = 8.0 × 10³ cal • Substitute the known quantities and solve the equation for heat in calories.

2 Calculating the energy of a phase change (4.6); example: determining the heat (in kcal) absorbed when 21 g of water is vaporized

1 Identify the original quantity and the desired quantity.	**2** Write out the conversion factors that place the unwanted units (g and cal) in the denominator.	**3** Solve the problem so that unwanted units cancel.
21 g ? kcal original quantity **desired** quantity • The heat of vaporization of water is 540 cal/g.	$\dfrac{540\ cal}{1\ g}$ $\dfrac{1\ kcal}{1000\ cal}$ **heat of vaporization** **cal–kcal** **conversion factor** **conversion factor** • The **heat of vaporization** is a conversion factor that relates the calories to grams.	$21\ g \times \dfrac{540\ cal}{1\ g} \times \dfrac{1\ kcal}{1000\ cal} = 11\ kcal\ absorbed$ • Substitute the known quantities and solve the equation for heat in kcal.

CHAPTER 4 SELF-TEST

The Self-Test consists of short-answer questions that test an understanding of definitions, equations, and other material encountered within the chapter. Answers to each question are provided at the end of the chapter. For any questions you are unable to answer correctly, revisit the appropriate section(s) of the text; then proceed to working the end-of-chapter problems to further reinforce your understanding of the chapter material.

1. Which term from "Key Terms" answers each question?
   a. What phase change involves conversion of a liquid to a solid?
   b. What term describes a process in which energy is absorbed?
   c. What term is used for the amount of energy needed to melt 1 g of a substance?
   d. What type of intermolecular force is due to the momentary changes in the electron density of a molecule?
   e. What physical property relates energy, mass, and temperature, and reflects the ability of a substance to absorb heat?
   f. What phase change is the opposite of deposition?

2. If a reaction releases 573 kJ of energy, how many kilocalories does this correspond to?

3. If a serving of crackers contains 2 g of protein, 5 g of fat, and 22 g of carbohydrates, estimate its number of Calories. Round off your answer to the tens place.

4. Which compounds are capable of intermolecular hydrogen bonding: (a) CH_4; (b) H_2O; (c) H_2S; (d) HCl; (e) C_3H_8?

5. Which compound in each pair has the higher boiling point: (a) HBr or HCl; (b) CO_2 or CH_3OH; (c) CH_3NH_2 or $H_2C=CH_2$?

6. How many grams of ice absorb 5.0×10^3 cal on melting? The heat of fusion of water is 79.7 cal/g.

7. If the specific heat of 2-propanol is larger than the specific heat of ethanol, which compound would have the higher temperature when 10 kcal of heat is added to each substance?

8. If the specific heat of copper is 0.0920 cal/(g • °C), what temperature change is observed when 20. cal of heat is added to 10.0 g of copper?

9. Label each process as endothermic or exothermic: (a) freezing 25 g of water; (b) subliming 25 g of carbon dioxide; (c) vaporizing 10 g of ethanol.

UNDERSTANDING KEY CONCEPTS

4.1 Which product has more Calories in a half-cup serving: Ben and Jerry's Cherry Garcia ice cream (13 g of fat, 23 g of carbohydrates, 4 g of protein) or Ben and Jerry's Cherry Garcia frozen yogurt (3 g of fat, 32 g of carbohydrates, 5 g of protein), which is lower in fat, but considerably higher in carbohydrate content? Round off each answer to the tens place.

Holly Curry/McGraw Hill

4.2 Using the nutrition labels for two different types of Caesar salad dressing (**A** and **B**), calculate the total Calories in one tablespoon of each product, and determine which label is for the "lite" variety of salad dressing. Round off each answer to the tens place.

A

	% Daily Values*
Total Fat 8.48g	13%
Saturated Fat 1.292g	6%
Polyunsaturated Fat 4.83g	
Monounsaturated Fat 1.985g	
Cholesterol 0mg	0%
Sodium 158mg	7%
Potassium 4mg	
Total Carbohydrate 0.46g	0%
Dietary Fiber 0g	0%
Sugars 0.19g	
Protein 0.18g	
Vitamin A 0% Vitamin C 0%	
Calcium 0% Iron 0%	

*Percent Daily Values are based on a 2000 Calorie diet. Your daily values may be higher or lower depending on your Calorie needs.

B

	% Daily Values*
Total Fat 0.66g	1%
Saturated Fat 0.105g	1%
Polyunsaturated Fat 0.355g	
Monounsaturated Fat 0.169g	
Cholesterol 0mg	0%
Sodium 162mg	7%
Potassium 4mg	
Total Carbohydrate 2.79g	1%
Dietary Fiber 0g	0%
Sugars 2.45g	
Protein 0.04g	
Vitamin A 0% Vitamin C 0%	
Calcium 0% Iron 0%	

*Percent Daily Values are based on a 2000 Calorie diet. Your daily values may be higher or lower depending on your Calorie needs.

4.3 What phase change is shown in the accompanying molecular art? Is energy absorbed or released during the process?

4.4 What phase change is shown in the accompanying molecular art? Is energy absorbed or released during the process?

4.5 Consider the cooling curve drawn below.

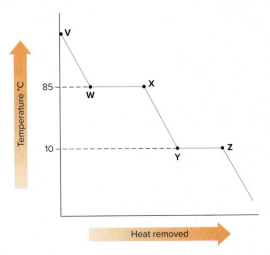

a. Which line segment corresponds to the following changes of state?

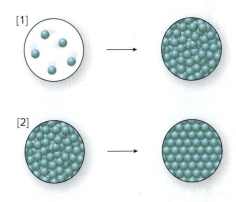

b. What is the melting point of the substance?

c. What is the boiling point of the substance?

4.6 Which line segments on the cooling curve in Problem 4.5 correspond to each of the following physical states?

4.7 Riding a bicycle at 12–13 miles per hour uses 563 Calories in an hour. Convert this value to (a) calories; (b) kilocalories; (c) joules; (d) kilojoules.

4.8 Estimate the number of Calories in two tablespoons of peanut butter, which contain 16 g of protein, 7 g of carbohydrates, and 16 g of fat. Round off your answer to the tens place.

4.9 What types of intermolecular forces are exhibited by each compound? Acetaldehyde is formed when ethanol, the alcohol in alcoholic beverages, is metabolized, and acetic acid gives vinegar its biting odor and taste.

a.   acetaldehyde   b.   acetic acid

4.10 Ethanol and dimethyl ether have the same molecular formula.

ethanol   dimethyl ether

a. What types of intermolecular forces are present in each compound?

b. Which compound has the higher boiling point?

4.11 Consider the two beakers containing the same mass of **X** and **Y** that were at the same initial temperature. Which compound, **X** or **Y**, has the higher specific heat if the same amount of heat was added to both substances?

4.12 Consider the following three containers (**A–C**) drawn below. (a) If the same amount of heat was added to all three samples, which sample has the highest temperature? (b) Which sample has the lowest temperature?

ADDITIONAL PROBLEMS

Energy

4.13 Carry out each of the following conversions.

a. 50 cal to kcal   c. 0.96 kJ to cal

b. 56 cal to kJ   d. 4,230 kJ to cal

4.14 Carry out each of the following conversions.

a. 5 kcal to cal   c. 1.22 kJ to cal

b. 2,560 cal to kJ   d. 4,230 J to kcal

4.15 Running at a rate of 6 mi/h uses 704 Calories in an hour. Convert this value to (a) calories; (b) kilocalories; (c) joules; (d) kilojoules.

4.16 Estimate the number of Calories in a serving of oatmeal that has 4 g of protein, 19 g of carbohydrates, and 2 g of fat. Round off your answer to the tens place.

4.17 A can of soda contains 120 Calories, and no protein or fat. How many grams of carbohydrates are present in each can?

4.18 Alcohol releases 29.7 kJ/g when it burns. Convert this value to the number of Calories per gram.

4.19 Use the number of grams of protein, carbohydrate, and fat in the accompanying label to estimate the number of Calories in a serving of whole milk. Round off your answer to the tens place.

Mark Dierker/McGraw Hill

4.20 Use the number of grams of protein, carbohydrate, and fat in the accompanying label to estimate the number of Calories in a serving of pizza. Round off your answer to the tens place.

Mark Dierker/McGraw Hill

Intermolecular Forces, Boiling Point, and Melting Point

4.21 Why is H_2O a liquid at room temperature, but H_2S, which has a higher molecular weight and a larger surface area, is a gas at room temperature?

4.22 Why is Cl_2 a gas, Br_2 a liquid, and I_2 a solid at room temperature?

4.23 What types of intermolecular forces are present in each compound?

a.   b.

4.24 What types of intermolecular forces are present in each compound?

a.   b.

4.25 What types of intermolecular forces are exhibited by each compound? Chloroethane is a local anesthetic and cyclopropane is a general anesthetic.

a. H—C—C—Cl

chloroethane

b. H—C—C—H

cyclopropane

4.26 Consider two compounds, ethylene and methanol.

ethylene   methanol

a. What types of intermolecular forces are present in each compound?
b. Which compound has the higher boiling point?

4.27 Which molecules are capable of intermolecular hydrogen bonding?

a. H—C≡C—H   b. CO_2 c. Br_2 d. H—C—N—H

4.28 Which molecules are capable of intermolecular hydrogen bonding?

a. N_2 b. H—C—F   c. HI   d. H—C—O—H

4.29 Which compound in each pair has the higher melting point?

a.   or

A   **B**

b.   or

C   **D**

4.30 Why is the melting point of NaCl (801 °C) much higher than the melting point of water (0 °C)?

4.31 Which compound, undecane or pentane, has the higher melting point? Explain.

undecane

pentane

4.32 Explain why the boiling point of **A** is higher than the boiling point of **B** despite the fact that **A** and **B** have the same chemical formula (C_3H_9N).

A   **B**

4.33 Consider two compounds, formaldehyde (H_2C=O) and ethylene (H_2C=CH_2).
a. Which compound exhibits the stronger intermolecular forces?
b. Which compound has the higher boiling point?
c. Which compound has the higher melting point?

4.34 Consider two compounds, formaldehyde (H_2C=O) and methanol (CH_3OH).
a. Which compound exhibits the weaker intermolecular forces?
b. Which compound has the lower boiling point?
c. Which compound has the lower melting point?

Specific Heat

4.35 How much energy is absorbed or lost in each of the following? Calculate your answer in both calories and joules.
a. the energy needed to heat 50. g of water from 15 °C to 50. °C
b. the energy lost when 250 g of aluminum is cooled from 125 °C to 50. °C

4.36 How many calories of heat are needed to increase the temperature of 55 g of ethanol from 18 °C to 48 °C?

4.37 Which of the following samples has the higher temperature?
a. 100. g of liquid water at 16.0 °C that absorbs 200. cal of heat
b. 50.0 g of liquid water at 16.0 °C that absorbs 350. J of heat

4.38 Which has the higher final temperature, 10.0 g of aluminum at 18 °C that absorbs 25.0 cal of heat or 12.0 g of iron at 22 °C that absorbs 65.0 J of heat?

4.39 If it takes 37.0 cal of heat to raise the temperature of 12.0 g of a substance by 8.5 °C, what is its specific heat?

4.40 Why does it take weeks for a lake to freeze in the winter even if the outdoor temperature is consistently below 0 °C?

Energy and Phase Changes

4.41 Classify each transformation as melting, freezing, vaporization, or condensation.
  a. Beads of water form on the glass of a cool drink in the summer.
  b. Wet clothes dry when hung on the clothesline in the sun.
  c. Water in a puddle on the sidewalk turns to ice when the temperature drops overnight.

4.42 Classify each transformation as melting, freezing, vaporization, or condensation.
  a. Fog forms on the mirror of the bathroom when a hot shower is taken.
  b. A puddle of water slowly disappears.
  c. A dish of ice cream becomes a bowl of liquid when left on the kitchen counter on a hot day.

4.43 Indicate whether heat is absorbed or released in each process.
  a. melting 100 g of ice
  b. freezing 25 g of water
  c. condensing 20 g of steam
  d. vaporizing 30 g of water

4.44 What is the difference between the heat of fusion and the heat of vaporization?

4.45 Which process requires more energy, melting 250 g of ice or vaporizing 50.0 g of water? The heat of fusion of water is 79.7 cal/g and the heat of vaporization is 540 cal/g.

4.46 How much energy in kilocalories is needed to vaporize 255 g of water? The heat of vaporization of water is 540 cal/g.

Heating and Cooling Curves

4.47 Draw the heating curve that is observed when octane is warmed from −70 °C to 130 °C. Octane, a component of gasoline, has a melting point of −57 °C and a boiling point of 126 °C.

4.48 Draw the heating curve that is observed when ice is warmed from −20 °C to 120 °C. Which sections of the curve correspond to the molecular art in **A, B,** and **C?**

A   B   C

Combined Energy Calculations

4.49 Use the following values to answer each part. The specific heat of water is 1.00 cal/(g · °C); the heat of fusion of water is 79.7 cal/g; and the heat of vaporization of water is 540 cal/g.
  a. How much energy (in calories) is needed to melt 45 g of ice at 0.0 °C and warm it to 55 °C?
  b. How much energy (in calories) is released when 45 g of water at 55 °C is cooled to 0.0 °C, and frozen to solid ice at 0.0 °C?
  c. How much energy (in kilocalories) is released when 35 g of steam at 100. °C is condensed to water, the water is cooled to 0.0 °C, and the water is frozen to solid ice at 0.0 °C?

4.50 Use the values in Problem 4.49 to solve each part.
  a. How much energy (in calories) is needed to heat 150 g of water from 35 °C to 100. °C and vaporize the water to steam at 100. °C?
  b. How much energy (in kilocalories) is released when 42 g of steam is condensed to water at 100. °C, the water is cooled to 0.0 °C, and the water is frozen to solid ice at 0.0 °C?

Applications

4.51 Explain why you feel cool when you get out of a swimming pool, even when the air temperature is quite warm. Then explain why the water feels warmer when you get back into the swimming pool.

4.52 To keep oranges from freezing when the outdoor temperature drops near 32 °F, an orchard is sprayed with water. Explain why this strategy is used.

4.53 A patient receives 2,000 mL of a glucose solution that contains 5 g of glucose in 100 mL. How many Calories does the glucose, a simple carbohydrate, contain?

4.54 Why does steam form when hot lava falls into the ocean?

4.55 Walking at a brisk pace burns off about 280 Cal/h. How long would you have to walk to burn off the Calories obtained from eating a cheeseburger that contained 32 g of protein, 29 g of fat, and 34 g of carbohydrates?

4.56 How many kilocalories does a runner expend when he runs for 4.5 h and uses 710 Cal/h? How many pieces of pizza that each contain 12 g of protein, 11 g of fat, and 30 g of carbohydrates could be eaten after the race to replenish these Calories?

CHALLENGE PROBLEMS

4.57 How much heat (in kcal) must be added to raise the temperature of the water in a 400.-gal hot tub from 60. °F to 110. °F? (Recall that the density of water is 1.00 g/mL.)

4.58 An energy bar contains 4 g of fat, 12 g of protein, and 24 g of carbohydrates. How many kilojoules of energy are obtained from eating two bars per day for a 30-day month? Write the answer in scientific notation using two significant figures.

BEYOND THE CLASSROOM

4.59 Some studies suggest that recycling one aluminum beverage can saves the energy equivalent of 0.5 gallons of gasoline. Estimate how many aluminum cans your household uses per week, and calculate how much gasoline would be saved by recycling these cans. If burning one gallon of gasoline releases about 3.1×10^4 kcal of energy, calculate how much energy is saved each week in recycling.

4.60 Obtain Calorie data from a fast-food restaurant and calculate how many Calories you ingest in a typical meal. How long would you have to walk or run to burn off those Calories? Assume that walking at a moderate pace expends 280 Cal/h and that running at a vigorous pace expends 590 Cal/h. Compare results for meals at different restaurants.

CHAPTER 4 ANSWERS

Concept Check Answers

4.1 a. gas < liquid < solid
b. solid < liquid < gas
c. gas < liquid < solid

4.2 The molecules of a gas are farthest apart, so they can be compressed into a smaller volume.

4.3 a. C_3H_8 b. Ne atoms

4.4 H–Cl   H–Cl
δ+ δ–   δ+ δ–

4.5 a. endothermic   b. exothermic

4.6 a. 40 °C   b. 77 °C   c. solid and liquid   d. liquid

4.7 a. gas   d. solid
b. solid and liquid   e. liquid and gas
c. liquid

4.8 a,b,c: liquid   d. solid   e. gas and liquid

Practice Problem Answers

4.1 a. 10. cal   b. 55,600 cal   c. 1,360 kJ   d. 107,000 J

4.2 48.1 kJ, 48,100 J

4.3 126 Cal, rounded to 130 Cal

4.4 128 Cal, rounded to 130 Cal

4.5 London dispersion: a–e; dipole–dipole: b,c,d; hydrogen bonding: c

4.6 a. H_2O b. HBr   c. H_2O d. C_2H_6

4.7 a. C_2H_6 b. CH_3OH c. HBr   d. CH_3Br

4.8 Water has stronger intermolecular forces and a higher boiling point because it can hydrogen bond.

4.9 a. **B**   b. **A**

4.10 Water has a high specific heat, so it can absorb or release a great deal of energy with only a small temperature change.

4.11 392 cal, 1,640 J

4.12 8,600 cal, 8.6 kcal

4.13 a. 3,990 cal   b. 2,790 cal   c. 2.79 kcal

4.14 a. 23,000 cal   b. 23,000 cal

4.15 condensation; exothermic

4.16 5,300 cal

4.17 18,000 cal

Self-Test Answers

1. a. freezing   d. London dispersion forces
b. endothermic   e. specific heat
c. heat of fusion   f. sublimation

2. 137 kcal

3. 141 Cal, rounded to 140 Cal

4. b.

5. a. HBr   b. CH_3OH c. CH_3NH_2

6. 63 g

7. ethanol

8. 22 °C

9. a. exothermic   b,c: endothermic

Odd-Numbered Problem Answers

4.1 ice cream: 225 Cal, rounded to 230 Cal
yogurt: 175 Cal, rounded to 180 Cal

4.3 vaporization; energy absorbed

4.5 a. [1] **W** → **X**; [2] **Y** → **Z**   b. 10 °C   c. 85 °C

4.7 a. 563,000 cal/h   c. 2.36×10^6 J/h
b. 563 kcal/h   d. 2,360 kJ/h

4.9 a. London forces and dipole–dipole
b. London forces, dipole–dipole, hydrogen bonding

4.11 **Y**

4.13 a. 0.05 kcal   c. 230 cal
b. 0.23 kJ   d. 1.01×10^6 cal

4.15 a. 704,000 cal   c. 2.95×10^6 J
b. 704 kcal   d. 2,950 kJ

4.17 30 g

4.19 148 Cal, rounded to 150 Cal

4.21 Water is capable of hydrogen bonding and H_2S is not.

4.23 a,b: London dispersion forces, dipole–dipole interactions

4.25 a. London forces, dipole–dipole
b. London forces only

4.27 d.

4.29 a. **A**   b. **D**

4.31 Undecane has the higher melting point because its size is larger, so its London forces are stronger.

4.33 a,b,c: formaldehyde

4.35 a. 1.8×10^3 cal, 7.5×10^3 J

b. 4.0×10^3 cal, 1.7×10^4 J

4.37 a.

4.39 0.36 cal/(g · °C)

4.41 a. condensation   b. vaporization   c. freezing

4.43 a,d: absorbed   b,c: released

4.45 Vaporizing 50.0 g of water takes more energy; 27,000 cal vs. 2.0×10^4 cal.

4.47

4.49 25 kcal

4.51 When you get out of a pool, the water on your body evaporates and this cools your skin. When you re-enter the water, the water feels warmer because the skin is cooler.

4.53 400 Calories

4.55 1.9 h, rounded to 2 h

4.57 4.2×10^4 kcal

Chemical Reactions

Aspirin, a widely used over-the-counter pain reliever, is one of the countless products synthesized by the chemical industry using chemical reactions. *Jill Braaten*

CHAPTER OUTLINE

CHAPTER GOALS

In this chapter you will learn how to:

1. Write and balance chemical equations
2. Classify chemical reactions as combination, decomposition, single replacement, or double replacement
3. Define oxidation and reduction and recognize the components of a redox reaction
4. Define a mole and use Avogadro's number in calculations
5. Calculate molar mass
6. Relate the mass of a substance to its number of moles
7. Carry out mole and mass calculations in chemical equations
8. Describe energy changes in reactions and classify a reaction as endothermic or exothermic
9. Understand the factors that affect the rate of a reaction
10. Describe the basic features of chemical equilibrium, and use Le Châtelier's principle to predict what happens when equilibrium is disturbed
11. Understand how temperature is regulated in the body

Why Study ...

Chemical Reactions?

Having learned about atoms, ionic compounds, and covalent molecules in Chapters 2 and 3, we now turn our attention to chemical reactions. Reactions are at the heart of chemistry. An understanding of chemical processes has made possible the conversion of natural substances into new compounds with different and sometimes superior properties. Aspirin, ibuprofen, and nylon are all products of chemical reactions utilizing substances derived from petroleum. Chemical reactions are not limited to industrial processes. The metabolism of food involves a series of reactions that both forms new compounds and also provides energy for the body's maintenance and growth. Burning gasoline, baking a cake, and photosynthesis involve chemical reactions. In Chapter 5 we learn the basic principles about chemical reactions.

5.1   Introduction to Chemical Reactions

Now that we have learned about compounds and the atoms that compose them, we can better understand the chemical changes first discussed in Section 1.2.

5.1A   General Features

- A chemical change—chemical reaction—converts one substance into another.

Chemical reactions involve breaking bonds in the starting materials, called *reactants*, and forming new bonds in the *products*. The combustion of methane (CH_4), the main constituent of natural gas, in the presence of oxygen (O_2) to form carbon dioxide (CO_2) and water (H_2O) is an example of a chemical reaction. The carbon–hydrogen bonds in methane and the oxygen–oxygen bond in elemental oxygen are broken, and new carbon–oxygen and hydrogen–oxygen bonds are formed in the products.

A chemical reaction is thus fundamentally different from a physical change such as melting or boiling discussed in Section 4.6. When water melts, for example, the highly organized water molecules in the solid phase become more disorganized in the liquid phase, but the bonds do *not* change.

Sample Problem 5.1	Identifying a Chemical Reaction and a Physical Change

Identify each process as a chemical reaction or a physical change.

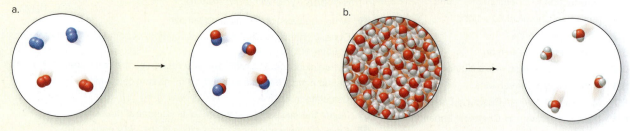

Analysis

- A chemical reaction occurs when the bonds in the reactants are broken and new bonds are formed in the products.
- A physical change occurs when the bonds in the reactants are the same as the bonds in the products.

Solution

Part (a) represents a chemical reaction—the reactants contain two N_2 molecules (with blue spheres joined) and two O_2 molecules (two red spheres joined), whereas the product contains four NO molecules (a red sphere joined to a blue sphere). Part (b) represents a physical change—boiling—because liquid H_2O molecules are converted to gaseous H_2O molecules and the bonds do not change.

| Practice Problem 5.1 | Use the molecular art to identify the process as a chemical reaction or a physical change, and explain your choice. |

| Practice Problem 5.2 | Use the molecular art to identify the process as a chemical reaction or a physical change, and explain your choice. |

More Practice: Try Problems 5.1, 5.2.

A chemical reaction may be accompanied by a visible change: two colorless reactants can form a colored product; a gas may be given off; two liquid reactants may yield a solid product. Sometimes heat is produced so that a reaction flask feels hot. A reaction having a characteristic visible change occurs when hydrogen peroxide (H_2O_2) is used to clean a bloody wound. An enzyme in the blood called catalase converts the H_2O_2 to water (H_2O) and oxygen (O_2), and bubbles of oxygen appear as a foam.

5.1B Writing Chemical Equations

- A *chemical equation* is an expression that uses chemical formulas and other symbols to illustrate what reactants constitute the starting materials in a reaction and what products are formed.

Chemical equations are written with the **reactants on the left** and the **products on the right,** separated by a horizontal arrow—a **reaction arrow**—that points from the reactants to the products. In the combustion of methane, methane (CH_4) and oxygen (O_2) are the reactants on the left side of the arrow, and carbon dioxide (CO_2) and water (H_2O) are the products on the right side.

The numbers written in front of any formula are called **coefficients. Coefficients show the number of molecules of a given element or compound that react or are formed.** When no number precedes a formula, the coefficient is assumed to be "1." In the combustion of methane, the coefficients tell us that one molecule of CH_4 reacts with two molecules of O_2 to form one molecule of CO_2 and two molecules of H_2O.

When a formula contains a subscript, **multiply its coefficient by the subscript** to give the total number of atoms of a given type in that formula.

2 O_2 molecules = **4** O atoms   2 H_2O molecules = **2** O atoms + **4** H atoms

Coefficients are used because all chemical reactions follow a fundamental principle of nature, the **law of conservation of matter,** which states:

> • **Atoms cannot be created or destroyed in a chemical reaction.**

Although bonds are broken and formed in reactions, the number of atoms of each element in the reactants must be the same as the number of atoms of each type in the products. **Coefficients are used to *balance* an equation,** making the number of atoms of each element the same on both sides of the equation.

$$CH_4 \quad + \quad 2\,O_2 \quad \longrightarrow \quad CO_2 \quad + \quad 2\,H_2O$$

Atoms in the reactants:
- 1 C atom
- 4 H atoms
- 4 O atoms

Atoms in the products:
- 1 C atom
- 4 H atoms
- 4 O atoms

Two other features are worthy of note. If heat is needed for a reaction to occur, the Greek letter delta (Δ) may be written over the arrow. The physical states of the reactants and products are sometimes indicated next to each formula—solid (s), liquid (l), or gas (g). If an aqueous solution is used—that is, if a reactant is dissolved in water—the symbol (aq) is used next to the reactant. When these features are added, the equation for the combustion of methane becomes:

Combustion of methane

$$CH_4(g) \quad + \quad 2\,O_2(g) \quad \xrightarrow{\Delta} \quad CO_2(g) \quad + \quad 2\,H_2O(g)$$

Table 5.1 **Symbols Used in Chemical Equations**

Symbol	Meaning
$\longrightarrow$	Reaction arrow
Δ	Heat
(s)	Solid
(l)	Liquid
(g)	Gas
(aq)	Aqueous solution

Concept Check 5.1

Write a chemical equation from the following description of a reaction: One molecule of gaseous methane (CH_4) is heated with four molecules of gaseous chlorine (Cl_2), forming one molecule of liquid carbon tetrachloride (CCl_4) and four molecules of gaseous hydrogen chloride (HCl).

The symbols used for chemical equations are summarized in Table 5.1.

Sample Problem 5.2

Determining the Number of Atoms Using a Coefficient and Subscript

Label the reactants and products, and indicate how many atoms of each type of element are present on each side of the equation.

$$C_2H_6O(l) + 3\,O_2(g) \longrightarrow 2\,CO_2(g) + 3\,H_2O(g)$$

Analysis

Reactants are on the left side of the arrow and products are on the right side in a chemical equation. When a formula contains a subscript, **multiply its coefficient by the subscript** to give the total number of atoms of a given type in the formula.

Solution

In this equation, the reactants are C_2H_6O and O_2, and the products are CO_2 and H_2O. If no coefficient is written, it is assumed to be "1." To determine the number of each type of atom when a formula has both a coefficient and a subscript, multiply the coefficient by the subscript.

1 C_2H_6O = **2** C's + **6** H's + 1 O

3 O_2 = 6 O's   Multiply the coefficient **3** by the subscript **2**.

2 CO_2 = 2 C's + 4 O's   Multiply the coefficient **2** by each subscript;
                                      2 × **1** C = **2** C's; **2** × **2** O's = **4** O's.

3 H_2O = 6 H's + 3 O's   Multiply the coefficient **3** by each subscript;
                                      3 × **2** H's = **6** H's; **3** × **1** O = **3** O's.

Add up the atoms on each side to determine the total number for each type of element.

$$C_2H_6O(l) \ + \ 3\,O_2(g) \longrightarrow 2\,CO_2(g) \ + \ 3\,H_2O(g)$$

Atoms in the reactants: Atoms in the products:
• 2 C's   6 H's   7 O's   • 2 C's   6 H's   7 O's

| **Practice Problem 5.3** | Label the reactants and products, and indicate how many atoms of each type of element are present on each side of the following equations. |

a. $2\,H_2O_2(aq) \longrightarrow 2\,H_2O(l) \ + \ O_2(g)$ b. $2\,C_8H_{18} \ + \ 25\,O_2 \longrightarrow 16\,CO_2 \ + \ 18\,H_2O$

| **Practice Problem 5.4** | Use the molecular art to write an equation for the given reaction. (The common element colors are shown in Appendix C.) |

More Practice: Try Problems 5.19, 5.20.

5.2 Balancing Chemical Equations

Sometimes a chemical equation is balanced as written and the coefficient of each formula is "1." For example, when charcoal is burned, the carbon (C) it contains reacts with oxygen (O_2) to form carbon dioxide (CO_2). One carbon atom reacts with one oxygen molecule to form one molecule of carbon dioxide.

$$C(s) \ + \ O_2(g) \longrightarrow CO_2(g)$$

More often, however, an equation must be balanced by adding coefficients in front of some formulas so that the **number of atoms of each element is equal on both sides of the equation.**

How To Balance a Chemical Equation

Example Write a balanced chemical equation for the reaction of propane (C_3H_8) with oxygen (O_2) to form carbon dioxide (CO_2) and water (H_2O).

Step [1] **Write the equation with the correct formulas.**

• Write the reactants on the left side and the products on the right side of the reaction arrow, and check if the equation is balanced without adding any coefficients.

$$C_3H_8 \quad + \quad O_2 \quad \longrightarrow \quad CO_2 \quad + \quad H_2O$$

• This equation is not balanced as written because none of the elements—carbon, hydrogen, and oxygen—has the same number of atoms on both sides of the equation. For example, there are 3 C's on the left and only 1 C on the right.

• **The subscripts in a formula can *never* be changed to balance an equation.** Changing a subscript changes the identity of the compound. For example, changing CO_2 to CO would balance oxygen (there would be 2 O's on both sides of the equation), but that would change CO_2 (carbon dioxide) into CO (carbon monoxide).

Step [2] **Balance the equation with coefficients one element at a time.**

• Begin with the most complex formula, and **start with an element that appears in only one formula on both sides of the equation.** Save the element found in multiple reactants or products for last. In this example, begin with either the C's or H's in C_3H_8. Because there are 3 C's on the left, place the coefficient **3** before CO_2 on the right.

$$C_3H_8 \quad + \quad O_2 \quad \longrightarrow \quad \mathbf{3}\,CO_2 \quad + \quad H_2O$$

3 C's on the left   Place a **3** to balance C's.

• To balance the 8 H's in C_3H_8, place the coefficient **4** before H_2O on the right.

$$C_3H_8 \quad + \quad O_2 \quad \longrightarrow \quad 3\,CO_2 \quad + \quad \mathbf{4}\,H_2O$$

8 H's on the left   Place a **4** to balance H's.
(4 × 2 H's in H_2O = **8** H's)

• The only element not balanced is oxygen, and at this point there are a total of 10 O's on the right—six from three CO_2 molecules and four from four H_2O molecules. To balance the 10 O's on the right, place the coefficient **5** before O_2 on the left.

$$C_3H_8 \quad + \quad \mathbf{5}\,O_2 \quad \longrightarrow \quad 3\,CO_2 \quad + \quad 4\,H_2O$$

Place a **5** to balance O's.   **10** O's on the right

Step [3] **Check to make sure that the smallest set of whole numbers is used.**

$$C_3H_8 \quad + \quad 5\,O_2 \quad \longrightarrow \quad 3\,CO_2 \quad + \quad 4\,H_2O$$

Atoms in the reactants:
• 3 C's   8 H's   10 O's (5 × 2 O's)

Atoms in the products:
• 3 C's   8 H's   10 O's [(3 × 2 O's) + (4 × 1 O)]

• This equation is balanced because the same number of C's, O's, and H's is present on both sides of the equation.

• Sometimes an equation is balanced but the lowest set of whole numbers is not used as coefficients. Say, for example, that balancing yielded the following equation:

$$2\,C_3H_8 + 10\,O_2 \longrightarrow 6\,CO_2 + 8\,H_2O$$

• This equation has the same number of C's, O's, and H's on both sides, but *each coefficient must be divided by two* to give the lowest set of whole numbers for the balanced equation, as drawn in the first equation in step [3].

Sample Problems 5.3–5.5 illustrate additional examples of balancing chemical equations. Sample Problem 5.4 shows how to balance an equation when there is an odd–even relationship of atoms in the reactants and products. Sample Problem 5.5 illustrates how to balance an equation that contains polyatomic ions.

Sample Problem 5.3	**Balancing a Chemical Equation**

Write a balanced equation for the reaction of glucose ($C_6H_{12}O_6$) with oxygen (O_2) to form carbon dioxide (CO_2) and water (H_2O).

Analysis

Balance an equation with coefficients, one element at a time, beginning with the most complex formula and starting with an element that appears in only one formula on both sides of the equation. Continue placing coefficients until the **number of atoms of each element is equal on both sides of the equation.**

Solution

[1] Write the equation with correct formulas.

$$C_6H_{12}O_6 + O_2 \longrightarrow CO_2 + H_2O$$
glucose

Bagels, pasta, bread, and rice are high in starch, which is hydrolyzed to the simple carbohydrate glucose after ingestion. The metabolism of glucose forms CO_2 and H_2O and provides energy for bodily functions (Sample Problem 5.3). *Scott Bauer/Agricultural Research Service/USDA*

- None of the elements is balanced in this equation. As an example, there are 6 C's on the left side, but only 1 C on the right side.

[2] Balance the equation with coefficients one element at a time.

- Begin with glucose, because its formula is most complex. Balance the 6 C's of glucose by placing the coefficient **6** before CO_2. Balance the 12 H's of glucose by placing the coefficient **6** before H_2O.

Place a **6** to balance C's.
$$C_6H_{12}O_6 + O_2 \longrightarrow 6\ CO_2 + 6\ H_2O$$
Place a **6** to balance H's.

- The right side of the equation now has 18 O's. Because glucose already has 6 O's on the left side, 12 additional O's are needed on the left side. The equation will be balanced if the coefficient **6** is placed before O_2.

$$C_6H_{12}O_6 + 6\ O_2 \longrightarrow 6\ CO_2 + 6\ H_2O$$

Place a **6** to balance O's.

[3] Check.

- The equation is balanced because the number of atoms of each element is the same on both sides.

Answer: $C_6H_{12}O_6 + 6\ O_2 \longrightarrow 6\ CO_2 + 6\ H_2O$

Atoms in the reactants:	Atoms in the products:
• 6 C's	• 6 C's (6 × 1 C)
• 12 H's	• 12 H's (6 × 2 H's)
• 18 O's (1 × 6 O's) + (6 × 2 O's)	• 18 O's (6 × 2 O's) + (6 × 1 O)

Practice Problem 5.5	Write a balanced equation for each reaction.

a. $H_2 + O_2 \longrightarrow H_2O$

b. $NO + O_2 \longrightarrow NO_2$

c. $Fe + O_2 \longrightarrow Fe_2O_3$

d. $CH_4 + Cl_2 \longrightarrow CH_2Cl_2 + HCl$

Practice Problem 5.6

Write a balanced equation for the following reaction, shown with molecular art.

More Practice: Try Problems 5.3–5.5; 5.7b; 5.8b; 5.11b; 5.23a, b; 5.24a, b; 5.25b; 5.26–5.28; 5.29b; 5.66b.

Sample Problem 5.4

Balancing a Chemical Equation with an Odd–Even Relationship of Atoms in the Reactants and Products

The airbag in an automobile inflates when ionic sodium azide (NaN_3), which is composed of Na^+ cations and the polyatomic anion, N_3^- (azide), rapidly decomposes to sodium (Na) and gaseous N_2 (Figure 5.1). Write a balanced equation for this reaction.

Analysis

Balance an equation with coefficients, one element at a time, beginning with the most complex formula and starting with an element that appears in only one formula on both sides of the equation. Continue placing coefficients until the **number of atoms of each element is equal on both sides of the equation.**

Solution

[1] Write the equation with correct formulas.

$$NaN_3 \longrightarrow Na + N_2$$
sodium azide

- The N atoms are not balanced because there are 3 N's on the left side and only 2 N's on the right.

[2] Balance the equation with coefficients.

- To balance the odd number of N atoms (3 N's) on the reactant side with the even number of N atoms on the product side (2 N's) requires the placement of *two* coefficients. Place the coefficient **2** on the left side (for a total of 6 N's in the reactants). Then place the coefficient **3** before N_2 (for a total of 6 N's in the product). **Placing two coefficients is necessary whenever there is an odd–even relationship of atoms in the reactants and products** (for any odd number other than one).

Place a **2** to give 6 N's on the left.

$$\textbf{2}\,NaN_3 \longrightarrow Na + \textbf{3}\,N_2$$

Place a **3** to give 6 N's on the right.

- The *subscript* of N_2 on the right side is numerically equal to the *coefficient* (**2**) before NaN_3 on the left side; similarly, the *subscript* of NaN_3 on the left side is numerically equal to the *coefficient* (**3**) before N_2 on the right side.
- Balance the 2 Na atoms on the left side by placing a **2** before the Na atoms on the right.

$$2\,NaN_3 \longrightarrow \textbf{2}\,Na + 3\,N_2$$

Place a **2** to balance Na's.

[3] Check.

- The equation is balanced because the number of atoms of each element is the same on both sides.

$$2\ NaN_3 \longrightarrow 2\ Na\ +\ 3\ N_2$$

Atoms in the reactants:	**Atoms in the products:**
• 2 Na atoms	• 2 Na atoms
• 6 N atoms	• 6 N atoms
(2 × 3 N's)	(3 × 2 N's)

Practice Problem 5.7

Write a balanced equation for the reaction of ethane (C_2H_6) with O_2 to form CO_2 and H_2O.

Practice Problem 5.8

The Haber process is an important industrial reaction that converts N_2 and H_2 to ammonia (NH_3), an agricultural fertilizer and starting material for the synthesis of nitrate fertilizers. Write a balanced equation for the Haber process.

More Practice: Try Problems 5.23c, 5.25a.

Figure 5.1   Chemistry of an Automobile Airbag

a. The chemical reaction that inflates an airbag

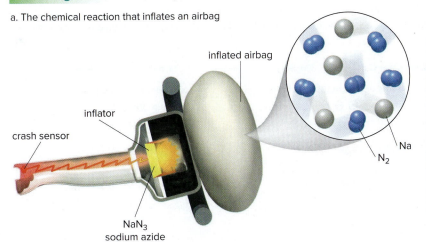

inflated airbag

inflator

crash sensor

Na

N_2

NaN_3
sodium azide

b. Airbags deployed in a head-on collision

Jack Sullivan/Alamy Stock Photo

A severe car crash triggers an airbag to deploy when an electric sensor causes sodium azide (NaN_3) to ignite, converting it to sodium (Na) and nitrogen gas (N_2). The nitrogen gas causes the bag to inflate fully in 40 milliseconds, helping to protect passengers from serious injury. The sodium atoms formed in this first reaction are hazardous and subsequently converted to a safe sodium salt. It took 30 years to develop a reliable airbag system for automobiles.

Sample Problem 5.5

Balancing an Equation with Polyatomic Ions

Balance the following equation.

$$Ca_3(PO_4)_2\ +\ H_2SO_4 \longrightarrow CaSO_4\ +\ H_3PO_4$$

calcium phosphate   sulfuric acid   calcium sulfate   phosphoric acid

Analysis

Balance an equation with coefficients, one element at a time, beginning with the most complex formula and starting with an element that appears in only one formula on both sides of the equation. Continue placing coefficients until the **number of atoms of each element is equal on both sides of the equation.**

Solution

[1] Write the equation with correct formulas.

- The correct formula for each compound is given in the problem statement. When the reactants and products contain polyatomic ions, PO_4^{3-} and SO_4^{2-} in this case, **balance each ion as a *unit*,** rather than balancing the individual atoms. Thus, phosphate is not balanced in the equation as written, because the left side has two PO_4^{3-} anions whereas the right side has only one.

[2] Balance the equation with coefficients.

- Begin with $Ca_3(PO_4)_2$. Balance the 3 Ca's by placing the coefficient **3** before $CaSO_4$. Balance the 2 PO_4^{3-} anions by placing the coefficient **2** before H_3PO_4.

Place a **3** to balance Ca's.

$$Ca_3(PO_4)_2 \; + \; H_2SO_4 \longrightarrow 3\,CaSO_4 \; + \; 2\,H_3PO_4$$

Place a **2** to balance PO_4^{3-}.

- Two components are still not balanced—H atoms and sulfate anions (SO_4^{2-}). Both can be balanced by placing the coefficient 3 before H_2SO_4 on the left.

6 H's   6 H's

$$Ca_3(PO_4)_2 \; + \; \mathbf{3}\,H_2SO_4 \longrightarrow 3\,CaSO_4 \; + \; 2\,H_3PO_4$$

3 SO_4^{2-} in both

Place a **3** to balance H and SO_4^{2-}.

[3] Check.

- The equation is balanced because the number of atoms and polyatomic anions is the same on both sides.

Answer: $Ca_3(PO_4)_2 \; + \; 3\,H_2SO_4 \longrightarrow 3\,CaSO_4 \; + \; 2\,H_3PO_4$

Atoms or ions in the reactants:	Atoms or ions in the products:
• 3 Ca's	• 3 Ca's
• 6 H's	• 6 H's
• 2 PO_4^{3-}	• 2 PO_4^{3-}
• 3 SO_4^{2-}	• 3 SO_4^{2-}

Practice Problem 5.9

Balance each chemical equation. Balance each polyatomic ion as a unit, rather than balancing the individual atoms.

a. $Al + H_2SO_4 \longrightarrow Al_2(SO_4)_3 + H_2$

b. $Na_2SO_3 + H_3PO_4 \longrightarrow H_2SO_3 + Na_3PO_4$

More Practice: Try Problems 5.6, 5.24c.

5.3 Types of Reactions

Although there is a great variety of known chemical reactions, many reactions fall into a few major categories. In this section we consider **combination, decomposition, single replacement, and double replacement reactions.** In Section 5.4, we will learn about reactions involving oxidation and reduction, and all of Chapter 8 is devoted to acid–base reactions.

5.3A Combination and Decomposition Reactions

- A *combination reaction* is the joining of two or more reactants to form a single product.

In a general combination reaction, reactant **A** combines with reactant **B** to form the product **AB.**

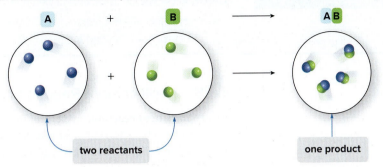

two reactants   one product

The reaction of hydrogen (H_2) with oxygen (O_2) to form water (H_2O) is an example of a combination reaction. **The reactants may be either elements or compounds, but the product is** *always* **a compound.**

$$2\,H_2 \qquad + \qquad O_2 \qquad \longrightarrow \qquad 2\,H_2O$$

reactant   reactant   one product

- A *decomposition reaction* is the conversion of a single reactant to two or more products.

In a general decomposition reaction, reactant **AB** forms product **A** and product **B.**

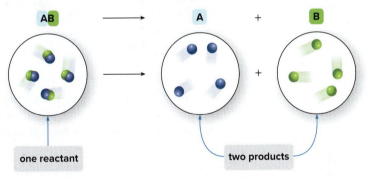

one reactant   two products

The conversion of hydrogen bromide (HBr) to hydrogen (H_2) and bromine (Br_2) is an example of a decomposition reaction. The conversion of NaN_3 to Na and N_2 in an automobile airbag (Sample Problem 5.4) is also a decomposition. **The products may be elements or compounds, but the reactant is** *always* **a compound.**

$$2\,HBr \qquad \longrightarrow \qquad H_2 \qquad + \qquad Br_2$$

one reactant   product   product

In comparing combination and decomposition:

- Combination reactions form only *one product,* whereas decomposition reactions begin with only *one reactant.*

Concept Check 5.2

Write a balanced equation for the reaction depicted in the molecular art, and classify the reaction as a combination or decomposition.

Classify each reaction as a combination or decomposition.

a. $Cu(s) + S(s) \longrightarrow CuS(s)$

b. $CuCO_3(s) \longrightarrow CuO(s) + CO_2(g)$

c. $H_2(g) + Cl_2(g) \longrightarrow 2\ HCl(g)$

d. $C_6H_{12}O_6(aq) \longrightarrow 2\ C_2H_6O(aq) + 2\ CO_2(g)$

5.3B   Replacement Reactions

In a **replacement reaction,** an element in a reactant is replaced by another element. Replacement reactions are characterized as **single replacement** or **double replacement,** depending on the particular reaction. **There are always** *two* **reactants and** *two* **products in replacement reactions.**

- A *single replacement* is a reaction in which one element replaces another element in a compound to form a different compound and element as products.

In a general single replacement reaction, reactants **A** and **BC** form products **B** and **AC.**

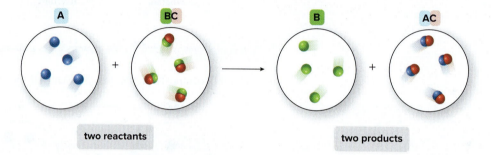

two reactants   two products

The reaction of zinc with $CuCl_2$ [copper(II) chloride] to form $ZnCl_2$ (zinc chloride) and copper is a single replacement reaction. The element Zn replaces Cu in the compound $CuCl_2$. In the most common single replacement reaction, **an element and compound are the reactants, and an element and compound are formed as products.**

- A *double replacement* is a reaction in which two compounds exchange "parts"—atoms or ions—to form two new compounds.

In a general double replacement reaction, reactants **AB** and **CD** form products **AD** and **CB** by the exchange of **A** for **C** in the reactants.

two reactants   two products

The reaction of $BaCl_2$ (barium chloride) with K_2CO_3 (potassium carbonate) to form $BaCO_3$ (barium carbonate) and KCl (potassium chloride) is a double replacement reaction. The cations

Ba^{2+} and K^+ exchange places to form two new ionic compounds. In double replacement, **two compounds are reactants and two compounds are products.**

Ba^{2+} and K^+ exchange places.

$$BaCl_2 \ + \ K_2CO_3 \longrightarrow BaCO_3 \ + \ 2\ KCl$$

compound   compound   compound   compound

Concept Check 5.4

Classify each reaction as a single replacement or double replacement reaction.

a. $Ni(NO_3)_2 + Mg \longrightarrow Ni + Mg(NO_3)_2$

b. $2\ KI + Sn(NO_3)_2 \longrightarrow SnI_2 + 2\ KNO_3$

c. $KCN + HCl \longrightarrow KCl + HCN$

d. $6\ Na + Al_2(SO_4)_3 \longrightarrow 3\ Na_2SO_4 + 2\ Al$

Table 5.2 summarizes the four reaction types.

Table 5.2 Summary of Reaction Types

	Type of Reaction	General Reaction			Example
①	Combination	A + B	⟶	AB	$Ca + Br_2 \longrightarrow CaBr_2$
②	Decomposition	AB ⟶	A	+ B	$2\ KClO_3 \longrightarrow 2\ KCl + 3\ O_2$
③	Single replacement	A + BC	⟶	B + AC	$3\ K + Al(NO_3)_3 \longrightarrow 3\ KNO_3 + Al$
④	Double replacement	AB + CD	⟶	AD + CB	$NaNO_3 + HCl \longrightarrow NaCl + HNO_3$

Sample Problem 5.6

Classifying a Reaction

Classify each reaction as a combination, decomposition, single replacement, or double replacement reaction.

a. $CuO + 2\ HCl \longrightarrow CuCl_2 + H_2O$ b. $2\ NO + O_2 \longrightarrow 2\ NO_2$

Analysis

- In a **combination** reaction, two or more reactants form one product.
- In a **decomposition** reaction, one reactant forms two or more products.
- In a **single replacement** reaction, **one element replaces another element** in a compound to form a new compound and new element.
- In a **double replacement** reaction, **two compounds exchange atoms or ions** to form two new compounds.

Solution

a. The reaction is a **double replacement** because the Cu^{2+} (in CuO) and H (in HCl) exchange to form two new compounds.

Cu^{2+} and H exchange.

$$CuO \ + \ 2\ HCl \longrightarrow CuCl_2 \ + \ H_2O$$

compound   compound   compound   compound

b. The reaction is a **combination** because two reactants (NO and O_2) combine to form one product (NO_2).

$$2\ NO \ + \ O_2 \longrightarrow 2\ NO_2$$

reactant   reactant   one product

Classify each reaction as a combination, decomposition, single replacement, or double replacement reaction.

a. $2\,HgO \longrightarrow 2\,Hg + O_2$

b. $Mg + ZnCl_2 \longrightarrow MgCl_2 + Zn$

c. $H_2C{=}CH_2 + HBr \longrightarrow CH_3CH_2Br$

d. $KNO_3 + HBr \longrightarrow KBr + HNO_3$

For the reaction depicted in the molecular art: (a) Use the common element colors shown in Appendix C to give the formula for each reactant and product. (b) Write a balanced equation. (c) Classify the reaction as combination, decomposition, single replacement, or double replacement.

More Practice: Try Problems 5.7c, 5.8c, 5.29c, 5.30–5.32.

5.4 Oxidation and Reduction

Another group of reactions—acid–base reactions—is discussed in Chapter 8.

Thus far we have examined features that are common to all types of chemical reactions. We now examine one class of reactions that involves electron transfer—oxidation–reduction reactions.

5.4A General Features of Oxidation–Reduction Reactions

A common type of chemical reaction involves the transfer of electrons from one element to another. When iron rusts, methane and wood burn, and a battery generates electricity, one element gains electrons and another loses them. These reactions involve **oxidation** and **reduction.**

- Oxidation is the *loss* of electrons from an atom.
- Reduction is the *gain* of electrons by an atom.

Oxidation and reduction are opposite processes, and both occur together in a single reaction called an **oxidation–reduction** or **redox reaction.** A redox reaction always has two components—one that is oxidized and one that is reduced.

- A redox reaction involves the transfer of electrons from one element to another.

An example of an oxidation–reduction reaction occurs when Zn metal reacts with Cu^{2+} cations, as shown in Figure 5.2.

Cu^{2+} gains two electrons.

$$Zn + Cu^{2+} \longrightarrow Zn^{2+} + Cu$$

Zn loses two electrons.

- **Zn** *loses* two electrons to form **Zn²⁺**, so **Zn** is *oxidized.*
- **Cu²⁺** *gains* two electrons to form **Cu** metal, so **Cu²⁺** is *reduced.*

Figure 5.2 A Redox Reaction—The Transfer of Electrons from Zn to Cu^{2+}

A redox reaction occurs when a strip of Zn metal is placed in a solution of Cu^{2+} ions. In this reaction, Zn loses two electrons to form Zn^{2+}, which goes into solution. Cu^{2+} gains two electrons to form Cu metal, which precipitates out of solution, forming a coating on the zinc strip.

Benzoyl peroxide ($C_{14}H_{10}O_4$) is the active ingredient in several acne medications. Benzoyl peroxide kills bacteria by oxidation reactions. *Jill Braaten*

Each of these processes can be written as individual reactions, called **half reactions,** to emphasize which electrons are gained and lost.

- A compound that gains electrons (is reduced) while causing another compound to be oxidized is called an *oxidizing agent.*
- A compound that loses electrons (is oxidized) while causing another compound to be reduced is called a *reducing agent.*

In this example, Zn loses electrons to Cu^{2+}. We can think of Zn as a **reducing agent** because it causes Cu^{2+} to gain electrons and become reduced. We can think of Cu^{2+} as an **oxidizing agent** because it causes Zn to lose electrons and become oxidized.

To draw the products of an oxidation–reduction reaction, we must decide which element or ion gains electrons and which element or ion loses electrons. Use the following guidelines.

- When considering neutral atoms, metals *lose* electrons and nonmetals *gain* electrons.
- When considering ions, cations tend to *gain* electrons and anions tend to *lose* electrons.

Thus, the metals sodium (Na) and magnesium (Mg) readily lose electrons to form the cations Na^+ and Mg^{2+}, respectively; that is, they are oxidized. The nonmetals O_2 and Cl_2 readily gain electrons to form $2 O^{2-}$ and $2 Cl^-$, respectively; that is, they are reduced. A positively charged ion like Cu^{2+} is reduced to Cu by gaining two electrons, whereas two negatively charged Cl^- anions are oxidized to Cl_2 by losing two electrons. These reactions and additional examples are shown in Figure 5.3.

Figure 5.3   Examples of Oxidation and Reduction Reactions

Oxidation Reactions	Reduction Reactions
$Na \longrightarrow Na^+ + e^-$	$Cl_2 + 2\,e^- \longrightarrow 2\,Cl^-$
$Mg \longrightarrow Mg^{2+} + 2\,e^-$	$O_2 + 4\,e^- \longrightarrow 2\,O^{2-}$
$2\,Cl^- \longrightarrow Cl_2 + 2\,e^-$	$Cu^{2+} + 2\,e^- \longrightarrow Cu$
$2\,O^{2-} \longrightarrow O_2 + 4\,e^-$	$Ag^+ + e^- \longrightarrow Ag$

Electrons are **lost.**   Electrons are **gained.**

Sample Problem 5.7

Identifying the Components of a Redox Reaction

Identify the species that is oxidized and the species that is reduced in the following reaction. Write out half reactions to show how many electrons are gained or lost by each species.

$$Mg(s) + 2\,H^+(aq) \longrightarrow Mg^{2+}(aq) + H_2(g)$$

Analysis

- **Metals and anions tend to lose electrons** and thus undergo oxidation.
- **Nonmetals and cations tend to gain electrons** and thus undergo reduction.

Solution

The metal Mg is oxidized to Mg^{2+}, thus losing two electrons. Two H^+ cations gain a total of two electrons, and so are reduced to the nonmetal H_2.

$$Mg(s) \longrightarrow Mg^{2+}(aq) \; + \; 2\,e^-$$

Mg is **oxidized.**

$$2\,H^+(aq) \; + \; 2\,e^- \longrightarrow H_2(g)$$

H^+ is **reduced.**   Two electrons are needed to balance charge.

We need enough electrons so that **the total charge is the same on both sides of the equation.** Because 2 H^+ cations have a +2 overall charge, this means that 2 e^- must be gained so that the total charge on both sides of the equation is zero.

Practice Problem 5.12

Identify the species that is oxidized and the species that is reduced in each reaction. Write out half reactions to show how many electrons are gained or lost by each species.

a. $Zn(s) + 2\,H^+(aq) \longrightarrow Zn^{2+}(aq) + H_2(g)$

b. $Fe^{3+}(aq) + Al(s) \longrightarrow Al^{3+}(aq) + Fe(s)$

c. $2\,I^- + Br_2 \longrightarrow I_2 + 2\,Br^-$

More Practice: Try Problems 5.13, 5.14, 5.33–5.36.

5.4B   Examples of Oxidation–Reduction Reactions

Many common processes involve oxidation and reduction. For example, common antiseptics like iodine (I_2) and hydrogen peroxide (H_2O_2) are oxidizing agents that clean wounds by oxidizing, thereby killing bacteria that might cause infection.

When iron (Fe) rusts, it is oxidized by the oxygen in air to form iron(III) oxide, Fe_2O_3. In this redox reaction, neutral iron atoms are oxidized to Fe^{3+} cations, and elemental O_2 is reduced to O^{2-} anions.

$$4\ Fe(s)\ +\ 3\ O_2(g)\ \longrightarrow\ 2\ Fe_2O_3(s)$$

neutral elements — Each Fe ion has a +3 charge.

Each O ion has a −2 charge.

Fe atom

Fe^{3+}

O^{2-}

Tony Cordoza/Alamy Stock Photo   Clair Dunn/Alamy Stock Photo

In some reactions it is much less apparent which reactant is oxidized and which is reduced. For example, in the combustion of methane (CH_4) with oxygen to form CO_2 and H_2O, there are no metals or cations that obviously lose or gain electrons, yet this is a redox reaction. In these instances, it is often best to count oxygen and hydrogen atoms.

- Oxidation results in the *gain* of oxygen atoms or the *loss* of hydrogen atoms.
- Reduction results in the *loss* of oxygen atoms or the *gain* of hydrogen atoms.

O_2 is **reduced.**

$$CH_4\ +\ 2\ O_2\ \longrightarrow\ CO_2\ +\ 2\ H_2O$$

4 H atoms   0 H atoms   2 O atoms   2 H atoms for each O atom

CH_4 is **oxidized.**

CH_4 is oxidized because it gains two oxygen atoms to form CO_2. O_2 is reduced because it gains two hydrogen atoms to form H_2O.

Concept Check 5.5	The following redox reaction occurs in mercury batteries for watches. Identify the species that is oxidized and the species that is reduced, and write out two half reactions to show how many electrons are gained or lost. In this reaction, oxide (O^{2-}) is unchanged; that is, it is neither oxidized nor reduced.

$$Zn\ +\ HgO\ \longrightarrow\ ZnO\ +\ Hg$$

Concept Check 5.6	Identify the species that is oxidized and the species that is reduced in the following redox reaction. Explain your choices.

$$C_2H_4O_2\ +\ 2\ H_2\ \longrightarrow\ C_2H_6O\ +\ H_2O$$

5.4C FOCUS ON HEALTH & MEDICINE
Pacemakers

A pacemaker is a small electrical device implanted in an individual's chest and used to maintain an adequate heart rate (Figure 5.4). A pacemaker contains a small, long-lasting battery that generates an electrical impulse by a redox reaction.

Most pacemakers used today contain a lithium–iodine battery. Each neutral lithium atom is oxidized to Li^+ by losing one electron. Each I_2 molecule is reduced by gaining two electrons and

Figure 5.4   The Lithium–Iodine Battery in a Pacemaker

Redox Reaction

Oxidation: $2\,Li \longrightarrow 2\,Li^+ + 2\,e^-$

Reduction: $I_2 + 2\,e^- \longrightarrow 2\,I^-$

A pacemaker generates a small electrical impulse that triggers the heart to beat. Today's pacemakers sense when the heart beats normally and provide an electrical signal only when the heart rate slows. Such devices are called "demand" pacemakers, and they quickly replaced earlier "fixed" rate models that continuously produced impulses to set the heart rate at a fixed value.

forming $2\,I^-$. Because the balanced equation contains two Li atoms for each I_2 molecule, the number of electrons lost by Li atoms equals the number of electrons gained by I_2.

• I_2 gains $2\,e^-$, forming $2\,I^-$.
• I_2 is **reduced.**

$$2\,Li \quad + \quad I_2 \quad \longrightarrow \quad 2\,LiI$$

• Each Li atom loses $1\,e^-$, forming Li^+.
• Li metal is **oxidized.**

Concept Check 5.7

Early pacemakers generated an electrical impulse by the following reaction. What species is the oxidizing agent and what species is the reducing agent in this reaction?

$$Zn + Hg^{2+} \longrightarrow Zn^{2+} + Hg$$

5.5   The Mole and Avogadro's Number

Although the chemical equations in Section 5.2 were discussed in terms of individual atoms and molecules, atoms are exceedingly small. It is more convenient to talk about larger quantities of atoms, and for this reason, scientists use the **mole.** A mole defines a quantity, much like a dozen items means 12, and a case of soda means 24 cans. The only difference is that a mole is much larger.

• A *mole* is a quantity that contains 6.022×10^{23} items—usually atoms, molecules, or ions.

The definition of a mole is based on the number of atoms contained in exactly 12 g of the carbon-12 isotope. This number is called **Avogadro's number,** after the Italian scientist Amadeo Avogadro,

Each sample contains one mole of the substance—water (H_2O molecules), salt (NaCl, one mole of Na^+ and one mole of Cl^-), and aspirin ($C_9H_8O_4$ molecules). Pictured is a mole of aspirin *molecules*, not a mole of aspirin *tablets*, which is a quantity too large to easily represent.
Anthony Arena/Jill Braaten/McGraw Hill

who first proposed the concept of a mole in the nineteenth century. One mole, abbreviated as **mol,** always contains an Avogadro's number of particles.

$$\textbf{1 mole} \text{ of C atoms} = \textbf{6.022} \times \textbf{10}^{23} \text{ C atoms}$$
$$\textbf{1 mole} \text{ of } H_2O \text{ molecules} = \textbf{6.022} \times \textbf{10}^{23} \ H_2O \text{ molecules}$$
$$\textbf{1 mole} \text{ of vitamin C molecules} = \textbf{6.022} \times \textbf{10}^{23} \text{ vitamin C molecules}$$

Concept Check 5.8

How many items are contained in one mole of (a) baseballs; (b) bicycles; (c) Cheerios; (d) CH_4 molecules?

We can use Avogadro's number as a conversion factor to relate the number of moles of a substance to the number of atoms or molecules it contains.

$$\text{Two possible conversion factors:} \quad \frac{1 \text{ mol}}{6.022 \times 10^{23} \text{ atoms}} \quad \text{or} \quad \frac{6.022 \times 10^{23} \text{ atoms}}{1 \text{ mol}}$$

These conversion factors allow us to determine how many atoms or molecules are contained in a given number of moles. To carry out calculations that contain numbers written in scientific notation, we must first learn how to multiply and divide numbers written in this form.

- To multiply two numbers in scientific notation, *multiply* the coefficients together and *add* the exponents in the powers of 10.

Add exponents.
$(5 + 2)$

$(3.0 \times 10^5) \quad \times \quad (2.0 \times 10^2) = 6.0 \times 10^7$

Multiply coefficients.
(3.0×2.0)

- To divide two numbers in scientific notation, *divide* the coefficients and *subtract* the exponents in the powers of 10.

For a number written in scientific notation as $y \times 10^x$, y is the coefficient and x is the exponent in the power of 10 (Section 1.6).

Divide coefficients. $\dfrac{6.0 \times 10^2}{2.0 \times 10^{20}}$ Subtract exponents. $= 3.0 \times 10^{-18}$
$(6.0 \div 2.0)$ $(2 - 20)$

Sample Problem 5.8 illustrates how to interconvert moles and molecules.

Sample Problem 5.8   Converting Moles to the Number of Molecules

How many molecules are contained in 5.0 moles of carbon dioxide (CO_2)?

Analysis and Solution

[1] Identify the original quantity and the desired quantity.

5.0 mol of CO_2	? number of molecules of CO_2
original quantity	desired quantity

[2] Write out the conversion factors.

- Choose the conversion factor that places the unwanted unit, mol, in the denominator, so that the units cancel.

$$\frac{1 \text{ mol}}{6.022 \times 10^{23} \text{ molecules}} \quad \text{or} \quad \frac{6.022 \times 10^{23} \text{ molecules}}{1 \text{ mol}}$$

Choose this conversion factor to cancel mol.

[3] Set up and solve the problem.

- Multiply the original quantity by the conversion factor to obtain the desired quantity.

Convert to a number between 1 and 10.

$$5.0 \text{ mol} \times \frac{6.022 \times 10^{23} \text{ molecules}}{1 \text{ mol}} = 30. \times 10^{23} \text{ molecules}$$

Moles cancel.

$$= 3.0 \times 10^{24} \text{ molecules of } CO_2$$

Answer

- Multiplication first gives an answer that is not written in scientific notation because the coefficient (30.) is greater than 10. Moving the decimal point one place to the *left* and *increasing* the exponent by one gives the answer written in the proper form.

Practice Problem 5.13

How many carbon atoms are contained in each of the following number of moles: (a) 2.00 mol; (b) 6.00 mol; (c) 0.500 mol; (d) 25.0 mol?

Practice Problem 5.14

How many molecules are contained in each of the following number of moles?

a. 2.5 mol of penicillin molecules

b. 0.25 mol of NH_3 molecules

c. 0.40 mol of sugar molecules

d. 55.3 mol of acetaminophen molecules

Practice Problem 5.15

How many moles of water contain each of the following number of molecules?

a. 6.02×10^{25} molecules   b. 3.01×10^{22} molecules   c. 9.0×10^{24} molecules

---

5.6   Mass to Mole Conversions

In Section 2.3, we learned that the *atomic weight* **is the average mass of an element,** reported in atomic mass units (amu). Thus, carbon has an atomic weight of 12.01 amu. We use atomic weights to calculate the mass of a compound.

- The *formula weight* is the sum of the atomic weights of all the atoms in a compound, reported in atomic mass units (amu).

The term "formula weight" is used for both ionic and covalent compounds. Often the term **"molecular weight"** is used in place of formula weight for covalent compounds, because they are composed of molecules, not ions. The formula weight of ionic sodium chloride (NaCl) is 58.44 amu, which is determined by adding up the atomic weights of Na (22.99 amu) and Cl (35.45 amu).

Formula weight of NaCl:

Atomic weight of 1 Na	=	22.99 amu
Atomic weight of 1 Cl	=	35.45 amu
Formula weight of NaCl	=	58.44 amu

Concept Check 5.9

Calculate the formula weight of each ionic compound.

a. $CaCO_3$, a common calcium supplement

b. KI, the essential nutrient added to NaCl to make iodized salt

---

5.6A   Molar Mass

When reactions are carried out in the laboratory, single atoms and molecules are much too small to measure out. Instead, substances are weighed on a balance and amounts are typically reported

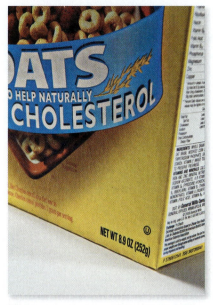

When a consumer product contains a great many lightweight small objects—for example, Cheerios—it is typically sold by weight, not by the number of objects. We buy Cheerios in an 8.9-oz box, not a box that contains 2,554 Cheerios.
Jill Braaten/McGraw Hill

in grams, not atomic mass units. To determine how many atoms or molecules are contained in a given mass, we use its **molar mass.**

- The *molar mass* is the mass of one mole of any substance, reported in grams per mole.

The value of the molar mass of an element in the periodic table (in grams per mole) is the same as the value of its atomic weight (in amu). Thus, the molar mass of carbon is 12.01 g/mol, because its atomic weight is 12.01 amu; that is, one mole of carbon atoms weighs 12.01 g.

- Carbon's **atomic weight** is 12.01 amu.
- Carbon's **molar mass** is 12.01 g/mol.
- **One mole of carbon atoms weighs 12.01 g.**

- The value of the molar mass of a compound in grams equals the value of its formula weight in amu.

Because the formula weight of NaCl is 58.44 amu, its molar mass is 58.44 g/mol. One mole of NaCl weighs 58.44 g. We use a compound's formula weight to calculate its molar mass, as shown in Sample Problem 5.9.

Sample Problem 5.9 **Calculating Molar Mass**

What is the molar mass of nicotine ($C_{10}H_{14}N_2$), the toxic and addictive stimulant in tobacco?

Analysis

Determine the number of atoms of each element from the subscripts in the chemical formula, multiply the number of atoms of each element by the atomic weight, and add up the results.

Solution

$$
\begin{aligned}
10 \text{ C atoms} \times 12.01 \text{ amu} &= 120.1 \text{ amu} \\
14 \text{ H atoms} \times 1.008 \text{ amu} &= 14.11 \text{ amu} \\
2 \text{ N atoms} \times 14.01 \text{ amu} &= \underline{28.02 \text{ amu}} \\
\text{Formula weight of nicotine} &= 162.23 \text{ amu rounded to } 162.2 \text{ amu}
\end{aligned}
$$

Answer: The formula weight of nicotine is 162.2 amu, so the molar mass of nicotine is 162.2 g/mol.

Practice Problem 5.16

What is the molar mass of each compound?
- a. Li_2CO_3 (lithium carbonate), a drug used to treat bipolar disorder
- b. C_2H_5Cl (ethyl chloride), a local anesthetic
- c. $C_{13}H_{21}NO_3$ (albuterol), a drug used to treat asthma

Practice Problem 5.17

The analgesic and narcotic effects of opium are largely due to morphine. What is the molar mass of morphine ($C_{17}H_{19}NO_3$)? Morphine is commonly used to treat severe pain, although it is highly addictive and tolerance develops with long-term use.

Morphine (Practice Problem 5.17) is obtained from the opium poppy.
Mafoto/Getty Images

More Practice: Try Problems 5.11a, 5.37, 5.38, 5.43b, 5.44b, 5.65a, 5.66a.

5.6B Relating Grams to Moles

The molar mass is a very useful quantity because it relates the number of *moles* to the number of *grams* of a substance. In this way, the molar mass can be used as a conversion factor. For example, because the molar mass of H_2O is 18.02 g/mol, two conversion factors can be written.

$$\frac{18.02 \text{ g } H_2O}{1 \text{ mol}} \quad \text{or} \quad \frac{1 \text{ mol}}{18.02 \text{ g } H_2O}$$

Using these conversion factors, we can convert a given number of moles of water to grams, or a specific number of grams of water to moles.

Sample Problem 5.10	Converting Moles to Mass

What is the mass of 0.25 moles of water?

Analysis and Solution

[1] Identify the original quantity and the desired quantity.

$$\begin{array}{cc} 0.25 \text{ mol of } H_2O & ? \text{ g of } H_2O \\ \text{original quantity} & \text{desired quantity} \end{array}$$

[2] Write out the conversion factors.

- Choose the conversion factor that places the unwanted unit, moles, in the denominator, so that the units cancel.

$$\frac{1 \text{ mol}}{18.02 \text{ g } H_2O} \quad \text{or} \quad \boxed{\frac{18.02 \text{ g } H_2O}{1 \text{ mol}}} \quad \text{Choose this conversion factor to cancel mol.}$$

[3] Set up and solve the problem.

- Multiply the original quantity by the conversion factor to obtain the desired quantity.

molar mass conversion factor

$$0.25 \text{ mol} \times \frac{18.02 \text{ g } H_2O}{1 \text{ mol}} = 4.5 \text{ g of } H_2O$$

Moles cancel. **Answer**

Practice Problem 5.18	Calculate the number of grams contained in each of the following number of moles.

a. 0.500 mol of NaCl   c. 3.60 mol of C_2H_4 (ethylene)

b. 2.00 mol of KI   d. 0.820 mol of CH_4O (methanol)

More Practice: Try Problems 5.39, 5.40.

Sample Problem 5.11	Converting Mass to Moles

How many moles are present in 100. g of aspirin ($C_9H_8O_4$, molar mass 180.15 g/mol)?

Analysis and Solution

[1] Identify the original quantity and the desired quantity.

$$\begin{array}{cc} 100. \text{ g of aspirin} & ? \text{ mol of aspirin} \\ \text{original quantity} & \text{desired quantity} \end{array}$$

[2] **Write out the conversion factors.**

- Choose the conversion factor that places the unwanted unit, grams, in the denominator, so that the units cancel.

$$\frac{180.15 \text{ g aspirin}}{1 \text{ mol}} \quad \text{or} \quad \frac{1 \text{ mol}}{180.15 \text{ g aspirin}}$$

Choose this conversion factor to cancel g.

[3] **Set up and solve the problem.**

- Multiply the original quantity by the conversion factor to obtain the desired quantity.

molar mass conversion factor

$$100. \, \cancel{g} \quad \times \quad \frac{1 \text{ mol}}{180.15 \, \cancel{g} \text{ aspirin}} \quad = \quad 0.555 \text{ mol of aspirin}$$

Grams cancel.   **Answer**

Vitamin C (Practice Problem 5.20) is needed in the formation of the protein collagen, which forms the connective tissue in blood vessels and muscle.
Denise Bush/Getty Images

Practice Problem 5.19

How many moles are contained in each of the following?

a. 100. g of NaCl

b. 25.5 g of CH_4

c. 0.250 g of aspirin ($C_9H_8O_4$)

d. 25.0 g of H_2O

Practice Problem 5.20

How many moles are contained in a 1,000.-mg dose of a vitamin C supplement (molar mass of vitamin C 176.12 g/mol)?

More Practice: Try Problems 5.41, 5.42, 5.43c, 5.44c, 5.67, 5.68a.

NO, nitrogen monoxide, is formed from N_2 and O_2 at very high temperature in automobile engines and coal-burning furnaces. NO is a reactive air pollutant that goes on to form other air pollutants, such as ozone (O_3) and nitric acid (HNO_3). HNO_3 is one component of acid rain that can devastate forests and acidify streams, making them unfit for fish and other wildlife.
Hisham F. Ibrahim/Photodisc/Getty Images

5.7   Mole Calculations in Chemical Equations

Having learned about moles and molar mass, we can now return to balanced chemical equations. As we learned in Section 5.2, the coefficients in a balanced chemical equation tell us the number of *molecules* of each compound that react or are formed in a given reaction.

- A balanced chemical equation also tells us the number of *moles* of each reactant that combine and the number of *moles* of each product formed.

$$\mathbf{1} \, N_2(g) \quad + \quad \mathbf{1} \, O_2(g) \quad \xrightarrow{\Delta} \quad \mathbf{2} \, NO(g)$$

one molecule of N_2	**one** molecule of O_2	**two** molecules of NO
one mole of N_2	**one** mole of O_2	**two** moles of NO

[The coefficient "1" has been written for emphasis.]

For example, the balanced chemical equation for the high temperature reaction of N_2 and O_2 to form nitrogen monoxide, NO, shows that one *molecule* of N_2 combines with one *molecule* of O_2 to form two *molecules* of NO. It also shows that one *mole* of N_2 combines with one *mole* of O_2 to form two *moles* of NO.

Coefficients are used to form mole ratios, which can serve as conversion factors. These ratios tell us the relative number of moles of reactants that combine in a reaction, as well as the relative number of moles of product formed from a given reactant, as shown in Sample Problem 5.12.

Mole ratios:

$$\frac{1 \text{ mol N}_2}{1 \text{ mol O}_2} \qquad \frac{1 \text{ mol N}_2}{2 \text{ mol NO}} \qquad \frac{1 \text{ mol O}_2}{2 \text{ mol NO}}$$

two reactants   reactant–product   reactant–product
N_2 and O_2 N_2 and NO O_2 and NO

- Use the mole ratio from the coefficients in the balanced equation to convert the number of moles of one compound (**A**) into the number of moles of another compound (**B**).

Moles of **A** $\longrightarrow$ Moles of **B**

**mole–mole
conversion factor**

Sample Problem 5.12 — Determining Moles of Product Using a Balanced Equation

Carbon monoxide (CO) is a poisonous gas that combines with hemoglobin in the blood, thus reducing the amount of oxygen that can be delivered to tissues. Under certain conditions, CO is formed when ethane (C_2H_6) in natural gas is burned in the presence of oxygen. Using the balanced equation, how many moles of CO are produced from 3.5 mol of C_2H_6?

$$2\,C_2H_6(g) \;+\; 5\,O_2(g) \;\xrightarrow{\;\Delta\;}\; 4\,CO(g) \;+\; 6\,H_2O(g)$$

Analysis and Solution

[1] **Identify the original quantity and the desired quantity.**

3.5 mol of C_2H_6 ? mol of CO
original quantity   desired quantity

[2] **Write out the conversion factors.**
- Use the coefficients in the balanced equation to write **mole–mole conversion factors** for the two compounds, C_2H_6 and CO. Choose the conversion factor that places the unwanted unit, moles of C_2H_6, in the denominator, so that the units cancel.

$$\frac{2 \text{ mol } C_2H_6}{4 \text{ mol CO}} \quad \text{or} \quad \boxed{\frac{4 \text{ mol CO}}{2 \text{ mol } C_2H_6}}$$

Choose this conversion factor to cancel mol C_2H_6.

[3] **Set up and solve the problem.**
- Multiply the original quantity by the conversion factor to obtain the desired quantity.

$$3.5 \text{ mol } C_2H_6 \;\times\; \frac{4 \text{ mol CO}}{2 \text{ mol } C_2H_6} \;=\; 7.0 \text{ mol CO}$$

Moles C_2H_6 cancel.   **Answer**

Practice Problem 5.21

Use the balanced equation for the reaction of N_2 and O_2 to form NO at the beginning of Section 5.7 to answer each question.

a. How many moles of NO are formed from 3.3 moles of N_2?

b. How many moles of NO are formed from 0.50 moles of O_2?

c. How many moles of O_2 are needed to completely react with 1.2 moles of N_2?

Practice Problem 5.22

Use the balanced equation in Sample Problem 5.12 to answer each question.

a. How many moles of O_2 are needed to react completely with 3.0 moles of C_2H_6?

b. How many moles of H_2O are formed from 0.50 moles of C_2H_6?

c. How many moles of C_2H_6 are needed to form 3.0 moles of CO?

More Practice: Try Problems 5.45, 5.46, 5.65b, 5.66c.

ENVIRONMENTAL NOTE

Lightning produces O_3 from O_2 during an electrical storm (*How To* p. 169). O_3 at the ground level is an unwanted pollutant. In the stratosphere, however, it protects us from harmful radiation from the sun.
Comstock Images/PunchStock

5.8   Mass Calculations in Chemical Equations

Because a mole represents an enormously large number of very small molecules, there is no way to directly count the number of moles or molecules used in a chemical reaction. Instead, we utilize a balance to measure the number of grams of a compound used and the number of grams of product formed. The number of grams of a substance and the number of moles it contains are related by the molar mass (Section 5.6).

5.8A   Converting Moles of Reactant to Grams of Product

To determine how many grams of product are expected from a given number of moles of reactant, two operations are necessary. First, we must determine how many moles of product to expect using the coefficients of the balanced chemical equation (Section 5.7). Then, we convert the number of moles of product to the number of grams using the molar mass (Section 5.6). Each step needs a conversion factor. The stepwise procedure is outlined in the accompanying *How To*, and then illustrated with an example in Sample Problem 5.13.

| Moles of reactant | —[1]→ | Moles of product | —[2]→ | Grams of product |
| | mole–mole conversion factor | | molar mass conversion factor | |

How To **Convert Moles of Reactant to Grams of Product**

Example   In the upper atmosphere, high-energy radiation from the sun converts oxygen (O_2) to ozone (O_3). Using the balanced equation, how many grams of O_3 are formed from 9.0 mol of O_2?

$$3\ O_2(g) \xrightarrow{\text{sunlight}} 2\ O_3(g)$$

Step [1] Convert the number of moles of reactant to the number of moles of product using a mole–mole conversion factor.

• Use the coefficients in the balanced chemical equation to write **mole–mole conversion factors.**

$$\frac{3\ \text{mol}\ O_2}{2\ \text{mol}\ O_3} \quad \text{or} \quad \frac{2\ \text{mol}\ O_3}{3\ \text{mol}\ O_2} \qquad \text{Choose this conversion factor to cancel mol } O_2.$$

• Multiply the number of moles of starting material (9.0 mol) by the conversion factor to give the number of moles of product. In this example, 6.0 mol of O_3 are formed.

| Moles of reactant | mole–mole conversion factor | Moles of product |

$$9.0\ \text{mol}\ O_2 \quad \times \quad \frac{2\ \text{mol}\ O_3}{3\ \text{mol}\ O_2} \quad = \quad 6.0\ \text{mol}\ O_3$$

Moles O_2 cancel.

—*Continued*

How To, continued . . .

Step [2] **Convert the number of moles of product to the number of grams of product using the product's molar mass.**

- Use the **molar mass** of the product (O_3) to write a conversion factor. The molar mass of O_3 is 48.00 g/mol (3 O atoms × 16.00 g/mol for each O atom = 48.00 g/mol).

$$\frac{1\ mol\ O_3}{48.00\ g\ O_3} \quad or \quad \boxed{\frac{48.00\ g\ O_3}{1\ mol\ O_3}} \quad \text{Choose this conversion factor to cancel mol.}$$

- Multiply the number of moles of product (from step [1]) by the conversion factor to give the number of grams of product.

$$6.0\ mol\ O_3 \quad \times \quad \frac{48.00\ g\ O_3}{1\ mol\ O_3} \quad = \quad 288\ g,\ rounded\ to\ 290\ g\ of\ O_3$$

Moles cancel.

Answer

It is also possible to combine the multiplication operations from steps [1] and [2] into a single operation using both conversion factors. This converts the moles of starting material to grams of product all at once. Both the one-step and stepwise approaches give the same overall result.

Moles of reactant		mole–mole conversion factor		molar mass conversion factor		Grams of product

$$9.0\ mol\ O_2 \quad \times \quad \frac{2\ mol\ O_3}{3\ mol\ O_2} \quad \times \quad \frac{48.00\ g\ O_3}{1\ mol\ O_3} \quad = \quad 290\ g\ of\ O_3$$

Moles O_2 cancel.   Moles O_3 cancel.

Answer

Sample Problem 5.13 — **Converting Moles of Reactant to Grams of Product**

Wine is produced by the fermentation of grapes. In fermentation, the carbohydrate glucose ($C_6H_{12}O_6$) is converted to ethanol and carbon dioxide according to the given balanced equation. How many grams of ethanol (C_2H_6O, molar mass 46.07 g/mol) are produced from 5.00 mol of glucose?

$$C_6H_{12}O_6(aq) \longrightarrow 2\ C_2H_6O(aq) + 2\ CO_2(g)$$
glucose   ethanol

Analysis and Solution

[1] **Convert the number of moles of reactant to the number of moles of product using a mole–mole conversion factor.**

- Use the coefficients in the balanced chemical equation to write **mole–mole conversion factors** for the two compounds—one mole of glucose ($C_6H_{12}O_6$) forms two moles of ethanol (C_2H_6O).
- Multiply the number of moles of reactant (glucose) by the conversion factor to give the number of moles of product (ethanol).

Moles of reactant		mole–mole conversion factor		Moles of product

$$5.00\ mol\ C_6H_{12}O_6 \quad \times \quad \frac{2\ mol\ C_2H_6O}{1\ mol\ C_6H_{12}O_6} \quad = \quad 10.0\ mol\ C_2H_6O$$

Moles $C_6H_{12}O_6$ cancel.

[2] Convert the number of moles of product to the number of grams of product using the product's molar mass.

- Use the **molar mass** of the product (C_2H_6O, molar mass 46.07 g/mol) to write a conversion factor.
- Multiply the number of moles of product (from step [1]) by the conversion factor to give the number of grams of product.

Moles of product		molar mass conversion factor		Grams of product
10.0 mol C_2H_6O	$\times$	$\dfrac{46.07 \text{ g } C_2H_6O}{1 \text{ mol } C_2H_6O}$	$=$	461 g C_2H_6O
		Moles cancel.		**Answer**

Practice Problem 5.23

Using the balanced equation for fermentation written in Sample Problem 5.13, answer the following questions.

a. How many grams of ethanol are formed from 0.55 mol of glucose?

b. How many grams of CO_2 are formed from 0.25 mol of glucose?

c. How many grams of glucose are needed to form 1.0 mol of ethanol?

Practice Problem 5.24

Using the balanced equation for the combustion of ethanol, answer the following questions.

$$C_2H_6O(l) + 3\,O_2(g) \longrightarrow 2\,CO_2(g) + 3\,H_2O(g)$$
ethanol

a. How many grams of CO_2 are formed from 0.50 mol of ethanol?

b. How many grams of H_2O are formed from 2.4 mol of ethanol?

c. How many grams of O_2 are needed to react with 0.25 mol of ethanol?

More Practice: Try Problems 5.47, 5.48.

5.8B   Converting Grams of Reactant to Grams of Product

The **coefficients in chemical equations tell us the ratio of the number of *molecules* or *moles* that are involved in a chemical reaction.** The coefficients do *not,* however, tell us directly about the number of grams. That's because the molar mass—the number of grams in one mole—of a substance depends on the identity of the elements that compose it. One mole of H_2O molecules weighs 18.02 g, one mole of NaCl weighs 58.44 g, and one mole of sugar molecules weighs 342.3 g (Figure 5.5).

In the laboratory, we measure out the number of grams of a reactant on a balance. This does not tell us directly the number of grams of a particular product that will form, because in all likelihood, the molar masses of the reactant and product are different. To carry out this type of calculation—grams of one compound (reactant) to grams of another compound (product)—three operations are necessary.

First, we must determine how many moles of reactant are contained in the given number of grams using the molar mass. Then, we can determine the number of moles of product expected using the coefficients of the balanced chemical equation. Finally, we convert the number of moles of product to the number of grams of product using its molar mass. Now there are three steps and three conversion factors. The stepwise procedure is illustrated in Sample Problem 5.14.

Grams of reactant	[1] molar mass conversion factor	Moles of reactant	[2] mole–mole conversion factor	Moles of product	[3] molar mass conversion factor	Grams of product

Figure 5.5   One Mole of Water, Table Salt, and Table Sugar

one mole of table sugar
C₁₂H₂₂O₁₁
342.30 g/mol

one mole of table salt
NaCl
58.44 g/mol

one mole of water molecules
H₂O
18.02 g/mol

One mole of each substance has the same number of units—6.022×10^{23} H_2O molecules, 6.022×10^{23} Na^+ and Cl^- ions, and 6.022×10^{23} sugar molecules. The molar mass of each substance is *different,* however, because they are each composed of *different* elements. *Jill Braaten/McGraw Hill*

Sample Problem 5.14	Converting Grams of Reactant to Grams of Product

Ethanol (C_2H_6O) is synthesized by reacting ethylene (C_2H_4) with water. How many grams of ethanol are formed from 14 g of ethylene?

$$C_2H_4 \ + \ H_2O \ \longrightarrow \ C_2H_6O$$
$$\text{ethylene} \qquad\qquad\qquad \text{ethanol}$$

Analysis and Solution

[1] **Convert the number of grams of reactant to the number of moles of reactant using the reactant's molar mass.**

- Use the **molar mass** of the reactant (C_2H_4, molar mass 28.05 g/mol) to write a conversion factor. Multiply the number of grams of reactant by the conversion factor to give the number of moles of reactant.

Grams of reactant		molar mass conversion factor		Moles of reactant
14 g C_2H_4	$\times$	$\dfrac{1 \text{ mol } C_2H_4}{28.05 \text{ g } C_2H_4}$	$=$	0.50 mol C_2H_4

Grams cancel.

[2] **Convert the number of moles of reactant to the number of moles of product using a mole–mole conversion factor.**

- Use the coefficients in the balanced chemical equation to write **mole–mole conversion factors.**
- Multiply the number of moles of reactant by the conversion factor to give the number of moles of product. In this example, 0.50 mol of C_2H_6O is formed.

Moles of reactant		mole–mole conversion factor		Moles of product
0.50 mol C_2H_4	$\times$	$\dfrac{1 \text{ mol } C_2H_6O}{1 \text{ mol } C_2H_4}$	$=$	0.50 mol C_2H_6O

Moles C_2H_4 cancel.

[3] **Convert the number of moles of product to the number of grams of product using the product's molar mass.**

- Use the **molar mass** of the product (C_2H_6O, 46.07 g/mol) to write a conversion factor. Multiply the number of moles of product (from step [2]) by the conversion factor to give the number of grams of product.

Moles of product		molar mass conversion factor		Grams of product
0.50 mol C_2H_6O	×	$\dfrac{46.07 \text{ g } C_2H_6O}{1 \text{ mol } C_2H_6O}$	=	23 g C_2H_6O
		Moles C_2H_6O cancel.		**Answer**

Practice Problem 5.25

Use the balanced equation, $N_2 + O_2 \longrightarrow 2\ NO$, to answer the following questions.

a. How many grams of NO are formed from 10.0 g of N_2?

b. How many grams of NO are formed from 10.0 g of O_2?

c. How many grams of O_2 are needed to react completely with 10.0 g of N_2?

More Practice: Try Problems 5.11d, 5.12, 5.65c, 5.66d.

It is also possible to combine the multiplication operations from steps [1], [2], and [3] of Sample Problem 5.14 into a single operation using all three conversion factors. This converts grams of starting material to grams of product all at once. Both the one-step and stepwise approaches give the same overall result.

Grams of reactant		molar mass conversion factor		mole–mole conversion factor		molar mass conversion factor		Grams of product
14 g C_2H_4	×	$\dfrac{1 \text{ mol } C_2H_4}{28.05 \text{ g } C_2H_4}$	×	$\dfrac{1 \text{ mol } C_2H_6O}{1 \text{ mol } C_2H_4}$	×	$\dfrac{46.07 \text{ g } C_2H_6O}{1 \text{ mol } C_2H_6O}$	=	23 g C_2H_6O
	Grams cancel.		Moles C_2H_4 cancel.		Moles C_2H_6O cancel.			**Answer**

5.9 Energy Changes in Reactions

When molecules come together and react, bonds are broken in the reactants and new bonds are formed in the products. **Breaking a bond requires energy, whereas forming a bond releases energy.**

5.9A Heat of Reaction

The energy absorbed or released in any reaction is called the **heat of reaction** or the **enthalpy change**, symbolized by ΔH. The heat of reaction is given a positive (+) or negative (–) sign depending on whether energy is absorbed or released.

- When energy is absorbed, the reaction is said to be *endothermic* and ΔH is positive (+).
- When energy is released, the reaction is said to be *exothermic* and ΔH is negative (–).

The heat of reaction measures the difference between the energy needed to break bonds in the reactants and the energy released from the bonds formed in the products. In other words, **ΔH indicates the relative strength of bonds broken and formed in a reaction.**

- When ΔH is negative, more energy is released in forming bonds than is needed to break bonds. The *products* are lower in energy than the reactants.

The CH_4 produced by decomposing waste material in large landfills is burned to produce energy for heating and generating electricity. *Digital Vision/Getty Images*

For example, when methane (CH_4) burns in the presence of oxygen (O_2) to form CO_2 and H_2O, 213 kcal/mol of energy is released in the form of heat.

Heat is **released.**

$$CH_4(g) \ + \ 2\,O_2(g) \ \longrightarrow \ CO_2(g) \ + \ 2\,H_2O(l) \quad \Delta H = -213 \text{ kcal/mol}$$

In this reaction energy is released, ΔH is negative (–), and the reaction is exothermic. **Because energy is released, the products are *lower* in energy than the reactants.**

- When ΔH is positive, more energy is needed to break bonds than is released in forming bonds. The *reactants* are lower in energy than the products.

For example, in the process of photosynthesis, green plants use chlorophyll to convert CO_2 and H_2O to glucose ($C_6H_{12}O_6$, a simple carbohydrate) and O_2 and 678 kcal of energy is absorbed.

$$6\,CO_2(g) \ + \ 6\,H_2O(l) \ \longrightarrow \ C_6H_{12}O_6(aq) \ + \ 6\,O_2(g) \quad \Delta H = +678 \text{ kcal/mol}$$

In this reaction energy is absorbed, ΔH is positive (+), and the reaction is endothermic. **Because energy is absorbed, the products are *higher* in energy than the reactants.**

Table 5.3 summarizes the characteristics of energy changes in reactions.

Photosynthesis is an endothermic reaction. Energy from sunlight is absorbed in the reaction and stored in the bonds of the products. *Comstock/Stockbyte/Getty Images*

Table 5.3 Endothermic and Exothermic Reactions

Endothermic Reaction	Exothermic Reaction
• Heat is **absorbed.**	• Heat is **released.**
• ΔH is **positive.**	• ΔH is **negative.**
• The products are *higher* in energy than the reactants.	• The products are *lower* in energy than the reactants.

Concept Check 5.10

Answer the following questions using the given equation and ΔH. (a) Is heat absorbed or released? (b) Are the reactants or products lower in energy? (c) Is the reaction endothermic or exothermic?

$$2\,NH_3(g) \ \longrightarrow \ 3\,H_2(g) \ + \ N_2(g) \quad \Delta H = +22.0 \text{ kcal/mol}$$

Concept Check 5.11

The ΔH for the combustion of propane (C_3H_8) with O_2 according to the given balanced chemical equation is –531 kcal/mol. (a) Is heat absorbed or released? (b) Are the reactants or products lower in energy? (c) Is the reaction endothermic or exothermic?

$$C_3H_8(g) \ + \ 5\,O_2(g) \ \longrightarrow \ 3\,CO_2(g) \ + \ 4\,H_2O(l) \quad \Delta H = -531 \text{ kcal/mol}$$
propane

5.9B Energy Diagrams

On a molecular level, what happens when a reaction occurs? In order for two molecules to react, they must collide, and in the collision, the kinetic energy they possess is used to break bonds.

The energy changes in a reaction are often illustrated on an **energy diagram,** which plots energy on the vertical axis, and the progress of the reaction—the **reaction coordinate**—on the horizontal axis. The reactants are written on the left side and the products on the right side, and a smooth curve that illustrates how energy changes with time connects them. Consider a general reaction between two starting materials, **A—B** and **C,** in which the **A—B** bond is broken and a new **B—C** bond is formed.

Let's assume that the products, **A** and **B—C,** are lower in energy than the reactants, **A—B** and **C.**

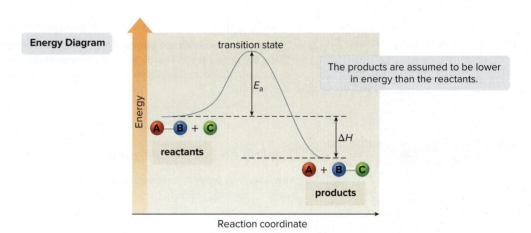

When the reactants **A—B** and **C** approach each other, their electron clouds feel some repulsion, causing an increase in energy until a maximum value is reached. This point is called the **transition state.** In the transition state, the bond between **A** and **B** is partially broken and the bond between **B** and **C** is partially formed.

At the transition state, the bond between **A** and **B** can re-form to regenerate reactants, or the bond between **B** and **C** can form to generate products. As the bond forms between **B** and **C,** the energy decreases until some stable energy minimum is reached. The products are drawn lower in energy than the reactants to reflect the initial assumption about their relative energies.

• The difference in energy between the reactants and the transition state is called the *energy of activation,* symbolized by E_a.

The energy of activation is the minimum amount of energy needed for a reaction to occur. The energy of activation is often called the **energy barrier** that must be crossed. The height of the energy barrier—the magnitude of the energy of activation—determines the **reaction rate, how fast the reaction occurs.**

• When the energy of activation is *high,* few molecules have enough energy to cross the energy barrier and the reaction is *slow.*
• When the energy of activation is *low,* many molecules have enough energy to cross the energy barrier and the reaction is *fast.*

The difference in energy between the reactants and products is the ΔH, which is also labeled on the energy diagram. When the products are *lower* in energy than the reactants, as is the case here, ΔH is *negative* (–) and the reaction is *exothermic.*

Energy diagrams can be drawn for any reaction. In the endothermic reaction shown in Figure 5.6, the products are higher in energy than the reactants.

Figure 5.6

Energy Diagram for an Endothermic
Reaction

- E_a **is the energy difference between the reactants and the transition state.**
- **ΔH is the difference in energy between the reactants and products.**
- Because the products are higher in energy than the reactants, ΔH is positive (+) and the reaction is endothermic.

Sample Problem 5.15   **Drawing an Energy Diagram**

Draw an energy diagram for a reaction with a low energy of activation and a ΔH of -10 kcal/mol. Label the axes, reactants, products, transition state, E_a, and ΔH.

Analysis

A low energy of activation means a low energy barrier and a small hill that separates reactants and products. When ΔH is $(-)$, the products are lower in energy than the reactants.

Solution

Practice Problem 5.26   Draw an energy diagram for a reaction with a high E_a and a $\Delta H = +20$ kcal/mol.

More Practice: Try Problems 5.15, 5.16, 5.55, 5.56.

5.10 Reaction Rates

Even though we may not realize it, the rate of chemical processes affects many facets of our lives. Aspirin is an effective pain reliever because it rapidly blocks the synthesis of pain-causing molecules. DDT is a persistent environmental pollutant because it does not react appreciably with water, oxygen, or any other chemical with which it comes into contact. These processes occur at different rates, resulting in beneficial or harmful effects.

The energy of activation, the minimum amount of energy needed for a reaction to occur, is a fundamental characteristic of a reaction. Some reactions are fast because they have low energies of activation. Other reactions are slow because the energy of activation is high.

5.10A How Concentration and Temperature Affect Reaction Rate

As we learned in Section 5.9, chemical reactions occur when molecules collide. How do changes in concentration and temperature affect the reaction rate?

- *Increasing* the concentration of the reactants *increases* the number of collisions, so the reaction rate *increases*.
- *Increasing* the temperature *increases* the reaction rate.

Increasing the temperature increases the kinetic energy, which increases the number of collisions. Increasing the temperature also increases the *average* kinetic energy of the reactants. Because the kinetic energy of colliding molecules is used for bond cleavage, more molecules have sufficient energy to cause bond breaking, and the reaction rate increases.

Concept Check 5.12

Consider the reaction of ozone (O_3) with nitrogen monoxide (NO), which occurs in smog. What effect would each of the following changes have on the rate of this reaction?

$$O_3(g) + NO(g) \longrightarrow O_2(g) + NO_2(g)$$

a. increasing the concentration of O_3
b. decreasing the concentration of NO
c. increasing the temperature
d. decreasing the temperature

5.10B Catalysts

Some reactions do not occur in a reasonable period of time unless a **catalyst** is added.

- A *catalyst* is a substance that speeds up the rate of a reaction. A catalyst is recovered unchanged in a reaction, and it does not appear in the product.

Catalysts accelerate a reaction by lowering the energy of activation (Figure 5.7). They have no effect on the energies of the reactants and products. Thus, the addition of a catalyst lowers E_a but does not affect ΔH.

We store food in a cold refrigerator to slow the reactions that cause food to spoil. *Jill Braaten/McGraw Hill*

Figure 5.7 The Effect of a Catalyst on a Reaction

A catalyst lowers the energy of activation, thus increasing the rate of the catalyzed reaction. The energy of the reactants and products is the same in both the uncatalyzed and catalyzed reactions.

Concept Check 5.13

The reaction of acetic acid ($C_2H_4O_2$) and ethanol (C_2H_6O) to form ethyl acetate ($C_4H_8O_2$) and water is catalyzed by sulfuric acid (H_2SO_4).

    a. What effect does H_2SO_4 have on the relative energies of the reactants and products?

    b. What effect does H_2SO_4 have on the energy of activation?

HEALTH NOTE

There is a direct link between the poor air quality in large metropolitan areas like Los Angeles and an increase in respiratory diseases. *Ingram Publishing/SuperStock*

5.10C FOCUS ON THE ENVIRONMENT
Catalytic Converters

The combustion of gasoline with oxygen provides a great deal of energy, and this energy is used to power vehicles. As the number of automobiles increased in the twentieth century, the air pollution they were responsible for became a major problem, especially in congested urban areas. In addition to CO_2 and H_2O formed during combustion, auto exhaust also contained unreacted gasoline molecules (general formula C_xH_y), the toxic gas carbon monoxide (CO, Section 5.7), and nitrogen monoxide (NO, Section 5.7, a contributing component of acid rain). **Catalytic converters** were devised to clean up these polluting automobile emissions.

The newest catalytic converters use a metal as a surface to catalyze three reactions, as shown in Figure 5.8. Both the unreacted gasoline molecules and carbon monoxide (CO) are oxidized to CO_2 and H_2O. Nitrogen monoxide is converted to oxygen and nitrogen. As a result, the only materials in the engine exhaust are CO_2, H_2O, N_2, and O_2.

Figure 5.8   How a Catalytic Converter Works

$$C_xH_y \; + \; O_2 \longrightarrow CO_2 \; + \; H_2O \text{ (unbalanced)}$$
$$2\,CO \; + \; O_2 \longrightarrow 2\,CO_2$$
$$2\,NO \longrightarrow N_2 \; + \; O_2$$

A catalytic converter uses a metal catalyst—rhodium, platinum, or palladium—to catalyze three reactions that clean up the exhaust from an auto engine.

Concept Check 5.14

Nitrogen dioxide, NO_2, also an undesired product formed during combustion, is converted to N_2 and O_2 in a catalytic converter. Write a balanced equation for this reaction.

5.11 Equilibrium

Thus far in discussing reactions we have assumed that the reactants are completely converted to products. A reaction of this sort is said to **go to completion.** Sometimes, however, a reaction is **reversible;** that is, reactants can come together and form products, and products can come together to re-form reactants.

> • A *reversible reaction* can occur in either direction, from reactants to products or from products to reactants.

5.11A Reversible Reactions

Consider the reversible reaction of carbon monoxide (CO) with water to form carbon dioxide (CO_2) and hydrogen. **Two full-headed arrows** ($\rightleftharpoons$) are used to show that the reaction can proceed from left to right and right to left as written.

The **forward** reaction proceeds to the **right.**

$$CO(g) \quad + \quad H_2O(g) \quad \rightleftharpoons \quad CO_2(g) \quad + \quad H_2(g)$$

The **reverse** reaction proceeds to the **left.**

> • The *forward* reaction proceeds from left to right as drawn.
> • The *reverse* reaction proceeds from right to left as drawn.

When CO and H_2O are mixed together they react to form CO_2 and H_2 by the forward reaction. Once CO_2 and H_2 are formed, they can react together to form CO and H_2O by the reverse reaction. The rate of the forward reaction is rapid at first, and then decreases as the concentration of reactants decreases. The rate of the reverse reaction is slow at first, but speeds up as the concentration of the products increases.

> • When the rate of the forward reaction equals the rate of the reverse reaction, the net concentrations of all species do not change and the system is at *equilibrium.*

The forward and reverse reactions do not stop once equilibrium has been reached. The reactants and products continue to react. Because the rates of the forward and reverse reactions are equal, however, the **net concentrations of all reactants and products do *not* change.** Molecular art in Figure 5.9 shows how the reversible reaction A $\rightleftharpoons$ B reaches equilibrium.

Figure 5.9   Attaining Equilibrium in a Reaction, **A** $\rightleftharpoons$ **B**

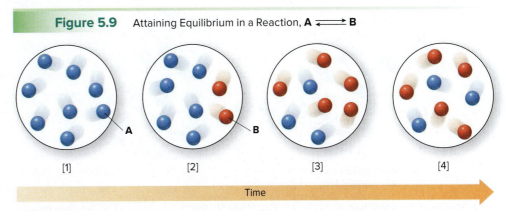

[1]   [2]   [3]   [4]

Time

The conversion of **A** (shown in blue spheres) to **B** (shown in red spheres) is a reversible reaction. Initially, the reaction contains only **A** (diagram [1]). As **A** reacts to form **B,** the amount of **B** increases (diagram [2]). As **B** is formed, it can be re-converted to **A.** At equilibrium (diagrams [3] and [4]), the amounts of **A** and **B** do not change, but the conversion of **A** to **B** and the conversion of **B** to **A** still occur.

Identify the forward and reverse reactions in each of the following reversible reactions.

a. $2 SO_2(g) + O_2(g) \rightleftharpoons 2 SO_3(g)$ b. $N_2(g) + O_2(g) \rightleftharpoons 2 NO(g)$

5.11B   Le Châtelier's Principle

What happens when a reaction is at equilibrium and something changes? For example, what happens when some additional reactant is added? **Le Châtelier's principle** is a general rule used to explain the effect of a change in reaction conditions on equilibrium. Le Châtelier's principle states:

> • If a chemical system at equilibrium is disturbed or stressed, the system will react in the direction that counteracts the disturbance or relieves the stress.

Consider the reaction of carbon monoxide (CO) with oxygen (O_2) to form carbon dioxide (CO_2).

$$2 CO(g) + O_2(g) \rightleftharpoons 2 CO_2(g)$$

If the reactants and products are at equilibrium, what happens if the concentration of CO is increased? Now the equilibrium is disturbed and, as a result, the rate of the forward reaction increases to produce more CO_2. We can think of **added reactant as driving the equilibrium to the** *right*.

Adding more reactant...

$2 CO(g) + O_2(g) \rightleftharpoons 2 CO_2(g)$

...drives the reaction to the *right*.

Figure 5.10 uses molecular art to illustrate the effect of added reactant in another reaction.

Figure 5.10   The Effect of Added Reactant on Equilibrium for the Reaction, $N_2 + 3 H_2 \rightleftharpoons 2 NH_3$

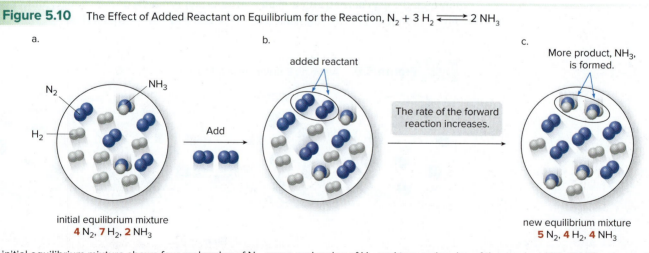

a.

NH$_3$
N$_2$
H$_2$

initial equilibrium mixture
4 N$_2$, **7** H$_2$, **2** NH$_3$

b.

added reactant

Add

The rate of the forward reaction increases.

c.

More product, NH$_3$, is formed.

new equilibrium mixture
5 N$_2$, **4** H$_2$, **4** NH$_3$

(a) The initial equilibrium mixture shows four molecules of N$_2$, seven molecules of H$_2$, and two molecules of the product NH$_3$. (b) When more N$_2$, a reactant, is added, the equilibrium is disturbed, and the rate of the forward reaction *increases* to form more product, NH$_3$. (c) In the new equilibrium mixture, the amount of NH$_3$ has increased (four molecules). Because H$_2$, a reactant, is used to form NH$_3$, the amount of H$_2$ has *decreased* (four molecules).

What happens, instead, if the concentration of CO_2 is increased? Now the equilibrium is disturbed but there is more product than there should be. As a result, the rate of the reverse reaction increases to produce more of both reactants, CO and O_2. We can think of **added product as driving the equilibrium to the** *left.*

Adding more product...

$$2\ CO(g)\ +\ O_2(g)\ \rightleftharpoons\ 2\ CO_2(g)$$

...drives the reaction to the ***left.***

Similar arguments can be made about the effect of *decreasing* the concentration of a reactant or product. For example, ethanol (C_2H_6O) can be converted to ethylene ($CH_2{=}CH_2$) and water in the presence of a small amount of acid, but equilibrium does not favor the products.

$$C_2H_6O\ \rightleftharpoons\ \underset{\text{ethanol}}{}\ \ \overset{H\quad\ H}{\underset{H\quad\ H}{C{=}C}}\ +\ H_2O$$

ethanol

Removing a product...

...drives the reaction to the ***right.***

In this case, water can be removed from the reaction as it is formed. A decrease in concentration of one product results in more of the forward reaction to form more product. **This process drives the equilibrium to the** *right.* If water is continuously removed, essentially all of the starting material can be converted to product. Table 5.4 summarizes the effects of concentration changes on the direction of equilibrium.

Table 5.4 The Effect of Concentration Changes on Equilibrium

Change	Effect on Equilibrium
Adding reactant	Equilibrium favors the **products.**
Removing reactant	Equilibrium favors the **reactants.**
Adding product	Equilibrium favors the **reactants.**
Removing product	Equilibrium favors the **products.**

Sample Problem 5.16

Determining the Effect of Concentration Changes on Equilibrium

In which direction is the equilibrium shifted with each of the following concentration changes for the given reaction: (a) increase [SO_2]; (b) increase [SO_3]; (c) decrease [O_2]; (d) decrease [SO_3]?

$$2\ SO_2(g)\ +\ O_2(g)\ \rightleftharpoons\ 2\ SO_3(g)$$

Analysis

Use Le Châtelier's principle to predict the effect of a change in concentration on equilibrium. Adding more reactant or removing product drives the equilibrium to the right. Adding more product or removing reactant drives the equilibrium to the left.

Solution

a. Increasing [SO_2], a reactant, drives the equilibrium to the right to form more product.
b. Increasing [SO_3], a product, drives the equilibrium to the left to form more reactants.
c. Decreasing [O_2], a reactant, drives the equilibrium to the left to form more reactants.
d. Decreasing [SO_3], a product, drives the equilibrium to the right to form more product.

| Practice Problem 5.27 | In which direction is the equilibrium shifted with each of the following concentration changes for the given reaction: (a) increase [H₂]; (b) increase [HCl]; (c) decrease [Cl₂]; (d) decrease [HCl]? |

In which direction is the equilibrium shifted with each of the following concentration changes for the given reaction: (a) increase $[H_2]$; (b) increase $[HCl]$; (c) decrease $[Cl_2]$; (d) decrease $[HCl]$?

$$H_2(g) + Cl_2(g) \rightleftharpoons 2\ HCl(g)$$

More Practice: Try Problems 5.17, 5.18, 5.61–5.64.

5.12 FOCUS ON THE HUMAN BODY
Body Temperature

The human body is an enormously complex organism that illustrates important features of energy and reaction rates. At any moment, millions of reactions occur in the body, when nutrients are metabolized and new cell materials are synthesized.

Normal body temperature, 37 °C, reflects a delicate balance between the amount of heat absorbed and released in all of the reactions and other processes. Because reaction rate increases with increasing temperature, it is crucial to maintain the right temperature for proper body function. When temperature increases, reactions proceed at a faster rate. An individual must breathe more rapidly and the heart must pump harder to supply oxygen for the faster metabolic processes. When temperature decreases, reactions slow down, less heat is generated in exothermic reactions, and it becomes harder and harder to maintain an adequate body temperature.

Thermoregulation—regulating temperature—is a complex mechanism that involves the brain, the circulatory system, and the skin (Figure 5.11). Temperature sensors in the skin and body core signal when there is a temperature change. The hypothalamus region of the brain, in turn, responds to changes in its environment.

Figure 5.11

Temperature Regulation in the Body

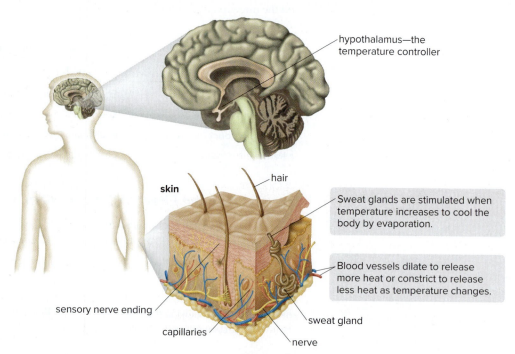

hypothalamus—the temperature controller

hair

skin

Sweat glands are stimulated when temperature increases to cool the body by evaporation.

Blood vessels dilate to release more heat or constrict to release less heat as temperature changes.

sensory nerve ending

capillaries

sweat gland

nerve

When the temperature in the environment around the body changes, the body works to counteract the change. The hypothalamus acts as a thermostat, which signals the body to respond to temperature changes. When the temperature increases, the body must dissipate excess heat by dilating blood vessels and sweating. When the temperature decreases, blood vessels constrict and the body shivers.

When the temperature increases, the body must somehow rid itself of excess heat. Blood vessels near the surface of the skin are dilated to release more heat. Sweat glands are stimulated, so the body can be cooled by the evaporation of water from the skin's surface.

When the temperature decreases, the body must generate more heat as well as slow down the loss of heat to the surroundings. Blood vessels constrict to reduce heat loss from the skin and muscles shiver to generate more heat.

An infection in the body is often accompanied by a fever; that is, the temperature in the body increases. A fever is part of the body's response to increase the rates of defensive reactions that kill bacteria. The respiratory rate and heart rate increase to supply more oxygen needed for faster reactions.

CHAPTER REVIEW

KEY TERMS

Avogadro's number (5.5)
Balanced chemical equation (5.2)
Catalyst (5.10)
Chemical equation (5.1)
Combination reaction (5.3)
Decomposition reaction (5.3)
Double replacement reaction (5.3)
Endothermic reaction (5.9)
Energy diagram (5.9)

Energy of activation (5.9)
Enthalpy change (5.9)
Equilibrium (5.11)
Exothermic reaction (5.9)
Formula weight (5.6)
Forward reaction (5.11)
Half reaction (5.4)
Heat of reaction (5.9)
Law of conservation of matter (5.1)

Le Châtelier's principle (5.11)
Molar mass (5.6)
Mole (5.5)
Molecular weight (5.6)
Oxidation (5.4)
Oxidizing agent (5.4)
Product (5.1)
Reactant (5.1)
Reaction rate (5.10)

Redox reaction (5.4)
Reducing agent (5.4)
Reduction (5.4)
Reverse reaction (5.11)
Reversible reaction (5.11)
Single replacement reaction (5.3)
Transition state (5.9)

KEY CONCEPTS

1 Types of reactions (5.3)

1 Combination	2 Decomposition	3 Single replacement	4 Double replacement
A + B → AB	AB → A + B	A + BC → B + AC	AB + CD → AD + CB
• The joining of two or more reactants to form a single product	• The conversion of a single reactant to two or more products	• A reaction in which one element replaces another element in a compound to form a different compound and element	• A reaction in which two compounds exchange atoms or ions to form two new compounds

2 Oxidation and reduction (5.4)

A **redox reaction** involves the *transfer* of electrons from one element to another.

1 Oxidation—loss of electrons	2 Reduction—gain of electrons
• **Metals** and **anions** tend to *lose* electrons in oxidation. • Oxidation may result in the **gain of O atoms** or the **loss of H atoms**.	• **Nonmetals** and **cations** tend to *gain* electrons in reduction. • Reduction may result in the **loss of O atoms** or the **gain of H atoms**.

3 The mole, formula weight, and molar mass; example: vitamin C ($C_6H_8O_6$)

1 The mole and Avogadro's number (5.5)	**2** Formula weight (5.6)	**3** Molar mass (5.6)

1 mole $C_6H_8O_6$ molecules = **6.022 × 10²³** molecules	6 C atoms × 12.01 amu = 72.06 amu
	8 H atoms × 1.008 amu = 8.064 amu
	6 O atoms × 16.00 amu = 96.00 amu
	                                      176.124 amu
	      Rounded to **176.12 amu**
		molar mass of $C_6H_8O_6$ = **176.12 g/mol**

• A quantity that contains 6.022 × 10²³ atoms, molecules, or ions • **6.022 × 10²³ = Avogadro's number**	• The sum of the atomic weights of all atoms in a compound reported in amu • The formula weight of $C_6H_8O_6$ is 176.12 amu.	• The mass of one mole of any substance, reported in **grams per mole** • The **molar mass** (in g/mol) is numerically equal to the **formula weight** (in amu).

4 Energy changes and reactions (5.9)

1 Bond breaking and bond making	**2** Endothermic reaction	**3** Exothermic reaction
• ΔH = **heat of reaction = enthalpy change =** the energy absorbed or released in a reaction • Bond breaking *requires* energy. • Bond formation *releases* energy.	• ΔH is positive (+). • Energy is **absorbed.** • The products are *higher* in energy than the reactants.	• ΔH is negative (–). • Energy is **released.** • The products are *lower* in energy than the reactants.

5 Energy diagrams (5.9)

1 A plot of energy changes versus the progress of the reaction	**2** Key features

	• The **transition state** is located at the top of the energy hill that separates reactants and products.
	• E_a—the energy of activation; the minimum amount of energy needed for a reaction to occur
	   ○ High E_a = slow reaction
	   ○ Low E_a = fast reaction
	• ΔH = energy difference between the reactants and products
	   ○ ΔH is (+) when the products are *higher* in energy than the reactants.
	   ○ ΔH is (–) when the products are *lower* in energy than the reactants.

6 Reaction rates (5.10)

1 Factors that affect reaction rate	**2** Factor that does not affect reaction rate
• **Energy of activation**—*Increasing* E_a *decreases* reaction rate. • **Concentration**—*Increasing* concentration *increases* reaction rate. • **Temperature**—*Increasing* temperature *increases* reaction rate. • **Catalyst**—*Increases* reaction rate	• The size of the enthalpy change (ΔH) does *not* affect reaction rate. The ΔH for a catalyzed and uncatalyzed reaction is the same, but the catalyzed reaction is faster because it has the lower E_a.

7 Using Le Châtelier's principle (5.11); example: predicting the effect of concentration change on the direction of equilibrium for the reaction, $C_2H_4 + H_2 \longrightarrow C_2H_6$, carried out with a Pd catalyst

Le Châtelier's principle: A system at equilibrium reacts in such a way as to counteract any disturbance to the equilibrium.

1 Increasing [H_2] or [C_2H_4]	**2** Increasing [C_2H_6]	**3** Decreasing [C_2H_6]	**4** Removing the catalyst
• Adding more reactant shifts the equilibrium to the *right*.	• Adding more product shifts the equilibrium to the *left*.	• Removing product shifts the equilibrium to the *right*.	• Catalysts increase the rate at which equilibrium is reached, but *do not change* the amount of any substance at equilibrium.

KEY SKILLS

1 Balancing a chemical equation (5.2); example: reaction of pentane (C_5H_{12}) with oxygen (O_2) to form carbon dioxide (CO_2) and water (H_2O)

1 Write the equation with correct formulas.	**2** Balance the equation with coefficients one element at a time.

2 Using molar mass as a conversion factor (5.6B); example: determining the number of grams in 0.25 mol aspirin (molar mass 180.15 g/mol)

1 Identify the original quantity and the desired quantity.	**2** Write out the molar mass conversion factor and place the unwanted unit (mol) in the denominator.	**3** Solve the problem so that the unwanted unit cancels.

3 Using the coefficients in a balanced equation to convert moles of one quantity into moles of another quantity (5.7); example: determining the number of moles of H_2 formed from 1.5 mol HCl in the given equation

1 Identify the original quantity and the desired quantity in the balanced equation.	**2** Write out the mole–mole conversion factor and place the unwanted unit in the denominator.	**3** Solve the problem so that the unwanted unit cancels.

4 Converting moles of reactant to grams of product (5.8); example: determining the number of grams of Fe_2O_3 formed from 3.0 mol Fe

1 Convert moles of reactant to moles of product using a mole–mole conversion factor.	**2** Convert moles of product to grams of product using a molar mass conversion factor (molar mass Fe_2O_3 = 159.70 g/mol).

- Multiply the number of moles of reactant (Fe) by the conversion factor to give the number of moles of product (Fe_2O_3).
- Multiply the number of moles of product (from step [1]) by the molar mass conversion factor to give the number of grams of product.

CHAPTER 5 SELF-TEST

The Self-Test consists of short-answer questions that test an understanding of definitions, equations, and other material encountered within the chapter. Answers to each question are provided at the end of the chapter. For any questions you are unable to answer correctly, revisit the appropriate section(s) of the text; then proceed to working the end-of-chapter problems to further reinforce your understanding of the chapter material.

1. Which term from "Key Terms" answers each question?
   a. What is the difference in energy between the reactants and the transition state called?
   b. What term is used for the mass of one mole of a substance, reported in grams per mole?
   c. In what type of reaction do two compounds exchange "parts"—atoms or ions—to form two new compounds?
   d. What general rule states that if a system at equilibrium is disturbed, the system will react in the direction that counteracts the disturbance?
   e. What two terms are used for the energy absorbed or released in a reaction?

2. Consider the following reaction: $CO + H_2O \rightarrow CO_2 + H_2$. In which direction is the equilibrium shifted with each of the following concentration changes: (a) increase $[H_2O]$; (b) increase $[CO_2]$; (c) decrease $[CO]$; (d) decrease $[H_2]$?

3. Consider a reaction in which $\Delta H = -88$ kcal/mol. (a) Are the reactants or products lower in energy? (b) Is the reaction endothermic or exothermic? (c) Are the bonds formed in the products stronger or weaker than the bonds broken in the reactants?

4. What is the molar mass of $C_6H_8O_6$?

5. How many molecules are contained in 8.00 mol of H_2O?

6. Consider the following reaction: $4\,Fe + 3\,O_2 \rightarrow 2\,Fe_2O_3$. How many grams of Fe_2O_3 (molar mass 159.70 g/mol) are produced from 1.0 mol of Fe?

7. Identify the species that is oxidized and the species that is reduced in the following reaction: $2\,Na + Cl_2 \rightarrow 2\,Na^+ + 2\,Cl^-$.

8. Fill in the boxes to balance the given equation.
   a. $Mg_3(PO_4)_2$ + b. $H_2SO_4 \longrightarrow$ c. $MgSO_4$ + d. H_3PO_4

9. Classify each reaction as a combination, decomposition, single replacement, or double replacement.
   a. $Fe + CuSO_4 \rightarrow FeSO_4 + Cu$
   b. $CH_3CH_2Cl \rightarrow C_2H_4 + HCl$
   c. $HCl + NaOH \rightarrow H_2O + NaCl$
   d. $N_2 + 3\,H_2 \rightarrow 2\,NH_3$

UNDERSTANDING KEY CONCEPTS

5.1 Use the molecular art to identify the process as a chemical reaction or a physical change. If a chemical reaction occurs, write the equation from the molecular art. If a physical change occurs, identify the phase change.

5.2 Diagram [1] represents the compound **AB,** which contains two elements, **A** shown with red spheres, and **B** shown with blue spheres. Label each transformation as a physical change or a chemical reaction: (a) [1] $\longrightarrow$ [2]; (b) [1] $\longrightarrow$ [3].

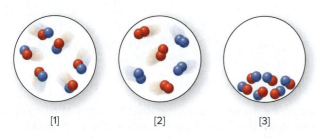

[1]   [2]   [3]

5.3 Use the molecular art to write a balanced equation for the given reaction.

5.4 Use the molecular art to write a balanced equation for the given reaction.

5.5 Some coal is high in sulfur (S) content, and when it burns, it forms sulfuric acid (H_2SO_4), a major component of acid rain, by a series of reactions. Balance the equation for the overall conversion drawn below.

$$S(s) + O_2(g) + H_2O(l) \longrightarrow H_2SO_4(l)$$

5.6 Balance the equation for the formation of magnesium hydroxide [$Mg(OH)_2$], one of the active ingredients in milk of magnesia.

$$MgCl_2 + NaOH \longrightarrow Mg(OH)_2 + NaCl$$

5.7 For the reaction depicted in the molecular art: (a) Use the common element colors shown in Appendix C to give the formula for each reactant and product. (b) Write a balanced equation. (c) Classify the reaction as combination, decomposition, single replacement, or double replacement.

5.8 Answer the questions in Problem 5.7 for the reaction depicted in the molecular art.

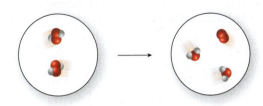

5.9 What is the mass in grams of each quantity of lactic acid ($C_3H_6O_3$, molar mass 90.08 g/mol), the compound responsible for the aching feeling of tired muscles during vigorous exercise?

    a. 3.60 mol   b. 7.3×10^{24} molecules

5.10 Spinach, cabbage, and broccoli are excellent sources of vitamin K (molar mass 450.7 g/mol), which is needed in adequate amounts for blood to clot. The recommended daily intake of vitamin K is 120 μg. How many molecules of vitamin K does this correspond to?

5.11 Fats, such as butter, and oils, such as corn oil, are formed from compounds called fatty acids, one of which is linolenic acid ($C_{18}H_{30}O_2$). Linolenic acid undergoes reactions with hydrogen and oxygen to form the products shown in each equation.

    [1] $C_{18}H_{30}O_2 + H_2 \longrightarrow C_{18}H_{36}O_2$
         linolenic acid

    [2] $C_{18}H_{30}O_2 + O_2 \longrightarrow CO_2 + H_2O$
         linolenic acid

    a. Calculate the molar mass of linolenic acid.
    b. Balance Equation [1], which shows the reaction with hydrogen.
    c. Balance Equation [2], which shows the reaction with oxygen.
    d. How many grams of product are formed from 10.0 g of linolenic acid in Equation [1]?

5.12 Iron, like most metals, does not occur naturally as the pure metal. Rather, it must be produced from iron ore, which contains iron(III) oxide, according to the given balanced equation.

$$Fe_2O_3(s) + 3 CO(g) \longrightarrow 2 Fe(s) + 3 CO_2(g)$$

    a. How many grams of Fe are formed from 10.0 g of Fe_2O_3?
    b. How many grams of Fe are formed from 25.0 g of Fe_2O_3?

5.13 Zinc–silver oxide batteries are used in cameras and hearing aids. Identify the species that is oxidized and the species that is reduced in the following redox reaction. Identify the oxidizing agent and the reducing agent. Oxide (O^{2-}) is unchanged in this reaction.

$$Zn + Ag_2O \longrightarrow ZnO + 2 Ag$$

5.14 Rechargeable nickel–cadmium batteries are used in appliances and power tools. Identify the species that is oxidized and the species that is reduced in the following redox reaction. Identify the oxidizing agent and the reducing agent.

$$Cd + Ni^{4+} \longrightarrow Cd^{2+} + Ni^{2+}$$

5.15 Consider the energy diagram drawn below.

    a. Which point on the graph corresponds to reactants?
    b. Which point on the graph corresponds to products?
    c. Which point on the graph corresponds to the transition state?
    d. The difference in energy between which two points equals the energy of activation?
    e. The difference in energy between which two points equals the ΔH?
    f. Which point is highest in energy?
    g. Which point is lowest in energy?

5.16 Compound **A** can be converted to either **B** or **C**. The energy diagrams for both processes are drawn on the graph below.

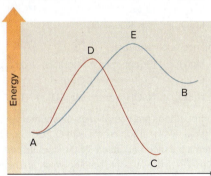

    a. Label each reaction as endothermic or exothermic.
    b. Which reaction is faster?
    c. Which reaction generates the product lower in energy?
    d. Which points on the graphs correspond to transition states?
    e. Label the energy of activation for each reaction.
    f. Label the ΔH for each reaction.

5.17 Consider the reversible reaction **A** + **B** ⇌ **AB,** shown in representation [1] at equilibrium. If representation [2] shows the system after the equilibrium has been disturbed, in which direction must the reaction be driven to achieve equilibrium again?

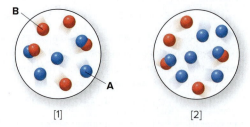

[1]   [2]

5.18 Consider the reversible reaction 2 **A** + **B**₂ ⇌ 2 **AB,** shown in representation [3] at equilibrium. If representation [4] shows the system after the equilibrium has been disturbed, in which direction must the reaction be driven to achieve equilibrium again?

[3]   [4]

ADDITIONAL PROBLEMS

Chemical Equations

5.19 How many atoms of each element are drawn on each side of the following equations? Label the equations as balanced or not balanced.

a. $2\ HCl(aq)\ +\ Ca(s)\ \longrightarrow\ CaCl_2(aq)\ +\ H_2(g)$

b. $Al(OH)_3\ +\ H_3PO_4\ \longrightarrow\ AlPO_4\ +\ 3\ H_2O$

5.20 How many atoms of each element are drawn on each side of the following equations? Label the equations as balanced or not balanced.

a. $3\ NO_2\ +\ H_2O\ \longrightarrow\ HNO_3\ +\ 2\ NO$

b. $Ca(OH)_2\ +\ 2\ HNO_3\ \longrightarrow\ 2\ H_2O\ +\ Ca(NO_3)_2$

5.21 Draw a representation using molecular art for the following equation: 2 **A** + 2 **B** ⟶ 2 **AB.** Use a red sphere for each **A** atom and a blue sphere for each **B** atom.

5.22 Draw a representation using molecular art for the following equation: **A**₂ + **B**₂ ⟶ 2 **AB.** Use a red sphere for each **A** atom and a blue sphere for each **B** atom.

5.23 Balance each equation.

a. $Ni(s)\ +\ HCl(aq)\ \longrightarrow\ NiCl_2(aq)\ +\ H_2(g)$

b. $CH_4(g)\ +\ Cl_2(g)\ \longrightarrow\ CCl_4(g)\ +\ HCl(g)$

c. $KClO_3\ \longrightarrow\ KCl\ +\ O_2$

5.24 Balance each equation.

a. $Mg(s)\ +\ HBr(aq)\ \longrightarrow\ MgBr_2(s)\ +\ H_2(g)$

b. $CO(g)\ +\ O_2(g)\ \longrightarrow\ CO_2(g)$

c. $H_2SO_4\ +\ NaOH\ \longrightarrow\ Na_2SO_4\ +\ H_2O$

5.25 Hydrocarbons are compounds that contain only C and H atoms. When a hydrocarbon reacts with O_2, CO_2 and H_2O are formed. Write a balanced equation for the combustion of each of the following hydrocarbons, both of which are high-octane components of gasoline.

a. C_6H_6 (benzene)

b. C_7H_8 (toluene)

5.26 MTBE ($C_5H_{12}O$) is a high-octane gasoline additive with a sweet, nauseating odor. Because small amounts of MTBE have contaminated the drinking water in some towns, it is now banned as a fuel additive in some states. MTBE reacts with O_2 to form CO_2 and H_2O. Write a balanced equation for the combustion of MTBE.

5.27 Consider the reaction, $O_3 + CO \longrightarrow O_2 + CO_2$. Molecular art is used to show the starting materials for this reaction. Fill in the molecules of the products using the balanced equation and following the law of conservation of matter.

5.28 Consider the reaction, $2\ NO + 2\ CO \longrightarrow N_2 + 2\ CO_2$. Molecular art is used to show the starting materials for this reaction. Fill in the molecules of the products using the balanced equation and following the law of conservation of matter.

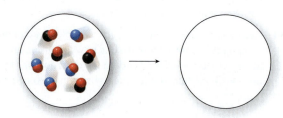

Types of Reactions

5.29 For the reaction depicted in the molecular art: (a) Use the common element colors shown in Appendix C to give a formula for each reactant and product. (b) Write a balanced equation. (c) Classify the reaction as a combination, decomposition, single replacement, or double replacement.

5.30 Classify the reaction depicted in the molecular art as combination, decomposition, single replacement, or double replacement.

5.31 Classify each reaction as combination, decomposition, single replacement, or double replacement.
a. $2\,Br{-}Cl \longrightarrow Br_2 + Cl_2$
b. $F_2 + 2\,NaBr \longrightarrow 2\,NaF + Br_2$
c. $4\,Fe + 3\,O_2 \longrightarrow 2\,Fe_2O_3$
d. $HNO_3 + KOH \longrightarrow KNO_3 + H_2O$

5.32 Fill in the needed reactant or product for each of the following reaction types, and balance the equation.
a. combination: $Na + \boxed{} \longrightarrow NaCl$
b. decomposition: $NI_3 \longrightarrow \boxed{} + \boxed{}$
c. single replacement: $Cl_2 + KI \longrightarrow \boxed{} + \boxed{}$
d. double replacement: $KOH + HI \longrightarrow \boxed{} + \boxed{}$

Oxidation–Reduction Reactions

5.33 Identify the species that is oxidized and the species that is reduced in each reaction. Write out two half reactions to show how many electrons are gained or lost by each species.
a. $Fe + Cu^{2+} \longrightarrow Fe^{2+} + Cu$
b. $Cl_2 + 2\,I^- \longrightarrow I_2 + 2\,Cl^-$
c. $2\,Na + Cl_2 \longrightarrow 2\,NaCl$

5.34 Identify the species that is oxidized and the species that is reduced in each reaction. Write out two half reactions to show how many electrons are gained or lost by each species.
a. $Mg + Fe^{2+} \longrightarrow Mg^{2+} + Fe$
b. $Cu^{2+} + Sn \longrightarrow Sn^{2+} + Cu$
c. $4\,Na + O_2 \longrightarrow 2\,Na_2O$

5.35 The reaction of magnesium metal (Mg) with oxygen (O_2) forms MgO. Write a balanced equation for this redox reaction. Write two half reactions to show how many electrons are gained or lost by each species.

5.36 When Cl_2 is used to disinfect drinking water, Cl^- is formed. Is Cl_2 oxidized or reduced in this process?

Formula Weight and Molar Mass

5.37 Calculate the formula weight and molar mass of each compound.
a. $NaNO_2$ (sodium nitrite), a preservative in hot dogs, ham, and other cured meats
b. C_2H_4 (ethylene), the industrial starting material for the plastic polyethylene
c. $C_9H_{13}NO_2$ (phenylephrine), a decongestant in Sudafed PE

5.38 L-Dopa is a drug used to treat Parkinson's disease.

L-dopa

a. What is the molecular formula of L-dopa?
b. What is the formula weight of L-dopa?
c. What is the molar mass of L-dopa?

Moles, Mass, and Avogadro's Number

5.39 How many grams are contained in 5.00 mol of each compound?
a. HCl   b. Na_2SO_4 c. C_2H_2

5.40 How many grams are contained in 0.50 mol of each compound?
a. NaOH   b. $CaSO_4$ c. C_3H_6

5.41 How many moles are contained in each number of grams of table sugar ($C_{12}H_{22}O_{11}$, molar mass 342.30 g/mol)?
a. 0.500 g   b. 5.00 g   c. 25.0 g

5.42 How many moles are contained in each number of grams of fructose ($C_6H_{12}O_6$, molar mass 180.15 g/mol), a carbohydrate that is about twice as sweet as table sugar? "Lite" food products use half as much fructose as table sugar to achieve the same sweet taste, but with fewer calories.
a. 0.500 g   b. 5.00 g   c. 25.0 g

5.43 Mescaline is a hallucinogen in peyote, a cactus native to the southwestern United States and Mexico.

mescaline

a. What is the chemical formula of mescaline?
b. Calculate its molar mass.
c. How many moles are contained in 7.50 g of mescaline?

5.44 (a) What is the chemical formula of **A**? (b) Calculate its molar mass. (c) How many moles are contained in 10.50 g of **A**?

A

Mass and Mole Calculations in Chemical Equations

5.45 Using the balanced equation for the combustion of acetylene, answer the following questions.

$$2\,H{-}C{\equiv}C{-}H + 5\,O_2 \longrightarrow 4\,CO_2 + 2\,H_2O$$
acetylene

a. How many moles of O_2 are needed to react completely with 5.00 mol of C_2H_2?
b. How many moles of CO_2 are formed from 6.0 mol of C_2H_2?
c. How many moles of H_2O are formed from 0.50 mol of C_2H_2?

5.46 Sodium metal (Na) reacts violently when added to water according to the following balanced equation.

$$2\ Na(s)\ +\ 2\ H_2O(l)\ \longrightarrow\ 2\ NaOH(aq)\ +\ H_2(g)$$

a. How many moles of H_2O are needed to react completely with 3.0 mol of Na?

b. How many moles of H_2 are formed from 0.38 mol of Na?

c. How many moles of H_2 are formed from 3.64 mol of H_2O?

5.47 Using the balanced equation for the combustion of acetylene in Problem 5.45, answer the following questions.

a. How many grams of CO_2 are formed from 2.5 mol of C_2H_2?

b. How many grams of CO_2 are formed from 0.50 mol of C_2H_2?

c. How many grams of H_2O are formed from 0.25 mol of C_2H_2?

5.48 Using the balanced equation for the reaction of Na with H_2O in Problem 5.46, answer the following questions.

a. How many grams of NaOH are formed from 3.0 mol of Na?

b. How many grams of H_2 are formed from 0.30 mol of Na?

c. How many grams of H_2O are needed to react completely with 0.20 mol of Na?

Energy Changes in Reactions

5.49 Do each of the following statements describe an endothermic or exothermic reaction?

a. ΔH is a negative value.

b. The energy of the reactants is lower than the energy of the products.

c. Energy is absorbed in the reaction.

5.50 Do each of the following statements describe an endothermic or exothermic reaction?

a. ΔH is a positive value.

b. The energy of the products is lower than the energy of the reactants.

c. Energy is released in the reaction.

5.51 The combustion of coal with oxygen forms CO_2 according to the given equation.

$$C(s)\ +\ O_2(g)\ \longrightarrow\ CO_2(g)\qquad \Delta H = -94\ kcal/mol$$

a. Is heat absorbed or released?

b. Are the reactants or products lower in energy?

c. Is the reaction endothermic or exothermic?

5.52 Ammonia (NH_3) is formed from hydrogen and nitrogen according to the given equation.

$$3\ H_2(g)\ +\ N_2(g)\ \longrightarrow\ 2\ NH_3(g)\qquad \Delta H = -22.0\ kcal/mol$$

a. Is heat absorbed or released?

b. Are the reactants or products lower in energy?

c. Is the reaction endothermic or exothermic?

5.53 The metabolism of glucose with oxygen forms CO_2 and H_2O and releases 678 kcal/mol of energy.

$$\underset{\text{glucose}}{C_6H_{12}O_6(aq)}\ +\ 6\ O_2(g)\ \longrightarrow\ 6\ CO_2(g)\ +\ 6\ H_2O(l)$$

a. Is ΔH for this reaction positive or negative?

b. Is the reaction endothermic or exothermic?

c. Are the reactants or products lower in energy?

5.54 Ethanol (C_2H_6O), a gasoline additive, is formed by the reaction of ethylene ($CH_2\!=\!CH_2$) with water. This reaction releases 9.0 kcal/mol of energy.

a. Is ΔH for this reaction positive or negative?

b. Is the reaction endothermic or exothermic?

c. Are the reactants or products lower in energy?

Energy Diagrams

5.55 Draw an energy diagram that fits each description.

a. an endothermic reaction with a high E_a

b. a reaction that has a low E_a and ΔH is negative

5.56 Draw an energy diagram for the following reaction in which $\Delta H = -12$ kcal/mol and $E_a = 5$ kcal: $A_2 + B_2 \longrightarrow 2\ AB$. Label the axes, reactants, products, transition state, E_a, and ΔH. Is the reaction endothermic or exothermic?

Reaction Rates

5.57 Which value (if any) in each pair corresponds to a faster reaction? Explain your choice.

a. $E_a = 10$ kcal or $E_a = 1$ kcal

b. $\Delta H = -2$ kcal/mol or $\Delta H = +2$ kcal/mol

5.58 Which of the following affect the rate of a reaction: (a) concentration; (b) ΔH; (c) energy difference between the reactants and the transition state?

5.59 Which of the following affect the rate of a reaction: (a) catalyst; (b) E_a; (c) temperature?

5.60 How does a catalyst affect each of the following: (a) reaction rate; (b) ΔH; (c) E_a; (d) relative energy of the reactants and products?

Equilibrium and Le Châtelier's Principle

5.61 Consider the reaction of $N_2(g) + O_2(g) \rightleftharpoons 2\ NO(g)$. What happens to the direction of equilibrium when (a) $[O_2]$ is increased; (b) [NO] is increased?

5.62 Consider the reaction of $H_2(g) + F_2(g) \rightleftharpoons 2\ HF(g)$. What happens to the direction of equilibrium when (a) $[H_2]$ is decreased; (b) [HF] is increased?

5.63 Consider the conversion of oxygen to ozone: $3\ O_2(g) \rightleftharpoons 2\ O_3(g)$. What effect does each of the following changes have on the direction of equilibrium?

a. decrease $[O_3]$

b. decrease $[O_2]$

c. increase $[O_3]$

5.64 Consider the reaction: $H_2(g) + I_2(g) \rightleftharpoons 2\ HI(g)$. What effect does each of the following changes have on the direction of equilibrium?

a. decrease [HI]

b. increase $[H_2]$

c. decrease $[I_2]$

General Questions and Applications

5.65 Answer the following questions about the conversion of the sucrose ($C_{12}H_{22}O_{11}$) in sugarcane to ethanol (C_2H_6O) and CO_2 according to the following balanced equation. In this way sugarcane is used as a renewable source of ethanol, which is used as a fuel additive in gasoline.

$$C_{12}H_{22}O_{11}(s) \ + \ H_2O(l) \ \longrightarrow \ 4 \ C_2H_6O(l) \ + \ 4 \ CO_2(g)$$
$$\text{sucrose} \qquad\qquad\qquad \text{ethanol}$$

a. What is the molar mass of sucrose?

b. How many moles of ethanol are formed from 2 mol of sucrose?

c. How many grams of ethanol are formed from 34.2 g of sucrose?

5.66 Answer the following questions about diethyl ether ($C_4H_{10}O$), the first widely used general anesthetic. Diethyl ether can be prepared from ethanol according to the following unbalanced equation.

$$C_2H_6O(l) \ \longrightarrow \ C_4H_{10}O(l) \ + \ H_2O(l)$$
$$\text{ethanol} \qquad \text{diethyl ether}$$

a. What is the molar mass of diethyl ether?

b. Balance the given equation.

c. How many moles of diethyl ether are formed from 2 mol of ethanol?

d. How many grams of diethyl ether are formed from 4.60 g of ethanol?

5.67 One dose of Maalox contains 500. mg each of $Mg(OH)_2$ and $Al(OH)_3$. How many moles of each compound are contained in a single dose?

Jill Braaten/McGraw Hill

5.68 A bottle of the pain reliever ibuprofen ($C_{13}H_{18}O_2$, molar mass 206.29 g/mol) contains 500 200.-mg tablets.

a. How many moles of ibuprofen does the bottle contain?

b. How many molecules of ibuprofen does the bottle contain?

Jill Braaten

CHALLENGE PROBLEMS

5.69 TCDD, also called dioxin ($C_{12}H_4Cl_4O_2$, molar mass 321.97 g/mol), is a potent poison. The average lethal dose in humans is estimated to be 3.0×10^{-2} mg per kg of body weight. (a) How many grams constitute a lethal dose for a 70.-kg individual? (b) How many molecules of TCDD does this correspond to?

5.70 The amount of energy released when a fuel burns is called its heat content. The heat content of fuels is often reported in kcal/g not kcal/mol so that fuels with different molar masses can be compared on a mass basis. The heat content of propane (C_3H_8), used as the fuel in gas grills, is 531 kcal/mol, whereas the heat content of butane (C_4H_{10}), used in lighters, is 688 kcal/mol. Show that the heat content of these two fuels is similar when converted to kcal/g.

BEYOND THE CLASSROOM

5.71 The combustion of gasoline is a redox reaction in which O_2 oxidizes the C—C and C—H bonds in the gasoline to form carbon dioxide (CO_2) and water (H_2O). One estimate suggests that driving an automobile 10,000 miles per year at 25 miles per gallon releases 10,000 lb of CO_2 into the atmosphere. Using these values as conversion factors, calculate how many pounds of CO_2 your family releases into the atmosphere given your car's gas mileage and the distance it is driven each year. Convert this value to the number of moles of CO_2, as well as the number of molecules of CO_2. Compare your result with others in the class whose driving habits are different from yours.

5.72 A 12-oz can of diet soda contains 180 mg of the artificial sweetener aspartame (trade names NutraSweet and Equal). When aspartame is ingested, it is metabolized to phenylalanine, aspartic acid, and methanol, according to the balanced equation:

$$C_{14}H_{18}N_2O_5 + 2 H_2O \longrightarrow C_9H_{11}NO_2 + C_4H_7NO_4 + CH_3OH$$

aspartame   phenylalanine aspartic   methanol
                                            acid

a. Using this equation, determine how much methanol is formed on metabolism of the aspartame in each 12-oz can. The molar mass of aspartame is 294.30 g/mol, and the molar mass of methanol is 32.04 g/mol.

b. Methanol is a toxic compound, with the average lethal dose in humans estimated at 428 mg/kg of body weight. How many cans of soda would you have to ingest to obtain a lethal dose of methanol? Compare your results with others in your class or family.

c. Estimate how many cans of diet soda you drink in a 30-day month, and determine how much methanol you consume during that interval.

CHAPTER 5 ANSWERS

Concept Check Answers

5.1 $CH_4(g) + 4 Cl_2(g) \longrightarrow CCl_4(l) + 4 HCl(g)$

5.2 decomposition: $2 NCl_3 \longrightarrow N_2 + 3 Cl_2$

5.3 a,c: combination   b,d: decomposition

5.4 a,d: single replacement   b,c: double replacement

5.5 Zn (oxidized), Hg^{2+} (reduced)

$$Zn \longrightarrow Zn^{2+} + 2 e^- \qquad Hg^{2+} + 2 e^- \longrightarrow Hg$$

5.6 $C_2H_4O_2$ reduced, H_2 oxidized

5.7 Zn reducing agent, Hg^{2+} oxidizing agent

5.8 a,b,c,d: 6.022×10^{23}

5.9 a. 100.09 amu   b. 166.00 amu

5.10 a. absorbed   b. reactants   c. endothermic

5.11 a. released   b. products   c. exothermic

5.12 a,c: increases   b,d: decreases

5.13 a. none   b. lowers E_a

5.14 $2 NO_2 \longrightarrow N_2 + 2 O_2$

5.15 a. forward: $2 SO_2(g) + O_2(g) \longrightarrow 2 SO_3(g)$

reverse: $2 SO_3(g) \longrightarrow 2 SO_2(g) + O_2(g)$

b. forward: $N_2(g) + O_2(g) \longrightarrow 2 NO(g)$

reverse: $2 NO(g) \longrightarrow N_2(g) + O_2(g)$

Practice Problem Answers

5.1 chemical reaction

5.2 physical change, freezing

5.3 a. reactant: $2 H_2O$, products: $2 H_2O + O_2$ (4 H, 4 O)

b. reactants: $C_8H_{18} + 25 O_2$, products $16 CO_2 + 18 H_2O$ (16 C, 50 O, 36 H)

5.4 $2 CH_4 + 3 O_2 \longrightarrow 2 CO + 4 H_2O$

5.5 a. $2 H_2 + O_2 \longrightarrow 2 H_2O$

b. $2 NO + O_2 \longrightarrow 2 NO_2$

c. $4 Fe + 3 O_2 \longrightarrow 2 Fe_2O_3$

d. $CH_4 + 2 Cl_2 \longrightarrow CH_2Cl_2 + 2 HCl$

5.6 $2 CO + O_2 \longrightarrow 2 CO_2$

5.7 $2 C_2H_6 + 7 O_2 \longrightarrow 4 CO_2 + 6 H_2O$

5.8 $N_2 + 3 H_2 \longrightarrow 2 NH_3$

5.9 a. $2 Al + 3 H_2SO_4 \longrightarrow Al_2(SO_4)_3 + 3 H_2$

b. $3 Na_2SO_3 + 2 H_3PO_4 \longrightarrow 3 H_2SO_3 + 2 Na_3PO_4$

5.10 a. decomposition   c. combination

b. single replacement   d. double replacement

5.11 a,b: $C_2H_6O \longrightarrow C_2H_4 + H_2O$ c. decomposition

5.12 a. Zn (oxidized), $2 H^+$ reduced

$$Zn \longrightarrow Zn^{+2} + 2 e^- \qquad 2 H^+ + 2 e^- \longrightarrow H_2$$

b. Fe^{3+} (reduced), Al (oxidized)

$$Al \longrightarrow Al^{3+} + 3 e^- \qquad Fe^{3+} + 3 e^- \longrightarrow Fe$$

c. $2 I^-$ (oxidized), Br_2 (reduced)

$$2 I^- \longrightarrow I_2 + 2 e^- \qquad Br_2 + 2 e^- \longrightarrow 2 Br^-$$

5.13 a. 1.20×10^{24} atoms   c. 3.01×10^{23} atoms

b. 3.61×10^{24} atoms   d. 1.51×10^{25} atoms

5.14 a. 1.5×10^{24} molecules   c. 2.4×10^{23} molecules

b. 1.5×10^{23} molecules   d. 3.33×10^{25} molecules

5.15 a. 100. mol   b. 0.0500 mol   c. 15 mol

5.16 a. 73.89 g/mol   b. 64.51 g/mol   c. 239.3 g/mol

5.17 285.33 g/mol

5.18 a. 29.2 g   b. 332 g   c. 101 g   d. 26.3 g

5.19 a. 1.71 mol   c. 1.39×10^{-3} mol

b. 1.59 mol   d. 1.39 mol

5.20 5.678×10^{-3} mol

5.21 a. 6.6 mol   b. 1.0 mol   c. 1.2 mol

5.22 a. 7.5 mol   b. 1.5 mol   c. 1.5 mol

5.23 a. 51 g   b. 22 g   c. 90. g

5.24 a. 44 g   b. 130 g   c. 24 g

5.25 a. 21.4 g   b. 18.8 g   c. 11.4 g

5.26

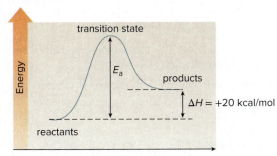

5.27 a,d: right   b,c: left

Self-Test Answers

1. a. energy of activation

   b. molar mass

   c. double replacement

   d. Le Châtelier's principle

   e. heat of reaction, enthalpy change

2. a,d: to the right   b,c: to the left

3. a. products   b. exothermic   c. stronger

4. 176.12 g/mol

5. 4.82×10^{24} molecules

6. 80. g

7. Na oxidized, Cl_2 reduced

8. a. 1   b. 3   c. 3   d. 2

9. a. single replacement   c. double replacement
   b. decomposition   d. combination

Odd-Numbered Problem Answers

5.1 $2\,CO + 2\,O_3 \longrightarrow 2\,CO_2 + 2\,O_2$ (not balanced)

5.3 $H_2 + Cl_2 \longrightarrow 2\,HCl$

5.5 $2\,S(s) + 3\,O_2(g) + 2\,H_2O(l) \longrightarrow 2\,H_2SO_4(l)$

5.7 a. reactants: 4 CO + 2 O_2, products: 4 CO_2
   b. $2\,CO + O_2 \longrightarrow 2\,CO_2$
   c. combination

5.9 a. 324 g   b. 1.1×10^3 g

5.11 a. 278.42 g/mol

   b. $C_{18}H_{30}O_2 + 3\,H_2 \longrightarrow C_{18}H_{36}O_2$

   c. $2\,C_{18}H_{30}O_2 + 49\,O_2 \longrightarrow 36\,CO_2 + 30\,H_2O$

   d. 10.2 g

5.13 Zn: oxidized, reducing agent; Ag^+: reduced, oxidizing agent

5.15 a. X   b. Z   c. Y   d. X,Y   e. X,Z   f. Y   g. Z

5.17 right

5.19 a. 2 H, 2 Cl, 1 Ca on both sides; therefore balanced

   b. 1 Al, 1 P, 7 O, 6 H on both sides; therefore balanced

5.21

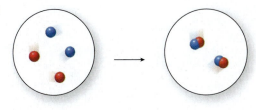

5.23 a. $Ni(s) + 2\,HCl(aq) \longrightarrow NiCl_2(aq) + H_2(g)$

   b. $CH_4(g) + 4\,Cl_2(g) \longrightarrow CCl_4(g) + 4\,HCl(g)$

   c. $2\,KClO_3 \longrightarrow 2\,KCl + 3\,O_2$

5.25 a. $2\,C_6H_6 + 15\,O_2 \longrightarrow 12\,CO_2 + 6\,H_2O$

   b. $C_7H_8 + 9\,O_2 \longrightarrow 7\,CO_2 + 4\,H_2O$

5.27

5.29 a. reactants: $2\,N_2 + 4\,O_2$, product: $4\,NO_2$
   b. $N_2 + 2\,O_2 \longrightarrow 2\,NO_2$
   c. combination

5.31 a. decomposition   c. combination
   b. single replacement   d. double replacement

5.33 a. Fe (oxidized), Cu^{2+} (reduced)

   $Fe \longrightarrow Fe^{2+} + 2\,e^-$ $Cu^{2+} + 2\,e^- \longrightarrow Cu$

   b. Cl_2 (reduced), $2\,I^-$ (oxidized)

   $2\,I^- \longrightarrow I_2 + 2\,e^-$ $Cl_2 + 2\,e^- \longrightarrow 2\,Cl^-$

   c. 2 Na (oxidized), Cl_2 (reduced)

   $2\,Na \longrightarrow 2\,Na^+ + 2\,e^-$ $Cl_2 + 2\,e^- \longrightarrow 2\,Cl^-$

5.35 $2\,Mg + O_2 \longrightarrow 2\,MgO$
   $2\,Mg \longrightarrow 2\,Mg^{2+} + 4\,e^-$ $O_2 + 4\,e^- \longrightarrow 2\,O^{2-}$

5.37 a. 69.00 amu, 69.00 g/mol   c. 167.2 amu, 167.2 g/mol

   b. 28.05 amu, 28.05 g/mol

5.39 a. 182 g   b. 710. g   c. 130. g

5.41 a. 1.46×10^{-3} mol   c. 0.0730 mol

      b. 0.0146 mol

5.43 a. $C_{11}H_{17}NO_3$ b. 211.26 g/mol   c. 0.0355 mol

5.45 a. 12.5 mol   b. 12 mol   c. 0.50 mol

5.47 a. 220 g   b. 44 g   c. 4.5 g

5.49 a. exothermic   b,c: endothermic

5.51 a. released   b. products   c. exothermic

5.53 a. negative   b. exothermic   c. products

5.55 a.

b.

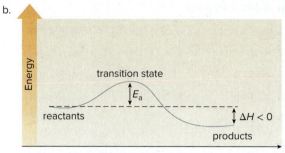

5.57 a. The reaction with $E_a = 1$ kcal will proceed faster because the energy of activation is lower.

      b. One cannot predict which reaction will proceed faster from the ΔH.

5.59 a, b, and c

5.61 a. proceeds to right   b. proceeds to left

5.63 a. proceeds to right   b,c: proceeds to left

5.65 a. 342.30 g/mol   b. 8 mol   c. 18.4 g

5.67 8.57×10^{-3} mol $Mg(OH)_2$ and 6.41×10^{-3} mol $Al(OH)_3$

5.69 a. 2.1×10^{-3} g   b. 3.9×10^{18} molecules

Gases

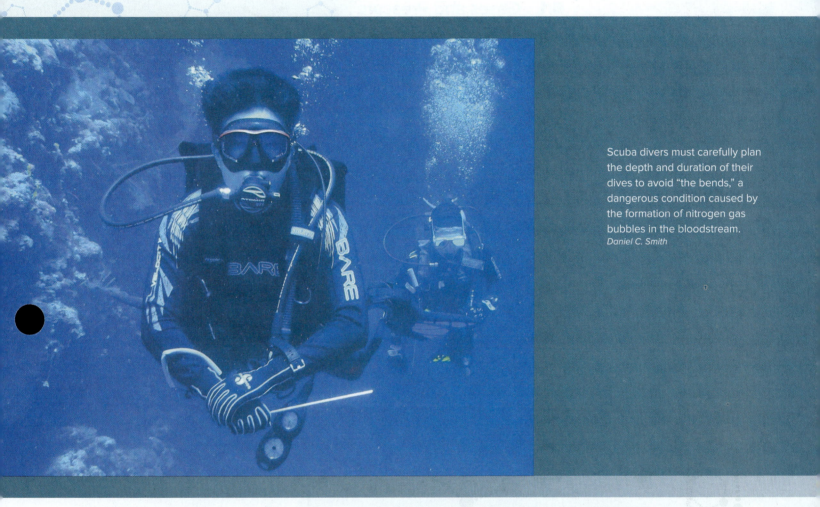

Scuba divers must carefully plan the depth and duration of their dives to avoid "the bends," a dangerous condition caused by the formation of nitrogen gas bubbles in the bloodstream.
Daniel C. Smith

CHAPTER OUTLINE

CHAPTER GOALS

In this chapter you will learn how to:

1. Measure pressure and convert one unit of pressure to another

2. Describe the relationship between the pressure, volume, and temperature of a gas using gas laws

3. Describe the relationship between the volume and number of moles of a gas

4. Write the equation for the ideal gas law and use it in calculations

5. Use Dalton's law to determine the partial pressure and total pressure of a gas mixture

6. Understand the importance of two minor components of the earth's atmosphere, ozone and carbon dioxide

Why Study ...

Gases?

In Chapter 6 we study the properties of gases. Why is air pulled into the lungs when we expand our rib cage and diaphragm? Why does a lid pop off a container of food when it is heated in the microwave? Why is a hyperbaric chamber used to treat a scuba diver suffering from the bends, a painful condition that may result from surfacing too quickly? To answer questions of this sort, we must understand the particulate properties of the gas state, as well as the laws that govern the behavior of all gases.

6.1 Gases and Pressure

Anyone who has ridden a bike against the wind knows that even though we can't see the gas molecules of the air, we can feel them as we move through them. Air is a mixture of 78% nitrogen (N_2), 21% oxygen (O_2), and 1% other gases, including carbon dioxide (CO_2), argon (Ar), water (H_2O), and ozone (O_3) (Figure 6.1).

Simple gases in the atmosphere—oxygen (O_2), carbon dioxide (CO_2), and ozone (O_3)—are vital to life. Oxygen, which constitutes 21% of the earth's atmosphere, is needed for metabolic processes that convert carbohydrates to energy. Green plants use carbon dioxide, a minor component of the atmosphere, to store the energy of the sun in the bonds of carbohydrate molecules during photosynthesis. Ozone forms a protective shield in the upper atmosphere to filter out harmful radiation from the sun, thus keeping it from the surface of the earth (Section 6.9).

6.1A Properties of Gases

Helium, a noble gas composed of He atoms, and oxygen, a gas composed of diatomic O_2 molecules, behave differently in chemical reactions. Many of their physical properties, however, and the properties of all gases, can be explained by the **kinetic–molecular theory of gases,** a set of principles based on the following assumptions:

- A gas consists of particles—atoms or molecules—that move randomly and rapidly.
- The size of gas particles is negligible compared to the space between the particles.
- Because the space between gas particles is large, gas particles exert no attractive forces on each other.
- The kinetic energy of gas particles increases with increasing temperature.
- When gas particles collide with each other, they rebound and travel in new directions. When gas particles collide with the walls of a container, they exert a pressure.

Because gas particles move rapidly, two gases mix together quickly. Moreover, when a gas is added to a container, the particles rapidly move to fill the entire container.

6.1B Gas Pressure

When many gas molecules strike a surface, they exert a measurable pressure. **Pressure (P) is the force (F) exerted per unit area (A) when gas particles hit a surface.**

$$\text{Pressure} = \frac{\text{Force}}{\text{Area}} = \frac{F}{A}$$

Gas pressure depends on the number of collisions of gas particles. Factors that *increase* the number of collisions *increase* pressure. All of the gases in the atmosphere collectively exert **atmospheric pressure** on the surface of the earth. The value of the atmospheric pressure varies with location, decreasing with increasing altitude. Atmospheric pressure also varies slightly from day to day, depending on the weather.

Atmospheric pressure is measured with a **barometer** (Figure 6.2). A barometer consists of a column of mercury (Hg) sealed at one end and inverted in a dish of mercury. The downward pressure exerted by the mercury in the column equals the atmospheric pressure on the mercury

Figure 6.1

Composition of the Atmosphere

N_2 78%

O_2 21%

1% $\begin{bmatrix} CO_2 \\ H_2O \\ Ar \\ O_3 \end{bmatrix}$

Figure 6.2

A Barometer—A Device for Measuring Atmospheric Pressure

vacuum

atmospheric pressure

760 mm

mercury-filled dish

A barometer measures atmospheric pressure. Air pressure on the Hg in the dish pushes Hg up a sealed glass tube to a height that equals the atmospheric pressure.

in the dish. Thus, the height of the mercury in the column measures the atmospheric pressure. Atmospheric pressure at sea level corresponds to a column of mercury 760. mm in height.

Many different units are used for pressure as shown in Table 6.1. The two most common units are the **atmosphere (atm)** and **millimeters of mercury (mm Hg)**, where **1 atm = 760. mm Hg.** One millimeter of mercury is also called one **torr.** In the United States, the common pressure unit is **pounds per square inch (psi)**, where **1 atm = 14.70 psi.** Pressure can also be measured in pascals (Pa), where 1 mm Hg = 133.32 Pa.

Table 6.1 Common Units to Measure Pressure

Unit	Abbreviation	Unit Equivalent to 1 atm
Atmosphere	atm	1 atm
Millimeters of mercury	mm Hg	760. mm Hg
Torr	torr	760. torr
Pounds per square inch	psi	14.70 psi
Pascal	Pa	101,325 Pa

To convert a value from one pressure unit to another, set up conversion factors and use the method in Sample Problem 6.1.

Sample Problem 6.1   Converting One Unit of Pressure to Another

A scuba diver typically begins a dive with a compressed air tank at 3,000. psi. Convert this value to (a) atmospheres; (b) mm Hg.

Analysis

To solve each part, set up conversion factors that relate the two units under consideration. Use conversion factors that place the unwanted unit, psi, in the denominator to cancel.

In part (a), the conversion factor must relate psi and atm:

psi–atm conversion factor

$$\frac{1 \text{ atm}}{14.70 \text{ psi}} \leftarrow \text{unwanted unit}$$

In part (b), the conversion factor must relate psi and mm Hg:

psi–mm Hg conversion factor

$$\frac{760. \text{ mm Hg}}{14.70 \text{ psi}} \leftarrow \text{unwanted unit}$$

Solution

a. Convert the original unit (3,000. psi) to the desired unit (atm) using the conversion factor:

$$3000. \text{ psi} \times \frac{1 \text{ atm}}{14.70 \text{ psi}} = 204.1 \text{ atm}$$

Psi cancels.

Answer

b. Convert the original unit (3,000. psi) to the desired unit (mm Hg) using the conversion factor:

$$3000. \text{ psi} \times \frac{760. \text{ mm Hg}}{14.70 \text{ psi}} = 155,100 \text{ mm Hg}$$

Psi cancels.

Answer

Practice Problem 6.1

Convert each pressure unit to the indicated unit.

   a. 3.0 atm to mm Hg   b. 720 mm Hg to psi   c. 424 mm Hg to atm

More Practice: Try Problems 6.1, 6.2, 6.15–6.18.

A scuba diver's pressure gauge shows the amount of air (usually measured in psi) in his tank before and during a dive.
Daniel C. Smith

6.1C FOCUS ON HEALTH & MEDICINE
Blood Pressure

Taking a patient's blood pressure is an important part of most physical examinations. Blood pressure measures the pressure in an artery of the upper arm using a device called a **sphygmomanometer.** A blood pressure reading consists of two numbers such as 120/80, where both values represent pressures in mm Hg. The higher number is the systolic pressure and refers to the maximum pressure in the artery right after the heart contracts. The lower number is the diastolic pressure and represents the minimum pressure when the heart muscle relaxes. A desirable systolic pressure is in the range of 100–120 mm Hg. A desirable diastolic pressure is in the range of 60–80 mm Hg. Figure 6.3 illustrates how a sphygmomanometer records pressure in a blood vessel.

Figure 6.3 Measuring Blood Pressure

(a) To measure blood pressure, a cuff is inflated around the upper arm and a stethoscope is used to listen to the sound of blood flowing through the brachial artery. When the pressure in the cuff is high, it constricts the artery, so that no blood can flow to the lower arm.
(b) Slowly the pressure in the cuff is decreased, and when it gets to the point that blood begins to spurt into the artery, a tapping sound is heard in the stethoscope. This value corresponds to the systolic blood pressure. (c) When the pressure in the cuff is further decreased, so that blood once again flows freely in the artery, the tapping sound disappears and the diastolic pressure is recorded.

When a patient's systolic pressure is routinely 140 mm Hg or greater or diastolic pressure is 90 mm Hg or greater, an individual is said to have **hypertension**—that is, high blood pressure. Consistently high blood pressure leads to increased risk of stroke and heart attacks.

Concept Check 6.1	Suppose blood pressure readings were reported in cm Hg rather than mm Hg. If this were the case, how would the pressure 140/90 be reported?

6.2 Boyle's Law—How the Pressure and Volume of a Gas Are Related

Four variables are important in discussing the behavior of gases—**pressure (P), volume (V), temperature (T),** and **number of moles (n).** The relationship of these variables is described by equations called **gas laws** that explain and predict the behavior of all gases as conditions change. Three gas laws illustrate the interrelationship of pressure, volume, and temperature.

- Boyle's law relates pressure and volume (Section 6.2).
- Charles's law relates volume and temperature (Section 6.3).
- Gay–Lussac's law relates pressure and temperature (Section 6.4).

Boyle's law describes how the volume of a gas changes as the pressure is changed.

- Boyle's law: For a fixed amount of gas at constant temperature, the pressure and volume of a gas are *inversely* related.

When two quantities are *inversely* related, one quantity *increases* as the other *decreases*. **The product of the two quantities, however, is a *constant*, symbolized by *k*.**

When pressure *increases...*

$$\text{Pressure} \quad \times \quad \text{Volume} \quad = \quad \text{constant}$$

...volume *decreases.*

$$PV \quad = \quad k$$

If the volume of a cylinder of gas is *decreased*, the particles of gas are more crowded together, more collisions occur, and pressure *increases*. Thus, if the volume of the cylinder is halved, the pressure of the gas inside the cylinder doubles. The same number of gas particles occupies *half* the volume and exerts *two* times the pressure.

$P = 1$ atm

Decrease volume.

$P = 2$ atm

Pressure increases.

$V = 2$ L

$V = 1$ L

If we know the pressure and volume under an initial set of conditions (P_1 and V_1), we can calculate the pressure or volume under a different set of conditions (P_2 and V_2), because the product of pressure and volume is a constant.

$$P_1V_1 \quad = \quad P_2V_2$$

initial conditions   new conditions

How To Use Boyle's Law to Calculate a New Gas Volume or Pressure

Example If a 4.0-L container of helium gas has a pressure of 10.0 atm, what pressure does the gas exert if the volume is increased to 6.0 L?

Step [1] Identify the known quantities and the desired quantity.

- To solve an equation using Boyle's law, we must know three quantities and solve for one quantity. In this case P_1, V_1, and V_2 are known and the final pressure, P_2, must be determined.

$P_1 = 10.0$ atm

$V_1 = 4.0$ L $V_2 = 6.0$ L $P_2 = ?$

known quantities   desired quantity

—*Continued*

How To, continued . . .

Step [2] Write the equation and rearrange it to isolate the desired quantity on one side.

- Rearrange the equation for Boyle's law so that the unknown quantity, P_2, is present alone on one side.

$$P_1V_1 \;=\; P_2V_2 \qquad \text{Solve for } P_2 \text{ by dividing both sides by } V_2.$$

$$\frac{P_1V_1}{V_2} \;=\; P_2$$

Step [3] Solve the problem.

- Substitute the known quantities into the equation and solve for P_2. Identical units must be used for two similar quantities (liters in this case) so that the units cancel.

$$P_2 \;=\; \frac{P_1V_1}{V_2} \;=\; \frac{(10.0 \text{ atm})(4.0 \, \cancel{L})}{6.0 \, \cancel{L}} \;=\; 6.7 \text{ atm}$$

Liters cancel. **Answer**

- In this example, the volume *increased* so the pressure *decreased*.

Sample Problem 6.2   **Determining Gas Volume Using Boyle's Law**

A tank of compressed air for scuba diving contains 8.5 L of gas at 204 atm pressure. What volume of air does this gas occupy at 1.0 atm?

Analysis

Boyle's law can be used to solve this problem because an initial pressure and volume (P_1 and V_1) and a final pressure (P_2) are known, and a final volume (V_2) must be determined.

Solution

[1] Identify the known quantities and the desired quantity.

$$P_1 = 204 \text{ atm} \qquad\qquad P_2 = 1.0 \text{ atm}$$
$$V_1 = 8.5 \text{ L} \qquad\qquad\qquad\qquad V_2 = \,?$$

known quantities   desired quantity

[2] Write the equation and rearrange it to isolate the desired quantity, V_2, on one side.

$$P_1V_1 \;=\; P_2V_2 \qquad \text{Solve for } V_2 \text{ by dividing both sides by } P_2.$$

$$\frac{P_1V_1}{P_2} \;=\; V_2$$

[3] Solve the problem.

- Substitute the three known quantities into the equation and solve for V_2.

$$V_2 \;=\; \frac{P_1V_1}{P_2} \;=\; \frac{(204 \, \cancel{\text{atm}})(8.5 \text{ L})}{1.0 \, \cancel{\text{atm}}} \;=\; 1{,}734 \text{ rounded to } 1{,}700 \text{ L}$$

Atm cancels.   **Answer**

- Thus, the volume *increased* because the pressure *decreased*.

Practice Problem 6.2   A sample of helium gas has a volume of 2.0 L at a pressure of 4.0 atm. What is the volume of gas at each of the following pressures?

a. 5.0 atm   b. 2.5 atm   c. 10.0 atm   d. 380 mm Hg

Practice Problem 6.3

A sample of nitrogen gas has a volume of 15.0 mL at a pressure of 0.50 atm. What is the pressure exerted by the gas if the volume is changed to each of the following values?

   a. 30.0 mL   b. 5.0 mL   c. 100. mL   d. 1.0 L

More Practice: Try Problems 6.5, 6.6, 6.19–6.24, 6.56, 6.57.

Boyle's law explains how air is brought into or expelled from the lungs as the rib cage and diaphragm expand and contract when we breathe (Figure 6.4).

Figure 6.4   Focus on the Human Body: Boyle's Law and Breathing

a.

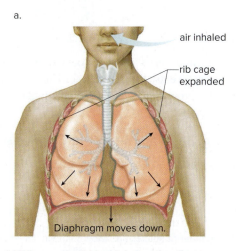

air inhaled

rib cage
expanded

Diaphragm moves down.

b.

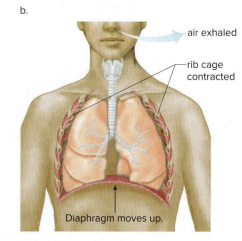

air exhaled

rib cage
contracted

Diaphragm moves up.

(a) When an individual inhales, the rib cage expands and the diaphragm is lowered, thus increasing the volume of the lungs. According to Boyle's law, *increasing* the volume of the lungs *decreases* the pressure inside the lungs. The decrease in pressure draws air into the lungs.

(b) When an individual exhales, the rib cage contracts and the diaphragm is raised, thus decreasing the volume of the lungs. Because the volume is now *decreased,* the pressure inside the lungs *increases,* causing air to be expelled into the surroundings.

A hot air balloon illustrates Charles's law. Heating the air inside the balloon causes it to expand and fill the balloon. When the air inside the balloon becomes less dense than the surrounding air, the balloon rises. *Open Door/Alamy Stock Photo*

6.3 Charles's Law—How the Volume and Temperature of a Gas Are Related

All gases expand when they are heated and contract when they are cooled. Charles's law describes how the volume of a gas changes as the Kelvin temperature is changed.

- Charles's law: For a fixed amount of gas at constant pressure, the volume of a gas is *proportional* to its Kelvin temperature.

Volume and temperature are *proportional;* that is, as one quantity *increases,* the other *increases* as well. Thus, **dividing volume by temperature is a constant (*k*).**

$$\frac{V}{T} = k$$

Increasing the temperature increases the kinetic energy of the gas particles, and they move faster and spread out, thus occupying a larger volume. Note that **Kelvin temperature** must be used in calculations involving gas laws. Any temperature reported in °C or °F must be converted to kelvins (K) prior to carrying out the calculation.

In molecular art, **longer "tails"** on spheres indicate **faster-moving particles.**

Increase temperature.

$V = 1\,L$

$T = 200\,K$

$V = 2\,L$

Volume increases.

$T = 400\,K$

Because dividing the volume of a gas by the temperature gives a constant, knowing the volume and temperature under an initial set of conditions (V_1 and T_1) means we can calculate the volume or temperature under another set of conditions (V_2 and T_2) when either volume or temperature is changed.

Equations for converting one temperature unit to another are given in Section 1.9.

$$\frac{V_1}{T_1} = \frac{V_2}{T_2}$$

initial conditions   new conditions

To solve a problem of this sort, we follow the same three steps listed in the *How To* outlined in Section 6.2, except we use the equation for Charles's law in step [2] in place of the equation for Boyle's law. This procedure is illustrated in Sample Problem 6.3.

Sample Problem 6.3

Determining Gas Volume Using Charles's Law

A balloon that contains 0.50 L of air at 25 °C is cooled to −196 °C. What volume does the balloon now occupy?

Analysis

Because this question deals with volume and temperature, Charles's law is used to determine a final volume because three quantities are known—the initial volume and temperature (V_1 and T_1), and the final temperature (T_2).

Solution

[1] **Identify the known quantities and the desired quantity.**

$V_1 = 0.50\,L$
$T_1 = 25\,°C$ $T_2 = -196\,°C$ $V_2 = ?$
        known quantities   desired quantity

- Both temperatures must be converted to Kelvin temperatures using the equation
  $T_K = T_C + 273.15.$
- $T_1 = 25\,°C + 273.15 = 298\,K$
- $T_2 = -196\,°C + 273.15 = 77\,K$

[2] **Write the equation and rearrange it to isolate the desired quantity, V_2, on one side.**

- Use Charles's law.

$$\frac{V_1}{T_1} = \frac{V_2}{T_2}$$ Solve for V_2 by multiplying both sides by T_2.

$$\frac{V_1 T_2}{T_1} = V_2$$

[3] **Solve the problem.**

- Substitute the three known quantities into the equation and solve for V_2.

$$V_2 = \frac{V_1 T_2}{T_1} = \frac{(0.50\,L)(77\,K)}{298\,K} = 0.13\,L$$

Answer

Kelvins cancel.

- Because the temperature has *decreased*, the volume of gas must *decrease* as well.

Practice Problem 6.4	A volume of 0.50 L of air at 37 °C is expelled from the lungs into cold surroundings at 0.0 °C. What volume does the expelled air occupy at this temperature?
Practice Problem 6.5	(a) A volume (25.0 L) of gas at 45 K is heated to 450 K. What volume does the gas now occupy? (b) A volume (50.0 mL) of gas at 400. °C is cooled to 50. °C. What volume does the gas now occupy?
Practice Problem 6.6	Calculate the Kelvin temperature to which 10.0 L of a gas at 27 °C would have to be heated to change the volume to 12.0 L.

More Practice: Try Problems 6.25–6.28.

Charles's law can be used to explain how wind currents form at the beach (Figure 6.5). The air above land heats up faster than the air above water. As the temperature of the air above the land increases, the volume that it occupies increases; that is, the air expands, and as a result, its density decreases. This warmer, less dense air then rises, and the cooler denser air above the water moves toward the land as wind, filling the space left vacant by the warm, rising air.

Figure 6.5 Focus on the Environment: How Charles's Law Explains Wind Currents

The same number of air molecules occupy a smaller volume, giving cooler air a higher density.

Warmer air rising

Heated air expands (Charles's law), so its density decreases. The hot air rises.

to take warm air's place

Cooler air moving

Land

Sea

6.4 Gay–Lussac's Law—How the Pressure and Temperature of a Gas Are Related

Gay–Lussac's law describes how the pressure of a gas changes as the Kelvin temperature is changed.

- **Gay–Lussac's law: For a fixed amount of gas at constant volume, the pressure of a gas is *proportional* to its Kelvin temperature.**

Pressure and temperature are *proportional;* that is, as one quantity *increases,* the other *increases.* Thus, **dividing the pressure by the temperature is a constant (k).**

$$\frac{P}{T} = k$$

Increasing the temperature increases the kinetic energy of the gas particles, thus increasing both the number of collisions and the force they exert. If the volume is kept constant, the pressure exerted by the particles increases.

$P = 1$ atm   Increase temperature. $P = 2$ atm

Pressure increases.

$V = 2$ L $V = 2$ L
$T = 200$ K $T = 400$ K

Because dividing the pressure of a gas by the temperature gives a constant, knowing the pressure and Kelvin temperature under an initial set of conditions (P_1 and T_1) means we can calculate the pressure or temperature under another set of conditions (P_2 and T_2) when either pressure or temperature is changed.

$$\frac{P_1}{T_1} = \frac{P_2}{T_2}$$

initial conditions   new conditions

How can you use Gay–Lussac's law to answer the question posed at the beginning of the chapter: Why does a lid pop off a container of food when it is heated in a microwave?

We solve this type of problem by following the same three steps in the *How To* in Section 6.2, using the equation for Gay–Lussac's law in step [2].

Sample Problem 6.4

Determining Pressure Using Gay–Lussac's Law

The tire on a bicycle stored in a cool garage at 18 °C had a pressure of 80. psi. What is the pressure inside the tire after riding the bike at 43 °C?

Analysis

Because this question deals with pressure and temperature, Gay–Lussac's law is used to determine a final pressure because three quantities are known—the initial pressure and temperature (P_1 and T_1), and the final temperature (T_2).

Solution

[1] **Identify the known quantities and the desired quantity.**

$P_1 = 80.$ psi
$T_1 = 18$ °C $T_2 = 43$ °C $P_2 = ?$
known quantities   desired quantity

- Both temperatures must be converted to Kelvin temperatures.
- $T_1 = T_c + 273.15 = 18\,°C + 273.15 = 291$ K
- $T_2 = T_c + 273.15 = 43\,°C + 273.15 = 316$ K

[2] **Write the equation and rearrange it to isolate the desired quantity, P_2, on one side.**
- Use Gay–Lussac's law. Because the initial pressure is reported in psi, the final pressure will be calculated in psi.

$$\frac{P_1}{T_1} = \frac{P_2}{T_2}$$ Solve for P_2 by multiplying both sides by T_2.

$$\frac{P_1 T_2}{T_1} = P_2$$

[3] **Solve the problem.**
- Substitute the three known quantities into the equation and solve for P_2.

$$P_2 = \frac{P_1 T_2}{T_1} = \frac{(80.\text{ psi})(316\text{ K})}{291\text{ K}} = 87\text{ psi}$$

Kelvins cancel. **Answer**

- Because the temperature has *increased,* the pressure of the gas must *increase* as well.

Practice Problem 6.7

A pressure cooker is used to cook food in a closed pot. By heating the contents of a pressure cooker at constant volume, the pressure increases. If the steam inside the pressure cooker is initially at 100. °C and 1.00 atm, what is the final temperature of the steam if the pressure is increased to 1.05 atm?

Practice Problem 6.8

The temperature of a 0.50-L gas sample at 25 °C and 1.00 atm is changed to each of the following temperatures. What is the final pressure of the system?
  a. 310. K   b. 150. K   c. 50. °C   d. 200. °C

More Practice: Try Problems 6.29–6.32.

6.5 The Combined Gas Law

All three gas laws—Boyle's, Charles's, and Gay–Lussac's laws—can be combined in a single equation, the **combined gas law,** that relates pressure, volume, and temperature.

$$\frac{P_1 V_1}{T_1} = \frac{P_2 V_2}{T_2}$$

initial conditions   new conditions

The combined gas law contains six terms that relate the pressure, volume, and temperature of an initial and final state of a gas. It can be used to calculate one quantity when the other five are known, as long as the amount of gas is constant. The combined gas law is used for determining the effect of changing two factors—such as pressure and temperature—on the third factor, volume.

We solve this type of problem by following the same three steps in the *How To* in Section 6.2, using the equation for the combined gas law in step [2]. Sample Problem 6.5 shows how this is done. Table 6.2 summarizes the equations for the gas laws presented in Sections 6.2–6.5.

Table 6.2 Summary of the Gas Laws That Relate Pressure, Volume, and Temperature

Law	Equation	Relationship
Boyle's law	$P_1V_1 = P_2V_2$	As P increases, V decreases for constant T and n.
Charles's law	$\dfrac{V_1}{T_1} = \dfrac{V_2}{T_2}$	As T increases, V increases for constant P and n.
Gay–Lussac's law	$\dfrac{P_1}{T_1} = \dfrac{P_2}{T_2}$	As T increases, P increases for constant V and n.
Combined gas law	$\dfrac{P_1V_1}{T_1} = \dfrac{P_2V_2}{T_2}$	The combined gas law shows the relationship of P, V, and T when two quantities are changed.

STUDY TIP

In solving a gas law problem that requires a specific equation:
1. Write down the equation.
2. Substitute the known quantities.
3. Solve the equation for the unknown quantity. **If there is more than one unknown, you are using the *wrong* equation.**

Sample Problem 6.5   Using the Combined Gas Law in a Calculation

A weather balloon contains 222 L of helium at 20. °C and 760 mm Hg. What is the volume of the balloon when it ascends to an altitude where the temperature is −40. °C and the pressure is 540 mm Hg?

Analysis

Because this question deals with pressure, volume, and temperature, the combined gas law is used to determine a final volume (V_2) because five quantities are known—the initial pressure, volume, and temperature (P_1, V_1, and T_1), and the final pressure and temperature (P_2 and T_2).

Solution

[1] Identify the known quantities and the desired quantity.

$$P_1 = 760 \text{ mm Hg} \qquad P_2 = 540 \text{ mm Hg}$$
$$T_1 = 20.\,°C \qquad\qquad T_2 = -40.\,°C$$
$$V_1 = 222 \text{ L}$$

known quantities $V_2 = ?$
                              desired quantity

- Both temperatures must be converted to Kelvin temperatures.
- $T_1 = T_c + 273.15 = 20.\,°C + 273.15 = 293 \text{ K}$
- $T_2 = T_c + 273.15 = -40.\,°C + 273.15 = 233 \text{ K}$

[2] Write the equation and rearrange it to isolate the desired quantity, V_2, on one side.

- Use the combined gas law.

$$\frac{P_1V_1}{T_1} = \frac{P_2V_2}{T_2}$$ Solve for V_2 by multiplying both sides by $\dfrac{T_2}{P_2}$.

$$\frac{P_1V_1T_2}{T_1P_2} = V_2$$

[3] Solve the problem.

- Substitute the five known quantities into the equation and solve for V_2.

$$V_2 = \frac{P_1V_1T_2}{T_1P_2} = \frac{(760 \text{ mm Hg})(222 \text{ L})(233 \text{ K})}{(293 \text{ K})(540 \text{ mm Hg})} = 248.5 \text{ L rounded to } 250 \text{ L}$$

Answer

Kelvins and mm Hg cancel.

Practice Problem 6.9

The pressure inside a 1.0-L balloon at 25 °C was 750 mm Hg. What is the pressure inside the balloon when it is cooled to −40. °C and expands to 2.0 L in volume?

More Practice: Try Problems 6.9, 6.33, 6.34.

6.6 Avogadro's Law—How the Volume and Moles of a Gas Are Related

Each equation in Table 6.2 was written for a constant amount of gas; that is, the number of moles (n) did not change. **Avogadro's law** describes the relationship between the number of moles of a gas and its volume.

> • Avogadro's law: When the pressure and temperature are held constant, the volume of a gas is *proportional* to the number of moles present.

As the number of moles of a gas *increases*, its volume *increases* as well. Thus, dividing the volume by the number of moles is a constant (k). **The value of k is the same regardless of the identity of the gas.**

$$\frac{V}{n} = k$$

Thus, if the pressure and temperature of a system are held constant, **increasing the number of moles increases the volume of a gas.**

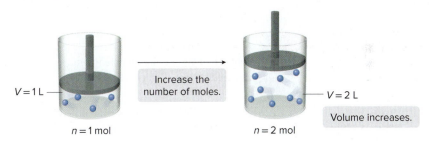

$V = 1\,L$ Increase the number of moles. $V = 2\,L$

Volume increases.

$n = 1\,mol$ $n = 2\,mol$

Because dividing the volume of a gas by the number of moles is a constant, knowing the volume and number of moles initially (V_1 and n_1) means we can calculate a new volume or number of moles (V_2 and n_2) when one of these quantities is changed.

$$\frac{V_1}{n_1} = \frac{V_2}{n_2}$$

initial conditions   new conditions

To solve a problem of this sort, we follow the same three steps listed in the *How To* outlined in Section 6.2, using Avogadro's law in step [2].

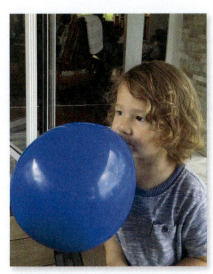

A balloon inflates when someone blows into it because the volume increases with an increased number of moles of air.
Daniel C. Smith

| Sample Problem 6.6 | Determining Moles Using Avogadro's Law |

The lungs of an average male hold 0.25 mol of air in a volume of 5.8 L. How many moles of air do the lungs of an average female hold if the volume is 4.6 L?

Analysis

This question deals with volume and number of moles, so Avogadro's law is used to determine a final number of moles when three quantities are known—the initial volume and number of moles (V_1 and n_1), and the final volume (V_2).

Solution

[1] Identify the known quantities and the desired quantity.

$V_1 = 5.8\,L$ $\qquad$ $V_2 = 4.6\,L$

$n_1 = 0.25\,mol$

known quantities $\qquad\qquad$ $n_2 = ?$

$\qquad\qquad\qquad\qquad$ desired quantity

[2] Write the equation and rearrange it to isolate the desired quantity, n_2, on one side.

- Use Avogadro's law. To solve for n_2, we must invert the numerator and denominator on *both* sides of the equation, and then multiply by V_2.

$$\frac{V_1}{n_1} = \frac{V_2}{n_2} \xrightarrow[\text{on both sides.}]{\text{Switch } V \text{ and } n} \quad \frac{n_1}{V_1} = \frac{n_2}{V_2} \qquad \text{Solve for } n_2 \text{ by multiplying both sides by } V_2.$$

$$\frac{n_1 V_2}{V_1} = n_2$$

[3] Solve the problem.

- Substitute the three known quantities into the equation and solve for n_2.

$$n_2 = \frac{n_1 V_2}{V_1} = \frac{(0.25 \text{ mol})(4.6 \text{ L})}{(5.8 \text{ L})} = 0.20 \text{ mol}$$

Liters cancel. **Answer**

Practice Problem 6.10	A balloon that contains 0.30 mol of helium in a volume of 6.4 L develops a leak so that its volume decreases to 3.85 L at constant temperature and pressure. How many moles of helium does the balloon now contain?
Practice Problem 6.11	A sample of nitrogen gas contains 5.0 mol in a volume of 3.5 L. Calculate the new volume of the container if the pressure and temperature are kept constant but the number of moles of nitrogen is changed to each of the following values: (a) 2.5 mol; (b) 3.65 mol; (c) 21.5 mol.

More Practice: Try Problems 6.4c, 6.8b.

Avogadro's law allows us to compare the amounts of any two gases by comparing their volumes. Often amounts of gas are compared at a set of **standard conditions of temperature and pressure,** abbreviated as **STP.**

- **STP conditions are:**   1 atm (760 mm Hg) for pressure
  273.15 K (0 °C) for temperature
- **At STP, one mole of any gas has the same volume, 22.4 L, called the *standard molar volume.***

Under STP conditions, one mole of nitrogen gas and one mole of helium gas each contain 6.022×10^{23} particles of gas and occupy a volume of 22.4 L at 0 °C and 1 atm pressure. Because the molar masses of nitrogen and helium are different (28.02 g for N_2 compared to 4.003 g for He), one mole of each substance has a *different* mass.

same volume
same number of particles

1 mol N_2 1 mol He

22.4 L   22.4 L
6.022×10^{23} particles 6.022×10^{23} particles
28.02 g   4.003 g

different masses

The standard molar volume can be used to set up conversion factors that relate the volume and number of moles of a gas at STP, as shown in the following stepwise procedure.

How To Convert Moles of Gas to Volume at STP

Example How many moles are contained in 2.0 L of N_2 at standard temperature and pressure?

Step [1] Identify the known quantities and the desired quantity.

2.0 L of N_2 ? moles of N_2

original quantity   desired quantity

Step [2] Write out the conversion factors.

- Set up conversion factors that relate the number of moles of a gas to volume at STP. Choose the conversion factor that places the unwanted unit, liters, in the denominator so that the units cancel.

$\dfrac{22.4\ L}{1\ mol}$ or $\boxed{\dfrac{1\ mol}{22.4\ L}}$ Choose this conversion factor to cancel L.

Step [3] Solve the problem.

- Multiply the original quantity by the conversion factor to obtain the desired quantity.

$2.0\ \cancel{L} \ \times \ \dfrac{1\ mol}{22.4\ \cancel{L}} \ = \ 0.089\ mol\ of\ N_2$

Liters cancel.   **Answer**

By using the molar mass of a gas, we can determine the volume of a gas from a given number of grams, as shown in Sample Problem 6.7.

Sample Problem 6.7   Converting Grams of Gas to Volume at STP

Burning one mole of propane in a gas grill adds 132.0 g of carbon dioxide (CO_2) to the atmosphere. What volume of CO_2 does this correspond to at STP?

Analysis

To solve this problem, we must convert the number of grams of CO_2 to moles using the molar mass. The number of moles of CO_2 can then be converted to its volume using a mole–volume conversion factor (1 mol/22.4 L).

Solution

[1] Identify the known quantities and the desired quantity.

132.0 g CO_2 ? L CO_2

known quantity   desired quantity

[2] Convert the number of grams of CO_2 to the number of moles of CO_2 using the molar mass.

molar mass
conversion factor

$132\ \cancel{g}\ CO_2 \ \times \ \dfrac{1\ mol\ CO_2}{44.01\ \cancel{g}\ CO_2} \ = \ 3.00\ mol\ CO_2$

Grams cancel.

[3] Convert the number of moles of CO_2 to the volume of CO_2 using a mole–volume conversion factor.

mole–volume
conversion factor

$3.00\ \cancel{mol}\ CO_2 \ \times \ \dfrac{22.4\ L}{1\ \cancel{mol}} \ = \ 67.2\ L\ CO_2$

Moles cancel.   **Answer**

| Practice Problem 6.12 | How many liters does each of the following quantities of O_2 occupy at STP: (a) 4.5 mol; (b) 0.35 mol; (c) 18.0 g? |

| Practice Problem 6.13 | How many moles are contained in the following volumes of air at STP: (a) 1.5 L; (b) 8.5 L; (c) 25 mL? |

More Practice: Try Problems 6.11, 6.12, 6.36–6.40.

6.7   The Ideal Gas Law

All four properties of gases—pressure, volume, temperature, and number of moles—can be combined into a single equation called the **ideal gas law.** The product of pressure and volume divided by the product of moles and Kelvin temperature is a constant, called the **universal gas constant** and symbolized by R.

$$\frac{PV}{nT} = R \quad \text{universal gas constant}$$

More often the equation is rearranged and written in the following way:

$$PV = nRT$$
Ideal gas law

For atm: $R = 0.08206 \dfrac{L \cdot atm}{mol \cdot K}$

For mm Hg: $R = 62.4 \dfrac{L \cdot mm\ Hg}{mol \cdot K}$

The value of the universal gas constant R depends on its units. The two most common values of R are given using atmospheres or mm Hg for pressure, liters for volume, and kelvins for temperature. **Be careful to use the correct value of R for the pressure units in the problem you are solving.**

The ideal gas law can be used to find any value—P, V, n, or T—as long as three of the quantities are known. Solving a problem using the ideal gas law is shown in the stepwise *How To* procedure and in Sample Problem 6.8. Although the ideal gas law gives exact answers only for a perfectly "ideal" gas, it gives a good approximation for most real gases, such as the oxygen and carbon dioxide in breathing, as well (Figure 6.6).

How To   Carry Out Calculations with the Ideal Gas Law

Example How many moles of gas are contained in a typical human breath that takes in 0.50 L of air at 1.0 atm pressure and 37 °C?

Step [1] Identify the known quantities and the desired quantity.

$$P = 1.0\ atm$$
$$V = 0.50\ L$$
$$T = 37\ °C \qquad\qquad n = ?\ mol$$

known quantities   desired quantity

Step [2] Convert all values to proper units and choose the value of R that contains these units.
- Convert T_C to T_K. $T_K = T_C + 273.15 = 37\ °C + 273.15 = 310.\ K$
- Use the value of R **in atm** because the pressure is given in atm; that is, $R = 0.08206\ L \cdot atm/(mol \cdot K)$.

—Continued

Step [3] Write the equation and rearrange it to isolate the desired quantity on one side.

- Use the ideal gas law and solve for *n* by dividing both sides by *RT*.

$$PV = nRT \qquad \text{Solve for } n \text{ by dividing both sides by } \mathbf{RT}.$$

$$\frac{PV}{RT} = n$$

Step [4] Solve the problem.

- Substitute the known quantities into the equation and solve for *n*.

$$n = \frac{PV}{RT} = \frac{(1.0 \text{ atm})(0.50 \text{ L})}{\left(0.08206 \dfrac{\text{L} \cdot \text{atm}}{\text{mol} \cdot \text{K}}\right)(310. \text{ K})} = 0.0197 \text{ rounded to } 0.020 \text{ mol}$$

Answer

Figure 6.6   Focus on the Human Body: The Lungs

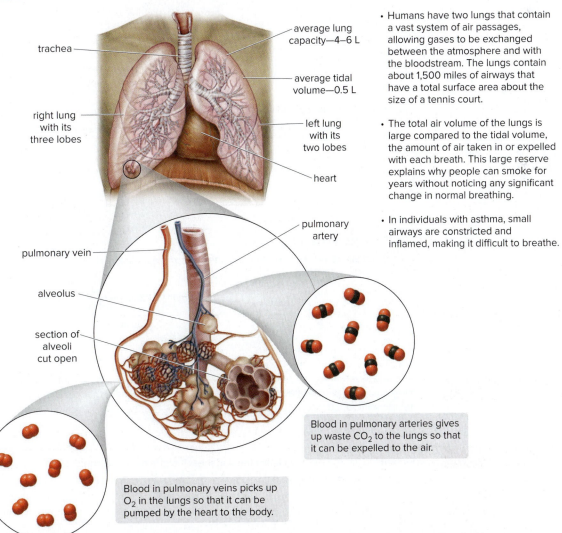

- trachea
- average lung capacity—4–6 L
- average tidal volume—0.5 L
- right lung with its three lobes
- left lung with its two lobes
- heart
- pulmonary artery
- pulmonary vein
- alveolus
- section of alveoli cut open

- Humans have two lungs that contain a vast system of air passages, allowing gases to be exchanged between the atmosphere and with the bloodstream. The lungs contain about 1,500 miles of airways that have a total surface area about the size of a tennis court.

- The total air volume of the lungs is large compared to the tidal volume, the amount of air taken in or expelled with each breath. This large reserve explains why people can smoke for years without noticing any significant change in normal breathing.

- In individuals with asthma, small airways are constricted and inflamed, making it difficult to breathe.

Blood in pulmonary arteries gives up waste CO_2 to the lungs so that it can be expelled to the air.

Blood in pulmonary veins picks up O_2 in the lungs so that it can be pumped by the heart to the body.

Sample Problem 6.8

Using the Ideal Gas Law in a Calculation

If a person exhales 25.0 g of CO_2 in an hour, what volume does this amount occupy at 1.00 atm and 37 °C?

Analysis

Use the ideal gas law to calculate V, because P and T are known and n can be determined by using the molar mass of CO_2 (44.01 g/mol).

Solution

[1] Identify the known quantities and the desired quantity.

$$P = 1.00 \text{ atm}$$
$$T = 37\,°C \qquad 25.0 \text{ g } CO_2 \qquad\qquad V = ? \text{ L}$$

known quantities   desired quantity

[2] Convert all values to proper units and choose the value of *R* that contains these units.

- Convert T_C to T_K. $T_K = T_C + 273.15 = 37\,°C + 273.15 = 310.\text{ K}$
- Use the value of **R** with atm because the pressure is given in atm; that is, $R = 0.08206 \text{ L} \cdot \text{atm/(mol} \cdot \text{K)}$.
- Convert the number of grams of CO_2 to the number of moles of CO_2 using the molar mass (44.01 g/mol).

molar mass
conversion factor

$$25.0 \text{ g } CO_2 \quad \times \quad \frac{1 \text{ mol } CO_2}{44.01 \text{ g } CO_2} \quad = \quad 0.568 \text{ mol } CO_2$$

Grams cancel.

[3] Write the equation and rearrange it to isolate the desired quantity, *V*, on one side.

- Use the ideal gas law and solve for V by dividing both sides by **P**.

$$PV \quad = \quad nRT \qquad \text{Solve for } V \text{ by dividing both sides by } \textbf{P}.$$

$$V \quad = \quad \frac{nRT}{P}$$

[4] Solve the problem.

- Substitute the three known quantities into the equation and solve for V.

$$V \;=\; \frac{nRT}{P} \;=\; \frac{(0.568 \text{ mol}) \left(0.08206 \dfrac{\text{L} \cdot \text{atm}}{\text{mol} \cdot \text{K}} \right)(310.\text{ K})}{1.00 \text{ atm}} \;=\; 14.4 \text{ L}$$

Answer

Practice Problem 6.14

How many moles of oxygen (O_2) are contained in a 5.0-L cylinder that has a pressure of 175 atm and a temperature of 20. °C?

Practice Problem 6.15

Determine the pressure of N_2 (in atm) under each of the following conditions.

   a. 0.45 mol at 25 °C in 10.0 L   b. 10.0 g at 20. °C in 5.0 L

More Practice: Try Problems 6.41–6.46.

Because the partial pressure of O_2 is low at very high altitudes, most mountain climbers use supplemental O_2 tanks above about 24,000 ft. *Vixit/Shutterstock*

6.8 Dalton's Law and Partial Pressures

Gas particles are very far apart compared to the size of an individual particle, so gas particles behave independently. As a result, the identity of the components of a gas mixture does not matter, and **a mixture of gases behaves like a pure gas.** Each component of a gas mixture is said to exert a pressure called its **partial pressure. Dalton's law** describes the relationship between the partial pressures of the components and the total pressure of a gas mixture.

> • **Dalton's law:** The *total pressure* (P_{total}) of a gas mixture is the sum of the partial pressures of its component gases.

Thus, if a mixture has two gases (**A** and **B**) with partial pressures P_A and P_B, respectively, the total pressure of the system (P_{total}) is the sum of the two partial pressures. The partial pressure of a component of a mixture is the same pressure that the gas would exert if it were a pure gas.

$$P_A = 5 \text{ atm} \qquad P_B = 3 \text{ atm} \qquad P_{total} = 8 \text{ atm}$$

P_A	+	P_B	=	P_{total}
partial pressure of **A**		partial pressure of **B**		total pressure

---

Sample Problem 6.9 — **Calculating Pressure Using Dalton's Law**

A sample of exhaled air from the lungs contains four gases with the following partial pressures: N_2 (563 mm Hg), O_2 (118 mm Hg), CO_2 (30. mm Hg), and H_2O (50. mm Hg). What is the total pressure of the sample?

Analysis

Using Dalton's law, the total pressure is the sum of the partial pressures.

Solution

Adding up the four partial pressures gives the total:

$$563 + 118 + 30. + 50. = 761 \text{ mm Hg (total pressure)}$$

Practice Problem 6.16 — CO_2 was added to a cylinder containing 2.5 atm of O_2 to give a total pressure of 4.0 atm of gas. What is the partial pressure of O_2 and CO_2 in the final mixture?

More Practice: Try Problems 6.13, 6.14.

We can also calculate the partial pressure of each gas in a mixture if two quantities are known—[1] the total pressure and [2] the percent of each component—as shown in Sample Problem 6.10.

Sample Problem 6.10   Determining Partial Pressures Using Dalton's Law

Air is a mixture of 21% O_2, 78% N_2, and 1% argon by volume. What is the partial pressure of each gas at sea level, where the total pressure is 760 mm Hg?

Analysis

- Convert each percent to a decimal by moving the decimal point two places to the left.
- Multiply each decimal by the total pressure to obtain the partial pressure for each component.

Solution

			Partial pressure
Fraction O_2:	21% = 0.21	0.21 × 760 mm Hg =	160 mm Hg (O_2)
Fraction N_2:	78% = 0.78	0.78 × 760 mm Hg =	590 mm Hg (N_2)
Fraction Ar:	1% = 0.01	0.01 × 760 mm Hg =	8 mm Hg (Ar)

758 rounded to 760 mm Hg

Practice Problem 6.17

A sample of natural gas at 750 mm Hg contains 85% methane, 10.% ethane, and 5.0% propane. What are the partial pressures of each gas in this mixture?

Practice Problem 6.18

Air contains 21% O_2 and 78% N_2. What are the partial pressures of N_2 and O_2 (in atm) in a hyperbaric chamber that contains air at 2.5 atm?

Practice Problem 6.19

At the summit of Kilimanjaro, the highest mountain in Africa with an elevation of 5,895 m, the partial pressure of oxygen is 78.5 mm Hg. What is the atmospheric pressure at the top of Mount Kilimanjaro?

Practice Problem 6.20

Nitrox is a gas mixture used in scuba diving that contains a higher-than-normal level of oxygen and a lower-than-normal level of nitrogen. A lower amount of nitrogen reduces the risk of "the bends."

a. What percent of N_2 and O_2 is present in the given nitrox tank?

b. What is the partial pressure of each gas (in psi) in the tank?

More Practice: Try Problems 6.47–6.50.

nitrox tank

3,500 psi

(Practice Problem 6.20)

The composition of the atmosphere does not change with location, even though the total atmospheric pressure decreases with increasing altitude. At high altitudes, therefore, the partial pressure of oxygen is much lower than it is at sea level, making breathing difficult. This is why mountain climbers use supplemental oxygen at altitudes above 8,000 meters.

In contrast, a hyperbaric chamber is a device that maintains air pressure two to three times higher than normal. Hyperbaric chambers have many uses. At this higher pressure the partial pressure of O_2 is higher. For burn patients, the higher pressure of O_2 increases the amount of O_2 in the blood, where it can be used by the body for reactions that fight infections.

When a scuba diver surfaces too quickly, the N_2 dissolved in the blood can form microscopic bubbles that cause pain in joints and can occlude small blood vessels, causing organ injury. This condition, called the bends, is treated by placing a diver in a hyperbaric chamber, where the elevated pressure decreases the size of the N_2 bubbles, which are then eliminated as N_2 gas from the lungs as the pressure is slowly decreased.

6.9 FOCUS ON THE ENVIRONMENT
Ozone and Carbon Dioxide in the Atmosphere

Although both ozone and carbon dioxide are present in only minor amounts in the earth's atmosphere, each plays an important role in the dynamics of the environment on the surface of the earth.

6.9A The Ozone Layer

Ozone (O_3) is a gas formed in the upper atmosphere (the stratosphere) by the reaction of oxygen molecules (O_2) with oxygen atoms (O). Stratospheric ozone acts as a shield that protects the earth by absorbing destructive ultraviolet radiation before it reaches the earth's surface (Figure 6.7). A decrease in ozone concentration in this protective layer would have some immediate negative consequences, including an increase in the incidence of skin cancer and eye cataracts. Other possible long-term effects include a reduced immune response, interference with photosynthesis in plants, and harmful effects on the growth of plankton, the mainstay of the ocean food chain.

Figure 6.7 Ozone in the Upper Atmosphere

ENVIRONMENTAL NOTE

O_3 destruction is most severe in the region of the South Pole, where an ozone hole is visible with satellite imaging. The declining levels of ozone-depleting compounds have had a positive effect on the size of the ozone hole in the southern hemisphere, but much more improvement is needed. *NOAA*

Research over the last 50 years has shown that **chlorofluorocarbons (CFCs),** simple compounds that contain the elements of carbon, chlorine, and fluorine, destroy ozone in the upper atmosphere. CFCs are synthetic compounds that were once widely used as refrigerants and aerosol propellants, but these findings led to a ban on the use of CFCs in aerosol propellants in the United States in 1978.

Concept Check 6.3

CF_2Cl_2 is a chlorofluorocarbon once used as an aerosol propellant. Draw a Lewis structure for CF_2Cl_2.

6.9B   Carbon Dioxide and Climate Change

Carbon dioxide (CO_2) is a very minor component of the earth's atmosphere; there are only approximately 411 CO_2 molecules found in every one million molecules of air. Carbon dioxide is produced when heating oil, natural gas, and gasoline are burned in the presence of oxygen, as we learned in Section 5.9. This reaction releases energy for heating homes, powering vehicles, and cooking food.

$$CH_4 + 2\,O_2 \xrightarrow{\text{flame}} CO_2 + 2\,H_2O + \text{energy}$$

methane
(natural gas)

The combustion of these fossil fuels adds a tremendous amount of CO_2 to the atmosphere each year. Quantitatively, data show a 30% increase in the atmospheric concentration of CO_2 in the last 62 years (Figure 6.8). Although the composition of the atmosphere has changed over the lifetime of the earth, this is likely the first time that the actions of humankind have altered that composition significantly and so quickly.

Figure 6.8

The Changing Concentration of CO_2 in the Atmosphere

The 2007 Nobel Peace Prize was awarded to former Vice President Al Gore and the Intergovernmental Panel on Climate Change for their roles in focusing attention on the potentially disastrous effects of rapid climate change caused by human activity.

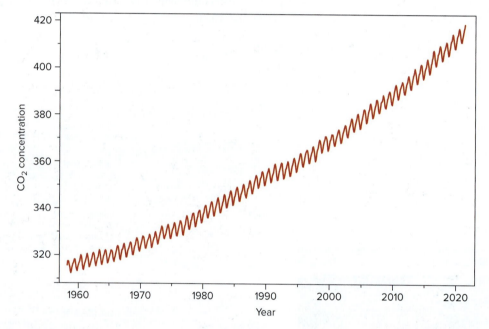

The concentration represents the number of CO_2 molecules in 1,000,000 molecules of air. The graph clearly shows the increasing level of CO_2 in the atmosphere (1958–2020). Two data points are recorded each year. The sawtooth nature of the graph is due to seasonal variation of CO_2 level with the seasonal variation in photosynthesis. (Data recorded at Mauna Loa, Hawaii.)

An increased CO_2 concentration in the atmosphere may have long-range and far-reaching effects. CO_2 is a **greenhouse gas** because it absorbs thermal energy that normally radiates from the earth's surface, and redirects it back to the surface. Higher levels of CO_2 may therefore contribute to an increase in the average temperature of the earth's atmosphere. The global climate change resulting from these effects may lead to the melting of polar ice caps, the rise in sea level, and many more unforeseen consequences.

Concept Check 6.4

Write a balanced equation for each combustion reaction. Propane and butane are minor components of natural gas. Balancing equations was described in Section 5.2.

a. $CH_3CH_2CH_3 + O_2 \xrightarrow{\text{flame}}$

propane

b. $CH_3CH_2CH_2CH_3 + O_2 \xrightarrow{\text{flame}}$

butane

CHAPTER REVIEW

KEY TERMS

Atmosphere (6.1)
Avogadro's law (6.6)
Barometer (6.1)
Boyle's law (6.2)
Charles's law (6.3)

Chlorofluorocarbons (CFCs, 6.9)
Combined gas law (6.5)
Dalton's law (6.8)
Gas laws (6.2)
Gay–Lussac's law (6.4)

Ideal gas law (6.7)
Kinetic–molecular theory (6.1)
Millimeters mercury (6.1)
Partial pressure (6.8)
Pressure (6.1)

Standard molar volume (6.6)
STP (6.6)
Universal gas constant (6.7)

KEY CONCEPTS

Gas pressure and partial pressures

① Definition (6.1)	**② Units (6.1)**	**③ Dalton's law (6.8)**
• Pressure (P)—the force (F) exerted per unit area (A) when gas particles strike a surface • Gas pressure *increases* as the number of collisions of gas particles *increases*. $$\text{Pressure} = \frac{\text{Force}}{\text{Area}} = \frac{F}{A}$$	• Pressure is recorded in atmospheres (atm), millimeters of mercury (mm Hg), or pounds per square inch (psi). See also Table 6.1. • 1 atm = 760. mm Hg = 14.70 psi	• The total pressure (P_{total}) of a gas mixture is the sum of the partial pressures of its component gases. • For two gases **A** and **B**: $$P_{total} = P_A + P_B$$

KEY EQUATIONS—THE GAS LAWS (6.2–6.7)

Relating pressure and volume (6.2)	Relating volume and temperature (6.3)	Relating pressure and temperature (6.4)	Relating pressure, volume, and temperature (6.5)	Relating volume and moles (6.6)	Relating pressure, volume, temperature, and moles (6.7)
$$P_1V_1 = P_2V_2$$ Boyle's law • P increases as V decreases.	$$\frac{V_1}{T_1} = \frac{V_2}{T_2}$$ Charles's law • V increases as T increases.	$$\frac{P_1}{T_1} = \frac{P_2}{T_2}$$ Gay–Lussac's law • P increases as T increases.	$$\frac{P_1V_1}{T_1} = \frac{P_2V_2}{T_2}$$ combined gas law	$$\frac{V_1}{n_1} = \frac{V_2}{n_2}$$ Avogadro's law • V increases as n increases.	$$PV = nRT$$ Ideal gas law • R = universal gas constant

KEY SKILLS

① Using Boyle's law (6.2); example: determining a new gas volume when an initial pressure (204 atm), initial volume (20. L), and final pressure (1.0 atm) are known

Boyle's law: The pressure and volume of a gas are *inversely* related. For example, when pressure *increases*, volume *decreases*.

① Identify the original quantities and the desired quantity.	**② Write the equation and rearrange it to isolate the desired quantity, V_2, on one side.**	**③ Solve the problem.**
P_1 = 204 atm V_1 = 20. L P_2 = 1.0 atom V_2 = ? original quantities **desired** quantity	Use Boyle's law: $P_1V_1 = P_2V_2$ Rearrange: $\dfrac{P_1V_1}{P_2} = V_2$	$V_2 = \dfrac{P_1V_1}{P_2} = \dfrac{(204\ \text{atm})(20.\ \text{L})}{1.0\ \text{atm}} = 4{,}100\ \text{L}$ • Substitute the three known quantities and solve for V_2.

② **Converting moles of gas to volume at STP (6.6); example: determining the number of moles contained in 4.0 L of O_2 at STP**

One mole of any gas has the same volume, 22.4 L, called the **standard molar volume,** at standard temperature and pressure **(STP).**

① **Identify the original quantity and the desired quantity.**	**②** **Write out the conversion factor that places the unwanted unit (L) in the denominator.**	**③** **Solve the problem so that the unwanted unit cancels.**
4.0 L ? mol original quantity **desired** quantity	$\dfrac{1\ \text{mol}}{22.4\ \text{L}}$ ← **desired** unit ← unwanted unit **conversion factor** • Use the **standard molar volume** at STP as a conversion factor.	**conversion factor** $4.0\ \cancel{L} \times \dfrac{1\ \text{mol}}{22.4\ \cancel{L}} = 0.18\ \textbf{mol}\ \textbf{O}_2$ └ Liters cancel. ┘

③ **Using the ideal gas law (6.7); example: determining the volume that 0.469 mol of O_2 occupies at 1.00 atm and 32 °C**

The ideal gas law can be used to calculate any one of the four variables (P, V, n, or T) for any gas, as long as the other three variables are known.

① **Identify the original quantities and the desired quantity. Convert all quantities to the proper units.**	**②** **Write the equation and rearrange it to isolate the desired quantity, V, on one side.**	**③** **Solve the problem.**
$P = 1.00$ atm $T = 32°$ C $n = 0.469$ mol $V = ?$ L original quantities **desired** quantity • Convert T_C to T_K: $T_K = T_C + 273.15 =$ 32 °C + 273.15 = 305 K. • Use the value of R with atm because the pressure is given in atm; that is, $R = 0.08206$ L · atm/(mol · K).	Ideal gas law: $PV = nRT$ Rearrange: $V = \dfrac{nRT}{P}$	$V = \dfrac{nRT}{P} = \dfrac{(0.469\ \cancel{\text{mol}})\left(0.08206\ \dfrac{\text{L} \cdot \cancel{\text{atm}}}{\cancel{\text{mol}} \cdot \cancel{K}}\right)(305\ \cancel{K})}{1.00\ \cancel{\text{atm}}} = 11.7\ \textbf{L}$ • Substitute the three known quantities and the proper value of R, then solve for **V**.

④ **Using Dalton's law to calculate partial pressure (6.8); example: determining partial pressures from the total pressure of a system (720 mm Hg)**

Dalton's law: The total pressure of a gas mixture is the sum of the partial pressures of its component gases.

① **Determine the fraction of each gas in the given mixture.**	**②** **Multiply each fraction by the total pressure to obtain the partial pressures.**

CHAPTER 6 SELF-TEST section below:

CHAPTER 6 SELF-TEST

The Self-Test consists of short-answer questions that test an understanding of definitions, equations, and other material encountered within the chapter. Answers to each question are provided at the end of the chapter. For any questions you are unable to answer correctly, revisit the appropriate section(s) of the text; then proceed to working the end-of-chapter problems to further reinforce your understanding of the chapter material.

1. Which term from "Key Terms" answers each question?
   a. Which gas law relates the pressure, volume, temperature, and number of moles of a gas?
   b. What device is used to measure atmospheric pressure?
   c. What term is used for the volume of one mole of any gas at STP?
   d. What simple compounds once used as aerosol propellants and refrigerants destroy ozone in the upper atmosphere?

2. In each scenario, decide if the atmospheric pressure is higher, lower, or equal to the pressure at sea level: (a) when scuba diving; (b) when flying in an airplane; (c) when hiking Mauna Loa, Hawaii (elevation 13,678 ft).

3. Convert 200. psi to mm Hg.

4. If a tank of compressed air contains 5.0. L of gas at 22.1 atm pressure, what volume does this gas occupy at 1.0 atm?

5. If the lungs of an average dog hold 0.15 mol of air in a volume of 3.6 L, how many moles of air do the lungs of an average cat hold if the volume is 1.2 L?

6. What is the total pressure of a sample composed of four gases with the following partial pressures: N_2 (530. mm Hg), O_2 (129 mm Hg), CO_2 (48 mm Hg), and H_2O (52 mm Hg)?

7. Container **A** contains 10 gas particles and container **B** contains six gas particles. (a) Which container has the higher pressure if the two containers have the same volume? (b) What happens to the pressure when **A** is heated and the volume is kept constant? (c) What happens to the pressure of **B** when its volume is halved and its temperature is kept constant?

8. Label each statement as True or False.
   a. As the temperature of a gas increases, the pressure increases if the volume and number of moles are kept constant.
   b. As the pressure of a gas increases, the volume increases if the temperature and number of moles are kept constant.
   c. One mole of nitrogen gas and one mole of oxygen gas have identical volumes but different masses at the same temperature.
   d. At 700 mm Hg and 298 K, one mole of a gas occupies 22.4 L.

UNDERSTANDING KEY CONCEPTS

6.1 (a) What is the pressure in psi (pounds per square inch) on the given pressure gauge used for scuba diving? (b) How many mm Hg does this correspond to?

6.2 (a) What is the pressure in psi (pounds per square inch) on the given pressure gauge used for scuba diving? (b) How many pascals (Pa) does this correspond to?

6.3 **X** consists of a flexible container with eight particles of a gas as shown. What happens to the pressure of the system when **X** is converted to the representations in parts (a), (b), and (c)?

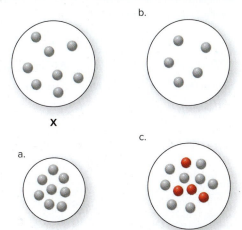

b.

X

a.

c.

6.4 Suppose **A** represents a balloon that can expand or contract. **A** contains 10 particles of a gas as shown. Draw a diagram that shows the volume of **A** when each change occurs.

A

a. The pressure is doubled but the temperature and number of particles remain the same.
b. The temperature is increased but the pressure and number of particles remain the same.
c. The number of particles is doubled but the pressure and temperature remain the same.

6.5 A gas syringe contains 50. mL of CO_2 at 1.0 atm pressure. What is the pressure inside the syringe when the plunger is depressed to 25 mL?

6.6 A cylinder contains 3.4 L of helium at 200. atm. How many 1.0-L balloons can be filled at 1 atm, assuming the temperature is kept constant?

6.7 Draw a picture that represents the given balloon when each of the following changes occurs.

a. The balloon is inflated outside on a cold winter day and then taken inside a building at 75 °F.
b. The balloon is taken to the top of Mauna Kea, Hawaii (elevation 13,796 ft). Assume the temperature is constant.
c. The balloon is taken inside an airplane pressurized at 0.8 atm.

6.8 Which representation ([1], [2], or [3]) shows what balloon **Z** resembles after each change occurs?

| Z | [1] | [2] | [3] |

   a. The balloon is cooled to a lower temperature.
   b. Some gas leaks out.
   c. The balloon is allowed to rise to a higher altitude.

6.9 If a compressed air cylinder for scuba diving contains 6.0 L of gas at 18 °C and 200. atm pressure, what volume does the gas occupy (in L) at 1.0 atm and 25 °C?

6.10 What happens to the pressure of a sample with each of the following changes?
   a. Double the volume and halve the Kelvin temperature.
   b. Double the volume and double the Kelvin temperature.
   c. Halve the volume and double the Kelvin temperature.

6.11 How many moles of helium are contained in each volume at STP: (a) 5.0 L; (b) 11.2 L; (c) 50.0 mL?

6.12 How many moles of argon are contained in each volume at STP: (a) 4.0 L; (b) 31.2 L; (c) 120 mL?

6.13 The partial pressure of N_2 in the air is 593 mm Hg at 1 atm. What is the partial pressure of N_2 in a bubble of air a scuba diver breathes when he is 66 ft below the surface of the water where the pressure is 3.00 atm?

6.14 If N_2 is added to a balloon that contains O_2 (partial pressure 450 mm Hg) and CO_2 (partial pressure 150 mm Hg) to give a total pressure of 850 mm Hg, what is the partial pressure of each gas in the final mixture?

ADDITIONAL PROBLEMS

Pressure

6.15 The highest atmospheric pressure ever measured is 814.3 mm Hg, recorded in Mongolia in December, 2001. Convert this value to atmospheres.

6.16 The lowest atmospheric pressure ever measured is 652.5 mm Hg, recorded during Typhoon Tip on October 12, 1979. Convert this value to atmospheres.

6.17 Convert each quantity to the indicated unit.
   a. 2.8 atm to psi
   b. 520 mm Hg to atm
   c. 20.0 atm to torr
   d. 100. mm Hg to Pa

6.18 The compressed air tank of a scuba diver reads 3,200 psi at the beginning of a dive and 825 psi at the end of a dive. Convert each of these values to atm and mm Hg.

Boyle's Law

6.19 An aluminum scuba tank holds 11.0 L of compressed air at 3,200 psi. What volume does this gas occupy at 1.0 atm pressure?

Daniel C. Smith

6.20 If you pack a bag of potato chips for a snack on a plane ride, the bag appears to have inflated when you take it out to open. If the initial volume of air in the bag was 250 mL at 760 mm Hg, and the plane is pressurized at 650 mm Hg, what is the final volume of the bag?

Martin Shields/Alamy Stock Photo

6.21 Assuming a fixed amount of gas at constant temperature, complete the following table.

	P_1	V_1	P_2	V_2
a.	2.0 atm	3.0 L	8.0 atm	?
b.	55 mm Hg	0.35 L	18 mm Hg	?
c.	705 mm Hg	215 mL	?	1.52 L

6.22 Assuming a fixed amount of gas at constant temperature, complete the following table.

	P_1	V_1	P_2	V_2
a.	2.5 atm	1.5 L	3.8 atm	?
b.	2.0 atm	350 mL	750 mm Hg	?
c.	75 mm Hg	9.1 mL	?	890 mL

6.23 If a scuba diver releases a 10.-mL air bubble below the surface where the pressure is 3.5 atm, what is the volume (in mL) of the bubble when it rises to the surface and the pressure is 1.0 atm?

6.24 If someone takes a breath and the lungs expand from 4.5 L to 5.6 L in volume, and the initial pressure was 756 mm Hg, what is the pressure inside the lungs (in mm Hg) before any additional air is pulled in?

Charles's Law

6.25 Assuming a fixed amount of gas at constant pressure, complete the following table.

	V_1	T_1	V_2	T_2
a.	5.0 L	310 K	?	250 K
b.	150 mL	45 K	?	45 °C
c.	60.0 L	0.0 °C	180 L	?

6.26 Assuming a fixed amount of gas at constant pressure, complete the following table.

	V_1	T_1	V_2	T_2
a.	10.0 mL	210 K	?	450 K
b.	255 mL	55 °C	?	150 K
c.	13 L	−150 °C	52 L	?

6.27 If a balloon containing 2.2 L of gas at 25 °C is cooled to −78 °C, what is its new volume (in L)?

6.28 How hot must the air in a balloon be heated if initially it has a volume of 750. L at 20 °C and the final volume must be 1,000. L?

Gay–Lussac's Law

6.29 Assuming a fixed amount of gas at constant volume, complete the following table.

	P_1	T_1	P_2	T_2
a.	3.25 atm	298 K	?	398 K
b.	550 mm Hg	273 K	?	−100. °C
c.	0.50 atm	250 °C	955 mm Hg	?

6.30 Assuming a fixed amount of gas at constant volume, complete the following table.

	P_1	T_1	P_2	T_2
a.	1.74 atm	120 °C	?	20. °C
b.	220 mm Hg	150 °C	?	300. K
c.	0.75 atm	198 °C	220 mm Hg	?

6.31 An autoclave is a pressurized container used to sterilize medical equipment by heating it to a high temperature under pressure. If an autoclave containing steam at 100. °C and 1.0 atm pressure is then heated to 150. °C, what is the pressure inside it (in atm)?

6.32 If a plastic container at 1.0 °C and 750. mm Hg is heated in a microwave oven to 80. °C, what is the pressure inside the container (in mm Hg)?

Combined Gas Law

6.33 Assuming a fixed amount of gas, complete the following table.

	P_1	V_1	T_1	P_2	V_2	T_2
a.	0.90 atm	4.0 L	265 K	?	3.0 L	310 K
b.	1.2 atm	75 L	5.0 °C	700. mm Hg	?	50 °C
c.	200. mm Hg	125 mL	298 K	100. mm Hg	0.62 L	?

6.34 Assuming a fixed amount of gas, complete the following table.

	P_1	V_1	T_1	P_2	V_2	T_2
a.	0.55 atm	1.1 L	340 K	?	3.0 L	298 K
b.	735 mm Hg	1.2 L	298 K	1.1 atm	?	0.0 °C
c.	7.5 atm	230 mL	−120 °C	15 atm	0.45 L	?

Avogadro's Law

6.35 What is the difference between STP and standard molar volume?

6.36 Given the same number of moles of two gases at STP conditions, how do the volumes of two gases compare? How do the masses of the two gas samples compare?

6.37 Calculate the volume of each substance at STP.
   a. 4.2 mol Ar
   b. 3.5 g CO_2
   c. 2.1 g N_2

6.38 Consider balloons **A** and **B,** which contain CH_4 and C_2H_6, respectively, at the same temperature and volume. (a) How does the pressure exerted by the gases in **A** and **B** compare? (b) How do the masses of the gases inside balloons **A** and **B** compare? (c) How do the densities of the gases inside balloons **A** and **B** compare?

6.39 What volume does 3.01×10^{21} molecules of N_2 occupy at STP?

6.40 What volume does 1.50×10^{24} molecules of CO_2 occupy at STP?

Ideal Gas Law

6.41 How many moles of gas are contained in a human breath that occupies 0.45 L and has a pressure of 747 mm Hg at 37 °C?

6.42 How many moles of gas are contained in a compressed air tank for scuba diving that has a volume of 7.0 L and a pressure of 210 atm at 25 °C?

6.43 How many moles of air are present in the lungs if they occupy a volume of 5.0 L at 37 °C and 740 mm Hg? How many molecules of air does this correspond to?

6.44 If a cylinder contains 10.0 g of CO_2 in 10.0 L at 325 K, what is the pressure?

6.45 Which sample contains more moles: 2.0 L of O_2 at 273 K and 500 mm Hg, or 1.5 L of N_2 at 298 K and 650 mm Hg? Which sample has the greater mass?

6.46 An unknown amount of gas occupies 30.0 L at 2.1 atm and 298 K. How many moles does the sample contain? What is the mass if the gas is helium? What is the mass if the gas is argon?

Dalton's Law and Partial Pressure

6.47 The molecular art shows a closed container with two gases, **A** (red spheres) and **B** (blue spheres), at a pressure of 630 mm Hg. What are the partial pressures of **A** and **B?**

6.48 If the pressure due to the red spheres in the molecular art is 480 mm Hg, what is the pressure due to the blue spheres? What is the total pressure of the system?

6.49 Air pressure on the top of Mauna Loa, a 13,000-ft mountain in Hawaii, is 460 mm Hg. What are the partial pressures of O_2 and N_2, which compose 21% and 78% of the atmosphere, respectively?

6.50 Nitrox is a gas mixture used in scuba diving that contains a higher-than-normal level of oxygen and a lower-than-normal level of nitrogen.

a. What percentages of N_2 and O_2 are present in the given nitrox tank?

b. What is the partial pressure of each gas (in psi) in the tank?

3,200 psi

General Problems and Applications

6.51 Assume that each of the following samples is at the same temperature. (a) Which diagram represents the sample at the highest pressure? (b) Which diagram represents the sample at the lowest pressure?

[1]   [2]   [3]

6.52 Use the diagrams in Problem 6.51 to answer the following questions.

a. What happens to the pressure when sample [1] is heated and the volume is kept constant?

b. What happens to the pressure of sample [2] when the volume is halved and the temperature is kept constant?

c. What happens to the pressure of sample [3] when its volume is halved and its Kelvin temperature is doubled?

6.53 A balloon is filled with helium at sea level. What happens to the volume of the balloon in each instance? Explain each answer.
a. The balloon floats to a higher altitude.
b. The balloon is placed in a bath of liquid nitrogen at −196 °C.
c. The balloon is placed inside a hyperbaric chamber at a pressure of 2.5 atm.
d. The balloon is heated inside a microwave.

6.54 Suppose you have a fixed amount of gas in a container with a movable piston, as drawn. Re-draw the container and piston to illustrate what it looks like after each of the following changes takes place.

— piston

a. The temperature is held constant and the pressure is doubled.
b. The pressure is held constant and the Kelvin temperature is doubled.
c. The pressure is halved and the Kelvin temperature is halved.

6.55 What is the difference between the systolic and diastolic blood pressure?

6.56 Why does a bubble at the bottom of a glass of a soft drink get larger as it rises to the surface?

6.57 If a scuba diver inhales 0.50 L of air at a depth of 100. ft and 4.0 atm pressure, what volume does this air occupy at the surface of the water, assuming air pressure is 1.0 atm? When a scuba diver must make a rapid ascent to the surface, he is told to exhale slowly as he ascends. How does your result support this recommendation?

6.58 What happens to the density of a gas if the temperature is increased but the pressure is held constant? Use this information to explain how wind currents arise.

6.59 A common laboratory test for a patient is to measure blood gases—that is, the partial pressures of O_2 and CO_2 in oxygenated blood. Normal values are 100 mm Hg for O_2 and 40 mm Hg for CO_2. A high or low level of one or both readings has some underlying cause. Explain the following. If a patient comes in agitated and hyperventilating—breathing very rapidly—the partial pressure of O_2 is normal but the partial pressure of CO_2 is 22 mm Hg.

6.60 Use the information in Problem 6.59 to explain the following. A patient with chronic lung disease has a partial pressure of O_2 of 60 mm Hg and a partial pressure of CO_2 of 60 mm Hg.

CHALLENGE PROBLEMS

6.61 A gas (4.0 g) occupies 11.2 L at 2 atm and 273 K. What is the molar mass of the gas? What is the identity of the gas?

6.62 As we learned in Chapter 5, an automobile airbag inflates when NaN_3 is converted to Na and N_2 according to the equation, $2 NaN_3 \longrightarrow 2 Na + 3 N_2$. What volume of N_2 would be produced if 100. g of NaN_3 completely reacted at STP?

BEYOND THE CLASSROOM

6.63 Several compounds besides carbon dioxide are classified as greenhouse gases, compounds that absorb thermal energy and redirect it back to the earth's surface. Research what other gases in the atmosphere fall into this category, and give their chemical structures. Is there a natural source of each gas, or is the gas produced solely by human activity? Is it known whether the concentration of the gas has changed significantly in the last 50–100 years?

6.64 When a major sporting event such as the World Cup or the Olympics occurs at a location that is considerably above sea level, the effects of higher altitude are often mentioned. Discuss some consequences of holding an athletic event at a venue with significantly lower atmospheric pressure and lower partial pressure of oxygen. Using these concepts, discuss what positive or negative effects might be observed on the athletes and the sport itself. Why do some athletes train at higher altitudes for an event that will be held at sea level? What types of sports might benefit from higher altitude and what types might be hindered? Give specific examples.

CHAPTER 6 ANSWERS

Concept Check Answers

6.1 14/9

6.2 When a sealed container is heated, the gases inside expand and increase the internal pressure, eventually popping the lid off.

6.3

$$:\overset{\displaystyle :\!\ddot{F}\!:}{\underset{\displaystyle :\!\ddot{F}\!:}{:\ddot{C}l-C-\ddot{C}l:}}$$

6.4
a. $CH_3CH_2CH_3 + 5 O_2 \longrightarrow 3 CO_2 + 4 H_2O$
b. $2 CH_3CH_2CH_2CH_3 + 13 O_2 \longrightarrow 8 CO_2 + 10 H_2O$

Practice Problem Answers

6.1 a. 2,300 mm Hg   b. 14 psi   c. 0.558 atm

6.2 a. 1.6 L   b. 3.2 L   c. 0.80 L   d. 16 L

6.3 a. 0.25 atm   b. 1.5 atm   c. 0.075 atm   d. 0.0075 atm

6.4 0.44 L

6.5 a. 250 L   b. 24.0 mL

6.6 360. K

6.7 392 K or 119 °C

6.8 a. 1.04 atm   b. 0.503 atm   c. 1.08 atm   d. 1.59 atm

6.9 290 mm Hg

6.10 0.18 mol

6.11 a. 1.8 L   b. 2.6 L   c. 15 L

6.12 a. 1.0×10^2 L   b. 7.8 L   c. 12.6 L

6.13 a. 0.067 mol   b. 0.38 mol   c. 1.1×10^{-3} mol

6.14 36 mol

6.15 a. 1.1 atm   b. 1.7 atm

6.16 O_2: 2.5 atm CO_2: 1.5 atm

6.17 methane: 640 mm Hg   ethane: 75 mm Hg
propane: 38 mm Hg

6.18 N_2: 2 atm O_2: 0.53 atm

6.19 370 mm Hg

6.20 a. N_2: 62.5% O_2: 37.5%
     b. N_2: 2,200 psi O_2: 1,300 psi

Self-Test Answers

1. a. ideal gas law   c. standard molar volume
     b. barometer   d. chlorofluorocarbons

2. a. higher   b,c: lower

3. 10,300 mm Hg

4. 110 L

5. 0.050 mol

6. 759 mm Hg

7. a. **A**   b,c: increases

8. a,c: True   b,d: False

Odd-Numbered Problem Answers

6.1 a. 2,600 psi   b. 1.3×10^5 mm Hg

6.3 a,c: increases   b. decreases

6.5 2.0 atm

6.7 a,b,c: Volume increases.

6.9 1,200 L

6.11 a. 0.22 mol   b. 0.500 mol   c. 0.002 23 mol

6.13 1,780 mm Hg

6.15 1.07 atm

6.17 a. 41 psi   b. 0.68 atm   c. 15,200 torr   d. 13,300 Pa

6.19 2,400 L

6.21 a. 0.75 L   b. 1.1 L   c. 99.7 mm Hg

6.23 35 mL

6.25 a. 4.0 L   b. 1.1 L   c. 820 K

6.27 1.4 L

6.29 a. 4.34 atm   b. 350 mm Hg   c. 1,300 K

6.31 1.1 atm

6.33 a. 1.4 atm   b. 110 L   c. 740 K

6.35 STP is "standard temperature and pressure," or 0 °C at
760 mm Hg. The standard molar volume is the volume that
one mole of a gas occupies at STP, or 22.4 liters.

6.37 a. 94 L   b. 1.8 L   c. 1.7 L

6.39 0.112 L or 112 mL

6.41 0.017 mol

6.43 0.19 mol or 1.1×10^{23} molecules

6.45 O_2 has more moles and also has the greater mass.

6.47 **A:** 210 mm Hg   **B:** 420 mm Hg

6.49 97 mm Hg for O_2 360 mm Hg for N_2

6.51 a. [3]   b. [2]

6.53 a. Volume increases as outside atmospheric pressure
     decreases.
    b. Volume decreases at the lower temperature.
    c. Volume decreases as external pressure increases.
    d. Volume increases as temperature increases.

6.55 The systolic pressure is the maximum pressure generated with
each heartbeat. The diastolic pressure is the lowest pressure
recorded between heartbeats.

6.57 The volume of air will increase from 0.50 L to 2.0 L as the scuba
diver ascends to the surface, so the scuba diver must exhale to
eliminate the excess volume of gas.

6.59 As a person breathes faster, he eliminates more CO_2 from the
lungs; therefore, the measured value of CO_2 is lower than the
normal value of 40 mm Hg.

6.61 The molar mass is 4.0 g/mol and the gas is helium.

Solutions

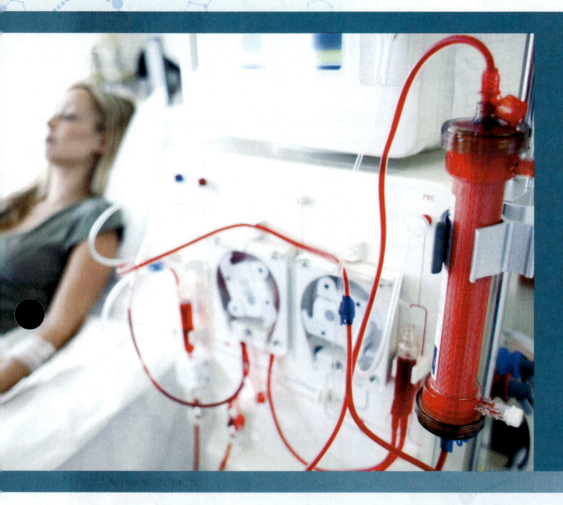

When an individual's kidneys no longer function, dialysis is used to remove waste products like urea from the blood.
Science Photo Library/Alamy Stock Photo

CHAPTER OUTLINE

CHAPTER GOALS

In this chapter you will learn how to:

1. Describe the fundamental properties of a solution and determine whether a mixture is a solution, colloid, or suspension

2. Classify a substance as an electrolyte or nonelectrolyte

3. Predict whether a substance is soluble in water or a nonpolar solvent

4. Predict the effect of temperature and pressure on solubility

5. Calculate the concentration of a solution

6. Prepare a dilute solution from a more concentrated solution

7. Describe the process of osmosis and how it relates to biological membranes and dialysis

In Chapter 7 we study solutions—homogeneous mixtures of two or more substances. Why are table salt (NaCl) and sugar (sucrose) soluble in water but vegetable oil and gasoline are not? How does a healthcare professional take a drug as supplied by the manufacturer and prepare a dilute solution to administer a proper dose to a patient? An understanding of solubility and concentration is needed to explain each of these phenomena.

7.1   Mixtures

Thus far we have concentrated primarily on **pure substances**—elements, covalent compounds, and ionic compounds. Most matter with which we come into contact, however, is a **mixture** composed of two or more pure substances. The air we breathe is composed of nitrogen and oxygen, together with small amounts of argon, water vapor, carbon dioxide, and other gases. Seawater is composed largely of sodium chloride and water. A mixture may be **heterogeneous** or **homogeneous.**

- A *heterogeneous* mixture does not have a uniform composition throughout a sample.
- A *homogeneous* mixture has a uniform composition throughout a sample.

7.1A   Solutions

The most common type of a homogeneous mixture is a solution.

- A *solution* is a homogeneous mixture that contains small particles. Liquid solutions are transparent.

A pepperoni pizza is a heterogeneous mixture, whereas a soft drink is a homogeneous solution. *Jill Braaten*

A cup of hot coffee, vinegar, and gasoline are solutions. Any phase of matter can form a solution (Figure 7.1). Air is a solution of gases. An intravenous saline solution contains solid sodium chloride (NaCl) in liquid water. A dental filling contains liquid mercury (Hg) in solid silver.

Figure 7.1   Three Different Types of Solutions

a. Air: Two (or more) gases

b. Saline solution: A solid in a liquid

c. Dental filling: A liquid in a solid

a. Air is a solution of gases, primarily N_2 and O_2. The lungs contain significant amounts of H_2O and CO_2 as well.

b. An IV saline solution contains solid sodium chloride (NaCl) dissolved in liquid water.

c. A dental filling contains a liquid, mercury (Hg), dissolved in solid silver (Ag).

(b): Janis Christie/Digital Vision/Alamy Stock Photo; (c): Keith Eng, 2008

When two substances form a solution, the substance present in the lesser amount is called the **solute,** and the substance present in the larger amount is the **solvent.** A solution with water as the solvent is called an **aqueous solution.**

Although a solution can be separated into its pure components, **one component of a solution cannot be filtered away from the other component,** because the particles of the solute are similar in size to the particles of the solvent. For a particular solute and solvent, solutions having different compositions are possible. For example, 1.0 g of NaCl can be mixed with 50.0 g of water or 10.0 g of NaCl can be mixed with 50.0 g of water.

7.1B Colloids and Suspensions

Colloids and suspensions are mixtures that contain larger particles than the particles in a solution.

- A *colloid* is a homogeneous mixture with larger particles, often having an opaque appearance.
- A *suspension* is a heterogeneous mixture that contains large particles suspended in a liquid.

Like solutions, the particles in a colloid generally cannot be filtered from its other components, and they do not settle out on standing. By definition, **a colloid contains particles that are 1 nm–1 μm in diameter.** Milk is a colloid. Homogenized milk is an opaque homogeneous mixture that contains large protein and fat molecules that do not dissolve in water. Homogenized milk does not have to be shaken before drinking because the particles of fat and protein do not settle out.

A suspension is a heterogeneous mixture that contains particles greater than 1 μm in diameter. The particles are so large that they do not dissolve in a liquid, and they can either be filtered away from the liquid, or separated using a centrifuge. The particles in a suspension settle out on standing and the suspension must be shaken to disperse the particles. Milk of magnesia, a common antacid, is a suspension that must be shaken before using because the active ingredients settle out.

Table 7.1 summarizes the properties of solutions, colloids, and suspensions.

Blood is a suspension containing blood cells that can be separated from liquid blood plasma by centrifugation.
Eugene_Sim/Getty Images

Table 7.1 Solutions, Colloids, and Suspensions

Mixture	Particle Size	Settling	Separation
Solution	Small, < 1 nm	No	The components cannot be separated by filtration.
Colloid	Larger, 1 nm–1 μm	No	The components cannot be separated by common filtration techniques.
Suspension	Large, > 1 μm	Yes	The components can be separated by filtration or centrifugation.

Concept Check 7.1

Classify each substance as a heterogeneous mixture, solution, or colloid: (a) Cherry Garcia ice cream (cherry ice cream + chocolate bits + cherries); (b) mayonnaise; (c) seltzer water; (d) nail polish remover; (e) brass (an alloy of Cu and Zn).

Concept Check 7.2

Use the appearance of each product to classify it as a solution, colloid, or suspension: (a) orange juice with pulp (label shown); (b) apple juice; (c) passion orange guava (POG) drink.

a.

b.

c.

Holly Curry/McGraw Hill   *Boarding1Now/iStock/Getty Images*   *Daniel C. Smith*

7.2 Electrolytes and Nonelectrolytes

A solute that dissolves in water to form ions conducts an electric current, whereas a solute that contains only neutral molecules does not.

- A substance that conducts an electric current in water is called an *electrolyte*.
- A substance that does not conduct an electric current in water is called a *nonelectrolyte*.

7.2A Classification

Consider the difference between an aqueous solution of sodium chloride, NaCl, and an aqueous solution of hydrogen peroxide, H_2O_2. A solution of sodium chloride contains Na^+ cations and Cl^- anions and conducts electricity. A solution of hydrogen peroxide contains only uncharged H_2O_2 molecules in H_2O, so it does not conduct electricity.

NaCl dissolved in water H_2O_2 dissolved in water

ions in H_2O Cl^- Na^+ neutral molecules in H_2O H_2O_2

electrolyte   nonelectrolyte

(left): Anthony Arena/Jill Braaten/McGraw Hill; (right): Jill Braaten/McGraw Hill

Electrolytes can be classified as strong or weak, depending on the extent that the compound dissociates (forms ions).

A **strong electrolyte dissociates completely to form ions** when it dissolves in water. NaCl is a strong electrolyte because all of the species present in solution are Na^+ cations and Cl^- anions.

A **weak electrolyte** dissolves in water to yield largely uncharged molecules, but a small fraction of the molecules form ions. When ammonia (NH_3) dissolves in water, the predominant species is uncharged NH_3 molecules, but some NH_3 molecules react with water to form NH_4^+ cations and OH^- anions. Because a weak electrolyte contains some ions, it conducts an electric current, but not as well as a strong electrolyte. Common weak electrolytes are the weak acids and bases that we will learn about in Chapter 8.

NH_3 is a weak electrolyte because it exists in H_2O largely as uncharged NH_3 molecules, together with a small number of ions (NH_4^+ and OH^-).

NH_3 dissolved in H_2O

NH_4^+
OH^-
NH_3

Jill Braaten/Anthony Arena/McGraw Hill

Table 7.2 summarizes the characteristics of strong electrolytes, weak electrolytes, and nonelectrolytes, and lists examples of each.

Table 7.2 Strong Electrolytes, Weak Electrolytes, and Nonelectrolytes

Solute	Species in Solution	Conductivity	Examples
Strong electrolyte	Ions	Conducts an electric current	NaCl, KOH, HCl, KBr
Weak electrolyte	Molecules with a few ions	Conducts an electric current	NH_3, CH_3CO_2H, HF
Nonelectrolyte	Molecules	Does not conduct an electric current	CH_3CH_2OH, H_2O_2

Concept Check 7.3

Consider the following diagrams for an aqueous solution of a compound **AB** (with **A** represented by red spheres and **B** represented by blue spheres). Label each diagram as a strong electrolyte, weak electrolyte, or nonelectrolyte.

a. b. c.

Concept Check 7.4

Classify each solution as an electrolyte or nonelectrolyte: (a) KCl in H_2O; (b) sucrose ($C_{12}H_{22}O_{11}$) in H_2O; (c) KI in H_2O.

Table 7.3 Moles and Equivalents

Ion	Equivalents
1 mol Na^+	**1** Eq Na^+
1 mol Ca^{2+}	**2** Eq Ca^{2+}
1 mol HCO_3^-	**1** Eq HCO_3^-
1 mol PO_4^{3-}	**3** Eq PO_4^{3-}

7.2B   Equivalents

Blood plasma contains electrolytes with dissolved ions such as Na^+, K^+, Cl^-, and HCO_3^-. An **equivalent (Eq)** is a unit used to express the amount of each ion in solution. An equivalent relates the amount of an ion to its charge.

> • An *equivalent* is the number of moles of charge that a mole of ions contributes to a solution.

Thus, one mole of K^+ equals one equivalent of K^+ ions because potassium bears a +1 charge. One mole of Ca^{2+} equals two equivalents of Ca^{2+} because calcium bears a +2 charge. **The number of equivalents per mole of an ion equals the charge on the ion,** as shown in Table 7.3.

Sample Problem 7.1

Calculating the Number of Equivalents of an Ion

How many equivalents of sulfate (SO_4^{2-}) are present in a solution that contains 3.2 moles of SO_4^{2-}?

Analysis

Use a conversion factor that relates the number of moles of the ion to its number of equivalents.

Solution

Because sulfate has a **−2** charge, there are **2** Eq of SO_4^{2-} ions per mole of SO_4^{2-}.

$$3.2 \; \cancel{mol} \; SO_4^{2-} \quad \times \quad \frac{2 \; Eq \; SO_4^{2-}}{1 \; \cancel{mol} \; SO_4^{2-}} \quad = \quad 6.4 \; Eq \; SO_4^{2-}$$

Answer

Practice Problem 7.1

Using the given number of moles, determine how many equivalents of each ion are present.

   a. 1 mol Na^+ b. 1 mol Mg^{2+} c. 0.5 mol K^+ d. 0.5 mol PO_4^{3-}

More Practice: Try Problem 7.17.

The amount of an electrolyte in blood plasma and IV solutions is often reported in the number of **milliequivalents per liter of solution (mEq/L),** where **1,000 mEq = 1 Eq.** In solutions that contain ions, **there must be a balance between the total positive charge and total negative charge of all the ions present.** Thus, if a solution of KCl contains 40 mEq/L of K^+ ions, it must also contain 40 mEq/L of Cl^- ions.

Sample Problem 7.2

Calculating Milliequivalents of an Ion

If an intravenous aqueous NaCl solution contains 154 mEq/L of Na^+ ions, how many milliequivalents of Na^+ is a patient given in 800. mL of solution?

Analysis and Solution

[1] Identify the known quantities and the desired quantity.

        154 mEq/L Na^+ in solution   ? mEq Na^+

        800. mL of solution

        known quantities   desired quantity

[2] Write out the conversion factors.

   • Set up a conversion factor that relates milliequivalents to volume. Because the volume of the solution is given in mL, a mL–L conversion factor is needed as well.

mEq–L conversion factors	L–mL conversion factors
$\dfrac{154 \; mEq \; Na^+}{1 \; L}$ or $\dfrac{1 \; L}{154 \; mEq \; Na^+}$	$\dfrac{1 \; L}{1000 \; mL}$ or $\dfrac{1000 \; mL}{1 \; L}$

Use the conversion factors with the unwanted units—L and mL—in the denominator.

[3] Solve the problem.

- Multiply the original quantity by the conversion factors to obtain the desired quantity.

L–mL conversion factor	mEq–L conversion factor

$$800. \, \cancel{mL} \quad \times \quad \frac{1 \, \cancel{L}}{1000 \, \cancel{mL}} \quad \times \quad \frac{154 \, mEq \, Na^+}{1 \, \cancel{L}} \quad = \quad 123 \, mEq \, Na^+$$

Answer

Practice Problem 7.2

If a KCl solution given intravenously to a patient that has a low potassium level contains 40. mEq of K^+ per liter, how many milliequivalents of K^+ are present in 550 mL of the solution?

Practice Problem 7.3

A solution contains the following ions: Na^+ (15 mEq/L), K^+ (10. mEq/L), and Ca^{2+} (4 mEq/L). If Cl^- is the only anion present in the solution, what is its concentration in mEq/L?

Practice Problem 7.4

If a solution contains 125 mEq of Na^+ per liter, how many milliliters of solution contain 25 mEq of Na^+?

More Practice: Try Problems 7.18–7.20.

7.3 Solubility—General Features

Solubility is the amount of solute that dissolves in a given amount of solvent, usually reported in grams of solute per 100 mL of solution (g/100 mL). A solution that has *less* than the maximum number of grams of solute is said to be **unsaturated.** A solution that has the *maximum* number of grams of solute that can dissolve is said to be **saturated.** If we added more solute to a saturated solution, the additional solute would remain undissolved in the flask.

What determines if a compound dissolves in a particular solvent? Whether a compound is soluble in a given solvent depends on the strength of the interactions between the compound and the solvent. As a result, compounds are soluble in solvents to which they are strongly attracted—that is, compounds are soluble in solvents with similar intermolecular forces. Solubility is often summed up in three words: **"Like dissolves like."**

- Most ionic and polar covalent compounds are soluble in water, a polar solvent.
- Nonpolar compounds are soluble in nonpolar solvents.

Water-soluble compounds are ionic or are small polar molecules that can hydrogen bond with the water solvent. For example, solid sodium chloride (NaCl) is held together by very strong electrostatic interactions of the oppositely charged ions. When it is mixed with water, the Na^+ and Cl^- ions are separated from each other and surrounded by polar water molecules (Figure 7.2). Each Na^+ is surrounded by water molecules arranged with their O atoms (which bear a partial negative charge) in close proximity to the positive charge of the cation. Each Cl^- is surrounded by water molecules arranged with their H atoms (which bear a partial positive charge) in close proximity to the negative charge of the anion.

- The attraction of an ion with a dipole in a molecule is called an ion–dipole interaction.

Figure 7.2   Dissolving Sodium Chloride in Water

When ionic NaCl dissolves in water, the Na⁺ and Cl⁻ interactions of the crystal are replaced by new interactions of Na⁺ and Cl⁻ ions with the solvent. Each ion is surrounded by a loose shell of water molecules arranged so that oppositely charged species are close to each other. *(left & right): Jill Braaten/McGraw Hill*

The ion–dipole interactions between Na⁺, Cl⁻, and water provide the energy needed to break apart the ions from the crystal lattice. The water molecules form a loose shell of solvent around each ion. The process of surrounding particles of a solute with solvent molecules is called **solvation.**

Small uncharged molecules that can hydrogen bond with water are also soluble. Thus, ethanol (C_2H_5OH), which is present in alcoholic beverages, dissolves in water because hydrogen bonding occurs between the OH group in ethanol and the OH group of water.

CH₃CH₂OH
ethanol

hydrogen bond

H₂O

Ethanol can hydrogen bond with water.

Foodcollection/StockFood

Water solubility for uncharged molecules occurs only with small polar molecules or those with many O or N atoms that can hydrogen bond to water. Thus, stearic acid ($C_{18}H_{36}O_2$), a component of animal fats, is *insoluble* in water because its nonpolar part (C—C and C—H bonds) is large compared to its polar part (C—O and O—H bonds). On the other hand, glucose ($C_6H_{12}O_6$), a simple carbohydrate, is *soluble* in water because it has many OH groups and thus many opportunities for hydrogen bonding with water.

Stearic acid–Water insoluble

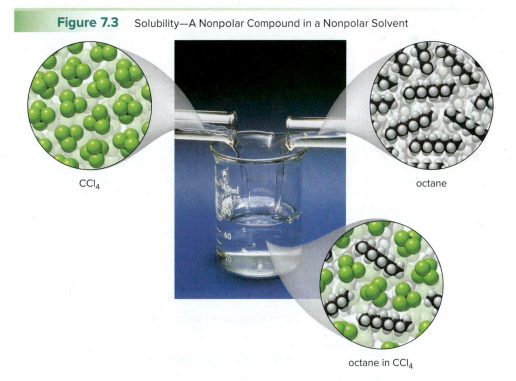

many nonpolar C–C and C–H bonds

Most of the molecule is nonpolar, so it is
not attracted to a polar solvent like H₂O.

polar C–O and
O–H bonds

Glucose–Water soluble

many O–H bonds for
hydrogen bonding to H₂O

Nonpolar compounds are soluble in nonpolar solvents. As a result, octane (C_8H_{18}), a component of gasoline, dissolves in the nonpolar solvent carbon tetrachloride (CCl_4), as shown in Figure 7.3. Animal fat and vegetable oils, which are composed largely of nonpolar C—C and C—H bonds, are soluble in CCl_4, but are insoluble in a polar solvent like water. These solubility properties explain why "oil and water don't mix."

Figure 7.3   Solubility—A Nonpolar Compound in a Nonpolar Solvent

CCl_4

octane

octane in CCl_4

Octane (C_8H_{18}) dissolves in CCl_4 because both are nonpolar liquids that exhibit only London dispersion forces. *Tom Pantages*

Keith Eng, 2008

Dissolving a solute in a solvent is a physical process that is accompanied by an energy change. Breaking up the particles of the solute requires energy, and forming new attractive forces between the solute and the solvent releases energy.

- When solvation *releases* more energy than that required to separate particles, the overall process is *exothermic* (heat is released).
- When the separation of particles *requires* more energy than is released during solvation, the process is *endothermic* (heat is absorbed).

These energy changes are used to an advantage in commercially available hot packs and cold packs. A hot pack, sometimes used for pain relief of sore muscles, contains calcium chloride ($CaCl_2$) or magnesium sulfate ($MgSO_4$) and water. Breaking the seal that separates them allows the salt to dissolve in the water, releasing heat, and the pouch gets warm. In contrast, ammonium nitrate (NH_4NO_3) absorbs heat on mixing with water, so this salt is found in instant cold packs used to reduce swelling.

Sample Problem 7.3	Predicting Water Solubility

Predict the water solubility of each compound: (a) KCl; (b) methanol (CH_3OH); (c) hexane (C_6H_{14}).

Analysis

Use the general solubility rule—"like dissolves like."

- **Generally, ionic and small polar compounds that can hydrogen bond are soluble in water.**
- **Nonpolar compounds are soluble in nonpolar solvents.**

Solution

a. KCl is an ionic compound, so it dissolves in water, a polar solvent.

b. CH_3OH is a small polar molecule that contains an OH group. As a result, it can hydrogen bond to water, making it soluble.

c. Hexane (C_6H_{14}) has only nonpolar C—C and C—H bonds, making it a nonpolar molecule that is therefore water insoluble.

Practice Problem 7.5	Which compounds are water soluble?

a. $NaNO_3$ b. CH_4 c. $HO-\overset{\overset{H}{|}}{\underset{\underset{H}{|}}{C}}-\overset{\overset{H}{|}}{\underset{\underset{H}{|}}{C}}-OH$ d. KBr   e. NH_2OH

Practice Problem 7.6	Which pairs of compounds will form a solution?

a. benzene (C_6H_6) and hexane (C_6H_{14})

b. Na_2SO_4 and H_2O

c. NaCl and hexane (C_6H_{14})

d. H_2O and CCl_4

More Practice: Try Problems 7.23–7.28.

7.4 Solubility—Effects of Temperature and Pressure

Both temperature and pressure can affect solubility.

7.4A Temperature Effects

For most ionic and molecular solids, solubility generally *increases* as temperature *increases*. Thus, sugar is much more soluble in a cup of hot coffee than in a glass of iced tea. If a solid is dissolved in a solvent at high temperature and then the solution is slowly cooled, the solubility of the solute decreases and it precipitates from the solution. Sometimes, however, if cooling is very slow, the solution becomes **supersaturated** with solute; that is, the solution contains more than the predicted maximum amount of solute at a given temperature. Such a solution is unstable, and when it is disturbed, the solute precipitates rapidly.

In contrast, **the solubility of gases *decreases* with increasing temperature.** Because increasing temperature increases the kinetic energy, more molecules escape into the gas phase and fewer remain in solution. Increasing temperature decreases the solubility of oxygen in lakes and streams. In cases where industrial plants operating near lakes or streams have raised water temperature, marine life dies from lack of sufficient oxygen in solution.

Scuba divers often hold onto a rope so that they ascend to the surface slowly after a dive, in order to avoid forming bubbles of nitrogen gas in joints and blood vessels. In a slow ascent, the external pressure around the diver slowly decreases and by Henry's law, the solubility of the nitrogen gas in the diver's blood slowly decreases as well.
Stephen Frink/Getty Images

> ### Concept Check 7.5
> Why does a soft drink become "flat" faster when it is left open at room temperature compared to when it is left open in the refrigerator?

7.4B   Pressure Effects

Pressure changes do not affect the solubility of liquids and solids, but pressure affects the solubility of gases a great deal. **Henry's law** describes the effect of pressure on gas solubility.

- Henry's law: The solubility of a gas in a liquid is proportional to the partial pressure of the gas above the liquid.

Thus, the *higher* the pressure, the *higher* the solubility of a gas in a solvent. A practical demonstration of Henry's law occurs whenever we open a carbonated soft drink. Soft drinks containing dissolved CO_2 are sealed under greater than 1 atm pressure. When a can is opened, the pressure above the liquid decreases to 1 atm, so the solubility of the CO_2 in the soda decreases as well and some of the dissolved CO_2 fizzes out of solution (Figure 7.4).

> ### Concept Check 7.6
> Predict the effect each change has on the solubility of [1] $Na_2CO_3(s)$; [2] $N_2(g)$.
>
> a. increasing the temperature   c. increasing the pressure
> b. decreasing the temperature   d. decreasing the pressure

Figure 7.4   Henry's Law and Carbonated Beverages

The gas pressure in a closed can of soda is approximately 2 atm. When the can is opened, the pressure above the liquid in the can decreases to 1 atm, so the CO_2 concentration in the soda decreases as well, and the gas fizzes from the soda. *(left & right): Jill Braaten/McGraw Hill*

Mouthwash, sore throat spray, and many other over-the-counter medications contain ingredients whose concentrations are reported in (w/v)%.
Suzie Ross/McGraw Hill

7.5 Concentration Units—Percent Concentration

In using a solution in the laboratory or in administering the proper dose of a liquid medication, we must know its *concentration*—**how much solute is dissolved in a given amount of solution.** Concentration can be measured in several different ways that use mass, volume, or moles. Two useful measures of concentration are reported as percentages—that is, the number of grams or milliliters of solute per 100 mL of solution.

7.5A Weight/Volume Percent

One of the most common measures of concentration is **weight/volume percent concentration, (w/v)%—that is, the number of grams of solute dissolved in 100 mL of solution.** Mathematically, weight/volume percent is calculated by dividing the number of grams of solute in a given number of milliliters of solution, and multiplying by 100%.

$$\text{Weight/volume percent concentration}$$
$$(w/v)\% = \frac{\text{mass of solute (g)}}{\text{volume of solution (mL)}} \times 100\%$$

For example, vinegar contains 5 g of acetic acid dissolved in 100 mL of solution, so the acetic acid concentration is 5% (w/v).

$$(w/v)\% = \frac{5 \text{ g acetic acid}}{100 \text{ mL vinegar solution}} \times 100\% = 5\% \text{ (w/v) acetic acid}$$

Note that the volume used to calculate concentration is the *final* volume of the solution, not the volume of solvent added to make the solution. A special flask called a **volumetric flask** is used to make a solution of a given concentration (Figure 7.5). The solute is placed in the flask and then enough solvent is added to dissolve the solute by mixing. Next, additional solvent is added until it reaches a calibrated line that measures the final volume of the solution.

Figure 7.5

Making a Solution with a Particular Concentration

a. Add the solute.

b. Add the solvent.

To make a solution of a given concentration: (a) add a measured number of grams of solute to a volumetric flask; (b) then add solvent to dissolve the solid, bringing the level of the solvent to the calibrated mark on the neck of the flask. *(a & b): Jill Braaten/McGraw Hill*

Sample Problem 7.4	**Calculating Weight/Volume Percent Concentration**

Chloraseptic sore throat spray contains 0.35 g of the antiseptic phenol dissolved in 25 mL of solution. What is the weight/volume percent concentration of phenol?

Analysis

Use the formula **(w/v)% = (grams of solute)/(mL of solution) × 100%.**

Solution

$$(w/v)\% = \frac{0.35 \text{ g phenol}}{25 \text{ mL solution}} \times 100\% = 1.4\% \text{ (w/v) phenol}$$
Answer

Practice Problem 7.7	Pepto-Bismol, an over-the-counter medication used for upset stomach and diarrhea, contains 525 mg of bismuth subsalicylate in each 15-mL tablespoon. What is the weight/volume percent concentration of bismuth subsalicylate?

Practice Problem 7.8	A commercial mouthwash contains 4.3 g of ethanol and 0.021 g of antiseptic in each 30.-mL portion. Calculate the weight/volume percent concentration of each component.

More Practice: Try Problems 7.8a, 7.9a, 7.10a, 7.31, 7.32, 7.43a, 7.61.

The alcohol (ethanol) content of wine, beer, and other alcoholic beverages is reported using volume/volume percent concentration (Sample Problem 7.5). Wines typically contain 10–13% (v/v) ethanol, whereas beer usually contains 3–5%. *Christian Delbert/Shutterstock*

7.5B   Volume/Volume Percent

When the solute in a solution is a liquid, its concentration is often reported using **volume/volume percent concentration, (v/v)%—that is, the number of milliliters of solute dissolved in 100 mL of solution.** Mathematically, volume/volume percent is calculated by dividing the number of milliliters of solute in a given number of milliliters of solution, and multiplying by 100%.

Volume/volume percent concentration

$$(v/v)\% = \frac{\text{volume of solute (mL)}}{\text{volume of solution (mL)}} \times 100\%$$

For example, a bottle of rubbing alcohol that contains 70 mL of 2-propanol in 100 mL of solution has a 70% (v/v) concentration of 2-propanol.

$$(v/v)\% = \frac{70 \text{ mL 2-propanol}}{100 \text{ mL rubbing alcohol}} \times 100\% = 70\% \text{ (v/v) 2-propanol}$$

Lemon extract (Practice Problem 7.9) used in cooking is prepared from lemon rind, ethanol, and water. *Sheila Fitzgerald/Shutterstock*

Sample Problem 7.5	**Calculating Volume/Volume Percent Concentration**

A 750-mL bottle of wine contains 101 mL of ethanol. What is the volume/volume percent concentration of ethanol?

Analysis

Use the formula **(v/v)% = (mL of solute)/(mL of solution) × 100%.**

Solution

$$(v/v)\% = \frac{101 \text{ mL ethanol}}{750 \text{ mL wine}} \times 100\% = 14\% \text{ (v/v) ethanol}$$
Answer

Practice Problem 7.9

A 1.0-oz bottle of lemon extract contains 24 mL of ethanol. What is the volume/volume percent concentration of ethanol?

More Practice: Try Problems 7.33, 7.34.

Ketamine (Sample Problem 7.6) is a widely used anesthetic in both human and veterinary medicine. It has been illegally used as a recreational drug because it can produce hallucinations. *Bob London/Alamy Stock Photo*

Alcohol-based hand sanitizers (Practice Problem 7.11) clean hands by killing many types of disease-causing bacteria and viruses. *Mark Dierker/McGraw Hill*

Mark Dierker/McGraw Hill
(Practice Problem 7.12)

7.5C   Using a Percent Concentration as a Conversion Factor

Percent concentration can be used as a conversion factor to relate the amount of solute (either grams or milliliters) to the amount of solution. For example, ketamine, an anesthetic especially useful for children, is supplied as a 5.0% (w/v) solution, meaning that 5.0 g of ketamine are present in 100 mL of solution. Two conversion factors derived from the percent concentration can be written.

$$5.0\% \text{ (w/v) ketamine} \qquad \frac{5.0 \text{ g ketamine}}{100 \text{ mL solution}} \quad \text{or} \quad \frac{100 \text{ mL solution}}{5.0 \text{ g ketamine}}$$

weight/volume
percent concentration

We can use these conversion factors to determine the amount of solute contained in a given volume of solution or to determine how much solution contains a given number of grams of solute (Sample Problem 7.6).

Sample Problem 7.6	Using Percent Concentration to Determine the Volume of Solution

What volume of a 5.0% (w/v) solution of ketamine contains 75 mg?

Analysis and Solution

[1]   Identify the known quantities and the desired quantity.

5.0% (w/v) ketamine solution

75 mg   ? mL ketamine
known quantities   desired quantity

[2]   Write out the conversion factors.

- Use the **weight/volume percent concentration** to set up conversion factors that relate grams of ketamine to mL of solution. Because percent concentration is expressed in grams, a mg–g conversion factor is needed as well.

mg–g conversion factors   **g–mL solution conversion factors**

$$\frac{1000 \text{ mg}}{1 \text{ g}} \quad \text{or} \quad \frac{1 \text{ g}}{1000 \text{ mg}} \qquad\qquad \frac{5.0 \text{ g ketamine}}{100 \text{ mL solution}} \quad \text{or} \quad \frac{100 \text{ mL solution}}{5.0 \text{ g ketamine}}$$

Choose the conversion factors with the unwanted units—mg and g—in the denominator.

[3]   Solve the problem.

- Multiply the original quantity by the conversion factors to obtain the desired quantity.

$$75 \text{ mg ketamine} \times \frac{1 \text{ g}}{1000 \text{ mg}} \times \frac{100 \text{ mL solution}}{5.0 \text{ g ketamine}} = 1.5 \text{ mL solution}$$

Answer

Practice Problem 7.10

A drink sold in a health food store contains 0.50% (w/v) of vitamin C. What volume would you have to ingest to obtain 1,000. mg of vitamin C?

Practice Problem 7.11

How many mL of ethanol are contained in an 8.0-oz bottle of hand sanitizer that has a concentration of 62% (v/v) of ethanol? (1 fl oz = 29.57 mL)

Practice Problem 7.12

The pain reliever ibuprofen is sold as an oral suspension for children 2–11 years of age. (a) What is the weight/volume percent concentration in the suspension of ibuprofen shown in the given label? (b) How many milligrams would you obtain in a 2.5-tsp dose? (1 tsp = 5 mL) (c) Why do the directions on the medicine bottle say, "Shake well before using"?

More Practice: Try Problems 7.9b, 7.10b, 7.45a, 7.60, 7.63a, 7.64a, 7.69, 7.70.

7.5D Parts Per Million

When a solution contains a very small concentration of solute, concentration is often expressed in **parts per million (ppm).** Whereas percent concentration is the number of "parts"—grams or milliliters—in 100 parts (100 mL) of solution, parts per million is the **number of "parts" in 1,000,000 parts of solution.** The "parts" may be expressed in either mass or volume units as long as the *same* unit is used for both the numerator and denominator.

$$\text{Parts per million}$$

$$\text{ppm} = \frac{\text{mass of solute (g)}}{\text{mass of solution (g)}} \times 10^6$$

or

$$\text{ppm} = \frac{\text{volume of solute (mL)}}{\text{volume of solution (mL)}} \times 10^6$$

A sample of seawater that contains 1.3 g of magnesium ions in 10^6 g of solution contains 1.3 ppm of magnesium.

$$\text{ppm} = \frac{1.3 \text{ g magnesium}}{10^6 \text{ g seawater}} \times 10^6 = 1.3 \text{ ppm magnesium}$$

Parts per million is used as a concentration unit for very dilute solutions. When water is the solvent, the density of the solution is close to the density of pure water, which is 1.0 g/mL at room temperature. In this case, the numerical value of the denominator is the same no matter if the unit is grams or milliliters. Thus, an aqueous solution that contains 2 ppm of MTBE, a gasoline additive and environmental pollutant, can be written in the following ways:

$$\frac{2 \text{ g MTBE}}{10^6 \text{ g solution}} \times 10^6 = \frac{2 \text{ g MTBE}}{10^6 \text{ mL solution}} \times 10^6 = 2 \text{ ppm MTBE}$$

10^6 mL has a mass of 10^6 g.

ENVIRONMENTAL NOTE

Seabirds such as osprey that feed on fish contaminated with the pesticide DDT (Sample Problem 7.7) accumulate an average of 25 parts per million of DDT in their fatty tissues. When DDT concentration is high, mother osprey produce eggs with very thin shells that are easily crushed, so fewer osprey chicks hatch. *Comstock/Getty Images*

Sample Problem 7.7   Calculating Parts Per Million

What is the concentration in parts per million of DDT in the tissues of a seabird that contains 50. mg of DDT in 1,900 g of tissue? DDT, a nonbiodegradable pesticide that is a persistent environmental pollutant, has been banned from use in the United States since 1973.

Analysis
Use the formula **ppm = (g of solute)/(g of solution) × 10^6.**

Solution

[1] **Convert milligrams of DDT to grams of DDT so that both the solute and solution have the same unit.**

$$50. \text{ mg DDT} \times \frac{1 \text{ g}}{1000 \text{ mg}} = 0.050 \text{ g DDT}$$

[2] **Use the formula to calculate parts per million.**

$$\frac{0.050 \text{ g DDT}}{1900 \text{ g tissue}} \times 10^6 = 26 \text{ ppm DDT}$$

Answer

Practice Problem 7.13

What is the concentration in parts per million of DDT in each of the following?

a. 0.042 mg in 1,400 g plankton

b. 5×10^{-4} g in 1.0 kg minnow tissue

c. 2.0 mg in 1.0 kg needlefish tissue

d. 225 µg in 1.0 kg breast milk

More Practice: Try Problems 7.47, 7.48.

7.6   Concentration Units—Molarity

The most common measure of concentration in the laboratory is *molarity*—**the number of moles of solute per liter of solution,** abbreviated as M.

$$\text{Molarity} \;=\; \mathbf{M} \;=\; \frac{\text{moles of solute (mol)}}{\text{liter of solution (L)}}$$

A solution that is formed from 1.00 mol (58.44 g) of NaCl in enough water to give 1.00 L of solution has a molarity of 1.00 M. A solution that is formed from 2.50 mol (146 g) of NaCl in enough water to give 2.50 L of solution is also a 1.00 M solution. Both solutions contain the *same number of moles per unit volume.*

$$\text{M} \;=\; \frac{\text{moles of solute (mol)}}{V\,(\text{L})} \;=\; \frac{1.00 \text{ mol NaCl}}{1.00 \text{ L solution}} \;=\; 1.00 \text{ M}$$

$$\text{M} \;=\; \frac{\text{moles of solute (mol)}}{V\,(\text{L})} \;=\; \frac{2.50 \text{ mol NaCl}}{2.50 \text{ L solution}} \;=\; 1.00 \text{ M}$$

same concentration
same number of moles per unit volume (*V*)

Because quantities in the laboratory are weighed on a balance, we must learn how to determine molarity beginning with a particular number of grams of a substance, as shown in the accompanying stepwise procedure.

How To   Calculate Molarity from a Given Number of Grams of Solute

Example   Calculate the molarity of a solution made from 20.0 g of NaOH in 250 mL of solution.

Step [1]   Identify the known quantities and the desired quantity.

20.0 g NaOH

250 mL solution   ? M (mol/L)

known quantities   desired quantity

Step [2]   Convert the number of grams of solute to the number of moles. Convert the volume of the solution to liters, if necessary.

- Use the **molar mass** to convert grams of NaOH to moles of NaOH (molar mass 40.00 g/mol).

molar mass
conversion factor

$$20.0 \text{ g NaOH} \;\times\; \frac{1 \text{ mol}}{40.00 \text{ g NaOH}} \;=\; 0.500 \text{ mol NaOH}$$

Grams cancel.

- Convert milliliters of solution to liters of solution using a **mL–L conversion factor.**

mL–L
conversion factor

$$250 \text{ mL solution} \;\times\; \frac{1 \text{ L}}{1000 \text{ mL}} \;=\; 0.25 \text{ L solution}$$

Milliliters cancel.

Step [3]   Divide the number of moles of solute by the number of liters of solution to obtain the molarity.

$$\mathbf{M} \;=\; \frac{\text{moles of solute (mol)}}{V\,(\text{L})} \;=\; \frac{0.500 \text{ mol NaOH}}{0.25 \text{ L solution}} \;=\; 2.0 \text{ M}$$

molarity   Answer

Sample Problem 7.8   Calculating the Molarity of a Solution

What is the molarity of an intravenous glucose solution prepared from 108 g of glucose in 2.0 L of solution?

Analysis and Solution

[1]   **Identify the known quantities and the desired quantity.**

108 g glucose
2.0 L solution   ? M (mol/L)
known quantities   desired quantity

[2]   **Convert the number of grams of glucose to the number of moles using the molar mass (180.16 g/mol).**

$$108 \text{ g glucose} \quad \times \quad \frac{1 \text{ mol}}{180.16 \text{ g}} \quad = \quad 0.599 \text{ mol glucose}$$

Grams cancel.

• Since the volume of the solution is given in liters, no conversion is necessary for volume.

[3]   **Divide the number of moles of solute by the number of liters of solution to obtain the molarity.**

$$M \; = \; \frac{\text{moles of solute (mol)}}{V \text{ (L)}} \; = \; \frac{0.599 \text{ mol glucose}}{2.0 \text{ L solution}} \; = \; 0.30 \text{ M}$$

molarity   **Answer**

Practice Problem 7.14

Calculate the molarity of each aqueous solution with the given amount of NaCl (molar mass 58.44 g/mol) and final volume.

a. 1.0 mol in 0.50 L   c. 0.050 mol in 5.0 mL
b. 2.0 mol in 250 mL   d. 12.0 g in 2.0 L

Practice Problem 7.15

What is the molarity of acetic acid in vinegar that contains 5.00 g of acetic acid (molar mass 60.05 g/mol) in 100. mL of solution?

More Practice: Try Problems 7.35, 7.36.

Molarity is a conversion factor that relates the number of moles of solute to the volume of solution it occupies. Thus, if we know the molarity and volume of a solution, we can calculate the number of moles it contains. If we know the molarity and number of moles, we can calculate the volume in liters.

To calculate the **moles** of solute...   ...rearrange the equation for molarity (M):

$$\frac{\text{moles of solute (mol)}}{V \text{ (L)}} \; = \; M$$ $$\boxed{\text{moles of solute (mol)} \; = \; M \; \times \; V \text{ (L)}}$$

To calculate the **volume** of solution...   ...rearrange the equation for molarity (M):

$$\frac{\text{moles of solute (mol)}}{V \text{ (L)}} \; = \; M$$ $$\boxed{V \text{ (L)} \; = \; \frac{\text{moles of solute (mol)}}{M}}$$

Sample Problem 7.9

Calculating Volume from Molarity and Moles

What volume in milliliters of a 0.30 M solution of glucose contains 0.025 mol of glucose?

Analysis

Use the equation, V = (moles of solute)/M, to find the volume in liters, and then convert the liters to milliliters.

Solution

[1] Identify the known quantities and the desired quantity.

$$0.30 \text{ M}$$
$$0.025 \text{ mol glucose}$$
known quantities

$$? \; V \text{ (L) solution}$$
desired quantity

[2] Divide the number of moles by molarity to obtain the volume in liters.

$$V \text{ (L)} = \frac{\text{moles of solute (mol)}}{M}$$

$$= \frac{0.025 \; \cancel{\text{mol}} \text{ glucose}}{0.30 \; \cancel{\text{mol}}/\text{L}} = 0.083 \text{ L solution}$$

[3] Use a mL–L conversion factor to convert liters to milliliters.

mL–L conversion factor

$$0.083 \; \cancel{\text{L}} \text{ solution} \quad \times \quad \frac{1000 \text{ mL}}{1 \; \cancel{\text{L}}} \quad = \quad 83 \text{ mL glucose solution}$$

Liters cancel.   **Answer**

Practice Problem 7.16

How many milliliters of a 1.5 M glucose solution contain each of the following number of moles?

a. 0.15 mol   b. 0.020 mol   c. 0.0030 mol   d. 3.0 mol

Practice Problem 7.17

How many moles of NaCl are contained in each volume of aqueous NaCl solution?

a. 2.0 L of a 2.0 M solution   c. 25 mL of a 2.0 M solution
b. 2.5 L of a 0.25 M solution   d. 250 mL of a 0.25 M solution

More Practice: Try Problems 7.39, 7.40.

Because the number of grams and moles of a substance is related by the molar mass, we can convert a given volume of solution to the number of grams of solute it contains by carrying out the stepwise calculation shown in Sample Problem 7.10.

Sample Problem 7.10

Determining the Number of Grams in a Solution

How many grams of aspirin are contained in 50.0 mL of a 0.050 M solution?

Analysis

Use the **molarity** to convert the volume of the solution to moles of solute. Then use the **molar mass** to convert moles to grams.

Solution

[1] Identify the known quantities and the desired quantity.

0.050 M   ? g aspirin

50.0 mL solution

known quantities   desired quantity

[2] Determine the number of moles of aspirin using the molarity.

$$50.0 \text{ mL solution} \times \frac{0.050 \text{ mol aspirin}}{1 \text{ L}} \times \frac{1 \text{ L}}{1000 \text{ mL}} = 0.0025 \text{ mol aspirin}$$

[3] Convert the number of moles of aspirin to grams using the molar mass (180.15 g/mol).

molar mass
conversion factor

$$0.0025 \text{ mol aspirin} \times \frac{180.15 \text{ g aspirin}}{1 \text{ mol aspirin}} = 0.45 \text{ g aspirin}$$

— Moles cancel. —

Answer

Practice Problem 7.18

How many grams of NaCl are contained in each of the following volumes of a 1.25 M solution?

 a. 0.10 L b. 2.0 L c. 0.55 L d. 50. mL

Practice Problem 7.19

How many milliliters of a 0.25 M sucrose solution contain each of the following number of grams? The molar mass of sucrose ($C_{12}H_{22}O_{11}$) is 342.30 g/mol.

 a. 0.500 g b. 2.0 g c. 1.25 g d. 50.0 mg

More Practice: Try Problems 7.41, 7.42.

7.7 Dilution

Sometimes a solution has a higher concentration than is needed. ***Dilution* is the addition of solvent to decrease the concentration of solute.** For example, a stock solution of a drug is often supplied in a concentrated form to take up less space on a pharmacy shelf, and then it is diluted so that it can be administered in a reasonable volume and lower concentration that allows for more accurate dosing.

A key fact to keep in mind is that the **amount of solute is *constant.*** Only the volume of the solution is changed by adding solvent.

Dilute with
more solvent.

The diluted solution contains the *same* number of molecules in a *larger* volume.

initial solution   diluted solution

Concept Check 7.7   If the solution of **A⁺B⁻** in **X** is diluted, which representation (**D–F**) represents the final solution?

X   D   E   F

In using molarity as a measure of concentration in Section 7.6, we learned that the number of moles of solute can be calculated from the molarity and volume of a solution.

$$\text{moles of solute} \;=\; \text{molarity} \;\times\; \text{volume}$$

$$\textbf{mol} \;=\; \textbf{M}V$$

Thus, if we have initial values for the molarity and volume (M_1 and V_1), we can calculate a new value for the molarity or volume (M_2 or V_2), because the product of the molarity and volume equals the number of moles, a constant.

$$\underset{\text{initial values}}{\textbf{M}_1\textbf{V}_1} \;=\; \underset{\text{final values}}{\textbf{M}_2\textbf{V}_2}$$

Although molarity is the most common concentration measure in the laboratory, the same facts hold in diluting solutions reported in other concentration units—percent concentration and parts per million—as well. In general, therefore, if we have initial values for the concentration and volume (C_1 and V_1), we can calculate a new value for the concentration or volume (C_2 or V_2), because the product of the concentration and volume is a constant.

$$\underset{\text{initial values}}{\textbf{C}_1\textbf{V}_1} \;=\; \underset{\text{final values}}{\textbf{C}_2\textbf{V}_2}$$

Sample Problem 7.11   **Calculating Concentration After Dilution**

What is the concentration of a solution formed by diluting 5.0 mL of a 3.2 M glucose solution to 40.0 mL?

Analysis

Because we know an initial molarity and volume (M_1 and V_1) and a final volume (V_2), we can calculate a new molarity (M_2) using the equation **$M_1V_1 = M_2V_2$.**

Solution

[1] Identify the known quantities and the desired quantity.

$$M_1 = 3.2 \text{ M}$$
$$V_1 = 5.0 \text{ mL} \qquad V_2 = 40.0 \text{ mL} \qquad\qquad M_2 = ?$$
$$\underset{\text{known quantities}}{\hspace{4cm}} \qquad \underset{\text{desired quantity}}{\hspace{2cm}}$$

[2] Write the equation and rearrange it to isolate the desired quantity, M_2, on one side.

$$M_1V_1 \;=\; M_2V_2 \qquad \text{Solve for } M_2 \text{ by dividing both sides by } \textbf{V}_2.$$

$$\frac{M_1V_1}{\textbf{V}_2} \;=\; M_2$$

[3] Solve the problem.

- Substitute the three known quantities into the equation and solve for M_2.

$$M_2 = \frac{M_1 V_1}{V_2} = \frac{(3.2\ M)(5.0\ \text{mL})}{(40.0\ \text{mL})} = 0.40\ M\ \text{glucose solution}$$

Answer

Practice Problem 7.20

What is the concentration of a solution formed by diluting 25.0 mL of a 3.8 M glucose solution to 275 mL?

More Practice: Try Problems 7.7, 7.8b, 7.51, 7.52.

Sample Problem 7.12

Determining the Volume Used in a Dilution

Dopamine is a potent drug administered intravenously to increase blood pressure in seriously ill patients. How many milliliters of a 4.0% (w/v) solution must be used to prepare 250 mL of a 0.080% (w/v) solution?

Analysis

Because we know an initial concentration (C_1), a final concentration (C_2), and a final volume (V_2), we can calculate the volume (V_1) of the initial solution that must be used with the equation, $C_1 V_1 = C_2 V_2$.

Solution

[1] Identify the known quantities and the desired quantity.

$$C_1 = 4.0\%\ \text{(w/v)} \qquad C_2 = 0.080\%\ \text{(w/v)}$$
$$V_2 = 250\ \text{mL} \qquad\qquad V_1 = ?$$

　　　　　known quantities　　　　　　desired quantity

[2] Write the equation and rearrange it to isolate the desired quantity, V_1, on one side.

$$C_1 V_1 = C_2 V_2 \qquad \text{Solve for } V_1 \text{ by dividing both sides by } C_1.$$

$$V_1 = \frac{C_2 V_2}{C_1}$$

[3] Solve the problem.

- Substitute the three known quantities into the equation and solve for V_1.

$$V_1 = \frac{(0.080\%)(250\ \text{mL})}{4.0\%} = 5.0\ \text{mL dopamine solution}$$

Answer

Practice Problem 7.21

How many milliliters of a 6.0 M NaOH solution would be needed to prepare each solution?

a. 525 mL of a 2.5 M solution
b. 750 mL of a 4.0 M solution
c. 450 mL of a 0.10 M solution
d. 25 mL of a 3.5 M solution

Practice Problem 7.22

Ketamine (Sample Problem 7.6) is supplied in a solution of 100. mg/mL. If 2.0 mL of this solution is diluted to a volume of 10.0 mL, how much of the diluted solution should be administered to supply a dose of 75 mg?

More Practice: Try Problems 7.53, 7.54.

7.8   Osmosis and Dialysis

The membrane that surrounds living cells is an example of a **semipermeable membrane**—a membrane that allows water (and some small molecules) to pass across, but ions and large molecules cannot.

semipermeable membrane

H_2O

solute

H_2O

solute + solvent   solvent only

- *Osmosis* is the passage of solvent, usually water, across a semipermeable membrane from a solution of low solute concentration to a solution of higher solute concentration.

7.8A   Osmotic Pressure

What happens when water and an aqueous glucose solution are separated by a semipermeable membrane? **Water flows back and forth across the membrane,** but more water flows from the side that has pure solvent toward the side that has dissolved glucose. This *decreases* the volume of pure solvent on one side of the membrane and *increases* the volume of the glucose solution on the other side.

semipermeable membrane

H_2O

glucose + H_2O solution

time

osmotic pressure

H_2O

glucose + H_2O solution

More H_2O molecules diffuse into the glucose solution, where the concentration of H_2O is *lower.*

H_2O

glucose

The increased weight of the glucose solution creates increased pressure on one side of the membrane. When the increased pressure gets to a certain point, it prevents more water movement to further dilute the glucose solution. Water continues to diffuse back and forth across the membrane, but the level of the two liquids does not change any further.

- *Osmotic pressure* is the pressure that prevents the flow of additional solvent into a solution on one side of a semipermeable membrane.

Osmotic pressure depends only on the number of particles in a solution. **The *greater* the number of dissolved particles, the *greater* the osmotic pressure.** A 0.1 M NaCl solution has twice the osmotic pressure as a 0.1 M glucose solution, because each NaCl is composed of two particles, Na^+ cations and Cl^- anions.

If, instead of having pure water on one side of the membrane, there were two solutions of different concentrations, **water would flow from the side of the *less* concentrated solution to dilute the *more* concentrated solution.**

Sample Problem 7.13	Predicting the Result of Osmosis

A 0.1 M glucose solution is separated from a 0.2 M glucose solution by a semipermeable membrane. (a) Which solution exerts the greater osmotic pressure? (b) In which direction will water flow between the two solutions? (c) Describe the level of the two solutions when equilibrium is reached.

Analysis

The solvent (water) flows from the *less* concentrated solution to the *more* concentrated solution.

Solution

a. The *greater* the number of dissolved particles, the *higher* the osmotic pressure, so the 0.2 M glucose solution exerts the *greater* pressure.

b. Water will flow from the *less* concentrated solution (0.1 M) to the *more* concentrated solution (0.2 M).

c. Because water flows into the 0.2 M solution, its height will *increase,* and the height of the 0.1 M glucose solution will *decrease.*

Practice Problem 7.23	Which solution in each pair exerts the greater osmotic pressure?

a. 1.0% sugar solution or 5.0% sugar solution

b. 3.0 M NaCl solution or a 4.0 M NaCl solution

c. 1.0 M glucose solution or a 0.75 M NaCl solution

Practice Problem 7.24	Consider the two aqueous solutions separated by a semipermeable membrane, shown in representation **A.** Which diagram (**B–D**) represents the final appearance of the liquids and dissolved solute?

More Practice: Try Problems 7.11, 7.12.

7.8B   FOCUS ON THE HUMAN BODY
Osmosis and Biological Membranes

Because cell membranes are semipermeable and biological fluids contain dissolved ions and molecules, osmosis is an ongoing phenomenon in living cells. Fluids on both sides of a cell membrane must have the same osmotic pressure to avoid pressure buildup inside or outside the cell. Any intravenous solution given to a patient, therefore, must have the same osmotic pressure as the fluids in the body.

- Two solutions with the same osmotic pressure are said to be *isotonic*.

Isotonic solutions used in hospitals include 0.92% (w/v) NaCl solution (or 0.15 M NaCl solution) and 5.0% (w/v) glucose solution. Although these solutions do not contain exactly the same ions or molecules present in body fluids, they exert the same osmotic pressure. **In osmosis, the *concentration* of dissolved particles is important, but the identity of those particles is not.**

If a red blood cell is placed in an isotonic NaCl solution, called **physiological saline solution,** the red blood cells retain their same size and shape because the osmotic pressure inside and out-side the cell is the same (Figure 7.6a). What happens if a red blood cell is placed in a solution having a different osmotic pressure?

- A *hypotonic* solution has a lower osmotic pressure than body fluids.
- A *hypertonic* solution has a higher osmotic pressure than body fluids.

In a **hypotonic solution,** the concentration of particles outside the cell is lower than the concen-tration of particles inside the cell. In other words, the concentration of water outside the cell is *higher* than the concentration of water inside the cell, so water diffuses *inside* (Figure 7.6b). As a result, the cell swells and eventually bursts. This swelling and rupture of red blood cells is called **hemolysis.**

In a **hypertonic solution,** the concentration of particles outside the cell is higher than the concentration of particles inside the cell. In other words, the concentration of water inside the cell is *higher* than the concentration of water outside the cell, so water diffuses *out* of the cell (Figure 7.6c). As a result, the cell shrinks. This process is called **crenation.**

Figure 7.6

The Effect of Osmotic Pressure Differences on Red Blood Cells

a.   b.   c.

isotonic
solution

hypotonic
solution

hypertonic
solution

(a) In an isotonic solution, the movement of water into and out of the red blood cell occurs to an equal extent and the red blood cell keeps its normal volume. (b) In a hypotonic solution, more water moves into the cell than diffuses out, so the cell swells and eventually it can rupture (hemolysis). (c) In a hypertonic solution, more water moves out of the cell than diffuses in, so the cell shrivels (crenation). *(a–c): Dennis Kunkel Microscopy/Science Source*

Concept Check 7.8

Which aqueous solution (**B–D**) is isotonic with solution **A**?

A B C D

Concept Check 7.9

Which diagram represents the appearance of a red blood cell when it is placed in each of the following solutions: (a) 3% (w/v) glucose solution; (b) 0.15 M KCl solution; (c) 0.15 M Na_2CO_3 solution?

A B C

7.8C FOCUS ON HEALTH & MEDICINE
Dialysis

Dialysis is a process similar to osmosis, with one important difference—the identity of the substances that can cross the semipermeable membrane, called a **dialyzing membrane.** In addition to water, **ions and small molecules like glucose and urea can cross a dialyzing membrane,** but large molecules like starch and proteins cannot.

Dialysis is the process by which blood is filtered through the kidneys in the human body (Figure 7.7). Each kidney contains over a million nephrons, tubelike structures with filtration membranes. These membranes filter small molecules—glucose, amino acids, urea, ions, and water—from the blood. Useful materials are then reabsorbed, but urea and other waste products are eliminated in urine.

Figure 7.7   Dialysis of Body Fluids by the Kidneys

filtration slits

kidney

fenestrae

Body fluids are dialyzed by passage through the kidneys, which contain more than a million nephrons that filter out small molecules and ions from the blood. Useful materials are then reabsorbed while urea and other waste products are eliminated in urine.

When an individual's kidneys are incapable of removing waste products from the blood, **hemodialysis** is used (Figure 7.8). A patient's blood flows through a long tube connected to a cellophane membrane suspended in an isotonic solution that contains NaCl, KCl, NaHCO$_3$, and glucose. Small molecules like urea cross the membrane into the solution, thus removing them from the blood. Red blood cells and large molecules are not removed from the blood because they are too big to cross the dialyzing membrane.

Figure 7.8   Hemodialysis

When a patient's kidneys no longer function properly, periodic dialysis treatments are used to remove waste products from the blood. Blood is passed through a dialyzer, which contains a membrane that allows small molecules to pass through, thus acting as an artificial kidney. Each treatment takes several hours. Patients usually require two to three treatments per week.

Concept Check 7.10	If a dialyzing bag containing a solution of NaCl, glucose, and protein was placed in a beaker of water, what substances would be found in the water outside the bag?
Concept Check 7.11	What is the difference between the semipermeable membrane of osmosis and the dialyzing membrane used in hemodialysis?

CHAPTER REVIEW

KEY TERMS

Aqueous solution (7.1)

Colloid (7.1)

Concentration (7.5)

Dialysis (7.8)

Dilution (7.7)

Electrolyte (7.2)

Equivalent (7.2)

Henry's law (7.4)

Heterogeneous mixture (7.1)

Homogeneous mixture (7.1)

Hypertonic solution (7.8)

Hypotonic solution (7.8)

Ion–dipole interaction (7.3)

Isotonic solution (7.8)

"Like dissolves like" (7.3)

Molarity (7.6)

Nonelectrolyte (7.2)

Osmosis (7.8)

Osmotic pressure (7.8)

Parts per million (7.5)

Saturated solution (7.3)

Semipermeable membrane (7.8)

Solubility (7.3)

Solute (7.1)

Solution (7.1)

Solvation (7.3)

Solvent (7.1)

Supersaturated solution (7.4)

Suspension (7.1)

Unsaturated solution (7.3)

Volume/volume percent
    concentration (7.5)

Weight/volume percent
    concentration (7.5)

KEY CONCEPTS

1 Comparing solutions, colloids, and suspensions (7.1)

1 Solution	**2** Colloid	**3** Suspension
solution	colloid	suspension
• A homogeneous mixture with **small** particles (< 1 nm) • Liquid solutions are **transparent.**	• A homogeneous mixture with **larger** particles (1 nm–1 μm) • Colloids are often **opaque.**	• A heterogeneous mixture that contains **large particles suspended in a liquid** • The particles **settle out** on standing. • A suspension **must be shaken** to disperse the particles.

2 Classifying a compound as a strong electrolyte, weak electrolyte, or nonelectrolyte (7.2)

1 Strong electrolyte	**2** Weak electrolyte	**3** Nonelectrolyte
• An aqueous solution of **AB** contains only **A⁺** and **B⁻** ions in water and conducts an electric current.	• An aqueous solution of **AB** contains mostly uncharged **AB** molecules and a small number of **A⁺** and **B⁻** ions, and weakly conducts an electric current.	• An aqueous solution of **AB** contains only neutral **AB** molecules and does not conduct an electric current.
A^+ B^-	AB	

3 Solubility

General rule: "Like dissolves like." (7.3)

1 Water-soluble compounds	**2** Water-insoluble compounds
• **Small polar molecules that can hydrogen bond** with water are *water soluble.* **Compounds with many N or O atoms** that can hydrogen bond with water are *water soluble.* • **Ionic compounds are generally water soluble.** CH_3OH Na^+Cl^- methanol sodium chloride **water soluble** **water soluble**	• **Nonpolar compounds** or those with **many nonpolar C–C and C–H bonds** are *water insoluble.* $C_{10}H_{22}$ decane **water insoluble** many C–C and C–H bonds

KEY EQUATIONS—CONCENTRATION

1 Percent concentration (7.5)

Weight/volume percent concentration	Volume/volume percent concentration
$$(w/v)\% = \frac{\text{mass of solute (g)}}{\text{volume of solution (mL)}} \times 100\%$$	$$(v/v)\% = \frac{\text{volume of solute (mL)}}{\text{volume of solution (mL)}} \times 100\%$$
• **Weight/volume percent concentration**—the number of grams of solute dissolved in 100 mL of solution	• **Volume/volume percent concentration**—the number of milliliters of solute dissolved in 100 mL of solution

② Parts per million (7.5D)	③ Molarity (7.6)

Parts per million

$$ppm = \frac{\text{parts of solute (g or mL)}}{\text{parts of solution (g or mL)}} \times 10^6$$

$$\text{Molarity} = M = \frac{\text{moles of solute (mol)}}{\text{liter of solution (L)}}$$

- **Parts per million**—the number of "parts" in 10^6 parts of solution, where "parts" is expressed in the same unit of mass or volume

- **Molarity**—the number of moles of solute per liter of solution

KEY SKILLS

① **Calculating the milliequivalents of an ion (7.2); example: determining the milliequivalents of Na$^+$ in 0.40 L of solution that contains 125 mEq/L of Na$^+$ ions**

① Identify the original quantities and the desired quantity.	**②** Write out the conversion factor.	**③** Solve the problem so that the unwanted unit (L) cancels.
125 mEq/L Na$^+$ in solution 0.40 L ? mEq Na$^+$ original quantities **desired** quantity	$\dfrac{125 \text{ mEq}}{1 \text{ L}}$ ← **desired** unit ← unwanted unit **conversion factor** • Use a conversion factor that relates milliequivalents to volume.	**conversion factor** $0.40 \text{ L} \times \dfrac{125 \text{ mEq}}{1 \text{ L}} = 50. \text{ mEq Na}^+$ Liters cancel.

② **Using percent concentration as a conversion factor (7.5); example: calculating the number of grams of X in 500. mL of a 0.47% (w/v) solution of X in water**

① Identify the original quantities and the desired quantity.	**②** Write out the conversion factor.	**③** Solve the problem so that the unwanted unit (mL) cancels.
0.47% (w/v) **X** solution 500. mL ? g **X** original quantities **desired** quantity	$\dfrac{0.47 \text{ g X}}{100 \text{ mL}}$ ← **desired** unit ← unwanted unit **conversion factor** • Use the weight/volume percent concentration as a conversion factor.	**conversion factor** $500. \text{ mL} \times \dfrac{0.47 \text{ g X}}{100 \text{ mL}} = 2.4 \text{ g X}$ mL cancel.

③ **Calculating molarity (7.6); example: calculating the molarity of a solution prepared from 4.2 g NaHCO$_3$ (molar mass 84.01 g/mol) in 0.050 L of solution**

① Identify the original quantities and the desired quantity.	**②** Use the molar mass to convert grams to moles.	**③** Substitute the quantities in the equation for molarity and solve.
4.2 g NaHCO$_3$ 0.050 L ? M NaHCO$_3$ original quantities **desired** quantity	**conversion factor** $4.2 \text{ g NaHCO}_3 \times \dfrac{1 \text{ mol}}{84.01 \text{ g}} = 0.050 \text{ mol NaHCO}_3$	$M = \dfrac{\text{mol}}{\text{L}} = \dfrac{0.050 \text{ mol NaHCO}_3}{0.050 \text{ L}}$ $= 1.0 \text{ M NaHCO}_3$ solution

④ **Calculating molarity after dilution (7.7); example: calculating the molarity of a solution formed by diluting 4.0 mL of a 1.2 M glucose solution to 60.0 mL**

① Identify the original quantities and the desired quantity.	**②** Write the equation and rearrange it to isolate the desired quantity, **M$_2$**, on one side.	**③** Solve the problem.
$M_1 = 1.2$ M $V_1 = 4.0$ mL $V_2 = 60.0$ mL $M_2 = ?$ original quantities **desired** quantity	$M_1 V_1 = \mathbf{M_2 V_2}$ Rearrange: $\dfrac{M_1 V_1}{V_2} = \mathbf{M_2}$ • In dilution, the number of moles (**mol = MV**) is constant.	$M_2 = \dfrac{M_1 V_1}{V_2} = \dfrac{(1.2 \text{ M})(4.0 \text{ mL})}{60.0 \text{ mL}} = 0.080 \text{ M glucose solution}$

5 **Predicting the result of osmosis (7.8); example: determining final liquid levels in two compartments with equal volumes of liquids A and B separated by a semipermeable membrane**

Osmosis—the passage of solvent, usually water, across a semipermeable membrane from a solution of low solute concentration to a solution of higher solute concentration

① Liquid A = 0.10 M glucose Liquid B = 0.20 M sucrose	② Liquid A = 2.0 M NaCl Liquid B = pure water	③ Liquid A = 2.0 M NaCl Liquid B = 4.0 M glucose
• Water flows from the less concentrated solution (0.10 M) to the more concentrated solution (0.20 M). • The height of **A** *decreases* and the height of **B** *increases*.	• Water flows from **B** to the more concentrated solution **A**. • The height of **A** *increases* and the height of **B** *decreases*.	• Because each NaCl gives two dissolved "particles," the concentration of dissolved particles is the same in both solutions. • The height of both liquids is *equal*.

CHAPTER 7 SELF-TEST

The Self-Test consists of short-answer questions that test an understanding of definitions, equations, and other material encountered within the chapter. Answers to each question are provided at the end of the chapter. For any questions you are unable to answer correctly, revisit the appropriate section(s) of the text; then proceed to working the end-of-chapter problems to further reinforce your understanding of the chapter material.

1. Which term from "Key Terms" answers each question?
   a. What type of homogeneous mixture is usually opaque and contains particles 1 nm–1 μm in size?
   b. What concentration unit equals the number of moles of solute per liter of solution?
   c. What type of solution contains the maximum number of grams of solute that can be dissolved?
   d. What term describes a solution that does not conduct an electric current?
   e. What process involves the passage of a solvent, usually water, across a semipermeable membrane from a solution of low solute concentration to a solution of higher solute concentration?

2. Label each compound as water soluble or water insoluble: (a) C_5H_{12}; (b) NaBr; (c) NH_3.

3. If a solution contains 2.5 mol Na^+ and 1.5 mol CO_3^{2-}, how many equivalents of each ion are present?

4. If a solution contains 0.80 g of NaCl dissolved in a total volume of 25 mL, what is the weight/volume percent concentration of NaCl?

5. What is the molarity of a sodium bicarbonate solution prepared from 8.4 g of sodium bicarbonate ($NaHCO_3$, molar mass 84.01 g/mol) in 50. mL of solution?

6. What is the concentration of a solution formed by diluting 15.0 mL of a 1.2 M glucose solution to 60.0 mL?

7. A flask contains two compartments (**A** and **B**) with equal volumes separated by a semipermeable membrane. Which compartment contains the larger final volume of liquid in each circumstance: (a) 0.10 M $CaCl_2$ solution (**A**) and 0.10 M NaCl solution (**B**); (b) pure water (**A**) and 0.20 M glucose solution (**B**)?

8. How many moles of solute are contained in 2.5 L of a 1.5 M HCl solution?

9. Label each substance as a solution, colloid, or suspension: (a) butter; (b) vinegar; (c) a bottle of juice with the label "Shake well."

UNDERSTANDING KEY CONCEPTS

7.1 Which representation of molecular art better shows a solution of KI dissolved in water? Explain your choice. Will this solution conduct an electric current?

A

B

7.2 Which representation of molecular art better shows a solution of NaF dissolved in water? Explain your choice. Will this solution conduct an electric current?

A

B

7.3   Using the ball-and-stick model for methanol (CH$_3$OH), draw a representation for the solution that results when methanol is dissolved in water.

7.4   Using the ball-and-stick model for dimethyl ether [(CH$_3$)$_2$O], draw a representation for the solution that results when dimethyl ether is dissolved in water.

7.5   The molecular art represents O$_2$ molecules dissolved in water at 25 °C. Draw a representation for the solution that results (a) after heating to 50 °C; (b) after cooling to 10 °C.

7.6   The molecular art represents NaCl dissolved in water at 25 °C. Assume that this solution is saturated in NaCl and that solid NaCl remains at the bottom of the flask in which the solution is stored. Draw a representation for the solution that results (a) after heating to 50 °C; (b) after cooling to 10 °C.

7.7   What is the weight/volume percent concentration of a 30.0% (w/v) solution of vitamin C after each of the following dilutions?
   a. 100. mL diluted to 200. mL
   b. 250. mL diluted to 1.5 L

7.8   One gram (1.00 g) of vitamin B$_3$ (niacin) is dissolved in water to give 10.0 mL of solution. (a) What is the weight/volume percent concentration of this solution? (b) What is the concentration of a solution formed by diluting 1.0 mL of this solution to each of the following volumes: [1] 10.0 mL; [2] 2.5 mL?

7.9   Cefdinir is an antibiotic used to treat infections of the middle ear, tonsils, and lungs. It is sold as a suspension in water.
   a. What is the weight/volume percent concentration in a suspension of cefdinir that contains 250 mg/5 mL?
   b. If the typical pediatric dose is 7.0 mg/kg every 12 hours, what volume of drug is given for a 21-kg child?

7.10   Amoxil is an antibiotic used to treat infections of the ear, nose, and throat. It is sold as a suspension in water.
   a. What is the weight/volume percent concentration in a suspension of Amoxil that contains 50 mg/mL?
   b. If the typical pediatric dose is 25 mg/kg per day, what volume of drug is given for a 15-kg child?

7.11   A flask contains two compartments (**A** and **B**) with equal volumes of solution separated by a semipermeable membrane. Describe the final level of the liquids when **A** and **B** contain each of the following solutions.

	A	**B**
a.	1% (w/v) glucose solution	pure water
b.	0.10 M glucose solution	0.20 M glucose solution
c.	0.10 M NaCl solution	0.10 M NaI solution
d.	0.20 M glucose solution	0.10 M NaCl solution

7.12   A flask contains two compartments (**A** and **B**) with equal volumes of solution separated by a semipermeable membrane. Which diagram represents the final level of the liquids when **A** and **B** contain each of the following solutions?

	A	**B**
a.	10% (w/v) glucose	20% (w/v) glucose
b.	0.20 M NaCl	0.30 M glucose
c.	pure water	5% (w/v) glucose
d.	3% (w/v) sucrose	1% (w/v) sucrose

ADDITIONAL PROBLEMS

Solutions, Colloids, and Suspensions

7.13 Classify each of the following as a solution, a colloid, or a suspension.
   a. bronze (an alloy of Sn and Cu)
   b. orange juice with pulp
   c. gasoline
   d. cream

7.14 Considering the size of the particles, classify each mixture as a solution, colloid, or suspension.

a.
1.5 × 10⁻⁷ m

b.
2.0 × 10⁻⁵ m

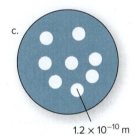
c.
1.2 × 10⁻¹⁰ m

Electrolytes and Equivalents

7.15 Label each diagram as a strong electrolyte, weak electrolyte, or nonelectrolyte.

a.

b.

7.16 Label each diagram as a strong electrolyte, weak electrolyte, or nonelectrolyte.

a.

b.

7.17 A solution contains 2.5 mol KCl and 2.5 mol CaSO₄. How many equivalents of each ion are present?

7.18 A dextrose solution contains 40. mEq/L Na⁺ and 15 mEq/L of HPO₄²⁻, in addition to other ions. How many milliequivalents of each ion are present in 320 mL of solution?

7.19 If a solution contains 132 mEq/L of Na⁺, how many milliequivalents of Na⁺ are present in each of the following volumes of solution: (a) 1.0 L; (b) 2.0 L; (c) 440 mL?

7.20 Ringer's solution is an intravenous solution used for the replacement of electrolytes when an individual is dehydrated. Ringer's solution contains the following cations: Na⁺ (147 mEq/L), K⁺ (4 mEq/L), and Ca²⁺ (4 mEq/L). (a) How many milliequivalents of Na⁺ are present in 125 mL of Ringer's solution? (b) If the only anion in Ringer's solution is Cl⁻, what is the concentration of Cl⁻ in mEq/L?

Solubility

7.21 If the solubility of KCl in 100 mL of H₂O is 34 g at 20 °C and 43 g at 50 °C, label each of the following solutions as unsaturated, saturated, or supersaturated. If more solid is added than can dissolve in the solvent, assume that undissolved solid remains at the bottom of the flask.
   a. adding 30 g to 100 mL of H₂O at 20 °C
   b. adding 65 g to 100 mL of H₂O at 50 °C
   c. adding 42 g to 100 mL of H₂O at 50 °C and slowly cooling to 20 °C to give a clear solution with no precipitate

7.22 If the solubility of sucrose in 100 mL of H₂O is 204 g at 20 °C and 260 g at 50 °C, label each of the following solutions as unsaturated, saturated, or supersaturated. If more solid is added than can dissolve in the solvent, assume that undissolved solid remains at the bottom of the flask.
   a. adding 200 g to 100 mL of H₂O at 20 °C
   b. adding 110 g to 50 mL of H₂O at 20 °C
   c. adding 220 g to 100 mL of H₂O at 50 °C and slowly cooling to 20 °C to give a clear solution with no precipitate

7.23 Which compounds are soluble in water?
   a. LiCl   b. C₇H₈   c. Na₃PO₄

7.24 Which compounds are soluble in water?
   a. C₅H₁₂   b. CaCl₂   c. CH₃Br

7.25 Consider a mixture of two substances shown in blue and red spheres in diagrams **A** and **B**. Which diagram best represents each type of mixture: (a) a polar solute and polar solvent; (b) a nonpolar solute and nonpolar solvent; (c) a nonpolar solute and a polar solvent; (d) a polar solute and nonpolar solvent?

A   B

7.26 Label each compound as water soluble or water insoluble.

a.

b.

pentane

sorbitol

7.27 Explain why cholesterol, a compound with molecular formula $C_{27}H_{46}O$ and one OH group, is soluble in CCl_4 but insoluble in water.

7.28 Which of the following pairs of compounds form a solution?
a. KCl and CCl_4
b. 1-propanol (C_3H_8O) and H_2O
c. pentane (C_5H_{12}) and hexane (C_6H_{14})

7.29 How is the solubility of solid NaCl in water affected by each of the following changes?
a. increasing the temperature from 25 °C to 50 °C
b. decreasing the temperature from 25 °C to 0 °C
c. increasing the pressure from 1 atm to 2 atm
d. decreasing the pressure from 5 atm to 1 atm

7.30 How is the solubility of helium gas in water affected by each of the following changes?
a. increasing the temperature from 25 °C to 50 °C
b. decreasing the temperature from 25 °C to 0 °C
c. increasing the pressure from 1 atm to 2 atm
d. decreasing the pressure from 5 atm to 1 atm

Concentration

7.31 What is the weight/volume percent concentration using the given amount of solute and total volume of solution?
a. 10.0 g of LiCl in 750 mL of solution
b. 25 g of $NaNO_3$ in 150 mL of solution

7.32 What is the weight/volume percent concentration using the given amount of solute and total volume of solution?
a. 5.5 g of LiCl in 550 mL of solution
b. 12.5 g of $NaNO_3$ in 250 mL of solution

7.33

Cinnamon sticks, powdered cinnamon, or cinnamon extract can be used for cinnamon flavoring in baking. How many milliliters of ethanol are contained in a 2.0-oz bottle of cinnamon extract that has a volume/volume percent concentration of 40.% ethanol?

Holly Curry/McGraw Hill

7.34

A 1.0-oz bottle of orange extract contains 23 mL of ethanol. What is the volume/volume percent concentration of ethanol?

Holly Curry/McGraw Hill

7.35 What is the molarity of a solution prepared using the given amount of solute and total volume of solution?
a. 3.5 mol of KCl in 1.50 L of solution
b. 25.0 g of NaCl in 650 mL of solution

7.36 What is the molarity of a solution prepared using the given amount of solute and total volume of solution?
a. 2.4 mol of NaOH in 1.50 L of solution
b. 25.0 g of KCl in 650 mL of solution

7.37 How would you use a 250-mL volumetric flask to prepare each of the following solutions?
a. 4.8% (w/v) acetic acid in water
b. 2.5 M NaCl solution

7.38 How would you use a 250-mL volumetric flask to prepare each of the following solutions?
a. 34% (v/v) ethanol in water
b. 4.0 M NaCl solution

7.39 How many moles of solute are contained in each solution?
a. 150 mL of a 0.25 M $NaNO_3$ solution
b. 45 mL of a 2.0 M HNO_3 solution
c. 2.5 L of a 1.5 M HCl solution

7.40 How many moles of solute are contained in each solution?
a. 250 mL of a 0.55 M $NaNO_3$ solution
b. 145 mL of a 4.0 M HNO_3 solution
c. 6.5 L of a 2.5 M HCl solution

7.41 How many grams of solute are contained in each solution in Problem 7.39?

7.42 How many grams of solute are contained in each solution in Problem 7.40?

7.43 The pain reliever acetaminophen is sold as an oral suspension for children 2–11 years of age.

a. What is the weight/volume percent concentration of acetaminophen in a suspension that contains 160 mg per 5.0 mL?
b. How many grams of acetaminophen are contained in a bottle that holds 4.0 fl oz?

Mark Dierker/McGraw Hill

7.44 What is the molarity of a 20.0% (v/v) aqueous ethanol solution? The density of ethanol (C_2H_6O, molar mass 46.07 g/mol) is 0.790 g/mL.

7.45 A 1.89-L bottle of vinegar contains 5.0% (w/v) of acetic acid ($C_2H_4O_2$, molar mass 60.05 g/mol) in water.
   a. How many grams of acetic acid are present in the container?
   b. How many moles of acetic acid are present in the container?
   c. Convert the weight/volume percent concentration to molarity.

7.46 What is the molarity of a 15% (w/v) glucose solution?

7.47 The maximum safe level of each compound in drinking water is given below. Convert each value to parts per million.
   a. chloroform ($CHCl_3$, a solvent), 80 µg/kg
   b. glyphosate (a pesticide), 700 µg/kg

7.48 The maximum safe level of each metal in drinking water is given below. Convert each value to parts per million.
   a. copper, 1,300 µg/kg
   b. arsenic, 10 µg/kg
   c. chromium, 100 µg/kg

Dilution

7.49 If the solution in **X** is diluted, which representation (**A–C**) represents the final solution?

X	**A**	**B**	**C**

7.50 Consider solutions **A, B,** and **C** in Problem 7.49. If the volumes of **A** and **B** are twice the volume of **C,** which solution is most concentrated? Which solution is least concentrated?

7.51 What is the concentration of a solution formed by diluting 125 mL of 12.0 M HCl solution to 850 mL?

7.52 What is the concentration of a solution formed by diluting 250 mL of 6.0 M NaOH solution to 0.45 L?

7.53 How many milliliters of a 2.5 M NaCl solution would be needed to prepare each solution?
   a. 25 mL of a 1.0 M solution
   b. 1.5 L of a 0.75 M solution
   c. 15 mL of a 0.25 M solution

7.54 (a) If **C** represents a 2.0 M aqueous NaCl solution, what is the concentration of **D,** a solution formed by dilution of **C** with water? (b) If the volume of **C** is 100. mL, how much water is added to form **D?**

C	**D**
	Add H_2O.

Osmosis

7.55 What is the difference between osmosis and osmotic pressure?

7.56 What is the difference between a hypotonic solution and an isotonic solution?

Applications

7.57 Explain why opening a warm can of soda causes a louder "whoosh" and more fizzing than opening a cold can of soda.

7.58 Explain why more sugar dissolves in a cup of hot coffee than a glass of iced coffee.

7.59 If the concentration of glucose in the blood is 90 mg/100 mL, what is the weight/volume percent concentration of glucose? What is the molarity of glucose (molar mass 180.16 g/mol) in the blood?

7.60 If the human body contains 5.0 L of blood, how many grams of glucose are present in the blood if the concentration is 90. mg/100. mL?

7.61 A sports drink contains 15 g of soluble complex carbohydrates in 8.0 oz (1 fl oz = 29.57 mL). What weight/volume percent concentration does this represent?

7.62 A sports drink contains 25 mg of magnesium in an 8.0-oz portion (1 fl oz = 29.57 mL). How many parts per million does this represent? Assume that the mass of 1.0 mL of the solution is 1.0 g.

7.63 Mannitol, a carbohydrate, is supplied as a 25% (w/v) solution. This hypertonic solution is given to patients who have sustained a head injury with associated brain swelling. (a) What volume should be given to provide a dose of 70. g? (b) How does the hypertonic mannitol benefit brain swelling?

7.64 A patient receives 750 mL of a 10.% (w/v) aqueous glucose solution. (a) How many grams of glucose does the patient receive? (b) How many moles of glucose (molar mass 180.16 g/mol) does the patient receive?

7.65 Explain why a cucumber placed in a concentrated salt solution shrivels.

7.66 Explain why a raisin placed in water swells.

7.67 Each day, the stomach produces 2.0 L of gastric juice that contains 0.10 M HCl. How many grams of HCl does this correspond to?

7.68 The molecular art illustrates a red blood cell in two different aqueous solutions that contain dissolved solute particles shown in blue spheres. Black arrows show the direction of movement of water across the cell membrane. In each diagram, is the red blood cell in a hypotonic, isotonic, or hypertonic solution?

A	**B**

7.69 An individual is legally intoxicated with a blood alcohol level of 0.08% (w/v) of ethanol. How many milligrams of ethanol are contained in 5.0 L of blood with this level?

7.70 A bottle of vodka labeled "80 proof" contains 40.% (v/v) ethanol in water. How many mL of ethanol are contained in 250 mL of vodka?

CHALLENGE PROBLEMS

7.71 The therapeutic concentration—the concentration needed to be effective—of acetaminophen ($C_8H_9NO_2$, molar mass 151.16 g/mol) is 10–20 μg/mL. Assume that the density of blood is 1.0 g/mL.
   a. If the concentration of acetaminophen in the blood was measured at 15 ppm, is this concentration in the therapeutic range?
   b. How many moles of acetaminophen are present at this concentration in 5.0 L of blood?

7.72 Very dilute solutions can be measured in parts per billion—that is, the number of parts in 1,000,000,000 parts of solution. To be effective, the concentration of digoxin, a drug used to treat congestive heart failure, must be 0.5–2.0 ng/mL. Convert both values to parts per billion (ppb).

BEYOND THE CLASSROOM

7.73 Pick two over-the-counter cough medicines that list ingredients as a percent concentration and are available in a local drug store or supermarket. Determine how many milligrams of each active ingredient are contained in a single dose. Determine how many milligrams (or grams) of each active ingredient are contained in the entire package. If you have chosen two cough medicines with the same active ingredients, is there a relationship between the cost and the quantity of the active ingredients?

7.74 Many sports drinks contain carbohydrates, sodium, and potassium, in addition to other dissolved minerals and vitamins. Pick a sports drink and calculate the number of grams or milligrams of carbohydrates, sodium, and potassium in 100 mL. Compare the values with sports drinks chosen by other members of the class. Is there any significant difference in commercial products? Is any difference reflected in price?

CHAPTER 7 ANSWERS

Concept Check Answers

7.1 a. heterogeneous mixture   b. colloid   c,d,e: solution

7.2 a. suspension   b. solution   c. colloid

7.3 a. weak electrolyte   c. strong electrolyte
   b. nonelectrolyte

7.4 a,c: electrolyte   b. nonelectrolyte

7.5 The solubility of gases in liquids decreases as temperature increases.

7.6 [1] a. increase   b. decrease   c,d: no change
   [2] a,d: decrease   b,c: increase

7.7 F

7.8 C

7.9 a. C   b. A   c. B

7.10 NaCl and glucose

7.11 In osmosis, water crosses the semipermeable membrane. In hemodialysis, water, ions, and small molecules like glucose and urea cross the dialyzing membrane.

Practice Problem Answers

7.1 a. 1 Eq Na^+ c. 0.5 Eq K^+
   b. 2 Eq Mg^{2+} d. 1.5 Eq PO_4^{3-}

7.2 22 mEq

7.3 29 mEq/L

7.4 2.0×10^2 mL

7.5 a,c,d,e: water soluble

7.6 a, b

7.7 3.5% (w/v)

7.8 14% (w/v) ethanol, 0.070% (w/v) antiseptic

7.9 81%

7.10 2.0×10^2 mL

7.11 150 mL

7.12 a. 2%
   b. 250 mg (rounded to 300 mg)
   c. The medicine is a suspension, so it settles out.

7.13 a. 0.030 ppm   b. 0.5 ppm   c. 2.0 ppm   d. 0.23 ppm

7.14 a. 2.0 M   b. 8.0 M   c. 10. M   d. 0.10 M

7.15 0.833 M

7.16 a. 1.0×10^2 mL   c. 2.0 mL
   b. 13 mL   d. 2.0×10^3 mL

7.17 a. 4.0 mol   b. 0.63 mol   c. 0.050 mol   d. 0.063 mol

7.18 a. 7.3 g   b. 150 g   c. 40. g   d. 3.7 g

7.19 a. 5.8 mL   b. 23 mL   c. 15 mL   d. 0.58 mL

7.20 0.35 M

7.21 a. 220 mL   b. 5.0×10^2 mL   c. 7.5 mL   d. 15 mL

7.22 3.8 mL

7.23 a. 5.0% sugar solution   b. 4.0 M NaCl   c. 0.75 M NaCl

7.24 C

Self-Test Answers

1.   a. colloid   d. nonelectrolyte
      b. molarity   e. osmosis
      c. saturated solution

2.   a. water insoluble   b,c: water soluble

3.   2.5 Eq Na^+, 3.0 Eq CO_3^{2-}

4.   3.2% (w/v) NaCl

5.   2.0 M

6.   0.30 M

7.   a. **A**   b. **B**

8.   3.8 mol

9.   a. colloid   b. solution   c. suspension

Odd-Numbered Problem Answers

7.1   **A.** The solution contains ions, so it conducts an electric current.

7.3

hydrogen bond

7.5   a.   b.

7.7   a. 15.0% (w/v)   b. 5.0% (w/v)

7.9   a. 5% (w/v)   b. 3 mL

7.11   a. **A** > **B**   b. **B** > **A**   c,d: no change

7.13   a,c: solution   b. suspension   d. colloid

7.15   a. weak electrolyte   b. strong electrolyte

7.17   2.5 Eq K^+ 5.0 Eq Ca^{2+}
            2.5 Eq Cl^- 5.0 Eq SO_4^{2-}

7.19   a. 130 mEq   b. 260 mEq   c. 58 mEq

7.21   a. unsaturated   b. saturated   c. supersaturated

7.23   a, c

7.25   a,b: **A**   c,d: **B**

7.27   Cholesterol is not water soluble because it is a large nonpolar molecule with a single OH group.

7.29   a. increased   b. decreased   c,d: no change

7.31   a. 1.3% (w/v)   b. 17% (w/v)

7.33   24 mL

7.35   a. 2.3 M   b. 0.66 M

7.37   a. Add 12 g of acetic acid to the flask and then water to bring the volume to 250 mL.
           b. Add 37 g of NaCl to the flask and then water to bring the volume to 250 mL.

7.39   a. 0.038 mol   b. 0.090 mol   c. 3.8 mol

7.41   a. 3.2 g   b. 5.7 g   c. 140 g

7.43   a. 3.2%   b. 3.8 g

7.45   a. 95 g   b. 1.6 mol   c. 0.85 M

7.47   a. 0.08 ppm   b. 0.7 ppm

7.49   **B**

7.51   1.8 M

7.53   a. 10. mL   b. 450 mL   c. 1.5 mL

7.55   Osmosis is the selective diffusion of solvent, usually water, across a semipermeable membrane. Osmotic pressure is the pressure that prevents the flow of additional solvent into a solution on one side of a semipermeable membrane.

7.57   At warmer temperatures, CO_2 is less soluble in water and more is in the gas phase and escapes as the can is opened and pressure is reduced.

7.59   0.09% (w/v)   0.005 M

7.61   6.3% (w/v)

7.63   a. 280 mL
           b. The hypertonic mannitol draws water out of swollen brain cells and thus reduces the pressure on the brain.

7.65   Water moves out of the cells of the cucumber to the hypertonic salt solution, so the cucumber shrivels.

7.67   7.3 g

7.69   4,000 mg

7.71   a. yes   b. 0.000 50 mol

Design Credits: DNA molecule structure background for Chapter Opener: Watchara/Shutterstock

8

Acids and Bases

Although sports drinks provide hydration and electrolyte replacement for athletes, their acidity may lead to dental erosion if they are heavily consumed. *Mark Dierker/McGraw Hill*

CHAPTER OUTLINE

8.1 Introduction to Acids and Bases

8.2 The Reaction of a Brønsted–Lowry Acid with a Brønsted–Lowry Base

8.3 Acid and Base Strength

8.4 Dissociation of Water

8.5 The pH Scale

8.6 Common Acid–Base Reactions

8.7 Titration

8.8 Buffers

8.9 FOCUS ON THE HUMAN BODY: Buffers in the Blood

CHAPTER GOALS

In this chapter you will learn how to:

1. Identify acids and bases and describe their characteristics

2. Write equations for acid–base reactions

3. Classify an acid or base as strong or weak

4. Define the ion–product of water and use it to calculate hydronium or hydroxide ion concentration

5. Calculate pH

6. Draw the products of common acid–base reactions

7. Use a titration to determine the concentration of an acid or a base

8. Describe the fundamental features of a buffer

9. Understand the importance of buffers in maintaining pH in the body

Chemical terms such as *anion* and *cation* may be unfamiliar to most nonscientists, but **acid** has found a place in everyday language. Commercials advertise the latest remedy for the heartburn caused by excess stomach *acid*. The nightly news may report the latest environmental impact of *acid* rain. Wine lovers often know that wine sours because its alcohol has turned to *acid*. *Acid* comes from the Latin word *acidus,* meaning sour, because when tasting compounds was a routine method of identification, these compounds were found to be sour. Acids commonly react with **bases,** and many products, including antacid tablets, glass cleaners, and drain cleaners, all contain bases. In Chapter 8 we learn about the characteristics of acids and bases and the reactions they undergo.

Why Study ...

Acids and Bases?

8.1   Introduction to Acids and Bases

The earliest definition of acids and bases was suggested by Swedish chemist Svante Arrhenius in the late nineteenth century. According to Arrhenius,

- An *acid* contains a hydrogen atom and dissolves in water to form a hydrogen ion, H^+.
- A *base* contains hydroxide and dissolves in water to form OH^-.

By the Arrhenius definition, HCl is an acid because it forms aqueous H^+ and Cl^- when it dissolves in water. Sodium hydroxide (NaOH) is a base because it contains OH^- and forms solvated Na^+ and OH^- ions when it dissolves in water.

H^+ is formed from HCl.

$$HCl(g) \longrightarrow H^+(aq) + Cl^-(aq)$$

acid

$$NaOH(s) \longrightarrow Na^+(aq) + OH^-(aq)$$

base

OH^- is formed from NaOH.

The names of common acids are derived from the anions formed when they dissolve in water.

- For acids derived from anions whose names end in the suffix *-ide,* add the prefix *hydro* and change the *-ide* ending to *-ic acid.*
- For acids derived from polyatomic anions whose names end in the suffix *-ate,* change the *-ate* ending to *-ic acid.*
- For acids derived from polyatomic anions whose names end in the suffix *-ite,* change the *-ite* ending to *-ous acid.*

Anion	Acid
Cl^- chlor**ide** -----→	HCl **hydro**chlor**ic acid**
SO_4^{2-} sulf**ate** -----→	H_2SO_4 sulfur**ic acid**
SO_3^{2-} sulf**ite** -----→	H_2SO_3 sulfur**ous acid**

Additional examples are listed in Table 8.1.

Table 8.1 Names of Common Acids

Anion	Name of Anion	Acid	Name of Acid
Br^-	Bromide	HBr	Hydrobromic acid
PO_4^{3-}	Phosphate	H_3PO_4	Phosphoric acid
CO_3^{2-}	Carbonate	H_2CO_3	Carbonic acid
$CH_3CO_2^-$	Acetate	CH_3CO_2H	Acetic acid
NO_2^-	Nitrite	HNO_2	Nitrous acid

Concept Check 8.1   Name each acid: (a) HF; (b) HNO_3; (c) HCN.

Concept Check 8.2   If the polyatomic anion ClO_2^- is called chlorite, what is the name for the acid $HClO_2$?

Common Arrhenius bases are formed from cations of Group 1A or Group 2A metals (such as Na^+, K^+, and Ca^{2+}) and the hydroxide anion (OH^-).

8.1A Brønsted–Lowry Definition

While the Arrhenius definition correctly predicts the behavior of many acids and bases, this definition is limited and sometimes inaccurate. We now know, for example, that the hydrogen ion, H^+, does *not* exist in water. H^+ is a naked proton with no electrons, and this concentrated positive charge reacts rapidly with a molecule of H_2O to form the **hydronium ion, H_3O^+.** Although $H^+(aq)$ will sometimes be written in an equation for emphasis, $H_3O^+(aq)$ is actually the reacting species.

$H^+(aq)$ and $H_3O^+(aq)$ are sometimes used interchangeably by chemists. Keep in mind, however, that $H^+(aq)$ does not really exist in aqueous solution.

Moreover, several compounds contain no hydroxide anions, yet they still exhibit the characteristic properties of a base. Examples include the neutral molecule ammonia (NH_3) and the salt sodium carbonate (Na_2CO_3). As a result, a more general definition of acids and bases, proposed by Johannes **Brønsted** and Thomas **Lowry** in the early twentieth century, is widely used today.

In the Brønsted–Lowry definition, acids and bases are classified according to whether they can donate or accept a **proton**—a positively charged hydrogen ion, H^+.

- A Brønsted–Lowry acid is a *proton donor.*
- A Brønsted–Lowry base is a *proton acceptor.*

Consider what happens when HCl is dissolved in water.

- HCl is a Brønsted–Lowry *acid* because it *donates* a proton to the solvent water.
- H_2O is a Brønsted–Lowry *base* because it *accepts* a proton from HCl.

Before we learn more about the details of this process, we must first learn about the characteristics of Brønsted–Lowry acids and bases.

8.1B Brønsted–Lowry Acids

A Brønsted–Lowry acid must contain a hydrogen atom. HCl is a Brønsted–Lowry acid because it *donates* a proton (H^+) to water when it dissolves, forming the hydronium ion (H_3O^+) and chloride (Cl^-).

Although hydrogen chloride, HCl, is a covalent molecule and a gas at room temperature, when it dissolves in water it reacts to form two ions, H_3O^+ and Cl^-. **Hydrochloric acid** is an aqueous solution of hydrogen chloride.

HCl gas   liquid H$_2$O   hydrochloric acid

(left & right): Jill Braaten/Anthony Arena/McGraw Hill

acetic acid

a **carboxylic acid**

Because a Brønsted–Lowry acid contains a hydrogen atom, a general Brønsted–Lowry acid is often written as **HA. A** can be a single atom such as Cl or Br. Thus, HCl and HBr are Brønsted–Lowry acids. **A** can also be a polyatomic ion. Sulfuric acid (H$_2$SO$_4$) and nitric acid (HNO$_3$) are Brønsted–Lowry acids, as well. **Carboxylic acids** are a group of Brønsted–Lowry acids that contain the atoms COOH arranged so that the carbon atom is doubly bonded to one O atom and singly bonded to another. Acetic acid, CH$_3$COOH, is a simple carboxylic acid. Although carboxylic acids may contain several hydrogen atoms, the **H atom of the OH group is the acidic proton that is donated.**

Although a Brønsted–Lowry acid must contain a hydrogen atom, **it may be a neutral molecule or contain a net positive or negative charge.** Thus, H$_3$O$^+$, HCl, and HSO$_4^-$ are all Brønsted–Lowry acids even though their net charges are +1, 0, and –1, respectively. Vinegar, citrus fruits, and carbonated soft drinks all contain Brønsted–Lowry acids, as shown in Figure 8.1.

Figure 8.1

Examples of Brønsted–Lowry Acids in Food Products

a.	b.	c.

acetic acid
CH$_3$COOH

citric acid
C$_6$H$_8$O$_7$

carbonic acid
H$_2$CO$_3$

a. Acetic acid is the sour-tasting component of vinegar. The air oxidation of ethanol to acetic acid is the process that makes "bad" wine taste sour.

b. Citric acid imparts a sour taste to oranges, lemons, and other citrus fruits.

c. Carbonated beverages contain carbonic acid, H$_2$CO$_3$.

(a): Jill Braaten; (b): Pikselstock/Shutterstock; (c): BananaStock/Getty Images

Sample Problem 8.1	**Identifying Brønsted–Lowry Acids**

Which of the following species can be Brønsted–Lowry acids: (a) HF; (b) HSO_3^-; (c) Cl_2?

Analysis

A Brønsted–Lowry acid must contain a hydrogen atom, but it may be neutral or contain a net positive or negative charge.

Solution

  a. HF is a Brønsted–Lowry acid because it contains a H.

  b. HSO_3^- is a Brønsted–Lowry acid because it contains a H.

  c. Cl_2 is not a Brønsted–Lowry acid because it does not contain a H.

Practice Problem 8.1	Which of the following species can be Brønsted–Lowry acids: (a) HI; (b) SO_4^{2-}; (c) $H_2PO_4^-$; (d) Cl^-?

More Practice: Try Problems 8.15, 8.16.

8.1C Brønsted–Lowry Bases

A Brønsted–Lowry base is a proton acceptor and as such, it must be able to form a bond to a proton. Because a proton has no electrons, **a base must contain a lone pair of electrons** that can be donated to form a new bond. Thus, ammonia (NH_3) is a Brønsted–Lowry base because it contains a nitrogen atom with a lone pair of electrons. When NH_3 is dissolved in water, its N atom accepts a proton from H_2O, forming an ammonium cation (NH_4^+) and hydroxide (OH^-).

$$NH_3(g) \ + \ H_2O(l) \longrightarrow NH_4^+(aq) \ + \ OH^-(aq)$$

This electron pair forms a new
bond to the H from H_2O.
Brønsted–Lowry base

A general Brønsted–Lowry base is often written as **B:** to emphasize that the base must contain a lone pair of electrons to bond to a proton. A base may be neutral or, more commonly, have a net negative charge. Hydroxide (OH^-) is the most common Brønsted–Lowry base. The source of hydroxide anions can be a variety of metal salts, including NaOH, KOH, $Mg(OH)_2$, and $Ca(OH)_2$. Ammonia (NH_3) and water (H_2O) are both Brønsted–Lowry bases because each contains an atom with a lone pair of electrons.

Many consumer products contain Brønsted–Lowry bases, as shown in Figure 8.2.

Figure 8.2   Examples of Brønsted–Lowry Bases in Consumer Products

a.

b.

c.

solid CaCO₃

OH⁻
NH₄⁺
NH₃
H₂O

CO₃²⁻
Ca²⁺

OH⁻
Na⁺

H₂O

solid NaOH

a. Calcium carbonate (CaCO₃), a base, is the active ingredient in the antacid Rolaids.

b. Windex and other household cleaners contain ammonia (NH₃) dissolved in water, forming NH₄⁺ cations and OH⁻ anions.

c. Drain cleaners contain pellets of solid sodium hydroxide (NaOH), which form Na⁺ cations and OH⁻ anions when mixed with water.

(a–c): Jill Braaten/McGraw Hill

Sample Problem 8.2

Identifying Brønsted–Lowry Bases

Which of the following species can be Brønsted–Lowry bases: (a) LiOH; (b) Cl⁻; (c) CH₄?

Analysis

A Brønsted–Lowry base must contain a lone pair of electrons, but it may be neutral or have a net negative charge.

Solution

a. LiOH is a base because it contains hydroxide, OH⁻, which has three lone pairs on its O atom.

b. Cl⁻ is a base because it has four lone pairs.

c. CH₄ is not a base because it has no lone pairs.

Practice Problem 8.2

Which of the following species can be Brønsted–Lowry bases: (a) Al(OH)₃; (b) Br⁻; (c) NH₄⁺; (d) CN⁻?

More Practice: Try Problems 8.17, 8.18.

Sample Problem 8.3

Identifying Brønsted–Lowry Acids and Bases

Classify each reactant as a Brønsted–Lowry acid or base.

a. $HF(g) + H_2O(l) \longrightarrow F^-(aq) + H_3O^+(aq)$

b. $SO_4^{2-}(aq) + H_2O(l) \longrightarrow HSO_4^-(aq) + OH^-(aq)$

Analysis

In each equation, the **Brønsted–Lowry acid loses a proton** and the **Brønsted–Lowry base gains a proton.**

Solution

a. HF is the acid because it loses a proton (H^+) to form F^-, and H_2O is the base because it gains a proton to form H_3O^+.

$$\text{HF}(g) \quad + \quad \text{H}_2\text{O}(l) \quad \longrightarrow \quad \text{F}^-(aq) \quad + \quad \text{H}_3\text{O}^+(aq)$$

b. H_2O is the acid because it loses a proton (H^+) to form OH^-, and SO_4^{2-} is the base because it gains a proton to form HSO_4^-.

$$\text{SO}_4^{2-}(aq) \quad + \quad \text{H}_2\text{O}(l) \quad \longrightarrow \quad \text{HSO}_4^-(aq) \quad + \quad \text{OH}^-(aq)$$

Practice Problem 8.3

Classify each reactant as a Brønsted–Lowry acid or base.

a. $HCl(g) + NH_3(g) \longrightarrow Cl^-(aq) + NH_4^+(aq)$

b. $CH_3COOH(l) + H_2O(l) \longrightarrow CH_3COO^-(aq) + H_3O^+(aq)$

c. $OH^-(aq) + HSO_4^-(aq) \longrightarrow H_2O(l) + SO_4^{2-}(aq)$

8.2   The Reaction of a Brønsted–Lowry Acid with a Brønsted–Lowry Base

When a Brønsted–Lowry acid reacts with a Brønsted–Lowry base, a proton is *transferred* from the acid to the base. **The Brønsted–Lowry acid donates a proton to the Brønsted–Lowry base, which accepts it.**

Consider, for example, the reaction of the general acid H—A with the general base B:. **In an acid–base reaction, one bond is broken and one bond is formed.** The electron pair of the base B: forms a new bond to the proton of the acid, forming H—B$^+$. The acid H—A loses a proton, leaving the electron pair in the H—A bond on A, forming A:$^-$.

Conjugate acid–base pair

HCl — acid → Cl$^-$ — conjugate base

- The product formed by loss of a proton from an acid is called its *conjugate base.*
- The product formed by gain of a proton by a base is called its *conjugate acid.*

Thus, the conjugate base of the acid HA is **A:$^-$.** The conjugate acid of the base B: is **HB$^+$.**

Conjugate acid–base pair

H_2O — base → H_3O^+ — conjugate acid

- Two species that differ by the presence of a proton are called a conjugate acid–base pair.

Thus, in an acid–base reaction, the acid and the base on the left side of the equation (HA and B:) form two products that are also an acid and a base (HB$^+$ and A:$^-$). Equilibrium arrows ($\rightleftharpoons$) are often used to separate reactants and products, because the reaction can proceed in either the forward or reverse directions. In some reactions, the products are greatly favored, as discussed in Section 8.3.

When HBr is dissolved in water, for example, the acid HBr loses a proton to form its conjugate base Br⁻, and the base H_2O gains a proton to form H_3O^+.

Thus, HBr and Br⁻ are a conjugate acid–base pair because these two species differ by the presence of a proton (H^+). H_2O and H_3O^+ are also a conjugate acid–base pair because these two species differ by the presence of a proton as well.

The net charge must be the same on both sides of the equation. In this example, the two reactants are neutral (zero net charge), and the sum of the −1 and +1 charges in the products is also zero.

Take particular note of what happens to the charges in each conjugate acid–base pair. **When a species gains a proton (H^+), it gains a +1 charge.** Thus, if a reactant is neutral to begin with, it ends up with a +1 charge. **When a species loses a proton (H^+), it effectively gains a −1 charge** because the product has one fewer proton (+1 charge) than it started with. Thus, if a reactant is neutral to begin with, it ends up with a −1 charge.

The reaction of ammonia (NH_3) with HCl is also a Brønsted–Lowry acid–base reaction. In this example, NH_3 is the base because it gains a proton to form its conjugate acid, NH_4^+. HCl is the acid because it donates a proton, forming its conjugate base, Cl⁻.

- A Brønsted–Lowry acid–base reaction is a *proton transfer reaction* because it always results in the transfer of a proton from an acid to a base.

The ability to identify and draw a conjugate acid or base from a given starting material is a necessary skill, illustrated in Sample Problems 8.4 and 8.5.

Sample Problem 8.4	Drawing the Conjugate Acid of a Base

Draw the conjugate acid of each base: (a) F⁻; (b) NO_3^-.

Analysis

Conjugate acid–base pairs differ by the presence of a proton. To draw a conjugate acid from a base, *add* a proton, H^+. This adds +1 to the charge of the base to give the charge on the conjugate acid.

Solution

a. F⁻ + H^+ gives HF as the conjugate acid.

b. NO_3^- + H^+ gives HNO_3 (nitric acid) as the conjugate acid.

Practice Problem 8.4

Draw the conjugate acid of each species: (a) H_2O; (b) I^-; (c) HCO_3^-.

More Practice: Try Problems 8.19, 8.20.

Sample Problem 8.5   **Drawing the Conjugate Base of an Acid**

Draw the conjugate base of each acid: (a) H_2O; (b) HCO_3^-.

Analysis

Conjugate acid–base pairs differ by the presence of a proton. To draw a conjugate base from an acid, *remove* a proton, H^+. **This adds −1 to the charge of the acid** to give the charge on the conjugate base.

Solution

    a. Remove H^+ from H_2O to form OH^-, the conjugate base.

    b. Remove H^+ from HCO_3^- to form CO_3^{2-}, the conjugate base. CO_3^{2-} has a −2 charge because −1 is added to an anion that had a −1 charge to begin with.

Practice Problem 8.5

Draw the conjugate base of each species: (a) H_2S; (b) HCN; (c) HSO_4^-.

More Practice: Try Problems 8.1, 8.2, 8.21, 8.22.

A compound that contains both a hydrogen atom and a lone pair of electrons can be either an acid or a base, depending on the particular reaction. Such a compound is said to be **amphoteric.** For example, when H_2O acts as a base it gains a proton, forming H_3O^+. Thus, H_2O and H_3O^+ are a conjugate acid–base pair. When H_2O acts as an acid it loses a proton, forming OH^-. H_2O and OH^- are also a conjugate acid–base pair.

Sample Problem 8.6   **Identifying the Acid, Base, Conjugate Acid, and Conjugate Base in a Reaction**

Label the acid and the base and the conjugate acid and the conjugate base in the following reaction.

$$NH_4^+(aq) + OH^-(aq) \longrightarrow NH_3(g) + H_2O(l)$$

Analysis

  • **The Brønsted–Lowry acid loses a proton to form its conjugate base.**
  • **The Brønsted–Lowry base gains a proton to form its conjugate acid.**

Solution

NH_4^+ is the acid because it loses a proton to form NH_3, its conjugate base. OH^- is the base because it gains a proton to form its conjugate acid, H_2O.

Practice Problem 8.6

Label the acid and the base and the conjugate acid and the conjugate base in each reaction.

a. $H_2O(l) + HI(g) \rightleftharpoons I^-(aq) + H_3O^+(aq)$

b. $CH_3COOH(l) + NH_3(g) \rightleftharpoons CH_3COO^-(aq) + NH_4^+(aq)$

c. $Br^-(aq) + HNO_3(aq) \rightleftharpoons HBr(aq) + NO_3^-(aq)$

Practice Problem 8.7

Ammonia, NH_3, is amphoteric. (a) Draw the conjugate acid of NH_3. (b) Draw the conjugate base of NH_3.

Practice Problem 8.8

Fill in the missing product in each acid–base equation.

a. $HSO_4^- + H_2O \rightleftharpoons$ [   ] $+ OH^-$

b. $HSO_4^- + H_2O \rightleftharpoons$ [   ] $+ H_3O^+$

More Practice: Try Problems 8.5, 8.6, 8.23, 8.24.

8.3   Acid and Base Strength

Although all Brønsted–Lowry acids contain protons, some acids readily donate protons whereas others do not. Similarly, some Brønsted–Lowry bases accept a proton much more readily than others. How readily proton transfer occurs is determined by the strength of the acid and base.

When a covalent acid dissolves in water, proton transfer forms H_3O^+ and an anion. **The splitting apart of a covalent molecule (or an ionic compound) into individual ions is called** *dissociation*. Acids differ in their tendency to donate a proton; that is, acids differ in the extent to which they *dissociate* in water.

- A strong acid readily donates a proton. When a strong acid dissolves in water, essentially 100% of the acid dissociates into ions.
- A weak acid less readily donates a proton. When a weak acid dissolves in water, only a small fraction of the acid dissociates into ions.

Common strong acids include **HI, HBr, HCl, H_2SO_4**, and **HNO_3** (Table 8.2). When each acid is dissolved in water, 100% of the acid dissociates, forming H_3O^+ and the conjugate base, as shown for HCl and H_2SO_4.

- Use a *single* reaction arrow.
- The **products** are greatly favored at equilibrium.

$HCl(g)$ + $H_2O(l)$ $\longrightarrow$ $H_3O^+(aq)$ + $Cl^-(aq)$
strong acid   conjugate base

$H_2SO_4(l)$ + $H_2O(l)$ $\longrightarrow$ $H_3O^+(aq)$ + $HSO_4^-(aq)$
strong acid   conjugate base

HCl, hydrochloric acid, is secreted by the stomach to digest food (Figure 8.3), and **H_2SO_4**, sulfuric acid, is an important industrial starting material in the synthesis of phosphate fertilizers. A single reaction arrow ($\longrightarrow$) is drawn to show that essentially all of the reactants are converted to products.

Acetic acid, **CH_3COOH**, is a weak acid. When acetic acid dissolves in water, only a small fraction of acetic acid molecules donate a proton to water to form H_3O^+ and the conjugate base, CH_3COO^-. The major species in solution is the undissociated acid, CH_3COOH. Two arrows that are unequal in length ($\rightleftharpoons$) are used between the reactants and products to show that both are present in solution. The longer arrow points toward the reactants, because few

Table 8.2 Relative Strength of Acids and Their Conjugate Bases

Acid		Conjugate Base	
Strong Acids			
Hydroiodic acid	HI	I⁻	Iodide ion
Hydrobromic acid	HBr	Br⁻	Bromide ion
Hydrochloric acid	HCl	Cl⁻	Chloride ion
Sulfuric acid	H_2SO_4	HSO_4^-	Hydrogen sulfate ion
Nitric acid	HNO_3	NO_3^-	Nitrate ion
Hydronium ion	H_3O^+	H_2O	**Water**
Weak Acids			
Phosphoric acid	H_3PO_4	$H_2PO_4^-$	Dihydrogen phosphate ion
Hydrofluoric acid	HF	F⁻	Fluoride ion
Acetic acid	CH_3COOH	CH_3COO^-	Acetate ion
Carbonic acid	H_2CO_3	HCO_3^-	Bicarbonate ion
Ammonium ion	NH_4^+	NH_3	Ammonia
Hydrocyanic acid	HCN	CN⁻	Cyanide ion
Water	H_2O	OH⁻	Hydroxide ion

Increasing acid strength (left arrow, upward)

Increasing base strength (right arrow, downward)

Figure 8.3

Focus on the Human Body: Hydrochloric Acid in the Stomach

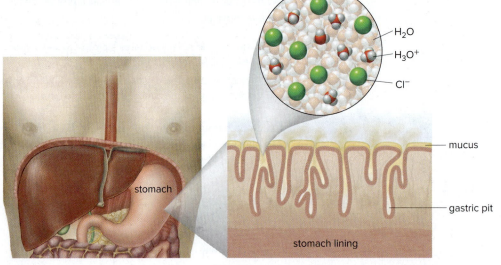

H₂O
H₃O⁺
Cl⁻

mucus

stomach

gastric pit

stomach lining

The thick mucous layer protects the stomach lining.

Although HCl is a corrosive acid secreted in the stomach, a thick layer of mucus covering the stomach wall protects it from damage by the strong acid. The strong acid HCl is completely dissociated to H_3O^+ and Cl⁻.

molecules of acetic acid dissociate. Other weak acids and their conjugate bases are listed in Table 8.2.

- Use **unequal** reaction arrows.
- The **reactants** are favored at equilibrium.

$$CH_3COOH(l) \; + \; H_2O(l) \; \rightleftharpoons \; H_3O^+(aq) \; + \; CH_3COO^-(aq)$$

weak acid   conjugate base

Figure 8.4 illustrates the difference between an aqueous solution of a strong acid that is completely dissociated and a weak acid that contains much undissociated acid. Sample Problem 8.7 illustrates how the extent of dissociation and acid strength are related.

Figure 8.4 A Strong and Weak Acid Dissolved in Water

hydrochloric acid

Cl⁻
Cl^-

H_3O^+

A strong acid is completely dissociated.

vinegar

CH_3COO^-

H_3O^+

CH_3COOH

A weak acid contains mostly undissociated acid, CH_3COOH.

- The **strong acid HCl completely dissociates** into H_3O^+ and Cl^- in water.
- Vinegar contains CH_3COOH dissolved in H_2O. The weak acid CH_3COOH is only slightly dissociated into H_3O^+ and CH_3COO^-, so **mostly CH_3COOH is present** in solution.

(left & right): Jill Braaten/Anthony Arena/McGraw Hill

| Sample Problem 8.7 | Using the Extent of Dissociation to Determine Relative Acid Strength |

Diagrams **A–C** represent three acids (HA) dissolved in water. Which representation shows the strongest acid? Which representation shows the weakest acid?

$\bigcirc$ = HA $\bullet$ = A⁻ $\bigcirc$ = H_3O^+

A   B   C

Analysis

The stronger the acid, the more readily it dissociates to form its conjugate base A⁻ and H_3O^+. In molecular art, the **strongest acid has the most A⁻ and H_3O^+ ions,** and the fewest molecules of undissociated HA.

Solution

B is the strongest acid because it contains the largest number of A^- and H_3O^+ ions (three each), and the smallest number (five) of undissociated HA molecules. **C** is the weakest acid because it contains the smallest number of A^- and H_3O^+ ions (one each), and the largest number (seven) of undissociated HA molecules.

Practice Problem 8.9

Diagrams **D–F** represent three acids (HA) dissolved in water. Rank the three acids (**D–F**) in order of increasing acidity.

D	E	F

More Practice: Try Problems 8.3, 8.4, 8.27, 8.28, 8.33.

Bases also differ in their ability to accept a proton.

- A strong base readily accepts a proton. When a strong base dissolves in water, 100% of the base dissociates into ions.
- A weak base less readily accepts a proton. When a weak base dissolves in water, only a small fraction of the base forms ions.

The most common strong base is hydroxide, **OH⁻,** used as a variety of metal salts, including NaOH and KOH. Solid NaOH dissolves in water to form solvated Na^+ cations and OH^- anions. In contrast, when **NH₃,** a weak base, dissolves in water, only a small fraction of NH_3 molecules react to form NH_4^+ and OH^-. The major species in solution is the undissociated molecule, NH_3. Figure 8.5 illustrates the difference between aqueous solutions of strong and weak bases. Table 8.2 lists common bases.

An inverse relationship exists between acid and base strength.

- A strong acid readily donates a proton, forming a *weak* conjugate base.
- A strong base readily accepts a proton, forming a *weak* conjugate acid.

Figure 8.5

A Strong and Weak Base Dissolved in Water

sodium hydroxide

Na⁺

OH⁻

A strong base is completely dissociated.

ammonia

NH_3

NH_4^+

OH^-

A weak base contains mostly undissociated base, NH_3.

- The **strong base NaOH completely dissociates** into Na⁺ and OH⁻ in water.
- The weak base NH_3 is only slightly dissociated into NH_4^+ and OH⁻, so **mostly NH_3 is present** in solution.

(left & right): Jill Braaten/Anthony Arena/McGraw Hill

Why does this inverse relationship exist? Because a strong acid readily donates a proton, it forms a conjugate base that has little ability to accept a proton. Because a strong base readily accepts a proton, it forms a conjugate acid that tightly holds onto its proton, making it a weak acid.

Thus, a *strong* acid like HCl forms a *weak* conjugate base (Cl⁻), and a *strong* base like OH⁻ forms a *weak* conjugate acid (H_2O). The entries in Table 8.2 are arranged in order of *decreasing* acid strength. This means that Table 8.2 is also arranged in order of *increasing* strength of the resulting conjugate bases. Knowing the relative strength of two acids makes it possible to predict the relative strength of their conjugate bases.

Sample Problem 8.8

Relating Acid and Base Strength

Using Table 8.2: (a) Is H_3PO_4 or HF the stronger acid? (b) Draw the conjugate base of each acid and predict which base is stronger.

Analysis

The stronger the acid, the weaker the conjugate base.

Solution

a. H_3PO_4 is located above HF in Table 8.2, making it the stronger acid.

b. **To draw each conjugate base, remove a proton (H⁺).** Because HF is the weaker acid, F⁻ is the stronger conjugate base.

$$H_3PO_4 \xrightarrow{\text{lose } H^+} H_2PO_4^-$$

stronger acid   **weaker** base

$$HF \xrightarrow{\text{lose } H^+} F^-$$

weaker acid   **stronger** base

lactic acid

Practice Problem 8.10

Label the stronger acid in each pair. Which acid has the stronger conjugate base?

a. H_2SO_4 or H_3PO_4 b. HF or HCl   c. H_2CO_3 or NH_4^+ d. HCN or HF

Practice Problem 8.11

If lactic acid ($C_3H_6O_3$) is a stronger acid than HCN, which compound forms the stronger conjugate base?

Practice Problem 8.12

(a) Draw the conjugate acids of NO_2^- and NO_3^-. (b) If NO_2^- is the stronger base, which acid is stronger?

More Practice: Try Problems 8.31, 8.32, 8.34, 8.35.

HEALTH NOTE

Lactic acid (Practice Problem 8.11) accumulates in tissues during vigorous exercise, making muscles feel tired and sore. The formation of lactic acid is discussed in greater detail in Section 18.5. *Digital Vision/SuperStock*

8.4 Dissociation of Water

In Section 8.2 we learned that water can behave as *both* a Brønsted–Lowry acid and a Brønsted–Lowry base. As a result, two molecules of water can react together in an acid–base reaction.

H_2O donates H^+. H_2O accepts H^+.

- One molecule of H_2O donates a proton (H^+), forming its conjugate base OH^-.
- One molecule of H_2O accepts a proton, forming its conjugate acid H_3O^+.

Because water is a very weak acid (Table 8.2), pure water contains an exceedingly low concentration of ions, H_3O^+ and OH^-. Because one H_3O^+ ion and one OH^- ion are formed in each reaction, the concentrations of H_3O^+ and OH^- are *equal* in pure water. Experimentally it can be shown that the $[H_3O^+] = [OH^-] = 1.0 \times 10^{-7}$ M at 25 °C.

- In pure water, $[H_3O^+] = [OH^-] = 1.0 \times 10^{-7}$ M.

Multiplying these concentrations together gives the **ion–product constant** for water, symbolized by K_w.

$$K_w = [H_3O^+][OH^-]$$

ion–product
constant

Substituting the concentrations for H_3O^+ and OH^- into the expression for K_w gives the following result.

$$K_w = [H_3O^+][OH^-]$$
$$K_w = (1.0 \times 10^{-7}) \times (1.0 \times 10^{-7})$$
$$\boldsymbol{K_w = 1.0 \times 10^{-14}}$$

- The product, $[H_3O^+][OH^-]$, is a constant, 1.0×10^{-14}, for all aqueous solutions at 25 °C.

Thus, **the value of K_w applies to any aqueous solution,** not just pure water. If we know the concentration of one ion, H_3O^+ or OH^-, we can find the concentration of the other by rearranging the expression for K_w.

To calculate [OH⁻] when [H₃O⁺] is known:

$$K_w = [H_3O^+][OH^-]$$

$$[OH^-] = \frac{K_w}{[H_3O^+]}$$

$$[OH^-] = \frac{1.0 \times 10^{-14}}{[H_3O^+]}$$

To calculate [H₃O⁺] when [OH⁻] is known:

$$K_w = [H_3O^+][OH^-]$$

$$[H_3O^+] = \frac{K_w}{[OH^-]}$$

$$[H_3O^+] = \frac{1.0 \times 10^{-14}}{[OH^-]}$$

How to write numbers in scientific notation was presented in Section 1.6. Useful mathematical concepts are presented in Appendix D.

Thus, if the concentration of H_3O^+ in a cup of coffee is 1.0×10^{-5} M, we can use this value to calculate $[OH^-]$.

$$[OH^-] = \frac{K_w}{[H_3O^+]} = \frac{1.0 \times 10^{-14}}{1.0 \times 10^{-5}}$$

$$[OH^-] = 1.0 \times 10^{-9} \text{ M}$$

hydroxide ion concentration
in a cup of coffee

In a cup of coffee, therefore, the concentration of H_3O^+ ions is greater than the concentration of OH^- ions, but the product of these concentrations, 1.0×10^{-14}, is a constant, K_w.

Pure water and any solution that has an equal concentration of H_3O^+ and OH^- ions (1.0×10^{-7}) is said to be *neutral.* **Other solutions are classified as acidic or basic,** depending on which ion is present in a higher concentration.

- In an *acidic* solution, $[H_3O^+] > [OH^-]$; thus, $[H_3O^+] > 10^{-7}$ M.
- In a *basic* solution, $[OH^-] > [H_3O^+]$; thus, $[OH^-] > 10^{-7}$ M.

In an acidic solution, the concentration of the acid H_3O^+ is greater than the concentration of the base OH^-. In a basic solution, the concentration of the base OH^- is greater than the concentration of the acid H_3O^+. Table 8.3 summarizes information about neutral, acidic, and basic solutions.

Coffee is an *acidic* solution because the concentration of H_3O^+ is *greater* than the concentration of OH^-.
Foodcollection/StockFood

Table 8.3 Neutral, Acidic, and Basic Solutions

Type	[H₃O⁺] and [OH⁻]	[H₃O⁺]	[OH⁻]
Neutral	$[H_3O^+] = [OH^-]$	10^{-7} M	10^{-7} M
Acidic	$[H_3O^+] > [OH^-]$	$> 10^{-7}$ M	$< 10^{-7}$ M
Basic	$[H_3O^+] < [OH^-]$	$< 10^{-7}$ M	$> 10^{-7}$ M

Sample Problem 8.9

Calculating [OH⁻] When [H₃O⁺] Is Known

If $[H_3O^+]$ in blood is 4.0×10^{-8} M, what is the value of $[OH^-]$? Is blood acidic, basic, or neutral?

Analysis

Use the equation $[OH^-] = K_w/[H_3O^+]$ to calculate the hydroxide ion concentration.

Solution

Substitute the given value of $[H_3O^+]$ in the equation to find $[OH^-]$.

$$[OH^-] = \frac{K_w}{[H_3O^+]} = \frac{1.0 \times 10^{-14}}{4.0 \times 10^{-8}} = 2.5 \times 10^{-7}\ M$$

hydroxide ion concentration
in the blood

Because $[OH^-] > [H_3O^+]$, blood is a basic solution.

Practice Problem 8.13

Calculate the value of $[OH^-]$ from the given $[H_3O^+]$ in each solution and label the solution as acidic or basic: (a) $[H_3O^+] = 10^{-3}$ M; (b) $[H_3O^+] = 10^{-11}$ M; (c) $[H_3O^+] = 2.8 \times 10^{-10}$ M; (d) $[H_3O^+] = 5.6 \times 10^{-4}$ M.

Practice Problem 8.14

Calculate the value of $[H_3O^+]$ from the given $[OH^-]$ in each solution and label the solution as acidic or basic: (a) $[OH^-] = 10^{-6}$ M; (b) $[OH^-] = 10^{-9}$ M; (c) $[OH^-] = 5.2 \times 10^{-11}$ M; (d) $[OH^-] = 7.3 \times 10^{-4}$ M.

More Practice: Try Problems 8.37–8.40.

Because a strong acid like HCl is completely dissociated in aqueous solution, the concentration of the acid tells us the concentration of hydronium ions present. Thus, a 0.1 M HCl solution completely dissociates, so the concentration of H_3O^+ is 0.1 M. This value can then be used to calculate the hydroxide ion concentration. Similarly, a strong base like NaOH completely dissociates, so the concentration of the base gives the concentration of hydroxide ions present. Thus, the concentration of OH^- in a 0.1 M NaOH solution is 0.1 M.

In **0.1 M** HCl solution: $[H_3O^+] = \mathbf{0.1\ M} = 1 \times 10^{-1}\ M$
strong acid

In **0.1 M** NaOH solution: $[OH^-] = \mathbf{0.1\ M} = 1 \times 10^{-1}\ M$
strong base

Sample Problem 8.10

Calculating $[H_3O^+]$ and $[OH^-]$ in a Solution of a Strong Base

Calculate the value of $[H_3O^+]$ and $[OH^-]$ in a 0.01 M NaOH solution.

Analysis

Because NaOH is a strong base that completely dissociates to form Na^+ and OH^-, the concentration of NaOH gives the concentration of OH^- ions. The $[OH^-]$ can then be used to calculate $[H_3O^+]$ from the expression for K_w.

Solution

The value of $[OH^-]$ in a 0.01 M NaOH solution is 0.01 M = 1×10^{-2} M.

$$[H_3O^+] = \frac{K_w}{[OH^-]} = \frac{1 \times 10^{-14}}{1 \times 10^{-2}} = 1 \times 10^{-12}\ M$$

concentration of H_3O^+

concentration of OH^-

Practice Problem 8.15

Calculate the value of $[H_3O^+]$ and $[OH^-]$ in each solution: (a) 0.001 M NaOH; (b) 0.001 M HCl; (c) 1.5 M HCl; (d) 0.30 M NaOH.

8.5 The pH Scale

Knowing the hydronium ion concentration is necessary in many different instances. The blood must have an H_3O^+ concentration in a very narrow range for an individual's good health. Plants thrive in soil that is not too acidic or too basic. The H_3O^+ concentration in a swimming pool must be measured and adjusted to keep the water clean and free from bacteria and algae.

8.5A Calculating pH

Because values for the hydronium ion concentration are very small, with negative powers of ten, the **pH scale** is used to more conveniently report $[H_3O^+]$. The pH of a solution is a number generally between 0 and 14, defined in terms of the *logarithm* (log) of the H_3O^+ concentration.

$$pH = -log\ [H_3O^+]$$

A logarithm is an exponent of a power of ten.

The log is the exponent.

$$log(10^5) = 5 \qquad log(10^{-10}) = -10 \qquad log(0.001) = log(10^{-3}) = -3$$

The log is the exponent.   Convert to scientific notation.

In calculating pH, first consider an H_3O^+ concentration that has a coefficient of *one* when the number is written in scientific notation. For example, the value of $[H_3O^+]$ in apple juice is about 1×10^{-4}, or 10^{-4} written without the coefficient. The pH of this solution is calculated as follows:

$$pH = -log\ [H_3O^+] = -log(10^{-4})$$
$$= -(-4) \qquad = 4$$

pH of apple juice

Because pH is defined as the *negative* logarithm of $[H_3O^+]$ and these concentrations have *negative* exponents (10^{-x}), pH values are *positive* numbers.

Whether a solution is acidic, neutral, or basic can now be defined in terms of its pH.

- **Acidic solution:** pH < 7 $\longrightarrow$ $[H_3O^+]$ > 1×10^{-7}
- **Neutral solution:** pH = 7 $\longrightarrow$ $[H_3O^+]$ = 1×10^{-7}
- **Basic solution:** pH > 7 $\longrightarrow$ $[H_3O^+]$ < 1×10^{-7}

Note the relationship between $[H_3O^+]$ and pH.

- The *lower* the pH, the *higher* the concentration of H_3O^+.

The pH of a solution can be measured using a pH meter as shown in Figure 8.6. Approximate pH values are determined using pH paper or indicators that turn different colors depending on the pH of the solution. The pH of various substances is shown in Figure 8.7.

Apple juice has a pH of about 4, so it is an acidic solution. *Wiktory/Shutterstock*

Figure 8.6   Measuring pH

a.

b.

c.

a. A pH meter is a small electronic device that measures pH when an electrode is dipped into a solution.

b. Paper strips called pH paper change color corresponding to a particular pH, when a drop of an aqueous solution is applied to them.

c. An acid–base indicator can be used to give an approximate pH. The indicator is a dye that changes color depending on the pH of the solution.

(a): Stephen Frisch/McGraw Hill; (b): JIPEN/Shutterstock; (c): Leslie Garland Picture Library/Alamy Stock Photo

Figure 8.7   The pH of Some Common Substances

The pH of many fruits is less than 7, making them acidic. Many cleaning agents, such as household ammonia and bleach, are basic (pH > 7).
(left to right): Humberto Olarte Cupas/Alamy Stock Photo; Ken Welsh/age fotostock; Brand X Pictures/Getty Images; Mitch Hrdlicka/Photodisc/Getty Images; Jill Braaten/McGraw Hill; Terry Wild Studio/McGraw Hill; Jill Braaten/McGraw Hill

Converting a given H_3O^+ concentration to a pH value is shown in Sample Problem 8.11. The reverse process, converting a pH value to an H_3O^+ concentration, is shown in Sample Problem 8.12.

Sample Problem 8.11

Calculating pH from [H_3O^+]

What is the pH of a urine sample that has an H_3O^+ concentration of 1×10^{-5} M? Classify the solution as acidic, basic, or neutral.

Analysis

Use the formula **pH = –log [H_3O^+].** When the coefficient of a number written in scientific notation is one, the **pH equals the value x in 10^{-x}.**

Solution

$$pH = -\log [H_3O^+] = -\log(10^{-5})$$
$$= -(-5) = 5 \quad \text{pH of urine sample}$$
Answer

The urine sample is acidic because the pH < 7.

Practice Problem 8.16

Convert each H_3O^+ concentration to a pH value.

   a. 1×10^{-6} M   b. 1×10^{-12} M   c. 0.000 01 M   d. 0.000 000 000 01 M

Sample Problem 8.12

Determining [H_3O^+] from pH

What is the H_3O^+ concentration in lemon juice that has a pH of about 2? Classify the solution as acidic, basic, or neutral.

Analysis

To find [H_3O^+] from a pH, which is logarithm, we must determine what number corresponds to the given logarithm. When the pH is a whole number x, the **value of x becomes the exponent in the expression $1 \times 10^{-x} = [H_3O^+]$.**

Solution

If the pH of lemon juice is **2**, $[H_3O^+] = 1 \times 10^{-2}$ M. Because the pH is less than 7, the lemon juice is acidic.

Practice Problem 8.17

What H_3O^+ concentration corresponds to each pH value: (a) 13; (b) 7; (c) 3? Label the solution as acidic, basic, or neutral.

More Practice: Try Problems 8.43a, b; 8.44a.

8.5B Calculating pH Using a Calculator

Determining logarithms and anti-logarithms using an electronic calculator is also discussed in Appendix D.

Because seawater contains dissolved salts, its pH is 8.50, making it slightly basic, not neutral like pure water. *Bogdan Bratosin/Moment/Getty Images*

To calculate the pH of a solution in which the hydronium ion concentration has a coefficient in scientific notation that is *not* equal to one—as in 2.0×10^{-3}—you need a calculator that has a log function. Described in this section are the steps that can be followed in calculations with *some* calculators. Consult your manual if these steps do *not* produce the stated result.

To calculate pH from a known $[H_3O^+]$, follow the steps in columns [1]–[3]. Two examples are shown.

$[H_3O^+]$	① Enter $[H_3O^+]$ in scientific notation[a]	② Display	③ Buttons to press	pH[b]
2.0×10^{-3} M	Press: 2.0 EE 3 Change sign (+ to –)	2.0^{-03}	LOG Change sign (+ to –)	2.70
7.3×10^{-8} M	Press: 7.3 EE 8 Change sign (+ to –)	7.3^{-08}	LOG Change sign (+ to –)	7.14

[a]Entering a number in scientific notation was described in Section 1.6B.
[b]The proper number of significant figures is explained before Sample Problem 8.13.

Similarly, when a reported pH is *not* a whole number—as in the pH = 8.50 for a sample of seawater—you need a calculator to calculate an *antilogarithm*—that is, the number that has a logarithm of 8.50.

To calculate $[H_3O^+]$ from a known pH, follow the steps in columns [1]–[3]. Two examples are shown.

pH	① Number to enter	② Buttons to press	③ Display	$[H_3O^+]$[a]
8.50	8.50	Change sign (+ to –) 2nd LOG	3.16227766017^{-09}	3.2×10^{-9} M
3.91	3.91	Change sign (+ to –) 2nd LOG	1.23026877081^{-04}	1.2×10^{-4} M

[a]The proper number of significant figures is explained before Sample Problem 8.13.

Care must be taken in keeping track of significant figures when using logarithms.

> • A logarithm has the same number of digits to the right of the decimal point as are contained in the coefficient of the original number.

$$[H_3O^+] = 3.2 \times 10^{-9} \text{ M} \qquad pH = 8.50 \qquad \textbf{two} \text{ digits after the decimal point}$$

two significant figures

Sample Problem 8.13

Determining pH from [H₃O⁺] Using a Calculator

What is the pH of wine that has an H_3O^+ concentration of 3.2×10^{-4} M?

Analysis

Use a calculator to determine the logarithm of a number that contains a coefficient in scientific notation that is not a whole number; **pH = –log $[H_3O^+]$.**

Solution

The order of the steps in using an electronic calculator, as well as the labels on the calculator buttons, vary. In some cases it is possible to calculate the pH by following three steps: **enter the number (H_3O^+ concentration); press the *log* button; and press the change sign key.** Consult your calculator manual if these steps do not give the desired value. Because the coefficient in the original number had two significant figures, the pH must have two digits to the right of the decimal point.

two significant figures

$$pH = -\log[H_3O^+] \quad = \quad -\log(3.2 \times 10^{-4})$$
$$= \quad -(-3.49) \quad = \quad 3.49$$

two digits to the right of the decimal point

Practice Problem 8.18

Convert each H_3O^+ concentration to a pH value.

a. 1.8×10^{-6} M   b. 9.21×10^{-12} M   c. 0.000 088 M   d. 0.000 000 000 076 2 M

More Practice: Try Problem 8.41a, b, d.

Sample Problem 8.14

Determining [H_3O^+] from pH Using a Calculator

What is the H_3O^+ concentration in sweat that has a pH of 5.8?

Analysis

Use a calculator to determine the antilogarithm of the negative of the pH; **[H_3O^+] = antilog(−pH).**

Solution

The order of the steps in using an electronic calculator, as well as the labels on the calculator buttons, vary. In some cases it is possible to calculate [H_3O^+] by the following steps: **enter the pH value; press the change sign key; and press the *2nd* + *log* buttons.** Because the pH has only one number to the right of the decimal point, the H_3O^+ concentration must have only one significant figure in its coefficient.

one digit to the right of the decimal point

$$[H_3O^+] \quad = \quad antilog(-pH) \quad = \quad antilog(-5.8)$$
$$[H_3O^+] \quad = \quad 2 \times 10^{-6} \text{ M}$$

one significant figure

Practice Problem 8.19

What H_3O^+ concentration corresponds to each pH value: (a) 10.2; (b) 7.8; (c) 4.3?

Practice Problem 8.20

What is the H_3O^+ concentration in a sports drink (mentioned in the chapter opener) that has a pH of 3.30?

More Practice: Try Problems 8.41c; 8.42; 8.43c; 8.44b, c; 8.45; 8.46.

8.5C FOCUS ON THE HUMAN BODY
The pH of Body Fluids

The human body contains fluids that vary in pH as shown in Figure 8.8. The gastric juice in the stomach has the lowest pH found in the body. The strongly acidic environment of the stomach aids in the digestion of food. It also kills many types of bacteria that might be inadvertently consumed along with food and drink. When food leaves the stomach, it passes to the basic environment of the small intestine. Bases in the small intestine react with acid from the stomach.

Figure 8.8   Variation in pH Values in the Human Body

Figure 8.8   Variation in pH Values in the Human Body

saliva
pH 5.8–7.1

esophagus

blood
pH 7.4

liver

stomach
pH 1.6–1.8

pancreas
pH 7.5–8.8

large intestine
pH 5.0–7.0

small intestine
pH 8.5

urine
pH 4.6–8.0

The pH of some body fluids must occupy a very narrow range. For example, a healthy individual has a blood pH in the range of 7.35–7.45. Maintaining this pH is accomplished by a complex mechanism described in Section 8.9. The pH of other fluids can be more variable. Urine has a pH anywhere from 4.6–8.0, depending on an individual's recent diet and exercise.

8.6   Common Acid–Base Reactions

Although we have already seen a variety of acid–base reactions in Sections 8.2–8.3, two common reactions deserve additional attention—reaction of acids with hydroxide bases (OH^-), and reaction of acids with bicarbonate (HCO_3^-) or carbonate (CO_3^{2-}).

8.6A   Reaction of Acids with Hydroxide Bases

The reaction of a Brønsted–Lowry acid (HA) with the metal salt of a hydroxide base (MOH) is an example of a *neutralization* reaction—**an acid–base reaction that produces a salt and water as products.**

$$HA(aq) + MOH(aq) \longrightarrow H{-}OH(l) + MA(aq)$$

$$\text{acid} \qquad \text{base} \qquad\qquad \text{water} \qquad \text{salt}$$

- The acid HA donates a proton (H^+) to the OH^- base to form H_2O.
- The anion A^- from the acid combines with the cation M^+ from the base to form the salt MA.

For example, hydrochloric acid, HCl, reacts with sodium hydroxide, NaOH, to form water and sodium chloride, NaCl.

$$HCl(aq) + NaOH(aq) \longrightarrow H{-}OH(l) + NaCl(aq)$$

$$\text{acid} \qquad \text{base} \qquad\qquad \text{water} \qquad \text{salt}$$

HEALTH NOTE

The antacid products Maalox and Mylanta both contain two bases—$Mg(OH)_2$ and $Al(OH)_3$—that react with excess stomach acid. A combination of bases is used so that the constipating effect of the aluminum salt is counteracted by the laxative effect of the magnesium salt. *Jill Braaten/McGraw Hill*

The important reacting species in this reaction are H^+ from the acid HCl and OH^- from the base NaOH. To more clearly see the acid–base reaction, we can write an equation that contains only the species that are actually involved in the reaction. Such an equation is called a **net ionic equation.**

- A *net ionic equation* contains only the species involved in a reaction.

To write a net ionic equation for an acid–base reaction, we first write the acid, base, and salt as individual ions in solution. This process is simplified if we use H^+ (not H_3O^+) as the reacting species of the acid, because it is the H^+ ion that is transferred to the base. The reaction of HCl with NaOH using individual ions is then drawn as:

$$H^+(aq) + Cl^-(aq) + Na^+(aq) + OH^-(aq) \longrightarrow H{-}OH(l) + Na^+(aq) + Cl^-(aq)$$

Writing the equation in this manner shows that the Na^+ and Cl^- ions are unchanged in the reaction. Ions that appear on both sides of an equation but undergo no change in a reaction are called **spectator ions.** Removing the spectator ions from the equation gives the net ionic equation.

$$H^+(aq) + Cl^-(aq) + Na^+(aq) + OH^-(aq) \longrightarrow H{-}OH(l) + Na^+(aq) + Cl^-(aq)$$

Omit the spectator ions.   Omit the spectator ions.

Net ionic equation $$H^+(aq) + OH^-(aq) \longrightarrow H{-}OH(l)$$

- Whenever a strong acid and strong base react, the net ionic equation is always the same—H^+ reacts with OH^- to form H_2O.

To draw the products of these neutralization reactions, keep in mind that **two products are always formed—water and a metal salt.** Balancing an acid–base equation can be done with the stepwise procedure for balancing a general reaction outlined in Section 5.2. The coefficients in a balanced chemical equation illustrate that one H^+ ion is always needed to react with each OH^- anion.

How To   Draw a Balanced Equation for a Neutralization Reaction Between HA and MOH

Example   Write a balanced equation for the reaction of $Mg(OH)_2$, an active ingredient in the antacid product Maalox, with the hydrochloric acid (HCl) in the stomach.

Step [1] Identify the acid and base in the reactants and draw H_2O as one product.

- HCl is the acid and $Mg(OH)_2$ is the base. H^+ from the acid reacts with OH^- from the base to form H_2O.

$$\underset{\text{acid}}{HCl(aq)} + \underset{\text{base}}{Mg(OH)_2(aq)} \longrightarrow H_2O(l) + \text{salt}$$

Step [2] Determine the structure of the salt formed as product.

- The salt is formed from the elements of the acid and base that are *not* used to form H_2O. **The anion of the salt comes from the acid and the cation of the salt comes from the base.**
- In this case, Cl^- (from HCl) and Mg^{2+} [from $Mg(OH)_2$] combine to form the salt $MgCl_2$.

Step [3] Balance the equation.

- Follow the procedure in Section 5.2 to balance an equation. The balanced equation shows that *two* moles of HCl are needed for *each* mole of $Mg(OH)_2$, because each mole of $Mg(OH)_2$ contains two moles of OH^-.

Place a **2** to balance H and O.

$$\mathbf{2}\,HCl(aq) + Mg(OH)_2(aq) \longrightarrow \mathbf{2}\,H_2O(l) + MgCl_2$$

Place a **2** to balance Cl.

Concept Check 8.3

Write a balanced equation for each acid–base reaction.

a. $HNO_3(aq) + NaOH(aq) \longrightarrow$ b. $H_2SO_4(aq) + KOH(aq) \longrightarrow$

Concept Check 8.4

Write the net ionic equation for each reaction in Concept Check 8.3.

8.6B Reaction of Acids with Bicarbonate and Carbonate

Acids react with the bases bicarbonate (HCO_3^-) and carbonate (CO_3^{2-}). A bicarbonate base reacts with *one* proton to form carbonic acid, H_2CO_3. A carbonate base reacts with *two* protons. The carbonic acid formed in these reactions is unstable and decomposes to form CO_2 and H_2O. Thus, when an acid reacts with either base, bubbles of CO_2 gas are given off.

$$H^+(aq) + HCO_3^-(aq) \longrightarrow [H_2CO_3(aq)] \longrightarrow H_2O(l) + CO_2(g)$$

1 H^+ needed   bicarbonate   bubbles of CO_2

$$2\,H^+(aq) + CO_3^{2-}(aq) \longrightarrow [H_2CO_3(aq)] \longrightarrow H_2O(l) + CO_2(g)$$

2 H^+ needed   carbonate

Sodium bicarbonate ($NaHCO_3$), an ingredient in the over-the-counter antacid Alka-Seltzer, is the metal salt of a bicarbonate base that reacts with excess stomach acid, releasing CO_2. Like the neutralization reactions in Section 8.6A, a salt, NaCl, is formed in which the cation (Na^+) comes from the base and the anion (Cl^-) comes from the acid.

$$HCl(aq) + NaHCO_3(aq) \longrightarrow NaCl(aq) + H_2CO_3(aq)$$

acid   base   salt

$$\longrightarrow H_2O(l) + CO_2(g)$$

Calcium carbonate ($CaCO_3$), a calcium supplement and antacid in Tums, also reacts with excess stomach acid with release of CO_2. Because each carbonate ion reacts with two protons, the balanced equation shows a 2:1 ratio of HCl to $CaCO_3$.

$$2\,HCl(aq) + CaCO_3(aq) \longrightarrow CaCl_2(aq) + H_2CO_3(aq)$$

acid   base   salt

$$\longrightarrow H_2O(l) + CO_2(g)$$

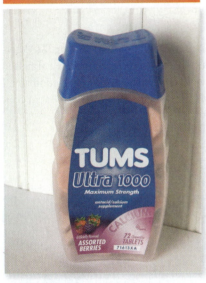

Like taking other over-the-counter medications, care must be exercised when using antacids. Ingestion of large amounts of $CaCO_3$ can increase the incidence of kidney stones.
Jill Braaten/McGraw Hill

Sample Problem 8.15

Writing a Balanced Acid–Base Equation with the Base NaHCO₃

Write a balanced equation for the reaction of H_2SO_4 with $NaHCO_3$.

Analysis

The acid and base react to form a salt and carbonic acid (H_2CO_3), which decomposes to CO_2 and H_2O.

Solution

H_2SO_4 is the acid and $NaHCO_3$ is the base. H^+ from the acid reacts with HCO_3^- from the base to give H_2CO_3, which decomposes to H_2O and CO_2. A salt (Na_2SO_4) is also formed from the cation of the base (Na^+) and the anion of the acid (SO_4^{2-}).

Unbalanced equation: $H_2SO_4(aq) + NaHCO_3(aq) \longrightarrow Na_2SO_4(aq) + H_2O(l) + CO_2(g)$

acid   base   salt

from H_2CO_3

To balance the equation, place coefficients so the number of atoms on both sides of the arrow is the same.

Place a **2** to balance Na...

$$H_2SO_4(aq) + \mathbf{2}\,NaHCO_3(aq) \longrightarrow Na_2SO_4(aq) + \mathbf{2}\,H_2O(l) + \mathbf{2}\,CO_2(g)$$

acid   base   salt

...then place **2**'s to balance C, H, and O.

Practice Problem 8.21	The acid in acid rain is generally sulfuric acid (H_2SO_4). When this rainwater falls on statues composed of marble ($CaCO_3$), the H_2SO_4 slowly dissolves the $CaCO_3$. Write a balanced equation for this acid–base reaction.
Practice Problem 8.22	Write a balanced equation for the reaction of nitric acid (HNO_3) with each base: (a) $NaHCO_3$; (b) $MgCO_3$.

More Practice: Try Problems 8.11, 8.49c, 8.50c.

8.7   Titration

Sometimes it is necessary to know the exact concentration of acid or base in a solution. To determine the molarity of a solution, we carry out a **titration.** A titration uses a *buret,* a calibrated tube with a stopcock at the bottom that allows a solution of known molarity to be added in small quantities to a solution of unknown molarity. The procedure for determining the total acid concentration of a solution of HCl is illustrated in Figure 8.9.

Figure 8.9   Titration of an Acid with a Base of Known Concentration

a.

b.

c.

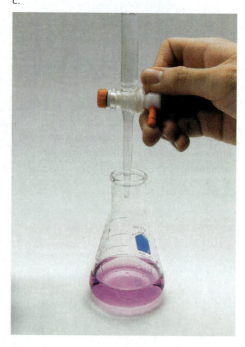

Steps in determining the molarity of a solution of HCl:

a. Add a measured volume of HCl solution to a flask. Add an acid–base indicator, often phenolphthalein, which is colorless in acid but turns bright pink in base.

b. Fill a buret with an NaOH solution of known molarity and slowly add it to the HCl solution.

c. Add NaOH solution until the *end point* is reached, the point at which the indicator changes color. At the end point, the **number of moles of NaOH added equals the number of moles of HCl** in the flask. In other words, all of the HCl has reacted with NaOH and the solution is no longer acidic. Read the volume of NaOH solution added from the buret. Using the known volume and molarity of the NaOH solution and the known volume of HCl solution, the molarity of the HCl solution can be calculated.

(a–c): Jill Braaten/Anthony Arena/McGraw Hill

How does a titration tell us the concentration of an HCl solution? A titration is based on the acid–base reaction that occurs between the acid in the flask (HCl) and the base that is added (NaOH).

When the number of moles of base added equals the number of moles of acid in the flask, the acid is *neutralized,* forming a salt and water.

At the end point, the number of moles of the acid (H⁺) and the base (OH⁻) are equal.

$$HCl(aq) \ + \ NaOH(aq) \longrightarrow NaCl(aq) \ + \ H_2O(l)$$

To determine an unknown molarity from titration data requires three operations.

First, we determine the number of moles of base added using its known molarity and volume. Then we use coefficients in the balanced acid–base equation to tell us the number of moles of acid that react with the base. Finally, we determine the molarity of the acid from the calculated number of moles and the known volume of the acid.

Sample Problem 8.16 — Determining the Molarity of an Acid from a Titration

What is the molarity of an HCl solution if 22.5 mL of a 0.100 M NaOH solution are needed to titrate 25.0 mL of the sample? The balanced equation for this acid–base reaction is given.

$$HCl(aq) + NaOH(aq) \longrightarrow NaCl(aq) + H_2O(l)$$

Analysis and Solution

[1] Determine the number of moles of base used to neutralize the acid.

- Convert milliliters to liters of base using a mL–L conversion factor. Use the molarity (M) and volume (V) of the base to calculate the number of moles **(mol = MV).**

volume of NaOH		mL–L conversion factor		M (mol/L) conversion factor		
22.5 mL NaOH	×	$\dfrac{1\,L}{1000\ mL}$	×	$\dfrac{0.100\ mol\ NaOH}{1\,L}$	=	0.002 25 mol NaOH

[2] Determine the number of moles of acid that react from the balanced chemical equation.

- One mole of HCl reacts with one mole of NaOH, so the number of moles of NaOH equals the number of moles of HCl at the end point.

$$HCl(aq) \ + \ NaOH(aq) \longrightarrow NaCl(aq) \ + \ H_2O(l)$$
$$\text{0.002 25 mol} \quad \text{0.002 25 mol}$$

[3] Determine the molarity of the acid from the number of moles and known volume.

				mL–L conversion factor		
M =	$\dfrac{mol}{L}$ =	$\dfrac{0.002\ 25\ mol\ HCl}{25.0\ mL\ solution}$	×	$\dfrac{1000\ mL}{1\ L}$	=	0.0900 M HCl

molarity

Practice Problem 8.23

What is the molarity of an HCl solution if 25.5 mL of a 0.24 M NaOH solution are needed to neutralize 15.0 mL of the sample?

Practice Problem 8.24

How many milliliters of 2.0 M NaOH are needed to neutralize 5.0 mL of a 6.0 M H_2SO_4 solution?

More Practice: Try Problems 8.51–8.54.

8.8   Buffers

A *buffer* is a solution whose pH changes very little when acid or base is added. Most buffers are solutions composed of approximately equal amounts of a weak acid and the salt of its conjugate base.

- The weak acid of the buffer reacts with added base, OH⁻.
- The conjugate base of the buffer reacts with added acid, H_3O^+.

8.8A   General Characteristics of a Buffer

The effect of a buffer can be illustrated by comparing the pH change that occurs when a small amount of strong acid or strong base is added to water, with the pH change that occurs when the same amount of strong acid or strong base is added to a buffer. As shown in Figure 8.10, when a small quantity of a strong acid or strong base is added to neutral water, the pH changes by over 5 pH units.

Figure 8.10

The Effect of a Buffer on pH Changes

a. The pH of pure water changes drastically when a small amount of strong acid or strong base is added. The pH changes from 7 to 1.7 when 0.020 mol HCl is added to the water. The pH changes from 7 to 12.3 when 0.020 mol NaOH is added.

b. The pH of a buffer changes very little when the same amount of strong acid or strong base is added. The pH changes from 4.74 to 4.70 when 0.020 mol HCl is added to the buffer. The pH changes from 4.74 to 4.77 when 0.020 mol NaOH is added.

In contrast, a buffer prepared from 0.50 M acetic acid (CH_3COOH) and 0.50 M sodium acetate ($NaCH_3COO$) has a pH of 4.74. When a small quantity of strong acid or strong base is added to the buffer, the pH changes by no more than 0.04 pH units!

Why is a buffer able to absorb acid or base with very little pH change? Let's use as an example a buffer that contains equal concentrations of acetic acid (CH_3COOH), and the sodium salt of its conjugate base, sodium acetate ($NaCH_3COO$).

$$CH_3COOH(aq) \;+\; H_2O(l) \;\rightleftharpoons\; H_3O^+(aq) \;+\; CH_3COO^-(aq)$$

approximately **equal** amounts

Suppose a small amount of strong acid is added to the buffer. Added H_3O^+ reacts with CH_3COO^- to form CH_3COOH, so that $[CH_3COO^-]$ decreases slightly and $[CH_3COOH]$ increases slightly, but the $[H_3O^+]$ and therefore the pH change only slightly.

On the other hand, if a small amount of strong base is added to the buffer, OH^- reacts with CH_3COOH to form CH_3COO^-, so that $[CH_3COOH]$ decreases slightly and $[CH_3COO^-]$ increases slightly but the $[H_3O^+]$ and therefore the pH change only slightly.

For a buffer to be effective, the amount of added acid or base must be small compared to the amount of buffer present. When a large amount of acid or base is added to a buffer, the H_3O^+ concentration and therefore the pH change a great deal.

Concept Check 8.5

Determine whether a solution containing each of the following substances is a buffer. Explain your reasoning.

    a. HBr and NaBr   b. HF and KF   c. CH_3COOH alone

Concept Check 8.6

Consider a buffer prepared from the weak acid HCO_3^- and its conjugate base CO_3^{2-}.

$$HCO_3^-(aq) + H_2O(l) \rightleftharpoons CO_3^{2-}(aq) + H_3O^+(aq)$$

    a. What happens to the concentrations of HCO_3^- and CO_3^{2-} when a small amount of acid is added to the buffer?

    b. What happens to the concentrations of HCO_3^- and CO_3^{2-} when a small amount of base is added to the buffer?

8.8B   FOCUS ON THE ENVIRONMENT
Acid Rain and a Naturally Buffered Lake

Unpolluted rainwater is not a neutral solution with a pH of 7; rather, because it contains dissolved carbon dioxide, it is slightly acidic with a pH of about 5.6.

$$CO_2(g) \;+\; 2\,H_2O(l) \;\rightleftharpoons\; H_3O^+(aq) \;+\; HCO_3^-(aq)$$

carbon dioxide
from the air

A low concentration of H_3O^+ gives rainwater a pH < 7.

Rainwater that contains dissolved H_2SO_4 (or HNO_3) from burning fossil fuels has a pH lower than 5.6. In some parts of the United States, rainwater often has a pH range of 4–5, and readings as low as pH = 1.8 have been recorded. When the rain in a region consistently has a lower-than-normal pH, this acid rain can have a devastating effect on plant and animal life.

The pH of some lakes changes drastically as the result of acid rain, whereas the pH of other lakes does not. In fact, the ability of some lakes to absorb acid rain without much pH change is entirely due to buffers (Figure 8.11). Lakes that are surrounded by limestone-rich soil are in contact with solid calcium carbonate, $CaCO_3$. As a result, the lake contains a natural carbonate/bicarbonate buffer. When acid precipitation falls on the lake, the dissolved carbonate (CO_3^{2-}) reacts with the acid to form bicarbonate (HCO_3^-).

The buffer reacts with added acid from rain.

$$CO_3^{2-}(aq) \;+\; H_3O^+(aq) \;\rightleftharpoons\; HCO_3^-(aq) \;+\; H_2O(l)$$

A lake surrounded by limestone contains a natural CO_3^{2-}/HCO_3^- buffer.

The carbonate/bicarbonate buffer thus allows the lake to resist large pH changes when acid is added. In some areas acidic lakes have been treated with limestone, thus adding calcium carbonate to neutralize the acid and restore the natural pH. This procedure is expensive and temporary because with time and more acid rain, the pH of the lakes decreases again.

Figure 8.11   Acid Rain and a Naturally Buffered Lake

Acid precipitation contains H_2SO_4 and HNO_3 dissolved in rain and snow.

CO_3^{2-}

HCO_3^-

lake surrounded by limestone ($CaCO_3$)

lake with no natural buffer

Acid rain adds H_3O^+ to the water, decreasing the pH.

The lake contains a natural buffer of CO_3^{2-} and HCO_3^-. H_3O^+ from rain reacts with dissolved CO_3^{2-}, forming more HCO_3^-. The concentrations of CO_3^{2-} and HCO_3^- do not change by much, so the pH changes only slightly.

Individuals with cystic fibrosis, the most common genetic disease in Caucasians, produce thick mucus in the lungs, resulting in a higher-than-normal level of CO_2 and respiratory acidosis.

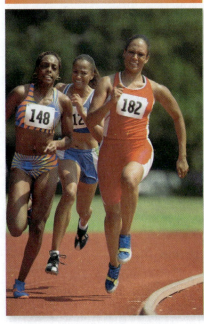

During strenuous exercise, the lungs expel more CO_2 than usual and the pH of the blood increases.
CS Productions/Jupiterimages/Brand X Pictures/Alamy Stock Photo

8.9 FOCUS ON THE HUMAN BODY
Buffers in the Blood

The normal blood pH of a healthy individual is in the range of 7.35 to 7.45. A pH above or below this range is generally indicative of an imbalance in respiratory or metabolic processes. The body is able to maintain a very stable pH because the blood and other tissues are buffered. The principal buffer in the blood is carbonic acid/bicarbonate (H_2CO_3/HCO_3^-).

In examining the carbonic acid/bicarbonate buffer system in the blood, two reactions are important. First of all, carbonic acid (H_2CO_3) is formed from CO_2 dissolved in the bloodstream (Section 8.6). Second, because carbonic acid is a weak acid, it is also dissociated in water to form its conjugate base, bicarbonate (HCO_3^-). Bicarbonate is also generated in the kidneys.

$$CO_2(g) + H_2O(l) \rightleftharpoons \underset{\text{carbonic acid}}{H_2CO_3(aq)} \overset{H_2O}{\rightleftharpoons} H_3O^+(aq) + \underset{\text{bicarbonate}}{HCO_3^-(aq)}$$

principal buffer in the blood

CO_2 is constantly produced by metabolic processes in the body and then transported to the lungs to be eliminated. Le Châtelier's principle explains the effect of increasing or decreasing the level of dissolved CO_2 on the pH of the blood. A higher-than-normal CO_2 concentration shifts the equilibrium to the right, increasing the H_3O^+ concentration and lowering the pH. **Respiratory acidosis** results when the body fails to eliminate adequate amounts of CO_2 through the lungs. This may occur in patients with advanced lung disease or respiratory failure.

A lower respiratory rate **increases [CO$_2$]**.

$$CO_2(g) + 2 H_2O(g) \rightleftharpoons H_3O^+(aq) + HCO_3^-(aq)$$

This drives the reaction to the right, **increasing [H$_3$O$^+$]**.
Blood has a higher [H$_3$O$^+$] $\longrightarrow$ **lower pH**

A lower-than-normal CO_2 concentration shifts the equilibrium to the left, decreasing the H_3O^+ concentration and raising the pH. **Respiratory alkalosis** is caused by hyperventilation, very rapid breathing that occurs when an individual experiences excitement or panic.

$$CO_2(g) + 2 H_2O(g) \rightleftharpoons H_3O^+(aq) + HCO_3^-(aq)$$

A faster respiratory rate **decreases [CO$_2$]**.

This drives the reaction to the left, **decreasing [H$_3$O$^+$]**.
Blood has a lower [H$_3$O$^+$] $\longrightarrow$ **higher pH**

The pH of the blood may also be altered when the metabolic processes of the body are not in balance. **Metabolic acidosis** results when excessive amounts of acid are produced and the blood pH falls. This may be observed in patients with severe infections (sepsis). It may also occur in poorly controlled diabetes. **Metabolic alkalosis** may occur when recurrent vomiting decreases the amount of acid in the stomach, thus causing a rise in pH.

CHAPTER REVIEW

KEY TERMS

Acid (8.1)

Acidic solution (8.4)

Amphoteric (8.2)

Base (8.1)

Basic solution (8.4)

Brønsted–Lowry acid (8.1)

Brønsted–Lowry base (8.1)

Buffer (8.8)

Conjugate acid (8.2)

Conjugate acid–base pair (8.2)

Conjugate base (8.2)

Dissociation (8.3)

Ion–product constant (8.4)

Net ionic equation (8.6)

Neutral solution (8.4)

Neutralization reaction (8.6)

pH scale (8.5)

Proton transfer reaction (8.2)

Spectator ion (8.6)

Titration (8.7)

KEY CONCEPTS

1 **General features of acids and bases (8.1, 8.3)**

1 **Brønsted–Lowry acid (HA)**

- A Brønsted–Lowry acid is a **proton donor.**
- A *strong* acid readily donates a proton, and 100% of the acid dissociates in water, forming H_3O^+.
- A *weak* acid less readily donates a proton, and only a small fraction of the acid dissociates in water.

strong acid
100% dissociated

weak acid
partially dissociated

2 **Brønsted–Lowry base (:B)**

- A Brønsted–Lowry base is a **proton acceptor.**
- A strong base readily accepts a proton, and 100% of the base dissociates in water, forming OH^-.
- A weak base less readily accepts a proton, and only a small fraction of the base forms ions in water.

strong base
100% dissociated

weak base
few ions in solution

2 **Acid–base reactions (8.2)**

1 **Drawing a conjugate base**

- **Remove a proton, H^+,** from an acid to form a conjugate base.
- **Add a –1 charge** to the acid to give the charge on the conjugate base.

acid   conjugate base

2 **Drawing a conjugate acid**

- **Add a proton, H^+,** to form a conjugate acid from a base.
- **Add a +1 charge** to the base to give the charge on the conjugate acid.

base   conjugate acid

3 **Acid–base reactions—Proton transfer reactions**

- In an acid–base reaction, **transfer a proton** from the acid to the base.
- A *strong* acid forms a *weak* conjugate base.

HA   +   B: A:⁻   +   HB⁺

acid   base   conjugate   conjugate
base   acid

3 **$[H_3O^+]$ and pH—The *higher* the concentration of H_3O^+, the *lower* the pH. (8.5)**

1 **Acidic solution**

- A solution is acidic when $[H_3O^+] > [OH^-]$.
- $[H_3O^+] > 1 \times 10^{-7}$.
- **pH < 7.**

2 **Neutral solution**

- A solution is neutral when $[H_3O^+] = [OH^-]$.
- $[H_3O^+] = 1 \times 10^{-7}$.
- **pH = 7.**

3 **Basic solution**

- A solution is basic when $[H_3O^+] < [OH^-]$.
- $[H_3O^+] < 1 \times 10^{-7}$.
- **pH > 7.**

4 **Buffers (8.8)**

1 **General features**

- A buffer is a solution whose pH changes little when acid or base is added.
- A buffer is composed of approximately equal amounts of a weak acid (HA) and its conjugate base (A⁻).

2 **When base is added**

- HA reacts with OH^- to increase [A⁻] a little.

[HA]   [A⁻]
decreases increases

3 **When acid is added**

- A⁻ reacts with H_3O^+ to increase [HA] a little.

[HA]   [A⁻]
increases   decreases

KEY EQUATIONS

① Ion–product constant for water, K_w (8.4)	② The logarithm of the H_3O^+ concentration, pH (8.5)
$$K_w = [H_3O^+][OH^-]$$	$$pH = -\log [H_3O^+]$$
• $K_w = 1.0 \times 10^{-14}$ at 25 °C.	

KEY SKILLS

① Calculating $[OH^-]$ when $[H_3O^+]$ is known (8.4); example: determining $[OH^-]$ when $[H_3O^+] = 2.5 \times 10^{-5}$

① Write the equation for K_w and rearrange.	② Solve the equation by substituting the known values.
$$K_w = [H_3O^+][OH^-] \qquad [OH^-] = \frac{K_w}{[H_3O^+]}$$ Ion–product constant for H_2O — Rearranged equation • Use the rearranged equation to solve for $[OH^-]$.	$$[OH^-] = \frac{K_w}{[H_3O^+]} = \frac{1.0 \times 10^{-14}}{2.5 \times 10^{-5}} = 4.0 \times 10^{-10}$$ concentration of OH^- • Use $K_w = 1.0 \times 10^{-14}$ at 25 °C.

② Interconverting $[H_3O^+]$ and pH (8.5)

① Determining pH from $[H_3O^+]$ when $[H_3O^+] = 5.3 \times 10^{-3}$	② Determining $[H_3O^+]$ from pH when pH = 4.82
• Use a calculator to determine the logarithm of a number that contains a coefficient in scientific notation that is not equal to one. $$\begin{aligned} pH &= -\log [H_3O^+] \\ &= -\log(5.3 \times 10^{-3}) \\ &= -(-2.28) = 2.28 \end{aligned}$$	• Use a calculator to determine the antilogarithm of the negative of the pH. $$\begin{aligned} [H_3O^+] &= \text{antilog}(-pH) \\ &= \text{antilog}(-4.82) \\ &= 1.5 \times 10^{-5} \end{aligned}$$

③ Determining the molarity of an acid from a titration (8.7); example: calculating the molarity of an HCl solution if 17.2 mL of 0.15 M NaOH are used to neutralize 5.00 mL of the sample

① Determine the number of moles of base used to neutralize the acid.	② Determine the number of moles of acid that react from the balanced chemical equation.	③ Determine the molarity of the acid from the number of moles and known volume.
$$17.2 \text{ mL NaOH} \times \frac{1 \text{ L}}{1000 \text{ mL}} \times \frac{0.15 \text{ mol NaOH}}{1 \text{ L}}$$ $$= 0.0026 \text{ mol NaOH}$$ • Use the molarity (M) and volume (V in L) of the base to calculate the number of moles (**mol = MV**).	$$NaOH + HCl \longrightarrow NaCl + H_2O$$ $$0.0026 \text{ mol} \quad 0.0026 \text{ mol}$$ • One mole of HCl reacts with one mole of NaOH, so 0.0026 mol of HCl reacts.	$$M = \frac{mol}{L} = \frac{0.0026 \text{ mol HCl}}{5.00 \text{ mL solution}} \times \frac{1000 \text{ mL}}{1 \text{ L}}$$ $$= 0.52 \text{ M HCl}$$ • The number of moles and known volume (in L) are used to give the molarity.

CHAPTER 8 SELF-TEST

The Self-Test consists of short-answer questions that test an understanding of definitions, equations, and other material encountered within the chapter. Answers to each question are provided at the end of the chapter. For any questions you are unable to answer correctly, revisit the appropriate section(s) of the text; then proceed to working the end-of-chapter problems to further reinforce your understanding of the chapter material.

1. Which term from "Key Terms" answers each question?
   a. What product is formed when a base gains a proton?
   b. What type of solution has a pH > 7?
   c. What term is used for the product of the concentrations of H_3O^+ and OH^- in water or an aqueous solution?
   d. What type of solution consists of an equal amount of a weak acid and its conjugate base, so its pH changes very little when an acid or base is added?
   e. What term describes a compound that can be both an acid and a base?

2. Label each compound as a Brønsted–Lowry acid (**A**), a Brønsted–Lowry base (**B**), or both: (a) $Ca(OH)_2$; (b) H_2O; (c) NO_2^-; (d) HBr.

3. (a) Draw the conjugate acid of HPO_4^{2-}. (b) Draw the conjugate base of H_2S.

4. Write a balanced equation for the following reaction:
$H_2SO_4(aq) + Mg(OH)_2(aq) \longrightarrow$.

5. Calculate $[H_3O^+]$ when $[OH^-] = 4.5 \times 10^{-9}$.

6. Calculate $[H_3O^+]$ when pH = 8.90.

7. How many milliliters of 1.0 M NaOH solution are needed to neutralize 10.0 mL of 2.5 M CH_3CO_2H solution?

8. Consider a buffer prepared from equal amounts of the acid HA and its conjugate base A^-. (a) What happens to the concentrations of HA and A^- when a small amount of acid is added? (b) What happens to the concentrations of HA and A^- when a small amount of base is added?

UNDERSTANDING KEY CONCEPTS

8.1 Draw the structure of the conjugate base of each acid.

a.

b.

8.2 Draw the structure of the conjugate base of each acid.

a.

b.

8.3 (a) Which of the following represents a strong acid HZ dissolved in water? (b) Which represents a weak acid HZ dissolved in water?

 = HZ   = Z⁻

A   B

8.4 (a) Using molecular art, draw a diagram that represents an aqueous solution of a strong acid H_2Z. (b) Using molecular art, draw a diagram that represents an aqueous solution of a weak acid H_2Z.

 = H_2Z = HZ⁻

8.5 Label the acid in the reactants and the conjugate acid in the products in each reaction.
a. $H_3PO_4(aq) + CN^-(aq) \longrightarrow H_2PO_4^-(aq) + HCN(aq)$
b. $Br^-(aq) + HSO_4^-(aq) \longrightarrow SO_4^{2-}(aq) + HBr(aq)$

8.6 Identify the acid, base, conjugate acid, and conjugate base in each diagram. Gray spheres correspond to H atoms.

a. + $\longrightarrow$ +

b. + $\longrightarrow$ +

8.7 Why is NH_3 a Brønsted–Lowry base but CH_4 is not?

8.8 Why is HCl a Brønsted–Lowry acid but NaCl is not?

8.9 Calculate the concentrations of H_3O^+ and OH^- in the water sample with the given pH.

8.10 Calculate the concentrations of H_3O^+ and OH^- in the sample of milk of magnesia with the given pH.

8.11 Marble statues, which are composed of calcium carbonate ($CaCO_3$), are slowly eaten away by the nitric acid (HNO_3) in acid rain. Write a balanced equation for the reaction of $CaCO_3$ with HNO_3.

8.12 Some liquid antacids contain suspensions of aluminum hydroxide [$Al(OH)_3$]. Write a balanced equation for the reaction of $Al(OH)_3$ with the HCl in stomach acid.

8.13 Consider a buffer prepared from the weak acid HNO_2 and its conjugate base NO_2^-. What happens to the concentrations of HNO_2 and NO_2^- when a small amount of acid is added to the buffer?

$$HNO_2(aq) + H_2O(l) \rightleftharpoons NO_2^-(aq) + H_3O^+(aq)$$

8.14 Referring to the equation in Problem 8.13, state what happens to the concentrations of HNO_2 and NO_2^- when a small amount of base is added to the buffer.

ADDITIONAL PROBLEMS

Acids and Bases

8.15 Which of the following species can be Brønsted–Lowry acids?
- a. HBr
- b. Br_2
- c. $AlCl_3$
- d. HCOOH
- e. NO_2^-
- f. HNO_2

8.16 Which of the following species can be Brønsted–Lowry acids?
- a. H_2O
- b. I^-
- c. HOCl
- d. $FeBr_3$
- e. CH_3CH_2COOH
- f. CO_2

8.17 Which of the following species can be Brønsted–Lowry bases?
- a. OH^-
- b. Ca^{2+}
- c. C_2H_6
- d. PO_4^{3-}
- e. OCl^-
- f. $MgCO_3$

8.18 Which of the following species can be Brønsted–Lowry bases?
- a. Cl^-
- b. BH_3
- c. H_2O
- d. Na^+
- e. $Ca(OH)_2$
- f. $HCOO^-$

8.19 Draw the conjugate acid of each base.
- a. HS^-
- b. CO_3^{2-}
- c. NO_2^-

8.20 Draw the conjugate acid of each base.
- a. Br^-
- b. HPO_4^{2-}
- c. CH_3COO^-

8.21 Draw the conjugate base of each acid.
- a. HNO_2
- b. NH_4^+
- c. H_2O_2

8.22 Draw the conjugate base of each acid.
- a. H_3O^+
- b. H_2Se
- c. HSO_4^-

8.23 Label the conjugate acid–base pairs in each equation.
- a. $HI(g) + NH_3(g) \longrightarrow NH_4^+(aq) + I^-(aq)$
- b. $HCOOH(l) + H_2O(l) \longrightarrow H_3O^+(aq) + HCOO^-(aq)$
- c. $HSO_4^-(aq) + H_2O(l) \longrightarrow H_2SO_4(aq) + OH^-(aq)$

8.24 Label the conjugate acid–base pairs in each equation.
- a. $Cl^-(aq) + HSO_4^-(aq) \longrightarrow HCl(aq) + SO_4^{2-}(aq)$
- b. $HPO_4^{2-}(aq) + OH^-(aq) \longrightarrow PO_4^{3-}(aq) + H_2O(l)$
- c. $NH_3(g) + HF(g) \longrightarrow NH_4^+(aq) + F^-(aq)$

8.25 Like H_2O, HCO_3^- is amphoteric. (a) Draw the conjugate acid of HCO_3^-. (b) Draw the conjugate base of HCO_3^-.

8.26 Like H_2O, $H_2PO_4^-$ is amphoteric. (a) Draw the conjugate acid of $H_2PO_4^-$. (b) Draw the conjugate base of $H_2PO_4^-$.

Acid and Base Strength

8.27 Which diagram represents an aqueous solution of HF and which represents HCl? Explain your choice.

A   B

8.28 Which diagram represents what happens when HCN dissolves in water? Explain your choice.

A   B

8.29 Use the data in Table 8.2 to label the stronger acid in each pair.
- a. H_2O or CH_3COOH
- b. H_3PO_4 or H_2SO_4

8.30 Use the data in Table 8.2 to label the stronger acid in each pair.
- a. H_2SO_4 or NH_4^+
- b. H_2O or HF

8.31 Which acid in each pair in Problem 8.29 has the stronger conjugate base?

8.32 Which acid in each pair in Problem 8.30 has the stronger conjugate base?

8.33 Which diagram illustrates the stronger acid?

 = HA   = A^- = H_3O^+

A   B

8.34 Which diagram in Problem 8.33 contains the stronger conjugate base? Explain your choice.

8.35 Which acid, **A** or **B**, is stronger in each part?
- a. **A** dissociates to a greater extent in water.
- b. The conjugate base of **A** is stronger than the conjugate base of **B**.

8.36 Fill in the missing terms (strong or weak) and formulas in the following table.

	Acid	Formula	Conjugate Base	Formula
a.	strong	H_2SO_4		
b.		HOBr	strong	
c.			weak	NO_3^-
d.	weak			HCO_2^-

Water and the pH Scale

8.37 Calculate the value of [OH⁻] from the given [H₃O⁺] and label the solution as acidic or basic.

   a. 10^{-8} M   c. 3.0×10^{-4} M

   b. 10^{-10} M   d. 2.5×10^{-11} M

8.38 Calculate the value of [H₃O⁺] from the given [OH⁻] and label the solution as acidic or basic.

   a. 10^{-2} M   c. 6.2×10^{-7} M

   b. 4.0×10^{-8} M   d. 8.5×10^{-13} M

8.39 Calculate the value of [H₃O⁺] from the given [OH⁻] in each liquid and label the solution as acidic or basic.

   a.   b.

       [OH⁻] = 1.1×10^{-9} M   [OH⁻] = 4.0×10^{-12} M

(a): Davydenko Yuliia/Shutterstock; (b): Ingram Publishing/SuperStock

8.40 Calculate the value of [OH⁻] from the given [H₃O⁺] in each liquid and label the solution as acidic or basic.

   a.   b.

      [H₃O⁺] = 5.0×10^{-7} [H₃O⁺] = 2.1×10^{-4}

(a): D. Hurst/Alamy Stock Photo; (b): Keith Homan/Shutterstock

8.41 Complete the following table with the needed [H₃O⁺], [OH⁻], and pH, and classify the solution as acidic, basic, or neutral.

	[H₃O⁺]	[OH⁻]	pH	Classification
a.	5.3×10^{-3}			
b.		2.0×10^{-8}		
c.			4.4	
d.		6.8×10^{-10}		

8.42 Determine the concentrations of H₃O⁺ and OH⁻ in each of the following drinks: (a) sparkling water (pH = 4.96); (b) pink lemonade (pH = 2.59); (c) Hawaiian Punch (pH = 2.87).

8.43 Calculate the H₃O⁺ concentration from each pH: (a) 12; (b) 1; (c) 1.80.

8.44 Calculate the H₃O⁺ concentration from each pH: (a) 4; (b) 2.60; (c) 11.30.

8.45 What are the concentrations of H₃O⁺ and OH⁻ in tomatoes that have a pH of 4.10?

8.46 What are the concentrations of H₃O⁺ and OH⁻ in a cola beverage that has a pH of 3.15?

8.47 Calculate the pH of each aqueous solution: (a) 0.0025 M HCl; (b) 0.015 M KOH.

8.48 Calculate the pH of each aqueous solution: (a) 0.015 M HNO₃; (b) 0.0025 M NaOH.

Acid–Base Reactions

8.49 Write a balanced equation for each reaction.

   a. $HBr(aq) + KOH(aq) \longrightarrow$

   b. $HNO_3(aq) + Ca(OH)_2(aq) \longrightarrow$

   c. $HCl(aq) + NaHCO_3(aq) \longrightarrow$

8.50 Write a balanced equation for each reaction.

   a. $HNO_3(aq) + LiOH(aq) \longrightarrow$

   b. $H_2SO_4(aq) + NaOH(aq) \longrightarrow$

   c. $K_2CO_3(aq) + HCl(aq) \longrightarrow$

Titration

8.51 What is the molarity of an HCl solution if 35.5 mL of 0.10 M NaOH are needed to neutralize 25.0 mL of the sample?

8.52 What is the molarity of an HCl solution if 17.2 mL of 0.15 M NaOH are needed to neutralize 5.00 mL of the sample?

8.53 What is the molarity of an acetic acid (CH₃COOH) solution if 15.5 mL of 0.20 M NaOH are needed to neutralize 25.0 mL of the sample?

8.54 What is the molarity of an H₂SO₄ solution if 18.5 mL of 0.18 M NaOH are needed to neutralize 25.0 mL of the sample?

Buffers

8.55 Consider a weak acid H₂A and its conjugate base HA⁻. Which diagram represents a buffer? Explain your choice.

= H₂A   = HA⁻

            A   **B**

8.56 Consider a weak acid H_2A and its conjugate base HA^-. Which diagram represents a buffer? Explain your choice.

  = H_2A = HA^-

           A   B

8.57 Consider a buffer prepared from the weak acid HF and its conjugate base F^-.

$$HF(aq) + H_2O(l) \rightleftharpoons F^-(aq) + H_3O^+(aq)$$

  a. What happens to the concentrations of HF and F^- when a small amount of acid is added to the buffer?

  b. What happens to the concentrations of HF and F^- when a small amount of base is added to the buffer?

8.58 Explain why both HF and F^- are needed to prepare the buffer in Problem 8.57.

CHALLENGE PROBLEMS

8.67 Calcium hypochlorite $[Ca(OCl)_2]$ is used to chlorinate swimming pools. $Ca(OCl)_2$ acts as a source of the weak acid hypochlorous acid, HOCl, a disinfectant that kills bacteria. Write the acid–base reaction that occurs when OCl^- dissolves in water and explain why this reaction makes a swimming pool more basic.

BEYOND THE CLASSROOM

8.69 Ascorbic acid (vitamin C) and pantothenic acid (vitamin B_5) are two acids needed by humans in the diet for normal cellular function. Pick one of these vitamins and draw its chemical structure. How much is required in the daily diet? What are some dietary sources of the vitamin? Why is the vitamin needed in the body and what symptoms result from its deficiency?

Applications

8.59 Why is the pH of unpolluted rainwater lower than the pH of pure water?

8.60 Why is the pH of acid rain lower than the pH of rainwater?

8.61 The optimum pH of a swimming pool is 7.50. Calculate the value of $[H_3O^+]$ and $[OH^-]$ at this pH.

8.62 A sample of rainwater has a pH of 4.18. (a) Calculate the H_3O^+ concentration in the sample. (b) Suggest a reason why this pH differs from the pH of unpolluted rainwater (5.6).

8.63 When an individual hyperventilates, he is told to blow into a paper bag held over his mouth. What effect should this process have on the CO_2 concentration and pH of the blood?

8.64 What is the difference between respiratory acidosis and respiratory alkalosis?

8.65 How is CO_2 concentration related to the pH of the blood?

8.66 Explain why a lake on a bed of limestone is naturally buffered against the effects of acid rain.

8.68 Most buffer solutions are prepared using a weak acid and a salt of its conjugate base. Explain how the following combination can also form a buffer solution: 0.20 M H_3PO_4 and 0.10 M NaOH.

8.70 Many conditions that cause an imbalance of respiratory or metabolic processes affect the pH of the blood. Research how the normal pH of the blood is altered by some or all of the following conditions: pneumonia, excessive alcohol consumption, emphysema, starvation, holding one's breath, hyperventilation, and the excessive use of antacids. Using the principles presented in Section 8.9, explain why the imbalance results. What other conditions affect the pH of the blood?

CHAPTER 8 ANSWERS

Concept Check Answers

8.1   a. hydrofluoric acid
      b. nitric acid
      c. hydrocyanic acid

8.2   chlorous acid

8.3   a. $HNO_3(aq) + NaOH(aq) \longrightarrow H_2O(l) + NaNO_3(aq)$
      b. $H_2SO_4(aq) + 2\,KOH(aq) \longrightarrow 2\,H_2O(l) + K_2SO_4(aq)$

8.4   a,b: $H^+(aq) + OH^-(aq) \longrightarrow H_2O(l)$

8.5   a,c: not a buffer
      b. buffer

8.6   a. $[HCO_3^-]$ increases, $[CO_3^{2-}]$ decreases
      b. $[HCO_3^-]$ decreases, $[CO_3^{2-}]$ increases

Practice Problem Answers

8.1   a,c

8.2   a,b,d

8.3 a. $HCl(g)$ + $NH_3(g)$ c. $OH^-(aq)$ + $HSO_4^-(aq)$
    acid   base   base   acid

b. $CH_3COOH(l)$ + $H_2O(l)$
    acid   base

8.4 a. H_3O^+ b. HI c. H_2CO_3

8.5 a. HS^- b. CN^- c. SO_4^{2-}

8.6 a. $H_2O(l)$ + $HI(g)$ $\rightleftharpoons$ $I^-(aq)$ + $H_3O^+(aq)$
    base   acid   conjugate conjugate
                            base   acid

b. $CH_3COOH(l)$ + $NH_3(g)$ $\rightleftharpoons$ $CH_3COO^-(aq)$ + $NH_4^+(aq)$
    acid   base   conjugate   conjugate
                                    base   acid

c. $Br^-(aq)$ + $HNO_3(aq)$ $\rightleftharpoons$ $HBr(aq)$ + $NO_3^-(aq)$
    base   acid   conjugate conjugate
                                  acid   base

8.7 a. NH_4^+ b. NH_2^-

8.8 a. H_2SO_4 b. SO_4^{2-}

8.9 D < F < E

8.10 a. H_2SO_4 is the stronger acid; H_3PO_4 has the stronger conjugate base.

b. HCl is the stronger acid; HF has the stronger conjugate base.

c. H_2CO_3 is the stronger acid; NH_4^+ has the stronger conjugate base.

d. HF is the stronger acid; HCN has the stronger conjugate base.

8.11 HCN

8.12 a. HNO_2 and HNO_3 b. HNO_3

8.13 a. 10^{-11} M acidic   c. 3.6×10^{-5} M basic

b. 10^{-3} M basic   d. 1.8×10^{-11} M acidic

8.14 a. 10^{-8} M basic   c. 1.9×10^{-4} M acidic

b. 10^{-5} M acidic   d. 1.4×10^{-11} M basic

8.15

	$[H_3O^+]$	$[OH^-]$
a.	10^{-11} M	10^{-3} M
b.	10^{-3} M	10^{-11} M
c.	1.5 M	6.7×10^{-15} M
d.	3.3×10^{-14} M	3.0×10^{-1} M

8.16 a. 6   b. 12   c. 5   d. 11

8.17 a. 1×10^{-13} M basic   c. 1×10^{-3} M acidic

b. 1×10^{-7} M neutral

8.18 a. 5.74   b. 11.036   c. 4.06   d. 10.118

8.19 a. 6×10^{-11} M   b. 2×10^{-8} M   c. 5×10^{-5} M

8.20 5.0×10^{-4} M

8.21 $H_2SO_4(aq)$ + $CaCO_3(s)$ $\longrightarrow$ $CaSO_4(aq)$ + $\underbrace{H_2O(l) + CO_2(g)}$
                                                                    from H_2CO_3

8.22 a. $HNO_3(aq)$ + $NaHCO_3(aq)$ $\longrightarrow$ $NaNO_3(aq)$ + $H_2O(l)$ + $CO_2(g)$

b. $2 HNO_3(aq)$ + $MgCO_3(s)$ $\longrightarrow$ $Mg(NO_3)_2(aq)$ + $H_2O(l)$ + $CO_2(g)$

8.23 0.41 M

8.24 30. mL

Self-Test Answers

1. a. conjugate acid   c. ion–product constant   e. amphoteric
    b. basic solution   d. buffer

2. a,c: **B**   b. both   d. **A**

3. a. $H_2PO_4^-$ b. HS^-

4. $H_2SO_4(aq)$ + $Mg(OH)_2(aq)$ $\longrightarrow$ $2 H_2O(l)$ + $MgSO_4(aq)$

5. 2.2×10^{-6} M

6. 1.3×10^{-9} M

7. 25 mL

8. a. [HA] increases, [A$^-$] decreases
    b. [HA] decreases, [A$^-$] increases

Odd-Numbered Problem Answers

8.1 a. NO_2^- b. CO_3^{2-}

8.3 a. **B**   b. **A**

8.5 a. H_3PO_4 acid, HCN conjugate acid
    b. HSO_4^- acid, HBr conjugate acid

8.7 NH_3 can accept a proton because the N atom has a lone pair, but CH_4 does not have a lone pair, so it cannot.

8.9 $[H_3O^+] = 3 \times 10^{-9}$ M, $[OH^-] = 3 \times 10^{-6}$ M

8.11 $2 HNO_3(aq)$ + $CaCO_3(s)$
                $\longrightarrow$ $Ca(NO_3)_2(aq)$ + $CO_2(g)$ + $H_2O(l)$

8.13 [HNO_2] increases, [NO_2^-] decreases

8.15 a,d,f

8.17 a,d,e,f

8.19 a. H_2S b. HCO_3^- c. HNO_2

8.21 a. NO_2^- b. NH_3 c. HO_2^-

8.23 a. $HI(g)$ acid $I^-(aq)$ conjugate base
    $NH_3(g)$ base $NH_4^+(aq)$ conjugate acid
    b. $HCOOH(l)$ acid $HCOO^-(aq)$ conjugate base
    $H_2O(l)$ base $H_3O^+(aq)$ conjugate acid
    c. $HSO_4^-(aq)$ base $H_2SO_4(aq)$ conjugate acid
    $H_2O(l)$ acid $OH^-(aq)$ conjugate base

8.25 a. H_2CO_3 b. CO_3^{2-}

8.27 **A** represents HCl and **B** represents HF.

8.29 a. CH_3COOH b. H_2SO_4

8.31 a. H_2O b. H_3PO_4

8.33 **B**

8.35 a. **A**   b. **B**

8.37 a. 10^{-6} M basic   c. 3.3×10^{-11} M acidic
    b. 10^{-4} M basic   d. 4.0×10^{-4} M basic

8.39 a. 9.1×10^{-6} acidic   b. 2.5×10^{-3} acidic

8.41

	$[H_3O^+]$	$[OH^-]$	pH	Classification
a.	5.3×10^{-3}	1.9×10^{-12}	2.28	acidic
b.	5.0×10^{-7}	2.0×10^{-8}	6.30	acidic
c.	4×10^{-5}	2.5×10^{-10}	4.4	acidic
d.	1.5×10^{-5}	6.8×10^{-10}	4.82	acidic

8.43 a. 1×10^{-12} M   c. 1.6×10^{-2} M

b. 1×10^{-1} M

8.45 $[H_3O^+] = 7.9 \times 10^{-5}$ M $[OH^-] = 1.3 \times 10^{-10}$ M

8.47 a. 2.60   b. 12.17

8.49 a. $HBr(aq) + KOH(aq) \longrightarrow KBr(aq) + H_2O(l)$

b. $2\,HNO_3(aq) + Ca(OH)_2(aq) \longrightarrow 2\,H_2O(l) + Ca(NO_3)_2(aq)$

c. $HCl(aq) + NaHCO_3(aq) \longrightarrow NaCl(aq) + H_2O(l) + CO_2(g)$

8.51 0.14 M

8.53 0.12 M

8.55 B

8.57 a. [HF] increases and $[F^-]$ decreases.

b. [HF] decreases and $[F^-]$ increases.

8.59 CO_2 combines with rainwater to form H_2CO_3, which is acidic, lowering the pH.

8.61 $[H_3O^+] = 3.2 \times 10^{-8}$ M $[OH^-] = 3.1 \times 10^{-7}$ M

8.63 By breathing into a bag, the individual breathes in air with a higher CO_2 concentration. Thus, the CO_2 concentration in the lungs and the blood increases, thereby lowering the pH.

8.65 A rise in CO_2 concentration leads to an increase in $H^+(aq)$ concentration by the following reactions:

$CO_2(g) + H_2O(l) \rightleftharpoons H_2CO_3(aq) \rightleftharpoons HCO_3^-(aq) + H^+(aq)$

8.67 $OCl^-(aq) + H_2O(l) \longrightarrow HOCl(aq) + OH^-(aq)$

OH^- is formed by this reaction, so the water is basic.

Nuclear Chemistry

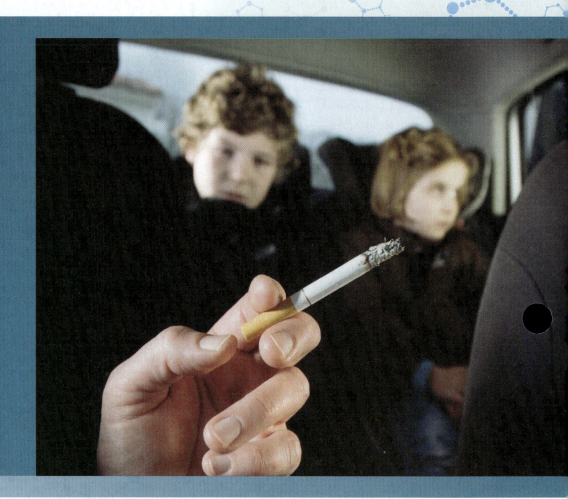

The Centers for Disease Control and Prevention warns consumers that radioactive substances in cigarette smoke harm both smokers and those exposed to secondhand smoke.
ClarkandCompany/Getty Images

CHAPTER OUTLINE

CHAPTER GOALS

In this chapter you will learn how to:

1. Describe the different types of radiation emitted by a radioactive nucleus
2. Write equations for nuclear reactions
3. Define half-life
4. Recognize the units used for measuring radioactivity
5. Give examples of common radioisotopes used in medical diagnosis and treatment
6. Describe the general features of nuclear fission and nuclear fusion
7. Describe the features of medical imaging techniques that do not use radioactivity

Thus far our study of reactions has concentrated on processes that involve the valence electrons of atoms. In these reactions, bonds that join atoms are broken and new bonds between atoms are formed, but the identity of the atoms does not change. In Chapter 9, we turn our attention to **nuclear reactions,** processes that involve changes in the nucleus of an atom. While certainly much less common than chemical reactions that occur with electrons, nuclear reactions form a useful group of processes with a wide range of applications. Nuclear medicine labs in hospitals use radioactive isotopes to diagnose disease, visualize organs, and treat tumors. Generating energy in nuclear power plants, dating archaeological objects using the isotope carbon-14, and designing a simple and reliable smoke detector all utilize the concepts of nuclear chemistry discussed here in Chapter 9.

9.1 Isotopes and Radioactivity

Although most reactions involve valence electrons, a small but significant group of reactions, **nuclear reactions,** involves the subatomic particles of the nucleus. To understand nuclear reactions we must first review facts presented in Chapter 2 regarding isotopes and the characteristics of the nucleus.

9.1A Isotopes

The nucleus of an atom is composed of protons and neutrons.

- The atomic number (Z) = the number of protons in the nucleus.
- The mass number (A) = the number of protons and neutrons in the nucleus.

Atoms of the same type of element have the same atomic number, but the number of neutrons may vary.

- Isotopes are atoms of the same element having a different number of neutrons.

As a result, **isotopes have the same atomic number (Z) but different mass numbers (A).** Carbon, for example, has three naturally occurring isotopes. Each isotope has six protons in the nucleus (i.e., $Z = 6$), but the number of neutrons may be six, seven, or eight. Thus, the mass numbers (A) of these isotopes are 12, 13, and 14, respectively. As we learned in Section 2.3, we can refer to these isotopes as carbon-12, carbon-13, and carbon-14. Isotopes are also written with the mass number to the upper left of the element symbol and the atomic number to the lower left.

p
n

6 protons
6 neutrons

6 protons
7 neutrons

6 protons
8 neutrons

mass number $\longrightarrow$ $^{12}_{6}$C
atomic number $\longrightarrow$
carbon-**12**

$^{13}_{6}$C
carbon-**13**

$^{14}_{6}$C
carbon-**14**

Many isotopes are stable, but a larger number are not.

- A *radioactive isotope,* called a *radioisotope,* is unstable and spontaneously emits energy to form a more stable nucleus.

Radioactivity **is the nuclear radiation emitted by a radioactive isotope.** Of the known isotopes of all the elements, 264 are stable and 300 are naturally occurring but unstable. An even larger number of radioactive isotopes, called **artificial isotopes,** have been produced in the laboratory. Both carbon-12 and carbon-13 are stable isotopes and occur in higher natural abundance than carbon-14, a radioactive isotope.

Using the isotope whose nucleus is shown in the accompanying diagram, give its: (a) atomic number; (b) mass number; (c) number of protons; (d) number of neutrons; (e) isotope symbol.

9.1B Types of Radiation

Different forms of radiation are emitted when a radioactive nucleus is converted to a more stable nucleus, including **alpha particles, beta particles, positrons,** and **gamma radiation.**

- An *alpha particle* is a high-energy particle that contains two protons and two neutrons.

$$\text{alpha particle:} \quad \alpha \quad \text{or} \quad {}_{2}^{4}\text{He}$$

An alpha particle, symbolized by the Greek letter **alpha (α)** or the element symbol for helium, has a +2 charge and a mass number of 4.

What is the difference between an α particle and a helium atom?

- A *beta particle* is a high-energy electron.

$$\text{beta particle:} \quad \beta \quad \text{or} \quad {}_{-1}^{0}\text{e}$$

An electron has a −1 charge and a negligible mass compared to a proton. A beta particle, symbolized by the **Greek letter beta (β),** is also drawn with the symbol for an electron, **e,** with a mass number of 0 in the upper left corner and a charge of −1 in the lower left corner. A β particle is formed when a neutron (n) is converted to a proton (p) and an electron.

$$\underset{\text{neutron}}{{}_{0}^{1}\text{n}} \longrightarrow \underset{\text{proton}}{{}_{1}^{1}\text{p}} + \underset{\boldsymbol{\beta \text{ particle}}}{{}_{-1}^{0}\text{e}}$$

- A *positron* is called an *antiparticle* of a β particle, because their charges are different but their masses are the same.

Thus, a **positron** has a negligible mass like a β particle, but is opposite in charge, +1. A positron, symbolized as $\boldsymbol{\beta^+}$, is also drawn with the symbol for an electron, **e,** with a mass number of 0 in the upper left corner and a charge of +1 in the lower left corner. A positron, which can be thought of as a "positive electron," is formed when a proton is converted to a neutron.

$$\textbf{Symbol:} \quad \underset{\text{positron}}{{}_{+1}^{0}\text{e}} \text{ or } \beta^+ \qquad \textbf{Formation:} \quad \underset{\text{proton}}{{}_{1}^{1}\text{p}} \longrightarrow \underset{\text{neutron}}{{}_{0}^{1}\text{n}} + \underset{\boldsymbol{\text{positron}}}{{}_{+1}^{0}\text{e}}$$

- *Gamma rays* are high-energy radiation released from a radioactive nucleus.

Gamma rays, symbolized by the **Greek letter gamma (γ),** are a form of energy and thus they have no mass or charge. Table 9.1 summarizes the properties of these different types of radiation.

gamma ray: γ

Table 9.1 Types of Radiation

Type of Radiation	Symbol	Charge	Mass Number
Alpha particle	α or ${}^{4}_{2}He$	+2	4
Beta particle	β or ${}^{0}_{-1}e$	−1	0
Positron	β^+ or ${}^{0}_{+1}e$	+1	0
Gamma ray	γ	0	0

Concept Check 9.3

What is the difference between an electron and a positron?

Concept Check 9.4

Identify Q in each of the following symbols.

a. ${}^{0}_{-1}Q$ b. ${}^{4}_{2}Q$ c. ${}^{0}_{+1}Q$

9.2 Nuclear Reactions

Radioactive decay **is the process by which an unstable radioactive nucleus emits radiation, forming a nucleus of new composition.** A nuclear equation can be written for this process, which contains the original nucleus, the new nucleus, and the radiation emitted. Unlike a chemical equation that balances atoms, in a nuclear equation the mass numbers and the atomic numbers of the nuclei must be balanced.

- The sum of the mass numbers (*A*) must be equal on both sides of a nuclear equation.
- The sum of the atomic numbers (*Z*) must be equal on both sides of a nuclear equation.

9.2A Alpha Emission

Alpha emission **is the decay of a nucleus by emitting an α particle.** For example, uranium-238 decays to form thorium-234 by loss of an α particle.

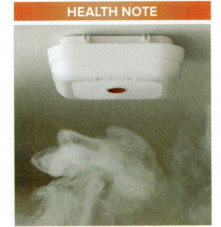

Americium-241 (*How To* p. 302) is a radioactive element contained in smoke detectors. The decay of α particles creates an electric current that is interrupted when smoke enters the detector, sounding an alarm.
Richard Lewisohn/Image Source

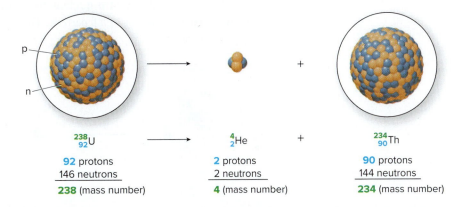

p

n

${}^{238}_{92}U$	$\longrightarrow$	${}^{4}_{2}He$	+	${}^{234}_{90}Th$
92 protons		**2** protons		**90** protons
146 neutrons		2 neutrons		144 neutrons
238 (mass number)		**4** (mass number)		**234** (mass number)

An α particle has two protons, so **the new nucleus has** *two fewer protons* **than the original nucleus.** Because it has a *different* number of protons, the new nucleus represents a *different* element. Uranium-238 has 92 protons, so loss of two forms the element thorium with 90 protons. The thorium nucleus has a mass number that is four fewer than the original—234—because it has been formed by loss of an α particle with a mass number of four.

As a result, the sum of the mass numbers is equal on both sides of the equation—238 = 4 + 234. The sum of the atomic numbers is also equal on both sides of the equation—92 = 2 + 90.

How To Balance an Equation for a Nuclear Reaction

Example Write a balanced nuclear equation showing how americium-241, a radioactive element used in smoke detectors, decays to form an α particle.

Step [1] Write an incomplete equation with the original nucleus on the left and the particle emitted on the right.
- Include the mass number and atomic number (from the periodic table) in the equation.

$$^{241}_{95}\text{Am} \longrightarrow {}^{4}_{2}\text{He} + ?$$

Step [2] Calculate the mass number and atomic number of the newly formed nucleus on the right.
- Mass number: The sum of the mass numbers must be equal on both sides of the equation. **Subtract the mass of an α particle (4) to obtain the mass of the new nucleus;** $241 - 4 = 237$.
- Atomic number: The sum of the atomic numbers must be equal on both sides of the equation. **Subtract the two protons of an α particle to obtain the atomic number of the new nucleus;** $95 - 2 = 93$.

Step [3] Use the atomic number to identify the new nucleus and complete the equation.
- From the periodic table, the element with an atomic number of 93 is neptunium, Np.
- Write the mass number and the atomic number with the element symbol to complete the equation.

$$241 = 4 + 237$$
$$95 = 2 + 93$$
$$^{241}_{95}\text{Am} \longrightarrow {}^{4}_{2}\text{He} + {}^{237}_{93}\text{Np}$$

Concept Check 9.5

Radon, a radioactive gas formed in the soil, can cause lung cancers when inhaled in high concentrations for a long period of time. Write a balanced nuclear equation for the decay of radon-222, which emits an α particle.

Concept Check 9.6

Write a balanced equation showing how each nucleus decays to form an α particle:
(a) polonium-218; (b) thorium-230.

9.2B Beta Emission

***Beta emission* is the decay of a nucleus by emitting a β particle.** For example, carbon-14 decays to nitrogen-14 by loss of a β particle. The decay of carbon-14 is used to date archaeological specimens (Section 9.3).

$$^{14}_{6}\text{C} \longrightarrow {}^{0}_{-1}\text{e} + {}^{14}_{7}\text{N}$$

6 protons	−1 charge	7 protons
8 neutrons	0 mass	7 neutrons
14 (mass number)		14 (mass number)

The mass number is constant.

In β emission, one neutron of the original nucleus decays to a β particle and a proton. As a result, the **new nucleus has *one more proton* and *one fewer neutron* than the original nucleus.** In this example, a carbon atom with six protons decays to a nitrogen atom with seven protons. Because the total number of particles in the nucleus does not change, the **mass number is constant.**

The subscripts that represent the atomic numbers are balanced because the β particle has a charge of –1. Seven protons on the right side plus a –1 charge for the β particle gives a total "charge" of +6, the atomic number of carbon on the left. The mass numbers are also balanced because a β particle has zero mass, and both the original nucleus and the new nucleus contain 14 subatomic particles (protons + neutrons).

Radioactive elements that emit β radiation are widely used in medicine. Iodine-131, a radioactive element that emits β radiation, is used to treat hyperthyroidism, a condition resulting from an overactive thyroid gland (Figure 9.1). Moreover, because β radiation is composed of high-energy electrons that penetrate tissue in a small, localized region, radioactive elements situated in close contact with tumor cells kill them. Although both healthy and diseased cells are destroyed by this internal radiation therapy, rapidly dividing tumor cells are more sensitive to its effects and therefore their growth and replication are affected the most.

Figure 9.1   The Use of Iodine-131 to Treat Hyperthyroidism

thyroid gland

Radioactive iodine-131 is incorporated into the four I atoms (in purple) of thyroxine.

thyroxine
$C_{15}H_{11}I_4NO_4$

Iodine-131 is incorporated into the thyroid hormone thyroxine. Beta radiation emitted by the radioactive isotope destroys nearby thyroid cells, thus decreasing the activity of the thyroid gland and bringing the disease under control.

Sample Problem 9.1

Balancing a Nuclear Equation Involving β Emission

Write a balanced nuclear equation for the β emission of phosphorus-32, a radioisotope used to treat leukemia and other blood disorders.

Analysis

- Balance the atomic numbers and mass numbers on both sides of a nuclear equation.
- With β emission, **treat the β particle as an electron with zero mass** in balancing mass numbers, and a **–1 charge** when balancing the atomic numbers.

Solution

[1] **Write an incomplete equation with the original nucleus on the left and the particle emitted on the right.**

- Use the identity of the element to determine the atomic number; phosphorus has an atomic number of 15.

$$^{32}_{15}P \longrightarrow \ ^{0}_{-1}e \ + \ ?$$

[2] **Calculate the mass number and the atomic number of the newly formed nucleus on the right.**

- Mass number: Because a β particle has no mass, the masses of the new particle and the original particle are the same, 32.
- Atomic number: Because β emission converts a neutron into a proton, the new nucleus has one more proton than the original nucleus; $15 = -1 + ?$. Thus the new nucleus has an atomic number of 16.

[3] **Use the atomic number to identify the new nucleus and complete the equation.**

- From the periodic table, the element with an atomic number of 16 is sulfur, S.
- Write the mass number and the atomic number with the element symbol to complete the equation.

$$^{32}_{15}\text{P} \longrightarrow \,^{0}_{-1}\text{e} + \,^{32}_{16}\text{S}$$

Practice Problem 9.1	Write a balanced nuclear equation for the β emission of iodine-131.

Practice Problem 9.2	Write a balanced nuclear equation for the β emission of each of the following isotopes.

    a. $^{20}_{9}\text{F}$ b. $^{92}_{38}\text{Sr}$ c. chromium-55

More Practice: Try Problems 9.8a, 9.17a, 9.18c, 9.19, 9.23b, 9.24a.

9.2C Positron Emission

***Positron emission* is the decay of a nucleus by emitting a positron (β⁺).** For example, carbon-11, an artificial radioactive isotope of carbon, decays to boron-11 by loss of a β⁺ particle. Positron emitters are used in positron emission tomography (PET), a diagnostic technique described in Section 9.5.

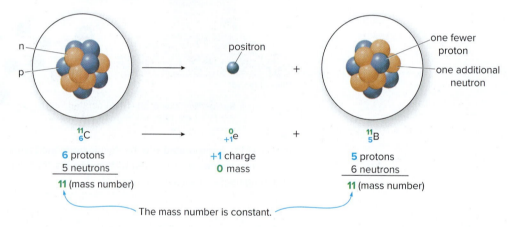

In positron emission, one proton of the original nucleus decays to a β⁺ particle and a neutron. As a result, the **new nucleus has *one fewer proton* and *one more neutron* than the original nucleus.** In this example, a carbon atom with six protons decays to a boron atom with five protons. Because the total number of particles in the nucleus does not change, the **mass number is constant.**

Sample Problem 9.2	**Balancing a Nuclear Equation Involving Positron Emission**

Write a balanced nuclear equation for the positron emission of fluorine-18, a radioisotope used for imaging in PET scans.

Analysis

- Balance the atomic numbers and mass numbers on both sides of a nuclear equation.
- With β⁺ emission, **treat the positron as a particle with zero mass** when balancing mass numbers, and a **+1 charge** when balancing the atomic numbers.

Solution

[1] **Write an incomplete equation with the original nucleus on the left and the particle emitted on the right.**

- Use the identity of the element to determine the atomic number; fluorine has an atomic number of 9.

$$^{18}_{9}\text{F} \longrightarrow \,^{0}_{+1}\text{e} + \,?$$

[2] Calculate the mass number and the atomic number of the newly formed nucleus on the right.

- Mass number: Because a β^+ particle has no mass, the masses of the new particle and the original particle are the same, 18.
- Atomic number: Because β^+ emission converts a proton into a neutron, the new nucleus has one fewer proton than the original nucleus; $9 - 1 = 8$. Thus, the new nucleus has an atomic number of 8.

[3] Use the atomic number to identify the new nucleus and complete the equation.

- From the periodic table, the element with an atomic number of 8 is oxygen, O.
- Write the mass number and the atomic number with the element symbol to complete the equation.

$$^{18}_{9}F \longrightarrow\ ^{\ 0}_{+1}e\ +\ ^{18}_{8}O$$

Practice Problem 9.3

Write a balanced nuclear equation for the positron emission of each of the following nuclei:
(a) arsenic-74; (b) oxygen-15.

More Practice: Try Problems 9.3, 9.7a, 9.17c, 9.18a, 9.23c, 9.24c.

9.2D   Gamma Emission

Gamma emission **is the decay of a nucleus by emitting γ radiation.** Because γ rays are simply a form of energy, their emission causes **no change in the atomic number or mass number** of a radioactive nucleus. Gamma emission sometimes occurs alone. For example, one form of technetium-99, written as technetium-99m, is an energetic form of the technetium nucleus that decays with emission of γ rays to technetium-99, a more stable but still radioactive element.

The *m* in technetium-99m stands for *metastable*. This designation is meant to indicate that the isotope decays to a more stable form of the same isotope.

$$^{99m}_{43}Tc \longrightarrow\ ^{99}_{43}Tc\ +\ \gamma$$

The mass number and atomic number are the same.

Technetium-99m is a widely used radioisotope in medical imaging. Because it emits high-energy γ rays but decays in a short period of time, it is used to image the brain, thyroid, lungs, liver, skeleton, and many other organs. It has also been used to detect ulcers in the gastrointestinal system, and combined with other compounds, it is used to map the circulatory system and gauge damage after a heart attack.

More commonly, γ emission accompanies α or β emission. For example, cobalt-60 decays with both β and γ emission. Because a β particle is formed, decay generates an element with the *same* mass but a *different* number of protons, and thus a new element, nickel-60.

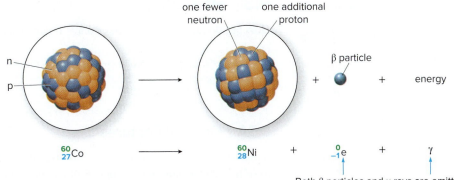

Cobalt-60 is used in external radiation treatment for cancer. Radiation generated by cobalt-60 decay is focused on a specific site in the body that contains cancerous cells (Figure 9.2). By directing the radiation on the tumor, damage to surrounding healthy tissues is minimized.

Figure 9.2 Focus on Health & Medicine: External Radiation Treatment for Tumors

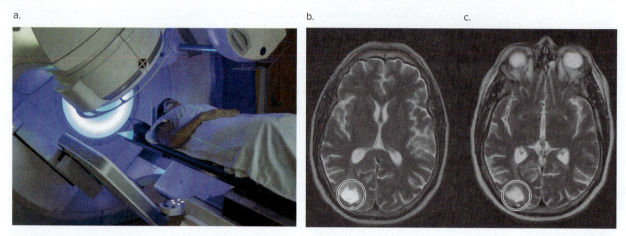

a. Gamma radiation from the decay of cobalt-60 is used to treat a variety of tumors, especially those that cannot be surgically removed.
b. A tumor (bright area in circle) before radiation treatment
c. A tumor (bright area in circle) that has decreased in size after six months of radiation treatment
(a): Mark Kostich/Getty Images; (b & c): STEVEN NEEDELL/Science Source

---

| Concept Check 9.7 | Complete the nuclear equation by drawing the nucleus of the missing atom. Give the symbol for each atom and the type of radiation. The blue spheres represent protons and the orange spheres represent neutrons. |

| Concept Check 9.8 | Write a nuclear equation for the decay of iridium-192 with β and γ emission. Iridium implants have been used to treat breast cancer. After the correct dose is administered, the iridium source is removed. |

9.3 Half-Life

How fast do radioactive isotopes decay? It depends on the isotope.

- The *half-life* ($t_{1/2}$) of a radioactive isotope is the time it takes for one-half of the sample to decay.

9.3A General Features

Suppose we have a sample that contains 16 g of phosphorus-32, a radioactive isotope that decays to sulfur-32 by β emission (Sample Problem 9.1). Phosphorus-32 has a half-life of approximately 14 days. Thus, after 14 days, the sample contains only half the amount of P-32—8.0 g. After another 14 days (a total of two half-lives), the 8.0 g of P-32 is again halved to 4.0 g. After another 14 days (a total of three half-lives), the 4.0 g of P-32 is halved to 2.0 g, and so on. Every 14 days, half of the P-32 decays.

three half-lives

Many naturally occurring isotopes have long half-lives. Examples include carbon-14 (5.70×10^3 years) and uranium-235 (7.03×10^8 years). Radioisotopes that are used for diagnosis and imaging in medicine have short half-lives so they do not linger in the body. Examples include technetium-99m (6.01 hours) and iodine-131 (8.02 days). The half-lives of several elements are given in Table 9.2.

Table 9.2 Half-Lives of Some Common Radioisotopes

Radioisotope	Symbol	Half-Life	Use
Carbon-14	$^{14}_{6}C$	5.70×10^3 years	Archaeological dating
Cobalt-60	$^{60}_{27}Co$	5.27 years	Cancer therapy
Iodine-131	$^{131}_{53}I$	8.02 days	Thyroid therapy
Potassium-40	$^{40}_{19}K$	1.25×10^9 years	Geological dating
Phosphorus-32	$^{32}_{15}P$	14.3 days	Leukemia treatment
Technetium-99m	$^{99m}_{43}Tc$	6.01 hours	Organ imaging
Uranium-235	$^{235}_{92}U$	7.03×10^8 years	Nuclear reactors

The half-life of a radioactive isotope is a property of a given isotope and is independent of the amount of sample, temperature, and pressure. Thus, if the half-life and amount of a sample are known, it is possible to predict how much of the radioactive isotope will remain after a period of time.

How To Use a Half-Life to Determine the Amount of Radioisotope Present

Example   If the half-life of iodine-131 is 8.02 days, how much of a 100. mg sample of iodine-131 remains after 32 days? Iodine-131 forms xenon-131 by β emission.

Step [1]   **Determine how many half-lives occur in the given amount of time.**
- Use the half-life of iodine-131 as a conversion factor to convert the number of days to the number of half-lives.

$$32 \text{ days} \times \frac{1 \text{ half-life}}{8.02 \text{ days}} = 4.0 \text{ half-lives}$$

Step [2]   **For each half-life, multiply the initial mass by one-half to obtain the final mass.**
- Because 32 days corresponds to *four* half-lives, multiply the initial mass by ½ *four* times to obtain the final mass. After four half-lives, 6.25 mg of iodine-131 remains.

four half-lives

$$100. \text{ mg} \times \frac{1}{2} \times \frac{1}{2} \times \frac{1}{2} \times \frac{1}{2} = 6.25 \text{ mg of iodine-131 remains.}$$

initial mass

The mass is halved four times.

| Sample Problem 9.3 | Using a Half-Life to Determine the Amount of Radioisotope Present |

A radioactive element, shown with blue spheres, forms a decay product with a half-life of 2.0 min. Draw a diagram that illustrates the sample composition after 6.0 min.

initial sample

Analysis and Solution

[1] Determine how many half-lives occur in the given amount of time.

- Use the half-life as a conversion factor to convert the number of minutes to the number of half-lives.

$$6.0 \ \text{min} \times \frac{1 \ \text{half-life}}{2.0 \ \text{min}} = 3.0 \ \text{half-lives}$$

[2] For each half-life, halve the number of particles of the radioactive element.

- Because three half-lives elapse, halve the number of particles three times to show the number of radioactive particles remaining and the number of new particles (in green) formed.

initial sample   2.0 min   2.0 min   2.0 min   final sample
                                                    Answer

| Practice Problem 9.4 | |

A radioactive element (shown with blue spheres) forms a decay product (shown with green spheres) with a half-life of 22 min. (a) How many half-lives have elapsed to form the sample in diagram **A**? (b) How many minutes have elapsed in the conversion of the initial sample to **A**?

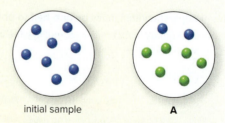

initial sample   A

| Practice Problem 9.5 | |

How much phosphorus-32 remains from a 1.00 g sample after each of the following number of half-lives: (a) 2; (b) 4; (c) 8?

More Practice: Try Problems 9.5, 9.6, 9.7b, 9.8b, 9.25–9.30.

9.3B Archaeological Dating

Archaeologists use the half-life of carbon-14 to determine the age of carbon-containing material derived from plants or animals. The technique, **radiocarbon dating,** is based on the fact that the ratio of radioactive carbon-14 to stable carbon-12 is a constant value in a living organism that is constantly taking in CO_2 and other carbon-containing nutrients from its surroundings. Once the organism dies, however, the radioactive isotope (C-14) decays (Section 9.2B) without being replenished, thus decreasing its concentration, while the stable isotope of carbon (C-12) remains at a constant value. By comparing the ratio of C-14 to C-12 in an artifact to the ratio of C-14 to C-12 in organisms today, the age of the artifact can be determined. Radiocarbon dating can be used to give the approximate age of wood, cloth, bone, charcoal, and many other substances that contain carbon.

The half-life of carbon-14 is 5.70×10^3 years, so half of the C-14 has decayed after about 6,000 years. Thus, a 6,000-year-old object has a ratio of C-14 to C-12 that has decreased by a factor of two, a 12,000-year-old object has a ratio of C-14 to C-12 that has decreased by a factor of four, and so forth.

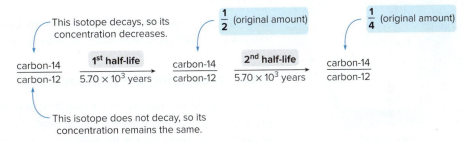

Using this technique, archaeologists have determined the age of the paintings on cave walls in France to be about 20,000 years old (Figure 9.3). Because the amount of carbon-14 decreases with time, artifacts older than about 50,000 years have too little carbon-14 to accurately estimate their age.

Concept Check 9.9

Estimate the age of an artifact that has 1/8 of the amount of C-14 (relative to C-12) compared to living organisms.

Figure 9.3

Radiocarbon Dating

Radiocarbon dating has been used to estimate the age of this cave painting in France at about 20,000 years. *Thipjang/Getty Images*

A Geiger counter is a device used to detect radiation. *Hank Morgan/Science Source*

Individuals who work with radioactivity wear badges to monitor radiation levels. *Landauer Corporation*

9.4   Detecting and Measuring Radioactivity

We all receive a miniscule daily dose of radiation from cosmic rays and radioactive substances in the soil. Additional radiation exposure comes from dental X-rays and other man-made sources. Moreover, we are still exposed to nuclear fallout, residual radiation resulting from the testing of nuclear weapons in the atmosphere decades ago.

Although this background radiation is unavoidable and minute, higher levels can be harmful and life-threatening, because radiation is composed of high-energy particles and waves that damage cells and disrupt key biological processes, often causing cell death. How can radiation be detected and measured when it can't be directly observed by any of the senses?

A **Geiger counter** is a small portable device used for measuring radioactivity. It consists of a tube filled with argon gas that is ionized when it comes into contact with nuclear radiation. This in turn generates an electric current that produces a clicking sound or registers on a meter. Geiger counters are used to locate a radiation source or a site that has become contaminated by radioactivity.

Individuals who work with radioactivity wear radiation badges. These badges are regularly monitored to assure that these individuals are not exposed to unhealthy levels of harmful radiation.

9.4A   Measuring the Radioactivity in a Sample

The amount of radioactivity in a sample is measured by the number of nuclei that decay per unit time—disintegrations per second. The most common unit is the **curie** (Ci), and smaller units derived from it, the **millicurie** (mCi) and the **microcurie** (μCi). One curie equals 3.7×10^{10} disintegrations/second, which corresponds to the decay rate of 1 g of the element radium.

The **becquerel** (Bq), an SI unit, is also used to measure radioactivity; 1 Bq = 1 disintegration/second. Because each nuclear decay corresponds to one becquerel, $1 \text{ Ci} = 3.7 \times 10^{10}$ Bq. Radioactivity units are summarized in Table 9.3.

Often a dose of radiation is measured in the number of millicuries that must be administered. For example, a diagnostic test for thyroid activity uses sodium iodide that contains iodine-131—that is, $Na^{131}I$. The radioisotope is purchased with a known amount of radioactivity per milliliter, such as 3.5 mCi/mL. By knowing the amount of radioactivity a patient must be given, as well as the concentration of radioactivity in the sample, one can calculate the volume of radioactive isotope that must be administered (Sample Problem 9.4).

Table 9.3 **Units Used to Measure Radioactivity**

1 Ci = 3.7×10^{10} disintegrations/s
1 Ci = 3.7×10^{10} Bq
1 Ci = 1,000 mCi
1 Ci = 1,000,000 μCi

The curie is named for Polish chemist Marie Skłodowska Curie who discovered the radioactive elements polonium and radium, and received Nobel Prizes for both Chemistry and Physics in the early twentieth century.

Sample Problem 9.4 | **Determining the Volume of a Dose of Radioisotope**

A patient must be given a 4.5-mCi dose of iodine-131, which is available as a solution that contains 3.5 mCi/mL. What volume of solution must be administered?

Analysis

Use the amount of radioactivity (**mCi/mL**) as a conversion factor to convert the dose of radioactivity from millicuries to a volume in milliliters.

Solution

The dose of radioactivity is known in millicuries, and the amount of radioactivity per unit volume (3.5 mCi/mL) is also known. Use 3.5 mCi/mL as a millicurie–milliliter conversion factor.

$$\text{4.5 mCi dose} \times \underbrace{\frac{1 \text{ mL}}{3.5 \text{ mCi}}}_{\text{mCi–mL conversion factor}} = \text{1.3 mL dose}$$

Millicuries cancel.   **Answer**

Practice Problem 9.6

To treat a thyroid tumor, a patient must be given a 110-mCi dose of iodine-131, supplied in a vial containing 25 mCi/mL. What volume of solution must be administered?

More Practice: Try Problems 9.7c, 9.8c, 9.35–9.38.

We can also use the half-life of a radioactive element to determine the amount of radioactivity remaining in a sample after a known time interval.

Sample Problem 9.5

Using Half-Life to Determine the Amount of Radioisotope Remaining

A patient is injected with a sample of technetium-99m ($t_{1/2}$ = 6.01 h), which has an activity of 32 mCi. What activity is observed after 12 h?

Analysis

First determine how many half-lives have occurred in the time interval. Then, multiply the initial activity by the number of half-lives to determine the remaining activity.

Solution

[1] Use the half-life of technetium-99m as a conversion factor to convert the number of hours to the number of half-lives.

$$12 \text{ h} \quad \times \quad \frac{1 \text{ half-life}}{6.01 \text{ h}} \quad = \quad 2.0 \text{ half-lives}$$

[2] For each half-life, multiply the initial activity by 1/2 to obtain the final activity.

$$\underbrace{32 \text{ mCi} \quad \times \quad \frac{1}{2} \quad \times \quad \frac{1}{2}}_{12 \text{ h = two half-lives}} \quad = \quad 8.0 \text{ mCi}$$

initial activity   **Answer**

Practice Problem 9.7

A sample of iodine-131 ($t_{1/2}$ = 8.02 days) has an activity of 240 mCi. What is the activity of the sample after 24 days?

More Practice: Try Problems 9.33, 9.34, 9.51c, 9.52c.

A lab worker must use protective equipment when working with radioactive substances. *AFP/Getty Images*

9.4B FOCUS ON HEALTH & MEDICINE
The Effects of Radioactivity

Radioactivity cannot be seen, smelled, tasted, heard, or felt, and yet it can have powerful effects. Because it is high in energy, nuclear radiation penetrates the surface of an object or living organism, where it can damage or kill cells. The cells that are most sensitive to radiation are those that undergo rapid cell division, such as those in bone marrow, reproductive organs, skin, and the intestinal tract. Because cancer cells also rapidly divide, they are also particularly sensitive to radiation, a fact that makes radiation an effective method of cancer treatment (Section 9.5).

Alpha (α) particles, β particles, and γ rays differ in the extent to which they can penetrate a surface. Alpha particles are the heaviest of the radioactive particles, and as a result they move the slowest and penetrate the least. Individuals who work with radioisotopes that emit α particles wear lab coats and gloves that provide a layer of sufficient protection. Beta particles move much faster because they have negligible mass, and they can penetrate into body tissue. Lab workers and health professionals must wear heavy lab coats and gloves when working with substances that give off β particles. Gamma rays travel the fastest and readily penetrate body tissue. Working with substances that emit γ rays is extremely hazardous, and a thick lead shield is required to halt their penetration.

CONSUMER NOTE

Strawberries that have been irradiated (on left) show no mold growth after two weeks, compared to strawberries that have not been irradiated (on right), which are moldy. *Cordelia Molloy/Science Source*

That γ rays kill cells is used to an advantage in the food industry. To decrease the incidence of harmful bacteria in foods, certain fruits and vegetables are irradiated with γ rays that kill any bacteria contained in them. Foods do not come into contact with radioisotopes and the food is not radioactive after radiation. Gamma rays merely penetrate the food and destroy any live organism, and often as a result, the food product has a considerably longer shelf life.

9.4C Measuring Human Exposure to Radioactivity

Several units are used to measure the amount of radiation *absorbed* by an organism.

- The **rad**—radiation absorbed dose—is the amount of radiation absorbed by one gram of a substance. The amount of energy absorbed varies with both the nature of the substance and the type of radiation.
- The **rem**—radiation equivalent for man—is the amount of radiation that also factors in its energy and potential to damage tissue. Using rem as a measure of radiation, one rem of any type of radiation produces the same amount of tissue damage.

Other units to measure absorbed radiation include the **gray** (1 Gy = 100 rad) and the **sievert** (1 Sv = 100 rem).

Although background radiation varies with location, the average radiation dose per year for an individual is estimated at 0.27 rem. Generally, no detectable biological effects are noticed when the dose of radiation is less than 25 rem. A single dose of 25–100 rem causes a temporary decrease in white blood cell count. The symptoms of radiation sickness—nausea, vomiting, fatigue, and prolonged decrease in white blood cell count—are visible at a dose of more than 100 rem.

Death results at still higher doses of radiation. The **LD$_{50}$—the lethal dose that kills 50% of a population**—is 500 rem in humans, and exposure to 600 rem of radiation is fatal for an entire population.

Concept Check 9.10

The unit millirem (1 rem = 1,000 mrem) is often used to measure the amount of radiation absorbed. (a) The average yearly dose of radiation from radon gas is 200 mrem. How many rem does this correspond to? (b) If a thyroid scan exposes a patient to 0.014 rem of radiation, how many mrem does this correspond to? (c) Which represents the larger dose?

9.5 FOCUS ON HEALTH & MEDICINE
Medical Uses of Radioisotopes

Radioactive isotopes are used for both diagnostic and therapeutic procedures in medicine. In a diagnostic test to measure the function of an organ or to locate a tumor, low doses of radioactivity are generally given. When the purpose of using radiation is therapeutic, such as to kill diseased cells or cancerous tissue, a much higher dose of radiation is required.

9.5A Radioisotopes Used in Diagnosis

Radioisotopes are routinely used to determine if an organ is functioning properly or to detect the presence of a tumor. The isotope is ingested or injected and the radiation it emits can be used to produce a scan. Sometimes the isotope is an atom or ion that is not part of a larger molecule. Examples include iodine-131, which is administered as the salt sodium iodide ($Na^{131}I$), and xenon-133, which is a gas containing radioactive xenon atoms. At other times the radioactive atom is bonded to a larger molecule that targets a specific organ. An organ that has increased or decreased uptake of the radioactive element can indicate disease, the presence of a tumor, or other conditions.

A HIDA scan (hepatobiliary iminodiacetic acid scan) uses a technetium-99m-labeled molecule to evaluate the functioning of the gallbladder and bile ducts (Figure 9.4). After injection, the technetium-99m travels through the bloodstream and into the liver, gallbladder, and bile ducts, where, in a healthy individual, the organs are all clearly visible on a scan. When the gallbladder is inflamed or the bile ducts are obstructed by gallstones, uptake of the radioisotope does not occur and these organs are not visualized because they do not contain the radioisotope.

Figure 9.4   HIDA Scan Using Technetium-99m

a.

b.

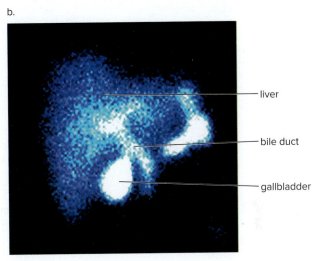

a. Schematic showing the location of the liver, gallbladder, and bile ducts
b. A scan using technetium-99m showing bright areas for the liver, gallbladder, and bile ducts, indicating normal function
(b): Living Art Enterprises, LLC/Science Source

HEALTH NOTE

Tiny seeds (shown next to a penny to illustrate their size) made of titanium with a small amount of iodine-125 are implanted near the prostate to treat prostate cancer. Because radiation is confined to a localized region, nearby organs such as the bladder and rectum receive only minimal radiation.
David A. Tietz/Editorial Image, LLC

Red blood cells tagged with technetium-99m are used to identify the site of internal bleeding in an individual. Bone scans performed with technetium-99m can show the location of metastatic cancer, so that specific sites can be targeted for radiation therapy.

Thallium-201 is used in stress tests to diagnose coronary artery disease. Thallium injected into a vein crosses cell membranes into normal heart muscle. Little radioactive thallium is found in areas of the heart that have a poor blood supply. This technique is used to identify individuals who may need a surgical procedure because of blocked coronary arteries.

Concept Check 9.11

The half-life of thallium-201 is three days. What fraction of thallium-201 is still present in an individual after nine days?

9.5B Radioisotopes Used in Treatment

The high-energy radiation emitted by radioisotopes can be used to kill rapidly dividing tumor cells. Two techniques are used. Sometimes the radiation source is external to the body. For example, a beam of radiation produced by decaying cobalt-60 can be focused at a tumor. Such a radiation source must have a much longer half-life—5.27 years in this case—than radioisotopes that are ingested for diagnostic purposes. With this method some destruction of healthy tissue often occurs, and a patient may experience some signs of radiation sickness, including vomiting, fatigue, and hair loss.

A more selective approach to cancer treatment involves brachytherapy, the implantation of a radioactive isotope internally at the site of the tumor within the body. Using iodine-131 to treat hyperthyroidism has already been discussed (Section 9.1). Other examples include using radioactive "seeds" or wire that can be implanted close to a tumor. Iodine-125 seeds are used to treat prostate cancer and iridium-192 wire is used to treat some cancers of the breast.

Figure 9.5 illustrates radioisotopes that are used for diagnosis or treatment.

Figure 9.5 Common Radioisotopes Used in Medicine

Xenon-133
lung function

Iodine-131
hyperthyroidism
and thyroid tumors

Technetium-99m
bone scan

Phosphorus-32
treating leukemia
and lymphomas

Technetium-99m
gallbladder function

Iridium-192
cancers of
the breast

Technetium-99m
visualizing gastrointestinal
bleeding

Thallium-201
heart function

Concept Check 9.12

Samarium-153, a radioactive isotope with a half-life of approximately 48 hours, is used to treat cancers lodged in the bone. (a) How many protons, neutrons, and electrons does samarium-153 contain? (b) Write a balanced nuclear equation for the emission of a β particle from samarium-153. (c) If a vial with an activity of 150 mCi is purchased, what is the activity of the solution after eight days?

9.5C Positron Emission Tomography—PET Scans

Positron emission tomography (PET) scans use radioisotopes that emit positrons when the nucleus decays. Once formed, a positron combines with an electron to form two γ rays, which create a scan of an organ.

Carbon-11, oxygen-15, nitrogen-13, and fluorine-18 are common radioactive isotopes used in PET scans. For example, a carbon-11 or fluorine-18 isotope can be incorporated in a glucose molecule. When this radioactive molecule is taken internally, its concentration becomes highest in areas in the body that continually use glucose. A healthy brain shows a high level of radioactivity from labeled glucose. When an individual suffers a stroke or has Alzheimer's disease, brain activity is significantly decreased and radioactivity levels are decreased, as shown in Figure 9.6a. Because Alzheimer's disease is characterized by insoluble deposits in the brain called amyloid plaques, PET scans are also used to visualize these regions when performed with a radioactive dye that binds to these plaques (Figure 9.6b).

Concept Check 9.13

Write a nuclear equation for the emission of a positron from nitrogen-13.

Figure 9.6 PET Scans and Alzheimer's Disease

a. PET scans with radioactive glucose

Normal | Mild cognitive impairment | Alzheimer's disease

- Reddish regions in the healthy brain indicate significant glucose metabolism, whereas in an Alzheimer's patient, there is diminished brain function and a decrease in glucose metabolism. *Science History Images/Photo Researchers/Alamy Stock Photo*

b. PET scans with a radioactive amyloid-binding agent

- The radioactive amyloid-binding agent shows no plaque buildup in the healthy brain on the left. The red regions in the PET scan on the right indicate regions of plaque buildup in the brain of an Alzheimer's patient. *Science Source*

9.6 Nuclear Fission and Nuclear Fusion

The nuclear reactions used in nuclear power plants occur by a process called *nuclear fission,* whereas the nuclear reactions that take place in the sun occur by a process called *nuclear fusion.*

- *Nuclear fission* is the splitting apart of a heavy nucleus into lighter nuclei and neutrons.
- *Nuclear fusion* is the joining together of two light nuclei to form a larger nucleus.

9.6A Nuclear Fission

When uranium-235 is bombarded by a neutron, it undergoes **nuclear fission** and splits apart into two lighter nuclei. Several different fission products have been identified. One common nuclear reaction is the fission of uranium-235 into krypton-91 and barium-142.

$$^{235}_{92}\text{U} \quad + \quad ^{1}_{0}\text{n} \quad \longrightarrow \quad ^{91}_{36}\text{Kr} \quad + \quad ^{142}_{56}\text{Ba} \quad + \quad 3\,^{1}_{0}\text{n}$$

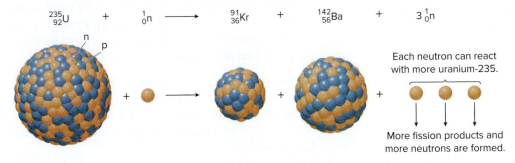

Each neutron can react with more uranium-235.

More fission products and more neutrons are formed.

Three high-energy neutrons are also produced in the reaction as well as a great deal of energy. Whereas burning 1 g of methane in natural gas releases 13 kcal of energy, fission of 1 g of uranium-235 releases 3.4×10^8 kcal. Each neutron produced during fission can go on to bombard three other uranium-235 nuclei to produce more nuclei and more neutrons. Such a process is called a **chain reaction.**

In order to sustain a chain reaction there must be a sufficient amount of uranium-235. When that amount—the **critical mass**—is present, the chain reaction occurs over and over again and an atomic explosion occurs. When less than the critical mass of uranium-235 is present, there is a more controlled production of energy, as is the case in a nuclear power plant.

A nuclear power plant utilizes the tremendous amount of energy produced by fission of the uranium-235 nucleus to heat water to steam, which powers a generator to produce electricity (Figure 9.7). While nuclear energy accounts for a small but significant fraction of the electricity needs in the United States, most of the electricity generated in some European countries comes from nuclear power.

Figure 9.7 A Nuclear Power Plant

a.

Kristin Smith/Shutterstock

b.

a. Nuclear power plant with steam rising from cooling towers
b. Fission occurs in a nuclear reactor core that is housed in a containment facility. Water surrounding the reactor is heated by the energy released during fission, and this energy drives a turbine, which produces electricity. Once the steam has been used to drive the turbine, it is cooled and re-circulated around the core of the reactor. To prevent the loss of any radioactive material to the environment, the water that surrounds the reactor core never leaves the containment building.

Two problems that surround nuclear power generation are the possibility of radiation leaks and the disposal of nuclear waste. Plants are designed and monitored to contain the radioactive materials within the nuclear reactor. The reactor core itself is located in a containment facility with thick walls, so that should a leak occur, the radiation should in principle be kept within the building. The nuclear reactor in Chernobyl, Russia, was built without a containment facility and in 1986 it exploded, releasing high levels of radioactivity to the immediate environment and sending a cloud of radioactivity over much of Europe.

The products of nuclear fission are radioactive nuclei with long half-lives, often hundreds or even thousands of years. As a result, nuclear fission generates radioactive waste that must be stored in a secure facility so that it does not pose a hazard to the immediate surroundings. Burying waste far underground is currently considered the best option, but this issue is still unresolved.

Sample Problem 9.6	Balancing a Nuclear Equation for a Fission Reaction

Write a nuclear equation for the fission of uranium-235 by neutron bombardment to form strontium-90, an isotope of xenon, and three neutrons.

Analysis

Balance the atomic numbers and mass numbers on both sides of the nuclear equation. In fission reactions, **include the neutron used for bombardment on the left and the high-energy neutrons produced on the right.** Each neutron has a mass of one and zero charge.

Solution

[1] Write an incomplete equation with the original nucleus and neutron used for bombardment on the left, and the particles formed on the right.

- Use the identity of each element to determine its atomic number. Uranium has an atomic number of 92, and xenon has an atomic number of 54.
- Include one neutron on the left side. In this reaction, three high-energy neutrons are formed, so include three neutrons on the right side.

$$^{235}_{92}U + ^{1}_{0}n \longrightarrow ^{90}_{38}Sr + ^{?}_{54}Xe + 3 \, ^{1}_{0}n$$

[2] Calculate the mass numbers and atomic numbers of all newly formed nuclei on the right.

- Atomic number: In this problem all atomic numbers are known from the identity of the elements.
- Mass number: **Balance mass numbers by taking into account the mass of the neutrons used or produced in the reaction.** On the left side, the total mass of the particles (the uranium nucleus and one neutron) is 236 (235 + 1). On the right side, the sum of the masses of Sr-90, xenon, and three neutrons (total mass of three) must also equal 236.

$$236 = 90 + ? + 3(1) \quad \text{← three neutrons}$$
$$236 = 93 + ? \quad \text{← one mass unit from each neutron}$$
$$143 = ?$$

The mass number of xenon is 143.

[3] Write the complete equation.

$$^{235}_{92}U + ^{1}_{0}n \longrightarrow ^{90}_{38}Sr + ^{143}_{54}Xe + 3 \, ^{1}_{0}n$$

Practice Problem 9.8

Write a nuclear equation for the fission of uranium-235 by neutron bombardment to form antimony-133, three neutrons, and one other isotope.

More Practice: Try Problems 9.45, 9.46.

9.6B Nuclear Fusion

Nuclear fusion occurs when two light nuclei join together to form a larger nucleus. For example, fusion of a deuterium nucleus with a tritium nucleus forms helium and a neutron. Recall from Section 2.3 that deuterium is an isotope of hydrogen that contains one proton and one neutron in its nucleus, whereas tritium is an isotope of hydrogen that contains one proton and two neutrons in its nucleus.

$$^{2}_{1}H + ^{3}_{1}H \longrightarrow ^{4}_{2}He + ^{1}_{0}n$$

deuterium   tritium   helium

Like fission, fusion also releases a great deal of energy—namely, 5.3×10^8 kcal/mol of helium produced. The light and heat of the sun and other stars result from nuclear fusion.

One limitation of using fusion to provide energy for mankind is the extreme experimental conditions needed to produce it. Because it takes a considerable amount of energy to overcome the repulsive forces of the like charges of two nuclei, fusion can only be accomplished at high temperatures (greater than 100,000,000 °C) and pressures (greater than 100,000 atm).

Because these conditions are not easily achieved, using controlled nuclear fusion as an energy source has yet to become a reality.

Controlled nuclear fusion has the potential of providing cheap and clean power. It is not plagued by the nuclear waste issues of fission reactors, and the needed reactants are readily available.

Concept Check 9.14	Nuclear fusion in the stars occurs by a series of reactions. Identify **X, Y,** and **Z** in the following nuclear reactions that ultimately convert hydrogen into helium.

$$\text{a. } {}^1_1\text{H} + \textbf{X} \longrightarrow {}^2_1\text{H} + {}^0_{+1}\text{e} \qquad \text{b. } {}^1_1\text{H} + {}^2_1\text{H} \longrightarrow \textbf{Y} \qquad \text{c. } {}^1_1\text{H} + {}^3_2\text{He} \longrightarrow {}^4_2\text{He} + \textbf{Z}$$

9.7 FOCUS ON HEALTH & MEDICINE
Medical Imaging Without Radioactivity

X-rays, CT scans, and **MRIs** are also techniques that provide an image of an organ or extremity that is used for diagnosis of a medical condition. Unlike PET scans and other procedures discussed thus far, however, **these procedures are *not* based on nuclear reactions and they do *not* utilize radioactivity.** In each technique, an energy source is directed toward a specific region in the body, and a scan is produced that is analyzed by a trained medical professional.

X-rays are a high-energy type of radiation called electromagnetic radiation. Tissues of different density interact differently with an X-ray beam, and so a map of bone and internal organs is created on an X-ray film. Dense bone is clearly visible in an X-ray, making it a good diagnostic technique for finding fractures (Figure 9.8a). Although X-rays are a form of high-energy radiation, they are lower in energy than the γ rays produced in nuclear reactions. Nonetheless, X-rays

Figure 9.8 Imaging the Human Body

a.

b.

c.

herniated disc

a. X-ray of a broken humerus in a patient's arm
b. A color-enhanced CT scan of the head showing the site of a stroke
c. MRI of the spinal cord showing spinal compression from a herniated disc
(a): Suttha Burawonk/Shutterstock; (b): Puwadol Jaturawutthichai/Shutterstock; (c): Daniel C. Smith

still cause adverse biological effects on the cells with which they come in contact, and the exposure of both the patient and X-ray technician must be limited.

CT (computed tomography) scans, which also use X-rays, provide high resolution images of "slices" of the body. Historically, CT images have shown a slice of tissue perpendicular to the long axis of the body. Modern CT scanners can now provide a three-dimensional view of the body's organs. CT scans of the head are used to diagnose bleeding and tumors in the brain (Figure 9.8b).

MRI (magnetic resonance imaging) uses low-energy radio waves to visualize internal organs. Unlike methods that use high-energy radiation, MRIs do not damage cells. An MRI is a good diagnostic method for visualizing soft tissue (Figure 9.8c), and thus it complements X-ray techniques.

CHAPTER REVIEW

KEY TERMS

Alpha (α) particle (9.1)

Becquerel (9.4)

Beta (β) particle (9.1)

Chain reaction (9.6)

Critical mass (9.6)

Curie (9.4)

Gamma (γ) ray (9.1)

Geiger counter (9.4)

Gray (9.4)

Half-life (9.3)

LD_{50} (9.4)

Nuclear fission (9.6)

Nuclear fusion (9.6)

Nuclear reaction (9.2)

Positron (9.1)

Rad (9.4)

Radioactive decay (9.2)

Radioactive isotope (9.1)

Radioactivity (9.1)

Radiocarbon dating (9.3)

Rem (9.4)

Sievert (9.4)

X-ray (9.7)

KEY CONCEPTS

① Types of nuclear radiation (9.1)

① Alpha particle	② Beta particle	③ Positron	④ Gamma ray
α or $^{4}_{2}He$	β or $^{0}_{-1}e$	β^{+} or $^{0}_{+1}e$	γ
• A high-energy nucleus that contains two protons and two neutrons	• A high-energy electron that has a −1 charge and negligible mass	• An antiparticle of a β particle that has a +1 charge and negligible mass	• High-energy radiation with no mass or charge

② Nuclear fission and nuclear fusion (9.6)

① Nuclear fission	② Nuclear fusion
$^{235}_{92}U + {}^{1}_{0}n \longrightarrow {}^{91}_{36}Kr + {}^{142}_{56}Ba + 3{}^{1}_{0}n$	$^{2}_{1}H + {}^{3}_{1}H \longrightarrow {}^{4}_{2}He + {}^{1}_{0}n$
• Nuclear fission is the splitting apart of a heavy nucleus into lighter nuclei and neutrons. • Nuclear fission releases a great deal of energy. • Fission is used in nuclear power plants to generate electricity.	• Nuclear fusion is the joining together of two light nuclei to form a larger nucleus. • Nuclear fusion releases a great deal of energy. • Nuclear fusion occurs in stars.

KEY SKILLS

1 Balancing an equation for a nuclear reaction (9.2); example: writing a balanced equation for the positron emission of rubidium-77

1 Write an incomplete equation with the original nucleus on the left and the particle emitted on the right.	**2** Calculate the mass number and the atomic number of the newly formed nucleus on the right.	**3** Use the atomic number to identify the new nucleus and complete the equation.
$$^{77}_{37}\text{Rb} \longrightarrow {}^{0}_{+1}\text{e} + ?$$ • Use the identity of the element to determine the atomic number; rubidium has an atomic number of 37.	• Mass number: **A β^+ particle has no mass,** so the masses of the new particle and the original particle are the *same*, 77. • Atomic number: Because β^+ emission converts a proton into a neutron, the new nucleus has **one fewer proton** than the original nucleus; $37 - 1 = 36$. Thus, the new nucleus has an atomic number of 36.	$$^{77}_{37}\text{Rb} \longrightarrow {}^{0}_{+1}\text{e} + {}^{77}_{36}\text{Kr}$$ • From the periodic table, the element with an atomic number of 36 is krypton, Kr. • Write the mass number and the atomic number with the element symbol to complete the equation.

2 Using a half-life to determine the amount of radioisotope present (9.3); example: calculating the amount of radioisotope ($t_{1/2}$ = 4.0 min) in a sample after 8.0 min

1 Determine how many half-lives occur in the given amount of time.	**2** For each half-life, halve the number of particles of the radioactive element.
$$8.0\ \cancel{\text{min}} \times \frac{1\ \text{half-life}}{4.0\ \cancel{\text{min}}} = 2.0\ \text{half-lives}$$ • Use the half-life as a conversion factor to convert the number of minutes to the number of half-lives.	 initial sample · after one half-life · after two half-lives • Because two half-lives have elapsed, halve the number of particles twice to show the number of radioactive particles remaining, and the number of new particles (in green) formed.

3 Determining the volume of a dose of a radioisotope (9.4); example: calculating the volume needed to administer a 7.5-mCi dose of iodine-131, which is available as a solution containing 2.5 mCi/mL

1 Identify the original quantity and the desired quantity.	**2** Write out the conversion factor that places the unwanted unit (mCi) in the denominator.	**3** Solve the problem so that the unwanted unit cancels.
7.5 mCi dose · ? mL original quantity · **desired** quantity	$$\frac{1\ \text{mL}}{2.5\ \text{mCi}} \leftarrow \textbf{desired unit}\\ \leftarrow \text{unwanted unit}$$ **conversion factor** • Use the amount of radioactivity per unit volume as a conversion factor.	**conversion factor** $$7.5\ \cancel{\text{mCi}} \times \frac{1\ \text{mL}}{2.5\ \cancel{\text{mCi}}} = 3.0\ \text{mL dose}$$ mCi cancel.

CHAPTER 9 SELF-TEST

The Self-Test consists of short-answer questions that test an understanding of definitions, equations, and other material encountered within the chapter. Answers to each question are provided at the end of the chapter. For any questions you are unable to answer correctly, revisit the appropriate section(s) of the text; then proceed to working the end-of-chapter problems to further reinforce your understanding of the chapter material.

1. Which term from "Key Terms" answers each question?

    a. What type of radiation has the same mass as a beta particle but is opposite in charge?

    b. What type of nuclear reaction involves the splitting apart of a heavy nucleus into lighter nuclei and neutrons?

    c. What unit used in measuring radioactivity equals 3.7×10^{10} disintegrations/second?

    d. What technique uses radioactivity to determine the age of carbon-containing material?

    e. What term is used for the amount of a radioactive element that is required to sustain a chain reaction?

2. For the isotope thallium-201, determine the: (a) atomic number; (b) mass number; (c) number of protons; (d) number of neutrons.

3. What new nucleus is formed when strontium-90 undergoes beta emission?

4. Write a balanced equation for the emission of an α particle from ^{252}Es.

5. If a sample of iodine-131 ($t_{1/2} = 8.02$ days) has an activity of 120 mCi, what is the activity of the sample after 16 days?

6. For which process does each statement apply—nuclear fission, nuclear fusion, or both fission and fusion?
    a. The reaction occurs in the sun.
    b. A neutron is used to bombard a nucleus.
    c. A large amount of energy is released.
    d. A very high temperature is required.

UNDERSTANDING KEY CONCEPTS

9.1 For each isotope of fluorine, whose nuclei are shown in the accompanying diagrams, give its (a) atomic number; (b) mass number; (c) number of protons; (d) number of neutrons; (e) isotope symbol. One isotope is a stable nucleus and one is a radioisotope used in PET scans.

A   B

9.2 For each isotope of nitrogen, whose nuclei are shown in the accompanying diagrams, give its (a) atomic number; (b) mass number; (c) number of protons; (d) number of neutrons; (e) isotope symbol. One isotope is a stable nucleus and one is a radioisotope used in PET scans.

A   B

9.3 Complete the nuclear equation by drawing the nucleus of the missing atom. Give the symbol for each atom and type of radiation. (The blue spheres represent protons and the orange spheres represent neutrons.)

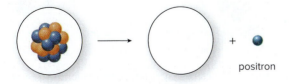

positron

9.4 Complete the nuclear equation by drawing the nucleus of the missing atom. Give the symbol for each atom and type of radiation. (The blue spheres represent protons and the orange spheres represent neutrons.)

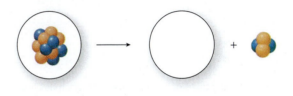

9.5 A radioactive element (shown with blue spheres) forms a decay product (shown with green spheres) with a half-life of three days. How many half-lives have elapsed in diagram **A**?

initial sample   A

9.6 A radioactive element (shown with blue spheres) forms a decay product (shown with green spheres) with a half-life of 10 min. How many minutes have elapsed to form the sample in diagram **B**?

initial sample   B

9.7 Arsenic-74 is a radioisotope used for locating brain tumors.

a. Write a balanced nuclear equation for the positron emission of arsenic-74.

b. If $t_{1/2}$ for As-74 is 18 days, how much of a 120-mg sample remains after 90. days?

c. If the radioactivity of a 2.0-mL vial of arsenic-74 is 10.0 mCi, what volume must be administered to give a 7.5-mCi dose?

9.8 Sodium-24 is a radioisotope used for examining circulation.

a. Write a balanced nuclear equation for the β decay of sodium-24.

b. If $t_{1/2}$ for Na-24 is 15 h, how much of an 84-mg sample remains after 2.5 days?

c. If the radioactivity of a 5.0-mL vial of sodium-24 is 10.0 mCi, what volume must be administered to give a 6.5-mCi dose?

ADDITIONAL PROBLEMS

Isotopes and Radiation

9.9 Complete the table of isotopes, each of which has found use in medicine.

	Atomic Number	Mass Number	Number of Protons	Number of Neutrons	Isotope Symbol
a. Chromium-51					
b.	46	103			
c.			19	23	
d.		133	54		

9.10 Complete the table of isotopes, each of which has found use in medicine.

	Atomic Number	Mass Number	Number of Protons	Number of Neutrons	Isotope Symbol
a. Sodium-24					
b.		89		51	
c.		59	26		
d. Samarium-153					

9.11 How much does the mass and charge of a nucleus change when each type of radiation is emitted: (a) α particle; (b) β particle; (c) γ ray; (d) positron?

9.12 Compare α particles, β particles, and γ rays with regard to each of the following: (a) speed the radiation travels; (b) penetrating power; (c) protective equipment that must be worn when handling.

9.13 What is the mass and charge of radiation that has each of the following symbols: (a) α; (b) n; (c) γ; (d) β?

9.14 What is the mass and charge of radiation that has each of the following symbols?

a. $_{-1}^{0}e$ b. $_{+1}^{0}e$ c. $_{2}^{4}He$ d. β^{+}

Nuclear Reactions

9.15 What particle is emitted when each of the following nuclear reactions occurs? Write a balanced nuclear equation for the process.

a. Cobalt-60 decays to nickel-60.

b. Cobalt-55 decays to iron-55.

c. Radon-220 decays to polonium-216.

9.16 What particle is emitted when each of the following nuclear reactions occurs? Write a balanced nuclear equation for the process.

a. Platinum-175 decays to osmium-171.

b. Magnesium-23 decays to sodium-23.

c. Sodium-22 decays to neon-22.

9.17 Complete each nuclear equation.

a. $_{26}^{59}Fe \longrightarrow ? + _{-1}^{0}e$ c. $_{80}^{178}Hg \longrightarrow ? + _{+1}^{0}e$

b. $_{78}^{190}Pt \longrightarrow ? + _{2}^{4}He$

9.18 Complete each nuclear equation.

a. $_{37}^{77}Rb \longrightarrow ? + _{+1}^{0}e$ c. $_{29}^{66}Cu \longrightarrow ? + _{-1}^{0}e$

b. $_{102}^{251}No \longrightarrow ? + _{2}^{4}He$

9.19 Complete the nuclear equation by drawing the nucleus of the missing atom. Give the symbol for each atom and type of radiation. The blue spheres represent protons and the orange spheres represent neutrons.

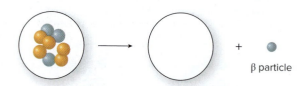

β particle

9.20 Complete the nuclear equation by drawing the nucleus of the missing atom. Give the symbol for each atom and the type of radiation. The blue spheres represent protons and the orange spheres represent neutrons.

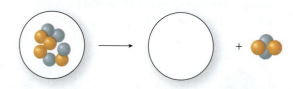

9.21 Bismuth-214 can decay to form either polonium-214 or thallium-210, depending on what type of radiation is emitted. Write a balanced nuclear equation for each process.

9.22 Lead-210 can be formed by the decay of either thallium-210 or polonium-214, depending on what type of radiation is emitted. Write a balanced nuclear equation for each process.

9.23 Write a balanced nuclear equation for each reaction.

    a. decay of thorium-232 by α emission

    b. decay of sodium-25 by β emission

    c. decay of xenon-118 by positron emission

9.24 Write a balanced nuclear equation for each reaction.

    a. decay of sulfur-35 by β emission

    b. decay of thorium-225 by α emission

    c. decay of rhodium-93 by positron emission

Half-Life

9.25 If the amount of a radioactive element decreases from 2.4 g to 0.30 g in 12 days, what is its half-life?

9.26 If the amount of a radioactive element decreases from 0.36 g to 90. mg in 22 min, what is its half-life?

9.27 Using the diagram in the *How To* in Section 9.3A as a guide, fill in the boxes in the following representation, which illustrates the decay of **A** to **B**. Indicate how many grams of **A** and **B** are present at each interval ($t_{1/2}$ = 2 days).

9.28 Using the diagram in the *How To* in Section 9.3A as a guide, fill in the boxes in the following representation, which illustrates the decay of **C** to **D**. Indicate how many grams of **C** and **D** are present at each interval.

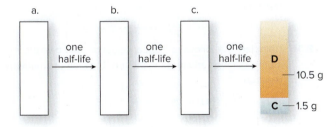

9.29 Radioactive iodine-131 ($t_{1/2}$ = 8.02 days) decays to form xenon-131 by emission of a β particle. How much of each isotope is present after each time interval if 64 mg of iodine-131 was present initially: (a) 8.0 days; (b) 16 days; (c) 24 days; (d) 32 days?

9.30 Radioactive phosphorus-32 decays to form sulfur-32 by emission of a β particle. Estimating the half-life to be 14 days, how much of each isotope is present after each time interval if 124 mg of phosphorus-32 was present initially: (a) 14 days; (b) 28 days; (c) 42 days; (d) 56 days?

9.31 If the half-life of an isotope is 24 hours, has all the isotope decayed in 48 hours?

9.32 Why can't radiocarbon dating be used to estimate the age of rocks?

9.33 A patient is injected with a sample of technetium-99m ($t_{1/2}$ = 6.01 h), which has an activity of 20 mCi. What activity is observed after each interval: (a) 6 h; (b) 12 h; (c) 24 h?

9.34 A sample of iodine-131 ($t_{1/2}$ = 8.02 days) has an activity of 200. mCi. What activity is observed after each interval: (a) 8.0 days; (b) 24 days; (c) 48 days?

Measuring Radioactivity

9.35 A patient must be administered a 28-mCi dose of technetium-99m, which is supplied in a vial containing a solution with an activity of 12 mCi/mL. What volume of solution must be given?

9.36 A radioactive isotope used for imaging is supplied in an 8.0-mL vial containing a solution with an activity of 108 mCi. What volume must be given to a patient who needs a 12-mCi dose?

9.37 Radioactive sodium-24, administered as $^{24}NaCl$, is given to treat leukemia. If a patient must receive 190 μCi/kg and the isotope is supplied as a solution that contains 5.0 mCi/mL, what volume is needed for a 68-kg patient?

9.38 Radioactive phosphorus-32, administered as sodium phosphate ($Na_3{}^{32}PO_4$), is used to treat chronic leukemia. The activity of an intravenous solution is 670 μCi/mL. What volume of solution must be used to supply a dose of 15 mCi?

9.39 The initial responders to the Chernobyl nuclear disaster were exposed to 20 Sv of radiation. Convert this value to rem. Did these individuals receive a fatal dose of radiation?

9.40 Many individuals who fought fires at the Chernobyl nuclear disaster site were exposed to 0.25 Sv of radiation. Convert this value to rem. Did these individuals receive a fatal dose of radiation? Would you expect any of these individuals to have shown ill health effects?

Nuclear Fission and Nuclear Fusion

9.41 What is the difference between nuclear fission and nuclear fusion?

9.42 For which process does each statement apply—nuclear fission, nuclear fusion, both fission and fusion?

    a. The reaction splits a nucleus into lighter nuclei.

    b. The reaction joins two lighter nuclei into a heavier nucleus.

    c. The reaction is used to generate energy in a nuclear power plant.

    d. The reaction generates radioactive waste with a long half-life.

9.43 Complete the nuclear fusion equation by drawing in the nucleus of the missing atom. Blue spheres represent protons and orange spheres represent neutrons.

9.44 Complete the nuclear fusion equation by drawing in the nucleus of the missing atom. Blue spheres represent protons and orange spheres represent neutrons.

9.45 Complete each nuclear fission equation.

a. $^{235}_{92}U + ^{1}_{0}n \longrightarrow ? + ^{97}_{42}Mo + 2\,^{1}_{0}n$

b. $^{235}_{92}U + ^{1}_{0}n \longrightarrow ? + ^{140}_{56}Ba + 3\,^{1}_{0}n$

9.46 Complete each nuclear fission equation.

a. $^{235}_{92}U + ^{1}_{0}n \longrightarrow ? + ^{139}_{57}La + 2\,^{1}_{0}n$

b. $^{235}_{92}U + ^{1}_{0}n \longrightarrow ? + ^{140}_{58}Ce + 2\,^{1}_{0}n + 6\,^{0}_{-1}e$

9.47 The fusion of two deuterium nuclei (hydrogen-2) forms a hydrogen nucleus (hydrogen-1) as one product. What other product is formed?

9.48 Fill in the missing product in the following nuclear fusion reaction.

$$^{3}_{2}He + ^{3}_{2}He \longrightarrow ? + 2\,^{1}_{1}H$$

9.49 Discuss two problems that surround the generation of electricity from a nuclear power plant.

9.50 Why are there as yet no nuclear power plants that use nuclear fusion to generate electricity?

General Questions

9.51 Answer the following questions about radioactive iridium-192.

a. Write a balanced nuclear equation for the decay of iridium-192, which emits both a β particle and a γ ray.

b. If $t_{1/2}$ for Ir-192 is 74 days, estimate how much of a 120-mg sample remains after five 30-day months.

c. If a sample of Ir-192 had an initial activity of 36 Ci, estimate how much activity remained in the sample after ten 30-day months.

9.52 Answer the following questions about radioactive samarium-153.

a. Write a balanced nuclear equation for the decay of samarium-153, which emits both a β particle and a γ ray.

b. If $t_{1/2}$ for Sm-153 is approximately 48 h, estimate how much of a 160-mg sample remains after four days.

c. If a sample of Sm-153 had an initial activity of 48 Ci, estimate how much activity remained in the sample after six days.

9.53 All nuclei with atomic numbers around 100 or larger do not exist naturally; rather, they have been synthesized by fusing two lighter-weight nuclei together. Using the principles for balancing nuclear reactions in Section 9.2, complete the nuclear equation by drawing the nucleus of the missing atom **X.** Give the element symbol, atomic number, and mass number of **X.**

$$^{238}_{92}U + \textbf{X} \longrightarrow ^{246}_{98}Cf + 4\,^{1}_{0}n$$

9.54 Using the principles for balancing nuclear reactions in Section 9.2, complete the nuclear equation by drawing the nucleus of the missing atom **Y.** Give the element symbol, atomic number, and mass number of **Y.**

$$^{252}_{98}Cf + \textbf{Y} \longrightarrow ^{257}_{103}Lr + 5\,^{1}_{0}n$$

9.55 Complete the following nuclear equation by giving the name, atomic number, and mass number of the element made by this reaction.

$$^{209}_{83}Bi + ^{58}_{26}Fe \longrightarrow ? + ^{1}_{0}n$$

9.56 Complete the following nuclear equation, and give the name, atomic number, and mass number of the element made by this reaction.

$$^{235}_{92}U + ^{14}_{7}N \longrightarrow ? + 5\,^{1}_{0}n$$

Applications

9.57 Explain how each isotope is used in medicine.

a. iodine-131   b. iridium-192   c. thallium-201

9.58 Explain how each isotope is used in medicine.

a. iodine-125   b. technetium-99m   c. cobalt-60

9.59 How does the half-life of each of the following isotopes of iodine affect the manner in which it is administered to a patient: (a) iodine-125, $t_{1/2} = 60$ days; (b) iodine-131, $t_{1/2} = 8.02$ days?

9.60 Explain why food is irradiated with γ rays.

9.61 A mammogram is an X-ray of the breast. Why does an X-ray technician leave the room or go behind a shield when a mammogram is performed on a patient?

9.62 A burnt piece of wood found at an archaeological site contained 0.0204 g of ^{14}C. If it was determined that the wood was about 11,500 years old, how much ^{14}C was present before it was burned?

9.63 One of the radioactive isotopes that contaminated the area around Chernobyl after the nuclear accident in 1986 was iodine-131. Suggest a reason why individuals in the affected region were given doses of NaI that contained the stable iodine-127 isotope.

9.64 The element strontium has similar properties to calcium. Suggest a reason why exposure to strontium-90, a product of nuclear testing in the atmosphere, is especially hazardous for children.

CHALLENGE PROBLEMS

9.65 An article states that the fission of 1.0 g of uranium-235 releases 3.4×10^8 kcal, the same amount of energy as burning one ton (2,000. lb) of coal. If this report is accurate, how much energy is released when 1.0 g of coal is burned?

9.66 Radioactive isotopes with high atomic numbers often decay to form isotopes that are themselves radioactive, and once formed, decay to form new isotopes. Sometimes a series of such decays occurs over many steps until a stable nucleus is formed. The following series of decays occurs: Polonium-218 decays with emission of an α particle to form **X**, which emits a β particle to form **Y**, which emits an α particle to form **Z**. Identify **X**, **Y**, and **Z**.

BEYOND THE CLASSROOM

9.67 The incident at the Three Mile Island Nuclear Generating Station in Pennsylvania in 1979 is considered the most serious accident in the history of the nuclear power industry in the United States. What factors contributed to the accident? How much radiation was released into the atmosphere and what radioactive isotopes were present? What improvements were made in employee training and reactor design as a result of this incident? Were any short- or long-term effects observed on the general health or cancer rate of individuals living near the plant?

9.68 Although the irradiation of certain food products provides an effective method for destroying harmful bacteria and disease-carrying pathogens, some consumer groups question the safety of irradiated food. Discuss the advantages and disadvantages of food irradiation. How might large-scale irradiation affect product availability? What products are typically irradiated, and how large of a radiation dose is used? Does your local market carry food that has been irradiated?

CHAPTER 9 ANSWERS

Concept Check Answers

9.1 a. 4   b. 11   c. 4   d. 7   e. $^{11}_{4}Be$

9.2 An α particle has no electrons around the nucleus and a +2 charge, whereas a helium atom has two electrons around the nucleus and is neutral.

9.3 An electron has a −1 charge and a positron has a +1 charge.

9.4 a. β particle   b. α particle   c. positron

9.5 $^{222}_{86}Rn \longrightarrow\ ^{4}_{2}He\ +\ ^{218}_{84}Po$

9.6 a. $^{218}_{84}Po \longrightarrow\ ^{4}_{2}He\ +\ ^{214}_{82}Pb$

   b. $^{230}_{90}Th \longrightarrow\ ^{4}_{2}He\ +\ ^{226}_{88}Ra$

9.7

9.8 $^{192}_{77}Ir \longrightarrow\ ^{192}_{78}Pt\ +\ ^{0}_{-1}e\ +\ \gamma$

9.9 17,200 years

9.10 a,c: 0.2 rem (larger dose)   b. 14 mrem

9.11 1/8

9.12 a. 62 protons, 91 neutrons, 62 electrons

   b. $^{153}_{62}Sm \longrightarrow\ ^{153}_{63}Eu\ +\ ^{0}_{-1}e$

   c. 9.4 mCi

9.13 $^{13}_{7}N \longrightarrow\ ^{0}_{+1}e\ +\ ^{13}_{6}C$

9.14 a. **X** $= ^{1}_{1}H$ b. **Y** $= ^{3}_{2}He$ c. **Z** $= ^{0}_{+1}e$

Practice Problem Answers

9.1 $^{131}_{53}I \longrightarrow\ ^{0}_{-1}e\ +\ ^{131}_{54}Xe$

9.2 a. $^{20}_{9}F \longrightarrow\ ^{0}_{-1}e\ +\ ^{20}_{10}Ne$

   b. $^{92}_{38}Sr \longrightarrow\ ^{0}_{-1}e\ +\ ^{92}_{39}Y$

   c. $^{55}_{24}Cr \longrightarrow\ ^{0}_{-1}e\ +\ ^{55}_{25}Mn$

9.3 a. $^{74}_{33}As \longrightarrow\ ^{0}_{+1}e\ +\ ^{74}_{32}Ge$

   b. $^{15}_{8}O \longrightarrow\ ^{0}_{+1}e\ +\ ^{15}_{7}N$

9.4 a. two   b. 44 min

9.5 a. 0.250 g   b. 0.0625 g   c. 0.003 91 g

9.6 4.4 mL

9.7 30. mCi

9.8 $^{235}_{92}U + ^{1}_{0}n \longrightarrow ^{133}_{51}Sb + ^{100}_{41}Nb + 3 ^{1}_{0}n$

Self-Test Answers

1. a. positron   c. curie   e. critical mass

     b. fission   d. radiocarbon dating

2. a. 81   b. 201   c. 81   d. 120

3. yttrium-90

4. $^{252}_{99}Es \longrightarrow ^{4}_{2}He + ^{248}_{97}Bk$

5. 30. mCi

6. a,d: fusion   b. fission   c. both

Odd-Numbered Problem Answers

9.1

	a. Atomic Number	b. Mass Number	c. Number of Protons	d. Number of Neutrons	Isotope Symbol
A	9	18	9	9	$^{18}_{9}F$
B	9	19	9	10	$^{19}_{9}F$

9.3

$^{13}_{7}N \longrightarrow ^{13}_{6}C + ^{0}_{+1}e$

9.5 two half-lives

9.7 a. $^{74}_{33}As \longrightarrow ^{0}_{+1}e + ^{74}_{32}Ge$

     b. 3.8 mg

     c. 1.5 mL

9.9

	Atomic Number	Mass Number	Number of Protons	Number of Neutrons	Isotope Symbol
a. Chromium-51	24	51	24	27	$^{51}_{24}Cr$
b. Palladium-103	46	103	46	57	$^{103}_{46}Pd$
c. Potassium-42	19	42	19	23	$^{42}_{19}K$
d. Xenon-133	54	133	54	79	$^{133}_{54}Xe$

9.11

	Change in Mass	Change in Charge
a. α particle	−4	−2
b. β particle	0	+1
c. γ ray	0	0
d. positron	0	−1

9.13

	Mass	Charge
a. α	4	+2
b. n	1	0
c. γ	0	0
d. β	0	−1

9.15 a. β particle, $^{60}_{27}Co \longrightarrow ^{60}_{28}Ni + ^{0}_{-1}e$

     b. positron, $^{55}_{27}Co \longrightarrow ^{55}_{26}Fe + ^{0}_{+1}e$

     c. α particle, $^{220}_{86}Rn \longrightarrow ^{216}_{84}Po + ^{4}_{2}He$

9.17 a. $^{59}_{26}Fe \longrightarrow ^{59}_{27}Co + ^{0}_{-1}e$

     b. $^{190}_{78}Pt \longrightarrow ^{186}_{76}Os + ^{4}_{2}He$

     c. $^{178}_{80}Hg \longrightarrow ^{178}_{79}Au + ^{0}_{+1}e$

9.19

$^{8}_{3}Li \longrightarrow ^{8}_{4}Be + ^{0}_{-1}e$

9.21 $^{214}_{83}Bi \longrightarrow ^{214}_{84}Po + ^{0}_{-1}e$

     $^{214}_{83}Bi \longrightarrow ^{210}_{81}Tl + ^{4}_{2}He$

9.23 a. $^{232}_{90}Th \longrightarrow ^{4}_{2}He + ^{228}_{88}Ra$

     b. $^{25}_{11}Na \longrightarrow ^{25}_{12}Mg + ^{0}_{-1}e$

     c. $^{118}_{54}Xe \longrightarrow ^{118}_{53}I + ^{0}_{+1}e$

9.25 4.0 days

9.27 a. 4.4 g **A**   c. 0.55 g **A**, 3.9 g **B**

     b. 2.2 g **A**, 2.2 g **B**

9.29

	Iodine-131	Xenon-131
a.	32 mg	32 mg
b.	16 mg	48 mg
c.	8.0 mg	56 mg
d.	4.0 mg	60. mg

9.31 No, 25% remains.

9.33 a. 10 mCi   b. 5 mCi   c. 1 mCi

9.35 2.3 mL

9.37 2.6 mL

9.39 20 Sv = 2,000 rem. This represents a fatal dose because 600 rem is uniformly fatal.

9.41 Nuclear fission refers to the splitting of nuclei and fusion refers to the joining of small nuclei to form larger ones.

9.43

9.45 a. $^{235}_{92}U + ^{1}_{0}n \longrightarrow ^{137}_{50}Sn + ^{97}_{42}Mo + 2\,^{1}_{0}n$
b. $^{235}_{92}U + ^{1}_{0}n \longrightarrow ^{93}_{36}Kr + ^{140}_{56}Ba + 3\,^{1}_{0}n$

9.47 tritium; $^{3}_{1}H$

9.49 The containment of radiation leaks and disposal of radioactive waste are two problems that are associated with nuclear power production.

9.51 a. $^{192}_{77}Ir \longrightarrow ^{192}_{78}Pt + ^{0}_{-1}e + \gamma$
b. 30. mg
c. 2.3 Ci

9.53 $^{12}_{6}C$

9.55 $^{209}_{83}Bi + ^{58}_{26}Fe \longrightarrow ^{266}_{109}Mt + ^{1}_{0}n$
meitnerium-266

9.57 a. used for the treatment and diagnosis of thyroid diseases
b. used for the treatment of breast cancer
c. used for the diagnosis of heart disease

9.59 a. Iodine-125, with its longer half-life, is used for the treatment of prostate cancers using implanted radioactive seeds.
b. Iodine-131, with its shorter half-life, is used for the diagnostic and therapeutic treatment of thyroid diseases and tumors. A patient is administered radioactive iodine-131, which is then incorporated into the thyroid hormone, thyroxine. Because its half-life is short, little remains after a month or so.

9.61 Radiology technicians must be shielded to avoid exposure to excessive and dangerous doses of radiation.

9.63 High doses of stable iodine will prevent the absorption and uptake of the radioactive iodine-131.

9.65 370 kcal

10 Introduction to Organic Molecules

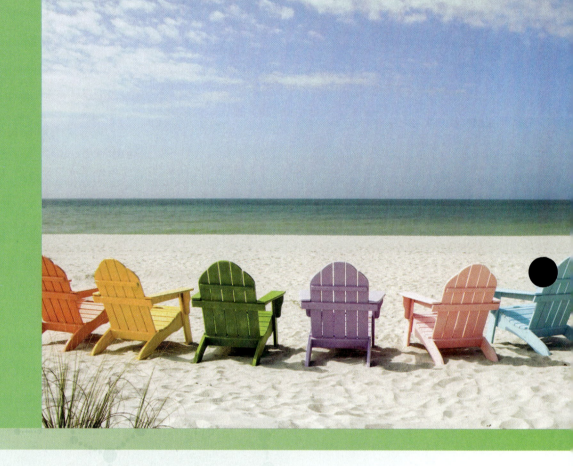

Many new varieties of lawn furniture are built from recycled polyethylene, a synthetic alkane routinely used in milk jugs and other plastic containers.
cmcderm1/Getty Images

CHAPTER OUTLINE

CHAPTER GOALS

In this chapter you will learn how to:

1. Distinguish organic compounds from ionic inorganic compounds
2. Recognize the characteristic features of organic compounds
3. Use shorthand methods to draw organic molecules
4. Recognize the common functional groups and understand their importance
5. Identify and draw acyclic alkanes and cycloalkanes
6. Identify constitutional isomers
7. Name alkanes using the IUPAC system of nomenclature
8. Predict the physical properties of alkanes
9. Determine the products of complete and incomplete combustion of alkanes

Consider for a moment the activities that occupied your past 24 hours. You likely showered with soap, drank a caffeinated beverage, ate at least one form of starch, and traveled in a vehicle that had rubber tires and was powered by fossil fuels. If you did any *one* of these, your life was touched by organic chemistry. In Chapter 10, we first examine the characteristic features of all organic molecules. Then, we study the alkanes, the major components of petroleum and the simplest organic molecules.

10.1   Introduction to Organic Chemistry

What is organic chemistry?

> • **Organic chemistry is the study of compounds that contain the element carbon.**

Some compounds that contain the element carbon are *not* organic compounds. Examples include carbon dioxide (CO_2), sodium carbonate (Na_2CO_3), and sodium bicarbonate ($NaHCO_3$).

Clothes, foods, medicines, gasoline, refrigerants, and soaps are composed almost solely of organic compounds. By studying the principles and concepts of organic chemistry, you can learn more about compounds present in these substances and how they affect the world around you. Figure 10.1 illustrates some common products of organic chemistry used in medicine.

Because organic compounds are composed of covalent bonds, their properties differ a great deal from those of ionic inorganic compounds.

The intermolecular forces in covalent compounds were discussed in Section 4.3.

> • **Organic compounds exist as discrete molecules with much weaker intermolecular forces—the forces that exist *between* molecules—than the very strong interactions of oppositely charged ions seen in ionic compounds.**

As a result, organic compounds resemble other covalent compounds in that they have much lower melting points and boiling points than ionic compounds. Whereas ionic compounds are generally solids at room temperature, many organic compounds are liquids and some are even gases. Table 10.1 compares these and other properties of a typical organic compound (butane, $CH_3CH_2CH_2CH_3$) and a typical ionic inorganic compound (sodium chloride, NaCl).

Figure 10.1   Some Common Products of Organic Chemistry Used in Medicine

a. Oral contraceptives

b. Plastic syringes

c. Antibiotics

d. Synthetic heart valves

• Organic chemistry has given us contraceptives, plastics, antibiotics, synthetic heart valves, and a myriad of other materials. Our lives would be vastly different today without these products of organic chemistry.

(a): Comstock Images/PictureQuest/Getty Images; (b): Tom Grill/Corbis Premium/Alamy Stock Photo; (c): Julian Claxton/Alamy Stock Photo; (d): Layne Kennedy/Corbis NX/Getty Images

Table 10.1 Comparing the Properties of an Organic Compound ($CH_3CH_2CH_2CH_3$) and an Ionic Inorganic Compound (NaCl)

or

$CH_3CH_2CH_2CH_3$
butane

discrete molecules
with covalent bonding   *Ingram Publishing*   *Fuse/Getty Images*   ionic bonding

Property	$CH_3CH_2CH_2CH_3$ (An Organic Compound)	NaCl (An Inorganic Compound)
Bonding	Covalent	Ionic
Physical state	Gas at room temperature	Solid at room temperature
Boiling point	Low (−0.5 °C)	High (1413 °C)
Melting point	Low (−138 °C)	High (801 °C)
Solubility in water	Insoluble	Soluble
Solubility in organic solvents	Soluble	Insoluble
Flammability	Flammable	Nonflammable

Concept Check 10.1

Which chemical formulas represent organic compounds and which represent inorganic compounds?

a. C_6H_{12} c. KI   e. CH_4O
b. H_2O d. $MgSO_4$ f. NaOH

10.2   Characteristic Features of Organic Compounds

What are the common features of organic compounds?

[1] **All organic compounds contain carbon atoms and most contain hydrogen atoms. Carbon always forms four covalent bonds, and hydrogen forms one covalent bond.**

Carbon is located in group 4A of the periodic table, so a carbon atom has four valence electrons available for bonding (Section 3.7). Because hydrogen has a single valence electron, methane (CH_4) consists of four single bonds, each formed from one electron from a hydrogen atom and one electron from carbon.

Methane, the main component of natural gas, burns in the presence of oxygen. The natural gas we use today was formed by the decomposition of organic material millions of years ago.
Ingram Publishing/SuperStock

$$H:\overset{\displaystyle H}{\underset{\displaystyle H}{C}}:H \quad = \quad H-\overset{\displaystyle H}{\underset{\displaystyle H}{C}}-H$$

← **two-electron bond**

methane

[2] **Carbon forms single, double, and triple bonds to other carbon atoms.**

Ethylene is an important starting material in the preparation of the plastic polyethylene. *Nextdoor Images/Creatas/ PunchStock*

Because **acetylene** produces a very hot flame on burning, it is often used in welding torches. *Phillip Spears/Getty Images*

When a compound contains two or more carbon atoms, the type of bonding is determined by the number of atoms around carbon. Consider the three compounds drawn below:

| **Each single bond contains two electrons.** | **A double bond contains four electrons.** | **A triple bond contains six electrons.** |

ethane   ethylene   acetylene

- **A C atom surrounded by four atoms forms four single bonds.** In ethane (C_2H_6), each carbon atom is bonded to three hydrogen atoms and one carbon atom. All bonds are single bonds.
- **A C atom surrounded by three atoms forms one double bond.** In ethylene (C_2H_4), each carbon atom is surrounded by three atoms (two hydrogens and one carbon); thus, each C forms a single bond to each hydrogen atom and a double bond to carbon.
- **A C atom surrounded by two atoms generally forms one triple bond.** In acetylene (C_2H_2), each carbon atom is surrounded by two atoms (one hydrogen and one carbon); thus, each C forms a single bond to hydrogen and a triple bond to carbon.

[3] Some compounds have chains of atoms and some compounds have rings.

For example, three carbon atoms can bond in a row to form propane, or form a ring called cyclopropane. Propane is the fuel burned in gas grills, and cyclopropane is an anesthetic.

propane
C_3H_8

cyclopropane
C_3H_6

[4] Organic compounds may also contain elements other than carbon and hydrogen. Any atom that is not carbon or hydrogen is called a *heteroatom*.

The most common heteroatoms are nitrogen, oxygen, and the halogens (F, Cl, Br, and I).

- Each heteroatom forms a characteristic number of bonds, determined by its location in the periodic table.
- The common heteroatoms also have nonbonding, lone pairs of electrons, so that each atom is surrounded by eight electrons.

The number of bonds formed by common elements was first discussed in Section 3.7.

Thus, nitrogen forms three bonds and has one lone pair of electrons, whereas oxygen forms two bonds and has two additional lone pairs. The halogens form one bond and have three additional lone pairs. Common bonding patterns for atoms in organic compounds are summarized in Table 10.2. Except for hydrogen, these common elements in organic compounds follow one rule in bonding:

| Number of bonds | + | Number of lone pairs | = | 4 |

Table 10.2 Common Bonding Patterns for Atoms in Organic Compounds

Oxygen and nitrogen form both single and multiple bonds to carbon. **The most common multiple bond between carbon and a heteroatom is a carbon–oxygen double bond (C=O).** The bonding patterns remain the same even when an atom is part of a multiple bond, as shown with methanol (CH_3OH) and formaldehyde ($H_2C=O$, a preservative).

| Sample Problem 10.1 | Drawing in H's and Lone Pairs in an Organic Compound |

Draw in all H's and lone pairs in each compound.

   a. C—C—Cl   b. C—C=O   c. C—C≡N

Analysis

Each C and heteroatom must be surrounded by eight electrons. Use the common bonding patterns in Table 10.2 to fill in the needed H's and lone pairs. C needs four bonds; Cl needs one bond and three lone pairs; O needs two bonds and two lone pairs; N needs three bonds and one lone pair.

Solution

a.

Cl has three lone pairs.

C needs **3 H's** to have four bonds.

b.

O has two lone pairs.

C needs **1 H** to have four bonds.

c.

N has one lone pair.

C already has **four bonds.**

Practice Problem 10.1

Fill in all H's and lone pairs in each compound.

a. C—C=C—C   b. C—C—C   c. C—C—C   d. C—C—N—C   e. (structure)

(b, c, d have C=O above middle carbon)

More Practice: Try Problems 10.15, 10.16.

10.3 Drawing Organic Molecules

Sections 10.1 and 10.2 illustrate that organic molecules can be represented in different ways. For example, all the atoms and bonds of pentane (C_5H_{12}) can be drawn (as in **A**), or a ball-and-stick model **B** can show the three-dimensional structure of pentane. Moreover, because organic molecules often contain many atoms, we use shorthand methods to simplify them. As a result, pentane can also be represented by the **condensed structure C** and the **skeletal structure D.**

pentane
complete structure
A

pentane
ball-and-stick model
B

$CH_3CH_2CH_2CH_2CH_3$
pentane
condensed structure
C

pentane
skeletal structure
D

10.3A Condensed Structures

Condensed structures are most often used for a compound having a chain of atoms bonded together, rather than a ring. The following conventions are used.

- All of the atoms are drawn in, but the two-electron bond lines are generally omitted.
- Lone pairs on heteroatoms are omitted.

To interpret a condensed formula, it is usually best to start at the *left side* of the molecule and remember that the *carbon atoms must have four bonds.*

- A carbon bonded to 3 H's becomes CH_3.
- A carbon bonded to 2 H's becomes CH_2.
- A carbon bonded to 1 H becomes CH.

Keep the vertical bond line to show where the CH_3 group is bonded.

Sometimes these structures are further simplified by using parentheses around like groups. Two CH_2 groups bonded together become $(CH_2)_2$. Two CH_3 groups bonded to the same carbon become $(CH_3)_2C$.

two CH_2 groups bonded together	two CH_3 groups bonded to the same C
$CH_3CH_2CH_2CH_3$ = $CH_3(CH_2)_2CH_3$	$CH_3CHCH_2CH_3$ = $(CH_3)_2CHCH_2CH_3$

Additional examples of drawing condensed structures are shown in Figure 10.2.

Figure 10.2   Examples of Condensed Structures

[1] $H-C-C-C-C-C-H$ = $CH_3CH_2CHCH_2CH_3$ or $CH_3CH_2CH(CH_3)CH_2CH_3$
                                                    CH_3

[2] $C=C$ = $CH_3CH=CHCH_3$

[3] $H-C-C-C-O-H$ = $(CH_3)_2CHCH_2OH$

- Entry [1]: Draw the H atom next to the C to which it is bonded, and use parentheses around CH_3 to show it is bonded to the carbon chain.
- Entry [2]: Keep the carbon–carbon double bond and draw the H atoms *after* each C to which they are bonded.
- Entry [3]: Omit the lone pairs on the O atom in the condensed structure.

Sample Problem 10.2   **Drawing a Condensed Structure**

Convert each compound into a condensed structure.

a. $H-C-C-C-C-C-H$

b. $H-C-C-C-O-C-Cl$

Analysis

- **Start at the left and proceed to the right, making sure that each carbon has four bonds.**
- **Omit lone pairs** on the heteroatoms O and Cl.
- When like groups are bonded together or bonded to the same atom, use **parentheses** to further simplify the structure.

Solution

	Condensed structure	Further simplified
a. $H-C-C-C-C-C-H$ =	$CH_3CHCH_2CHCH_3$ CH_3 CH_3	$(CH_3)_2CHCH_2CH(CH_3)_2$ Use parentheses to show 2 CH_3 groups on one C.

b. $H-\underset{\underset{H}{|}}{\overset{\overset{H}{|}}{C}}-\underset{\underset{H}{|}}{\overset{\overset{H}{|}}{C}}-\underset{\underset{H}{|}}{\overset{\overset{H}{|}}{C}}-\ddot{\overset{..}{O}}-\underset{\underset{H}{|}}{\overset{\overset{H}{|}}{C}}-\ddot{\overset{..}{Cl}}:$ = $CH_3CH_2CH_2OCH_2Cl$ = $CH_3(CH_2)_2OCH_2Cl$

Practice Problem 10.2

Convert each compound to a condensed formula.

a. $:\ddot{Br}-\underset{\underset{H}{|}}{\overset{\overset{H}{|}}{C}}-\underset{\underset{H}{|}}{\overset{\overset{H}{|}}{C}}-\ddot{Br}:$

b. $H-\underset{\underset{H}{|}}{\overset{\overset{H}{|}}{C}}-\underset{\underset{H}{|}}{\overset{\overset{H}{|}}{C}}-\ddot{\overset{..}{O}}-\underset{\underset{H}{|}}{\overset{\overset{H}{|}}{C}}-\underset{\underset{H}{|}}{\overset{\overset{H}{|}}{C}}-\ddot{\overset{..}{O}}-H$

c. $H-\overset{\overset{\displaystyle H-\overset{H}{\overset{|}{C}}-H}{|}}{\underset{\underset{\displaystyle H-\overset{|}{\underset{|}{C}}-H}{|}}{C}}$ ... structure

---

Practice Problem 10.3

Convert each condensed formula to a complete structure with lone pairs on heteroatoms.

a. $CH_3(CH_2)_4OH$ c. $CH_3(CH_2)_4CH(CH_3)_2$

b. CH_3CCl_3 d. $(CH_3)_2CHCH_2NH_2$

More Practice: Try Problems 10.3, 10.4, 10.21, 10.22, 10.25a, 10.26a.

10.3B Skeletal Structures

Skeletal structures are used for organic compounds containing both rings and chains of atoms. Three important rules are used in drawing them.

- Assume there is a carbon atom at the junction of any two lines or at the end of any line.
- Assume there are enough hydrogens around each carbon to give it four bonds.
- Draw in all heteroatoms and the hydrogens directly bonded to them.

As shown for hexane, carbon chains are drawn in a zigzag fashion in skeletal structures to resemble the zigzag arrangement in the ball-and-stick model.

hexane   =   hexane
                            skeletal structure   =   hexane
                                                                    ball-and-stick model

---

Concept Check 10.2

Give the molecular formula for each molecule drawn in a skeletal structure.

a.   b. OH   c.

Rings are drawn as polygons with a carbon atom "understood" at each vertex, as shown for cyclohexane and cyclopentanol. All carbons and hydrogens in these molecules are understood, except for H's bonded to heteroatoms.

Each C is bonded to 2 H's.   Draw in the **H** on O.

cyclohexane   skeletal structure   cyclopentanol   skeletal structure

This C is bonded to only 1 H.

Figure 10.3 illustrates the stepwise conversion of a skeletal structure with a ring to a complete structure.

Figure 10.3 Example of a Skeletal Structure with a Ring

- C labeled in blue has 3 H's, so it has four bonds.
- C labeled in red has 1 H, so it has four bonds.
- C labeled in green already has four bonds, so it has no H's bonded to it.
- All the other C's have 2 H's, so each C has four bonds.

Sample Problem 10.3   Interpreting the Skeletal Structure of a Cyclic Compound

Convert each skeletal structure to a complete structure with all C's and H's drawn in.

a.   b. —Cl

Analysis

To draw each complete structure, place a C atom at the corner of each polygon and add H's to give carbon four bonds. Add lone pairs to a heteroatom to give an octet.

Solution

a.   Each C needs 2 H's.

b.   This C needs only **1 H**.

Add three lone pairs to Cl.

Practice Problem 10.4

Convert each skeletal structure to a complete structure with all C's and H's drawn in. Add lone pairs on all heteroatoms.

a.   cyclooctane

b. Cl   lindane (an insecticide)

c. CH_3— —$CH(CH_3)_2$ menthol (isolated from peppermint)

Practice Problem 10.5

Using the skeletal structure, determine the molecular formula of sitagliptin, a drug used to treat type 2 diabetes.

sitagliptin

More Practice: Try Problems 10.23, 10.24, 10.25b, 10.26b, 10.27.

10.4 Functional Groups

In addition to strong C—C and C—H bonds, organic molecules may have other structural features as well. Although over 50 million organic compounds are currently known, only a limited number of common structural features, called **functional groups,** are found in these molecules.

- A *functional group* is an atom or a group of atoms with characteristic chemical and physical properties.
- A functional group contains a heteroatom, a multiple bond, or sometimes both a heteroatom *and* a multiple bond.

A functional group determines a molecule's shape, properties, and the type of reactions it undergoes. A functional group behaves the same whether it is bonded to a carbon backbone having as few as two or as many as 20 carbons. For this reason, we often abbreviate the carbon and hydrogen portion of the molecule by a capital letter **R,** and draw the **R** bonded to a particular functional group.

Ethanol (CH_3CH_2OH), for example, has two carbons and five hydrogens in its carbon backbone, as well as an OH group, a functional group called a **hydroxyl group.** The hydroxyl group determines the physical properties of ethanol as well as the type of reactions it undergoes. Moreover, any organic molecule containing a hydroxyl group has properties similar to ethanol. Compounds that contain a hydroxyl group are called **alcohols.**

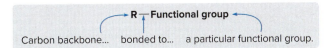

ethanol

The most common functional groups can be subdivided into three types.

- Hydrocarbons
- Compounds containing a single bond to a heteroatom
- Compounds containing a C═O group

10.4A Hydrocarbons

Hydrocarbons **are compounds that contain only the elements of carbon and hydrogen,** as shown in Table 10.3.

- *Alkanes* have only C—C single bonds and no functional group. Ethane, CH_3CH_3, is a simple alkane.
- *Alkenes* have a C—C double bond as their functional group. Ethylene, $CH_2{=}CH_2$, is a simple alkene.
- *Alkynes* have a C—C triple bond as their functional group. Acetylene, HC≡CH, is a simple alkyne.
- *Aromatic hydrocarbons* contain a benzene ring, a six-membered ring with three double bonds.

All hydrocarbons other than alkanes contain multiple bonds. Alkanes, which have no functional groups and therefore no reactive sites, are notoriously unreactive except under very drastic conditions. For example, **polyethylene** is a synthetic plastic and high-molecular-weight alkane, consisting of long chains of —CH_2— groups bonded together, hundreds or even thousands of

Ethanol, the alcohol present in wine and other alcoholic beverages, is formed by the fermentation of sugar. Ethanol can also be made in the lab by a totally different process. **Ethanol produced in the lab is identical to the ethanol produced by fermentation.**
Goodshoot/PunchStock

Table 10.3 Hydrocarbons

Type of Compound	General Structure	Example	3-D Structure	Functional Group
Alkane	R—H	CH_3CH_3		—
Alkene	C=C	H, H / C=C / H, H		Carbon–carbon double bond
Alkyne	—C≡C—	H—C≡C—H		Carbon–carbon triple bond
Aromatic compound				Benzene ring

Polyethylene was first produced in the 1930s and initially used as insulating material for radar during World War II. It is now a plastic in milk containers, sandwich bags, and plastic wrapping. Over 100 billion pounds of polyethylene are manufactured each year.
Jill Braaten/McGraw Hill Education

Tetrahydrocannabinol (THC) is the primary active constituent in marijuana. THC can be used legally in some states for medical and recreational purposes.
Adam Bailleaux/Atomazul/123RF

atoms long. Because it has no reactive sites, it is a very stable compound that does not readily degrade and thus persists for years in landfills.

polyethylene
The chain continues in both directions.

=

10.4B   Compounds Containing a Single Bond to a Heteroatom

Several types of functional groups contain a carbon atom singly bonded to a heteroatom. Common examples include alkyl halides, alcohols, ethers, amines, and thiols, as shown in Table 10.4.

Molecules containing these functional groups may be simple or very complex. It doesn't matter what else is present in other parts of the molecule. **Always dissect it into small pieces to identify the functional groups.** For example, diethyl ether, the first general anesthetic, is an ether because it has an O atom bonded to two C's. Tetrahydrocannabinol (THC), the active component in marijuana, is also an ether because it contains an O atom bonded to two carbon atoms. In this case, the O atom is also part of a ring.

ether

CH_3CH_2—Ö—CH_2CH_3

or

diethyl ether

ether

OH

ether

tetrahydrocannabinol
(THC)

Table 10.4 Compounds Containing a Carbon–Heteroatom Single Bond

Type of Compound	General Structure	Example	3-D Structure	Functional Group
Alkyl halide	R—Ẍ: (X = F, Cl, Br, I)	CH₃—B̈r:		—X halo group
Alcohol	R—ÖH	CH₃—ÖH		—OH hydroxyl group
Ether	R—Ö—R	CH₃—Ö—CH₃		—OR alkoxy group
Amine	R—N̈H₂ or R₂N̈H or R₃N̈	CH₃—N̈H₂		—NH₂ amino group
Thiol	R—S̈H	CH₃—S̈H		—SH sulfhydryl group

Concept Check 10.3

Identify the functional groups in each compound. A compound may contain more than one functional group.

        OH
        |
a.   CH₃CHCH₃

2-propanol
(disinfectant in
rubbing alcohol)

b. ⌬—CH=CH₂

styrene
(starting material used
to synthesize Styrofoam)

c. H₂NCH₂CH₂CH₂CH₂NH₂

putrescine
(putrid odor of rotting fish)

10.4C Compounds Containing a C=O Group

carbonyl group

Many different kinds of compounds contain a carbon–oxygen double bond (**C=O, carbonyl group**), as shown in Table 10.5. Carbonyl compounds include aldehydes, ketones, carboxylic acids, esters, and amides. The type of atom bonded to the carbonyl carbon—hydrogen, carbon, or a heteroatom—determines the specific class of carbonyl compound.

Take special note of the condensed structures used to draw aldehydes, carboxylic acids, and esters.

- **An aldehyde has a hydrogen atom bonded directly to the carbonyl carbon.**

= CH₃**CHO** C is bonded to both H and O.

acetaldehyde   condensed structure

- **A carboxylic acid contains an OH group bonded directly to the carbonyl carbon.**

= CH₃**COOH** or CH₃**CO₂H** C is bonded to both O atoms.

acetic acid   condensed structures

Table 10.5 Compounds Containing a C=O Group

Type of Compound	General Structure	Example	Condensed Structure	3-D Structure	Functional Group
Aldehyde	R–C(=O)–H	CH₃–C(=O)–H	CH₃CHO		C(=O)–H
Ketone	R–C(=O)–R	CH₃–C(=O)–CH₃	(CH₃)₂CO		C=O
Carboxylic acid	R–C(=O)–OH	CH₃–C(=O)–OH	CH₃CO₂H		C(=O)–OH carboxyl group
Ester	R–C(=O)–OR	CH₃–C(=O)–OCH₃	CH₃CO₂CH₃		C(=O)–OR
Amide	R–C(=O)–N(H (or R))(H (or R))	CH₃–C(=O)–NH₂	CH₃CONH₂		C(=O)–N

The characteristic odor of many fruits is due to low-molecular-weight esters.

Jill Braaten

CH₃COOCH₂CH₂CH(CH₃)₂

isoamyl acetate
odor of banana

• An ester contains an OR group bonded directly to the carbonyl carbon.

CH₃–C(=O)–OCH₂CH₃
ethyl acetate

= CH₃**COOCH₂CH₃** or CH₃**CO₂CH₂CH₃**

condensed structures

C is bonded to both O atoms.

Sample Problem 10.4 — Identifying Functional Groups

Identify the functional group in each compound.

a.

H   CH₃
 C=C
H   CH₃

A

b.

⬡—OH

B

c.

⬠=O

C

Analysis

Concentrate on the multiple bonds and heteroatoms and refer to Tables 10.3, 10.4, and 10.5.

Solution

a.

H   CH₃
 C=C
H   CH₃

A

b.

⬡—OH

B

c.

⬠=O

C

C's on both sides of the C=O

A is a hydrocarbon with a carbon–carbon double bond, making it an alkene.

B has a carbon atom bonded to a hydroxyl group (OH), making it an alcohol.

C contains a C=O. Because the carbonyl carbon is bonded to two other carbons in the ring, **C** is a ketone.

Practice Problem 10.6

For each compound: [1] Identify the functional group; [2] draw out the complete compound, including lone pairs on heteroatoms.

a. ⬡—CHO

c. ⬡=O

e. (cyclopentane)—C(=O)—NHCH₃

b. $CH_3CH_2CH_2CO_2H$

d. $CH_3CH_2CO_2CH_2CH_3$

Practice Problem 10.7

Label each of the following condensed structures as an ether or a ketone.

a. $(CH_3)_2O$ b. $(CH_3)_2CO$ c. $(CH_3CH_2)_2CO$ d. $CH_3CH_2OCH(CH_3)_2$

More Practice: Try Problems 10.29–10.31, 10.35, 10.36.

Many useful organic compounds contain complex structures with two or more functional groups. Sample Problem 10.5 illustrates an example of identifying several functional groups in a single molecule. Sample Problem 10.6 shows examples of identifying functional groups in simple and complex molecules beginning with ball-and-stick models.

Sample Problem 10.5

Identifying Functional Groups in a Complex Molecule

HEALTH NOTE

Tamiflu (Sample Problem 10.5) is the trade name for oseltamivir, an antiviral drug used to treat influenza. *Jill Braaten*

Tamiflu is an antiviral drug effective against avian influenza. Identify all of the functional groups in Tamiflu.

(structure of Tamiflu with labels: $(CH_3CH_2)_2CH$, O, $CO_2CH_2CH_3$, HN, NH_2, CH_3, C=O)

Tamiflu

Analysis

To identify functional groups, **look for multiple bonds and heteroatoms.** With functional groups that contain O atoms, look at what is bonded to the O's to decide if the group is an alcohol, ether, or other group. With carbonyl-containing groups, look at what is bonded to the carbonyl carbon.

Solution

Re-draw Tamiflu to further clarify the functional groups:

(re-drawn structure with labels: double bond—alkene, ether, ester, amide, amine, $(CH_3CH_2)_2$ CH, O, C, OCH_2CH_3, HN, NH_2, CH_3)

- Tamiflu contains a total of five functional groups: a carbon–carbon double bond (an alkene), ether, ester, amide, and amine.
- Note the difference between an amine and an amide. **An amine contains a N atom but no C═O. An amide contains a N atom bonded directly to a C═O.**

Practice Problem 10.8

HEALTH NOTE

Atenolol (Practice Problem 10.8) is one of the many drugs available to treat patients with high blood pressure. *Russell Illig/Photodisc/Getty Images*

Identify all of the functional groups in atenolol, an antihypertensive agent; that is, it is drug used to treat high blood pressure.

(structure of atenolol with labels: O, C, H_2N, CH_2, O, $CH_2CHCH_2NHCH(CH_3)_2$, OH)

atenolol

More Practice: Try Problems 10.32, 10.34.

Sample Problem 10.6	Converting a Ball-and-Stick Model to a Condensed or Skeletal Structure

Convert each ball-and-stick model to a condensed or skeletal structure and identify the functional groups.

a.

cis-2-butene

b.

methyl anthranilate

Analysis

Use the common element colors shown in Appendix C to convert the ball-and-stick model to a shorthand representation. **Look for multiple bonds and heteroatoms** to locate the functional groups, and refer to Tables 10.3, 10.4, and 10.5.

Solution

a.

cis-2-butene   carbon–carbon double bond
(an alkene)

b.

methyl anthranilate

ester   ester
benzene ring amine

Practice Problem 10.9	Convert each ball-and-stick model to a condensed or skeletal structure and identify the functional groups.

a.   b.

More Practice: Try Problems 10.1, 10.2, 10.12c, 10.33.

10.5 Alkanes

10.5A Introduction

Now that we have learned some general features of organic compounds, we can examine the alkanes, the simplest organic molecules, in more detail. *Alkanes* **are hydrocarbons having only C—C and C—H single bonds.** The carbons of an alkane can be joined together to form chains or rings of atoms.

The prefix *a-* means *not*, so an **a**cyclic alkane is *not* cyclic.

- **Alkanes that contain chains of carbon atoms but no rings are called *acyclic alkanes.*** An acyclic alkane has the molecular formula C_nH_{2n+2}, where *n* is the number of carbons it contains. Acyclic alkanes are also called **saturated hydrocarbons** because they have the maximum number of hydrogen atoms per carbon.
- *Cycloalkanes* contain carbons joined in one or more rings. Because a cycloalkane has two fewer H's than an acyclic alkane with the same number of carbons, its general formula is C_nH_{2n}.

Undecane and cyclohexane are examples of two naturally occurring alkanes. Undecane is an acyclic alkane with molecular formula $C_{11}H_{24}$. Undecane is a *pheromone,* **a chemical substance used for communication** in a specific animal species, most commonly an insect population. Secretion of undecane by a cockroach causes other members of the species to aggregate. Cyclohexane, a cycloalkane with molecular formula C_6H_{12}, is one component of the mango, the most widely consumed fruit in the world.

$CH_3CH_2CH_2CH_2CH_2CH_2CH_2CH_2CH_2CH_2CH_3$

undecane

cyclohexane

(left): God of Insects; (right): Dinodia/Pixtal/age fotostock

---

Concept Check 10.4

How many hydrogen atoms are present in each compound?

a. an acyclic alkane with three carbons

b. a cycloalkane with four carbons

c. a cycloalkane with nine carbons

d. an acyclic alkane with seven carbons

Concept Check 10.5

Which formulas represent acyclic alkanes and which represent cycloalkanes?

a. C_5H_{12} b. C_4H_8 c. $C_{12}H_{24}$ d. $C_{10}H_{22}$

10.5B Acyclic Alkanes Having Fewer Than Five Carbons

The structures for the two simplest acyclic alkanes were given in Section 10.2.

- **Methane, CH_4,** has a single carbon atom surrounded by four hydrogens to give it four bonds.
- **Ethane, CH_3CH_3,** has two carbon atoms joined together by a single bond. Each carbon is also bonded to three hydrogens to give it four bonds total.

Recall how to draw a tetrahedron from Section 3.10: Place two bonds in the plane of the page (on solid lines), one bond in front (on a wedge), and one bond behind the plane (on a dashed wedge).

The shape around atoms in organic molecules is determined by counting groups using the principles of VSEPR theory (Section 3.10). Because each carbon in an alkane is surrounded by four atoms, each carbon is **tetrahedral,** and all bond angles are 109.5°.

CH_4 methane

3-D representation   ball-and-stick model

CH_3CH_3 ethane

To draw a three-carbon alkane, draw three carbons joined together with single bonds and add enough hydrogens to give each carbon four bonds. This forms **propane, $CH_3CH_2CH_3$.**

$CH_3CH_2CH_3$ propane

3-D representation   ball-and-stick model   skeletal structure

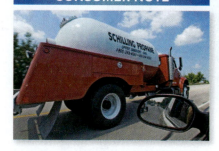

The carbon skeleton in propane and other alkanes can be drawn in a variety of different ways and still represent the same molecule. For example, the three carbons of propane can be drawn in a horizontal row or with a bend. *These representations are equivalent.* **If you follow the carbon chain from one end to the other, you move across the *same* three carbon atoms in both representations.**

3 C's in a row   3 C's with a bend

Similarly, as long as a skeletal structure has three carbons in a row, it represents propane.

$CH_3CH_2CH_3$ $CH_3CH_2CH_3$ $CH_3CH_2CH_3$ $CH_3CH_2CH_3$

All of these skeletal structures represent propane, C_3H_8.

- The bends in a carbon chain don't matter when it comes to identifying different compounds.

There are two different ways to arrange four carbons, giving two compounds with molecular formula C_4H_{10}.

- Butane, $CH_3CH_2CH_2CH_3$, has four carbon atoms in a row. Butane is a **straight-chain alkane**, an alkane that has all of its carbons in one continuous chain.
- Isobutane, $(CH_3)_3CH$, has three carbon atoms in a row and one carbon bonded to the middle carbon. Isobutane is a **branched-chain alkane**, an alkane that contains one or more carbon branches bonded to a carbon chain.

Propane is the principal alkane in LPG (liquefied petroleum gas), a fuel used for vehicles and cooking. LPG has also been used as an aerosol propellant, replacing the ozone-depleting chlorofluorocarbons (Section 6.9). *Gary Gardiner/Bloomberg/Getty Images*

CONSUMER NOTE

butane = $CH_3CH_2CH_2CH_3$

| butane | 4 C's in a row | skeletal structure | **straight-chain alkane** |

isobutane =

isobutane | 3 C's with a one-carbon branch | skeletal structure | **branched-chain alkane**

Butane and isobutane are *isomers,* **two different compounds with the same molecular formula.** They belong to one of the two major classes of isomers called **constitutional isomers.**

Constitutional isomers are also called structural isomers.

• **Constitutional isomers** differ in the way the atoms are connected to each other.

Concept Check 10.6

Label each representation as butane or isobutane (2-methylpropane).

a.   H—C—C—C—H (with H, H, H on top; H, H, CH₃ on bottom)

b. $HC(CH_3)_3$

c.   (branched structure)

d.   H—C—C—H (with CH₃, CH₃ on top; H, H on bottom)

Constitutional isomers like butane and isobutane belong to the same family of compounds: they are both **alkanes.** This is not always the case. For example, there are two different arrangements of atoms for a compound of molecular formula C_2H_6O.

CH_3CH_2OH

ethanol

CH_3OCH_3

dimethyl ether

Ethanol (CH_3CH_2OH) and dimethyl ether (CH_3OCH_3) are constitutional isomers with different functional groups: CH_3CH_2OH is an **alcohol** and CH_3OCH_3 is an **ether.**

Sample Problem 10.7   Identifying Constitutional Isomers

Are the compounds in each pair constitutional isomers or are they not constitutional isomers of each other?

a. $CH_3CH_2CH_3$ and   (C=C structure with H, H, H, CH₃)

b. $CH_3CH_2CH_2NH_2$ and $CH_3—N—CH_3$ (with CH₃ below N)

Analysis

- **Compare molecular formulas;** two compounds are isomers only if they have the same molecular formula.
- **Check how the atoms are connected to each other.** Constitutional isomers have atoms bonded to *different* atoms.

Solution

a. $CH_3CH_2CH_3$ and

molecular formula
C_3H_8

molecular formula
C_3H_6

The two compounds have the same number of C's but a *different* number of H's, so they have *different* molecular formulas. Thus, they are *not* isomers of each other.

b. $CH_3CH_2CH_2NH_2$ and $CH_3{-}N{-}CH_3$ with CH_3

Both compounds have molecular formula C_3H_9N. Because one compound has C–C bonds and the other does not, the atoms are connected differently. These compounds are *constitutional isomers.*

Practice Problem 10.10

Are the compounds in each pair constitutional isomers or are they not constitutional isomers of each other?

a. $CH_3CH_2CH_2CH_3$ and $CH_3CH_2CH_3$

b. $CH_3CH_2CH_2OH$ and $CH_3OCH_2CH_3$

c. and —CH_3

More Practice: Try Problems 10.39, 10.40.

10.5C   Acyclic Alkanes Having Five or More Carbons

As the number of carbon atoms in an alkane increases, so does the number of isomers. There are three constitutional isomers for the five-carbon alkane, each having molecular formula C_5H_{12}: **pentane, isopentane** (or 2-methylbutane), and **neopentane** (or 2,2-dimethylpropane).

$CH_3CH_2CH_2CH_2CH_3$

pentane

$CH_3{-}\overset{\displaystyle CH_3}{\underset{\displaystyle H}{C}}{-}CH_2CH_3$

isopentane
(2-methylbutane)

$CH_3{-}\overset{\displaystyle CH_3}{\underset{\displaystyle CH_3}{C}}{-}CH_3$

neopentane
(2,2-dimethylpropane)

With alkanes having five or more carbons, the names of the straight-chain isomers are derived from Greek roots: *pent*ane for **five** carbons, *hex*ane for **six**, and so on. Table 10.6 lists the names and structures for the straight-chain alkanes having up to 10 carbons. The suffix -*ane* identifies a molecule as an alk*ane.* The remainder of the name—meth-, eth-, prop-, and so forth—indicates the number of carbons in the long chain.

The suffix -*ane* identifies a molecule as an alk*ane.*

Table 10.6 Straight-Chain Alkanes

Number of C's	Molecular Formula	Structure	Name
1	CH_4	CH_4	*methane*
2	C_2H_6	CH_3CH_3	*ethane*
3	C_3H_8	$CH_3CH_2CH_3$	*propane*
4	C_4H_{10}	$CH_3CH_2CH_2CH_3$	*butane*
5	C_5H_{12}	$CH_3CH_2CH_2CH_2CH_3$	*pentane*
6	C_6H_{14}	$CH_3CH_2CH_2CH_2CH_2CH_3$	*hexane*
7	C_7H_{16}	$CH_3CH_2CH_2CH_2CH_2CH_2CH_3$	*heptane*
8	C_8H_{18}	$CH_3CH_2CH_2CH_2CH_2CH_2CH_2CH_3$	*octane*
9	C_9H_{20}	$CH_3CH_2CH_2CH_2CH_2CH_2CH_2CH_2CH_3$	*nonane*
10	$C_{10}H_{22}$	$CH_3CH_2CH_2CH_2CH_2CH_2CH_2CH_2CH_2CH_3$	*decane*

Sample Problem 10.8 **Drawing Isomers**

Draw two isomers with molecular formula C_6H_{14} that have five carbon atoms in the longest chain and a one-carbon branch coming off the chain.

Analysis

Because isomers are different compounds with the same molecular formula, we must add a one-carbon branch to two different carbons to form two different products. Then add enough H's to give each C four bonds.

Solution

Compounds **A** and **B** are isomers because the CH$_3$ group is bonded to different atoms in the five-carbon chain. Note, too, that we cannot add the one-carbon branch to an *end* carbon because that creates a continuous six-carbon chain. Remember that **bends in the chain don't matter.**

Practice Problem 10.11 Draw two isomers with molecular formula C_6H_{14} that have four carbon atoms in the longest chain and two one-carbon branches coming off the chain.

More Practice: Try Problems 10.41–10.44, 10.67b, 10.68b.

allicin
(odor of garlic)

Garlic has been used in Chinese herbal medicine for over 4,000 years. Today it is sold as a dietary supplement because of its reported health benefits. **Allicin,** the molecule responsible for garlic's odor, is not stored in the garlic bulb, but rather is produced by the action of enzymes when the bulb is crushed or bruised.
Photodisc/Getty Images

10.6   Alkane Nomenclature

10.6A   The IUPAC System of Nomenclature

How are organic compounds named? Long ago, the name of a compound was often based on the plant or animal source from which it was obtained. For example, the name allicin, the principal component of the odor of garlic, is derived from the botanical name for garlic, *Allium sativum.*

With the isolation and preparation of thousands of new organic compounds it became obvious that each organic compound must have an unambiguous name. A systematic method of naming compounds (a system of **nomenclature**) was developed by the *I*nternational *U*nion of *P*ure and *A*pplied *C*hemistry. It is referred to as the **IUPAC** system of nomenclature.

10.6B   The Basic Features of Alkane Nomenclature

Although the names of the straight-chain alkanes having 10 carbons or fewer were already given in Table 10.6, we must also learn how to name alkanes that have carbon branches, called **substituents,** bonded to a long chain. The names of these organic molecules have three parts.

- The **parent name** indicates the number of carbons in the longest continuous carbon chain in the molecule.
- The **suffix** indicates what functional group is present.
- The **prefix** tells us the identity, location, and number of substituents attached to the carbon chain.

The names of the straight-chain alkanes in Table 10.6 consist of two parts. The suffix **-ane** indicates that the compounds are alkanes. The remainder of the name is the parent name, which indicates the number of carbon atoms in the longest carbon chain. The parent name for **one carbon is meth-,** for **two carbons is eth-,** and so on. Thus, we are already familiar with two parts of the name of an organic compound.

To determine the third part of a name, the prefix, we must learn how to name the substituents that are bonded to the longest carbon chain.

10.6C   Naming Substituents

Carbon substituents bonded to a long carbon chain are called *alkyl groups.*

- An *alkyl group* is formed by removing one hydrogen from an alkane.

An alkyl group is a part of a molecule that is now able to bond to another atom or a functional group. **To name an alkyl group, change the -ane ending of the parent alkane to -yl.** Thus, methane (CH_4) becomes methyl (CH_3-) and ethane (CH_3CH_3) becomes ethyl (CH_3CH_2-).

Remove 1 **H.**

methane ⟶ methyl   = CH_3- The line means that the C can bond to something else.

Remove 1 **H.**

ethane ⟶ ethyl   = CH_3CH_2-

Removing one hydrogen from an *end* carbon in any straight-chain alkane forms other alkyl groups named in a similar fashion. Thus, propane (CH₃CH₂CH₃) becomes propyl (CH₃CH₂CH₂–) and butane (CH₃CH₂CH₂CH₃) becomes butyl (CH₃CH₂CH₂CH₂–). The names of alkyl groups having six carbons or fewer are summarized in Table 10.7.

Table 10.7 Some Common Alkyl Groups

Number of C's	Structure	Name
1	CH₃–	methyl
2	CH₃CH₂–	ethyl
3	CH₃CH₂CH₂–	propyl
4	CH₃CH₂CH₂CH₂–	butyl
5	CH₃CH₂CH₂CH₂CH₂–	pentyl
6	CH₃CH₂CH₂CH₂CH₂CH₂–	hexyl

10.6D Naming an Acyclic Alkane

Four steps are needed to name an alkane.

How To Name an Alkane Using the IUPAC System

Step [1] Find the parent carbon chain and add the suffix.

- **Find the longest continuous carbon chain,** and name the molecule using the parent name for that number of carbons, given in Table 10.6. To the name of the parent, add the suffix **-ane** for an alkane. Each functional group has its own suffix.

$$CH_3-\overset{\overset{\displaystyle CH_3}{|}}{CH}-CH_2-CH_2-CH_2-CH_3$$

6 C's in the longest chain

6 C's - - -→ **hexane**

- The longest chain may not be written horizontally across the page. Remember that **it does not matter if the chain is *straight* or has *bends.*** All of the following representations are equivalent.

6 C's in the longest chain of each structure

Step [2] Number the atoms in the carbon chain to give the first substituent the lower number.

CORRECT   Start numbering here.

$$CH_3-\overset{\overset{\displaystyle CH_3}{|}}{CH}-CH_2-CH_2-CH_2-CH_3$$

1   2   3   4   5   6

first substituent at C2

INCORRECT

$$CH_3-\overset{\overset{\displaystyle CH_3}{|}}{CH}-CH_2-CH_2-CH_2-CH_3$$

6   5   4   3   2   1

first substituent at C5

Step [3] Name and number the substituents.

- Name the substituents as alkyl groups, and use the numbers from step [2] to designate their location.

methyl at C2

$$CH_3-\overset{\overset{\displaystyle CH_3}{|}}{CH}-CH_2-CH_2-CH_2-CH_3$$

←— longest chain (6 C's) **hexane**

1   2

- Every carbon belongs to *either* the longest chain or a substituent, but *not both.*
- Each substituent needs its *own* number.

—Continued

How To, continued . . .

- If two or more identical substituents are bonded to the longest chain, use **prefixes** to indicate how many: **di-** for two groups, **tri-** for three groups, **tetra-** for four groups, and so forth. The following compound has two methyl groups, so its name contains the prefix di- before methyl → *di*methyl.

methyl at C2

CH₃—CH—CH—CH₂—CH₂—CH₃   Use the prefix *di-*.

1   2 3   methyl at C3

Step [4] Combine substituent names and numbers + parent + suffix.

- Precede the name of the parent by the names of the substituents. **Alphabetize** the names of the substituents, ignoring any prefixes like *di-*. For example, triethyl precedes dimethyl because the e of **e**thyl comes before the m of **m**ethyl in the alphabet.
- Precede the name of each substituent by the number that indicates its location. There must be **one number for each substituent.**
- Separate numbers by commas and separate numbers from letters by dashes.

To help identify which carbons belong to the longest chain and which are substituents, always box in the atoms of the long chain. Every other carbon atom then becomes a substituent that needs its own name as an alkyl group.

methyl at C2

CH₃—CH—CH₂—CH₂—CH₂—CH₃

1   2   longest chain (6 C's)
hexane

Answer: 2-methylhexane

methyl at C2

CH₃—CH—CH—CH₂—CH₂—CH₃   ← **hexane**

1   2 3   methyl at C3

- Two numbers are needed, one for each methyl group.

Answer: 2,3-dimethylhexane

Concept Check 10.7

How many carbons are in the longest chain of each molecule?

a. CH₃CH₂CHCH₂CH₂CHCH₃
(with CH₃ up and CH₃ down)

b. (CH₃)₂CH(CH₂)₃CH₃

c.

Sample Problem 10.9

Naming an Alkane

Give the IUPAC name for the following compound.

CH₃—C—CH₂CH₂—C—C—CH₂CH₃
(with CH₃ and H on left carbon; CH₃CH₂, H, H, CH₃ on right carbons)

Analysis and Solution

[1] Name the parent and use the suffix -*ane* because the molecule is an alkane.

CH₃—C—CH₂CH₂—C—C—CH₂CH₃

8 C's in the longest chain   - - - →   **octane**

- Box in the atoms of the longest chain to clearly show which carbons are part of the longest chain and which carbons are substituents.

[2] Number the chain to give the first substituent the lower number.

• Numbering from left to right puts the first substituent at C2.

[3] Name and number the substituents.

methyl at C2   ethyl at C5

methyl at C6

• This compound has three substituents: two methyl groups at C2 and C6 and an ethyl group at C5.

[4] Combine the parts.

• Write the name as one word and use the prefix di- before methyl because there are two methyl groups.
• Alphabetize the *e* of **e**thyl before the *m* of **m**ethyl. The prefix di- is ignored when alphabetizing.

methyl at C2   ethyl at C5   ← octane

methyl at C6

Answer: 5-ethyl-2,6-dimethyloctane

Practice Problem 10.12

Give the IUPAC name for each compound.

a. CH₃CH₂CHCH₂CH₃ (with CH₃ substituent)

b. H—C—CH₂—CHCH₃ (with CH₂CH₃, CH₃, CH₃ substituents)

c.

Practice Problem 10.13

Give the IUPAC name for each compound.

a. (CH₃)₂CHCH(CH₃)₂

b. CH₃CH₂CH₂CH₂—C—C—CH₂CH₃ (with CH₃, H, CH₂, CH₃, CH₃ substituents)

c.

More Practice: Try Problems 10.45, 10.46.

You must also know how to derive a structure from a given name. Sample Problem 10.10 demonstrates a stepwise method.

Sample Problem 10.10   **Deriving a Structure from a Name**

Give the structure with the following IUPAC name: 2,3,3-trimethylpentane.

Analysis

To derive a structure from a name, first **look at the end of the name to find the parent name and suffix.** From the parent we know the number of C's in the longest chain, and the suffix tells us the functional group; the suffix **-*ane* = an alkane.** Then, number the carbon chain from either end and add the substituents. Finally, add enough H's to give each C four bonds.

Solution

2,3,3-Trimethyl**pentane** has **pentane** (5 C's) as the longest chain and three methyl groups at carbons 2, 3, and 3.

Draw 5 C's and
number the chain:

C—C—C—C—C   ------→   C—C—C—C—C   ------→   CH₃—C—C—CH₂—CH₃

1 2 3 4 5   Add 3 CH₃ groups   2   3   Add H's

$$CH_3-\overset{\overset{CH_3}{|}}{\underset{\underset{H}{|}}{C}}-\overset{\overset{CH_3}{|}}{\underset{\underset{CH_3}{|}}{C}}-CH_2-CH_3$$

Answer:

2,3,3-trimethylpentane

skeletal structure

Give the structure corresponding to each IUPAC name.

a. 3-methylhexane

b. 3,3-dimethylpentane

c. 3,5,5-trimethyloctane

d. 3-ethyl-4-methylhexane

Give the structure corresponding to each IUPAC name.

a. 2,2-dimethylbutane

b. 6-butyl-3-methyldecane

c. 4,4,5,5-tetramethylnonane

d. 3-ethyl-5-propylnonane

More Practice: Try Problems 10.51a, b, c; 10.52a, b.

10.6E FOCUS ON HEALTH & MEDICINE
Naming New Drugs

Naming organic compounds has become big business for drug companies. The IUPAC name of an organic compound can be long and complex. As a result, most drugs have three names:

- **Systematic:** The systematic name follows the accepted rules of nomenclature; this is the IUPAC name.
- **Generic:** The generic name is the official, internationally approved name for the drug.
- **Trade:** The trade name for a drug is assigned by the company that manufactures it. Trade names are often "catchy" and easy to remember. Companies hope that the public will continue to purchase a drug with an easily recalled trade name long after a cheaper generic version becomes available.

Consider the world of over-the-counter anti-inflammatory agents. The compound a chemist calls 2-[4-(2-methylpropyl)phenyl]propanoic acid has the generic name ibuprofen. It is marketed under a variety of trade names, including Motrin and Advil.

Jill Braaten

Systematic name: 2-[4-(2-methylpropyl)phenyl]propanoic acid

Generic name: ibuprofen

Trade name: Motrin or Advil

10.7 Cycloalkanes

Cycloalkanes contain carbon atoms arranged in a ring. Think of a cycloalkane as being formed by removing two H's from the end carbons of a chain, and then bonding the two carbons together.

10.7A Simple Cycloalkanes

Simple cycloalkanes are named by adding the prefix *cyclo-* to the name of the acyclic alkane having the same number of carbons. Cycloalkanes having three to six carbon atoms are shown in the accompanying figure. They are drawn using polygons in skeletal representations (Section 10.3). Each corner of the polygon has a carbon atom with two hydrogen atoms to give it four bonds.

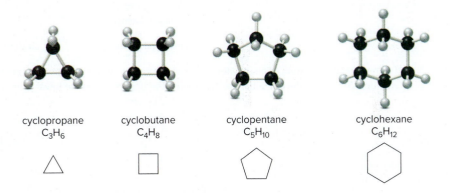

cyclopropane
C_3H_6

cyclobutane
C_4H_8

cyclopentane
C_5H_{10}

cyclohexane
C_6H_{12}

Although we draw cycloalkanes as flat polygons, in reality cycloalkanes with more than three carbons are not planar molecules. Cyclohexane, for example, adopts a puckered arrangement called the **chair** form, in which all bond angles are 109.5°.

cyclohexane

skeletal structure

nonplanar cyclohexane
chair form

ball-and-stick model

10.7B Naming Cycloalkanes

Cycloalkanes are named using the rules in Section 10.6, but the prefix *cyclo-* immediately precedes the name of the parent.

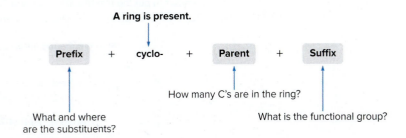

A ring is present.

Prefix + **cyclo-** + **Parent** + **Suffix**

What and where
are the substituents?

How many C's are in the ring?

What is the functional group?

How To Name a Cycloalkane Using the IUPAC System

Step [1] **Find the parent cycloalkane.**

- Count the number of carbon atoms in the ring and use the parent name for that number of carbons. Add the prefix *cyclo-* and the suffix *-ane* to the parent name.

6 C's in the ring

cyclohexane

Step [2] **Name and number the substituents.**

- No number is needed to indicate the location of a **single** substituent.

methyl

butyl

methylcyclohexane   butylcyclopentane

- For rings with more than one substituent, begin numbering at one substituent, and then give the **second substituent** the lower number. With two **different** substituents, number the ring to assign the lower number to the substituents **alphabetically.**

• Place CH₃ groups at C1 and C3.   Earlier letter – – – → lower number

1,3-dimethylcyclohexane   • **ethyl** group at **C1**
(*not* 1,5-dimethylcyclohexane)   • **methyl** group at **C3**

1-ethyl-3-methylcyclohexane
(*not* 3-ethyl-1-methylcyclohexane)

Sample Problem 10.11   ## Naming a Cycloalkane

Give the IUPAC name for the following compound.

Analysis and Solution

[1] **Name the ring.** The ring has 5 C's so the molecule is named as a cyclopentane.

[2] **Name and number the substituents.**

- There are two substituents: CH₃CH₂– is an ethyl group and CH₃CH₂CH₂– is a propyl group.
- Number to put the two groups at C1 and C2 (not C1 and C5).
- Place the ethyl group at C1 because the *e* of **ethyl** comes before the *p* of **propyl** in the alphabet.

Start numbering here.

ethyl at C1

5 C's – – – → cyclopentane

propyl at C2 (not C5)

Answer: 1-ethyl-2-propylcyclopentane

Practice Problem 10.16

Give the IUPAC name for each compound.

a.   b.   c.   d.

Practice Problem 10.17

Give the structure corresponding to each IUPAC name.

a. propylcyclopentane

b. 1,2-dimethylcyclobutane

c. 1,1,2-trimethylcyclopropane

d. 4-ethyl-1,2-dimethylcyclohexane

More Practice: Try Problems 10.47–10.50; 10.51d; 10.52c, d; 10.67a; 10.68a.

A significant source of atmospheric methane, the simplest **alkane,** comes from flooded rice fields. *Daniel C. Smith*

CONSUMER NOTE

Natural gas is odorless. The smell observed in a gas leak is due to minute amounts of a sulfur additive such as methanethiol, CH_3SH, which provides an odor for easy detection and safety.

10.8 FOCUS ON THE ENVIRONMENT
Methane and Other Fossil Fuels

Many alkanes occur in nature. **Methane** in the atmosphere comes from natural and humanmade sources. As global temperatures increase, methane trapped in permafrost and glaciers is released with melting. Microorganisms in the gut of ruminant animals produce methane that is released during defecation and belching. The microorganisms in wetlands and flooded rice fields decompose organic material to form methane when no oxygen is present. Methane is a greenhouse gas with significant global warming potential, and its concentration has increased significantly in the last 200 years.

Natural gas is composed largely of methane (60–80% depending on its source), with lesser amounts of ethane, propane, and butane. These organic compounds burn in the presence of oxygen, releasing energy for cooking and heating (Section 10.10).

Petroleum is a complex mixture of compounds, most of which are hydrocarbons containing 1–40 carbon atoms. Distilling crude petroleum, a process called **refining,** separates it into usable fractions that differ in boiling point (Figure 10.4). Most products of petroleum refining provide fuel for home heating, automobiles, diesel engines, and airplanes. Each fuel type has a different composition of hydrocarbons: gasoline (C_5H_{12}–$C_{12}H_{26}$), kerosene ($C_{12}H_{26}$–$C_{16}H_{34}$), and diesel fuel ($C_{15}H_{32}$–$C_{18}H_{38}$).

Figure 10.4

Refining Crude Petroleum into Usable Fuel and Other Petroleum Products

gases → boiling point range below 20 °C

gasoline → 20–200 °C

kerosene → 175–275 °C

fuel oil → 250–400 °C

lubricating oil → above 350 °C

Crude oil and vapor are preheated.

residue (asphalt)

(a) **An oil refinery.** At an oil refinery, crude petroleum is separated into fractions of similar boiling point. *Ingram Publishing*

(b) **A refinery tower.** As crude petroleum is heated, the lower-boiling components come off at the top of the tower, followed by fractions of higher boiling point.

Petroleum provides more than fuel. About 3% of crude oil is used to make plastics and other synthetic compounds, including drugs, fabrics, dyes, and pesticides. Imagine what life would be like without air-conditioning, refrigeration, anesthetics, and pain relievers, all products of the petroleum industry.

Barrel of crude oil

petroleum starting materials for chemical synthesis (1.3 gal)

asphalt and road oil (1.3 gal)

boiler oil (2.9 gal)

lubricants, waxes, solvents (4.2 gal)

jet fuel (4.2 gal)

diesel and home heating oil (8.4 gal)

gasoline (19.7 gal)

1 barrel = 42 gal
McGraw Hill

products made from petroleum
Jill Braaten/McGraw Hill

Energy from petroleum is *nonrenewable,* and the remaining known oil reserves are limited. Given our dependence on petroleum, not only for fuel, but also for the many necessities of modern society, it becomes obvious that we must both conserve what we have and find alternate energy sources.

10.9 Physical Properties

Organic molecules resemble other covalent compounds in that three types of intermolecular forces are possible, depending on the functional group. Intermolecular forces were first discussed in Section 4.3.

- London dispersion forces are weak intermolecular forces present in all covalent compounds, and the only forces present in nonpolar molecules.
- Dipole–dipole interactions are stronger intermolecular forces due to the attraction of permanent dipoles in polar molecules.
- Hydrogen bonding is the strongest intermolecular force that occurs when a hydrogen atom bonded to O, N, or F is electrostatically attracted to an O, N, or F atom in another molecule.

Because **alkanes contain only nonpolar C—C and C—H bonds,** they cannot exhibit dipole–dipole interactions or hydrogen bonding. **Alkanes are nonpolar molecules that exhibit only weak London dispersion forces.** Because the strength of the intermolecular forces determines the boiling point and melting point of compounds, **alkanes have low boiling points and low melting points.** Low-molecular-weight alkanes are gases at room temperature, and alkanes used in gasoline are all liquids.

The melting points and boiling points of alkanes increase as the number of carbons increases. Increased surface area increases the force of attraction between molecules, thus raising the boiling point and melting point. This is seen in comparing the boiling points of three straight-chain alkanes.

ENVIRONMENTAL NOTE

Crude oil that leaks into the sea forms an insoluble layer on the surface.
Tigergallery/Shutterstock

$CH_3CH_2CH_2CH_3$
butane

$CH_3CH_2CH_2CH_2CH_3$
pentane

$CH_3CH_2CH_2CH_2CH_2CH_3$
hexane

bp = −0.5 °C

bp = 36 °C

bp = 69 °C

Increasing surface area
Increasing boiling point

HEALTH NOTE

Vaseline petroleum jelly (Concept Check 10.9), a skin lubricant often used for chapped lips and diaper rash, is a mixture of high-molecular-weight alkanes. *Jill Braaten*

Because nonpolar alkanes are water insoluble and less dense than water, crude petroleum spilled into the sea from a ruptured oil tanker creates an insoluble oil slick on the surface. The insoluble hydrocarbon oil poses a special threat to birds whose feathers are coated with natural nonpolar oils for insulation. Because these oils dissolve in the crude petroleum, birds lose their layer of natural protection and many die.

Concept Check 10.8

Answer the following questions about pentane (C_5H_{12}), heptane (C_7H_{16}), and decane ($C_{10}H_{22}$).

   a. Which compound has the highest boiling point?
   b. Which compound has the lowest boiling point?
   c. Which compound has the highest melting point?
   d. Which compound has the lowest melting point?

Concept Check 10.9

Explain why Vaseline, a complex mixture of hydrocarbons used as a skin lubricant and ointment, is insoluble in water and soluble in dichloromethane (CH_2Cl_2).

Concept Check 10.10

The coating of apple skins contains an alkane with 27 carbon atoms. (a) How many hydrogen atoms does this alkane contain? (b) What purpose does this alkane serve?

10.10 FOCUS ON THE ENVIRONMENT
Combustion

Combustion was first discussed in Sections 5.4B and 6.9B.

CONSUMER NOTE

A spark or a flame is needed to initiate combustion. Gasoline, which is composed largely of alkanes, can be safely handled and stored in the air, but the presence of a spark or open flame causes immediate and violent combustion. *Comstock Images/Getty Images*

HEALTH NOTE

Meters that measure CO levels can be purchased, as described in Section 5.7. When wood is burned in a poorly ventilated fireplace situated in a well-insulated room, the CO concentration can reach an unhealthy level.

Alkanes are the only family of organic molecules that has no functional group, so **alkanes undergo few reactions.** In this chapter, we consider only one reaction of alkanes—**combustion**—that is, **alkanes burn in the presence of oxygen to form carbon dioxide (CO_2) and water.** This is a practical example of oxidation. Every C—H and C—C bond in the starting material is converted to a C—O bond in the product.

$$2\ (CH_3)_3CCH_2CH(CH_3)_2\ +\ 25\ O_2\ \xrightarrow{\text{flame}}\ 16\ CO_2\ +\ 18\ H_2O\ +\ \text{energy}$$
isooctane
(high-octane component
of gasoline)

Note that the products, **$CO_2 + H_2O$,** are the same regardless of the identity of the starting material. Combustion of alkanes in the form of natural gas, gasoline, or heating oil releases energy for heating homes, powering vehicles, and cooking food.

When there is not enough oxygen available to completely burn a hydrocarbon, **incomplete combustion** may occur and **carbon monoxide** (CO) is formed instead of carbon dioxide (CO_2).

Incomplete combustion
$$2\ CH_4\ +\ 3\ O_2\ \xrightarrow{\text{flame}}\ 2\ CO\ +\ 4\ H_2O\ +\ \text{energy}$$
methane   carbon monoxide

Carbon monoxide is a poisonous gas that binds to hemoglobin in the blood, thus reducing the amount of oxygen that can be transported through the bloodstream to cells. CO can be formed whenever hydrocarbons burn. When an automobile engine burns gasoline, unwanted carbon monoxide can be produced. Yearly car inspections measure CO and other pollutant levels and are designed to prevent cars from emitting potentially hazardous substances into ambient air. Carbon monoxide is also formed when cigarettes burn, so heavy smokers have an unhealthy concentration of CO in their bloodstream.

Concept Check 10.11

(a) What products are formed when the ethane (CH_3CH_3) in natural gas undergoes combustion in the presence of a flame? (b) Write a balanced equation for the incomplete combustion of ethane (CH_3CH_3) to form carbon monoxide as one product.

CHAPTER REVIEW

KEY TERMS

Acyclic alkane (10.5)

Alcohol (10.4)

Aldehyde (10.4)

Alkane (10.4)

Alkene (10.4)

Alkyl group (10.6)

Alkyl halide (10.4)

Alkyne (10.4)

Amide (10.4)

Amine (10.4)

Amino group (10.4)

Aromatic compound (10.4)

Branched-chain alkane (10.5)

Carbonyl group (10.4)

Carboxyl group (10.4)

Carboxylic acid (10.4)

Combustion (10.10)

Condensed structure (10.3)

Constitutional isomer (10.5)

Cycloalkane (10.5)

Ester (10.4)

Ether (10.4)

Functional group (10.4)

Heteroatom (10.2)

Hydrocarbon (10.4)

Hydroxyl group (10.4)

Incomplete combustion (10.10)

Isomer (10.5)

IUPAC nomenclature (10.6)

Ketone (10.4)

Organic chemistry (10.1)

Parent name (10.6)

Pheromone (10.5)

Saturated hydrocarbon (10.5)

Skeletal structure (10.3)

Straight-chain alkane (10.5)

Thiol (10.4)

KEY CONCEPTS

1 **Characteristic features of organic compounds (10.1)**

1 **General features**	**2** **Properties**
• Organic compounds contain carbon atoms and most contain hydrogen atoms. • Carbon forms four bonds, including single, double, and triple bonds to itself and other atoms. • Carbon atoms can bond to form chains or rings. • Organic compounds often contain heteroatoms, commonly N, O, and the halogens.	• Organic compounds consist of **discrete molecules** held together by covalent bonds. • Organic compounds have weaker intermolecular forces than ionic compounds, resulting in **lower melting points** and **boiling points.**

2 **Identifying functional groups (10.4)**

A functional group is an atom or a group of atoms with characteristic chemical and physical properties. To locate functional groups, concentrate on multiple bonds and heteroatoms and refer to Tables 10.3, 10.4, and 10.5.

1 **Hydrocarbons (10.4A)**	**2** **Compounds with a C–Z single bond (10.4B)**	**3** **Compounds with a C=O group (10.4C)**
$CH_3CH_2CH_3$ alkane $CH_2{=}CH_2$ alkene $CH_3{-}C{\equiv}CH$ alkyne aromatic compound	$CH_3CH_2CH_2\mathbf{Cl}$ alkyl halide $CH_3CH_2CH_2\mathbf{SH}$ thiol $CH_3CH_2CH_2\mathbf{OH}$ alcohol $CH_3CH_2CH_2\mathbf{NH_2}$ amine $CH_3CH_2\mathbf{O}CH_3$ ether	$CH_3\overset{O}{\underset{H}{C}}$ aldehyde $CH_3\overset{O}{\underset{OH}{C}}$ carboxylic acid $CH_3\overset{O}{\underset{NH_2}{C}}$ amide $CH_3\overset{O}{\underset{CH_3}{C}}$ ketone $CH_3\overset{O}{\underset{OCH_3}{C}}$ ester • With carbonyl-containing functional groups, look at what is bonded to the **carbonyl carbon.**

3 **Facts about alkanes**

1 **General molecular formulas (10.5)**		**2** **Intermolecular forces and physical properties (10.9)**
Acyclic alkanes	**Cycloalkanes**	• Alkanes are **nonpolar** with weak intermolecular forces. • Alkanes have **low melting points** and **low boiling points** and are water **insoluble.** • Melting point and boiling point *increase* as the number of carbons *increases* due to increased surface area.
• C_nH_{2n+2} • saturated hydrocarbons $CH_3CH_2CH_3$ propane	• C_nH_{2n} • two fewer H atoms than acyclic alkanes △ cyclopropane	

KEY REACTIONS

① **Complete combustion (10.10)**	② **Incomplete combustion (10.10)**
$CH_4 + 2 O_2 \xrightarrow{\text{flame}} CO_2 + 2 H_2O$	$2 CH_4 + 3 O_2 \xrightarrow{\text{flame}} 2 CO + 4 H_2O$
• Alkanes burn in the presence of oxygen to form carbon dioxide (CO_2) and water.	• When oxygen supply is limited, alkanes react with oxygen to form carbon monoxide (CO) and water.

KEY SKILLS

①   **Converting a compound to a condensed structure (10.3A); example: isopropyl chloroacetate**

① **Proceed from left to right, making sure each C has four bonds.**	② **Omit lone pairs and use parentheses around like groups.**
 isopropyl chloroacetate	
• A C bonded to 3 H's becomes **CH₃**; a C bonded to 2 H's becomes **CH₂**; a C bonded to 1 H becomes **CH**.	

②   **Converting a skeletal structure to a complete structure (10.3B); example: 2-amino-4-ethyl-1-hexanol**

① **Place a C at the intersection of any two lines and at the end of any line.**	② **Add enough H's to make each C tetravalent.**	③ **Add lone pairs to give each heteroatom an octet.**
 2-amino-4-ethyl-1-hexanol		
• This molecule has eight carbons, labeled in red, green, and blue.	• Each C labeled in blue needs 3 H's to be tetravalent. • Each C in red needs 2 H's. • Each C in green needs 1 H.	• Add one lone pair to N and two lone pairs to O to complete the structure.

③   **Determining whether compounds are constitutional isomers (10.5)**

Constitutional isomers are compounds with the same molecular formula, which differ in the way the atoms are connected to each other.

① **Compare molecular formulas.**	② **Check how the atoms are connected to each other.**
 cyclopentane 1,2-dimethylcyclobutane **C₅H₁₀** **C₆H₁₂**	 1,2-dimethylcyclobutane methylcyclopentane **C₆H₁₂** **C₆H₁₂**
• Two compounds are isomers only if they have the *same* molecular formula. • Cyclopentane and 1,2-dimethylcyclobutane are *not* isomers of each other because they have *different* molecular formulas.	• Constitutional isomers have atoms bonded to *different* atoms. • 1,2-Dimethylcyclobutane and methylcyclopentane are constitutional isomers because they have the same molecular formula, but the connectivity of the atoms is different. One compound has a 4-C ring and one compound has a 5-C ring.

4 **Naming an alkane using the IUPAC system (10.6)**

1 Name the parent.	**2** Number the chain.	**3** Name and number the substituents.	**4** Combine the parts.
• Count the number of carbons in the longest chain. • Use the suffix -ane.	• Number to give the first substituent the lower number. • Numbering from left to right puts the first substituent at C**2**.	• This compound has three substituents—two methyl groups at C2 and an ethyl group at C4.	• Alphabetize the e of **ethyl** before the m of **methyl**. • Ignore the prefix **di-** in alphabetizing. **Answer:** 4-ethyl-2,2-dimethylhexane

6 C's in the chain ⟶ **hexane**

two methyls at C2
Use the prefix **di-**.
dimethyl

ethyl at C4

5 **Naming a cycloalkane (10.7)**

1 Name the ring.	**2** Name and number the substituents.
• The ring has 6 C's, so the molecule is named as a cyclohexane.	• There are two substituents: a 3-C **propyl** group and a 4-C **butyl** group. • Number to put the two groups at C1 and C4. • Place the butyl group at C1 because the b of **b**utyl comes before the p of **p**ropyl in the alphabet.

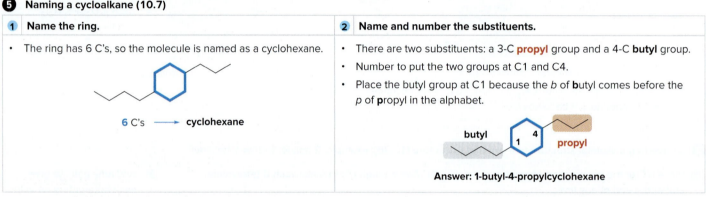

6 C's ⟶ **cyclohexane**

butyl

propyl

Answer: 1-butyl-4-propylcyclohexane

CHAPTER 10 SELF-TEST

The Self-Test consists of short-answer questions that test an understanding of definitions, equations, and other material encountered within the chapter. Answers to each question are provided at the end of the chapter. For any questions you are unable to answer correctly, revisit the appropriate section(s) of the text; then proceed to working the end-of-chapter problems to further reinforce your understanding of the chapter material.

1. Which term from "Key Terms" answers each question?
    a. What functional group contains an OH group bonded to a carbonyl carbon?
    b. What class of compounds contains only the elements of carbon and hydrogen?
    c. What type of shorthand structure for an organic compound is used for acyclic compounds but not for cyclic compounds?
    d. What class of organic compounds contains a sulfhydryl group (SH)?
    e. What term is used for any atom in an organic compound that is not carbon or hydrogen?

2. Identify each labeled functional group.

3. How many H atoms are bonded to each labeled carbon?

4. Which compound is *not* a representation for 2,5-dimethylheptane?

    a.

    c.

    b. $(CH_3)_2CHCH_2CH_2CH_2CH(CH_3)_2$

5. Label each pair of compounds as constitutional isomers, identical molecules, or not isomers of each other: (a) **A** and **B**; (b) **A** and **C**; (c) **A** and **D**; (d) **B** and **D**.

    $CH_3CH_2CH_2CO_2CH_3$ $(CH_3)_2CHOCH_2CH_2OCH_3$

    A   **C**

    $CH_3CH_2CO_2CH_2CH_3$

    B   **D**

6. Give the IUPAC name for the following compound:
    $(CH_3)_3CCH_2CH_2CH_2CH_2CH(CH_3)CH_2CH_3$.

UNDERSTANDING KEY CONCEPTS

10.1   Identify the functional group in each ball-and-stick model.

a.   b.

10.2   GHB is an addictive, illegal recreational drug that depresses the central nervous system and results in intoxication. Identify the functional groups in GHB.

GHB

10.3   Convert each shorthand structure to a complete structure with all atoms and lone pairs drawn in.
   a. $(CH_3)_2CH(CH_2)_6CH_3$ b. $(CH_3)_3COH$

10.4   Convert each shorthand structure to a complete structure with all atoms and lone pairs drawn in.
   a. $(CH_3)_2CHO(CH_2)_4CH_3$ b. $(CH_3)_3C(CH_2)_3CBr_3$

10.5   The waxy coating that covers tobacco leaves contains a straight-chain alkane having 31 carbons. How many hydrogens does this alkane contain?

Zbigniew Guzowski/Shutterstock

10.6   The waxy coating on the surface of apple skins contains an alkane with 27 carbon atoms. What is the molecular formula of this alkane?

lynx/iconotec/Glow Images

10.7   Draw three constitutional isomers having molecular formula C_7H_{14} that contain a five-membered ring and two CH_3 groups bonded to the ring. Give the IUPAC name for each isomer.

10.8   Draw four constitutional isomers having molecular formula C_6H_{12} that contain a four-membered ring. Give the IUPAC name for each isomer.

10.9   Answer the following questions about the alkane drawn below.

a. Give the IUPAC name.
b. Draw one constitutional isomer.
c. Predict the solubility in water.
d. Predict the solubility in an organic solvent.
e. Write a balanced equation for complete combustion.

10.10   Answer the questions in Problem 10.9 for the alkane drawn below.

10.11   Methyl salicylate is responsible for the characteristic odor of the oil wintergreen.

a. Give the molecular formula for methyl salicylate.
b. Draw in all lone pairs on heteroatoms using a skeletal structure.
c. Predict the water solubility of methyl salicylate.

methyl salicylate

10.12   Procaine (trade name Novocain) is a local anesthetic once commonly used in medicine and dentistry.

a. What is the molecular formula of procaine?
b. Draw in all the lone pairs on heteroatoms using a skeletal structure.
c. Identify the functional groups.

procaine

ADDITIONAL PROBLEMS

General Characteristics of Organic Molecules

10.13 Which molecular formulas represent organic compounds and which represent inorganic compounds: (a) H_2SO_4; (b) Br_2; (c) C_5H_{12}?

10.14 Which chemical formulas represent organic compounds and which represent inorganic compounds: (a) LiBr; (b) HCl; (c) CH_5N?

10.15 Complete each structure by filling in all H's and lone pairs.

a. C—C≡C—C≡C

c. C≡C—C—C
              |
              Cl

b. C≡C—C—O

d. C≡C—C—C—N
                    ‖
                    O

10.16 Complete the structure of mepivacaine by filling in all H's and lone pairs. Mepivacaine, sold under the trade name Carbocaine, is a fast-acting local anesthetic.

mepivacaine

Zoetis Services LLC

Properties of Organic Compounds

10.17 You are given two unlabeled bottles of solids, one containing sodium chloride (NaCl) and one containing cholesterol ($C_{27}H_{46}O$). You are also given two labeled bottles of liquids, one containing water and one containing dichloromethane (CH_2Cl_2). How can you determine which solid is sodium chloride and which solid is cholesterol?

10.18 State how potassium iodide (KI) and pentane ($CH_3CH_2CH_2CH_2CH_3$) differ in each of the following properties: (a) type of bonding; (b) solubility in water; (c) solubility in an organic solvent; (d) melting point; (e) boiling point.

10.19 Spermaceti wax [$CH_3(CH_2)_{14}CO_2(CH_2)_{15}CH_3$], a compound isolated from sperm whales, was once a common ingredient in cosmetics. Its use is now banned to help protect whales. (a) Identify the functional group in spermaceti wax. (b) Predict its solubility properties in water and organic solvents and explain your reasoning.

10.20 Acetic acid (CH_3CO_2H) and palmitic acid [$CH_3(CH_2)_{14}CO_2H$] are both carboxylic acids. Acetic acid is the water-soluble carboxylic acid in vinegar, whereas palmitic acid is a water-insoluble fatty acid. Give a reason for this solubility difference.

Drawing Organic Molecules

10.21 Convert each compound to a condensed structure.

a.

c.

b.

10.22 Convert each compound to a condensed structure.

a.

c.

b.

10.23 Convert each compound to a skeletal structure.

a.

b.

10.24 Convert each compound to a skeletal structure.

a.

b.

10.25 Convert each shorthand structure to a complete structure with all atoms and lone pairs drawn in.

a. $CH_3CO_2(CH_2)_3CH_3$

b.

10.26 Convert each shorthand structure to a complete structure with all atoms and lone pairs drawn in.

a. $(CH_3)_2CHCONH_2$

b.

10.27 Albuterol (trade names: Proventil and Ventolin) is a bronchodilator, a drug that widens airways, thus making it an effective treatment for individuals suffering from asthma. Draw a complete structure for albuterol with all atoms and lone pairs.

albuterol

10.28 Naproxen is the anti-inflammatory agent in Aleve and Naprosyn. Draw a complete structure for naproxen including all atoms and lone pairs.

naproxen

Functional Groups

10.29 Identify the functional groups in each molecule.

a.
isoprene
(emitted by plants)

b. propyl acetate
(from pears)

10.30 Identify the functional groups in each molecule.

a. tartaric acid
(from grapes)

b. butanedione
(component of butter flavor)

10.31 Identify the functional groups in each molecule.

a. $CH_2\!=\!CHCHCH_2\!-\!C\!\equiv\!C\!-\!C\!\equiv\!C\!-\!CH_2CH\!=\!CH(CH_2)_5CH_3$
           |
           OH

carotatoxin
(neurotoxin from carrots)

b.
vanillin
(isolated from vanilla beans)

10.32 (a) Identify the functional groups in donepezil, trade name Aricept, used to treat dementia and Alzheimer's disease. (b) What is the molecular formula of donepezil?

donepezil

Jill Braaten

10.33 Convert the ball-and-stick model of neral, a compound isolated from lemons and limes, to a condensed or skeletal structure and identify the functional groups.

neral

10.34 (a) Identify the functional groups in venlafaxine, trade name Effexor, an antidepressant. (b) What is the molecular formula of venlafaxine?

venlafaxine

Jb Reed/Bloomberg/Getty Images

10.35 Draw an organic compound that fits each of the following criteria.
  a. a hydrocarbon having molecular formula C_3H_4 that contains a triple bond
  b. an aldehyde containing three carbons
  c. a ketone having molecular formula C_4H_8O

10.36 Draw an organic compound that fits each of the following criteria.
  a. an amine containing two CH_3 groups bonded to the N atom
  b. an alkene that has the double bond in a ring
  c. an amide that has molecular formula C_3H_7NO

Alkanes and Isomers

10.37 Draw the structure of an alkane, an alkene, and an alkyne, each having five carbon atoms. What is the molecular formula for each compound?

10.38 An alkane has 20 hydrogen atoms. How many carbon atoms would it contain if it were (a) a straight-chain alkane; (b) a branched-chain alkane; (c) a cycloalkane?

10.39 Label each pair of compounds as constitutional isomers or identical molecules.

a. $CH_3CHCHCH_3$ and $CH_3CH_2CHCH_2CH(CH_3)_2$

b. $CH_3CH_2CHCHCH_2CH_3$ and $CH_3CHCHCH_3$

c. and

10.40 Label each pair of compounds as constitutional isomers, identical molecules, or not isomers of each other.
  a. $CH_3CH_2CH_2CH_3$ and $CH_3CH_2CH_2CH_2CH_3$

  b. and

  c. $CH_3\!-\!C$ and $CH_3CH_2\!-\!C$

10.41 Draw structures that fit the following descriptions:
  a. an ether and an alcohol that are constitutional isomers with molecular formula $C_5H_{12}O$
  b. two constitutional isomers of molecular formula C_3H_7Cl

10.42 Draw the five constitutional isomers having molecular formula C_6H_{14}.

10.43 Draw three constitutional isomers with molecular formula C_3H_6O. Draw the structure of one alcohol, one ketone, and one cyclic ether.

10.44 Draw one constitutional isomer of each compound.

a.   b.

Alkane Nomenclature

10.45 Give the IUPAC name for each compound.

a. $CH_3CH_2CHCH_2CH_2CH_2CH_2CH_2CH_3$
       |
       CH_3

b. $CH_3CH_2CHCH_2CHCH_2CH_2CH_3$
       | |
       CH_3 CH_3

c. $CH_3CH_2CH_2C(CH_2CH_3)_3$

d.

10.46 Give the IUPAC name for each compound.

a.
```
      H   CH3
      |   |
CH3—C—CHCH3
      |
      CH2CH3
```

b.

c.
```
        CH2CH2CH2CH3
        |
CH3CH2CHCH2CH2CH2CH3
```

d. $(CH_3CH_2)_2CHCH(CH_2CH_3)_2$

10.47 Give the IUPAC name for each compound.

a.   b.

10.48 Give the IUPAC name for each compound.

a.   b.

10.49 Give the IUPAC name for each cycloalkane.

a. (octagon)

b. (triangle)—CH₃

c.
```
        CH3
        |
      —CH3
        |
      —CH3
        |
      CH3
```

d. (cyclohexane with substituents)

10.50 Give the IUPAC name for each cycloalkane.

a. (cycloheptane with ethyl group)

b. (cyclohexane with substituents)

c. (cyclopentane with substituents)

d. (cyclohexane with substituents)

10.51 Give the structure corresponding to each IUPAC name.
a. 3-ethylhexane
b. 3-ethyl-3-methyloctane
c. 2,3,4,5-tetramethyldecane
d. 1,1,3-trimethylcyclohexane

10.52 Give the structure corresponding to each IUPAC name.
a. 3-ethyl-3-methylhexane
b. 4-ethyl-2,2-dimethyloctane
c. 1,3,5-triethylcycloheptane
d. 3-ethyl-3,4-dimethylnonane

10.53 Each of the following IUPAC names is incorrect. Explain why it is incorrect and give the correct IUPAC name.
a. 3-methylbutane   c. 1,3-dimethylbutane
b. 1-methylcyclopentane   d. 5-ethyl-2-methylhexane

10.54 Each of the following IUPAC names is incorrect. Explain why it is incorrect and give the correct IUPAC name.
a. 4-methylpentane   c. 1,3-dimethylpentane
b. 2,3,3-trimethylbutane   d. 3-methyl-5-ethylhexane

Physical Properties

10.55 Which compound has the higher melting point?

10.56 Which compound has the higher boiling point?

10.57 Explain why hexane is more soluble in dichloromethane (CH_2Cl_2) than in water.

10.58 Mineral oil and Vaseline are both mixtures of alkanes, but mineral oil is a liquid at room temperature and Vaseline is a solid. Which product is composed of alkanes that contain a larger number of carbon atoms? Explain your choice.

Reactions

10.59 Write a balanced equation for the combustion of each alkane: (a) CH_3CH_3; (b) $(CH_3)_2CHCH_2CH_3$.

10.60 Write a balanced equation for the combustion of each cycloalkane.

a. (hexagon)   b. (pentagon)

10.61 Write a balanced equation for the incomplete combustion of each alkane: (a) $CH_3CH_2CH_3$; (b) $CH_3CH_2CH_2CH_3$.

10.62 Benzene (C_6H_6) is a fuel additive sometimes used to make gasoline burn more efficiently. Write a balanced equation for the incomplete combustion of benzene.

Applications and General Problems

10.63 The gasoline industry seasonally changes the composition of gasoline in locations where it gets very hot in the summer and very cold in the winter. Gasoline is refined to contain a larger fraction of higher-molecular-weight alkanes in warmer weather. In colder weather, it is refined to contain a larger fraction of lower-molecular-weight alkanes. What is the purpose of producing different types of gasoline for different temperatures?

10.64 Polyethylene (Section 10.4) is a high-molecular-weight alkane that contains hundreds or even thousands of carbon atoms, bonded together in long carbon chains. When a new home is built, the concrete foundation is often wrapped with polyethylene (sold under the trade name of Tyvek). What purpose does the polyethylene serve?

10.65 Cabbage leaves are coated with a hydrocarbon of molecular formula $C_{29}H_{60}$. What purpose might this hydrocarbon coating serve?

10.66 Skin moisturizers come in two types. (a) One type of moisturizer is composed mainly of hydrocarbon material. Suggest a reason as to how a moisturizer of this sort helps to keep the skin from drying out. (b) A second type of moisturizer is composed mainly of propylene glycol [$CH_3CH(OH)CH_2OH$]. Suggest a reason as to how a moisturizer of this sort helps to keep the skin from drying out.

10.67 Answer the following questions for the cycloalkane depicted in the given ball-and-stick model.

a. Give the IUPAC name.
b. Draw one constitutional isomer.
c. Predict the solubility in water.
d. Predict the solubility in an organic solvent.
e. Draw a skeletal structure.

10.68 Answer the questions in Problem 10.67 for the cycloalkane depicted in the given ball-and-stick model.

CHALLENGE PROBLEMS

10.69 THC is the active component in marijuana (Section 10.4), and ethanol (CH_3CH_2OH) is the alcohol in alcoholic beverages. Which compound is more water soluble? Explain why drug screenings are able to detect the presence of THC but not ethanol weeks after these substances have been introduced into the body.

10.70 Cocaine is a widely abused, addicting drug. Cocaine is usually obtained as its hydrochloride salt (cocaine hydrochloride, an ionic salt). This salt can be converted to crack (a neutral molecule) by treatment with base.

cocaine (crack)
neutral organic molecule

cocaine hydrochloride
a salt

a. Identify the functional groups in cocaine.
b. Given what you have learned about ionic and covalent bonding, which of the two compounds—crack or cocaine hydrochloride—has a higher boiling point?
c. Which compound is more soluble in water?
d. Can you use the relative solubility to explain why crack is usually smoked but cocaine hydrochloride is injected directly into the bloodstream?

BEYOND THE CLASSROOM

10.71 Research one of the following drugs: propoxyphene, meperidine, salmeterol, sertraline, donepezil, or fexofenadine (or choose another drug). Give its trade name, chemical formula, and chemical structure. Identify all of the functional groups. What is the drug used for? Is its mechanism of action known?

10.72 Several localities now have waste-to-energy facilities—that is, plants that convert garbage to energy. What are the pros and cons of a garbage-burning facility? Research the nearest waste-burning plant near your home and determine how much garbage is combusted annually. If 27,000 lb of garbage provide the energy equivalent of 21 barrels of crude oil, how much oil does this translate into? What happens to the garbage that is not combustible as well as the ash that is produced after combustion? How much impact does a waste-burning facility have on the local landfill?

CHAPTER 10 ANSWERS

Concept Check Answers

10.1 a,e: organic   b,c,d,f: inorganic

10.2 a. C_4H_{10} b. C_3H_8O c. $C_4H_{10}O$

10.3 a. hydroxyl   b. alkene, aromatic ring   c. two amines

10.4 a. 8   b. 8   c. 18   d. 16

10.5 a,d: acyclic   b,c: cyclic

10.6 a,c,d: butane   b. isobutane

10.7 a. 7   b. 6   c. 10

10.8 a,c: decane   b,d: pentane

10.9 Because Vaseline is composed of nonpolar hydrocarbons, it is insoluble in a polar solvent like water and soluble in a weakly polar solvent like CH_2Cl_2.

10.10 a. 56   b. The nonpolar alkane keeps water inside.

10.11 a. CO_2 and H_2O
b. $2\ CH_3CH_3 + 5\ O_2 \longrightarrow 4\ CO + 6\ H_2O$

Practice Problem Answers

10.1 a.   d.

b.   e.

c.

10.2 a. $Br(CH_2)_2Br$
b. $CH_3CH_2O(CH_2)_2OH$
c. $CH_3CH_2C(CH_3)_3$

10.3 a.   c.

b.   d.

10.4 a.   c.

b.

10.5 $C_{16}H_{15}F_6N_5O$

10.6 a.   d.
aldehyde   ester

b.   e.
carboxylic acid   amide

c.
ketone

10.7 a,d: ether   b,c: ketone

10.8 amide, aromatic ring, ether, hydroxyl, amine

10.9 a. $CH_3CH_2OCH_2CH_3$, ether
b. $CH_3CH_2CH{=}CHCH_2CHO$, alkene, aldehyde

10.10 a. not isomers   b,c: isomers

10.11 $CH_3CHCHCH_3$ $CH_3CCH_2CH_3$

10.12 a. 3-methylpentane   c. 3-methylheptane
b. 2,4-dimethylhexane

10.13 a. 2,3-dimethylbutane
b. 4-ethyl-3,4-dimethyloctane
c. 4-ethyl-3-methylnonane

10.14 a. $CH_3CH_2CHCH_2CH_2CH_3$ c. $CH_3CH_2CHCH_2CCH_2CH_2CH_3$

b. $CH_3CH_2CCH_2CH_3$ d. $CH_3CH_2CHCHCH_2CH_3$

10.15 a. $CH_3{-}CCH_2CH_3$

b. $CH_3CH_2CHCH_2CH_2CH(CH_2)_3CH_3$

c. $CH_3CH_2CH_2C{-}CCH_2CH_2CH_2CH_3$

d. $CH_3CH_2CHCH_2CH(CH_2)_3CH_3$

10.16 a. methylcyclobutane
   b. 1,1-dimethylcyclohexane
   c. 1-ethyl-3-propylcyclopentane
   d. 1-ethyl-4-methylcyclohexane

10.17 a.   c.

   b.   d.

Self-Test Answers

1. a. carboxylic acid   c. condensed structure   e. heteroatom
   b. hydrocarbons   d. thiol

2. a. ether   c. ester   e. amide
   b. ketone   d. hydroxyl group

3. a. 2   b. 1   c. 0   d. 3

4. b

5. a,c: constitutional isomers   b. not isomers   d. identical

6. 2,2,8-trimethyldecane

Odd-Numbered Problem Answers

10.1 a. aldehyde   b. amine

10.3 a.   b.

10.5 64

10.7

1,2-dimethylcyclopentane   1,1-dimethylcyclopentane

1,3-dimethylcyclopentane

10.9 a. 3-ethyl-4,4-dimethylheptane
   b. $CH_3CH_2CH{-}C{-}C{-}CH_2CH_3$ with CH_3, CH_3, CH_2CH_3 groups
   c. not water soluble
   d. soluble in organic solvents
   e. $C_{11}H_{24} + 17\ O_2 \longrightarrow 11\ CO_2 + 12\ H_2O$

10.11 a. $C_8H_8O_3$ c. water soluble
   b.

10.13 a,b: inorganic   c. organic

10.15 a.   c.

   b.   d.

10.17 NaCl dissolves in water but not in dichloromethane, and cholesterol dissolves in dichloromethane but not in water.

10.19 a. Spermaceti wax is an ester.
   b. The very long hydrocarbon chains would make it insoluble in water and soluble in organic solvents.

10.21 a. $CH_3(CH_2)_2C(CH_3)_3$ c. $CH_3(CH_2)_2CHO$
   b. $(CH_3)_2CHCH_2O(CH_2)_2CH_3$

10.23 a.   b.

10.25 a.   b.

10.27

10.29 a. two alkenes   b. ester

10.31 a. two alkenes, two alkynes, one hydroxyl
   b. hydroxyl, ether, aldehyde, aromatic ring

10.33 $(CH_3)_2C{=}CH(CH_2)_2C{=}CHCHO$ with CH_3

   alkenes aldehyde

10.35 a. $HC{\equiv}C{-}CH_3$ b. CH_3CH_2CHO c. $CH_3CH_2CCH_3$ (with C=O)

10.37 alkane $CH_3CH_2CH_2CH_2CH_3$ C_5H_{12}
   alkene $CH_3CH_2CH{=}CHCH_3$ C_5H_{10}
   alkyne $CH_3CH_2C{\equiv}CCH_3$ C_5H_8

10.39 a,c: constitutional isomers   b. identical

10.41 a. $CH_3CH_2CH_2CH_2CH_2OH$ $CH_3OCH_2CH_2CH_2CH_3$
   b. $CH_3CH_2CH_2Cl$ CH_3CHCH_3 with Cl

10.43 �─OH CH_3CCH_3 (with C=O)   ─CH₃ (epoxide with CH₃)

10.45 a. 3-methyldecane   c. 3,3-diethylhexane
   b. 3,5-dimethyloctane   d. 4-ethyl-2,6-dimethyloctane

10.47 a. 4-methylheptane   b. ethylcyclopentane

10.49 a. cyclooctane

b. methylcyclopropane

c. 1,1,2,2-tetramethylcyclobutane

d. 1,3-diethylcyclohexane

10.51 a. $CH_3CH_2CHCH_2CH_2CH_3$
$\quad\quad\quad\quad\quad\quad\;|$
$\quad\quad\quad\quad\quad CH_2CH_3$

c.
$\quad\quad\quad\quad\quad CH_3\;\; CH_3$
$\quad\quad\quad\quad\quad\;|\quad\;\;|$
$\quad CH_3CHCHCHCHCH_2CH_2CH_2CH_2CH_3$
$\quad\quad\quad\quad\quad\quad\;|\quad\;\;|$
$\quad\quad\quad\quad\quad\quad CH_3\;\; CH_3$

b.
$\quad\quad\quad\quad\;\; CH_3$
$\quad\quad\quad\quad\;\;\;|$
$\quad CH_3CH_2CCH_2CH_2CH_2CH_2CH_3$
$\quad\quad\quad\quad\;\;\;|$
$\quad\quad\quad\quad CH_2CH_3$

d.

10.53 a. 2-methylbutane: Number to give CH_3 the lower number, 2 not 3.

b. methylcyclopentane: no number assigned if only one substituent

c. 2-methylpentane: five-carbon chain

d. 2,5-dimethylheptane: longest chain not chosen

10.55 $CH_3CH_2CH_2CH_3$

10.57 Hexane is a nonpolar hydrocarbon and is soluble in organic solvents but not in water.

10.59 a. $2\,CH_3CH_3 + 7\,O_2 \longrightarrow 4\,CO_2 + 6\,H_2O$

b. $(CH_3)_2CHCH_2CH_3 + 8\,O_2 \longrightarrow 5\,CO_2 + 6\,H_2O$

10.61 a. $2\,CH_3CH_2CH_3 + 7\,O_2 \longrightarrow 6\,CO + 8\,H_2O$

b. $2\,CH_3CH_2CH_2CH_3 + 9\,O_2 \longrightarrow 8\,CO + 10\,H_2O$

10.63 Higher-molecular-weight alkanes in warmer weather means less evaporation. Lower-molecular-weight alkanes in colder weather means the gasoline won't freeze.

10.65 The waxy coating will prevent loss of water from leaves and will keep leaves crisp.

10.67 a,e:

propylcyclopentane

b. $-CH_2CH_3$ (one possibility)
$\quad\quad CH_3$

c. insoluble

d. soluble

10.69 Ethanol is more water soluble, so it will be quickly excreted in the urine, which is mostly water. THC has many nonpolar C—C and C—H bonds that make it soluble in fatty tissues; therefore, it will persist in tissues for an extended period of time.

Design Credits: DNA molecule structure background for Chapter Opener: Watchara/Shutterstock

Unsaturated Hydrocarbons

11

Vinyl chloride is an alkene used to prepare poly(vinyl chloride), a high-molecular-weight organic compound used in blood bags, tubing, and materials needed for hemodialysis.
Hin255/iStock/Getty Images

CHAPTER GOALS

In this chapter you will learn how to:

1. Identify the three major types of unsaturated hydrocarbons—alkenes, alkynes, and aromatic compounds

2. Name alkenes, alkynes, and substituted benzenes

3. Recognize the difference between constitutional isomers and stereoisomers, as well as identify cis and trans isomers

4. Identify saturated and unsaturated fatty acids and predict their relative melting points

5. Draw the products of addition reactions of alkenes

6. Explain what products are formed when a vegetable oil is partially hydrogenated

7. Draw the structure of polymers formed from alkene monomers

In Chapter 11 we continue our study of hydrocarbons by examining three families of compounds that contain carbon–carbon multiple bonds. **Alkenes** contain a double bond and **alkynes** contain a triple bond. **Aromatic hydrocarbons** contain a benzene ring, a six-membered ring with three double bonds. These compounds differ from the alkanes of Chapter 10 because they each have a functional group, making them much more reactive. Thousands of biologically active molecules contain these functional groups, and many useful synthetic products result from their reactions.

11.1 Alkenes and Alkynes

Alkenes and alkynes are two families of organic molecules that contain multiple bonds.

- Alkenes are compounds that contain a carbon–carbon double bond.

The general molecular formula of an alkene is C_nH_{2n}, so an alkene has **two** fewer hydrogens than an acyclic alkane, which has a general molecular formula of C_nH_{2n+2}. Ethylene (C_2H_4) is the simplest alkene. Because each carbon of ethylene is surrounded by three atoms, each carbon is **trigonal planar.** All six atoms of ethylene lie in the same plane, and all bond angles are **120°.**

- Alkynes are compounds that contain a carbon–carbon triple bond.

The general molecular formula for an alkyne is C_nH_{2n-2}, so an alkyne has **four** fewer hydrogens than an acyclic alkane. Acetylene (C_2H_2) is the simplest alkyne. Each carbon of acetylene is surrounded by two atoms, making each carbon **linear** with bond angles of **180°.**

Because alkenes and alkynes are composed of nonpolar carbon–carbon and carbon–hydrogen bonds, their physical properties are similar to other hydrocarbons. Like alkanes:

- Alkenes and alkynes have low melting points and boiling points and are insoluble in water.

Recall from Chapter 10 that acyclic alkanes are called saturated hydrocarbons, because they contain the maximum number of hydrogen atoms per carbon. In contrast, **alkenes and alkynes are called *unsaturated* hydrocarbons.**

- Unsaturated hydrocarbons are compounds that contain fewer than the maximum number of hydrogen atoms per carbon.

CONSUMER NOTE

A ripe banana speeds up the ripening of green tomatoes because the banana gives off ethylene, a plant growth hormone. Fruit grown in faraway countries can be picked green, and then sprayed with ethylene when ripening is desired upon arrival at its destination.
Jill Braaten/McGraw Hill

Concept Check 11.1

Determine whether each molecular formula corresponds to a saturated hydrocarbon, an alkene, or an alkyne.

a. C_3H_6 b. C_5H_{12} c. C_8H_{14} d. C_6H_{12}

Concept Check 11.2

Give the molecular formula for each of the following compounds.

a. an alkene that has four carbons

b. a saturated hydrocarbon that has six carbons

c. an alkyne that has seven carbons

d. an alkene that has five carbons

The multiple bond of an alkene or alkyne is always drawn in a condensed structure. To translate a condensed structure to a complete structure with all bond lines drawn in, make sure that each carbon of a double bond has three atoms around it, and each carbon of a triple bond has two atoms around it, as shown in Sample Problem 11.1.

Sample Problem 11.1   **Interpreting a Condensed Structure of an Alkene and an Alkyne**

Draw a complete structure for each alkene or alkyne.

a. $CH_2\!\!=\!\!CHCH_2CH_3$

b. $CH_3C\!\!\equiv\!\!CCH_2CH_3$

Analysis

First, draw the multiple bond in each structure. Draw an alkene so that each C of the double bond has *three* atoms around it. Draw an alkyne so that each C of the triple bond has *two* atoms around it. All other C's have four single bonds.

Solution

a. The C labeled in red has single bonds to 2 H's. The C labeled in blue has single bonds to 1 H and a CH_2CH_3 group.

b. The C labeled in red has a single bond to a CH_3 group. The C labeled in blue has a single bond to a CH_2CH_3 group.

Practice Problem 11.1

Convert each condensed structure to a complete structure with all atoms, bond lines, and lone pairs drawn in.

a. $CH_2\!\!=\!\!CHCH_2OH$

b. $(CH_3)_2C\!\!=\!\!CH(CH_2)_2CH_3$

c. $(CH_3)_2CHC\!\!\equiv\!\!CCH_2C(CH_3)_3$

11.2 Nomenclature of Alkenes and Alkynes

Whenever we encounter a new functional group, we must learn how to use the IUPAC system to name it. In the IUPAC system:

- An alkene is identified by the suffix *-ene*.
- An alkyne is identified by the suffix *-yne*.

How To Name an Alkene or an Alkyne

Example Give the IUPAC name of each alkene and alkyne.

a. $CH_2=CHCHCH_3$ (with CH_3 group)

b. $CH_3CH_2CHC\equiv CCH_3$ (with CH_2CH_3 group)

Step [1] Find the longest chain that contains *both* carbon atoms of the double or triple bond.

a. $CH_2=CHCHCH_3$ (with CH_3 group)

4 C's in the longest chain

but**ane** ---→ but**ene**

- The compound is an alkene, so change the *-ane* ending of the parent alkane to *-ene*.

b. $CH_3CH_2CHC\equiv CCH_3$ (with CH_2CH_3 group)

6 C's in the longest chain

hex**ane** ---→ hex**yne**

- The compound is an alkyne, so change the *-ane* ending of the parent alkane to *-yne*.

Step [2] Number the carbon chain from the end that gives the multiple bond the lower number.

For each compound, number the chain and name the compound using the *first* number assigned to the multiple bond.

Start numbering here.

a. $CH_2=CHCHCH_3$ (with CH_3 group)

1   2 3 4

1-butene

- Numbering the chain from left to right puts the double bond at **C1** (not C3). The alkene is named using the *first* number assigned to the double bond, making it **1**-butene.

Start numbering here.

b. $CH_3CH_2CHC\equiv CCH_3$ (with CH_2CH_3 group)

6   5   4 3   2 1

2-hexyne

- Numbering the chain from right to left puts the triple bond at **C2** (not C4). The alkyne is named using the *first* number assigned to the triple bond, making it **2**-hexyne.

Step [3] Number and name the substituents, and write the name.

a. methyl group at C3

$CH_2=CHCHCH_3$

1   3

Answer: 3-methyl-1-butene

b. ethyl group at C4

$CH_3CH_2CHC\equiv CCH_3$

4   2

Answer: 4-ethyl-2-hexyne

A few simple alkenes and alkynes have names that do not follow the IUPAC system. The simplest alkene, $CH_2=CH_2$, is called *ethene* in the IUPAC system, but it is commonly called **ethylene**. The simplest alkyne, $HC\equiv CH$, is called *ethyne* in the IUPAC system, but it is commonly named **acetylene.** We will use these common names because they are more widely used than their systematic IUPAC names.

Sample Problem 11.2 | Naming an Alkene

Give the IUPAC name for each compound.

a. CH₃CH₂   CH₃
         \\   /
          C=C
         /   \\
     CH₃   CH₃

b.

Analysis and Solution

[1] Find the longest chain containing both carbon atoms of the multiple bond.

a. CH₃CH₂   CH₃
         \\   /
          C=C
         /   \\
     CH₃   CH₃

5 C's in the longest chain − − − → **pentene**

b.

7 C's in the longest chain − − − → **heptene**

[2] Number the chain to give the double bond the lower number.

a. CH₃CH₂   CH₃
   5   4 \\   /
            C=C
     CH₃ 3 2 CH₃
          1

- Numbering from right to left is preferred because the double bond begins at C2 (not C3). The molecule is named as a **2-pentene.**

b.

3-heptene

- Numbering from left to right places the double bond at C3 (not C4), so the molecule is named as a **3-heptene.**

[3] Name and number the substituents and write the complete name.

a. Use the prefix di- before methyl →
   2,3-**di**methyl.

CH₃CH₂   CH₃ methyl at C2
         \\   /
          C=C
     CH₃ 3 2 CH₃

methyl at C3

Answer: 2,3-dimethyl-2-pentene

b.
ethyl at C3

Answer: 3-ethyl-3-heptene

Practice Problem 11.2

Give the IUPAC name for each alkene.

a. CH₂=CHCHCH₂CH₃
           |
           CH₃

b. (CH₃CH₂)₂C=CHCH₂CH₂CH₃

c.

d.

Practice Problem 11.3

Give the IUPAC name for each alkyne.

a. CH₃CH₂CH₂CH₂CH₂C≡CCH(CH₃)₂

b. CH₃CH₂—C≡C—CH₂—C—CH₃
                        |
                        CH₂CH₃
                        |
                        CH₃

Practice Problem 11.4

Give the structure corresponding to each name.

a. 4-methyl-1-hexene

b. 5-ethyl-2-methyl-2-heptene

c. 2,5-dimethyl-3-hexyne

d. 4-ethyl-1-decyne

More Practice: Try Problems 11.3; 11.4; 11.5a; 11.6a; 11.17–11.19; 11.20a, b; 11.21a, b, c; 11.22a, c, d.

11.3   Cis–Trans Isomers

As we learned in Section 10.5 on alkanes, constitutional isomers are possible for alkenes of a given molecular formula. For example, there are three constitutional isomers for an alkene of molecular formula C_4H_8—1-butene, 2-butene, and 2-methylpropene.

$CH_2{=}CHCH_2CH_3$	$CH_3CH{=}CHCH_3$	$CH_2{=}CCH_3$ (with CH_3 above)
1-butene	2-butene	2-methylpropene

11.3A   Stereoisomers—A New Class of Isomer

2-Butene illustrates another important aspect about alkenes. There is **restricted rotation** around the carbon atoms of a double bond. As a result, the groups on one side of the double bond *cannot* rotate to the other side.

With 2-butene, there are two ways to arrange the atoms on the double bond. The two CH_3 groups can be on the *same side* of the double bond or they can be on *opposite sides* of the double bond. These molecules are *different* compounds with the same molecular formula; that is, they are **isomers.**

$CH_3CH{=}CHCH_3$

2-butene

cis isomer

skeletal structure

The two CH_3 groups are on the **same side** of the line that joins the doubly bonded C's.

trans isomer

The two CH_3 groups are on **opposite sides** of the line that joins the doubly bonded C's.

skeletal structure

- When the two CH_3 groups are on the *same side* of the double bond, the compound is called the *cis* isomer.
- When the two CH_3 groups are on *opposite sides* of the double bond, the compound is called the *trans* isomer.

Thus, one isomer of 2-butene is called ***cis*-2-butene,** and the other isomer is called ***trans*-2-butene.**

two CH_3 groups on the **same** side

two CH_3 groups on **opposite** sides

cis-**2-butene**   *trans*-**2-butene**

different compounds

Concept Check 11.3

Bombykol is secreted by the female silkworm moth (*Bombyx mori*) to attract mates. Bombykol contains two double bonds, and each double bond must have a particular three-dimensional arrangement of groups around it to be biologically active. Label the double bonds of bombykol as cis or trans.

bombykol

The cis and trans isomers of 2-butene are a specific example of a general class of isomer that occurs at carbon–carbon double bonds. **Whenever the two groups on *each* end of a C=C are *different from each other*, two isomers are possible.**

These two groups must be **different** from each other... ...*and* these two groups must also be **different** from each other.

When the two groups on one end of the double bond are identical, there is still restricted rotation, but no cis and trans isomers are possible. With 1-butene, $CH_2=CHCH_2CH_3$, one end of the double bond has two hydrogens, so the ethyl group (CH_2CH_3) is always cis to a hydrogen, no matter how the molecule is drawn.

two identical groups

1-butene   identical

No cis and trans isomers are possible.

Sample Problem 11.3

Drawing Cis and Trans Isomers

Draw *cis-* and *trans-*3-hexene.

Analysis

First, use the parent name to draw the carbon skeleton, and place the double bond at the correct carbon; 3-hexene indicates a 6-C chain with the double bond beginning at C3. Then use the definitions of cis and trans to draw the isomers.

Solution

Each C of the double bond is bonded to a CH_3CH_2 group and a hydrogen. A cis isomer has the CH_3CH_2 groups bonded to the *same* side of the double bond. A trans isomer has the two CH_3CH_2 groups bonded to the *opposite* sides of the double bond.

$CH_3CH_2CH=CHCH_2CH_3$
1 2 3   4 5 6

3-hexene   cis-3-hexene   trans-3-hexene

Practice Problem 11.5

Draw the structure of each compound: (a) *cis-*2-octene; (b) *trans-*3-heptene; (c) *trans-*4-methyl-2-pentene.

More Practice: Try Problems 11.1c, 11.2c, 11.20c, 11.21d, 11.22b, 11.23, 11.25–11.28.

Cis and trans compounds are isomers, but they are *not* constitutional isomers. Each carbon atom of *cis-* and *trans-*2-butene is bonded to the same atoms. The only difference is the three-dimensional arrangement of the groups around the double bond. Isomers of this sort are called **stereoisomers.**

- *Stereoisomers* are isomers that differ *only* in the 3-D arrangement of atoms.

Cis–trans isomers are also called **geometric isomers.**

Thus, *cis-*2-butene and *trans-*2-butene are a specific type of stereoisomer, but each of these compounds is a constitutional isomer of 1-butene, as shown in Figure 11.1. We will learn about another type of stereoisomer in Section 12.10.

Figure 11.1

Comparing Three Isomers: 1-Butene, *cis*-2-Butene, and *trans*-2-Butene

We have now learned the two major classes of isomers.

- Constitutional isomers differ in the way the atoms are bonded to each other.
- Stereoisomers differ only in the three-dimensional arrangement of atoms.

Concept Check 11.4

Label each pair of compounds as stereoisomers or constitutional isomers: (a) **A** and **B**; (b) **A** and **C**.

A   **B**   **C**

Concept Check 11.5

Double bonds in rings can also be characterized as cis or trans. Label each double bond in the following compounds as cis or trans.

a.   b.

11.3B FOCUS ON HEALTH & MEDICINE
Saturated and Unsaturated Fatty Acids

Naturally occurring animal fats and vegetable oils are formed from fatty acids. **Fatty acids are carboxylic acids (RCOOH) with long carbon chains of 12–20 carbon atoms.** Because a fatty acid has many nonpolar C—C and C—H bonds and few polar bonds, fatty acids are insoluble in water. There are two types of fatty acids.

- *Saturated* fatty acids have no double bonds in their long hydrocarbon chains.
- *Unsaturated* fatty acids have one or more double bonds in their long hydrocarbon chains.

Table 11.1 lists the structure and melting point of four fatty acids containing 18 carbon atoms. Stearic acid is one of the two most common saturated fatty acids, whereas oleic and linoleic acids are the most common unsaturated ones. One structural feature of unsaturated fatty acids is especially noteworthy.

- Generally, double bonds in naturally occurring fatty acids are cis.

The presence of cis double bonds affects the melting point of these fatty acids greatly.

- As the number of double bonds in the fatty acid *increases,* the melting point *decreases*.

Table 11.1 Common Saturated and Unsaturated Fatty Acids

Name	Structure	Mp (°C)
Stearic acid (0 C=C)		71
Oleic acid (1 C=C)		16
Linoleic acid (2 C=C)		−5
Linolenic acid (3 C=C)		−11

The cis double bonds introduce kinks in the long hydrocarbon chain, as shown in Figure 11.2. This makes it difficult for the molecules to pack closely together in a solid. **The *larger* the number of cis double bonds, the *more* kinks in the hydrocarbon chain, and the *lower* the melting point.**

Fats and oils are organic molecules synthesized in plant and animal cells from fatty acids. Fats and oils have different physical properties.

- Fats are solids at room temperature. Fats are generally formed from fatty acids having few double bonds.
- Oils are liquids at room temperature. Oils are generally formed from fatty acids having a larger number of double bonds.

Figure 11.2   The Three-Dimensional Structure of Four Fatty Acids

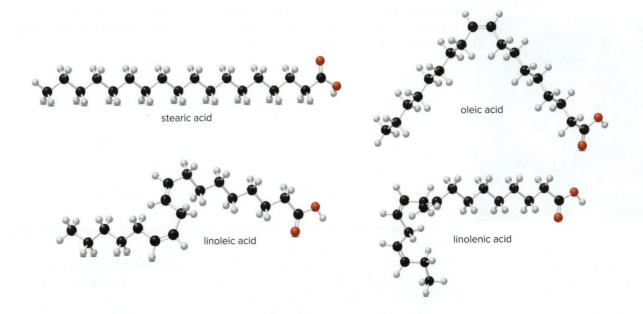

stearic acid

oleic acid

linoleic acid

linolenic acid

Saturated fats are typically obtained from animal sources, whereas unsaturated oils are common in vegetable sources. Thus, butter and lard are formed from saturated fatty acids, and olive oil and safflower oil are formed from unsaturated fatty acids. An exception to this generalization is coconut oil, which is composed largely of saturated fatty acids.

Considerable evidence suggests that an elevated cholesterol level is linked to increased risk of heart disease. Saturated fats stimulate cholesterol synthesis in the liver, resulting in an increase in cholesterol concentration in the blood. We will learn more about fats and oils in Section 11.6 and Chapter 15.

Concept Check 11.6

You have two fatty acids, one with a melting point of 63 °C, and one with a melting point of 1 °C. Which structure corresponds to each melting point?

palmitic acid   palmitoleic acid

11.4 FOCUS ON HEALTH & MEDICINE
Oral Contraceptives

The development of synthetic oral contraceptives in the 1960s revolutionized the ability to control fertility. Prior to that time, women ingested all sorts of substances—iron rust, gunpowder, tree bark, sheep's urine, elephant dung, and others—in the hope of preventing pregnancy.

Synthetic birth control pills are similar in structure to the female sex hormones **estradiol** and **progesterone,** but they also contain a carbon–carbon triple bond. Most oral contraceptives contain two synthetic hormones that are more potent than these natural hormones, so they can be administered in lower doses.

estradiol   progesterone

Two common components of birth control pills are **ethynylestradiol** and **norethindrone.** Ethynylestradiol is a synthetic estrogen that resembles the structure and biological activity of estradiol. Norethindrone is a synthetic progesterone that is similar to the natural hormone progesterone. These compounds act by artificially elevating hormone levels in a woman, and this prevents pregnancy, as illustrated in Figure 11.3.

Figure 11.3 How Oral Contraceptives Work

[The carbon–carbon triple bond is drawn in red.]

Monthly cycles of hormones from the pituitary gland cause ovulation, the release of an egg from an ovary. To prevent pregnancy, the two synthetic hormones in many oral contraceptives have different effects on the female reproductive system. **A:** The elevated level of **ethynylestradiol,** a synthetic estrogen, "fools" the pituitary gland into thinking a woman is pregnant, so ovulation does not occur. **B:** The elevated level of **norethindrone,** a synthetic progesterone, stimulates the formation of a thick layer of mucus in the cervix, making it difficult for sperm to reach the uterus.

11.5 Reactions of Alkenes

Most families of organic compounds undergo a characteristic type of reaction. **Alkenes undergo addition reactions.** In an addition reaction, new groups X and Y are added to a starting material. One bond of the double bond is broken and two new single bonds are formed.

- **Addition is a reaction in which elements are added to a compound.**

Addition

$$\text{C=C} \quad + \quad \text{X—Y} \quad \longrightarrow \quad \overset{|}{\underset{|}{\text{C}}}-\overset{|}{\underset{|}{\text{C}}}$$
$$\qquad\qquad\qquad\qquad\qquad\qquad\qquad\qquad \text{X} \quad \text{Y}$$

One bond is broken.   Two single bonds are formed.

Why does addition occur? A double bond is composed of one strong bond and one weak bond. In an addition reaction, **the weak bond is broken and two new strong single bonds are formed.** For example, alkenes react with hydrogen (H_2, Section 11.5A) and water (H_2O, Section 11.5B).

11.5A Addition of Hydrogen—Hydrogenation

Hydrogenation is the addition of hydrogen (H_2) to an alkene. Two bonds are broken—one bond of the carbon–carbon double bond and the H—H bond—and two new C—H bonds are formed.

Hydrogenation

$$\text{C=C} \quad + \quad \text{H—H} \quad \xrightarrow[\text{catalyst}]{\text{metal}} \quad \overset{|}{\underset{|}{\text{C}}}-\overset{|}{\underset{|}{\text{C}}}$$
$$\text{alkene}\qquad\qquad\qquad\qquad\qquad \text{H} \quad \text{H} \longleftarrow H_2 \text{ is added.}$$

alkane

The addition of H_2 occurs only in the presence of a **metal catalyst** such as palladium (Pd). The metal provides a surface that binds both the alkene and H_2, and this speeds up the rate of reaction. Hydrogenation of an alkene forms an **alkane** because the product has only C—C single bonds.

ethylene   ethane H_2 is added.

Sample Problem 11.4   **Drawing the Product of Hydrogenation**

Draw the product of the following reaction.

$$CH_3CH_2CH_2CH{=}CH_2 \;+\; H_2 \xrightarrow{\;Pd\;}$$

Analysis

To draw the product of a hydrogenation reaction:

- Locate the C=C and mentally **break one bond in the double bond.**
- Mentally **break the H—H bond** of the reagent.
- **Add one H atom to each C of the C=C,** thereby forming two new C—H single bonds.

Solution

Practice Problem 11.6   What product is formed when each alkene is treated with H_2 and a Pd catalyst?

a. $CH_3CH_2CH{=}CHCH_2CH_3$ b.   c.

More Practice: Try Problems 11.5b, 11.6b, 11.33, 11.35a, 11.36a.

11.5B Addition of Water—Hydration

Hydration is the addition of water to an alkene. Two bonds are broken—one bond of the carbon–carbon double bond and the H—OH bond—and new C—H and C—OH bonds are formed.

Hydration   alkene   alcohol H_2O is added.

Hydration occurs only if a strong acid such as H_2SO_4 is added to the reaction mixture. The product of hydration is an **alcohol.** For example, hydration of ethylene forms ethanol.

ethylene   ethanol

Gasohol contains 10% ethanol.
NoDerog/Getty Images

Ethanol is used as a solvent in many reactions in the laboratory. **Ethanol is also used as a gasoline additive** because, like alkanes, it burns in the presence of oxygen to form CO_2 and H_2O with the release of a great deal of energy. Although ethanol can also be formed by the fermentation of carbohydrates in grains and potatoes, much of the ethanol currently used in gasoline and solvent comes from the hydration of ethylene.

There is one important difference in this addition reaction compared to the addition of H_2. In this case, addition puts different groups—H and OH—on the two carbons of the double bond. As a result, H_2O can add to the double bond to give two constitutional isomers when an unsymmetrical alkene is used as starting material.

For example, the addition of H_2O to propene could in theory form two products. If H adds to the end carbon (labeled C1) and OH adds to the middle carbon (C2), 2-propanol is formed. If OH adds to the end carbon (C1) and H adds to the middle carbon (C2), 1-propanol is formed. In fact, **addition forms *only* 2-propanol.** This is a specific example of a general trend called **Markovnikov's rule,** named for the Russian chemist who first observed the selectivity in alkene additions.

- In the addition of H_2O to an unsymmetrical alkene, the H atom bonds to the less substituted carbon atom—that is, the carbon that has more H's to begin with.

The end carbon (C1) of propene has two hydrogens whereas the middle carbon (C2) has just one hydrogen. Addition puts the H of H_2O on C1 because it had more hydrogens (two versus one) to begin with.

Addition of H_2O to propene forms 2-propanol [$(CH_3)_2CHOH$], the main component of rubbing alcohol. *McGraw Hill*

This C has only 1 **H**, so the **OH** bonds here. propene. This C has 2 **H**'s, so the **H** bonds here. 2-propanol

Sample Problem 11.5 — **Drawing the Product of Hydration**

What product is formed when 1-methylcyclohexene is treated with H_2O in the presence of H_2SO_4?

1-methylcyclohexene + H_2O →(H_2SO_4)

Analysis

Alkenes undergo addition reactions, so the elements of H and OH must be added to the double bond. Because the alkene is unsymmetrical, the **H atom of H_2O bonds to the carbon that has more H's to begin with.**

Solution

This C has no H's, so the **OH** bonds here.

This C has 1 H, so the **H** bonds here.

Practice Problem 11.7

What product is formed when each alkene is treated with H_2O in the presence of H_2SO_4?

a. $CH_3CH=CHCH_3$

b. (structure with CH₃ groups on cyclohexene)

c. $(CH_3)_2C=CHCH_3$

More Practice: Try Problems 11.5c, 11.6c, 11.34, 11.35b, 11.36b, 11.37, 11.38.

11.6 FOCUS ON HEALTH & MEDICINE
Margarine or Butter?

One addition reaction of alkenes, hydrogenation, is especially important in the food industry. It lies at the heart of the debate over which product, butter or margarine, is better for the consumer.

As we learned in Section 11.3, butter is derived from saturated fatty acids like stearic acid $[CH_3(CH_2)_{16}COOH]$, compounds with long carbon chains that contain only carbon–carbon single bonds. As a result, butter is a *solid* at room temperature.

A portion of a fat molecule in butter

all C—C single bonds in the carbon chain

Fuse/Corbis/Getty Images

Margarine, on the other hand, is a synthetic product that mimics the taste and texture of butter. It is prepared from vegetable oils derived from unsaturated fatty acids like linoleic acid $[CH_3(CH_2)_4CH=CHCH_2CH=CH(CH_2)_7COOH]$. Margarine is composed mainly of *partially hydrogenated* vegetable oils formed by adding hydrogen to the double bonds in the carbon chain derived from unsaturated fatty acids.

When an unsaturated liquid vegetable oil is treated with hydrogen, some (or all) of the double bonds add H_2, as shown in Figure 11.4. This *increases* the melting point of the oil, thus giving it a *semi-solid* consistency that more closely resembles butter.

As we will learn in Section 15.4, unsaturated oils with carbon–carbon double bonds are healthier than saturated fats with no double bonds. Why, then, does the food industry partially hydrogenate oils, thus reducing the number of double bonds? The reasons relate to texture and shelf life. Consumers prefer the semi-solid consistency of margarine to a liquid oil. Imagine pouring vegetable oil on a piece of toast or pancakes!

Figure 11.4 Partial Hydrogenation of the Double Bonds in a Vegetable Oil

Unsaturated vegetable oil
- two C=C's
- **lower melting**
- **liquid** at room temperature

Add H₂ to one
C=C only.

H₂, Pd

Partially hydrogenated oil in margarine
- one C=C
- **higher melting**
- **semi-solid** at room temperature

- When an oil is *partially* hydrogenated, some double bonds react with H₂, whereas some double bonds remain in the product. Because the product has fewer double bonds, it has a higher melting point. Thus, a liquid oil is converted to a semi-solid.
(top & bottom): Jill Braaten/McGraw Hill

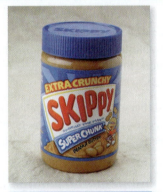
Furthermore, unsaturated oils are more susceptible than saturated fats to oxidation, which makes the oil rancid and inedible. Hydrogenating the double bonds reduces the likelihood of oxidation, thereby increasing the shelf life of the food product. This process reflects a delicate balance between providing consumers with healthier food products, while maximizing shelf life to prevent spoilage.

One other fact is worthy of note. During hydrogenation, some of the cis double bonds in vegetable oils are converted to trans double bonds, forming so-called "trans fats." The shape of the resulting fatty acid chain is very different, closely resembling the shape of a *saturated* fatty acid chain. As a result, trans fats are thought to have the same negative effects on blood cholesterol levels as saturated fats; that is, trans fats stimulate cholesterol synthesis in the liver, thus increasing blood cholesterol levels, a factor linked to increased risk of heart disease.

A "trans" fat and a saturated fat have a similar 3-D shape.

trans double bond

a trans fatty acid chain

similar shape

a saturated fatty acid chain

Candlenuts, known as kukui nuts in Hawaii, are rich in oils derived from unsaturated fatty acids, such as linoleic acid and linolenic acid (Concept Check 11.7).
inga spence / Alamy Stock Photo

Concept Check 11.7

Oil from the kukui nut contains linolenic acid (Table 11.1). (a) When linolenic acid is partially hydrogenated with one equivalent of H_2 in the presence of a Pd catalyst, three constitutional isomers are formed. One of them is linoleic acid. Draw the structures of the other two possible products. (b) What product is formed when linolenic acid is completely hydrogenated with three equivalents of H_2?

Concept Check 11.8

Conjugated linoleic acids (CLAs) are a group of unsaturated fatty acids that contain 18 carbons and two carbon–carbon double bonds separated by one single bond. Conjugated linoleic acids are found in meat and dairy products obtained from grass-fed sheep and cows. Some dietary supplements purported to reduce body fat contain CLAs. (a) Label the double bonds in conjugated linoleic acid **A** as cis or trans. (b) Would **A** have a higher or lower melting point than linoleic acid? (c) Is **A** a stereoisomer or constitutional isomer of linoleic acid?

A

11.7 Polymers—The Fabric of Modern Society

***Polymers* are large molecules made up of repeating units of smaller molecules—called *monomers*—covalently bonded together.** Polymers include the naturally occurring proteins that compose hair, tendons, and fingernails. They also include such industrially important plastics as polyethylene, poly(vinyl chloride) (PVC), and polystyrene. Since 1976, the U.S. production of synthetic polymers has exceeded its steel production.

11.7A Synthetic Polymers

Many synthetic polymers—that is, those synthesized in the lab—are among the most widely used organic compounds in modern society. Soft drink bottles, plastic bags, food wrap, compact discs, Teflon, and Styrofoam are all made of synthetic polymers. Figure 11.5 illustrates several consumer products and the polymers from which they are made.

To form a polymer from an alkene monomer, the weak bond that joins the two carbons of the double bond is broken and new strong carbon–carbon single bonds join the monomers together. For example, joining **ethylene monomers** together forms the polymer **polyethylene,** a plastic used in milk containers and sandwich bags.

HDPE (high-density polyethylene) and **LDPE** (low-density polyethylene) are common types of polyethylene prepared under different reaction conditions and having different physical properties. HDPE is opaque and rigid, and used in milk containers and water jugs. LDPE is less opaque and more flexible, and used in plastic bags and electrical insulation.
Jill Braaten

three monomer units joined together

[New bonds are shown in red.]

- *Polymerization* is the joining together of monomers to make polymers.

Figure 11.5 Polymers in Some Common Products

poly(vinyl chloride)
PVC

polyacrylonitrile

polyisobutylene

polystyrene
(Styrofoam)

(top left): John Thoeming/McGraw Hill; (top right): Dynamicgraphics/JupiterImages; (bottom left): Fernando Bengoechea/Getty Images; (bottom right): John Thoeming/McGraw Hill

Many ethylene derivatives having the general structure $CH_2{=}CHZ$ are also used as monomers for polymerization. Polymerization of $CH_2{=}CHZ$ usually yields polymers with the Z groups on every other carbon atom in the chain.

Repeating unit: shorthand structure

Many monomers are joined together.

Polymer structures are often abbreviated by placing the atoms in the repeating unit in brackets, as shown. Table 11.2 lists two common monomers and polymers used in medicine and dentistry.

Table 11.2 Common Monomers and Polymers Used in Medicine and Dentistry

Monomer ⟶	Polymer	Product
propene	polypropylene	polypropylene syringes
tetrafluoroethylene	polytetrafluoroethylene Teflon	dental floss

(top): Steve Allen/Shutterstock; (bottom): Jill Braaten

Sample Problem 11.6 — Drawing the Product of Polymerization

What polymer is formed when $CH_2{=}CHCl$ (vinyl chloride, mentioned in the chapter opener) is polymerized?

Analysis

Draw three or more alkene molecules and arrange the carbons of the double bonds next to each other. **Break one bond of each double bond,** and join the alkenes together with single bonds. With unsymmetrical alkenes, **substituents are bonded to every other carbon.**

Solution

New bonds are drawn in red.

Practice Problem 11.8

What polymer is formed when each compound is polymerized?

a. $CH_2{=}C$ with CH_3 and CH_2CH_3

b. $CH_2{=}C$ with CH_3 and CN

c. $CH_2{=}CH$ with aryl group bearing Cl

Practice Problem 11.9

Draw the structure of poly(acrylic acid), the polymer formed by polymerizing acrylic acid $(CH_2{=}CHCOOH)$.

More Practice: Try Problems 11.7, 11.8, 11.41–11.46.

HDPE

LDPE

11.7B FOCUS ON THE ENVIRONMENT
Polymer Recycling

The same desirable characteristics that make polymers popular materials for consumer products—durability, strength, and lack of reactivity—also contribute to environmental problems. Polymers do not degrade readily, and as a result, billions of pounds of polymers end up in landfills every year. Recycling existing polymer types to make new materials is one solution to the waste problem created by polymers.

Although thousands of different synthetic polymers have now been prepared, six compounds account for the bulk of the synthetic polymers produced in the United States each year. Each polymer is assigned a recycling code (1–6) that indicates its ease of recycling; **the lower the number, the easier it is to recycle.** Table 11.3 lists these six most common polymers, as well as the type of products made from each recycled polymer.

Of the most common polymers, only the polyethylene terephthalate (PET) in soft drink bottles and the high-density polyethylene (HDPE) in milk jugs and juice bottles are recycled to any great extent. Recycled HDPE is converted to Tyvek, an insulating wrap used in new housing construction, and outdoor furniture. Recycled PET is used to make fibers for fleece clothing and carpeting.

Table 11.3 Recyclable Polymers

Recycling Code	Polymer Name	Shorthand Structure	Recycled Product
1	PET polyethylene terephthalate	$\left[CH_2CH_2-O-\overset{O}{\underset{\parallel}{C}}-\bigcirc-\overset{O}{\underset{\parallel}{C}}-O \right]_n$	fleece jackets carpeting plastic bottles
2	HDPE high-density polyethylene	$\left[CH_2CH_2 \right]_n$	Tyvek insulation outdoor furniture sports clothing
3	PVC poly(vinyl chloride)	$\left[CH_2CH \atop Cl \right]_n$	floor mats
4	LDPE low-density polyethylene	$\left[CH_2CH_2 \right]_n$	trash bags
5	PP polypropylene	$\left[CH_2CH \atop CH_3 \right]_n$	furniture
6	PS polystyrene	$\left[CH_2CH \atop C_6H_5 \right]_n$	molded trays trash cans

11.8 Aromatic Compounds

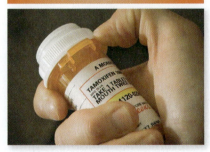

Tamoxifen, a potent anticancer drug sold under the trade name of Novaldex, contains three benzene rings.
Sergio Piumatti

OCH₂CH₂N(CH₃)₂ — OCH$_2$CH$_2$N(CH$_3$)$_2$

tamoxifen

CH₃CH₂ — CH$_3$CH$_2$

Aromatic compounds represent another example of unsaturated hydrocarbons. Aromatic compounds were originally named because many simple compounds in this family have characteristic odors. Today, the word **aromatic refers to compounds that contain a benzene ring,** or rings that react in a similar fashion to benzene.

Benzene, the simplest and most widely known aromatic compound, contains a six-membered ring and three double bonds. Because each carbon of the ring is also bonded to a hydrogen atom, the molecular formula for benzene is C_6H_6. Each carbon is surrounded by three groups, making it trigonal planar. Thus, **benzene is a planar molecule,** and all bond angles are **120°.**

benzene
C_6H_6

120°

planar molecule

Although benzene is drawn with a six-membered ring and three double bonds, there are two different ways to arrange the double bonds so that they alternate with single bonds around the ring. Each of these representations is equivalent.

This bond is a single bond in one structure and a double bond in the second structure.

The physical properties of aromatic hydrocarbons are similar to other hydrocarbons—they have low melting points and boiling points and are water insoluble.

11.9 Nomenclature of Benzene Derivatives

Many organic molecules contain a benzene ring with one or more substituents, so we must learn how to name them.

11.9A Monosubstituted Benzenes

To name a benzene ring with one substituent, **name the substituent and add the word** *benzene.* Carbon substituents are named as alkyl groups. When a halogen is a substituent, name the halogen by changing the *-ine* ending of the name of the halogen to the suffix *-o;* for example, chlor*ine* → chlor*o.*

—CH_2CH_3
ethyl group
ethylbenzene

—$CH_2CH_2CH_2CH_3$
butyl group
butylbenzene

—Cl
chloro group
chlorobenzene

Many monosubstituted benzenes, such as those with methyl (CH_3–), hydroxyl (–OH), and amino (–NH_2) groups, have common names that you must learn, too.

—CH_3
toluene
(methylbenzene)

—OH
phenol
(hydroxybenzene)

—NH_2
aniline
(aminobenzene)

11.9B Disubstituted Benzenes

There are three different ways that two groups can be attached to a benzene ring, so a prefix— **ortho, meta,** or **para**—is used to designate the relative position of the two substituents. Ortho, meta, and para are generally abbreviated as *o, m,* and *p,* respectively.

The pain reliever acetaminophen (trade name Tylenol) contains a para-disubstituted benzene ring.

Jill Braaten/McGraw Hill

acetaminophen
(Trade name: Tylenol)

1,2-Disubstituted benzene **ortho** isomer	**1,3-Disubstituted benzene** **meta** isomer	**1,4-Disubstituted benzene** **para** isomer
CH_2CH_3 CH_2CH_3	CH_2CH_3 CH_2CH_3	CH_2CH_3 CH_2CH_3
***o*-diethylbenzene** or **1,2-diethylbenzene**	***m*-diethylbenzene** or **1,3-diethylbenzene**	***p*-diethylbenzene** or **1,4-diethylbenzene**

If the two groups on the benzene ring are different, **alphabetize the name of the substituents** preceding the word benzene. If one of the substituents is part of a **common root,** name the **molecule as a derivative of that monosubstituted benzene.**

Alphabetize two different substituent names: **Use a common root name:**

o-**b**romo**c**hloro-benzene   m-**e**thyl**f**luoro-benzene   p-bromo**toluene**   o-chloro**phenol**

11.9C Polysubstituted Benzenes

For three or more substituents on a benzene ring:

> 1. Number to give the lowest possible numbers around the ring.
> 2. Alphabetize the substituent names.
> 3. When substituents are part of common roots, name the molecule as a derivative of that monosubstituted benzene. The substituent that comprises the common root is located at C1, but the "1" is omitted from the name.

- Assign the lowest set of numbers.
- Alphabetize the names of all the substituents.

4-**c**hloro-**1**-**e**thyl-**2**-**p**ropylbenzene

- Name the molecule as a derivative of the common root **aniline.**
- Designate the position of the NH_2 group as "1," and then assign the lowest possible set of numbers to the other substituents.

2,5-dichloro**aniline**

Naming Aromatic Compounds

Name each of the following aromatic compounds.

a.   b.

Analysis

- Name the substituents on the benzene ring.
- With two groups, alphabetize the substituent names and use the prefix **ortho, meta,** or **para** to indicate their location.
- With three substituents, alphabetize the substituent names, and **number to give the lowest set of numbers.**

Solution

a.

CH₃CH₂— ... —CH₂CH₂CH₃

ethyl group   propyl group

toluene

- The two substituents are located 1,3- or **meta** to each other.
- Alphabetize the e of **e**thyl before the p of **p**ropyl.

Answer: *m*-ethylpropylbenzene

b.

CH₃— ring with Cl (3), Br (4), position 1

- Because a CH₃– group is bonded to the ring, name the molecule as a derivative of toluene.
- Place the CH₃ group at the "1" position, and number to give the lowest set of numbers.

Answer: 4-bromo-3-chlorotoluene

Practice Problem 11.10

Give the IUPAC name of each compound.

a. (benzene ring with CH₂CH₂CH₃)

b. (benzene ring with CH₂CH₃ and I para)

c. (benzene ring with OH and CH₂CH₂CH₂CH₃)

d. (benzene ring with CH₃, Br, Cl)

Practice Problem 11.11

Draw the structure corresponding to each name.

a. pentylbenzene
b. *o*-dichlorobenzene

c. *m*-bromoaniline
d. 4-chloro-1,2-diethylbenzene

More Practice: Try Problems 11.9, 11.10, 11.47–11.54.

HEALTH NOTE

Commercial sunscreens are given an **SPF** rating (sun protection factor), according to the amount of sunscreen present. The higher the number, the greater the protection. *Jill Braaten/McGraw Hill*

11.10 FOCUS ON HEALTH & MEDICINE
Sunscreens and Antioxidants

11.10A Sunscreens

Many commercial sunscreens like oxybenzone contain a benzene ring. Oxybenzone effectively filters a broad spectrum of harmful ultraviolet radiation, but it can be washed off while swimming. Recent research at the University of Hawaii suggests that a minute amount of oxybenzone causes coral bleaching. Bleaching is a phenomenon that occurs when an external stress causes corals to expel symbiotic algae from their tissues, causing the coral to turn white. Because the algae provide the coral with most of its nutrients, an extended period of bleaching leads to starvation and decay. For this reason, the state of Hawaii now prohibits the sale of sunscreens that contain oxybenzone.

(structure of oxybenzone: OH, O, CH₃O)

oxybenzone

Concept Check 11.9

Which of the following compounds might be an ingredient in a commercial sunscreen? Explain why or why not.

a.

b.

11.10B   Phenols as Antioxidants

A wide variety of **phenols,** compounds that contain a hydroxyl group bonded to a benzene ring, occur in nature. **Vanillin** from the vanilla bean is a phenol, as is **curcumin,** a yellow pigment isolated from turmeric, a tropical perennial in the ginger family and a principal ingredient in curry powder. Curcumin has long been used as an anti-inflammatory agent in traditional eastern medicine.

vanilla bean

turmeric

vanillin

curcumin

(left): MonikaMaria/Shutterstock; (right): FLPA/Alamy Stock Photo

Many **phenols are antioxidants,** compounds that prevent unwanted oxidation reactions from occurring. Two examples are naturally occurring vitamin E and synthetic BHT. **The OH group on the benzene ring is the key functional group that prevents oxidation reactions from taking place.**

vitamin E

BHT
(**b**utylated **h**ydroxy **t**oluene)

Nuts are an excellent source of vitamin E. *Stockbyte/Corbis/Getty Images*

Vitamin E is a natural antioxidant found in fish oil, peanut oil, wheat germ, and leafy greens. Although the molecular details of its function remain obscure, it is thought that vitamin E prevents the unwanted oxidation of unsaturated fatty acid residues in cell membranes. In this way, vitamin E helps retard the aging process.

Synthetic antioxidants such as **BHT**—**b**utylated **h**ydroxy **t**oluene—are added to packaged and prepared foods to prevent oxidation and spoilage. BHT has been used as an additive in breakfast cereals.

Concept Check 11.10

Which of the following compounds might be antioxidants?

a. CH₃—⬡—CH(CH₃)₂

OH

menthol

b. curcumin

c. CH₃—⬡—CH₃

p-xylene
(gasoline additive)

CHAPTER REVIEW

KEY TERMS

Addition reaction (11.5)	Fat (11.3)	Monomer (11.7)	Polymer (11.7)
Alkene (11.1)	Fatty acid (11.3)	Oil (11.3)	Polymerization (11.7)
Alkyne (11.1)	Hydration (11.5)	Ortho isomer (11.9)	Stereoisomer (11.3)
Antioxidant (11.10)	Hydrogenation (11.5)	Para isomer (11.9)	Trans isomer (11.3)
Aromatic compound (11.8)	Markovnikov's rule (11.5)	Partial hydrogenation (11.6)	Unsaturated hydrocarbon (11.1)
Cis isomer (11.3)	Meta isomer (11.9)	Phenol (11.10)	

KEY CONCEPTS

① **Types of unsaturated hydrocarbons**

Unsaturated hydrocarbons are compounds that contain fewer than the maximum number of hydrogen atoms per carbon.

① **Alkenes (11.1)**	**②** **Alkynes (11.1)**	**③** **Aromatic compounds (11.8)**
• Alkenes contain a carbon–carbon double bond and have molecular formula C_nH_{2n}. • Each carbon of the double bond is trigonal planar.	• Alkynes contain a carbon–carbon triple bond and have molecular formula C_nH_{2n-2}. • Each carbon of the triple bond is linear.	• **Benzene**, molecular formula C_6H_6, is the most common aromatic hydrocarbon. • Benzene contains a six-membered ring and three double bonds. • Each carbon of benzene is trigonal planar.
H H C=C H H ethylene	H—C≡C—H acetylene	⬡ benzene

② **Constitutional isomers and stereoisomers compared (11.3)**

① **Constitutional isomers**	**②** **Stereoisomers**
• Constitutional isomers have atoms bonded to *different* atoms. • 1-Pentene and *cis*-2-pentene are constitutional isomers because they have the same molecular formula (C_5H_{10}), but the double bond is in a *different* location in the carbon chain.	• Stereoisomers differ only in the three-dimensional arrangement of atoms. • Cis and trans isomers are one type of stereoisomer. • *cis*-2-Pentene has two alkyl groups on the same side of the double bond, whereas *trans*-2-pentene has two alkyl groups on opposite sides.
  H CH₂CH₂CH₃ CH₃ CH₂CH₃ C=C C=C H H H H **1**-pentene *cis*-**2**-pentene	 CH₃ CH₂CH₃ H CH₂CH₃ C=C C=C H H CH₃ H *cis*-2-pentene *trans*-2-pentene two groups on the **same** side two groups on **opposite** sides

KEY REACTIONS

Addition reactions of alkenes (11.5)

Addition is a reaction in which elements are *added* to a compound. In addition, the weak bond of a double bond is broken and two new strong single bonds are formed.

KEY SKILLS

1 **Naming an alkyne using the IUPAC system (11.2)**

1 **Name the parent.**	**2** **Number the chain.**	**3** **Name and number the substituent and combine the parts.**

• Find the longest chain that contains both atoms of the multiple bond.
• Use the suffix *-yne*.
 	
$CH_3-C{\equiv}C-CH\,CH_3$
    $\|$
    $CH_2CH_2CH_3$
 	
7 C's in the longest chain ⟶ **heptyne**
	• Number to give the multiple bond the lower number.
	• Numbering from left to right puts the first carbon of the triple bond at C**2**.
	
	  1 **2** 3 4
	$CH_3-C{\equiv}C-CH\,CH_3$
	    $\|$
	    $CH_2CH_2CH_3$
	       7
	
	2-heptyne
		• This compound has one substituent, a methyl group at C4.
		
		         4
		$CH_3-C{\equiv}C-CH\,CH_3$ methyl at C4
		    $\|$
		    $CH_2CH_2CH_3$
		
		Answer: 4-methyl-2-heptyne

2 **Drawing the product of hydration using Markovnikov's rule (11.5B); example: reaction of $(CH_3)_2C{=}CHCH_2CH_3$ with H_2O, H_2SO_4**

3 **Drawing the polymer formed from a monomer (11.7); example: using $CH_2{=}CHCN$ in polymerization**

4   **Naming substituted benzenes (11.9)**

① **Naming a disubstituted benzene**	**②** **Naming a trisubstituted benzene**

①   **Naming a disubstituted benzene**

- Use the prefixes **ortho** (1,2-), **meta** (1,3-), or **para** (1,4-) to designate the position of the two groups.
- Alphabetize the names of the substituents.

Answer:
m-chloroethylbenzene

chloro group   **ethyl** group

two groups **meta**
(C1 and C3)

②   **Naming a trisubstituted benzene**

- Number to give the lowest possible numbers around the ring.
- Alphabetize the substituent names.
- When a substituent is part of a common root, name the molecule as a derivative of that monosubstituted benzene and place the substituent of the common root at C1.

Answer:
2-bromo-5-chlorophenol

phenol

CHAPTER 11 SELF-TEST

The Self-Test consists of short-answer questions that test an understanding of definitions, equations, and other material encountered within the chapter. Answers to each question are provided at the end of the chapter. For any questions you are unable to answer correctly, revisit the appropriate section(s) of the text; then proceed to working the end-of-chapter problems to further reinforce your understanding of the chapter material.

1. Which term from "Key Terms" answers each question?

    a. What type of hydrocarbon contains fewer than the maximum number of hydrogen atoms per carbon?

    b. What type of compound contains a hydroxyl group bonded to a benzene ring?

    c. What type of compound consists of repeating units of smaller molecules covalently bonded together?

    d. What type of compound contains a carboxyl group (CO_2H) bonded to a long chain of carbon atoms?

    e. What type of isomer differs only in the three-dimensional arrangement of atoms?

2. Which label—cis, trans, ortho, meta, or para—can be used for each of the following compounds?

    a. HO—⟨⟩—OH

    b. (structure)

    c. (structure with Br and CH_3)

    d. (structure with Cl, Cl, H, H)

3. What is the molecular formula for each compound: (a) an alkene with eight carbons; (b) an alkyne with four carbons; (c) an aromatic compound with two methyl groups bonded to a benzene ring?

4. Give the IUPAC name for each compound: (a) $CH_2{=}C(CH_2CH_2CH_3)_2$; (b) $CH_3CH_2C{\equiv}CCH_2CH_2CH(CH_2CH_3)_2$; (c) a benzene ring that contains two CH_3 groups bonded to adjacent carbons.

5. What product is formed when $CH_2{=}CHCH_2C(CH_3)_3$ is treated with each reagent: (a) H_2, Pd; (b) H_2O, H_2SO_4?

6. Draw the monomer used to form the following polymer:

(polymer structure with CH_3, O groups)

UNDERSTANDING KEY CONCEPTS

11.1 Anethole, the major constituent of anise oil, is used in licorice-flavored sweets and flavored brandy. Answer the following questions using the ball-and-stick model of anethole.

anethole

    a. What is the molecular formula of anethole?

    b. Identify the functional groups.

    c. Label the carbon–carbon double bond as cis or trans.

    d. How many trigonal planar carbons does anethole contain?

11.2   Methyl cinnamate, isolated from red clover flowers, is used in perfumery and as a flavoring agent because of its strawberry-like taste and odor. Answer the following questions using the ball-and-stick model of methyl cinnamate.

methyl cinnamate

a. What is the molecular formula of methyl cinnamate?

b. Identify the functional groups.

c. Label the carbon–carbon double bond as cis or trans.

d. Draw the structure of methyl cinnamate and label each carbon as trigonal planar or tetrahedral.

11.3   Give the IUPAC name for each compound.

a. $CH_2=CHCH_2CH_2C(CH_3)_3$ b. $CH_3C\equiv CCH_2C(CH_3)_3$

11.4   Give the IUPAC name for each compound.

a. $CH_2=CHCH_2CHCH_2CH_3$
                            |
                           CH_3

b. $CH_3C\equiv CCHCH_2CH_3$
                  |
          $CH_2CH_2CH_2CH_3$

11.5   Answer the following questions about compound **A**, represented in the given ball-and-stick model.

A

a. Convert **A** to a condensed structure and give its IUPAC name.

b. What product is formed when **A** is treated with H_2 in the presence of a metal catalyst?

c. What product is formed when **A** is treated with H_2O in the presence of H_2SO_4?

11.6   Answer the following questions about compound **B**, represented in the given ball-and-stick model.

B

a. Convert **B** to a condensed structure and give its IUPAC name.

b. What product is formed when **B** is treated with H_2 in the presence of a metal catalyst?

c. What product is formed when **B** is treated with H_2O in the presence of H_2SO_4?

11.7   What polymer is formed when $CH_2=CCl_2$ is polymerized?

11.8   What polymer is formed when $CH_2=CHNHCOCH_3$ is polymerized?

11.9   Draw the structure of the three constitutional isomers that have a Cl atom and an NH_2 group bonded to a benzene ring. Name each compound using the IUPAC system.

11.10   Draw the structure of 2,4,6-trichlorotoluene.

ADDITIONAL PROBLEMS

Alkene, Alkyne, and Benzene Structure

11.11   What is the molecular formula for a hydrocarbon with 10 carbons that is (a) completely saturated; (b) an alkene; (c) an alkyne?

11.12   Draw the structure of a hydrocarbon with molecular formula C_6H_{10} that also contains: (a) a carbon–carbon triple bond; (b) two carbon–carbon double bonds; (c) one ring and one C=C.

11.13   Draw structures for the three alkynes having molecular formula C_5H_8.

11.14   Draw the structures of the five constitutional isomers of molecular formula C_5H_{10} that contain a double bond.

11.15   Label each carbon in the following molecules as tetrahedral, trigonal planar, or linear.

a. [structure]   b. [structure]—$C\equiv CH$

11.16   Falcarinol is a natural pesticide found in carrots that protects them from fungal diseases. Predict the indicated bond angles in falcarinol.

falcarinol

Nomenclature of Alkenes and Alkynes

11.17   Give the IUPAC name for each molecule depicted in the ball-and-stick models.

a.   b.

11.18   Give the IUPAC name for each molecule depicted in the ball-and-stick models.

a.   b.

11.19 Give the IUPAC name for each compound.

a. (CH₃CH₂)₂C=CHCHCH₂CHCH₃ (with CH₃ CH₃ substituents)

c. CH₃C≡C−CH₂−CH−C−CH₂CH₃ (with CH₃ CH₃ and CH₂CH₃ substituents)

b.

11.20 Give the IUPAC name for each compound.

a. (CH₃)₂C=CHCH₂CHCH₂CH₃ (with CH₂CH₂CH₃ substituent)

c.

b. (CH₃)₃CC≡CC(CH₃)₃

11.21 Give the structure corresponding to each IUPAC name.
   a. 3-methyl-1-octene
   b. 2-methyl-3-hexyne
   c. 3,5-diethyl-2-methyl-3-heptene
   d. *cis*-7-methyl-2-octene

11.22 Give the structure corresponding to each IUPAC name.
   a. 6-ethyl-2-octyne
   b. *trans*-5-methyl-2-hexene
   c. 5,6-dimethyl-2-heptyne
   d. 3,4,5,6-tetramethyl-1-decyne

Isomers

11.23 Leukotriene C₄ is a key compound that causes the constriction of small airways, and in this way contributes to the asthmatic response. Label each double bond in leukotriene C₄ as cis or trans.

leukotriene C₄

11.24 Ichthyothereol is a toxic compound isolated from several Central and South American plants of the genus *Ichthyothere*. Identify the functional groups in ichthyothereol and label any alkene as cis or trans.

Mauricio Mercadante

ichthyothereol

11.25 Label the carbon–carbon double bond as cis or trans, and give the IUPAC name for the alkene depicted in the ball-and-stick model.

11.26 Label the carbon–carbon double bond as cis or trans, and give the IUPAC name for the alkene depicted in the ball-and-stick model.

11.27 Draw the cis and trans isomers for 2-methyl-3-heptene.

11.28 Draw the cis and trans isomers for 4,4-dimethyl-2-hexene.

11.29 How are the compounds in each pair related? Choose from constitutional isomers, stereoisomers, or identical.

a. [structure] and [structure]

b. [structure] and [structure]

11.30 Consider alkenes **A**, **B**, and **C**. How are the compounds in each pair related? Choose from constitutional isomers, stereoisomers, or identical: (a) **A** and **B**; (b) **A** and **C**; (c) **B** and **C**.

A   **B**   **C**

11.31 How are the compounds in each pair related? Choose from constitutional isomers, stereoisomers, or identical.

a. [ball-and-stick] and [ball-and-stick]

b. [ball-and-stick] and [ball-and-stick]

11.32 Consider compounds **A**, **B**, and **C**. Label each pair of compounds as constitutional isomers, stereoisomers, or not isomers of each other: (a) **A** and **B**; (b) **A** and **C**; (c) **B** and **C**.

A   **B**   **C**

Reactions of Alkenes

11.33 What alkane is formed when each alkene is treated with H_2 in the presence of a Pd catalyst?

a. $CH_2\!\!=\!\!CHCH_2CH_2CH_2CH_3$

c.

b. $(CH_3)_2C\!\!=\!\!CHCH_2CH_2CH_3$

d.

11.34 What alcohol is formed when each alkene is treated with H_2O and H_2SO_4?

a. (square)

c. $CH_2\!\!=\!\!CHCH_2CH(CH_3)_2$

b. $(CH_3)_2C\!\!=\!\!C(CH_3)_2$

d. (cyclohexane =CH_2)

11.35 Draw the product formed when 1-ethylcyclohexene is treated with each reagent: (a) H_2, Pd; (b) H_2O, H_2SO_4.

1-ethylcyclohexene

11.36 Draw the products formed in each reaction.

a. (cyclopentene) + H_2 →(Pd)→

b. (dimethyl-methylcyclopentene) + H_2O →(H_2SO_4)→

11.37 The hydration of 2-pentene ($CH_3CH\!\!=\!\!CHCH_2CH_3$) with H_2O and H_2SO_4 forms two alcohols. Draw the structure of both products and explain why more than one product is formed.

11.38 When myrcene is treated with three equivalents of H_2O in the presence of H_2SO_4, a single addition product of molecular formula $C_{10}H_{22}O_3$ is formed. Draw the structure of the product.

(myrcene structure)

myrcene

11.39 What alkene is needed as a starting material to prepare each of the following alcohols?

a. CH_3CH_2OH

b. (cyclohexanol with OH)

11.40 2-Butanol can be formed as the only product of the addition of H_2O to two different alkenes. In contrast, 2-pentanol can be formed as the *only* product of the addition of H_2O to just one alkene. Draw the structures of the alkene starting materials and explain the observed results.

$CH_3CHCH_2CH_3$ $CH_3CHCH_2CH_2CH_3$
| |
OH   OH

2-butanol   2-pentanol

Polymers

11.41 What is the structure of polychlorotrifluoroethylene, the polymer formed by polymerizing chlorotrifluoroethylene ($F_2C\!\!=\!\!CFCl$)? Polychlorotrifluoroethylene is used in packaging materials for some medications.

11.42 What polymer is formed when methyl α-methylacrylate [$CH_2\!\!=\!\!C(CH_3)CO_2CH_3$] is polymerized? This polymer is used in Lucite and Plexiglas, transparent materials that are lighter but more impact resistant than glass.

11.43 Draw the structure of each compound shown in the ball-and-stick models. Then draw the polymer that is formed when each compound is polymerized.

a. (ball-and-stick model)   b. (ball-and-stick model)

11.44 Draw the structure of each compound shown in the ball-and-stick models. Then draw the polymer that is formed when each compound is polymerized.

a. (ball-and-stick model)   b. (ball-and-stick model)

11.45 What monomer is used to form the following polymer?

11.46 What monomer is used to form the following polymer?

Nomenclature of Benzene

11.47 Name each aromatic compound.

a. (ball-and-stick model)   b. (ball-and-stick model)

11.48 Name each aromatic compound.

a. (ball-and-stick model)   b. (ball-and-stick model)

11.49 Give the IUPAC name for each substituted benzene.

a.

b.

c.

d.

11.50 Give the IUPAC name for each substituted benzene.

a.

b.

c.

d.

11.51 Give the structure corresponding to each IUPAC name.
   a. *m*-dibutylbenzene
   b. *o*-iodophenol
   c. 2-bromo-4-chlorotoluene
   d. 2-chloro-6-iodoaniline

11.52 Give the structure corresponding to each IUPAC name.
   a. *o*-difluorobenzene
   b. *p*-bromotoluene
   c. 1,3,5-tributylbenzene
   d. 2,4-dibromophenol

11.53 Each of the following IUPAC names is incorrect. Explain why it is incorrect and give the correct IUPAC name: (a) 5,6-dichlorophenol; (b) *m*-dibromoaniline.

11.54 Each of the following IUPAC names is incorrect. Explain why it is incorrect and give the correct IUPAC name: (a) 1,5-dichlorobenzene; (b) 1,3-dibromotoluene.

Applications

11.55 Rosemary contains the antioxidant rosmarinic acid. Identify each functional group that contains an OH group as an alcohol, carboxylic acid, or phenol.

rosmarinic acid

Pixtal/SuperStock

11.56 Why has BHA been used as an ingredient in some breakfast cereals and other packaged foods?

BHA
(**b**utylated **h**ydroxy **a**nisole)

11.57 Although nonpolar compounds tend to dissolve and remain in fatty tissues, polar substances are more water soluble, and more readily excreted into an environment where they may be degraded by other organisms. Explain why methoxychlor is more biodegradable than DDT (Problem 11.58).

CH_3O — — CCl_3 — — OCH_3

methoxychlor

11.58 Explain why the pesticide DDT is insoluble in water, but the herbicide 2,4-D is water soluble. 2,4-D is one component of the defoliant Agent Orange used extensively during the Vietnam War.

DDT   2,4-D

11.59 Macadamia nuts have a high concentration of unsaturated oils formed from palmitoleic acid [$CH_3(CH_2)_5CH=CH(CH_2)_7COOH$]. (a) Draw the structure of the naturally occurring fatty acid with a cis double bond. (b) Draw a stereoisomer of palmitoleic acid. (c) Draw a constitutional isomer of palmitoleic acid.

11.60 Eleostearic acid is an unsaturated fatty acid found in tung oil, obtained from the seeds of the tung oil tree (*Aleurites fordii*), a deciduous tree native to China. Eleostearic acid is unusual in that the double bond at C9 is cis, but the other two double bonds are trans. (a) Draw the structure of eleostearic acid, showing the arrangement of groups around each double bond. (b) Draw a stereoisomer of eleostearic acid in which all of the double bonds are trans. (c) Which compound, eleostearic acid or its all-trans isomer, has the higher melting point? Explain your reasoning.

$CH_3(CH_2)_3CH=CHCH=CHCH=CH(CH_2)_7CO_2H$
C9

eleostearic acid

11.61 Which of the following compounds might be an antioxidant?

a. —$CH_2CH_2CH_2OH$

b. —OCH_3

c. HO— —OCH_3

11.62 Which of the following compounds might be an ingredient in a commercial sunscreen?

a.

CH_3O—

b.

$CH_2CH_2CH_3$

CHALLENGE PROBLEMS

11.63 Are *cis*-2-hexene and *trans*-3-hexene constitutional isomers or stereoisomers? Explain.

11.64 Some polymers are copolymers, formed from two different alkene monomers joined together. An example is Saran, the polymer used in the well known plastic food wrap. What two alkene monomers combine to form Saran?

Saran

BEYOND THE CLASSROOM

11.65 Research the subject of using paper or plastic bags for supermarket purchases. What are the advantages and disadvantages of each? Consider the volume that the bags occupy and the raw materials used to make the bags. Also consider whether the bags decompose in a landfill, or whether they can be recycled. Draw the chemical structures for the principal organic compound in each bag type. In your opinion, which has the lesser negative environmental impact, using paper or plastic bags?

11.66 Pick a spice used routinely in your household. Possibilities might include vanilla, basil, hot pepper, thyme, mint, caraway, or cloves. Determine the major organic compound in the spice, give its name, and draw its chemical structure. Identify all the functional groups. Has the spice been used in any therapeutic or medicinal preparations in any societies in the world, and if it has, what are its purported health benefits? Have there been any experimental studies to support these claims?

CHAPTER 11 ANSWERS

Concept Check Answers

11.1 a,d: alkene   b. saturated hydrocarbon   c. alkyne

11.2 a. C_4H_8 b. C_6H_{14} c. C_7H_{12} d. C_5H_{10}

11.3 trans (left C=C), cis (right C=C)

11.4 a. constitutional isomers   b. stereoisomers

11.5 a. cis   b. trans (upper C=C), cis (right C=C)

11.6 palmitic acid, mp = 63 °C
palmitoleic acid, mp = 1 °C

11.7 a. $CH_3CH_2CH=CH(CH_2)_4CH=CH(CH_2)_7CO_2H$
$CH_3CH_2CH=CHCH_2CH=CH(CH_2)_{10}CO_2H$
b. $CH_3(CH_2)_{16}CO_2H$

11.8 a. trans, cis (closer to CO_2H)
b. higher mp
c. constitutional isomer

11.9 a.

11.10 b.

Practice Problem Answers

11.1 a.   c.

b.

11.2 a. 3-methyl-1-pentene   c. 4,5-dimethyl-1-octene
b. 3-ethyl-3-heptene   d. 2,3-dimethyl-2-hexene

11.3 a. 2-methyl-3-nonyne   b. 6,6-dimethyl-3-octyne

11.4 a. $CH_2=CHCH_2CH(CH_3)CH_2CH_3$
b. $(CH_3)_2C=CHCH_2CH(CH_2CH_3)_2$
c. $(CH_3)_2CHC\equiv CCH(CH_3)_2$
d. $HC\equiv CCH_2CH(CH_2CH_3)(CH_2)_5CH_3$

11.5 a.   c.

b.

11.6 a. $CH_3CH_2CH_2CH_2CH_2CH_3$ c.
b. $(CH_3)_2CHCH_2CH_2CH(CH_3)_2$

11.7 a.   c.

b.

11.8 a.

b.

c.

11.9

11.10 a. propylbenzene

b. *p*-ethyliodobenzene

c. *m*-butylphenol

d. 2-bromo-5-chlorotoluene

11.11 a.

b.

c.

d.

Self-Test Answers

1. a. unsaturated hydrocarbon   d. fatty acid

b. phenol   e. stereoisomer

c. polymer

2. a. para   b. trans   c. ortho   d. cis

3. a. C_8H_{16} b. C_4H_6 c. C_8H_{10}

4. a. 2-propyl-1-pentene   c. *o*-dimethylbenzene

b. 7-ethyl-3-nonyne

5. a. $CH_3CH_2CH_2C(CH_3)_3$ b. $CH_3CH(OH)CH_2C(CH_3)_3$

6. $CH_2{=}CHOCH_3$

Odd-Numbered Problem Answers

11.1 a. $C_{10}H_{12}O$ c. trans

b. aromatic ring, alkene, ether   d. eight

11.3 a. 5,5-dimethyl-1-hexene   b. 5,5-dimethyl-2-hexyne

11.5 a. $CH_2{=}CHCH_2CH_2CH_2CH_2CH_3$
       1-heptene

b. $CH_3(CH_2)_5CH_3$

c. $CH_3CH(OH)CH_2CH_2CH_2CH_2CH_3$

11.7

11.9

o-chloroaniline   m-chloroaniline   p-chloroaniline

11.11 a. $C_{10}H_{22}$ b. $C_{10}H_{20}$ c. $C_{10}H_{18}$

11.13 $HC{\equiv}CCH_2CH_2CH_3$ $CH_3C{\equiv}CCH_2CH_3$ $HC{\equiv}C{-}CH(CH_3)_2$

11.15 a.

b.

= trigonal planar   = tetrahedral   = linear

11.17 a. 2-ethyl-1-butene   b. 2-hexyne

11.19 a. 3-ethyl-5,7-dimethyl-3-octene

b. 3,4-diethyl-5-methyl-3-octene

c. 6-ethyl-5,6-dimethyl-2-octyne

11.21 a. $CH_2{=}CHCH(CH_3)(CH_2)_4CH_3$ c. $(CH_3CH_2)_2CHCH{=}C(CH_2CH_3)CH(CH_3)_2$

b. $CH_3CH_2C{\equiv}CCH(CH_3)_2$ d.

11.23 Left to right: two cis C=C, two trans C=C

11.25 *cis*-4-methyl-2-pentene

11.27

cis-2-methyl-3-heptene   trans-2-methyl-3-heptene

11.29 a. identical   b. stereoisomers

11.31 a. identical   b. constitutional isomers

11.33 a. $CH_3CH_2CH_2CH_2CH_2CH_3$ c.

b. $(CH_3)_2CHCH_2CH_2CH_2CH_3$ d.

11.35 a.

b.

11.37 $CH_3CH_2CHCH_2CH_3$ (OH) $CH_3CHCH_2CH_2CH_3$ (OH)

OH adds to either side of the C=C because both carbons have an equal number of H's.

11.39 a. $CH_2{=}CH_2$ b.

11.41

11.43 a. $CH_2=CHCH_2CH_3$

b.

11.45

11.47 a. *p*-chloroethylbenzene   b. *o*-bromofluorobenzene

11.49 a. *m*-bromochlorobenzene   c. *o*-butylethylbenzene

b. *o*-bromoaniline   d. 2,5-dichlorophenol

11.51 a.   c.

b.   d.

11.53 a. Assign the lowest numbers: 2,3-dichlorophenol.

b. You can only use ortho, meta, para for a disubstituted benzene: 3,5-dibromoaniline.

11.55 four phenols (OH's bonded to a benzene ring), one carboxylic acid (OH bonded to a C=O)

11.57 Methoxychlor is more water soluble. The OCH_3 groups can hydrogen bond to water. This increase in water solubility makes methoxychlor more biodegradable.

11.59 a.   c.

b.

11.61 c.

11.63 They are constitutional isomers because the double bond is located in a different place on the carbon chain.

12

Organic Compounds That Contain Oxygen, Halogen, or Sulfur

Ethanol, the alcohol in beer, wine, and other alcoholic beverages, is formed by the fermentation of carbohydrates in grapes, grains, and potatoes.
John A. Rizzo/Getty Images

CHAPTER OUTLINE

CHAPTER GOALS

In this chapter you will learn how to:

1. Identify alcohols, ethers, alkyl halides, thiols, aldehydes, and ketones

2. Determine the properties of each functional group

3. Name alcohols, ethers, alkyl halides, aldehydes, and ketones

4. Determine the products of alcohol dehydration and oxidation

5. Convert thiols to disulfides

6. Draw the products of oxidation reactions of aldehydes

7. Identify chirality centers and recognize when a molecule is chiral or achiral

8. Draw Fischer projection formulas

Chapter 12 concentrates on six families of compounds. **Alcohols (ROH), ethers (ROR), alkyl halides (RX, X = F, Cl, Br, or I),** and **thiols (RSH)** contain a carbon atom singly bonded to a heteroatom, whereas **aldehydes (RCHO)** and **ketones (RCOR)** contain a carbonyl group (C=O). Alcohols such as ethanol (CH_3CH_2OH) are widely occurring, and ethers are the most common anesthetics in use today. Alkyl halides are often used as solvents, and the —SH group of thiols plays an important role in protein chemistry. Many naturally occurring compounds are aldehydes and ketones, and these carbonyl-containing molecules serve as useful starting materials and solvents in industrial processes. In Chapter 12, we learn about the properties of these families of compounds.

12.1 Introduction

Chapter 12 focuses on families of organic compounds that contain the heteroatoms oxygen, halogen, or sulfur—alcohols, ethers, alkyl halides, thiols, aldehydes, and ketones.

Alcohols (ROH), ethers (ROR), alkyl halides (RX), and **thiols (RSH)** are four families of compounds that contain a carbon atom singly bonded to an oxygen, halogen, or sulfur.

| Alcohol | Ether | Alkyl halide | Thiol |

$CH_3CH_2—\ddot{O}—H$ $\quad$ $CH_3CH_2—\ddot{O}—CH_2CH_3$ $\quad$ $CH_3CH_2—\ddot{C}\ddot{l}:$ $\quad$ $CH_3CH_2—\ddot{S}—H$

ethanol $\qquad$ diethyl ether $\qquad$ chloroethane $\qquad$ ethanethiol

The characteristic odor of coffee is due to 2-mercaptomethylfuran, a compound that contains both an ether and a thiol (labeled in red). *Foodcollection*

- Alcohols contain a hydroxyl group (OH) bonded to a tetrahedral carbon.
- Ethers contain two alkyl groups bonded to an oxygen atom.
- Alkyl halides contain a halogen atom (X = F, Cl, Br, or I) bonded to a tetrahedral carbon.
- Thiols contain a sulfhydryl group (SH) bonded to a tetrahedral carbon.

2-mercaptomethylfuran

The oxygen atom in alcohols and ethers and the sulfur atom in thiols have two lone pairs of electrons, so each heteroatom is surrounded by eight electrons. The halogen atom in alkyl halides has three lone pairs of electrons.

Aldehydes (RCHO) and **ketones (RCOR)** contain a **carbonyl group (C=O)** with the carbonyl carbon bonded to carbon or hydrogen atoms.

| Aldehyde | Ketone |

acetaldehyde $\qquad\qquad$ acetone

Octanal [$CH_3(CH_2)_6CHO$] and decanal [$CH_3(CH_2)_8CHO$] are two aldehydes that contribute to the flavor and odor of an orange. *Ekaterina Semenets/Shutterstock*

- An aldehyde has at least one H atom bonded to the carbonyl carbon.
- A ketone has two alkyl groups bonded to the carbonyl carbon.

The double bond of a carbonyl group is usually omitted in shorthand structures. Acetaldehyde, for example, is written as CH_3CHO. Remember that the **H atom is bonded to the carbon atom,** not the oxygen. Likewise, acetone is written as CH_3COCH_3 or $(CH_3)_2CO$. Remember that **each compound contains a C=O.**

Concept Check 12.1

a. Label the hydroxyl groups, thiols, halogens, and ether oxygens in each compound.

b. Which –OH group in salmeterol (**C**) is *not* part of an alcohol? Explain.

A
chondrochole A
(isolated from red seaweed,
Chondrococcus hornemanni)

B
trans-2-butene-1-thiol
(constituent of skunk odor)

C
salmeterol
Trade name: Serevent
(dilates airways; used to treat asthma)

Concept Check 12.2

Draw out each compound to clearly show what groups are bonded to the carbonyl carbon. Label each compound as a ketone or aldehyde.

a. CH_3CH_2CHO b. $CH_3CH_2COCH_3$ c. $(CH_3)_3CCOCH_3$ d. $(CH_3CH_2)_2CHCHO$

12.2 Structure and Properties of Alcohols

Alcohols (ROH) are classified as **primary (1°), secondary (2°),** or **tertiary (3°)** based on the number of carbon atoms bonded to the carbon with the OH group.

- A primary (1°) alcohol has an OH group on a carbon bonded to one carbon.
- A secondary (2°) alcohol has an OH group on a carbon bonded to two carbons.
- A tertiary (3°) alcohol has an OH group on a carbon bonded to three carbons.

Sample Problem 12.1

Classifying an Alcohol as 1°, 2°, or 3°

Classify each alcohol as 1°, 2°, or 3°.

a. ⬡—CH₂CH₂OH   b. ⬡—OH

Analysis

To determine whether an alcohol is 1°, 2°, or 3°, locate the C with the OH group and **count the number of C's bonded to it.** A 1° alcohol has the OH group on a C bonded to one C, and so forth.

Solution

Draw out the structure or add H's to the skeletal structure to clearly see how many C's are bonded to the C bearing the OH group.

a.

This C is bonded to 1 C.
1° alcohol

b.

This C is bonded to 2 C's in the ring.
2° alcohol

Linalool and lavandulol (Practice Problem 12.1) are two of the over 300 compounds that determine the characteristic fragrance of lavender. *Zakharova Tatiana/Shutterstock*

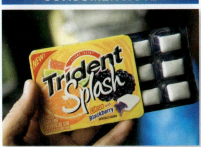

Sorbitol (Practice Problem 12.2) occurs naturally in some berries and fruits. It is used as a substitute sweetener in sugar-free—that is, sucrose-free—candy and gum. *Jill Braaten/McGraw Hill*

Practice Problem 12.1

Classify each alcohol as 1°, 2°, or 3°.

a. $(CH_3)_2C=CHCH_2CH_2-\overset{\overset{\displaystyle CH_3}{|}}{\underset{\underset{\displaystyle OH}{|}}{C}}-CH=CH_2$

linalool
(isolated from lavender)

b. $(CH_3)_2C=CHCH_2-\overset{\overset{\displaystyle CH_2OH}{|}}{\underset{\underset{\displaystyle H}{|}}{C}}-C(CH_3)=CH_2$

lavandulol
(isolated from lavender)

Practice Problem 12.2

Classify each hydroxyl group in sorbitol as 1°, 2°, or 3°.

sorbitol

More Practice: Try Problems 12.1, 12.3b, 12.4c, 12.7b, 12.8b, 12.13, 12.34a.

12.2A Physical Properties of Alcohols

An alcohol contains an oxygen atom with a **bent** shape like H_2O. The C—O—H bond angle is similar to the tetrahedral bond angle of 109.5°. **Alcohols are capable of intermolecular hydrogen bonding,** because they possess a hydrogen atom bonded to an oxygen. This gives alcohols much stronger intermolecular forces than the hydrocarbons of Chapters 10 and 11.

Alcohols exhibit intermolecular hydrogen bonding.

$CH_3-\ddot{\overset{\displaystyle ..}{O}}:$ $\underset{H}{}$ = hydrogen bond = $H-\overset{\displaystyle ..}{\underset{\displaystyle ..}{O}}{\,}_{CH_3}$

As a result:

- Alcohols have *higher* boiling points and melting points than hydrocarbons of comparable molecular weight and shape.

$CH_3CH_2CH_2CH_3$ $CH_3CH_2CH_2OH$ ← **stronger intermolecular forces higher boiling point and melting point**

butane   1-propanol

melting point: −138 °C   melting point: −127 °C
boiling point: −0.5 °C   boiling point: 97 °C

The general rule governing solubility—**"like dissolves like"**—explains the solubility properties of alcohols.

- Alcohols are soluble in organic solvents.
- Low-molecular-weight alcohols (those having less than six carbons) are soluble in water.
- Higher-molecular-weight alcohols (those having six carbons or more) are not soluble in water.

Thus, both ethanol (CH_3CH_2OH) and 1-octanol [$CH_3(CH_2)_7OH$] are soluble in organic solvents, but ethanol is water soluble and 1-octanol is not.

CH_3CH_2OH $CH_3CH_2CH_2CH_2CH_2CH_2CH_2CH_2OH$

only 2 C's in the chain   many nonpolar C–C and C–H bonds
water soluble   **water insoluble**

Concept Check 12.3

Which compound in each pair has the higher boiling point?

a. [structure with OH]   or   [structure with CH_3]   b. $(CH_3)_3C{-}OH$ or $(CH_3)_4C$

Concept Check 12.4

Label each compound as water soluble or water insoluble.

a. [bicyclic structure]   b. $(CH_3)_3C{-}OH$ c. $CH_2{=}CHCH_2CH_3$ d. [bicyclic structure with OH]

12.2B Nomenclature of Alcohols

In the IUPAC system, alcohols are identified by the suffix *-ol.* To name an alcohol:

- Find the longest carbon chain containing the carbon bonded to the OH group.
- Number the carbon chain to give the OH group the lower number, and apply all other rules of nomenclature.

When an OH group is bonded to a ring, the **ring is numbered beginning with the carbon bonded to the OH group,** and the "1" is usually omitted from the name. The ring is then numbered in a clockwise or counterclockwise fashion to give the next substituent the lower number.

[structure: CH_3 ... OH; positions 3 and 1]   [structure: OH; positions 1 and 2, CH_2CH_3]

3-methylcyclohexanol   **2-ethylcyclopentanol**

⎡ The OH group is at C1; the second substituent ⎤   ⎡ The OH group is at C1; the second substituent ⎤
⎣ (CH_3) gets the lower number.   ⎦   ⎣ (CH_3CH_2) gets the lower number.   ⎦

Sample Problem 12.2   **Naming an Alcohol**

Give the IUPAC name of the following alcohol.

$$CH_3$$
$$CH_3CH_2CCH_2CH_2OH$$
$$CH_3$$

Analysis and Solution

[1] Find the longest carbon chain that contains the carbon bonded to the OH group.

$$CH_3CH_2CCH_2CH_2{-}OH$$
(with CH₃ above and CH₃ below the third carbon)

- Change the **-e** ending of the parent alkane to the suffix **-ol.**

5 C's in the longest chain ----→ **pentanol**

[2] Number the carbon chain to give the OH group the lower number, and apply all other rules of nomenclature.

a. **Number** the chain.

$$CH_3CH_2CCH_2CH_2{-}OH$$
(CH₃ above; 5   3   CH₃   1 below)

- Number the chain to put the OH group at C**1**, not C5.

1-pentanol

b. **Name** and **number** the substituents.

two methyls at C3 ⟶ dimethyl

$$CH_3CH_2CCH_2CH_2{-}OH$$
(CH₃ above; 3   CH₃   1 below)

Answer: 3,3-dimethyl-1-pentanol

Practice Problem 12.3

Give the IUPAC name for each compound.

a. $CH_3CH(CH_2)_4CH_3$ (with OH above the second carbon)

b. $(CH_3CH_2)_2CHCHCH_2CH_3$ (with OH above)

c. cyclohexane with CH₃ and OH substituents

Practice Problem 12.4

Give the structure corresponding to each name.

a. 7,7-dimethyl-4-octanol

b. 5-methyl-4-propyl-3-heptanol

c. 2-ethyl-3-methylcyclohexanol

More Practice: Try Problems 12.3c; 12.7a; 12.8a; 12.17a; 12.18b; 12.19; 12.20; 12.21a, c; 12.22a, d; 12.23.

12.3 Structure and Properties of Ethers

Ethers **(ROR) are organic compounds that have two alkyl groups bonded to an oxygen atom.** The oxygen atom of an ether is surrounded by two carbon atoms and two lone pairs of electrons, giving it a **bent** shape like the oxygen in H_2O. The C—O—C bond angle is similar to the tetrahedral bond angle of 109.5°.

$$CH_3{-}\overset{..}{\underset{..}{O}}{-}CH_3 =$$ 109.5°

dimethyl ether   bent shape

Simple ethers are usually assigned common names. To do so, **name both alkyl groups** bonded to the oxygen, arrange these names alphabetically, and add the word **ether.** For ethers with identical alkyl groups, name the alkyl group and add the prefix **di-.**

$$CH_3CH_2CH_2{-}O{-}CH_2CH_3$$

propyl   ethyl

ethyl propyl ether

$$CH_3CH_2{-}O{-}CH_2CH_3$$

ethyl   ethyl

diethyl ether

Name each ether.

a. CH_3—O—$CH_2CH_2CH_2CH_3$ b. $CH_3CH_2CH_2$—O—$CH_2CH_2CH_3$

Draw the structure of the three constitutional isomers of molecular formula $C_4H_{10}O$ that contain an ether.

Because oxygen is more electronegative than carbon, the C—O bonds of an ether are both polar. An ether contains two polar bonds and a bent shape, so it has a **net dipole.**

Ethers do not contain a hydrogen atom bonded to oxygen, so unlike alcohols, two ether molecules *cannot* intermolecularly hydrogen bond to each other. **This gives ethers stronger intermolecular forces than alkanes but weaker intermolecular forces than alcohols.** As a result:

- Ethers have higher melting points and boiling points than hydrocarbons of comparable molecular weight and shape.
- Ethers have lower melting points and boiling points than alcohols of comparable molecular weight and shape.

$CH_3CH_2CH_2CH_3$	$CH_3OCH_2CH_3$	$CH_3CH_2CH_2OH$
butane	ethyl methyl ether	1-propanol
boiling point −0.5 °C	boiling point 11 °C	boiling point 97 °C

Increasing boiling point →

All ethers are soluble in organic solvents. Like alcohols, **low-molecular-weight ethers are water soluble,** because the oxygen atom of the ether can hydrogen bond to one of the hydrogens of water. When the alkyl groups of the ether have more than a total of five carbons, the nonpolar portion of the molecule is too large, so the ether is water insoluble.

Determining the Relative Boiling Point of an Alcohol, Ether, and Alkane

Rank the following compounds in order of increasing boiling point:

Analysis

Look at the functional groups to determine the strength of the intermolecular forces—**the *stronger* the forces, the *higher* the boiling point.**

Solution

B is an alkane with nonpolar C—C and C—H bonds, so it has the weakest intermolecular forces and therefore the lowest boiling point. **C** is an alcohol capable of intermolecular hydrogen bonding, so it has the strongest intermolecular forces and the highest boiling point. **A** is an ether, so it contains a net dipole but is incapable of intermolecular hydrogen bonding. **A**, therefore, has intermolecular forces of intermediate strength and has a boiling point between the boiling points of **B** and **C.** Thus, the relative boiling points are **B < A < C.**

Practice Problem 12.5	Which compound in each pair has the higher boiling point?

a. $CH_3(CH_2)_6CH_3$ or $CH_3(CH_2)_5OCH_3$ c. $CH_3(CH_2)_6OH$ or $CH_3(CH_2)_5OCH_3$

b. [hexane ring] or [oxane ring with O]   d. $CH_3(CH_2)_5OCH_3$ or CH_3OCH_3

More Practice: Try Problems 12.9, 12.10, 12.27.

12.4   Interesting Alcohols and Ethers

12.4A   Simple Alcohols

The most well known alcohol is ethanol, CH_3CH_2OH. **Ethanol** (Figure 12.1), formed by the fermentation of carbohydrates in grains and grapes, is the alcohol present in alcoholic beverages. Fermentation requires yeast, which provides the needed enzymes for the conversion. Ethanol is likely the first organic compound synthesized by humans, because alcohol has been produced for at least 4,000 years. Two other simple alcohols are listed in Figure 12.2.

Figure 12.1   Ethanol—The Alcohol in Alcoholic Beverages

Fermentation

$$C_6H_{12}O_6 \xrightarrow{\text{yeast}} 2\ CH_3CH_2OH\ +\ 2\ CO_2$$
glucose   ethanol

CH_3CH_2OH
ethanol

Ethanol is the alcohol in red wine, obtained by the fermentation of grapes. All alcoholic beverages are mixtures of ethanol and water in various proportions. Beer has 3–8% ethanol, wines have 10–14% ethanol, and other liquors have 35–90% ethanol. *(left): Dynamic Graphics Group/Getty Images; (middle): Goodshoot/Getty Images; (right): Isadora Getty Buyou/Image Source*

Figure 12.2 Two Simple Alcohols

(CH₃)₂CHOH

• **2-Propanol [(CH₃)₂CHOH, isopropyl alcohol]** is the major component of rubbing alcohol. 2-Propanol is used to clean skin before medical procedures and to sterilize medical instruments.

HOCH₂CHCH₂OH
|
OH

• **Glycerol [(HOCH₂)₂CHOH]** is used in lotions, liquid soap, and shaving cream. It is sweet tasting and nontoxic, so it is also an additive in candy and some prepared foods. Its three OH groups readily hydrogen bond to water, so it helps to retain moisture in these products.

12.4B FOCUS ON HEALTH & MEDICINE
Ethers as Anesthetics

A general anesthetic is a drug that interferes with nerve transmission in the brain, resulting in a loss of consciousness and the sensation of pain. The discovery that **diethyl ether** (CH₃CH₂OCH₂CH₃) is a general anesthetic dramatically changed surgery in the nineteenth century. Diethyl ether is an imperfect anesthetic, but considering the alternatives at the time, it was considered revolutionary. It is safe and easy to administer with low patient mortality, but it is highly flammable, and it causes nausea in many patients.

For these reasons, alternatives to diethyl ether are now widely used. Many of these newer general anesthetics, which cause little patient discomfort, are also ethers. These include isoflurane, enflurane, and methoxyflurane. Replacing some of the hydrogen atoms in the ether by halogens results in compounds with similar anesthetic properties but decreased flammability.

isoflurane
(Trade name: Forane)

enflurane
(Trade name: Ethrane)

methoxyflurane
(Trade name: Penthrane)

Concept Check 12.7

desflurane

(a) Translate the ball-and-stick model of desflurane, a potent general anesthetic, into a condensed structure. (b) The solubility properties of desflurane resemble other general anesthetics, in that it readily crosses the nonpolar medium of the blood–brain barrier. Explain why desflurane is soluble in organic solvents and other organic media.

12.5 Reactions of Alcohols

Alcohols undergo two useful reactions—**dehydration** and **oxidation.**

12.5A Dehydration

When an alcohol is treated with a strong acid such as H_2SO_4, the elements of water are lost and an alkene is formed as product. **Loss of H_2O from a starting material is called** *dehydration.* Dehydration takes place by breaking bonds on two adjacent atoms—the C—OH bond and an adjacent C—H bond.

Dehydration is an example of a general type of organic reaction called an **elimination reaction.**

- *Elimination* is a reaction in which elements of the starting material are "lost" and a new multiple bond is formed.

For example, dehydration of ethanol (CH_3CH_2OH) with H_2SO_4 forms ethylene (CH_2=CH_2), as shown. To draw the product of any dehydration, **remove the elements of H and OH from two adjacent atoms and draw a carbon–carbon double bond in the product.**

Sample Problem 12.4	Drawing the Product of Dehydration

Draw the product formed when cyclohexanol is dehydrated with H_2SO_4.

cyclohexanol

Analysis

To draw the products of dehydration, find the carbon bonded to the OH group. **Remove the elements of H and OH from two adjacent C's, and draw a double bond between these C's in the product.**

Solution

Elimination of H and OH from two adjacent atoms forms cyclohexene.

cyclohexene

Practice Problem 12.6

Draw the product formed when each alcohol is dehydrated with H_2SO_4.

More Practice: Try Problems 12.7c, 12.8c, 12.29, 12.30, 12.35a, 12.36a.

12.5B Oxidation

To determine if an organic compound has been oxidized, we compare the relative number of C—H and C—O bonds in the starting material and product.

- Oxidation results in an *increase* in the number of C—O bonds or a *decrease* in the number of C—H bonds.

Thus, an organic compound like CH_4 can be oxidized by replacing C—H bonds with C—O bonds, as shown in Figure 12.3. Organic chemists use the symbol [O] to indicate oxidation.

Figure 12.3 A General Scheme for Oxidizing an Organic Compound

4 C—H bonds   **3** C—H bonds   **2** C—H bonds   **1** C—H bond   **0** C—H bonds
0 C—O bonds   **1** C—O bond   **2** C—O bonds   **3** C—O bonds   **4** C—O bonds

Increasing number of C—O bonds

Alcohols can be oxidized to a variety of compounds, depending on the type of alcohol and the reagent. **Oxidation occurs by replacing the C—H bonds *on the carbon bearing the OH group* by C—O bonds.** All oxidation products from alcohol starting materials contain a **C=O, a carbonyl group.**

A common reagent for alcohol oxidation is **potassium dichromate, $K_2Cr_2O_7$,** a red-orange solid. Oxidation with this chromium reagent is characterized by a color change, as the **red-orange reagent** is reduced to a **green Cr^{3+} product.** Because a variety of reagents can be used, we will often write oxidations with the general symbol [O].

- Primary (1°) alcohols are first oxidized to aldehydes (RCHO), which are further oxidized to carboxylic acids (RCOOH) by replacing one and then two C—H bonds by C—O bonds.

ethanol   acetaldehyde   acetic acid

Oxidation of one C—H bond of ethanol forms acetaldehyde. Because acetaldehyde still contains a hydrogen atom on the carbonyl carbon, converting this C—H bond to a C—O bond forms acetic acid, a carboxylic acid with three C—O bonds.

- Secondary (2°) alcohols are oxidized to ketones (R₂CO), by replacing one C—H bond by one C—O bond.

2-propanol　　　　　　　　acetone

Because 2° alcohols have only one hydrogen atom bonded to the carbon with the OH group, they can be oxidized to only one type of compound, a ketone. Thus, 2-propanol is oxidized to acetone.

- Tertiary 3° alcohols have no H atoms on the carbon with the OH group, so they are *not* oxidized.

no C—H bond

3° alcohol

Sample Problem 12.5　　**Drawing the Product of Alcohol Oxidation**

Draw the carbonyl products formed when each alcohol is oxidized.

a. 　　b. $CH_3CH_2CH_2CH_2OH$

　cyclohexanol　　　　　　　1-butanol

Analysis

Classify the alcohol as 1°, 2°, or 3° by drawing in all of the H atoms on the C with the OH. Then **concentrate on the C with the OH group** and replace H atoms by bonds to O. Keep in mind:

- **RCH$_2$OH** (1° alcohols) are oxidized to **RCHO**, which are then oxidized to **RCOOH**.
- **R$_2$CHOH** (2° alcohols) are oxidized to **R$_2$CO**.

Solution

a. Because cyclohexanol is a 2° alcohol with only one H atom on the C bonded to the OH group, it is oxidized to a ketone, cyclohexanone.

only 1 H atom

cyclohexanol　　　　　　　　cyclohexanone
2° alcohol　　　　　　　　　　ketone

b. 1-Butanol is a 1° alcohol with two H atoms on the C bonded to the OH group. Thus, it is first oxidized to an aldehyde and then to a carboxylic acid.

2 H atoms

1-butanol　　　　　　　butanal　　　　　　　butanoic acid
1° alcohol　　　　　　　aldehyde　　　　　　carboxylic acid

More Practice: Try Problems 12.3d, 12.7d, 12.8d, 12.31–12.34, 12.35b, 12.36b, 12.67c.

Practice Problem 12.7	What products are formed when each alcohol is oxidized? In some cases, no reaction occurs.

a. $CH_3CHCH_2CH_2CH_3$
       |
      OH

b. $(CH_3)_3CCH_2CH_2OH$

c. ⬠—OH

d. ⬡ with CH₃ and OH

While alcohol use is socially acceptable, alcohol-related traffic fatalities are common with irresponsible alcohol consumption. Data from the National Highway Traffic Safety Administration from 2017 show that the leading cause of death in car crashes on U.S. highways is drunk driving. *Source: Gwinnet County Police Department/Centers for Disease Control and Prevention*

12.5C FOCUS ON HEALTH & MEDICINE
The Metabolism of Ethanol

Throughout history, humans have ingested alcoholic beverages for their pleasant taste and the feeling of euphoria they impart. When ethanol is consumed, it is quickly absorbed in the stomach and small intestine and then rapidly transported in the bloodstream to other organs. Ethanol is metabolized in the liver by a two-step oxidation sequence. The body does not use chromium reagents as oxidants. Instead, high-molecular-weight enzymes, alcohol dehydrogenase and aldehyde dehydrogenase, and a small molecule called a **coenzyme** carry out these oxidations.

The products of the biological oxidation of ethanol are the same as the products formed in the laboratory. When ethanol (CH_3CH_2OH, a 1° alcohol) is ingested, it is oxidized in the liver first to CH_3CHO (acetaldehyde), and then to CH_3COOH (acetic acid).

$$CH_3CH_2{-}OH \xrightarrow[\text{dehydrogenase}]{[O] \atop \text{alcohol}} \underset{\substack{\text{acetaldehyde}}}{CH_3{-}\overset{\displaystyle O}{\overset{\|}{C}}{-}H} \xrightarrow[\text{dehydrogenase}]{[O] \atop \text{aldehyde}} \underset{\substack{\text{acetic acid}}}{CH_3{-}\overset{\displaystyle O}{\overset{\|}{C}}{-}OH}$$

ethanol

If more ethanol is ingested than can be metabolized in a given time period, the concentration of acetaldehyde accumulates. This toxic compound is responsible for the feelings associated with a hangover.

12.6 Alkyl Halides

Alkyl halides have the general molecular formula $C_nH_{2n+1}X$, and are formally derived from an alkane by replacing a hydrogen atom with a halogen.

Alkyl halides are organic molecules containing a halogen atom X (X = F, Cl, Br, I) bonded to a tetrahedral carbon atom. Alkyl halides are classified as **primary (1°), secondary (2°), or tertiary (3°)** depending on the number of carbons bonded to the carbon with the halogen.

- A primary (1°) alkyl halide has a halogen on a carbon bonded to one carbon.
- A secondary (2°) alkyl halide has a halogen on a carbon bonded to two carbons.
- A tertiary (3°) alkyl halide has a halogen on a carbon bonded to three carbons.

Dexamethasone (Concept Check 12.9) relieves inflammation and is used to treat some forms of arthritis, skin conditions, and asthma. Dexamethasone has recently been effectively used to treat patients who are critically ill with COVID-19. *Jill Braaten*

Concept Check 12.8

Classify each alkyl halide as 1°, 2°, or 3°.

a. $CH_3CH_2CH_2CH_2CH_2$—Br

b. [cyclohexane ring with CH₃ and F on same carbon]

c. CH_3—C(CH_3)(CH_3)—CHCH$_3$ with Cl

Concept Check 12.9

Dexamethasone is a halogen-containing steroid. (a) Classify the alkyl halide in dexamethasone as 1°, 2°, or 3°. (b) Classify the hydroxyl groups as 1°, 2°, or 3°.

dexamethasone

12.6A Physical Properties

Alkyl halides contain a polar C—X bond, but because they have all of their hydrogens bonded to carbon, they are incapable of intermolecular hydrogen bonding. Alkyl halides with one halogen are polar molecules, therefore, because they contain a net dipole. As a result, they have higher melting points and boiling points than alkanes with the same number of carbons.

polar bond

$X = F, Cl, Br, I$

net dipole

The size of the alkyl groups and the halogen also affects the physical properties of an alkyl halide.

- The boiling point and melting point of an alkyl halide *increase* with the size of the alkyl group because of increased surface area.
- The boiling point and melting point of an alkyl halide *increase* with the size of the halogen.

Thus, $CH_3CH_2CH_2Cl$ has a higher boiling point than CH_3CH_2Cl because it has one more carbon, giving it a larger surface area. $CH_3CH_2CH_2Br$ has a higher boiling point than $CH_3CH_2CH_2Cl$ because Br is further down the column of the periodic table, making it larger than Cl.

CH_3CH_2Cl	$CH_3CH_2CH_2Cl$	$CH_3CH_2CH_2Br$
chloroethane	1-chloropropane	1-bromopropane
bp = 12 °C	bp = 47 °C	bp = 71 °C

Increasing boiling point
Increasing size of the alkyl group
Increasing size of the halogen

Because alkyl halides are incapable of hydrogen bonding, they are insoluble in water regardless of size.

Rank the compounds in each group in order of increasing boiling point.

a. $CH_3CH_2CH_2I$, $CH_3CH_2CH_2Cl$, $CH_3CH_2CH_2F$

b. $CH_3(CH_2)_4CH_3$, $CH_3(CH_2)_5Br$, $CH_3(CH_2)_5Cl$

12.6B   Nomenclature

In the IUPAC system, an alkyl halide is named as an alkane with a halogen substituent—that is, as a **halo alkane.** To name a halogen substituent, change the **-ine** ending of the name of the halogen to the suffix **-o** (e.g., chlor*ine* → chlor*o*).

How To   Name an Alkyl Halide Using the IUPAC System

Example: Give the IUPAC name of the following alkyl halide.

$$\underset{\underset{}{\overset{\overset{CH_3}{|}}{}}{CH_3CH_2}\underset{}{CH}CH_2CH_2\underset{\underset{}{\overset{\overset{Cl}{|}}{}}{CH}CH_3}$$

Step [1]   Find the parent carbon chain containing the carbon bonded to the halogen.

$$\underset{\underset{}{\overset{\overset{CH_3}{|}}{}}{CH_3CH_2}CHCH_2CH_2\underset{\underset{}{\overset{\overset{Cl}{|}}{}}{CH}CH_3}$$

7 C's in the longest chain

7 C's ----→ **heptane**

• Name the parent chain as an **alkane,** with the halogen as a substituent bonded to the longest chain.

Step [2]   Apply all other rules of nomenclature.

a. **Number** the chain.

$$\underset{\underset{}{\overset{\overset{CH_3}{|}}{}}{CH_3CH_2}CHCH_2CH_2\underset{\underset{}{\overset{\overset{Cl}{|}}{}}{CH}CH_3}$$

7 6   5 4   3 **2 1**

• Begin at the end nearest the first substituent, either alkyl or halogen.

b. **Name** and **number** the substituents.

methyl at C5   chloro at C2

$$\underset{\underset{}{\overset{\overset{CH_3}{|}}{}}{CH_3CH_2}CHCH_2CH_2\underset{\underset{}{\overset{\overset{Cl}{|}}{}}{CH}CH_3}$$

5   2

c. Alphabetize: **c** for **c**hloro, then **m** for **m**ethyl.

Answer: 2-chloro-5-methylheptane

Common names for alkyl halides are used only for simple alkyl halides. To assign a common name, name the carbon atoms as an alkyl group. Then name the halogen by changing the **-ine** ending of the halogen name to the suffix **-ide;** for example, **brom*ine*** → **brom*ide.***

$$CH_3CH_2-Cl \quad = \quad$$ chlor*ine* → chlor*ide*

ethyl
group

Common name:
ethyl chloride

Give the structure corresponding to each name.

a. 3-chloro-2-methylhexane

b. 4-ethyl-5-iodo-2,2-dimethyloctane

c. 1,1,3-tribromohexane

d. propyl chloride

12.6C Interesting Alkyl Halides

Many simple alkyl halides make excellent solvents because they are not flammable and they dissolve a wide variety of organic compounds. Two compounds in this category include $CHCl_3$ (chloroform or trichloromethane) and CCl_4 (carbon tetrachloride or tetrachloromethane). Large quantities of these solvents are produced industrially each year, but like many chlorinated organic compounds, both chloroform and carbon tetrachloride are toxic if inhaled or ingested. Two other simple alkyl halides are given in Figure 12.4.

Giant kelp forests in the ocean release chloromethane (CH_3Cl) into the atmosphere. *Source: Claire Fackler/ CINMS/NOAA*

Figure 12.4   Two Simple Alkyl Halides

CH_3Cl

- **Chloromethane (CH_3Cl)** is produced by giant kelp and algae and is also found in emissions from volcanoes such as Hawaii's Kilauea. Almost all of the atmospheric chloromethane results from these natural sources.

CH_2Cl_2

- **Dichloromethane (or methylene chloride, CH_2Cl_2)** is an important solvent, once used to decaffeinate coffee. Coffee is now decaffeinated using liquid CO_2 due to concerns over the possible ill effects of trace amounts of residual CH_2Cl_2 in the coffee. Subsequent studies on rats have shown, however, that no cancers occurred when animals ingested the equivalent of over 100,000 cups of decaffeinated coffee per day.

HEALTH NOTE

Fluticasone (Concept Check 12.12) is used as an aerosol inhalation to treat seasonal and recurrent allergies and to combat chronic lung inflammation in asthma. *Mark Dierker/McGraw Hill*

Concept Check 12.12

Classify each of the labeled functional groups in the drug fluticasone as 1°, 2°, or 3°.

fluticasone

12.7 Thiols

Thiols are organic compounds that contain a sulfhydryl group (SH group) bonded to a tetrahedral carbon. Because sulfur is directly below oxygen in the periodic table, thiols can be considered sulfur analogues of alcohols.

Thiol

sulfhydryl group

CH_3CH_2—$\overset{..}{\underset{..}{S}}$—H   =   =

ethanethiol

bent shape around S

Thiols differ from alcohols in one important way. They contain no O—H bonds, so they are incapable of intermolecular hydrogen bonding. **This gives thiols *lower* boiling points and melting points compared to alcohols having the same size and shape.**

Hydrogen bonding can occur. stronger intermolecular forces **higher boiling point**	→ CH_3CH_2—OH	CH_3CH_2—SH ←	Hydrogen bonding is impossible. weaker intermolecular forces **lower boiling point**
	ethanol	ethanethiol	
	bp 78 °C	bp 35 °C	

1-Propanethiol ($CH_3CH_2CH_2SH$) is partly responsible for the characteristic odor of onions. *Creative-i/Shutterstock*

The most obvious physical property of thiols is their distinctive foul odor. 3-Methyl-1-butanethiol [$(CH_3)_2CHCH_2CH_2SH$] is one of the main components of the defensive spray of skunks.

Thiols undergo one important reaction: **thiols are oxidized to disulfides,** compounds that contain a sulfur–sulfur bond. This is an **oxidation** reaction because two hydrogen atoms are removed in forming the disulfide.

$$\text{Oxidation} \qquad 2\ CH_3CH_2\text{—}S\text{—}H \xrightarrow{[O]} CH_3CH_2\text{—}S\text{—}S\text{—}CH_2CH_3$$

new S—S bond

thiol   disulfide

Disulfides can also be converted back to thiols with a **reducing agent.** The symbol for a general reducing agent is **[H],** because hydrogen atoms are often added to a molecule during reduction.

$$\text{Reduction} \qquad CH_3CH_2\text{—}S\text{—}S\text{—}CH_2CH_3 \xrightarrow{[H]} 2\ CH_3CH_2\text{—}S\text{—}H$$

new S—H bond

disulfide   thiol

The chemistry of thiols and disulfides plays an important role in determining the properties and shape of some proteins. For example, α-keratin, the protein in hair, contains many disulfide bonds. Straight hair can be made curly by cleaving the disulfide bonds in α-keratin, then rearranging and re-forming them, as shown schematically in Figure 12.5.

Concept Check 12.13

(a) Draw the disulfide formed when $CH_3CH_2CH_2SH$ is oxidized. (b) Draw the products formed when the following disulfide is reduced: $CH_3CH_2CH_2CH_2SSCH_2CH_3$.

Figure 12.5 Focus on the Human Body: Making Straight Hair Curly

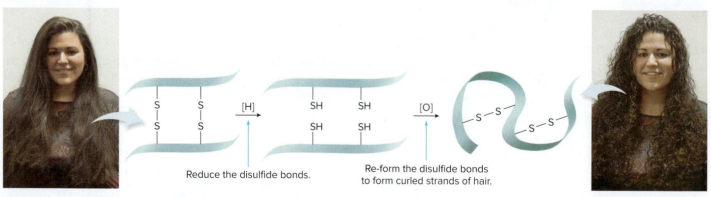

To make straight hair curly, the disulfide bonds holding the protein chains together are reduced. This forms free SH groups. The hair is turned around curlers and then an oxidizing agent is applied. This re-forms the disulfide bonds to the hair, now giving it a curly appearance. *(left & right): Suzie Ross/McGraw Hill*

Diallyl disulfide (Concept Check 12.14), which is formed by the decomposition of allicin (Section 10.6) when garlic cloves are cut or crushed, is principally responsible for the pungent odor of garlic oil. *Mark Dierker/McGraw Hill*

formaldehyde
$CH_2{=}O$

acetone
$(CH_3)_2C{=}O$

Concept Check 12.14

What product is formed when diallyl disulfide, a major component of garlic oil, is reduced?

diallyl disulfide

12.8 Structure and Properties of Aldehydes and Ketones

Aldehydes (RCHO) and ketones **(RCOR or R$_2$CO)** are two families of compounds that contain a carbonyl group. Two structural features dominate the properties and chemistry of the carbonyl group.

- The carbonyl carbon atom is trigonal planar, and all bond angles are 120°.
- Because oxygen is more electronegative than carbon, a carbonyl group is polar. The carbonyl carbon is electron poor (δ+) and the oxygen is electron rich (δ−).

12.8A Naming Aldehydes

- In IUPAC nomenclature, aldehydes are identified by the suffix *-al*.

To name an aldehyde using the IUPAC system:

- Find the longest chain containing the CHO group, and change the -e ending of the parent alkane to the suffix *-al*.
- Number the chain or ring to put the CHO group at C1, but omit this number from the name. Apply all of the other usual rules of nomenclature.

Simple aldehydes have common names that are virtually always used instead of their IUPAC names. Common names all contain the suffix *-aldehyde.*

formaldehyde
(methanal)

acetaldehyde
(ethanal)

benzaldehyde
(benzenecarbaldehyde)

(IUPAC names are in parentheses.)

Sample Problem 12.6   **Naming an Aldehyde**

Give the IUPAC name for the following aldehyde.

Analysis and Solution

[1] Find and name the longest chain containing the CHO.	[2] Number and name substituents, making sure the CHO group is at C1.

butane - - - → butan*al*
(4 C's)

Answer: 2,3-dimethylbutanal

Practice Problem 12.8

Give the IUPAC name for each aldehyde.

a. $(CH_3)_2CHCH_2CH_2CH_2CHO$ b. $(CH_3)_3CC(CH_3)_2CH_2CHO$ c.

Practice Problem 12.9

Give the structure corresponding to each IUPAC name.

a. 2-chloropropanal   c. 3,6-diethylnonanal

b. 3,4,5-triethylheptanal   d. *o*-ethylbenzaldehyde

Practice Problem 12.10

Give the IUPAC name for each aldehyde depicted in the ball-and-stick models. Both aldehydes are isolated from yuzu, a citrus fruit extensively cultivated in Japan.

Yuzu is a small citrus fruit used widely in Japanese cooking. Its unusual aroma is partly due to the aldehydes depicted in Practice Problem 12.10. *Toshihiko Watanabe/Aflo Co., Ltd./Alamy Stock Photo*

More Practice: Try Problems 12.5c; 12.6c; 12.39a; 12.40a; 12.41a, b; 12.42a, b; 12.43a, b; 12.44a, b; 12.45.

12.8B   Naming Ketones

- In the IUPAC system, ketones are identified by the suffix *-one.*

To name an acyclic ketone using IUPAC rules:

- Find the longest chain containing the carbonyl group, and change the *-e* ending of the parent alkane to the suffix *-one.*
- Number the carbon chain to give the carbonyl carbon the lower number. Apply all of the other usual rules of nomenclature.

With cyclic ketones, numbering always begins at the carbonyl carbon, but the "1" is usually omitted from the name. The ring is then numbered clockwise or counterclockwise to give the first substituent the lower number.

Three simple ketones have widely used common names.

acetone   acetophenone   benzophenone

Sample Problem 12.7	**Naming a Ketone**

Give the IUPAC name for the following ketone.

Analysis and Solution

[1] Name the ring.

cyclohexane – – –→ cyclohexan*one*
(6 C's)

[2] Number and name substituents, making sure the carbonyl carbon is at C1.

Answer: 3-ethyl-4-methylcyclohexanone

Practice Problem 12.11	Give the IUPAC name for each ketone.

a. $CH_3CH_2CCHCH_2CH_2CH_3$
          CH_3

b.

c.

More Practice: Try Problems 12.5d; 12.6d; 12.39b; 12.40b; 12.41c, d; 12.42c, d; 12.43c, d; 12.44c, d; 12.46.

12.8C   Physical Properties

Aldehydes and ketones have a polar carbonyl group, so they are **polar molecules** with stronger intermolecular forces than the hydrocarbons of Chapters 10 and 11. Because they have no O—H bond, two molecules of RCHO or RCOR are incapable of intermolecular hydrogen bonding, giving them weaker intermolecular forces than alcohols.

As a result:

- **Aldehydes and ketones have *higher* boiling points than hydrocarbons of comparable molecular weight.**
- **Aldehydes and ketones have *lower* boiling points than alcohols of comparable molecular weight.**

CH₃CH₂CH₂CH₂CH₃   CH₃CH₂CH₂CHO   CH₃CH₂CH₂CH₂OH

pentane   butanal   1-butanol

bp 36 °C   bp 76 °C   bp 118 °C

Increasing strength of intermolecular forces
Increasing boiling point

Based on the general rule governing solubility (i.e., **"like dissolves like"**), aldehydes and ketones are soluble in organic solvents. Moreover, because aldehydes and ketones contain an oxygen atom with an available lone pair, they can intermolecularly hydrogen bond to water.

acetone   hydrogen bond

As a result:

- Low-molecular-weight aldehydes and ketones (those having less than six carbons) are soluble in both organic solvents and water.
- Higher-molecular-weight aldehydes and ketones (those having six carbons or more) are soluble in organic solvents, but insoluble in water.

Concept Check 12.15

Which compound in each pair has the higher boiling point?

a. or   c. or

b. (CH₃CH₂)₂CO or (CH₃CH₂)₂C=CH₂   d. CH₃(CH₂)₆CH₃ or CH₃(CH₂)₅CHO

Concept Check 12.16

Acetone and progesterone are two ketones that occur naturally in the human body. Discuss the solubility properties of both compounds in water and organic solvents.

acetone

progesterone

12.9 FOCUS ON HEALTH & MEDICINE
Interesting Aldehydes and Ketones

Many aldehydes with characteristic odors occur in nature, including vanillin from vanilla beans and cinnamaldehyde from cinnamon bark.

vanillin
(flavoring agent from
vanilla beans)

cinnamaldehyde
(odor of cinnamon)

Jill Braaten/McGraw Hill

Acetone [$(CH_3)_2C{=}O$, the simplest ketone] is produced naturally in cells during the breakdown of fatty acids. In diabetes, a disease where normal metabolic processes are altered because of the inadequate secretion of insulin, individuals often have unusually high levels of acetone in the bloodstream. The characteristic odor of acetone can be detected on the breath of diabetic patients when their disease is poorly controlled.

Ketones play an important role in the tanning industry. Dihydroxyacetone is the active ingredient in commercial tanning agents that produce sunless tans. Dihydroxyacetone reacts with proteins in the skin, producing a complex colored pigment that gives the skin a brown hue. In addition, many commercial sunscreens are ketones. Examples include avobenzone, oxybenzone (Section 11.10A), and dioxybenzone.

dihydroxyacetone

avobenzone

oxybenzone

dioxybenzone

sunscreen

sunless tanning agent

(left & right): Jill Braaten/McGraw Hill

[Ketone carbonyls are shown in red.]

| **Concept Check 12.17** | Which sunscreen—avobenzone, oxybenzone, or dioxybenzone—is probably most soluble in water, and therefore most readily washed off when an individual goes swimming? Explain your choice. |

12.10 Oxidation of Aldehydes

Because aldehydes contain a hydrogen atom bonded directly to the carbonyl carbon, they can be oxidized to carboxylic acids; that is, the aldehyde C—H bond can be converted to a C—OH bond. Ketones have no hydrogen atom bonded to the carbonyl group, so they are not oxidized under similar reaction conditions.

Replace 1 C—H bond by 1 C—O bond.

Oxidation

aldehyde $\xrightarrow{[O]}$ carboxylic acid

A common reagent for this oxidation is potassium dichromate, $K_2Cr_2O_7$, the oxidizing agent introduced in Section 12.5B, which oxidizes 1° and 2° alcohols as well.

$CH_3CH_2CH_2$—CHO (butanal) $\xrightarrow{K_2Cr_2O_7}$ $CH_3CH_2CH_2$—COOH (butanoic acid)

no C—H bond

CH_3CH_2—CO—CH_3 (2-butanone) $\xrightarrow{K_2Cr_2O_7}$ No reaction

Aldehydes can be oxidized *selectively* in the presence of other functional groups using **silver(I) oxide (Ag₂O) in aqueous ammonium hydroxide (NH₄OH).** This is called **Tollens reagent.** Only aldehydes react with Tollens reagent; all other functional groups are inert. Oxidation with Tollens reagent provides a distinct color change because the Ag^+ reagent is converted to silver metal (Ag), which precipitates out of the reaction mixture as a silver mirror.

$\xrightarrow[NH_4OH]{Ag_2O}$ + Ag

silver mirror

Only the aldehyde is oxidized.

Aldehydes are said to give a *positive* Tollens test; that is, they react with Ag^+ to form RCOOH and Ag. When the reaction is carried out in a glass flask, a silver mirror is formed on its walls. Other functional groups give a *negative* Tollens test; that is, no silver mirror forms.
Charles D. Winters/McGraw Hill

Sample Problem 12.8	Drawing the Product of Oxidation with Tollens Reagent

What product is formed when each compound is treated with Tollens reagent (Ag₂O, NH₄OH)?

a. CH_3—CHO

b. $HOCH_2$—cyclohexane—CHO

Analysis

Only aldehydes (RCHO) react with Tollens reagent. Ketones and alcohols are inert to oxidation.

Solution

The aldehyde in both compounds is oxidized to RCO$_2$H, but the 1° alcohol in part (b) does not react with Tollens reagent.

a.

CH$_3$ — C(=O) — H →(Ag$_2$O, NH$_4$OH)→ CH$_3$ — C(=O) — OH

> Replace 1 C—H bond by 1 C—O bond.

b. HOCH$_2$ — [cyclohexane] — C(=O) — H →(Ag$_2$O, NH$_4$OH)→ HOCH$_2$ — [cyclohexane] — C(=O) — OH

> Only the aldehyde is oxidized.

Practice Problem 12.12

What product is formed when each compound is treated with Tollens reagent (Ag$_2$O, NH$_4$OH)? In some cases, no reaction occurs.

a. CH$_3$(CH$_2$)$_6$CHO

b. [cyclopentane]=O

c. [cyclopentane]—CHO

d. [cyclohexane]—OH

More Practice: Try Problems 12.4d, 12.49, 12.50, 12.51b, 12.52c.

12.11 Looking Glass Chemistry—Molecules and Their Mirror Images

The remainder of Chapter 12 concentrates on **stereochemistry**—the three-dimensional structure of compounds. The principles presented here apply to all of the families of molecules in this chapter, as well as the hydrocarbons we encountered in Chapters 10 and 11. To understand stereochemistry, recall the definition of stereoisomers learned in Section 11.3.

> • **Stereoisomers** are isomers that differ *only* in the three-dimensional arrangement of atoms.

cis-2-butene

The cis and trans isomers of 2-butene are one example of stereoisomers. The cis isomer has the two CH$_3$ groups on the same side of the double bond, whereas the trans isomer has the two CH$_3$ groups on opposite sides.

trans-2-butene

Another type of stereoisomer occurs at tetrahedral carbons. To learn more about this type of stereoisomer, we must turn our attention to molecules and their mirror images. **Everything, including molecules, has a mirror image.** What's important in chemistry is **whether a molecule is *identical* to or *different* from its mirror image.**

12.11A What It Means to Be Chiral or Achiral

Some molecules are like hands. **Left and right hands are mirror images of each other, but they are *not* identical.** If you try to mentally place one hand inside the other hand you can never superimpose either all the fingers, or the tops and palms. To *superimpose* an object on its mirror

The dominance of right-handedness over left-handedness occurs in all races and cultures. Despite this fact, even identical twins can exhibit differences in hand preference. Pictured scuba diving are Matthew (right-handed) and Zachary (left-handed), identical twin sons of the author. *Daniel C. Smith*

The adjective *chiral* (pronounced ky-rel) comes from the Greek *cheir,* meaning *hand.* Left and right hands are **chiral:** they are mirror images that do not superimpose on each other.

image means to align *all* parts of the object with its mirror image. With molecules, this means aligning all atoms and all bonds.

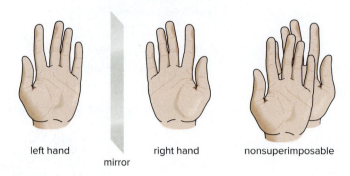

left hand   right hand   nonsuperimposable

mirror

- A molecule (or object) that is *not* superimposable on its mirror image is said to be *chiral.*

Other molecules are like socks. **Two socks from a pair are mirror images that *are* superimposable.** One sock can fit inside another, aligning toes and heels, and tops and bottoms. A sock and its mirror image are *identical.*

mirror   superimposable

- A molecule (or object) that *is* superimposable on its mirror image is said to be *achiral.*

Classify each object as chiral or achiral.

a.   b.   c.   d.

When is a molecule chiral or achiral? To determine whether a molecule is chiral or achiral, we must examine what groups are bonded to each carbon atom.

- A chiral molecule has at least one carbon atom bonded to four *different* groups.
- An achiral molecule does *not* contain a carbon atom bonded to four different groups.

Using these definitions, we can classify CH_2BrCl and $CHBrClF$ as chiral or achiral molecules.

CH_2BrCl is an achiral molecule. Its single carbon atom is bonded to two H atoms. Ball-and-stick models illustrate that this molecule is superimposable on its mirror image.

A and B are **stereoisomers** because they are isomers differing only in the three-dimensional arrangement of substituents. They represent a new type of stereoisomer that occurs at tetrahedral carbon atoms. These stereoisomers are called *enantiomers.*

- **Enantiomers are mirror images that are not superimposable.**

Thus, CH_2BrCl is an achiral molecule with no chirality center, whereas CHBrClF is a chiral molecule with one chirality center.

12.11B Locating Chirality Centers

A necessary skill in studying the three-dimensional structure of molecules is the ability to locate chirality centers—carbon atoms bonded to four different groups. To locate a chirality center, examine each **tetrahedral** carbon atom in a molecule, and look at the four *groups*—not the four *atoms*—bonded to it. CBrClFI has one chirality center because its carbon atom is bonded to four different elements—Br, Cl, F, and I. 3-Bromohexane also has one chirality center because one carbon is bonded to H, Br, CH_2CH_3, and $CH_2CH_2CH_3$. We consider all atoms in a group as a *whole unit,* not just the atom bonded directly to the carbon in question.

chirality center

chirality center

3-bromohexane

Many types of snails possess a chiral, right-handed helical shell.
Jill Braaten/McGraw Hill

Naming a carbon atom with four different groups is a topic that has no firm agreement among organic chemists. IUPAC recommends the term *chirality center,* and so this is the term used in this text. Other terms in common use are chiral center, chiral carbon, asymmetric carbon, stereocenter, and stereogenic center.

With CHBrClF, the result is different. CHBrClF contains a carbon atom bonded to four different groups—H, Br, Cl, and F. **A carbon bonded to four different groups is called a *chirality center.*** The molecule (labeled **A**) and its mirror image (labeled **B**) are not superimposable. No matter how you rotate **A** and **B**, all of the atoms never align. **CHBrClF is thus a chiral molecule,** and **A** and **B** are different compounds.

Keep in mind the following:

> • CH$_2$ and CH$_3$ groups have more than one H atom bonded to the same carbon, so each of these carbons is *never* a chirality center.
> • A carbon that is part of a multiple bond does not have four groups around it, so it can *never* be a chirality center.

Sample Problem 12.9   Locating Chirality Centers

Locate the chirality center in each drug.

a.
$$
\begin{array}{c}
O \overset{\displaystyle \diagup OH}{\underset{C}{\big\|}} \\
H_2N-\overset{|}{C}-H \\
CH_3-\overset{|}{C}-CH_3 \\
\overset{|}{S H}
\end{array}
$$
penicillamine

b. HO—⟨ring⟩—$\overset{OH}{\underset{H}{\overset{|}{C}}}$—CH$_2$NHC(CH$_3$)$_3$

HOCH$_2$

albuterol

Analysis

In compounds with many C's, look at each C individually and eliminate those C's that can't be chirality centers. Thus, **omit all CH$_2$ and CH$_3$ groups and all multiply bonded C's.** Check all remaining C's to see if they are bonded to four different groups.

Solution

a. **Penicillamine** is used to treat Wilson's disease, a genetic disorder that leads to a buildup of copper in the liver, kidneys, and the brain. The CH$_3$ groups and COOH group in penicillamine are not chirality centers. One C is bonded to two CH$_3$ groups, so it can be eliminated from consideration as well. This leaves one C with four different groups bonded to it.

b. **Albuterol** is a bronchodilator—that is, it widens air passages—so it is used to treat asthma. Omit all CH$_2$ and CH$_3$ groups and all C's that are part of a double bond. Also omit from consideration the one C bonded to three CH$_3$ groups. This leaves one C with four different groups bonded to it.

penicillamine   albuterol

Practice Problem 12.13

Label the chirality center in the given molecules. The compounds contain zero or one chirality center.

a. CH$_3$CHCH$_2$CH$_3$ b. (CH$_3$)$_3$CH   c. CH$_2$=CHCHCH$_3$ d. ⟨ring⟩—$\overset{H}{\underset{Br}{\overset{|}{C}}}$—CH$_3$
   $\;\;\;\;\;$| |
   $\;\;\;\;$Cl   OH

Practice Problem 12.14

Label the chirality center in each drug.

a. Br—⟨ring⟩—$\overset{H}{\underset{CH_2CH_2N(CH_3)_2}{\overset{|}{C}}}$—⟨pyridine, N⟩

brompheniramine
(antihistamine)

b. ⟨ring⟩—$\overset{H}{\underset{CH_2CH_2NHCH_3}{\overset{|}{C}}}$—O—⟨ring⟩—CF$_3$

fluoxetine
Trade name: Prozac
(antidepressant)

More Practice: Try Problems 12.11a, 12.12a, 12.57–12.60, 12.67a, 12.68a.

Larger organic molecules can have two, three, or even hundreds of chirality centers. Tartaric acid, isolated from grapes, and ephedrine, isolated from the herb ma huang (used to treat respiratory ailments in traditional Chinese medicine), each have two chirality centers. Once a popular drug to promote weight loss and enhance athletic performance, ephedrine use has now been linked to episodes of sudden death, heart attack, and stroke.

tartaric acid

ephedrine
(bronchodilator, decongestant)

[Chirality centers are labeled in blue.]

Foodcollection

WILDLIFE GmbH/Alamy Stock Photo

Concept Check 12.19

Label the chirality centers in each molecule. Compounds may have one or two chirality centers.

a. $CH_3CH_2CH_2 - \overset{\displaystyle H}{\underset{\displaystyle OH}{C}} - CH_3$

b. $H_2N - \overset{\displaystyle COOH}{\underset{\displaystyle CH_3 - \overset{\displaystyle |}{\underset{\displaystyle CH_2CH_3}{C} - H}}{C} - H}$

c. $CH_3CH_2 - \overset{\displaystyle Br}{\underset{\displaystyle H}{C}} - CH_2CH_2 - \overset{\displaystyle Cl}{\underset{\displaystyle H}{C}} - CH_3$

In locating chirality centers in skeletal structures, always remember that **each carbon has enough hydrogens to make it tetravalent.** 2-Chloropentane has one chirality center, because C2 is bonded to four different groups—H, CH_3, $CH_2CH_2CH_3$, and Cl.

Cl — chirality center

2-chloropentane

H Cl

C2 is bonded to four different groups.

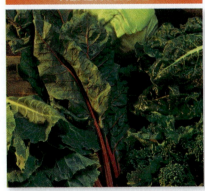

Lettuce, spinach, broccoli, and cabbage are all excellent sources of vitamin K (Concept Check 12.20). *Comstock/Getty Images*

Concept Check 12.20

Translate the ball-and-stick model to a skeletal structure, and label the two chirality centers in vitamin K. Vitamin K is essential to blood clot formation. A deficiency of this vitamin may lead to excessive and often fatal bleeding.

vitamin K

12.11C Fischer Projections

The chirality centers in some organic compounds, most notably carbohydrates (Chapter 14), are often drawn using the following convention.

Instead of drawing a tetrahedron with two bonds in the plane, one in front of the plane, and one behind it, the **tetrahedron is tipped so that both horizontal bonds come forward (drawn on** wedges) and both vertical bonds go behind (on dashed wedges). This structure is then abbreviated by a **cross formula,** also called a **Fischer projection formula.** In a Fischer projection formula, therefore,

Draw a tetrahedron as: Abbreviate it as a cross formula:

Horizontal bonds come forward, on wedges.
Vertical bonds go back, on dashed wedges.

chirality center
Fischer projection formula

- A carbon atom is located at the intersection of the two lines of the cross.
- The horizontal bonds come forward, on wedges.
- The vertical bonds go back, on dashed wedges.

For example, to draw the chirality center of 2-butanol [$CH_3CH(OH)CH_2CH_3$] using this convention, **draw the tetrahedron with horizontal bonds on wedges and vertical bonds on dashed wedges.** Then, **replace the chirality center with a cross** to draw the Fischer projection. The carbon skeleton is generally arranged vertically.

$$= \quad HO \blacktriangleright \overset{CH_3}{\underset{CH_2CH_3}{C}} \blacktriangleleft H \qquad H \blacktriangleright \overset{CH_3}{\underset{CH_2CH_3}{C}} \blacktriangleleft OH \quad =$$

one enantiomer of 2-butanol   mirror image

Replace the chirality center with a cross.

$$HO \overline{\underset{CH_2CH_3}{\overset{CH_3}{|}}} H \qquad\qquad H \overline{\underset{CH_2CH_3}{\overset{CH_3}{|}}} OH$$

Fischer projection formulas for both enantiomers of 2-butanol

Sample Problem 12.10 illustrates another example of drawing two enantiomers of a carbohydrate using this convention. The carbonyl group of a carbohydrate is placed on the top vertical bond in the Fischer projection.

Sample Problem 12.10   Drawing Fischer Projections

Draw both enantiomers of glyceraldehyde, a simple carbohydrate, using Fischer projection formulas.

$$HOCH_2CHCHO$$
$$\underset{OH}{|}$$

glyceraldehyde

Analysis

- **Draw the tetrahedron of one enantiomer with the horizontal bonds on wedges, and the vertical bonds on dashed wedges.** Arrange the four groups on the chirality center—H, OH, CHO, and CH_2OH—with the CHO group at the top and the carbon skeleton drawn vertically.
- Draw the second enantiomer by arranging the substituents in the mirror image so they are a **reflection** of the groups in the first molecule.
- Replace the chirality center with a cross to draw the Fischer projections.

Solution

chirality center

HOCH₂—C—CHO
    |
    H
    |
    OH

glyceraldehyde

Draw the compound.

CHO
|
H►C◄OH
|
CH₂OH

- **Horizontal bonds come forward.**
- **Vertical bonds go back.**

CHO
|
H——OH
|
CH₂OH

Fischer projection of
one enantiomer

Then, draw the mirror image.

CHO
|
HO►C◄H
|
CH₂OH

CHO
|
HO——H
|
CH₂OH

Fischer projection of the
second enantiomer

Practice Problem 12.15

Convert each molecule into a Fischer projection formula.

a. CH₃
    |
    H►C◄Br
    |
    Cl

b. COOH
    |
    H►C◄OH
    |
    CH₂Cl

Practice Problem 12.16

Draw Fischer projections of both enantiomers for each compound.

a. CH₃CHCH₂CH₂CH₃
         |
         OH

b. CH₃CH₂CHCH₂Cl
              |
              Cl

Practice Problem 12.17

Although fresh human sweat has no odor, enzymes in skin bacteria form a variety of compounds with distinctive, pungent odors. The two enantiomers of 3-methyl-3-sulfanyl-1-hexanol (**A**), which smell somewhat differently, contribute to the characteristic foul odor of sweat. Locate the chirality center in **A** and draw both enantiomers.

CH₃
|
HOCH₂CH₂—C—CH₂CH₂CH₃
|
SH

A

More Practice: Try Problems 12.11b, 12.12b, 12.65, 12.66, 12.67b, 12.68b.

The foul odor of human sweat is due in part to the presence of unequal amounts of both enantiomers of 3-methyl-3-sulfanyl-1-hexanol (Practice Problem 12.17). *Daniel C. Smith*

12.12   FOCUS ON HEALTH & MEDICINE
Chiral Drugs

A living organism is a sea of chiral molecules. Many drugs are chiral, and often they must inter-act with a chiral receptor to be effective. One enantiomer of a drug may be effective in treating a disease, whereas its mirror image may be ineffective.

Ibuprofen is commonly used to relieve headaches and muscle and joint pain.
Jill Braaten/McGraw Hill

Why should such a difference in biological activity be observed? One enantiomer may "fit" the chiral receptor and evoke a specific response. Its mirror image may not fit the same receptor, making it ineffective; or if it "fits" another receptor, it can evoke a totally different response. Figure 12.6 schematically illustrates this difference in binding between two enantiomers and a chiral receptor.

Ibuprofen and L-dopa are two drugs that illustrate how two enantiomers can have different biological activities.

Ibuprofen is the generic name for the pain relievers known as Motrin and Advil. Ibuprofen has one chirality center, and thus exists as a pair of enantiomers. Only one enantiomer, labeled **A,** is an active anti-inflammatory agent. Its enantiomer **B** is inactive. **B,** however, is slowly converted to **A** in the body. Ibuprofen is sold as a mixture of both enantiomers.

ibuprofen
active anti-inflammatory agent
A

inactive enantiomer
B

Figure 12.6 The Interaction of Two Enantiomers with a Chiral Receptor

chiral receptor surface   chiral receptor surface

One enantiomer of the amino acid phenylalanine, labeled **A,** has three groups that can interact with the appropriate binding sites of the chiral receptor. The groups around the chirality center in the mirror image **B,** however, can never be arranged so that all three groups can bind to these same three binding sites. Thus, enantiomer **B** does not "fit" the same receptor, so it does not evoke the same response.

Concept Check 12.21

Pregabalin, sold under the trade name Lyrica, is a drug used to treat chronic pain.
(a) Locate the chirality center in pregabalin. (b) Draw both enantiomers of pregabalin.

pregabalin

The broad bean *Vicia faba* produces the drug L-dopa. To meet the large demand for L-dopa today, the commercially available drug is synthesized in the laboratory. *Nic Miller/Organics image library/Alamy Stock Photo*

L-Dopa was first isolated from the seeds of the broad bean *Vicia faba,* and since 1967 it has been the drug of choice for the treatment of Parkinson's disease. L-Dopa is a chiral molecule, which is sold as a single enantiomer. Like many chiral drugs, only one enantiomer is active against Parkinson's disease. The inactive enantiomer is also responsible for neutropenia, a decrease in certain white blood cells that help to fight infection.

L-dopa
active against
Parkinson's disease

D-dopa
enantiomer responsible for
undesired side effects

Parkinson's disease, which afflicts 1.5 million individuals in the United States, results from the degeneration of neurons that produce the neurotransmitter dopamine in the brain. When the level of dopamine drops, the loss of motor control symptomatic of Parkinson's disease results.

L-Dopa is an oral medication that is transported to the brain by the bloodstream. In the brain it is converted to dopamine (Figure 12.7). Dopamine itself cannot be given as a medication because it cannot pass from the bloodstream into the brain; that is, it does not cross the blood–brain barrier.

Figure 12.7   L-Dopa, Dopamine, and Parkinson's Disease

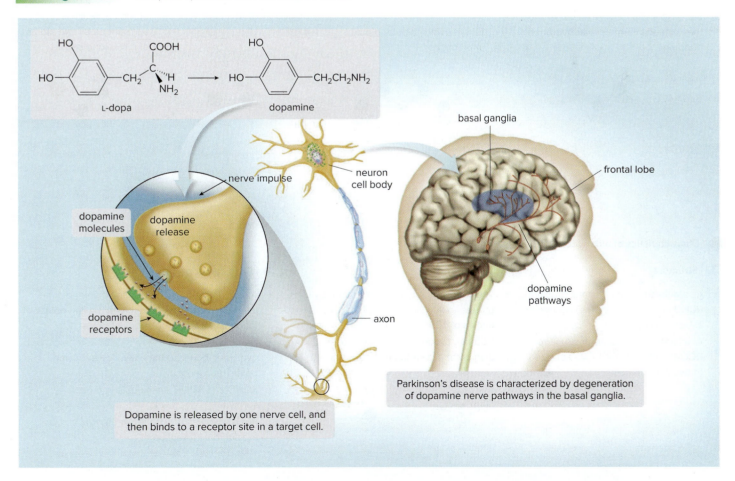

L-dopa

dopamine

basal ganglia

frontal lobe

nerve impulse

neuron
cell body

dopamine
molecules

dopamine
release

dopamine
pathways

dopamine
receptors

axon

Dopamine is released by one nerve cell, and then binds to a receptor site in a target cell.

Parkinson's disease is characterized by degeneration of dopamine nerve pathways in the basal ganglia.

CHAPTER REVIEW

KEY TERMS

Achiral (12.11)

Alcohol (12.1)

Aldehyde (12.1)

Alkyl halide (12.1)

Carbonyl group (12.1)

Chiral (12.11)

Chirality center (12.11)

Cross formula (12.11)

Dehydration (12.5)

Disulfide (12.7)

Elimination (12.5)

Enantiomer (12.11)

Ether (12.1)

Fischer projection formula (12.11)

Hydroxyl group (12.1)

Ketone (12.1)

Oxidation (12.5)

Primary (1°) alcohol (12.2)

Primary (1°) alkyl halide (12.6)

Secondary (2°) alcohol (12.2)

Secondary (2°) alkyl halide (12.6)

Stereochemistry (12.11)

Stereoisomer (12.11)

Sulfhydryl group (12.1)

Tertiary (3°) alcohol (12.2)

Tertiary (3°) alkyl halide (12.6)

Thiol (12.1)

Tollens reagent (12.10)

KEY CONCEPTS

1 **Characteristics of alcohols, ethers, alkyl halides, and thiols**

1 Alcohols (12.2)	**2** Ethers (12.3)	**3** Alkyl halides (12.6)	**4** Thiols (12.7)
• Alcohols (**ROH**) contain a hydroxyl group (OH) bonded to a tetrahedral carbon. • Alcohols have a bent shape around O, which is surrounded by two atoms and two lone pairs. • Because alcohols have an OH group, they exhibit intermolecular hydrogen bonding.	• Ethers (**ROR**) have two alkyl groups bonded to O. • Ethers have a bent shape around O, which is surrounded by two atoms and two lone pairs. • Ethers exhibit dipole–dipole interactions.	• Alkyl halides (**RX**) contain a halogen (X = F, Cl, Br, I) bonded to a tetrahedral carbon. • Alkyl halides have one polar bond and a net dipole.	• Thiols (**RSH**) contain a sulfhydryl group (SH) bonded to a tetrahedral carbon. • Thiols have a bent shape around S, which is surrounded by two atoms and two lone pairs.
 CH_3CH_2OH ethanol	 $CH_3CH_2OCH_2CH_3$ diethyl ether	 CH_3Cl chloromethane	 CH_3CH_2SH ethanethiol

2 **Characteristics of aldehydes and ketones (12.8)**

1 Structure	**2** Shape	**3** Physical properties
• An **aldehyde** has the general structure **RCHO** and contains a C=O bonded to at least one hydrogen. • A **ketone** has the general structure **RCOR'** and contains a C=O bonded to two carbon atoms.	• The carbonyl carbon is surrounded by three groups, making it trigonal planar, with bond angles of 120°. • Because oxygen is more electronegative than carbon, a carbonyl group is polar.	• Aldehydes and ketones exhibit dipole–dipole interactions, so they have higher boiling points than nonpolar hydrocarbons of similar size. • Two molecules of an aldehyde or ketone cannot hydrogen bond, so they have lower boiling points than alcohols. • RCHO and RCOR' are soluble in organic solvents. • RCHO and RCOR' with fewer than six carbons are soluble in water, too.
aldehyde ketone		

3 Chirality and enantiomers (12.11)

1 Chiral versus achiral molecules	2 Enantiomers
• An **achiral molecule** is superimposable on its mirror image. H_2O and CH_2Cl_2 are achiral molecules. • A **chiral molecule** is *not* superimposable on its mirror image.	• Enantiomers are mirror images that are not superimposable. Enantiomers are one type of stereoisomer. • **A** and **B** are two enantiomers of 2-chlorobutane.

chiral molecule   **chiral** molecule
mirror
not superimposable

2-chlorobutane   2-chlorobutane
A   **B**

KEY REACTIONS

1 Reactions of alcohols (12.5)

1

H_2SO_4
dehydration
+ H_2O

3

[O]
oxidation
2° alcohol   ketone

2

[O]
oxidation
1° alcohol   aldehyde
[O]
oxidation
carboxylic acid

4

[O]
oxidation
3° alcohol   **No reaction**

2 Reactions of sulfur compounds (12.7)

1

2   —SH   [O]
oxidation
—S—S—

2

—S—S—   [H]
reduction
2   —SH

3 Oxidation of aldehydes (12.10)

aldehyde $K_2Cr_2O_7$ or Ag_2O, NH_4OH carboxylic acid
oxidation

KEY SKILLS

1 Classifying alcohols and alkyl halides as 1°, 2°, or 3° (12.2, 12.6)

1 Locate the C bonded to OH (or X).	2 Count the number of C's bonded to the C with the OH (or X).
• Draw out the structure or add H's in a skeletal structure to clearly see how many C's are bonded to the C with the functional group.	• Use the definitions: a 1° ROH or RX has the functional group on a C bonded to one C, and so forth.

1° bromide—Br on C bonded to **1** C
2° chloride—Cl on C bonded to **2** C's
3° alcohol—OH on C bonded to **3** C's

2 Naming an alcohol using the IUPAC system (12.2B)

1 Name the parent.	2 Number the chain.	3 Name and number the substituents and combine the parts.
• Find the longest chain that contains the carbon bonded to the OH group. • Use the suffix -ol.	• Number to give the OH group the lower number. • Numbering from right to left puts the OH group on C1.	• This compound has one substituent, a methyl group at C2.

4 C's in the longest chain → **butanol**

1-butanol

methyl at C2

Answer: 2-methyl-1-butanol

3 Naming ethers and alkyl halides

1 Naming an ether (12.3)	2 Naming an alkyl halide (12.6)
• For ethers with different alkyl groups bonded to oxygen, name both alkyl groups, arrange them alphabetically, and add the word **ether**. • For ethers with identical alkyl groups, name the alkyl group and add the prefix **di-**.	• Name an alkyl halide as an alkane with a halogen substituent. • Change the -ine ending of the name of the halogen to the suffix **-o**.

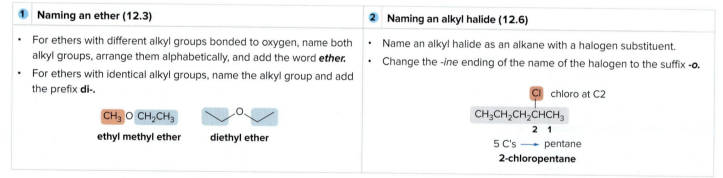

CH_3 O CH_2CH_3

ethyl methyl ether   **diethyl ether**

Cl   chloro at C2

$CH_3CH_2CH_2CHCH_3$

5 C's → pentane

2-chloropentane

4 Determining relative boiling points by identifying functional groups (12.3)

1 Identify the functional groups.	2 Determine the type and strength of intermolecular forces.
	• The *stronger* the intermolecular forces, the *higher* the boiling point. • **C** is an alkane with nonpolar C—C and C—H bonds, so it has the weakest intermolecular forces. • **B** is an ether with two polar C—O bonds, so it has dipole–dipole interactions. • **A** is an alcohol with an OH group, so it is capable of intermolecular hydrogen bonding, the strongest force. • In order of increasing boiling point: **C < B < A**.

A   **B**   **C**
alcohol   ether   alkane

5 Drawing the products formed when alcohols undergo dehydration and oxidation (12.5); example: cyclohexanol

1 Dehydration (12.5A)	**2** Oxidation (12.5B)
• Treatment of an alcohol with H_2SO_4 results in **dehydration,** elimination of the elements of H and OH from two adjacent carbons to form a C=C.	• Oxidation of an alcohol occurs by replacing the C—H bonds on the **carbon bearing the OH group** by C—O bonds.

6 Naming an aldehyde using the IUPAC system (12.8A)

1 Name the parent.	**2** Number the chain.	**3** Name and number the substituents and combine the parts.
• Find the longest chain that contains the carbonyl carbon. • Use the suffix *-al*. $CH_3CH_2CH_2CH_2$ — C — C — H with H, O above and CH_2CH_3 below **6** C's ⟶ **hexan*al***	• Number to put the C=O at C**1**, but omit this number from the name. **6** ... $CH_3CH_2CH_2CH_2$ — C — C — H numbered 2, 1 CH_2CH_3	• This compound has one substituent, an ethyl group at C2. $CH_3CH_2CH_2CH_2$ — C — C — H CH_2CH_3 ethyl at C2 **Answer: 2-ethylhexanal**

7 Naming a ketone using the IUPAC system (12.8B)

1 Name the parent.	**2** Number the ring.	**3** Name and number the substituents and combine the parts.
• Name the ring. • Use the suffix *-one*. CH_3 on cyclopentane ring with =O cyclopentane ⟶ cyclopentan*one*	• Number to put the C=O at C**1**, but omit this number from the name. CH_3 on ring numbered 3, 2, 1 with =O	• This compound has one substituent, a methyl group at C3. CH_3 on ring position 3 with =O methyl at C3 **Answer: 3-methylcyclopentanone**

8 Drawing an oxidation product of an aldehyde (12.10)

1 Basic principles	**2** Draw the product
• Locate the H atom bonded to the C=O. • RCHO are oxidized to **RCO₂H** with $K_2Cr_2O_7$ or Tollens reagent. • RCOR' are *not* oxidized with either reagent. ketone no reaction ⟶ $K_2Cr_2O_7$ or Ag_2O, NH_4OH	• Replace the C–H bond of the aldehyde with C–OH.

9 Locating chirality centers in a molecule (12.11)

1 Omit the C atoms that can't be chirality centers.	**2 Check remaining C's to see if they are bonded to four different groups.**
• Omit CH_2 and CH_3 groups. • Omit C's that are part of a multiple bond.	• A **chirality center** is a carbon bonded to four different groups. • Two carbons in this molecule are each bonded to four different groups, making them chirality centers.
 Omit C's labeled in gray.	 Two chirality centers are labeled in blue.

10 Drawing two enantiomers using Fischer projections (12.11); example: $HOCH_2CHBrCl$

1 Draw the chirality center of each enantiomer with the horizontal bonds on wedges and the vertical bonds on dashed wedges.	**2 Replace each chirality center with a cross.**
• Arrange the four groups on the chirality center—H, Br, Cl, and CH_2OH—arbitrarily in the first enantiomer. • Draw the second enantiomer by arranging the substituents in the mirror image so they are a reflection of the groups in the first molecule.	• In a Fischer projection, a carbon atom is located at the intersection of the two lines of the cross. • The **horizontal lines** represent bonds that come *forward,* on wedges. • The **vertical lines** represent bonds that go *back,* on dashed wedges.
 <center>enantiomers</center>	 <center>Fischer projection formulas of enantiomers A and B</center>

CHAPTER 12 SELF-TEST

The Self-Test consists of short-answer questions that test an understanding of definitions, equations, and other material encountered within the chapter. Answers to each question are provided at the end of the chapter. For any questions you are unable to answer correctly, revisit the appropriate section(s) of the text; then proceed to working the end-of-chapter problems to further reinforce your understanding of the chapter material.

1. Which term from "Key Terms" answers each question?
    a. What type of reaction results in an increase in the number of C—O bonds or a decrease in the number of C—H bonds?
    b. What type of compound contains a sulfur–sulfur bond?
    c. What term is used for a carbon bonded to four different groups?
    d. What term is used for a molecule that is superimposable on its mirror image?
    e. What type of reaction results in the loss of water from a starting material?

2. How many chirality centers are located in each compound: (a) $(CH_3)_2CHCH(Cl)C(CH_3)_3$; (b) $CH_3CH_2CH(OH)CH(OH)CH_3$; (c) $(CH_3)_2C(Br)CH_2CH_3$?

3. Give an acceptable name for each compound:
    (a) $(CH_3)_2CHCH_2CH_2CH_2OH$; (b) $CH_3CH_2OCH_2CH_2CH_2CH_3$;
    (c) $CH_3CH_2CH(CH_3)CH(CH_3)CH_2CHO$.

4. Which compound(s) are oxidized with $K_2Cr_2O_7$:
    (a) $CH_3CH_2CH_2OH$; (b) $(CH_3CH_2)_2C=O$; (c) $CH_3CH_2CH(CH_3)CHO$?

5. Rank the following compounds in order of increasing boiling point: $CH_3CH_2CH_2CH_2CH(CH_3)_2$ (**A**), $CH_3CH_2CH_2CH(CH_3)CH_2OH$ (**B**), $CH_3CH_2CH(CH_3)CH_2OCH_3$ (**C**).

6. Classify each labeled alcohol and alkyl halide as 1°, 2°, or 3°.

7. (a) Which structure (**B–D**) is identical to Fischer projection **A**? (b) Which structure is the enantiomer of **A**? (c) Which structure is a constitutional isomer of **A**?

UNDERSTANDING KEY CONCEPTS

12.1   Locate the chirality center in the neurotransmitter norepinephrine. Classify the hydroxyl group of the alcohol (not the phenols) as 1°, 2°, or 3°.

norepinephrine

12.2   Locate the chirality center in the general anesthetic isoflurane. Why does isoflurane have a bent arrangement around its O atom?

isoflurane

12.3   Consider the following ball-and-stick model of an alcohol.

    a. Locate the chirality center.
    b. Classify the alcohol as 1°, 2°, or 3°.
    c. Name the alcohol.
    d. Draw the structure of the product formed when the alcohol is oxidized.

12.4   Consider the following ball-and-stick model.

    a. Locate the chirality center.
    b. Is the carbonyl group part of an aldehyde or a ketone?
    c. Classify each hydroxyl group as 1°, 2°, or 3°.
    d. What product is formed when this compound is treated with Tollens reagent?

12.5   Name each compound.

    a. $CH_3CHCH_2CH_2CH_3$ with F substituent

    b. $CH_3CH_2OCH_2CH_2CH_2CH_3$

    c.

    d. (structure)

12.6   Name each compound.

    a. $CH_3CH_2CHCH_2CH_2CH(CH_3)_2$ with Cl substituent

    b. $CH_3OCH_2CH_2CH_3$

    c. (structure)

    d. (structure)

12.7   Answer the following questions about alcohol **A**.

A

    a. Give the IUPAC name.
    b. Classify the alcohol as 1°, 2°, or 3°.
    c. Draw the products formed when **A** is dehydrated with H_2SO_4.
    d. What product is formed when **A** is oxidized with $K_2Cr_2O_7$?
    e. Draw a constitutional isomer of **A** that contains an OH group.
    f. Draw a constitutional isomer of **A** that contains an ether.

12.8   Answer the questions in Problem 12.7 for alcohol **B**.

B

12.9   Rank the following compounds in order of increasing melting point.

    A   **B**   **C**

12.10   Menthone and menthol are both isolated from mint. Explain why menthol is a solid at room temperature but menthone is a liquid.

    menthone   menthol

12.11   Hydroxydihydrocitronellal is an example of a compound that has two enantiomers that smell different. One enantiomer smells minty, and the other smells like lily of the valley.

$$CH_3CCH_2CH_2CH_2CHCH_2-C{\scriptstyle H}^{\scriptstyle O}$$

with OH, CH_3, CH_3 groups

hydroxydihydrocitronellal

    a. Locate the chirality center in this compound.
    b. Draw Fischer projection formulas for both enantiomers.

12.12   Two enantiomers can sometimes taste very different. L-Leucine, a naturally occurring amino acid used in protein synthesis, tastes bitter, but its enantiomer, D-leucine, tastes sweet.

$$H_2N-\underset{CH_2CH(CH_3)_2}{\overset{COOH}{\underset{|}{\overset{|}{C}}}}-H$$

leucine

    a. Locate the chirality center in leucine.
    b. Draw Fischer projection formulas for both enantiomers.

ADDITIONAL PROBLEMS

Alcohols, Ethers, Alkyl Halides, and Thiols

12.13 Classify each alcohol as 1°, 2°, or 3°.

a. $CH_3CH_2CH_2OH$

b.

12.14 Label each halide and alcohol in plocamenol A, a marine natural product, as 1°, 2°, or 3°.

plocamenol A

12.15 Draw the structure of a molecule that fits each description:
  a. a 2° alcohol of molecular formula $C_6H_{14}O$
  b. an ether with molecular formula $C_6H_{14}O$ that has a methyl group bonded to oxygen
  c. a thiol that contains four carbons

12.16 Draw the structure of a molecule that fits each description:
  a. a 2° alcohol of molecular formula $C_6H_{12}O$
  b. a thiol of molecular formula C_3H_8S
  c. a 1° alkyl iodide with molecular formula $C_6H_{13}I$

Nomenclature of Alcohols, Ethers, and Alkyl Halides

12.17 Name each molecule depicted in the ball-and-stick models.

a.

b.

12.18 Name each molecule depicted in the ball-and-stick models.

a.

b.

12.19 Give the IUPAC name for each alcohol.

a. $CH_3CHCCH_2CH_2CH_3$ with HO and CH_3 above, CH_3 below

c.

b. $(CH_3)_2CHCH_2CHCH_2CH_3$ with $CH_2CH_2CH_2OH$ below

12.20 Give the IUPAC name for each alcohol.

a. $CH_3CH_2CH_2CHCHCH_2CH_3$ with CH_2OH above

b. $CH_3(CH_2)_3CHCH_3$ with CH_2CH_2OH below

c. (cyclopentane structure with OH and ethyl and methyl)

12.21 Give the structure corresponding to each name.
  a. 3-hexanol
  b. dibutyl ether
  c. 2-methylcyclopropanol
  d. 1-chloro-3,5-dimethylheptane

12.22 Give the structure corresponding to each name.
  a. 3-methyl-3-pentanol
  b. ethyl propyl ether
  c. 2-bromo-2,4-dimethylhexane
  d. 3,5-dimethylcyclohexanol

12.23 Draw the structures and give the IUPAC names for all constitutional isomers of molecular formula $C_7H_{16}O$ that contain an OH group and have seven carbons in the longest chain.

12.24 Draw structures for the four constitutional isomers of molecular formula C_4H_9Br. Give the IUPAC name for each alkyl halide.

Physical Properties

12.25 Explain the following observation. Dimethyl ether [$(CH_3)_2O$] and ethanol (CH_3CH_2OH) are both water-soluble compounds. The boiling point of ethanol (78 °C), however, is much higher than the boiling point of dimethyl ether (–24 °C).

12.26 Explain why two four-carbon organic molecules have very different solubility properties: 1-butanol ($CH_3CH_2CH_2CH_2OH$) is water soluble but 1-chlorobutane ($CH_3CH_2CH_2CH_2Cl$) is water insoluble.

12.27 Which compound in each pair has the higher boiling point?

  a. $(CH_3)_3CCH_2CH_2CH_3$ or $(CH_3)_3CCH_2CHO$

  b. (cyclopentane)—$COCH_3$ or (cyclopentane)—CH_2CH_2OH
      X   **Y**

12.28 Which compound in each pair is more water soluble?

  a. $CH_3(CH_2)_6CHO$ or $CH_3(CH_2)_7OH$

  b. (benzene)—C(=O)—CH_3 or (benzene)—$CH(CH_3)_2$
      Z   **W**

Reactions of Alcohols and Thiols

12.29 Draw the product formed when each alcohol is dehydrated with H_2SO_4.

  a. (cyclobutane)—OH

  b. (benzene ring with $CH(OH)CH_2CH_3$)

12.30 Draw the product formed when each alcohol is dehydrated with H_2SO_4.

  a. $(CH_3)_2CHCH_2CH_2CH_2OH$
  b. $(CH_3CH_2)_3COH$

12.31 Draw the product formed when each alcohol is oxidized with $K_2Cr_2O_7$. In some cases, no reaction occurs.

a. $CH_3(CH_2)_6CH_2OH$

b. $(CH_3CH_2)_2CHOH$

c.

12.32 Draw the product formed when each alcohol is oxidized with $K_2Cr_2O_7$. In some cases, no reaction occurs.

a.

b. $CH_3(CH_2)_8CH_2OH$

c. $(CH_3CH_2)_3COH$

12.33 Cortisol is an anti-inflammatory agent that also regulates carbohydrate metabolism. What oxidation product is formed when cortisol is treated with $K_2Cr_2O_7$?

cortisol

12.34 Xylitol is a nontoxic compound as sweet as table sugar but with only one-third the calories, so it is often used as a sweetening agent in gum and hard candy.

xylitol

a. Classify the OH groups as 1°, 2°, or 3°.

b. What product is formed when only the 1° OH groups are oxidized?

c. What product is formed when all of the OH groups are oxidized?

12.35 What products are formed when **X** is treated with each reagent: (a) H_2SO_4; (b) $K_2Cr_2O_7$?

X

12.36 What products are formed when **Y** is treated with each reagent: (a) H_2SO_4; (b) $K_2Cr_2O_7$?

Y

12.37 What disulfide is formed when each thiol is oxidized?

a. —SH

b. $CH_3(CH_2)_4SH$

12.38 The smell of fried onions is in part determined by the two disulfides below. What thiols are formed when each disulfide is reduced?

a. $CH_3S—SCH_2CH_2CH_3$

b. $CH_2{=}CHCH_2S—SCH_2CH_2CH_3$

Structure and Nomenclature of Aldehydes and Ketones

12.39 Give an acceptable name for each carbonyl compound depicted in the ball-and-stick models.

a.

b.

12.40 Give an acceptable name for each carbonyl compound depicted in the ball-and-stick models.

a.

b.

12.41 Give an acceptable name for each aldehyde or ketone.

a. $CH_3CH_2CHCH_2CHCH_2CHO$ (with CH_3 and CH_3 substituents)

b. $CH_3CH_2CH_2CHCH_2CH_2CH_3$ (with CH_2CHO substituent)

c.

d. $(CH_3)_2CH$—C(=O)—$CHCH_2CH_3$ (with CH_3 substituent)

12.42 Give an acceptable name for each aldehyde or ketone.

a. $CH_3CH_2CH_2CH_2CHCHCH_3$ (with CH_3 and CH_2CHO substituents)

b. $CH_3CH_2CH_2$—C—$CH_2CH_2CHCH_2CHO$ (with CH_3, CH_3, CH_2CH_3 substituents)

c. CH_3CH_2—C(=O)—$(CH_2)_5CH(CH_3)_2$

d.

12.43 Draw the structure corresponding to each name.

a. 3,4-dimethylhexanal

b. 4-hydroxyheptanal

c. 3,3-dimethyl-2-hexanone

d. 2,4,5-triethylcyclohexanone

12.44 Draw the structure corresponding to each name.

a. 2-propylheptanal

b. 3,4-dihydroxynonanal

c. 1-chloro-3-pentanone

d. 3-hydroxycyclopentanone

12.45 Draw the structure of the four constitutional isomers of molecular formula $C_6H_{12}O$ that contain an aldehyde and four carbons in the longest chain. Give the IUPAC name for each aldehyde.

12.46 Draw the structure of the three isomeric ketones of molecular formula $C_5H_{10}O$. Give the IUPAC name for each ketone.

Oxidation Reactions

12.47 What product is formed when each compound is treated with $K_2Cr_2O_7$?

a. $CH_3(CH_2)_4CHO$

b. —CH_2CHO

c. $\sim\sim\sim$OH

12.48 What product is formed when each compound is treated with $K_2Cr_2O_7$?

a.

c.

b. $CH_3(CH_2)_8CHO$

12.49 What product is formed when each compound in Problem 12.47 is treated with Tollens reagent (Ag_2O, NH_4OH)? With some compounds, no reaction occurs.

12.50 What product is formed when each compound in Problem 12.48 is treated with Tollens reagent (Ag_2O, NH_4OH)? With some compounds, no reaction occurs.

12.51 Benzaldehyde is the compound principally responsible for the odor of almonds. What products are formed when benzaldehyde is treated with each reagent?

a. $K_2Cr_2O_7$
b. Ag_2O, NH_4OH

benzaldehyde

12.52 Answer each question about phenylacetaldehyde, depicted in the ball-and-stick model and one of the components responsible for the fragrance of the plumeria flower.

Don Mammoser/Shutterstock

phenylacetaldehyde

a. Draw a skeletal structure for phenylacetaldehyde.
b. How many trigonal planar carbons are present in phenylacetaldehyde?
c. What product is formed when phenylacetaldehyde is treated with Tollens reagent?

Chiral Compounds, Chirality Centers, and Enantiomers

12.53 Classify each object as chiral or achiral.

a.   b.   c.

(a): Glow Images; (b): Mark Dierker/McGraw Hill; (c): Sylvain Sonnet/Photographer's Choice RF/Getty Images

12.54 Classify each object as chiral or achiral.

a.   b.   c.

(a): Photodisc/Getty Images; (b): Electronic Publishing Services, Inc., NY/McGraw Hill; (c): India Images/Dinodia Photos/Alamy Stock Photo

12.55 Label each compound as chiral or achiral.

a. halothane (general anesthetic)

b. dichloromethane (common solvent)

12.56 Label each compound as chiral or achiral.

a. CH_3CH_2—O—CH_2CH_3
diethyl ether (general anesthetic)

b. glycerol (triol used in lotions)

12.57 Label the chirality center (if one exists) in each compound. Compounds contain zero or one chirality center.

a. CH_3CH_2CHCl with Cl

c. CH_3CHCH_3 with Cl

b. $CH_3CHCHCl_2$ with Cl

d. $CH_3CCH_2CHCH_3$ with CH_3, Cl, H

12.58 Label the chirality center (if one exists) in each compound. Compounds contain zero or one chirality center.

a. CH_3—C(=O)—CH_2CH_3

c. CH_3CHCHO with CH_3

b. CH_3—C(=O)—$CHCH_2CH_3$ with CH_3

d. CH_3CHCHO with Cl

12.59 How many chirality centers are present in each compound?

a.

b.

12.60 Locate the chirality centers in aspartame, the synthetic sweetener sold under the trade name Equal.

Jill Braaten/McGraw Hill

aspartame

Isomers and Fischer Projections

12.61 Consider the ball-and-stick model of **A**, and label **B** and **C** as either identical to **A** or an enantiomer of **A**.

A   **B**   **C**

12.62 Consider the ball-and-stick model of **D**, and label **E** and **F** as either identical to **D** or an enantiomer of **D**.

D	**E**	**F**

12.63 How are the compounds in each pair related? Are they identical molecules or enantiomers?

a.   CH₃CH₂CH₂—C—CH₃   and   CH₃—C—CH₂CH₂CH₃
with Br above and Br below on each

b.   (3-D structure with Br, CH₃, Cl, H)   and   (3-D structure with Br, Cl, CH₃, H)

12.64 How are the compounds in each pair related? Are they identical molecules or enantiomers?

a.   HOCH₂—C—OH   and   HO—C—CH₂OH
with COOH above and CH₃ below

b.   CH₃O—C—CHO   and   OHC—C—OCH₃
with OCH₃ above and CH₃ below

12.65 Convert each three-dimensional representation into a Fischer projection.

a.   H—C—OH
with CHO above and CH₃ below

b.   H—C—NH₂
with COCH₃ above and CH₂CH₃ below

12.66 Convert each three-dimensional representation into a Fischer projection.

a.   CH₃—C—H
with OCH₂CH₃ above and N(CH₃)₂ below

b.   HO—C—CH₃
with CHO above and CH(CH₃)₂ below

Applications

12.67 Lactic acid is a product of glucose metabolism. During periods of strenuous exercise, lactic acid forms faster than it can be oxidized, resulting in the aching feeling of tired muscles.

CH₃CHCO₂H
with OH below

lactic acid

a. Locate the chirality center in lactic acid.
b. Draw Fischer projection formulas for both enantiomers of lactic acid.
c. What product is formed when lactic acid is oxidized by $K_2Cr_2O_7$?

12.68 Plavix, the trade name for the generic drug clopidogrel, is used in the treatment of coronary artery disease. Like many newer drugs, Plavix is sold as a single enantiomer.

clopidogrel
Trade name: Plavix

a. Locate the chirality center.
b. Draw Fischer projection formulas of both enantiomers.

12.69 Diethyl ether [$(CH_3CH_2)_2O$] is an imperfect anesthetic in part because it readily combusts in air to form CO_2 and H_2O. Write a balanced equation for the complete combustion of diethyl ether.

12.70 (a) Translate the ball-and-stick model of sevoflurane, an inhaled, sweet-smelling general anesthetic, into a complete structure with all atoms, bond lines, and lone pairs. (b) Give two reasons why halogenated ethers such as sevoflurane are now used in place of diethyl ether as anesthetics.

sevoflurane

CHALLENGE PROBLEMS

12.71 Identify the nine chirality centers in sucrose, the sweet-tasting carbohydrate we use as table sugar.

sucrose

12.72 Dehydration of alcohol **C** forms two products of molecular formula $C_{14}H_{12}$ that are isomers, but they are not constitutional isomers. Draw the structures of these two products.

C

BEYOND THE CLASSROOM

12.73 Thalidomide and naproxen are two drugs that contain a chirality center. Research the structure of each drug and label the chirality center. Each drug has two enantiomers, one of which has beneficial therapeutic effects, whereas the other is harmful. How does the biological activity of the enantiomers of each drug differ? Can the drug be used effectively now that the difference between the enantiomers is understood? See if you can find other examples of drugs whose enantiomers have very different properties.

12.74 Many alcohols and ketones derived from plant sources have pleasant odors. Fragrant alcohols include linalool, sclareol, patchoulol, and nerolidol, while fragrant ketones include camphor, carvone, beta-ionone, and *cis*-jasmone. Pick one alcohol and one ketone from these lists or use other compounds of your choosing that contain these functional groups. Draw the chemical structure. Where does each compound occur in nature? Is the molecule used as a flavor or fragrance in commercial products? Does the compound have any useful medicinal activity? Is the nature-made compound used in commercial products, or is a synthetic compound that has been prepared in the laboratory used?

CHAPTER 12 ANSWERS

Concept Check Answers

12.1 a. **A:** two halides (one chloride, one bromide), ether in ring
       B: thiol
       C: three hydroxyl groups, ether
    b. The OH on the benzene ring of **C** is a phenol.

12.2 a,d: aldehyde   b,c: ketone

12.3 a.   b. $(CH_3)_3COH$

12.4 a,c: insoluble   b,d: soluble

12.5 a. butyl methyl ether   b. dipropyl ether

12.6 $CH_3OCH_2CH_2CH_3$, $(CH_3CH_2)_2O$, $(CH_3)_2CHOCH_3$

12.7 a. $CF_3CH(F)OCHF_2$
    b. By the rule "like dissolves like," the weakly polar organic compound (desflurane) is soluble in an organic solvent.

12.8 a. 1°   b. 3°   c. 2°

12.9 a. 3°
    b. 2° OH on six-membered ring, 3° OH on five-membered ring, 1° OH on CH_2 group

12.10 a. $CH_3CH_2CH_2F < CH_3 CH_2CH_2Cl < CH_3CH_2CH_2I$
    b. $CH_3(CH_2)_4CH_3 < CH_3(CH_2)_5Cl < CH_3(CH_2)_5Br$

12.11 a. $(CH_3)_2CHCH(Cl)CH_2CH_2CH_3$
    b. $(CH_3)_3CCH_2CH(CH_2CH_3)CH(I)CH_2CH_2CH_3$
    c. $Br_2CHCH_2CH(Br)CH_2CH_2CH_3$
    d. $CH_3CH_2CH_2Cl$

12.12 2° OH, 2° and 3° fluoride

12.13 a. $CH_3CH_2CH_2SSCH_2CH_2CH_3$
    b. $CH_3(CH_2)_3SH$ and CH_3CH_2SH

12.14 $2\ CH_2{=}CHCH_2SH$

12.15 a.   c.
    b. $(CH_3CH_2)_2CO$ d. $CH_3(CH_2)_5CHO$

12.16 Acetone is soluble in water and organic solvents. Progesterone is soluble only in organic solvents.

12.17 Dioxybenzone has two OH groups, making it the most water soluble and most easily washed off.

12.18 a,d: achiral   b,c: chiral

12.19 a. $CH_3CH_2CH_2CH(OH)CH_3$
    b. $HOOCCH(NH_2)CH(CH_3)CH_2CH_3$
    c. $CH_3CH_2CH(Br)CH_2CH_2CH(Cl)CH_3$

12.20

two chirality centers

12.21 a,b:

Practice Problem Answers

12.1 a. 3°   b. 1°

12.2 two 1° OH, four 2° OH

12.3 a. 2-heptanol   c. 2-methylcyclohexanol
    b. 4-ethyl-3-hexanol

12.4 a. $CH_3CH_2CH_2CH(OH)CH_2CH_2C(CH_3)_3$
    b. $CH_3CH_2CH(OH)CH(CH_2CH_2CH_3)CH(CH_3)CH_2CH_3$
    c.

12.5 a. $CH_3(CH_2)_5OCH_3$ c. $CH_3(CH_2)_6OH$
    b.   d. $CH_3(CH_2)_5OCH_3$

12.6 a. $CH_3CH{=}CH_2$ b.   c.

12.7 a. $CH_3CCH_2CH_2CH_3$ (with =O below)

c. (cyclopentanone structure with =O)

b. $(CH_3)_3CCH_2CO_2H$

d. no reaction

12.8 a. 5-methylhexanal

c. 2,5,6-trimethyloctanal

b. 3,3,4,4-tetramethylpentanal

12.9 a. $CH_3CH(Cl)CHO$

b. $(CH_3CH_2)_2CHCH(CH_2CH_3)CH(CH_2CH_3)CH_2CHO$

c. $CH_3CH_2CH_2CH(CH_2CH_3)CH_2CH_2CH(CH_2CH_3)CH_2CHO$

d. (benzene ring with CHO and CH_2CH_3 substituents)

12.10 a. $(CH_3)_2CH(CH_2)_6CHO$

b. $CH_3CH_2CH(CH_3)(CH_2)_6CHO$

12.11 a. 4-methyl-3-heptanone

c. 2,2-dimethyl-3-heptanone

b. 2-methylcyclopentanone

12.12 a. $CH_3(CH_2)_6CO_2H$

c. (cyclopentane with C(=O)OH group)

b,d: no reaction

12.13 a. $CH_3CH(Cl)CH_2CH_3$

c. $CH_2=CHCH(OH)CH_3$

b. none

d. (benzene ring with $CH(Br)CH_3$ group)

12.14 a. C bonded to two six-membered rings

b. C bonded to benzene ring and O atom

12.15 a. (Fischer projection: CH_3 top, H—left, Br—right, Cl bottom)

b. (Fischer projection: COOH top, H—left, OH—right, CH_2Cl bottom)

12.16 a. (structure: CH_3—C with H top, $CH_2CH_2CH_3$ right, OH bottom) $CH_3CH_2CH_2$—C with H top, CH_3 right, OH bottom

b. CH_3CH_2—C with H top, CH_2Cl right, Cl bottom; $ClCH_2$—C with H top, CH_2CH_3 right, Cl bottom

12.17 $HOCH_2CH_2$—C (with CH_3 top, $CH_2CH_2CH_3$ wedge, SH bottom) ; $CH_3CH_2CH_2$—C (with CH_3 top, CH_2CH_2OH wedge, SH bottom)

Self-Test Answers

1. a. oxidation   c. chirality center   e. dehydration

   b. disulfide   d. achiral

2. a. one   b. two   c. zero

3. a. 4-methyl-1-pentanol

   b. butyl ethyl ether

   c. 3,4-dimethylhexanal

4. a, c

5. **A < C < B**

6. a. 1°   b. 2°   c. 3°

7. a. **B**   b. **D**   c. **C**

Odd-Numbered Problem Answers

12.1 chirality center: C bonded to benzene, OH, H, CH_2NH_2; 2° alcohol

12.3 a. $CH_3CH(OH)CH_2CH_3$ c. 2-butanol

   b. 2° alcohol   d. $CH_3COCH_2CH_3$

12.5 a. 2-fluoropentane   c. 3-methylhexanal

   b. butyl ethyl ether   d. 4-methyl-2-pentanone

12.7 a. 3-ethylcyclopentanol

d. (cyclopentanone with CH_2CH_3 substituent)

   b. 2°

e. (cyclopentane with CH_2CH_3 and OH substituents)

   c. (cyclopentene with CH_2CH_3) and (cyclopentene with CH_2CH_3)

f. (cyclopentane with OCH_2CH_3)

12.9 **C < A < B**

12.11 a. $(CH_3)_2C(OH)(CH_2)_3CH(CH_3)CH_2CHO$

   b. CH_3—C with CH_2CHO top, H right, $(CH_2)_3C(CH_3)_2OH$ bottom ; H—C with CH_2CHO top, CH_3 right, $(CH_2)_3C(CH_3)_2OH$ bottom

12.13 a. 1°   b. 2°

12.15 a. $CH_3CH_2CH(OH)CH_2CH_2CH_3$

   b. CH_3—O—$CH_2CH_2CH_2CH_2CH_3$

   c. $CH_3CH_2CH_2CH_2SH$

12.17 a. 2-methyl-3-pentanol

   b. 1-chloro-2-methylcyclopentane

12.19 a. 3,3-dimethyl-2-hexanol

   b. 4-ethyl-6-methyl-1-heptanol

   c. 4-butylcyclohexanol

12.21 a. $CH_3CH_2CH(OH)CH_2CH_2CH_3$

   b. $CH_3CH_2CH_2CH_2OCH_2CH_2CH_2CH_3$

   c. (cyclopropane with OH and CH_3 substituents)

   d. $ClCH_2CH_2CH(CH_3)CH_2CH(CH_3)CH_2CH_3$

12.23 $CH_3(CH_2)_6OH$ $CH_3(CH_2)_4CH(OH)CH_3$

   1-heptanol   2-heptanol

   $CH_3(CH_2)_3CH(OH)CH_2CH_3$ $(CH_3CH_2CH_2)_2CHOH$

   3-heptanol   4-heptanol

12.25 Ethanol can hydrogen bond with itself and water. The stronger intermolecular forces make its boiling point higher. Both ethanol and dimethyl ether have only two carbons and can hydrogen bond to H_2O, so both are water soluble.

12.27 a. $(CH_3)_3CCH_2CHO$ b. **Y**

12.29 a. (square/cyclobutane)   b. (benzene ring with $CH=CHCH_3$)

12.31 a. $CH_3(CH_2)_6CO_2H$ b. $(CH_3CH_2)_2CO$ c. no reaction

12.33

12.35 a. $CH_3CH_2CH_2CH=CHCH_2CH_3$ b. $(CH_3CH_2CH_2)_2CO$

12.37 a.

b. $CH_3(CH_2)_4S-S(CH_2)_4CH_3$

12.39 a. 2-methylpentanal   b. 3-ethylcyclohexanone

12.41 a. 3,5-dimethylheptanal   c. 2,6-dimethylcyclohexanone
b. 3-propylhexanal   d. 2,4-dimethyl-3-hexanone

12.43 a. $CH_3CH_2CH(CH_3)CH(CH_3)CH_2CHO$ c. $CH_3CH_2CH_2C(CH_3)_2COCH_3$

b. $CH_3CH_2CH_2CH(OH)CH_2CH_2CHO$ d.

12.45 $CH_3CH_2C(CH_3)_2CHO$ $(CH_3)_3CCH_2CHO$
2,2-dimethylbutanal   3,3-dimethylbutanal

$(CH_3)_2CHCH(CH_3)CHO$ $(CH_3CH_2)_2CHCHO$
2,3-dimethylbutanal   2-ethylbutanal

12.47 a. $CH_3(CH_2)_4COOH$ c. $CH_3(CH_2)_4COOH$

b. $-CH_2COOH$

12.49 a. $CH_3(CH_2)_4COOH$ c. no reaction

b. $-CH_2COOH$

12.51 a,b: $-CO_2H$

12.53 a,c: achiral   b. chiral

12.55 a. chiral   b. achiral

12.57 a,c: none
b. $CH_3CH(Cl)CHCl_2$ d. $(CH_3)_2CHCH_2CH(Cl)CH_3$

12.59 a. three   b. two

12.61 **A** and **B** are enantiomers.
A and **C** are identical.

12.63 a. identical   b. enantiomers

12.65 a.

b.

12.67 a. $CH_3CH(OH)CO_2H$

b.

c. CH_3COCO_2H

12.69 $2 (CH_3CH_2)_2O + 13 O_2 \longrightarrow 8 CO_2 + 10 H_2O$

12.71

* = chirality center

Design Credits: DNA molecule structure background for Chapter Opener: Watchara/Shutterstock

Carboxylic Acids, Esters, Amines, and Amides

13

Capsaicin is the amide responsible for the characteristic spiciness of hot peppers. *Daniel C. Smith*

CHAPTER OUTLINE

CHAPTER GOALS

In this chapter you will learn how to:

1 Identify the characteristics of carboxylic acids, esters, amines, and amides

2 Name carboxylic acids, esters, amines, and amides

3 Draw the products of acid–base reactions of carboxylic acids

4 Explain how soap cleans away dirt

5 Draw the products of reactions involving carboxylic acids, esters, and amides

6 Draw the products of acid–base reactions of amines

7 Identify and name ammonium salts

8 Explain how penicillin works

hapter 13 discusses the chemistry and properties of **carboxylic acids (RCOOH), esters (RCOOR), amines (RNH₂, R₂NH, or R₃N),** and **amides (RCONR₂).** All four families of organic compounds occur widely in nature, and many useful synthetic compounds have been prepared as well. Simple carboxylic acids like acetic acid (CH_3COOH) have a sour taste and a biting odor, whereas simple esters have easily recognized fragrances. Common amines include the caffeine in coffee and soft drinks and the nicotine in tobacco products. Several useful drugs are amides, and all proteins, such as those that form hair, muscle, and connective tissue, contain many amide units.

13.1 Introduction

Carboxylic acids and **esters** are two families of organic molecules that contain a carbonyl group (C=O) singly bonded to an oxygen atom.

Carboxylic acid		Ester	
CH_3—C—ÖH	=	CH_3—C—ÖCH₃	=
acetic acid	CH_3CO_2H	methyl acetate	$CH_3CO_2CH_3$

- A carboxylic acid contains a carboxyl group (COOH).
- An ester contains an alkoxy group (OR) bonded to the carbonyl carbon.

The structure of a carboxylic acid is often abbreviated as RCOOH or RCO_2H, whereas the structure of an ester is abbreviated as RCOOR or RCO_2R. Keep in mind that the **central carbon in both functional groups has a double bond to one oxygen and a single bond to another.**

Amines and **amides** are nitrogen-containing organic molecules.

Amine		Amide	
H—C—ṄH₂	=	CH_3—C—ṄH₂	=
methylamine	CH_3NH_2	acetamide	CH_3CONH_2

- An amine is an organic nitrogen compound formed by replacing one or more hydrogen atoms of ammonia (NH_3) with alkyl groups.
- An amide is a carbonyl compound that contains a nitrogen atom bonded to the carbonyl carbon.

Each oxygen has two lone pairs and each nitrogen has one lone pair so each atom is surrounded by eight electrons.

Concept Check 13.1

Label each compound as a carboxylic acid, ester, or amide.

a. $CH_3CH_2CO_2CH_2CH_3$ b. $CH_3CONHCH_3$ c. $(CH_3)_3CCO_2H$ d. $(CH_3)_2CHCON(CH_3)_2$

Concept Check 13.2

Label each compound as an amine or amide.

a. $CH_3CH_2CONHCH_2CH_3$ c. $CH_3COCH_2CH_2NH_2$

b. $(CH_3)_3N$ d. $(CH_3)_2NCH_2CH_2OCH_3$

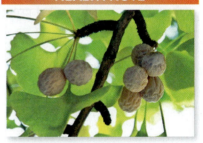
Concept Check 13.3

Identify each functional group in lisinopril, a drug used to treat high blood pressure.

lisinopril

13.2 Nomenclature of Carboxylic Acids and Esters

To name carboxylic acids and esters, we must learn the suffix that identifies each functional group in the IUPAC system.

13.2A Naming a Carboxylic Acid—RCOOH

- In the IUPAC system, carboxylic acids are identified by the suffix *-oic acid.*

To name a carboxylic acid using the IUPAC system:

1. Find the longest chain containing the COOH group, and change the *-e* ending of the parent alkane to the suffix *-oic acid.*
2. Number the carbon chain to put the COOH group at C1, but omit this number from the name. Apply all of the other usual rules of nomenclature.

Many simple carboxylic acids are often referred to by their common names. A common name uses the suffix *-ic acid.* Three common names are generally used.

formic acid
(methanoic acid)

acetic acid
(ethanoic acid)

benzoic acid
(benzenecarboxylic acid)

[IUPAC names are given in parentheses, and are rarely used.]

Sample Problem 13.1   Naming a Carboxylic Acid

Give the IUPAC name for the following carboxylic acid.

$$CH_3CHCHCH_2CH_2COOH$$
(with CH₃ substituents)

Analysis and Solution

[1] Find and name the longest chain containing COOH.

hexane ----→ hexan*oic acid*
(6 C's)

The COOH contributes one C to the longest chain.

[2] Number and name the substituents, making sure the COOH group is at C1.

two methyl substituents on C4 and C5

Answer: 4,5-dimethylhexanoic acid

Give the IUPAC name for each compound.

a. CH₃CH₂CH₂C(CH₃)₂CH₂COOH with CH₃ above and CH₃ below

b. CH₃CHCH₂CH₂COOH with Cl below

c. (CH₃CH₂)₂CHCH₂CHCOOH with CH₂CH₃ above

a. $CH_3CH_2CH_2\underset{\underset{CH_3}{|}}{\overset{\overset{CH_3}{|}}{C}}CH_2COOH$ b. $CH_3\underset{\underset{Cl}{|}}{CH}CH_2CH_2COOH$ c. $(CH_3CH_2)_2CHCH_2\underset{}{\overset{\overset{CH_2CH_3}{|}}{CH}}COOH$

Give the structure corresponding to each IUPAC name.

a. 2-bromobutanoic acid

b. 2,3-dimethylpentanoic acid

c. 2-ethyl-5,5-dimethyloctanoic acid

d. 3,4,5,6-tetraethyldecanoic acid

More Practice: Try Problems 13.5a; 13.6a; 13.17b; 13.19a, b; 13.20a, b; 13.25a, b; 13.26a, b.

13.2B Naming an Ester—RCOOR'

The names of esters are derived from the names of the parent carboxylic acids. Keep in mind that the common names **formic acid, acetic acid,** and **benzoic acid** are used for the parent acid, so these common parent names are used for their derivatives as well.

An ester has two parts to its structure, each of which must be named separately: the **RCO– group** that contains the carbonyl group and is derived from a carboxylic acid, and an **alkyl group** (designated as **R'**) bonded to the oxygen atom.

Ethyl butanoate [CH₃(CH₂)₂CO₂CH₂CH₃] is an ester isolated from mangoes.
JIANG HONGYAN/Shutterstock

an alkyl group

from RCO₂H

- In the IUPAC system, esters are identified by the suffix *-ate.*

To name an ester:

Esters are often written as RCOOR', where the alkyl group (R') is written *last.* When an ester is named, however, the R' group appears *first* in the name.

- Name the R' group bonded to the oxygen atom as an alkyl group.
- Name the RCO– group by changing the *-ic acid* ending of the parent carboxylic acid to the suffix *-ate.*

Naming an Ester

Give the IUPAC name of the following ester.

$$CH_3CH_2CH_2-\overset{\overset{\displaystyle O}{\|}}{C}-OCH_3$$

Analysis and Solution

[1] Name the alkyl group on the O atom.

$$CH_3CH_2CH_2-\overset{\overset{\displaystyle O}{\|}}{C}-OCH_3$$

methyl group

The word *methyl* becomes the first part of the name.

[2] To name the RCO– group, find and name the longest chain containing the carbonyl group, placing the C=O at C1.

$$CH_3CH_2CH_2-\overset{\overset{\displaystyle O}{\|}}{C}-OCH_3$$

butano*ic acid* - - - → butano*ate*
(4 C's)

Answer: methyl butanoate

Practice Problem 13.3

Give an acceptable name for each ester.

a. $CH_3(CH_2)_4CO_2CH_3$ b.   c. $CH_3CH_2CH_2CH_2COOCH_2CH_2CH_3$

Practice Problem 13.4

Give the structure corresponding to each name.

a. propyl propanoate   b. butyl acetate   c. ethyl hexanoate   d. methyl benzoate

---

Practice Problem 13.5

Draw the structure of ethyl formate, a compound with a rum-like odor that is partly responsible for the flavor of raspberries.

More Practice: Try Problems 13.18a; 13.19d; 13.20c; 13.25c, d; 13.26c, d.

Ethyl formate (Practice Problem 13.5) is one of the many compounds that contribute to the characteristic flavor of raspberries. *StockFood/Foodcollection*

hydrogen bond

hydrogen bond
carboxylic acid dimer

13.3   Physical Properties of Carboxylic Acids and Esters

Carboxylic acids and esters are polar compounds because they possess a polar carbonyl group. Carboxylic acids also exhibit intermolecular **hydrogen bonding** because they possess a hydrogen atom bonded to an electronegative oxygen atom. Carboxylic acids often exist as **dimers,** held together by *two* intermolecular hydrogen bonds. The carbonyl oxygen atom of one molecule hydrogen bonds to the hydrogen atom of another molecule. As a result:

- Carboxylic acids have stronger intermolecular forces than esters, giving them *higher* boiling points and melting points, when comparing compounds of similar molecular weight.

$$CH_3-C{\overset{O}{\underset{OCH_3}{}}}$$
bp 58 °C

no hydrogen bonding possible

$$CH_3CH_2-C{\overset{O}{\underset{OH}{}}}$$ hydrogen bonding possible
bp 141 °C

stronger intermolecular forces
higher boiling point

Carboxylic acids also have stronger intermolecular forces than alcohols, even though both functional groups exhibit hydrogen bonding; two molecules of a carboxylic acid are held together by *two* intermolecular hydrogen bonds, whereas only *one* hydrogen bond is possible between two molecules of an alcohol (Section 12.2). As a result:

- Carboxylic acids have *higher* boiling points and melting points than alcohols of comparable molecular weight.

$CH_3CH_2CH_2$**OH**

1-propanol
bp 97 °C

$$CH_3-C{\overset{O}{\underset{OH}{}}}$$

acetic acid
bp 118 °C

more hydrogen-bonding interactions possible
higher boiling point

Like other oxygen-containing compounds, carboxylic acids and esters having fewer than six carbons are soluble in water. Higher-molecular-weight compounds are insoluble in water because the nonpolar portion of the molecule, the C—C and C—H bonds, gets larger than the polar carbonyl group.

Concept Check 13.4

Rank the following compounds in order of increasing boiling point. Which compound is the most water soluble? Which compound is the least water soluble?

CH₂COOH — **A**
COOCH₃ — **B**
CH₂CH₂CH₂CH₃ — **C**

HCO₂H

CH₃CO₂H

13.4 Interesting Carboxylic Acids in Consumer Products and Medicines

Simple carboxylic acids have biting or foul odors. **Formic acid** (HCO₂H) is responsible for the sting of some types of ants. **Acetic acid** (CH₃CO₂H) is the sour-tasting component of vinegar. Air oxidation of ethanol (CH₃CH₂OH) to acetic acid makes "bad" wine taste sour.

13.4A FOCUS ON HEALTH & MEDICINE
Skin Care Products

Several skin care products that purportedly smooth fine lines and improve skin texture contain α-hydroxy acids (alpha-hydroxy acids). **α-Hydroxy acids contain a hydroxyl group on the carbon bonded to the carboxyl group.** Two common α-hydroxy acids are glycolic acid and lactic acid. Glycolic acid occurs naturally in sugarcane, and lactic acid gives sour milk its distinctive taste.

General structure

α-hydroxy acid   glycolic acid   lactic acid

Do products that contain α-hydroxy acids make the skin look younger? α-Hydroxy acids react with the outer layer of skin cells, causing them to loosen and flake off. Underneath is a layer of healthier looking skin that has not been exposed to the sun. In this way, these skin products do not actually reverse the aging process; rather, they remove a layer of old skin that may be less pliant or contain small age spots. They can, however, give the appearance of younger skin for a time.

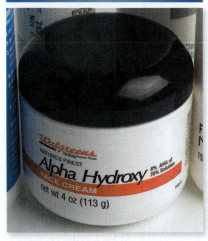

Concept Check 13.5

Which compounds are α-hydroxy acids?

a.
tartaric acid
(isolated from grapes)

b. CH₃CHCH₂—C
3-hydroxybutanoic acid
(starting material for
polymer synthesis)

c.
salicylic acid
(isolated from meadowsweet)

13.4B FOCUS ON HEALTH & MEDICINE
Aspirin and Anti-Inflammatory Agents

Aspirin and ibuprofen are common pain relievers and anti-inflammatory agents that contain a carboxyl group.

aspirin

ibuprofen

How does aspirin relieve pain and reduce inflammation? In the 1970s, it was shown that aspirin blocks the synthesis of **prostaglandins,** carboxylic acids containing 20 carbons that are responsible for mediating pain, inflammation, and a wide variety of other biological functions.

Prostaglandins are not stored in cells. Rather, they are synthesized on an as-needed basis from arachidonic acid, an unsaturated fatty acid with four cis double bonds. **Aspirin relieves pain and decreases inflammation because it prevents the synthesis of prostaglandins, the compounds responsible for both of these physiological responses, from arachidonic acid.**

arachidonic acid

Aspirin blocks this process.

PGF$_{2\alpha}$ and compounds like it cause pain and inflammation.

PGF$_{2\alpha}$
a prostaglandin

Aspirin was first used in medicine for its analgesic (pain-relieving), antipyretic (fever-reducing), and anti-inflammatory properties. Today it is also commonly used to prevent blood clots from forming in arteries. In this way, it is used to treat and prevent heart attacks and strokes.

13.5 The Acidity of Carboxylic Acids

As their name implies, **carboxylic acids are acids;** that is, **they are *proton donors*.** When a carboxylic acid is dissolved in water, an acid–base reaction occurs: the carboxylic acid donates a proton to H_2O, forming its conjugate base, a **carboxylate anion,** and water gains a proton, forming its conjugate acid, H_3O^+.

General reaction

carboxylic acid   This proton is removed.   carboxylate anion

While carboxylic acids are more acidic than other families of organic compounds, they are weak acids compared to inorganic acids like HCl or H_2SO_4. Thus, only a small percentage of a carboxylic acid is ionized in aqueous solution.

13.5A   Reaction with Bases

Carboxylic acids react with bases such as NaOH to form water-soluble salts. In this reaction, essentially all of the carboxylic acid is converted to its carboxylate anion.

This proton is transferred from the acid to the base.

- A proton is removed from acetic acid (CH_3COOH) to form its conjugate base, the acetate anion (CH_3COO^-), which is present in solution as its sodium salt, sodium acetate.
- Hydroxide (OH^-) gains a proton to form neutral H_2O.

Similar acid–base reactions occur with other hydroxide bases (KOH), sodium bicarbonate ($NaHCO_3$), and sodium carbonate (Na_2CO_3). **In each reaction, a proton is transferred from the acid (RCOOH) to the base.**

Sample Problem 13.3	Drawing the Products Formed When a Carboxylic Acid Reacts with a Base

What products are formed when propanoic acid (CH_3CH_2COOH) reacts with potassium hydroxide (KOH)?

Analysis

In any acid–base reaction with a carboxylic acid:
- **Remove a proton from the carboxyl group (COOH)** and form the carboxylate anion ($RCOO^-$).
- **Add a proton to the base.**
- Balance the charge of the carboxylate anion by drawing it as a salt with a metal cation.

Solution

This proton is transferred from the acid to the base.

The carboxylate anion is formed as a potassium salt.

Thus, CH_3CH_2COOH loses a proton to form $CH_3CH_2COO^-$, which is present in solution as its potassium salt, $CH_3CH_2COO^-$ K^+. Hydroxide (OH^-) gains a proton to form H_2O.

Practice Problem 13.6	Draw the products of each acid–base reaction.

a. [cyclohexane-C(=O)-OH] + NaOH ⟶   b. [$CH_3CH_2CH_2$-C(=O)-OH] + Na_2CO_3 ⟶

More Practice: Try Problems 13.5d, 13.6d, 13.33, 13.34.

To name the metal salts of carboxylate anions, three parts must be put together: the name of the metal cation, the parent name indicating the number of carbons in the parent carboxylic acid, and the suffix indicating that the species is a salt. The suffix is added by changing the *-ic acid* ending of the parent carboxylic acid to the suffix *-ate*.

| name of the metal cation | + | parent | + | *-ate* (suffix) |

Sample Problem 13.4 — Naming a Salt of a Carboxylate Anion

Give an acceptable name for each salt: (a) $CH_3CO_2^-Na^+$; (b) $CH_3CH_2CO_2^-K^+$.

Analysis

Name the carboxylate salt by putting three parts together:

- the name of the **metal cation**
- the **parent name** that indicates the number of carbons in the parent chain
- the suffix, *-ate*

Solution

a.

sodium cation

parent + suffix
acet-   -ate

- The first part of the name is the metal cation, sodium.
- The parent name is derived from the common name, acetic acid. Change the *-ic acid* ending to *-ate;* acetic acid → acetate.

Answer: sodium acetate

---

b.

potassium cation

parent + suffix
propano-   -ate

- The first part of the name is the metal cation, potassium.
- The parent name is derived from the IUPAC name, propanoic acid. Change the *-ic acid* ending to *-ate;* propanoic acid → propanoate.

Answer: potassium propanoate

Practice Problem 13.7

Name each salt of a carboxylic acid.

a. $CH_3CH_2CH_2CO_2^-Na^+$

b. ⟨benzene ring⟩—COO^-Li^+

More Practice: Try Problems 13.19c, 13.20d.

The salts of carboxylic acids that are formed by acid–base reactions are water-soluble ionic solids. Thus, a water-*insoluble* carboxylic acid like octanoic acid can be converted to its water-*soluble* sodium salt by reaction with NaOH.

$$CH_3(CH_2)_6\text{—}C(=O)\text{—O—H} \quad + \quad Na^+ \, OH^- \quad \longrightarrow \quad CH_3(CH_2)_6\text{—}C(=O)\text{—O}^- \, Na^+ \quad + \quad H\text{—O—H}$$

octanoic acid   base   sodium octanoate

| The acid has more than 5 C's, so it is **insoluble** in water. |

| The ionic salt is **soluble** in water. |

Salts of carboxylic acids are commonly used as preservatives. **Sodium benzoate,** which inhibits the growth of fungus, is a preservative used in soft drinks, and potassium sorbate is an additive that prolongs the shelf-life of baked goods and other foods. These salts do not kill bacteria or fungus. They increase the pH of the product, thus preventing further growth of microorganisms.

sodium benzoate

$CH_3CH=CHCH=CH\text{—}C(=O)\text{—O}^- \, K^+$

potassium sorbate

13.5B   How Does Soap Clean Away Dirt?

Soap has been used by humankind for some 2,000 years. **Soaps are salts of carboxylic acids that have many carbon atoms in a long hydrocarbon chain.** A soap has two parts.

- The ionic end is called the *polar head.*
- The carbon chain of nonpolar C—C and C—H bonds is called the *nonpolar tail.*

Dissolving soap in water forms **spherical droplets having the ionic heads on the surface and the nonpolar tails packed together in the interior.** These spherical droplets are called *micelles* and are illustrated in Figure 13.1. In this arrangement, the ionic heads can interact with the polar solvent water, and this brings the nonpolar, "greasy" hydrocarbon portion of the soap into solution.

Figure 13.1

Dissolving Soap in Water

When soap is dissolved in H$_2$O, the molecules form spherical droplets with the nonpolar tails in the interior and the polar heads on the surface.

All soaps are metal salts of carboxylate anions. The main difference between brands is the addition of other ingredients that do not alter their cleaning properties: dyes for color, scents for a pleasing odor, and oils for lubrication. *Jill Braaten/McGraw Hill*

How does soap dissolve grease and oil? The polar solvent water alone cannot dissolve dirt, which is composed largely of nonpolar hydrocarbons. When soap is mixed with water, however, the nonpolar hydrocarbon tails dissolve the dirt in the interior of the micelle. The polar head of the soap remains on the surface to interact with water. The nonpolar tails of the soap molecules are so well sealed off from the water by the polar head groups that the micelles are water soluble, so they can separate from the fibers of our clothes and be washed down the drain with water. In this way, soaps do a seemingly impossible task: they remove nonpolar hydrocarbon material from skin and clothes by dissolving it in the polar solvent water.

Cross-section of a soap micelle with a grease particle dissolved in the interior

Concept Check 13.6

Draw the structure of a soap that has a potassium cation and a carboxylate anion containing 16 carbons.

13.6 Reactions Involving Carboxylic Acids and Esters

Ethyl acetate, a common organic solvent with a very characteristic odor, is used in nail polish remover. *Jill Braaten*

Carboxylic acids and esters undergo a common type of reaction—*substitution*. When a carboxylic acid or ester (RCOZ) undergoes substitution, the group Z (OH or OR') bonded to the carbonyl carbon is *replaced* by another group of atoms Y (OR' or OH).

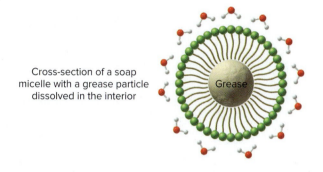

13.6A Ester Formation

Treatment of a carboxylic acid (RCOOH) with an alcohol (R'OH) in the presence of an acid catalyst forms an ester (RCOOR'). This reaction is called a **Fischer esterification.** Esterification is a substitution because the OR' group of an alcohol replaces the OH group of the starting carboxylic acid.

Sample Problem 13.5   Drawing the Product of Fischer Esterification

What ester is formed when propanoic acid (CH_3CH_2COOH) is treated with methanol (CH_3OH) in the presence of H_2SO_4?

Analysis

To draw the products of esterification, replace the **OH** group of the carboxylic acid by the **OR'** group of the alcohol, forming a new C—O bond at the carbonyl carbon.

Remove **OH** and H to form H_2O.   new C—O bond

$$\underset{\text{carboxylic acid}}{\overset{O}{\underset{OH}{||}}} \quad + \quad H\text{—OR'} \quad \longrightarrow \quad \underset{\text{ester}}{\overset{O}{\underset{OR'}{||}}} \quad + \quad H\text{—OH}$$

Solution

Replace the **OH** group of propanoic acid by the **OCH₃** group of methanol to form the ester.

new C—O bond

$$\underset{CH_3CH_2}{\overset{O}{||}}\text{—OH} \quad + \quad H\text{—OCH}_3 \quad \underset{\longleftarrow}{\overset{H_2SO_4}{\longrightarrow}} \quad \underset{CH_3CH_2}{\overset{O}{||}}\text{—OCH}_3 \quad + \quad H\text{—OH}$$

OCH₃ replaces **OH**.

Practice Problem 13.8

What ester is formed when each carboxylic acid is treated with ethanol (CH_3CH_2OH) in the presence of H_2SO_4?

a.
$$\underset{(CH_3)_2CH}{\overset{O}{||}}\text{—OH}$$

b. HCO_2H

c.
$$\underset{CH_3(CH_2)_6}{\overset{O}{||}}\text{—OH}$$

d.
⬡—CO_2H

Practice Problem 13.9

What ester **A** is formed in the following reaction? **A** was converted in one step to blattellaquinone, the sex pheromone of the female German cockroach, *Blattella germanica*.

$$\text{(structure)} + \text{(structure)} \xrightarrow{H_2SO_4} \mathbf{A} \longrightarrow \text{blattellaquinone}$$

More Practice: Try Problems 13.5e, 13.6e, 13.35, 13.36, 13.43a.

13.6B   Ester Hydrolysis

Esters are hydrolyzed with water in the presence of acid or base. **Treatment of an ester (RCOOR') with water in the presence of an acid catalyst forms a carboxylic acid (RCOOH) and a molecule of alcohol (R'OH).** This reaction is a **hydrolysis, because bonds are cleaved by reaction with water.**

$$\underset{\text{ester}}{\underset{CH_3}{\overset{O}{||}}\text{—OCH}_2CH_3} \quad + \quad H\text{—OH} \quad \underset{\longleftarrow}{\overset{H_2SO_4}{\longrightarrow}} \quad \underset{\textbf{carboxylic acid}}{\underset{CH_3}{\overset{O}{||}}\text{—OH}} \quad + \quad \underset{\text{alcohol}}{H\text{—OCH}_2CH_3}$$

Sample Problem 13.6	**Drawing the Products of Ester Hydrolysis in Acid**

What products are formed when ethyl propanoate ($CH_3CH_2CO_2CH_2CH_3$) is hydrolyzed with water in the presence of H_2SO_4?

Analysis

To draw the products of hydrolysis in acid, replace the **OR'** group of the ester by an **OH** group from water, forming a new C—O bond at the carbonyl carbon. A molecule of alcohol (R'OH) is also formed from the alkoxy group (OR') of the ester.

Solution

Replace the **OCH₂CH₃** group of ethyl propanoate by the **OH** group of water to form propanoic acid ($CH_3CH_2CO_2H$) and ethanol (CH_3CH_2OH).

Practice Problem 13.10	

What products are formed when each ester is treated with H_2O and H_2SO_4?

a. $CH_3(CH_2)_8$ —C(=O)— OCH_3

b. CH_3CHCH_2 —C(=O)— OCH_2CH_3 (with CH_3 branch)

c.   (cyclohexyl)— $CO_2CH_2CH_2CH_3$

Practice Problem 13.11	

Aspirin cannot be sold as a liquid solution for children because it slowly undergoes hydrolysis in water. What products are formed when aspirin is hydrolyzed?

aspirin

More Practice: Try Problems 13.7, 13.37, 13.41, 13.43b.

Esters are hydrolyzed in aqueous base to form carboxylate anions and a molecule of alcohol. Basic hydrolysis of an ester is called **saponification.**

ester   **carboxylate anion**   alcohol

Concept Check 13.7	What products are formed when each ester in Practice Problem 13.10 is treated with H_2O and NaOH?

13.6C   Lipid Hydrolysis

The most prevalent naturally occurring esters are the **triacylglycerols.** *Triacylglycerols* **contain three ester groups, each having a long carbon chain (abbreviated as R, R', and R'') bonded to the carbonyl group.** Triacylglycerols are **lipids;** that is, they are water-insoluble organic compounds found in biological systems. Animal fats and vegetable oils are composed of triacylglycerols.

R groups have 11–19 C's.

[Three ester groups are labeled in red.]

triacylglycerol

Animals store energy in the form of triacylglycerols kept in a layer of fat cells below the surface of the skin. This fat serves to insulate the organism, as well as provide energy for its metabolic needs for long periods of time. The first step in the metabolism of a triacylglycerol is hydrolysis of the ester bonds to form glycerol and three fatty acids—long-chain carboxylic acids. **This reaction is simply ester hydrolysis.** In cells, this reaction is carried out with enzymes called **lipases.**

$$3\,H\!-\!OH \xrightarrow{\text{lipase}}$$

Three carboxylic acids containing 12–20 C's are formed as products.

triacylglycerol   glycerol

The fatty acids produced on hydrolysis are then oxidized, yielding CO_2 and H_2O, as well as a great deal of energy.

Concept Check 13.8

What products are formed when the following triacylglycerol is hydrolyzed with water and H_2SO_4?

13.7   Amines

Amines **are organic nitrogen compounds,** formed by replacing one or more hydrogen atoms of ammonia (NH_3) with alkyl groups.

13.7A   Structure and Classification

Amines are classified as 1°, 2°, or 3° by the number of alkyl groups bonded to the *nitrogen* atom.

1° amine   2° amine   3° amine

- A primary (1°) amine has one C—N bond and the general structure RNH_2.
- A secondary (2°) amine has two C—N bonds and the general structure R_2NH.
- A tertiary (3°) amine has three C—N bonds and the general structure R_3N.

Like ammonia, **the amine nitrogen atom has a lone pair of electrons,** which is generally omitted in condensed structures. An amine nitrogen atom is surrounded by three atoms and one nonbonded electron pair, making it trigonal pyramidal in shape, with bond angles of approximately 109.5°.

$$CH_3NH_2 = \overset{\ddot{N}}{CH_3{\diagdown}{}^{\cdots\cdots}H} = \text{(model)}$$
methylamine　　109.5°　　trigonal pyramidal

The amine nitrogen can also be part of a ring. Figure 13.2 illustrates the structures of cocaine and atropine, two **alkaloids—naturally occurring amines derived from plant sources.** Each alkaloid contains a nitrogen atom in a ring.

Figure 13.2　　Biologically Active Alkaloids

coca plant　　　　　　cocaine　　　　　　atropine　　　　　　deadly nightshade plant

- Cocaine is an addictive stimulant obtained from the leaves of the coca plant.
- Atropine is isolated from *Atropa belladonna,* the deadly nightshade plant. Atropine dilates pupils, increases heart rate, and relaxes smooth muscles.
(left): Likit Supasai/Shutterstock; (right): Westend61/Shutterstock

---

Sample Problem 13.7　　　**Classifying Amines**

Classify each amine in the following compounds as 1°, 2°, or 3°. Putrescine is partly responsible for the foul odor of decaying fish. MDMA is the illegal stimulant commonly called "Ecstasy."

a.　$H_2N(CH_2)_4NH_2$

putrescine

b.

MDMA
"Ecstasy"

Analysis

To determine whether an amine is 1°, 2°, or 3°, **count the number of carbons bonded to the nitrogen atom.** A 1° amine has one C—N bond, and so forth.

Solution

Draw out the structure or add H's to the skeletal structure to clearly see how many C—N bonds the amine contains.

a. H—N—CH$_2$CH$_2$CH$_2$CH$_2$—N—H
  | |
  H   H
         putrescine

Each N is bonded to only 1 C.
Both amines are **1°.**

b.

MDMA
"Ecstasy"

This N is bonded to **2 C's,**
making it a **2° amine.**

Practice Problem 13.12

Classify each amine in the following compounds as 1°, 2°, or 3°.

a. H$_2$N(CH$_2$)$_3$NH(CH$_2$)$_4$NH(CH$_2$)$_3$NH$_2$

spermine
(isolated from semen)

b.

meperidine
(Trade name: Demerol)

Practice Problem 13.13

Label the amine and hydroxyl group in scopolamine, a drug used to treat motion sickness, as 1°, 2°, or 3°. Scopolamine is obtained from angel's trumpets, ornamental plants native to South America.

scopolamine

More Practice: Try Problems 13.15, 13.16, 13.50a, 13.53b, 13.54b.

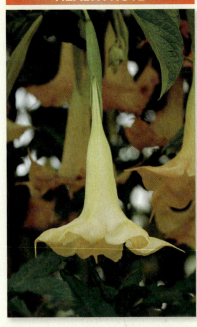

HEALTH NOTE

Angel's trumpets are ornamental plants that contain the toxic amine scopolamine (Practice Problem 13.13). Transdermal patches that release a minute dose of scopolamine are used to treat the nausea and vomiting associated with motion sickness. *James Forte/Getty Images*

13.7B Nomenclature

To name a primary (1°) amine, name the alkyl group bonded to the nitrogen atom and add the suffix **-amine,** forming a single word. For 2° and 3° amines with different alkyl groups, alphabetize the names of the alkyl groups. Secondary (2°) and 3° amines having identical alkyl groups are named by using the prefix **di-** or **tri-** with the name of the primary amine.

CH$_3$NH$_2$

methylamine

           CH$_2$CH$_3$
            |
CH$_3$CH$_2$—N—CH$_2$CH$_3$

triethylamine

           H
            |
CH$_3$CH$_2$CH$_2$—N—CH$_2$CH$_3$

ethylpropylamine

Sample Problem 13.8	**Naming Amines**

Name each amine: (a) $CH_3CH_2CH_2CH_2CH_2NH_2$; (b) $CH_3NHCH_2CH_3$.

Analysis and Solution

a. For a 1° amine, name the alkyl group and add the suffix -*amine*.

b. For a 2° amine, name each alkyl group, alphabetize the names, and add the suffix -*amine*.

$CH_3CH_2CH_2CH_2CH_2NH_2$

5 C's

Answer: pentylamine

a methyl and ethyl group on N

Answer: ethylmethylamine

Practice Problem 13.14	

Name each amine.

a. $CH_3CH_2CH_2NH_2$ 　　b. $CH_3CH_2CH_2NHCH_3$ 　　c. $(CH_3CH_2CH_2CH_2)_2NH$

More Practice: Try Problems 13.23a, 13.24a, 13.27b, 13.28a.

Aromatic amines, amines having a nitrogen atom bonded directly to a benzene ring, are named as derivatives of aniline. Use the prefix *N*- before any alkyl group bonded to the amine nitrogen.

aniline 　　　　　 *N*-ethylaniline 　　　　　 *o*-ethylaniline

Concept Check 13.9	

Draw a structure corresponding to each name: (a) *N*-methylaniline; (b) *m*-ethylaniline; (c) 3,5-diethylaniline; (d) *N,N*-diethylaniline.

13.7C Physical Properties

Many low-molecular-weight amines have *very* foul odors. **Trimethylamine** [$(CH_3)_3N$], formed when enzymes break down certain fish proteins, has the characteristic odor of rotting fish. **Cadaverine** ($NH_2CH_2CH_2CH_2CH_2CH_2NH_2$) is a poisonous diamine with a putrid odor also present in rotting fish, and partly responsible for the odor of semen, urine, and bad breath.

intermolecular hydrogen bonding

Because nitrogen is much more electronegative than carbon or hydrogen, amines contain polar C—N and N—H bonds. **Primary (1°) and 2° amines are also capable of intermolecular hydrogen bonding,** because they contain N—H bonds.

Because nitrogen is less electronegative than oxygen, however, intermolecular hydrogen bonds between N and H are *weaker* than those between O and H. As a result:

- In comparing compounds of similar molecular weight, 1° and 2° amines have *higher* boiling points than compounds incapable of hydrogen bonding, but *lower* boiling points than alcohols that have stronger intermolecular hydrogen bonds.

$CH_3CH_2OCH_2CH_3$ 　　　 $CH_3CH_2CH_2CH_2NH_2$ 　　　 $CH_3CH_2CH_2CH_2OH$

diethyl ether 　　　　　 butylamine 　　　　　 1-butanol

bp = 38 °C 　　　　　 bp = 78 °C 　　　　　 bp = 118 °C

Increasing intermolecular forces
Increasing boiling point

- Tertiary (3°) amines have lower boiling points than 1° and 2° amines of comparable molecular weight, because they have no N—H bonds.

$$CH_3CH_2N(CH_3)_2$$

3 C's bonded to N
3° amine

no N—H bond
bp = 38 °C

$$CH_3CH_2—N(H)—CH_2CH_3$$

one N—H bond
bp = 56 °C

2 C's bonded to N
2° amine

higher boiling point

Amines are soluble in organic solvents regardless of size. Amines with fewer than six carbons are water soluble because they can hydrogen bond with water. Larger amines are water insoluble because the nonpolar alkyl portion is too large to dissolve in the polar water solvent.

Sample Problem 13.9

Determining Relative Boiling Points

Which compound in each pair has the higher boiling point: (a) $CH_3CH_2NHCH_3$ or $CH_3CH_2OCH_3$; (b) $(CH_3)_3N$ or $CH_3CH_2CH_2NH_2$?

Analysis

Keep in mind the general rule: For compounds of comparable molecular weight, **the *stronger* the intermolecular forces, the *higher* the boiling point.** Compounds that can hydrogen bond have higher boiling points than compounds that are polar but cannot hydrogen bond. Polar compounds have higher boiling points than nonpolar compounds.

Solution

a. The 2° amine ($CH_3CH_2NHCH_3$) has an N—H bond, so intermolecular hydrogen bonding is possible. The ether ($CH_3CH_2OCH_3$) has only C—H bonds, so there is no possibility of intermolecular hydrogen bonding. $CH_3CH_2NHCH_3$ has a higher boiling point because it has stronger intermolecular forces.

$$CH_3CH_2—N(H)—CH_3$$

a 2° amine with an N—H bond
intermolecular hydrogen bonding
higher boiling point

$$CH_3CH_2—O—CH_3$$

an ether with only C—H bonds

b. The 1° amine ($CH_3CH_2CH_2NH_2$) has N—H bonds, so intermolecular hydrogen bonding is possible. The 3° amine [$(CH_3)_3N$] has only C—H bonds, so there is no possibility of intermolecular hydrogen bonding. $CH_3CH_2CH_2NH_2$ has a higher boiling point because it has stronger intermolecular forces.

$$CH_3CH_2CH_2NH_2$$

a 1° amine with N—H bonds
intermolecular hydrogen bonding
higher boiling point

$$CH_3—N(CH_3)—CH_3$$

a 3° amine with only C—H bonds

Practice Problem 13.15

Which compound in each pair has the higher boiling point?

a. $CH_3—C(=O)—CH_2CH_3$ or $(CH_3)_2CHCH_2NH_2$

b. $(CH_3)_2CHCH_2NH_2$ or $(CH_3)_2CHCH_2OH$

c. ⬡—NH_2 or   ⬡—CH_3

More Practice: Try Problems 13.11, 13.29–13.32.

13.8 Amines as Bases

Like ammonia (NH_3), **amines are bases;** that is, **they are *proton acceptors*.** When an amine is dissolved in water, an acid–base reaction occurs: the amine accepts a proton from H_2O, forming its conjugate acid, an **ammonium ion,** and water loses a proton, forming hydroxide, OH^-.

This acid–base reaction occurs with 1°, 2°, and 3° amines. While amines are more basic than other families of organic compounds, they are weak bases compared to inorganic bases like NaOH.

13.8A Reaction of Amines with Acids

Amines also react with acids such as HCl to form water-soluble salts. The lone pair of electrons from the amine nitrogen atom is always used to form a new bond to a proton from the acid.

- The amine [$(CH_3)_3N$] gains a proton to form its conjugate acid, an ammonium cation [$(CH_3)_3NH^+$].
- A proton is removed from the acid (HCl) to form its conjugate base, the chloride anion (Cl^-).

Similar acid–base reactions occur with other inorganic acids (H_2SO_4), and with organic acids like CH_3COOH, as well.

- In an acid–base reaction of an amine, the amine nitrogen always forms a new bond to a proton, forming an ammonium ion.

Sample Problem 13.10 **Drawing the Products When an Amine Reacts with Acid**

What products are formed when methamphetamine reacts with HCl?

methamphetamine

Analysis

In any acid–base reaction with an amine:

- **Locate the N atom of the amine and add a proton to it.**
- **Remove a proton from the acid (HCl)** and form its conjugate base (Cl^-).

Solution

Transfer a proton from the acid to the base. Use the lone pair on the N atom to form the new bond to the proton of the acid.

methamphetamine

This proton is transferred from the acid to the amine base.

Thus, HCl loses a proton to form Cl⁻, and the N atom of methamphetamine gains a proton to form an ammonium cation.

What products are formed when each of the following amines is treated with HCl: (a) $CH_3CH_2NH_2$; (b) $(CH_3CH_2)_2NH$; (c) $(CH_3CH_2)_3N$?

More Practice: Try Problems 13.9d, 13.10d, 13.45–13.49, 13.50b.

13.8B   Ammonium Salts

When an amine reacts with an acid, the product is an *ammonium salt:* **the amine forms its conjugate acid, a positively charged ammonium ion, and the acid forms its conjugate base, an anion.**

positively charged
ammonium ion   chloride anion

ammonium salt

To name an ammonium salt, change the suffix *-amine* of the parent amine from which the salt is formed to the suffix **-ammonium.** Then add the name of the anion as a separate word.

Sample Problem 13.11   **Naming an Ammonium Salt**

Name the ammonium salt: $[(CH_3)_3NH]^+$ CH_3COO^-.

Analysis

To name an ammonium salt, identify the four groups bonded to the N atom. Remove one hydrogen from the N atom to draw the structure of the parent amine. Then put the two parts of the name together.

• Name the ammonium ion by changing the suffix *-amine* of the parent amine to the suffix **-ammonium.**
• Add the name of the anion.

Solution

• Change the name trimethyl*amine* to trimethyl*ammonium.*
• Add the name of the anion, acetate.

Answer: trimethylammonium acetate

More Practice: Try Problems 13.27c, 13.28c.

Practice Problem 13.17	Name each ammonium salt.

a. $(CH_3NH_3)^+$ Cl^- b. $[(CH_3CH_2CH_2)_2NH_2]^+$ Br^- c. $[(CH_3)_2NHCH_2CH_3]^+$ CH_3COO^-

Ammonium salts are ionic compounds, and as a result:

- **Ammonium salts are water-soluble solids.**

In this way, the solubility properties of an amine can be changed by treatment with acid. For example, octylamine has eight carbons, making it water *insoluble*. Reaction with HCl forms octylammonium chloride. This ionic solid is now *soluble* in water.

$$CH_3CH_2CH_2CH_2CH_2CH_2CH_2CH_2-\underset{\underset{H}{|}}{\overset{}{N}}-H \;+\; H-Cl \longrightarrow \left[CH_3CH_2CH_2CH_2CH_2CH_2CH_2CH_2-\underset{\underset{H}{|}}{\overset{\overset{H}{|}}{N}}-H\right]^+ \; Cl^-$$

octylamine   octylammonium chloride

water-insoluble amine -------------------→ **water-soluble salt**

The solubility properties change.

Concept Check 13.10	Label each compound as water soluble or water insoluble.

a. $(CH_3CH_2)_3N$ b. $[(CH_3CH_2)_3NH]^+$ Br^- c. $CH_3CH_2NH_2$ d. $\left[\text{(ring)} \underset{\underset{H}{|}}{\overset{\overset{H}{|}}{N}} \right]^+$ Cl^-

Ammonium salts can be re-converted to amines by treatment with base. Base removes a proton from the nitrogen atom of the ammonium ion, regenerating the neutral amine.

This proton is removed by the base.

$$\left[CH_3CH_2CH_2CH_2CH_2CH_2CH_2CH_2-\underset{\underset{H}{|}}{\overset{\overset{H}{|}}{N}}-H\right]^+ \; Cl^- \;+\; Na^+OH^- \longrightarrow CH_3CH_2CH_2CH_2CH_2CH_2CH_2CH_2-\underset{\underset{H}{|}}{\overset{}{N}}-H \;+\; H-OH \;+\; NaCl$$

octylammonium chloride
water-soluble salt

octylamine
water-insoluble amine

Concept Check 13.11	Draw the products formed when each ammonium salt is treated with NaOH.

a. $[(CH_3CH_2)_3NH]^+$ Br^- b. $(CH_3CH_2NH_3)^+$ HSO_4^-

13.8C FOCUS ON HEALTH & MEDICINE
Ammonium Salts as Useful Drugs

Many amines with useful medicinal properties are sold as their ammonium salts. Because the ammonium salts are more water soluble than the parent amine, they are easily transported through the body in the aqueous medium of the blood.

Many antihistamines and decongestants are sold as their hydrochloride salts.
Jill Braaten/McGraw Hill

Sudafed PE is an over-the-counter decongestant that contains phenylephrine hydrochloride (Concept Check 13.12).
Jill Braaten/McGraw Hill

For example, diphenhydramine is a 3° amine that is sold as its ammonium salt under the name of Benadryl. Benadryl, formed by treating diphenhydramine with HCl, is an over-the-counter antihistamine that is used to relieve the itch and irritation of skin rashes and hives.

diphenhydramine   +   HCl   ⟶   ammonium salt
Benadryl

The names of medicines sold as ammonium salts are often derived from the names of the amine and the acid used to form them. Because Benadryl is formed from diphenhydramine and HCl, it is called **diphenhydramine hydrochloride.**

Concept Check 13.12

Phenylephrine hydrochloride is the decongestant in Sudafed PE. Draw the amine and acid used to form phenylephrine hydrochloride.

phenylephrine hydrochloride
(decongestant in Sudafed PE,
Theraflu, and DayQuil)

13.9 Amides

Amides contain a carbonyl group bonded to a nitrogen atom.

General structure

| amide
R' = H or alkyl | 1° amide
1 C—N bond | 2° amide
2 C—N bonds | 3° amide
3 C—N bonds |

The N atom of an amide may be bonded to other hydrogen atoms or alkyl groups. Amides are classified as **1°, 2°,** or **3°** depending on the number of carbon atoms bonded directly to the *nitrogen* atom.

- A primary (1°) amide contains one C—N bond. A 1° amide has the structure $RCONH_2$.
- A secondary (2°) amide contains two C—N bonds. A 2° amide has the structure RCONHR'.
- A tertiary (3°) amide contains three C—N bonds. A 3° amide has the structure $RCONR'_2$.

For example, **capsaicin,** the compound responsible for the characteristic spiciness of hot peppers mentioned in the chapter opener, is a 2° amide. Capsaicin is the active ingredient in several topical creams for pain relief.

Jill Braaten/McGraw Hill

Concept Check 13.13

Draw the structure of a 1°, 2°, and 3° amide, each having the molecular formula C_4H_9NO.

13.9A Naming an Amide

- In the IUPAC system, amides are identified by the suffix *-amide.*

All 1° amides are named by replacing the *-oic acid* ending (or *-ic acid* ending of a common name) with the suffix *-amide.*

Naming 1° amides

derived from
acet*ic acid* ----→ **acet*amide***

derived from
benz*oic acid* ----→ **benz*amide***

A 2° or 3° amide has two parts to its structure: the **RCO– group** that contains the carbonyl and one or two **alkyl groups** bonded to the nitrogen atom. To name a 2° or 3° amide:

- Name the alkyl group (or groups) bonded to the N atom of the amide. Use the prefix **"N-"** preceding the name of each alkyl group.
- Name the RCO– group with the suffix *-amide.*

Sample Problem 13.12

Naming a 2° or 3° Amide

Name the following amide: $HCONHCH_2CH_3$.

Analysis and Solution

[1] **Name the alkyl group on the N atom, and precede its name with N-.**

ethyl group

[2] **Name the RCO– group. Change the *-ic acid* ending of the parent carboxylic acid to *-amide*, and combine the parts together.**

derived from
form*ic acid* ----→ **form*amide***

Answer: *N*-ethylformamide

Practice Problem 13.18

Name each amide.

a. [structure: pentanamide chain ending in C(=O)NH₂]

b. [structure: benzene ring with C(=O)NHCH₃]

c. [structure: H—C(=O)—N(CH₂CH₂CH₃)₂]

Practice Problem 13.19

Draw the structure of each amide.

a. propanamide   b. *N*-ethylhexanamide   c. *N,N*-dimethylacetamide

More Practice: Try Problems 13.17a; 13.18b; 13.23b, c; 13.24b, c; 13.25e, f; 13.26e, f.

13.9B   Physical Properties

Primary (1°) and 2° amides contain N—H bonds, so hydrogen bonding is possible between two molecules of the amide. This gives 1° and 2° amides stronger intermolecular forces than 3° amides and esters, which can't intermolecularly hydrogen bond. As a result:

- Primary (1°) and 2° amides have higher boiling points and melting points than esters and 3° amides of comparable molecular weight.

CH_3—$\overset{\displaystyle O}{\underset{\displaystyle OCH_3}{C}}$

ester
bp 58 °C

H—$\overset{\displaystyle O}{\underset{\displaystyle N(CH_3)_2}{C}}$

3° amide
bp 153 °C

CH_3CH_2—$\overset{\displaystyle O}{\underset{\displaystyle NH_2}{C}}$ hydrogen bonding possible

1° amide
bp 213 °C

strongest intermolecular forces
highest boiling point

Amides having fewer than six carbons are soluble in water. Higher-molecular-weight amides are insoluble in water because the nonpolar portion of the molecule, the C—C and C—H bonds, gets larger than the polar carbonyl group.

Concept Check 13.14

Why is the boiling point of CH_3CONH_2 (221 °C) higher than the boiling point of $CH_3CON(CH_3)_2$ (166 °C), even though the latter compound has a higher molecular weight and more surface area?

13.9C   Amide Hydrolysis

Like esters, amides undergo hydrolysis, but amides are much less reactive than esters. Nonetheless, under forcing conditions amides can be hydrolyzed with water in the presence of acid or base. **Treatment of an amide (RCONHR') with water in the presence of an acid catalyst (HCl) forms a carboxylic acid (RCOOH) and an ammonium salt (R'NH₃⁺ Cl⁻).**

CH_3CH_2—$\overset{\displaystyle O}{\underset{}{C}}$—$NHCH_3$ +   H—OH $\xrightarrow{\text{HCl}}$ CH_3CH_2—$\overset{\displaystyle O}{\underset{}{C}}$—$OH$ + $\left[\text{H}-\overset{\displaystyle H}{\underset{\displaystyle H}{\text{N}}}-CH_3 \right]^+$ Cl⁻

amide

carboxylic acid   amine salt

Amides are also hydrolyzed in aqueous base to form carboxylate anions and a molecule of ammonia (NH_3) or amine.

| amide | | carboxylate anion | ammonia |

The relative lack of reactivity of the amide bond is important in the proteins in the body. Proteins are polymers connected by amide linkages, as we will learn in Chapter 16. Proteins are stable in water in the absence of acid or base, so they can perform their various functions in the cell without breaking down. The hydrolysis of the amide bonds in proteins requires a variety of specific enzymes.

Sample Problem 13.13

Drawing the Products of Amide Hydrolysis in Base

What products are formed when N-methylacetamide ($CH_3CONHCH_3$) is hydrolyzed with water in the presence of NaOH?

Analysis

To draw the products of amide hydrolysis in base, replace the **NHR'** group of the amide by an oxygen anion (**O⁻**), forming a new C—O bond at the carbonyl carbon. A molecule of amine (R'NH$_2$) is also formed from the nitrogen group (NHR') of the amide.

Solution

Replace the **NHCH₃** group of N-methylacetamide by a negatively charged oxygen atom (**O⁻**) to form sodium acetate ($CH_3CO_2^-$ Na⁺) and methylamine (CH_3NH_2).

Practice Problem 13.20

What products are formed when each amide is treated with H_2O and H_2SO_4?

a. $CH_3(CH_2)_8$ —C(=O)— NH_2

b. CH_3CHCH_2 —C(=O)— $NHCH_3$ with CH_3 substituent

c. cyclohexane carbonyl with N-dipropyl

Practice Problem 13.21

What products are formed when each amide in Practice Problem 13.20 is treated with H_2O and NaOH?

More Practice: Try Problems 13.39, 13.40, 13.44.

13.10   Interesting Amines and Amides

Caffeine and **nicotine** are widely used stimulants of the central nervous system that contain nitrogen atoms in rings. Caffeine and nicotine, like the amines in Figure 13.2, are *alkaloids,* **naturally occurring amines derived from plant sources.**

coffee "cherries"
Jessica Byrne

caffeine

nicotine

tobacco plant
Kai4107/Shutterstock

13.10A   FOCUS ON THE HUMAN BODY
Epinephrine and Related Compounds

Epinephrine, or **adrenaline** as it is commonly called, is an amine synthesized in the adrenal glands from norepinephrine (noradrenaline).

norepinephrine
(noradrenaline)

epinephrine
(adrenaline)

When an individual senses danger or is confronted by stress, the hypothalamus region of the brain signals the adrenal glands to synthesize and release epinephrine, which enters the bloodstream and then stimulates a response in many organs (Figure 13.3). Stored carbohydrates are metabolized in the liver to form glucose, which is further metabolized to provide an energy boost. Heart rate and blood pressure increase, and lung passages are dilated. These physiological changes are commonly referred to as a "rush of adrenaline," and they prepare an individual for "fight or flight."

The search for drugs that were structurally related to epinephrine but exhibited only some components of its wide range of biological activities led to the discovery of some useful medications. Both **albuterol** and **salmeterol** dilate lung passages; that is, they are *bronchodilators.* They do not, however, stimulate the heart. This makes both compounds useful for the treatment of asthma. Albuterol is a short-acting drug used to relieve the wheezing associated with asthma. Salmeterol is much longer acting, and thus it is often used before bedtime to keep an individual symptom free overnight.

albuterol
(Trade names: Ventolin, Proventil)

salmeterol
(Trade name: Serevent)

Concept Check 13.15	Classify the amines and the alcohols (not the phenols) in albuterol and salmeterol as 1°, 2°, or 3°.

Figure 13.3

Epinephrine—The "Fight or Flight" Hormone

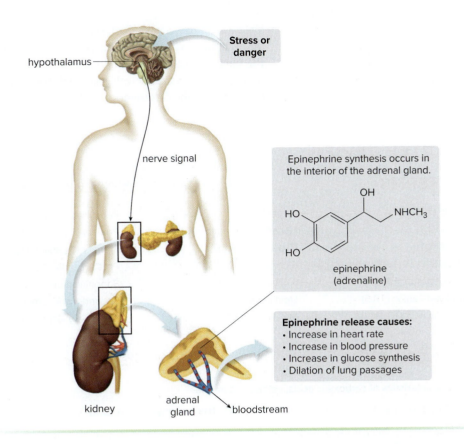

Stress or danger

hypothalamus

nerve signal

Epinephrine synthesis occurs in the interior of the adrenal gland.

epinephrine (adrenaline)

Epinephrine release causes:
• Increase in heart rate
• Increase in blood pressure
• Increase in glucose synthesis
• Dilation of lung passages

kidney

adrenal gland

bloodstream

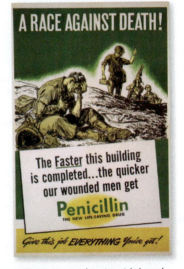

Penicillin was used to treat injured soldiers in World War II before its structure had been conclusively determined. *Source: Library of Congress Prints and Photographs Division LC-USZC4-1986*

13.10B FOCUS ON HEALTH & MEDICINE
Penicillin

In the twenty-first century it is hard to imagine that an infected cut or scrape could be life-threatening. Before antibiotics were discovered in the early twentieth century, however, that was indeed the case.

The antibiotic properties of **penicillin** were first discovered in 1928 by Sir Alexander Fleming, who noticed that a mold of the genus *Penicillium* inhibited the growth of certain bacteria. After years of experimentation, penicillin was first used to treat a female patient who had developed a streptococcal infection in 1942. By 1944, penicillin production was given high priority by the United States government, because it was needed to treat the many injured soldiers in World War II.

The penicillins are a group of related antibiotics. All penicillins contain two amide units. One amide is part of a four-membered ring called a **β-lactam.** The second amide is bonded to the four-membered ring. Particular penicillins differ in the identity of the R group in the amide side chain. The first penicillin to be discovered was penicillin G. Amoxicillin is another example in common use today.

General structure Penicillin

α carbon

β-lactam

penicillin G

amoxicillin

Unlike mammalian cells, bacterial cells are surrounded by a fairly rigid cell wall, which allows the bacterium to live in many different environments. **Penicillin interferes with the synthesis of the bacterial cell wall.** The β-lactam ring of the penicillin molecule reacts with an enzyme needed to synthesize the cell wall, and this deactivates the enzyme. Cell wall construction is halted, killing the bacterium.

CHAPTER REVIEW

KEY TERMS

Alkaloid (13.7)

Amide (13.1)

Amine (13.1)

Ammonium salt (13.8)

Carboxyl group (13.1)

Carboxylate anion (13.5)

Carboxylic acid (13.1)

Ester (13.1)

Fischer esterification (13.6)

Hydrolysis (13.6)

α-Hydroxy acid (13.4)

β-Lactam (13.10)

Lipid (13.6)

Micelle (13.5)

Penicillin (13.10)

Primary (1°) amide (13.9)

Primary (1°) amine (13.7)

Prostaglandin (13.4)

Saponification (13.6)

Secondary (2°) amide (13.9)

Secondary (2°) amine (13.7)

Soap (13.5)

Tertiary (3°) amide (13.9)

Tertiary (3°) amine (13.7)

Triacylglycerol (13.6)

KEY CONCEPTS

1 Characteristics of carboxylic acids, esters, and amides

1 Structure (13.1)	**2** The carbonyl group (13.3)	**3** Physical properties (13.3)
• A **carboxylic acid** has the general structure **RCOOH** (or RCO_2H). • An **ester** has the general structure **RCOOR'** (or RCO_2R'). • An **amide** has the general structure **RCONR'$_2$,** where R' = H or alkyl. Classification: $RCONH_2$ (1° amide), RCONHR' (2° amide), and RCONR'$_2$ (3° amide).	• Because oxygen is more electronegative than carbon, a carbonyl group is polar. 	• RCOOH, $RCONH_2$, and RCONHR' exhibit intermolecular hydrogen bonding. • RCOOH, $RCONH_2$, and RCONHR' have higher boiling points and melting points than RCOOR' and RCONR'$_2$, which are incapable of hydrogen bonding. • Compounds with fewer than six carbons are water soluble.

2 Characteristics of amines (13.7)

1 General features	**2** Classification	**3** Shape

• Amines are organic nitrogen compounds, formed by replacing one or more H atoms of NH_3 by alkyl groups.
 	
 • The amine nitrogen has a **lone pair of electrons,** and this feature dominates its chemistry.
	• Amines are classified as **1°, 2°,** or **3°** by the number of alkyl groups bonded to the nitrogen atom.
	
	 • A primary amine has one C—N bond, and so forth.
	
	   RNH$_2$ **R$_2$NH** **R$_3$N**
	   1° amine **2° amine** **3° amine**
		• An amine nitrogen is surrounded by three atoms and one lone pair, making it **trigonal pyramidal** in shape.
		
		 CH_3—N···H = (trigonal pyramidal)
		 109.5°

KEY REACTIONS

1 Acid–base reactions

2 Esterification of carboxylic acids

$$\text{carboxylic acid} + CH_3OH \xrightarrow[\text{Fischer esterification}]{H_2SO_4} \text{ester} + H-OH$$

(13.6A)

3 Hydrolysis of esters and amides

1
$$\text{ester} + H-OH \xrightarrow[\text{hydrolysis in acid}]{H_2SO_4} \text{carboxylic acid} + CH_3OH$$
(13.6B)

3
$$\text{amide} + H-OH \xrightarrow[\text{hydrolysis in acid}]{HCl} \text{carboxylic acid} + \left[CH_3NH_2\ |\ H \right]^+ Cl^-$$
(13.9C)

2
$$\text{ester} + H-OH \xrightarrow[\text{hydrolysis in base}]{OH^-} \text{carboxylate} + CH_3OH$$
(13.6B)

4
$$\text{amide} + H-OH \xrightarrow[\text{hydrolysis in base}]{OH^-} \text{carboxylate} + CH_3NH\ |\ H$$
(13.9C)

KEY SKILLS

1 Naming a carboxylic acid and an ester using the IUPAC system (13.2)

1 Naming a carboxylic acid (13.2A)

- Find the longest chain that contains the carbonyl carbon.
- Use the suffix *-oic acid.*
- Number to put the C=O at C**1**, but omit this number from the name.
- This compound has a methyl group on C3 and a chloro group on C4.

chloro on C4

$$CH_3CHCHCH_2-C\underset{CH_3}{\overset{}{\underset{1}{\|}}}OH$$

methyl on C3

Answer: 4-chloro-3-methylpentanoic acid

2 Naming an ester (13.2B)

- Name the alkyl group bonded to the ester O.
- Name the RCO– group by finding the longest chain that contains the carbonyl carbon.
- Change the *-ic acid* ending of the parent carboxylic acid to the suffix *-ate.*

$$CH_3CH_2O-\overset{O}{\overset{\|}{C}}-CH_2CH_2CH_2CH_3$$

ethyl group   pentan*oic acid* → pentan*oate*
                           5 C's

Answer: ethyl pentanoate

2 Naming an amine and amide

1 Naming an amine (13.7B)

- For a 1° amine, name the alkyl group on nitrogen and add the suffix *-amine.*
- For a 2° or 3° amine with different alkyl groups, alphabetize the names of the alkyl groups.

$$CH_3CH_2CH_2NH_2$$

1° amine   2° amine
propylamine   ethylpropylamine

2 Naming an amide (13.9A)

- Name the alkyl group (or groups) bonded to the N atom, and use the prefix *N-* preceding the name of each alkyl group.
- Name the RCO– group with the suffix *-amide.*

$$(CH_3CH_2)_2N \qquad$$ propan*oic acid* → propan*amide*
                                          3 C's
two ethyl groups
N,N-diethyl

Answer: *N,N*-diethylpropanamide

3 Drawing and naming the product formed when a carboxylic acid is treated with base (13.5)

1 Transfer a proton from the carboxylic acid to the base.	**2** Name the metal salt of the resulting carboxylate anion.
• Remove a proton from the carboxyl group (COOH) to form a carboxylate anion (RCOO⁻). • Add a proton to the base and adjust the charge. • Draw the carboxylate anion as a salt with a metal cation.	• To name the metal salt of a carboxylate anion, put three parts together: the name of the metal cation, the parent name, and the suffix (**-ate**).

carboxylic acid + Na⁺OH⁻ (base) ⟶ sodium salt of the carboxylate anion + H—OH

Answer: sodium benzoate

sodium cation

parent + suffix
benzo- -ate

4 Drawing hydrolysis products from an ester and an amide

1 Hydrolysis of an ester in acid (13.6B)	**2** Hydrolysis of an amide in base (13.9C)
• To draw the products of hydrolysis in acid, replace the OR' group of the ester by OH. • The alkoxy group of the ester becomes an alcohol R'OH.	• To draw the products of hydrolysis in base, replace the NH_2 group of the amide by O⁻. • The NH_2 group of the amide becomes NH_3.

ethyl acetate + H—OH →(H_2SO_4)→ acetic acid + H—OCH_2CH_3 ethanol

acetamide + H—OH →(NaOH)→ sodium acetate + H—NH_2 ammonia

5 Comparing the physical properties of amines with other compounds (13.7C); example: determining the relative boiling points of $CH_3CH_2N(CH_3)_2$, $CH_3CH_2CH_2CH_2NH_2$, and $CH_3CH_2CH_2CH_2OH$

1 Determine the types of intermolecular forces present.	**2** Rank the compounds in order of increasing boiling point.
• Amines have polar C–N and N–H bonds. • Primary (1°) and 2° amines have N–H bonds, so they are capable of intermolecular **hydrogen bonding.** • Because N is *less* electronegative than O, hydrogen bonds between N and H are *weaker* than those between O and H.	• The *stronger* the intermolecular forces, the *higher* the boiling point.

no N–H bond / no hydrogen bonding — **weakest forces**

hydrogen bonding with N–H bond

hydrogen bonding with O–H bond — **strongest forces**

increasing strength of intermolecular forces
increasing boiling point

6 Drawing and naming the product formed when an amine is treated with acid (13.8)

1 Transfer a proton from the acid to the amine base.	**2** Name the resulting salt.
• Locate the N atom of the amine and add a proton to it. Because the amine nitrogen is neutral to begin with, adding a proton gives it a (+1) charge. This forms an ammonium ion. • Remove a proton from the acid (HCl) and form its conjugate base (Cl⁻).	• To name the ammonium salt, change the suffix *-amine* of the parent amine to *-ammonium.* Add the name of the anion.

cyclohexylamine (base) + H—Cl (acid) ⟶ ammonium ion chloride
ammonium salt

chloride

cyclohexyl**amine** ⟶ cyclohexyl**ammonium**

Answer: cyclohexylammonium chloride

CHAPTER 13 SELF-TEST

The Self-Test consists of short-answer questions that test an understanding of definitions, equations, and other material encountered within the chapter. Answers to each question are provided at the end of the chapter. For any questions you are unable to answer correctly, revisit the appropriate section(s) of the text; then proceed to working the end-of-chapter problems to further reinforce your understanding of the chapter material.

1. Which term from "Key Terms" answers each question?
   a. What type of reaction results in basic hydrolysis of an ester?
   b. What term is used for the spherical droplets containing ionic heads on the surface and nonpolar tails in the interior, which are formed when soap is dissolved in water?
   c. What type of compound contains three C–N bonds and the general structure R_3N?
   d. What type of compound contains an amide as part of a four-membered ring?
   e. What type of lipid contains three ester groups, each having a long carbon chain bonded to the carbonyl group?

2. Name each of the following compounds: (a) $(CH_3)_2CHCH_2CO_2CH_3$; (b) $CH_3CH_2CH_2CON(CH_3)_2$; (c) $CH_3CH_2CH_2CO_2^-K^+$; (d) $(CH_3CH_2)_2NH$.

3. Rank the following compounds in order of increasing boiling point: $CH_3CH_2CO_2H$ (**A**); $CH_3CO_2CH_3$ (**B**); $CH_3(CH_2)_3OH$ (**C**).

4. Draw the products formed in each acid–base reaction.
   a. $(CH_3)_3N + HBr \longrightarrow$
   b. $CH_3CH_2CO_2H + KOH \longrightarrow$

5. What products are formed when $(CH_3)_3CCO_2H$ is treated with CH_3OH in the presence of H_2SO_4?

6. What products are formed when $CH_3CH_2CON(CH_3)_2$ is treated with aqueous NaOH?

7. Classify each amide and amine as 1°, 2°, or 3°.

a.   b.   c.

UNDERSTANDING KEY CONCEPTS

13.1 Label each nitrogen-containing functional group in lidocaine, a local anesthetic, as an amine or amide, and classify it as 1°, 2°, or 3°.

lidocaine

13.2 Label each nitrogen-containing functional group in fentanyl, a narcotic pain reliever, as an amine or amide, and classify it as 1°, 2°, or 3°.

fentanyl

13.3 Which compound(s) can hydrogen bond to another molecule like itself? Which compound(s) can hydrogen bond to water? (a) HCO_2CH_3; (b) $CH_3CH_2CO_2H$.

13.4 Which compound(s) can hydrogen bond to another molecule like itself? Which compound(s) can hydrogen bond to water? (a) $CH_3CONHCH_3$; (b) $HCON(CH_3)_2$.

13.5 Answer the following questions about **A**, depicted in the ball-and-stick model.

A

a. What is the IUPAC name for **A**?
b. Draw an isomer of **A** that has the same functional group.
c. Draw an isomer of **A** that has a different functional group.
d. What products are formed when **A** is treated with NaOH?
e. What product is formed when **A** is treated with CH_3CH_2OH and H_2SO_4?

13.6 Answer the questions in Problem 13.5 for **B**, depicted in the ball-and-stick model.

B

13.7 Sunscreens that contain an ester can undergo hydrolysis. What products are formed when octyl salicylate, a commercial sunscreen, is hydrolyzed with water?

octyl salicylate

13.8 What products are formed when the pain reliever phenacetin is hydrolyzed with water and NaOH?

phenacetin

13.9 Benzphetamine (trade name: Didrex) is a habit-forming diet pill.

benzphetamine

a. Label the amine as 1°, 2°, or 3°.
b. Draw a constitutional isomer that contains a 1° amine.
c. Draw a constitutional isomer that contains a 3° amine.
d. What products are formed when benzphetamine is treated with acetic acid, CH_3COOH?

13.10 Phentermine is one component of the banned diet drug fen–phen.

phentermine

a. Label the amine as 1°, 2°, or 3°.
b. Draw a constitutional isomer that contains a 1° amine.
c. Draw a constitutional isomer that contains a 2° amine.
d. What products are formed when phentermine is treated with benzoic acid, C_6H_5COOH?

13.11 Explain why a 1° amine and a 3° amine having the same number of carbons are soluble in water to a similar extent, but the 1° amine has a higher boiling point.

13.12 Which compound has the higher water solubility: $CH_3(CH_2)_5NH_2$ or $CH_3(CH_2)_5NH_3^+ Cl^-$? Explain your choice.

ADDITIONAL PROBLEMS

Structure and Bonding

13.13 Draw the structure of a compound that fits each description:
a. a carboxylic acid of molecular formula $C_8H_{16}O_2$ that has six carbons in its longest chain
b. an ester of molecular formula $C_6H_{12}O_2$ that contains a methoxy group (OCH_3) bonded to the carbonyl group
c. an ester of molecular formula $C_6H_{10}O_2$ that contains a ring

13.14 Match the description to compounds **E** and **F**.
a. a compound that contains a 1° amine and a 2° amide
b. a compound that contains a 2° amine and a 3° amide

E   **F**

13.15 Classify each amine as 1°, 2°, or 3°.

a. $CH_3CHCH_2CH_2NHCH_3$
      |
      CH_3

b.

13.16 Draw the structure of a compound of molecular formula $C_4H_{11}NO$ that fits each description:
a. a compound that contains a 1° amine and a 1° alcohol
b. a compound that contains a 2° amine and a 2° alcohol
c. a compound that contains a 1° amine and a 3° alcohol

Nomenclature

13.17 Give an acceptable name for each compound.

13.18 Give an acceptable name for each compound.

13.19 Give an acceptable name for each compound.

a. $(CH_3)_2CHCH_2CH_2COOH$

b. $CH_3CH_2CH_2CHCH CH_2CH_2COOH$
             | |
          CH_2CH_3 CH_2CH_3

c.

d. $\bigcirc\!\!-CO_2(CH_2)_3CH_3$

13.20 Give an acceptable name for each compound.
a. $(CH_3)_2CHCH_2CH_2CH_2CH_2COOH$ c. $CH_3CO_2(CH_2)_4CH_3$

b. $CH_3CHCHCHCH_2COOH$
        | | |
       CH_3 CH_3 CH_3

d. $CH_3(CH_2)_5COO^- K^+$

13.21 Give an acceptable name for the following amine.

13.22 Give an acceptable name for the following amine.

13.23 Give an acceptable name for each amine or amide.

a. $CH_3CH_2-\overset{\overset{H}{|}}{N}-CH_2CH_3$

c. $HCONHCH_2CH_2CH_2CH_3$

b. $CH_3(CH_2)_4CONH_2$

13.24 Give an acceptable name for each amine or amide.

a. $CH_3(CH_2)_6NH_2$

c.

b. $CH_3(CH_2)_6CONH_2$

13.25 Draw the structure corresponding to each name.
a. 2-hydroxyheptanoic acid
b. 4-chlorononanoic acid
c. butyl butanoate
d. heptyl benzoate
e. N-ethylhexanamide
f. N-ethyl-N-methylheptanamide

13.26 Draw the structure corresponding to each name.
a. 3-methylhexanoic acid
b. 3-hydroxy-4-methylheptanoic acid
c. hexyl pentanoate
d. propyl hexanoate
e. N-butylbenzamide
f. N,N-dimethyloctanamide

13.27 Draw the structure of each amine or ammonium salt.
a. p-bromoaniline   c. butylammonium bromide
b. ethylhexylamine

13.28 Draw the structure of each amine or ammonium salt.
a. butylamine   c. triethylammonium iodide
b. N-pentylaniline

Physical Properties and Intermolecular Forces

13.29 Rank the following compounds in order of increasing boiling point: $CH_3CH_2CH(CH_3)_2$, $CH_3CH_2CO_2H$, and $CH_3CH_2COCH_3$.

13.30 Which compound in each pair is more water soluble? Which compound in each pair is more soluble in an organic solvent: (a) $CH_3CH_2CH_2CH_3$ or $CH_3CH_2CH_2CH_2CO_2H$; (b) $CH_3(CH_2)_4COOH$ or $CH_3(CH_2)_4COO^- Na^+$?

13.31 Which compound in each pair has the higher boiling point?
a. $CH_3(CH_2)_6OH$ or $CH_3(CH_2)_6NH_2$

b. —$CH_2N(CH_3)_2$ or —$CH_2CH_2CH_2NH_2$

13.32 Which compound in each pair has the higher boiling point?

a. ⬡ or ⬡NH

b. $CH_3CHCH_2CHCH_3$ (with Cl and NH_2) or $[CH_3CH_2CH_2CHCH_3$ with $NH_3]^+$ Cl^-

Acidity

13.33 Draw the products of each acid–base reaction.

a. $CH_3(CH_2)_3$—C(=O)—OH + KOH ⟶

b. $(CH_3)_2CHCH_2CH_2COOH$ + Na_2CO_3 ⟶

13.34 Draw the products of each acid–base reaction.

a. CH_3—⬡—C(=O)—OH + NaOH ⟶

b. Cl_3CCOOH + $NaHCO_3$ ⟶

Esterification and Hydrolysis

13.35 What ester is formed when butanoic acid ($CH_3CH_2CH_2CO_2H$) is treated with each of the following alcohols in the presence of H_2SO_4: (a) CH_3OH; (b) $CH_3CH_2CH_2OH$?

13.36 What ester is formed when each of the following carboxylic acids is treated with 2-propanol [$(CH_3)_2CHOH$] in the presence of H_2SO_4: (a) CH_3CH_2COOH; (b) HCO_2H?

13.37 What products are formed when each ester is hydrolyzed with water and H_2SO_4?

a. $CH_3CH_2CH_2$—C(=O)—$OCH(CH_3)_2$

b. cyclopentyl—O—C(=O)—$CH_2CH_2CH_3$

13.38 What products are formed when each ester in Problem 13.37 is hydrolyzed with water and NaOH?

13.39 What products are formed when each amide is hydrolyzed with water and HCl?
a. $(CH_3)_3CCON(CH_3)_2$ b. $HO(CH_2)_4CONHCH_3$

13.40 What products are formed when each amide in Problem 13.39 is hydrolyzed with water and NaOH?

13.41 Ethyl phenylacetate ($C_6H_5CH_2CO_2CH_2CH_3$) is a naturally occurring ester in honey. What hydrolysis products are formed when this ester is treated with water and H_2SO_4?

Emilio Ereza/Alamy Stock Photo

13.42 Benzyl acetate ($CH_3CO_2CH_2C_6H_5$) is a naturally occurring ester in peaches. What hydrolysis products are formed when this ester is treated with water and NaOH?

Pixtal/SuperStock

13.43 Draw the products formed in each reaction.

a. $CH_3\overset{\displaystyle O}{\underset{}{C}}OH$ + CH_3OH $\xrightarrow{H_2SO_4}$

b. $(CH_3)_2CHO\overset{\displaystyle O}{\underset{}{C}}CH_3$ + H_2O $\xrightarrow{H_2SO_4}$

13.44 Draw the products formed in each reaction.

a. $(CH_3)_2CHCH_2NH\overset{\displaystyle O}{\underset{}{C}}CH_3$ + H_2O $\xrightarrow{HCl}$

b. $(CH_3)_2CHNH\overset{\displaystyle O}{\underset{}{C}}CH_2CH_3$ + H_2O $\xrightarrow{NaOH}$

Acid–Base Reactions of Amines

13.45 Draw the acid–base reaction that occurs when each amine dissolves in water: (a) $CH_3CH_2NH_2$; (b) $(CH_3CH_2)_2NH$; (c) $(CH_3CH_2)_3N$.

13.46 What ammonium salt is formed when each amine is treated with HCl?

a. [benzene ring]–NH_2 b. [benzene ring]–CH_2NHCH_3

13.47 Draw the products of each acid–base reaction.

a. $CH_3CH_2CH_2N(CH_3)_2$ + HCl $\longrightarrow$

b. $CH_3CH_2\overset{\overset{\displaystyle NH_2}{|}}{CH}CH_2CH_3$ + H_2SO_4 $\longrightarrow$

13.48 Draw the products of each acid–base reaction.

a. $CH_3CH_2\overset{\underset{\displaystyle NHCH_3}{|}}{CH}CH_2CH_3$ + H_2SO_4 $\longrightarrow$

b. CH_3NH_2 + CH_3COOH $\longrightarrow$

13.49 What ammonium salt is formed when coniine, an alkaloid from hemlock, is treated with HCl?

[structure: piperidine ring with N–H and $CH_2CH_2CH_3$ substituent]

coniine

13.50 Oxycodone (trade name: OxyContin) is an addictive, narcotic pain reliever prescribed for chronic or severe pain. (a) Identify the functional groups in oxycodone and label any alcohol or amine as 1°, 2°, or 3°. (b) What hydrochloride salt is formed when oxycodone reacts with HCl?

oxycodone

PureRadiancePhoto/Shutterstock

Applications

13.51 Which of the following structures represent soaps? Explain your answers.

a. $CH_3CO_2^-\ Na^+$ c. $CH_3(CH_2)_{12}COOH$
b. $CH_3(CH_2)_{14}CO_2^-\ Na^+$

13.52 Explain how soap is able to dissolve nonpolar hydrocarbons in a polar solvent like H_2O.

13.53 Ritalin is the trade name for methylphenidate, a drug used to treat attention deficit hyperactivity disorder (ADHD).

[structure of methylphenidate with benzene ring, CH, COOCH₃, and piperidine ring with HN]

methylphenidate
(Trade name: Ritalin)

a. Identify the functional groups.
b. Label the amine as 1°, 2°, or 3°.
c. Draw the structure of methylphenidate hydrochloride.

13.54 Seldane is the trade name for terfenadine, an antihistamine once used in the United States but withdrawn from the market because of cardiac side effects observed in some patients.

[structure of terfenadine]

terfenadine
(Trade name: Seldane)

a. Identify the functional groups.
b. Label the amine as 1°, 2°, or 3°.
c. Draw the structure of terfenadine hydrochloride.

CHALLENGE PROBLEMS

13.55 Draw the products formed by acidic hydrolysis of **aspartame,** the artificial sweetener used in Equal and many diet beverages. One of the hydrolysis products of this reaction is the naturally occurring amino acid phenylalanine. Infants afflicted with phenylketonuria cannot metabolize this amino acid, so it accumulates, causing mental retardation. When identified early, a diet limiting the consumption of phenylalanine (and compounds like aspartame that are converted to it) can allow for a normal life.

aspartame

13.56 Polyhydroxybutyrate (PHB) is a biodegradable polyester; that is, the polymer is degraded by microorganisms that naturally occur in the environment. Such biodegradable polymers are attractive alternatives to the commonly used polymers that persist in landfills for years. What is the single product formed when PHB is hydrolyzed with acid and water?

PHB

BEYOND THE CLASSROOM

13.57 Norepinephrine, dopamine, and serotonin are three neurotransmitters, molecules that transmit nerve impulses from one nerve cell to another. All three compounds contain an amine, in addition to other functional groups. Determine the structure of one or more of these neurotransmitters and research why proper levels are needed for an individual's mental health. What conditions result if the level is too high or too low? What medications are available to treat these conditions?

13.58 Research the nicotine content (in mg) in different cigarettes. Are there significant variations among different types of cigarettes: regular cigarettes (70 mm in length), king size varieties (100 mm in length), those with filters, or menthols? How does the nicotine content of cigarettes compare with smokeless tobacco, e-cigarettes, or cigars? Ask family members or classmates who smoke to estimate how many cigarettes they smoke in a day. Calculate how many grams of nicotine the smoker would inhale in a year with various cigarette types. Research how nicotine can be both a stimulant and a relaxant, two apparently conflicting effects.

CHAPTER 13 ANSWERS

Concept Check Answers

13.1   a. ester   b,d: amide   c. carboxylic acid

13.2   a. amide   b,c,d: amine

13.3   two carboxylic acids, amide, two amines, aromatic ring

13.4   boiling point: **C < B < A**
A most water soluble, **C** least water soluble

13.5   a.

13.6 $CH_3(CH_2)_{14}CO_2^-K^+$

13.7   a. $CH_3(CH_2)_8CO_2^-Na^+ + CH_3OH$
b. $(CH_3)_2CHCH_2CO_2^-Na^+ + CH_3CH_2OH$

c.

13.8 $(HOCH_2)_2CHOH + CH_3(CH_2)_{10}CO_2H + CH_3(CH_2)_{12}CO_2H + CH_3(CH_2)_{14}CO_2H$

13.9   a.

b.

c.

d.

13.10   a. insoluble   b,c,d: soluble

13.11   a. $(CH_3CH_2)_3N + H_2O + NaBr$
b. $CH_3CH_2NH_2 + H_2O + NaHSO_4$

13.12

13.13 $CH_3CH_2CH_2CONH_2$ (1°), $CH_3CH_2CONHCH_3$ (2°), $CH_3CON(CH_3)_2$ (3°)

13.14 CH_3CONH_2 can intermolecularly hydrogen bond because it has 2 N–H bonds, and $CH_3CON(CH_3)_2$ cannot.

13.15   both compounds: 2° amine, 1° and 2° alcohols

Practice Problem Answers

13.1   a. 3,3-dimethylhexanoic acid
b. 4-chloropentanoic acid
c. 2,4-diethylhexanoic acid

13.2   a. $CH_3CH_2CH(Br)CO_2H$
b. $CH_3CH_2CH(CH_3)CH(CH_3)CO_2H$
c. $CH_3CH_2CH_2C(CH_3)_2CH_2CH_2CH(CH_2CH_3)CO_2H$
d. $CH_3(CH_2)_3CH(CH_2CH_3)CH(CH_2CH_3)CH(CH_2CH_3)CH(CH_2CH_3)CH_2CO_2H$

13.3   a. methyl hexanoate   c. propyl pentanoate
b. ethyl benzoate

13.4 a. $CH_3CH_2CO_2(CH_2)_2CH_3$ c. $CH_3(CH_2)_4CO_2CH_2CH_3$

b. $CH_3CO_2(CH_2)_3CH_3$ d. —CO_2CH_3

13.5 $HCO_2CH_2CH_3$

13.6 a. —$CO_2^- Na^+$ + H_2O

b. $CH_3CH_2CH_2CO_2^- Na^+$ + $Na^+HCO_3^-$

13.7 a. sodium butanoate   b. lithium benzoate

13.8 a. $(CH_3)_2CHCO_2CH_2CH_3$ c. $CH_3(CH_2)_6CO_2CH_2CH_3$

b. $HCO_2CH_2CH_3$ d. —$CO_2CH_2CH_3$

13.9

A

13.10 a. $CH_3(CH_2)_8CO_2H$ + CH_3OH

b. $(CH_3)_2CHCH_2CO_2H$ + CH_3CH_2OH

c. —CO_2H + $CH_3CH_2CH_2OH$

13.11   + CH_3CO_2H

13.12 a. two 1° amines (on ends), two 2° amines

b. 3° amine

13.13 3° amine, 1° alcohol

13.14 a. propylamine   c. dibutylamine

b. methylpropylamine

13.15 a. $(CH_3)_2CHCH_2NH_2$ c. —NH_2

b. $(CH_3)_2CHCH_2OH$

13.16 a. $(CH_3CH_2NH_3)^+$ + Cl^- c. $[(CH_3CH_2)_3NH]^+$ + Cl^-

b. $[(CH_3CH_2)_2NH_2]^+$ + Cl^-

13.17 a. methylammonium chloride

b. dipropylammonium bromide

c. ethyldimethylammonium acetate

13.18 a. pentanamide   c. N,N-dipropylformamide

b. N-methylbenzamide

13.19 a. $CH_3CH_2CONH_2$ c. $CH_3CON(CH_3)_2$

b. $CH_3(CH_2)_4CONHCH_2CH_3$

13.20 a. $CH_3(CH_2)_8CO_2H$ + $NH_4^+ HSO_4^-$

b. $(CH_3)_2CHCH_2CO_2H$ + $(CH_3NH_3)^+ HSO_4^-$

c. —CO_2H + $[(CH_3CH_2CH_2)_2NH_2]^+ HSO_4^-$

13.21 a. $CH_3(CH_2)_8COO^- Na^+$ + NH_3

b. $(CH_3)_2CHCH_2COO^- Na^+$ + CH_3NH_2

c. —$CO_2^- Na^+$ + $(CH_3CH_2CH_2)_2NH$

Self-Test Answers

1. a. saponification   d. β-lactam

b. micelles   e. triacylglycerol

c. tertiary (3°) amine

2. a. methyl 3-methylbutanoate   c. potassium butanoate

b. N,N-dimethylbutanamide   d. diethylamine

3. **B < C < A**

4. a. $[(CH_3)_3NH]^+ Br^-$ b. $CH_3CH_2CO_2^- K^+$ + H_2O

5. $(CH_3)_3CCO_2CH_3$ + H_2O

6. $CH_3CH_2CO_2^- Na^+$ + $HN(CH_3)_2$

7. a. 3°   b. 2°   c. 1°

Odd-Numbered Problem Answers

13.1 2° amide, 3° amine

13.3 (a) HCO_2CH_3 can hydrogen bond to water.
(b) CH_3CH_2COOH can hydrogen bond to itself and to water.

13.5 a. 5-methylhexanoic acid

b. $CH_3CH_2CH(CH_3)CH_2CH_2CO_2H$

c. $CH_3(CH_2)_4CO_2CH_3$

d. $(CH_3)_2CH(CH_2)_3CO_2^- Na^+$ + H_2O

e. $(CH_3)_2CH(CH_2)_3CO_2CH_2CH_3$

13.7   + $HO(CH_2)_7CH_3$

13.9 a. 3° amine

b.

c.

d. + CH_3COO^-

13.11 Primary amines can hydrogen bond to each other, whereas 3° amines cannot, giving 1° amines a higher boiling point. Any amine can hydrogen bond to water, so both 1° and 3° amines have similar solubility properties.

13.13 a. $(CH_3)_3C(CH_2)_3CO_2H$ c.

b. $CH_3(CH_2)_3CO_2CH_3$

13.15 a. 2°   b. 3°

13.17 a. 4-methylpentanamide   b. o-chlorobenzoic acid

13.19 a. 4-methylpentanoic acid   c. lithium formate

b. 4,5-diethyloctanoic acid   d. butyl benzoate

13.21 N-ethyl-N-methylaniline

13.23 a. diethylamine   c. N-butylformamide

b. hexanamide

13.25 a. $CH_3(CH_2)_4CH(OH)CO_2H$

b. $CH_3(CH_2)_4CH(Cl)CH_2CH_2CO_2H$

c. $CH_3CH_2CH_2CO_2(CH_2)_3CH_3$

d. —$CO_2(CH_2)_6CH_3$

e. $CH_3(CH_2)_4CONHCH_2CH_3$

f. $CH_3(CH_2)_5CON(CH_3)CH_2CH_3$

13.27 a.

c. $(CH_3CH_2CH_2CH_2NH_3)^+ \ Br^-$

b. $CH_3CH_2NH(CH_2)_5CH_3$

13.29 $CH_3CH_2CH(CH_3)_2 < CH_3CH_2COCH_3 < CH_3CH_2CO_2H$

13.31 a. $CH_3(CH_2)_6OH$

b. —$CH_2CH_2CH_2NH_2$

13.33 a. $CH_3(CH_2)_3CO_2^- \ K^+ + H_2O$

b. $(CH_3)_2CHCH_2CH_2CO_2^-Na^+ \ + \ NaHCO_3$

13.35 a. $CH_3CH_2CH_2CO_2CH_3$

b. $CH_3CH_2CH_2CO_2CH_2CH_2CH_3$

13.37 a. $CH_3CH_2CH_2CO_2H \ + \ HOCH(CH_3)_2$

b. —OH $+ \ CH_3CH_2CH_2CO_2H$

13.39 a. $(CH_3)_3CCO_2H + [(CH_3)_2NH_2]^+ \ Cl^-$

b. $HO(CH_2)_4CO_2H + (CH_3NH_3)^+ \ Cl^-$

13.41 —$CH_2CO_2H \ + \ CH_3CH_2OH$

13.43 a. $CH_3CO_2CH_3$

b. $(CH_3)_2CHOH + CH_3CO_2H$

13.45 a. $CH_3CH_2NH_2 + H_2O \longrightarrow (CH_3CH_2NH_3)^+ + OH^-$

b. $(CH_3CH_2)_2NH + H_2O \longrightarrow [(CH_3CH_2)_2NH_2]^+ + OH^-$

c. $(CH_3CH_2)_3N + H_2O \longrightarrow [(CH_3CH_2)_3NH]^+ + OH^-$

13.47 a. $[CH_3CH_2CH_2NH(CH_3)_2]^+ \ + \ Cl^-$

b. $[(CH_3CH_2)_2CHNH_3]^+ \ + \ HSO_4^-$

13.49 Cl^-

13.51 b. This is a sodium salt of a long-chain carboxylic acid.

13.53 a,b: aromatic ring, ester, 2° amine

c. Cl^-

13.55 $HO_2CCH_2CH(NH_2)CO_2H \ +$ —$CH_2CH(NH_2)CO_2H \ + \ CH_3OH$

14 Carbohydrates

Table sugar contains **sucrose,** a sweet-tasting carbohydrate obtained from sugarcane.
MaraZe/Shutterstock

CHAPTER GOALS

In this chapter you will learn how to:

1. Identify the three major types of carbohydrates

2. Recognize the major structural features of monosaccharides

3. Draw the cyclic forms of monosaccharides and classify them as α or β isomers

4. Draw reduction and oxidation products of monosaccharides

5. Recognize the major structural features of disaccharides

6. Describe the characteristics of cellulose, starch, and glycogen

7. Describe the role that carbohydrates play in determining blood type

Why Study ...

Carbohydrates?

Chapter 14 is the first of four chapters that deal with the chemistry of ***biomolecules,*** **organic molecules found in biological systems.** Chapter 14 discusses carbohydrates, the largest group of biomolecules in nature, whereas Chapter 15 focuses on lipids, biomolecules that contain many carbon–carbon and carbon–hydrogen bonds, making them soluble in organic solvents and insoluble in water. Chapter 16 focuses on proteins and the amino acids that compose them. Finally, the properties of DNA, the polymer responsible for the storage of genetic information in the chromosomes of cells, is presented in Chapter 17. These compounds are all organic molecules, so many of the principles and chemical reactions that you have already learned will be examined once again. But, as you will see, each class of compound has its own unique features that we will discuss as well.

14.1 Introduction

Carbohydrates, commonly referred to as sugars and starches, are polyhydroxy aldehydes and ketones, or compounds that can be hydrolyzed to them. Carbohydrates can be simple or complex, having as few as three or as many as thousands of carbon atoms. They are the largest group of organic molecules in nature, comprising approximately 50% of the earth's biomass.

Carbohydrates are classified into three groups:

- Monosaccharides (Sections 14.1–14.4)
- Disaccharides (Section 14.5)
- Polysaccharides (Sections 14.6–14.7)

Monosaccharides or simple sugars are the simplest carbohydrates. Glucose and fructose, the two major constituents of honey, are monosaccharides. Glucose contains an aldehyde at one end of a six-carbon chain, and fructose contains a ketone. Every other carbon atom has a hydroxyl group bonded to it. Monosaccharides cannot be converted to simpler compounds by hydrolysis.

honey

Ninikas/Getty Images

common monosaccharides

Disaccharides are composed of two monosaccharides joined together. Lactose, the principal carbohydrate in milk, is a disaccharide. Although disaccharides contain no carbonyl groups, they are hydrolyzed to simple monosaccharides that contain an aldehyde or ketone, as we will learn in Section 14.5.

lactose
a common disaccharide

Image Source/Getty Images

Polysaccharides have three or more monosaccharides joined together. Starch, the main carbohydrate found in the seeds and roots of plants, is a polysaccharide composed of hundreds of glucose molecules joined together. Like disaccharides, polysaccharides are hydrolyzed to simple monosaccharides that contain carbonyl groups. Pasta, bread, rice, and potatoes are foods that contain a great deal of starch.

one form of starch
a polysaccharide

James Noble/The Image Bank/Getty Images

Carbohydrates are storehouses of chemical energy. Carbohydrates are synthesized in green plants and algae through **photosynthesis,** a process that uses the energy from the sun to convert carbon dioxide and water into glucose and oxygen. Plants store glucose in the form of polysaccharides like starch and cellulose (Section 14.6).

Energy is stored in **photosynthesis.**

$$6\ CO_2 \ + \ 6\ H_2O \ \xrightarrow[\text{chlorophyll}]{\text{sunlight}} \ \underset{\text{glucose}}{C_6H_{12}O_6} \ + \ 6\ O_2$$

Energy is released in **metabolism.**

Chlorophyll in green leaves converts CO_2 and H_2O to glucose and O_2 during photosynthesis. *Daniel C. Smith*

The energy stored in glucose bonds is released when glucose is metabolized. The oxidation of glucose is a multistep process that forms carbon dioxide, water, and a great deal of energy. Although the metabolism of lipids provides more energy per gram than the metabolism of carbohydrates, glucose is the preferred source when a burst of energy is needed during exercise. Glucose is water soluble, so it can be quickly and easily transported through the bloodstream to tissues.

14.2 Monosaccharides

Monosaccharides, the simplest carbohydrates, generally have three to six carbon atoms in a chain, with a **carbonyl group** at either the terminal carbon, numbered C1, or the carbon adjacent to it, numbered C2. In most carbohydrates, each of the remaining carbon atoms has a **hydroxyl group.** Monosaccharides are drawn vertically, with the carbonyl group at (or near) the top.

monosaccharide

- Monosaccharides with a carbonyl group at C1 are aldehydes called *aldoses.*
- Monosaccharides with a carbonyl group at C2 are ketones called *ketoses.*

Glyceraldehyde is the simplest aldose and dihydroxyacetone is the simplest ketose. Glyceraldehyde and dihydroxyacetone both have molecular formula $C_3H_6O_3$, so they are **constitutional isomers;** that is, they have the same molecular formula but a different arrangement of atoms. Glucose is the most prevalent aldose.

glyceraldehyde   dihydroxyacetone   glucose

A monosaccharide is characterized by the number of carbons in its chain.

- A triose has three carbons.
- A tetrose has four carbons.
- A pentose has five carbons.
- A hexose has six carbons.

These terms are then combined with the words *aldose* and *ketose* to indicate both the number of carbon atoms in the monosaccharide and whether it contains an aldehyde or ketone. Thus, glyceraldehyde is an **aldotriose** (three carbons and an aldehyde), dihydroxyacetone is a **ketotriose** (three carbons and a ketone), and glucose is an **aldohexose** (six carbons and an aldehyde).

CONSUMER NOTE

Dihydroxyacetone (DHA) is the active ingredient in many artificial tanning agents. *Elite Images/McGraw Hill*

| Sample Problem 14.1 | Classifying a Monosaccharide |

Classify each monosaccharide by the type of carbonyl group and the number of carbons in the chain.

a.
```
        CHO
         |
    H—C—OH
         |
    H—C—OH
         |
    H—C—OH
         |
       CH2OH
```
ribose

b.
```
       CH2OH
         |
       C=O
         |
   HO—C—H
         |
    H—C—OH
         |
    H—C—OH
         |
       CH2OH
```
fructose

Analysis

Identify the type of carbonyl group to label the monosaccharide as an aldose or ketose. **An aldose has the C=O at C1** so that a hydrogen atom is bonded to the carbonyl carbon. **A ketose has two carbons bonded to the carbonyl carbon.** Count the number of carbons in the chain to determine the suffix— namely, *-triose, -tetrose,* and so forth.

Solution

a.

$$
\begin{array}{c}
\text{H}\diagdown \overset{\displaystyle O}{\underset{\displaystyle C}{\parallel}} \\
\mid \\
\text{H—C—OH} \\
\mid \\
\text{H—C—OH} \\
\mid \\
\text{H—C—OH} \\
\mid \\
\text{CH}_2\text{OH}
\end{array}
\quad \text{aldehyde}
$$

ribose
5 C's in the chain
Answer: aldopentose

b.

$$
\begin{array}{c}
\text{CH}_2\text{OH} \\
\mid \\
\text{C=O} \\
\mid \\
\text{HO—C—H} \\
\mid \\
\text{H—C—OH} \\
\mid \\
\text{H—C—OH} \\
\mid \\
\text{CH}_2\text{OH}
\end{array}
\quad \text{ketone}
$$

fructose
6 C's in the chain
Answer: ketohexose

Practice Problem 14.1

Classify each monosaccharide by the type of carbonyl group and the number of carbons in the chain.

a.
$$
\begin{array}{c}
\text{CHO} \\
\mid \\
\text{HO—C—H} \\
\mid \\
\text{H—C—OH} \\
\mid \\
\text{H—C—OH} \\
\mid \\
\text{CH}_2\text{OH}
\end{array}
$$
arabinose

b.
$$
\begin{array}{c}
\text{CHO} \\
\mid \\
\text{HO—C—H} \\
\mid \\
\text{H—C—OH} \\
\mid \\
\text{CH}_2\text{OH}
\end{array}
$$
threose

c.
$$
\begin{array}{c}
\text{CH}_2\text{OH} \\
\mid \\
\text{C=O} \\
\mid \\
\text{H—C—OH} \\
\mid \\
\text{CH}_2\text{OH}
\end{array}
$$
erythrulose

Practice Problem 14.2

Draw the structure of (a) an aldotetrose; (b) a ketopentose; (c) an aldohexose.

More Practice: Try Problems 14.1a, 14.2a, 14.3, 14.4, 14.5b, 14.6b, 14.11, 14.12.

Monosaccharides are all sweet tasting, but their relative sweetness varies a great deal. Monosaccharides are polar compounds with high melting points. The presence of so many polar functional groups capable of hydrogen bonding makes them very water soluble.

Concept Check 14.1

Rank the following compounds in order of increasing water solubility: glucose, hexane [$CH_3(CH_2)_4CH_3$], and 1-decanol [$CH_3(CH_2)_9OH$]. Explain your choice.

14.2A Fischer Projection Formulas

A striking feature of carbohydrate structure is the presence of chirality centers. **All carbohydrates except for dihydroxyacetone contain one or more chirality centers.** The simplest aldose, glyceraldehyde, has one chirality center—one carbon atom bonded to four different groups. Thus, there are two possible enantiomers—mirror images that are not superimposable.

Two enantiomers of glyceraldehyde

chirality center

naturally occurring isomer
D-glyceraldehyde

OH on **right**

chirality center

unnatural isomer
L-glyceraldehyde

OH on **left**

Chirality centers and enantiomers were first encountered in Section 12.10.

Only one enantiomer of glyceraldehyde occurs in nature. When the carbon chain is drawn vertically with the aldehyde at the top, the naturally occurring enantiomer has the OH group drawn on the right side of the carbon chain. **To distinguish the two enantiomers, the prefixes D and L precede the name.** Thus, the naturally occurring enantiomer is labeled D-glyceraldehyde, whereas the unnatural isomer is L-glyceraldehyde.

Fischer projection formulas are commonly used to depict the chirality centers in monosaccharides. Recall from Section 12.10 that a Fischer projection formula uses a cross to represent a tetrahedral carbon. In a Fischer projection formula:

- A carbon atom is located at the intersection of the two lines of the cross.
- The horizontal bonds come forward, on wedges.
- The vertical bonds go back, on dashed wedges.

In a carbohydrate, the carbon skeleton is drawn vertically, with the carbonyl group at the top. Using a Fischer projection formula, D-glyceraldehyde becomes:

D-glyceraldehyde

Fischer projection formula

A 5% intravenous glucose (dextrose) solution provides a patient with calories and hydration. *Janis Christie/Digital Vision/Alamy Stock Photo*

Concept Check 14.2

Draw L-glyceraldehyde using a Fischer projection formula.

14.2B Monosaccharides with More Than One Chirality Center

Fischer projection formulas are also used for compounds like aldohexoses that contain several chirality centers. Glucose, for example, contains four chirality centers labeled in the structure below. To convert the molecule to a Fischer projection, the molecule is drawn with a vertical carbon skeleton with the aldehyde at the top, and the horizontal bonds are assumed to come forward (on wedges). In the Fischer projection, each chirality center is replaced by a cross.

Replace each chirality center (in blue) with a cross.

glucose

Fischer projection formula

The letters D and L are used to label all monosaccharides, even those with many chirality centers. **The configuration of the chirality center** *farthest* **from the carbonyl group determines whether a monosaccharide is D or L.**

- A D monosaccharide has the OH group on the chirality center farthest from the carbonyl on the right (like D-glyceraldehyde).
- An L monosaccharide has the OH group on the chirality center farthest from the carbonyl on the left (like L-glyceraldehyde).

Glucose and all other naturally occurring sugars are D sugars. L-Glucose, a compound that does not occur in nature, is the enantiomer of D-glucose. L-Glucose has the opposite configuration at *every* chirality center.

Determining the Stereochemical Features of a Monosaccharide

Consider the aldopentose ribose. (a) Label all chirality centers. (b) Classify ribose as a D or L monosaccharide. (c) Draw the enantiomer.

CHO

H————OH

H————OH

H————OH

CH₂OH

ribose

Analysis

- A chirality center has four different groups around a carbon atom.
- The labels D and L are determined by the position of the OH group on the chirality center farthest from the carbonyl group: **a D sugar has the OH group on the right and an L sugar has the OH group on the left.**
- To draw an enantiomer, draw the mirror image so that each group is a reflection of the group in the original compound.

Solution

a. The three carbons that contain both H and OH groups in ribose are chirality centers.

H—C=O
H——OH
H——OH
H——OH
CH₂OH

b. Ribose is a D sugar because the OH group on the chirality center farthest from the carbonyl is on the **right**.

H—C=O
H——OH
H——OH
H——**OH**
CH₂OH

D sugar

c. The enantiomer of D-ribose, L-ribose, has **all three OH groups on the left side** of the carbon chain.

H—C=O
HO——H
HO——H
HO——H
CH₂OH

enantiomer

Practice Problem 14.3

For each monosaccharide: [1] label all chirality centers; [2] classify the monosaccharide as D or L; [3] draw the enantiomer.

a.
CHO
H——OH
H——OH
CH₂OH

b.
CHO
HO——H
H——OH
HO——H
CH₂OH

c.
CHO
HO——H
HO——H
H——OH
CH₂OH

Practice Problem 14.4

For each type of monosaccharide, [1] give an example of a D sugar; [2] label each chirality center: (a) an aldopentose; (b) a ketohexose; (c) a ketotetrose.

More Practice: Try Problems 14.5a, c, d; 14.6a, c, d; 14.13; 14.14; 14.17; 14.18.

14.2C   Common Monosaccharides

The most common monosaccharides in nature are the aldohexoses L-glucose and D-galactose, and the ketohexose D-fructose.

CHO
H—C—OH
HO—C—H
H—C—OH
H—C—OH
CH₂OH

D-glucose

CHO
H—C—OH
HO—C—H
HO—C—H
H—C—OH
CH₂OH

D-galactose

CH₂OH
C=O
HO—C—H
H—C—OH
H—C—OH
CH₂OH

D-fructose

Glucose, also called dextrose, is the sugar referred to when blood sugar is measured. It is the most abundant monosaccharide. Glucose is the building block for the polysaccharides starch and cellulose. Glucose, the carbohydrate that is transported in the bloodstream, provides energy for cells when it is metabolized. Normal blood glucose levels are in the range of 70–110 mg/dL.

An individual with galactosemia must avoid cow's milk and all products derived from cow's milk (Section 14.2C).
Comstock/PunchStock/Getty Images

Insulin, a protein produced in the pancreas, regulates blood glucose levels. When glucose concentration increases after eating, insulin stimulates the uptake of glucose in tissues and its conversion to glycogen. Patients with diabetes produce insufficient insulin to adequately regulate blood glucose levels, and the concentration of glucose rises. With close attention to diet and daily insulin injections or other medications, a normal level of glucose can be maintained in most diabetic patients. Individuals with poorly controlled diabetes can develop many other significant complications, including cardiovascular disease, chronic renal failure, and blindness.

Galactose is one of the two monosaccharides that form the disaccharide lactose (Section 14.5). Individuals with galactosemia, a rare inherited disease, lack an enzyme needed to metabolize galactose. Galactose accumulates, causing a variety of physical problems, including cataracts, cirrhosis, and mental retardation. Galactosemia can be detected in newborn screening, and affected infants must be given soy-based formula to avoid all milk products with lactose.

Fructose is one of two monosaccharides that form the disaccharide sucrose (Section 14.5). Fructose is a ketohexose found in honey and is almost twice as sweet as normal table sugar with about the same number of calories per gram.

Some "lite" food products have fewer calories because they use only half as much fructose as sucrose for the same level of sweetness (Section 14.2C).
Matt Meadows/McGraw Hill Education

Concept Check 14.3

(a) Draw a Fischer projection formula for D-galactose. (b) Draw a Fischer projection formula for the enantiomer of D-galactose.

14.3 The Cyclic Forms of Monosaccharides

Although the monosaccharides in Section 14.2 were drawn as acyclic carbonyl compounds, the hydroxyl and carbonyl groups can react together to form a ring. Let's illustrate the process with D-glucose, and then learn a general method for drawing the cyclic forms of any aldohexose.

Which of the five OH groups reacts with the aldehyde carbonyl? In glucose, the OH group on C5 reacts with the carbonyl carbon to form a six-membered ring.

$$
\begin{array}{c}
\overset{1}{H}\diagdown \!\!\! C \!\!\! \diagup\!\!\! O \\
\overset{2}{H-C-OH} \\
\overset{3}{HO-C-H} \\
\overset{4}{H-C-OH} \\
\overset{5}{H-C-\textbf{OH}} \\
\overset{6}{CH_2OH}
\end{array}
$$

This OH group is the right distance away from the carbonyl for cyclization to a six-membered ring.

D-glucose

To convert this acyclic form (labeled **A**) into a cyclic monosaccharide, rotate the carbon skeleton clockwise 90°, and twist the chain to put the OH group on C5 close to the aldehyde carbonyl, forming **B.** In this process, the CH_2OH group at the end of the chain ends up *above* the carbon skeleton. Reaction of the OH group on C5 (in red) with the aldehyde carbonyl forms a six-membered ring with a new chirality center. Cyclization yields two isomers, because the OH group on the new chirality center can be located above or below the six-membered ring.

D-glucose

Draw the CH₂OH up.

Rotate and twist.

A

B

α-D-glucose

β-D-glucose

- The α isomer, called α-D-glucose, has the OH group on the new chirality center drawn down (shown in red).
- The β isomer, called β-D-glucose, has the OH group on the new chirality center drawn up (shown in blue).

These flat, six-membered rings used to represent the cyclic forms of glucose and other sugars are called **Haworth projections.**

Thus, D-glucose really exists in three different forms—an acyclic aldehyde and two cyclic compounds. The mixture has 37% of the α isomer, 63% of the β isomer, and only a trace amount of the acyclic aldehyde (Figure 14.1).

Figure 14.1　　The Three Forms of D-Glucose

α-D-glucose
37%

acyclic D-glucose
trace

β-D-glucose
63%

14.3A　Haworth Projections

All aldohexoses exist primarily as cyclic compounds typically drawn in Haworth projections. Sample Problem 14.3 shows how to convert an acyclic monosaccharide to a Haworth projection.

Sample Problem 14.3　　**Drawing a Haworth Projection from an Acyclic Aldohexose**

Draw the α isomer of the cyclic form of D-galactose.

CHO
H—C—OH
HO—C—H
HO—C—H
H—C—OH
CH₂OH

D-galactose

Analysis and Solution

[1] Draw a hexagon with an O atom in the upper right corner. Add the CH₂OH above the ring on the first carbon to the left of the O atom.

D-galactose

[2] Draw the new chirality center on the first carbon clockwise from the O atom.

- The α isomer has the OH group drawn down, whereas a β isomer has the OH group drawn up.

α isomer

[3] Add the OH groups and H atoms to the three remaining carbons (C2–C4).

- Groups on the *right* side in the acyclic form are drawn *down,* below the six-membered ring, and groups on the *left* side in the acyclic form are drawn *up,* above the six-membered ring.

α isomer of cyclic D-galactose

Practice Problem 14.5

Convert each aldohexose to the indicated isomer using a Haworth projection.

a.
CHO
H——OH
H——OH
H——OH
H——OH
CH₂OH

α isomer

b.
CHO
HO——H
HO——H
HO——H
H——OH
CH₂OH

β isomer

c.
CHO
HO——H
H——OH
H——OH
H——OH
CH₂OH

α isomer

More Practice: Try Problems 14.5e, 14.6e, 14.19, 14.20, 14.21b, 14.22b, 14.23, 14.24.

14.3B   The Cyclic Forms of Fructose, a Ketohexose

Certain monosaccharides—notably aldopentoses and ketohexoses—form five-membered rings, *not* six-membered rings, in solution. The same principles apply to drawing these structures as for drawing six-membered rings, except the ring size is one atom smaller.

> • Cyclization forms two isomers. For a D sugar, the OH group is drawn down on the new chirality center in the α isomer and up in the β isomer.

For example, D-fructose forms a five-membered ring when it cyclizes because the carbonyl group is a ketone at C2, instead of an aldehyde at C1.

acyclic D-fructose

This OH group is the right distance away from the carbonyl group for cyclization to a five-membered ring.

re-draw

α isomer

α-D-fructose

β isomer

β-D-fructose

new chirality center

The two cyclic forms of D-fructose contain a five-membered ring.

Concept Check 14.4

Label each compound as an α or β isomer.

14.4   Reactions of Monosaccharides

The aldehyde carbonyl group of a monosaccharide undergoes two common reactions—**reduction to an alcohol** and **oxidation to a carboxylic acid.**

14.4A   Reduction of the Aldehyde Carbonyl Group

Like the double bond in an alkene, the double bond in the carbonyl of an aldose reacts with hydrogen (H_2) in the presence of a palladium (Pd) metal catalyst (Section 11.5). The product, an alcohol called an **alditol,** is sometimes referred to as a "sugar alcohol." For example, reduction of D-glucose with H_2 and Pd yields glucitol, commonly called sorbitol. This reaction is a **reduction,** because the number of carbon–oxygen bonds decreases.

H₂ adds to the C=O.

D-glucose
(D-sorbose)

glucitol
(sorbitol)

The alditol contains an OH group on every C.

Sorbitol is 60% as sweet as table sugar (sucrose) and contains two-thirds the calories per gram. Sorbitol is used as a sweetening agent in a variety of sugar-free candies and gum.

Sample Problem 14.4 — Drawing the Product from the Reduction of an Aldose

Draw the structure of xylitol, the product formed when D-xylose is treated with H₂ in the presence of a Pd catalyst.

$$
\begin{array}{c}
CHO \\
H-C-OH \\
HO-C-H \\
H-C-OH \\
CH_2OH
\end{array}
$$

D-xylose

CONSUMER NOTE

Xylitol (Sample Problem 14.4) is used as a substitute sweetener in sugar-free—that is, sucrose-free—products.
Mark Dierker/McGraw Hill

Analysis

- Locate the C=O and mentally break one bond in the double bond.
- Mentally break the H—H bond of the reagent and add one H atom to each atom of the C=O, forming new C—H and O—H bonds.

Solution

Break one bond.

+ H—H

Break the single bond.

Pd

xylose

xylitol

Practice Problem 14.6

What compound is formed when each aldose is treated with H₂ in the presence of a Pd catalyst?

a.
$$
\begin{array}{c}
CHO \\
H-C-OH \\
H-C-OH \\
H-C-OH \\
CH_2OH
\end{array}
$$

b.
$$
\begin{array}{c}
CHO \\
H-C-OH \\
H-C-OH \\
CH_2OH
\end{array}
$$

More Practice: Try Problems 14.5g, 14.6g, 14.27[1], 14.28[1].

14.4B   Oxidation of the Aldehyde Carbonyl Group

The aldehyde carbonyl of an aldose is easily oxidized with a variety of reagents to form a carboxyl group, yielding an **aldonic acid.** For example, D-glucose is oxidized with a Cu^{2+} reagent called **Benedict's reagent** to form gluconic acid. In the process, a characteristic color change occurs as the blue Cu^{2+} is reduced to Cu^{+}, forming brick-red Cu_2O.

D-glucose   gluconic acid

This reaction is an oxidation because the product carboxylic acid has one more C—O bond than the starting aldehyde. Carbohydrates that are oxidized with Benedict's reagent are called **reducing sugars,** because the Cu^{2+} in Benedict's reagent is reduced to Cu^{+} during the reaction. Those that do not react with Benedict's reagent are called **nonreducing sugars. All aldoses are reducing sugars.**

Sample Problem 14.5

Drawing the Product from the Oxidation of an Aldose

Draw the product formed when L-threose is oxidized with Benedict's reagent.

CHO
H—C—OH
HO—C—H
CH₂OH

L-threose

Analysis

To draw the oxidation product of an aldose, **convert the CHO group to COOH.**

Solution

L-threose

Practice Problem 14.7

What aldonic acid is formed by oxidation of each monosaccharide?

a.   CHO
    H—C—OH
    H—C—OH
    H—C—OH
    CH₂OH

b.   CHO
    H—C—OH
    H—C—OH
    CH₂OH

Practice Problem 14.8

Practice Problem 14.8

What product is formed when D-arabinose is treated with each reagent: (a) H_2, Pd; (b) Benedict's reagent?

CHO
HO—C—H
H—C—OH
H—C—OH
CH₂OH

D-arabinose

More Practice: Try Problems 14.5f, 14.6f, 14.27[2], 14.28[2], 14.29, 14.30.

14.4C   FOCUS ON HEALTH & MEDICINE
Monitoring Glucose Levels

Test strips that contain glucose oxidase are used to measure glucose concentration in urine. *Jill Braaten/McGraw Hill*

In order to make sure that their blood glucose levels are in the proper range, individuals with diabetes frequently measure the concentration of glucose in their blood. A common method for carrying out this procedure today involves the oxidation of glucose to gluconic acid using the enzyme glucose oxidase.

D-glucose   gluconic acid

In the presence of glucose oxidase, oxygen (O_2) in the air oxidizes the aldehyde of glucose to a carboxyl group. The O_2, in turn, is reduced to hydrogen peroxide, H_2O_2. In the first generation of meters for glucose monitoring, the H_2O_2 produced in this reaction was allowed to react with another organic compound to produce a colored product. The intensity of the colored product was then correlated to the amount of glucose in the blood. Test strips used for measuring glucose concentration in the urine are still based on this technology.

Modern glucose meters are electronic devices that measure the amount of oxidizing agent that reacts with a known amount of blood (Figure 14.2). This value is correlated with blood glucose concentration and the result is displayed digitally. A high blood glucose level may mean that an individual needs more insulin, whereas a low level may mean that it is time to ingest some calories.

Figure 14.2

Monitoring Blood Glucose Levels

glucose meter

test strip

small pin prick for a blood sample

A small drop of blood is placed on a disposable test strip that is inserted in an electronic blood glucose meter, and the glucose concentration is read on a digital display. *Jill Braaten/McGraw Hill*

14.5 Disaccharides

Disaccharides are carbohydrates composed of two monosaccharides. A disaccharide is formed when a hydroxyl group of one monosaccharide reacts with a hydroxyl group of a second monosaccharide. The new C—O bond that joins the two rings together is called a **glycosidic linkage.** The carbon in a glycosidic linkage (labeled in blue) is bonded to two O atoms—one O atom is part of a ring, and the other O atom joins the two rings together.

For example, maltose is a disaccharide formed from two molecules of glucose. Maltose, which is found in grains such as barley, is a product of the hydrolysis of starch. Each ring in maltose is numbered beginning at the carbon bonded to two oxygen atoms. In maltose, the glycosidic linkage joins C1 of one ring to C4 of the other ring.

Maltose gets its name from malt, the liquid obtained from barley used in the brewing of beer. *Josef Aichner/Glow Images*

The glycosidic linkage that joins the two monosaccharides in a disaccharide can be oriented in two different ways, shown with Haworth projections in structures **A** and **B**. (Several OH groups are omitted for clarity.)

α glycoside

The glycoside bond is **down.**

1→4-α-glycosidic linkage

A

β glycoside

The glycoside bond is **up.**

1→4-β-glycosidic linkage

B

- An α glycoside has the glycosidic linkage oriented down.
- A β glycoside has the glycosidic linkage oriented up.

Numbers are used to designate which ring atoms are joined in the disaccharide. Disaccharide **A** has a **1→4-α-glycosidic linkage** because the glycoside bond is oriented down and joins C1 of one ring to C4 of the other. Disaccharide **B** has a **1→4-β-glycosidic linkage** because the glycoside bond is oriented up and joins C1 of one ring to C4 of the other.

Sample Problem 14.6

Locating and Classifying Glycosidic Linkages

(a) Locate the glycosidic linkage in the following disaccharide. (b) Number the carbon atoms in both rings. (c) Classify the glycosidic linkage as α or β, and use numbers to designate its location.

Analysis and Solution

a. and b. The glycosidic linkage labeled in red contains a carbon bonded to two oxygens—one ring oxygen as well as the oxygen that joins the two rings together. Each ring is numbered beginning at the carbon bonded to two oxygen atoms.

c. The disaccharide has an α glycosidic linkage because the C—O bond is drawn down. The glycosidic linkage joins C1 of one ring to C4 of the other ring, so the disaccharide contains a 1→4-α-glycosidic linkage.

Practice Problem 14.9

(a) Locate the glycosidic linkage in cellobiose. (b) Number the carbon atoms in both rings. (c) Classify the glycosidic linkage as α or β, and use numbers to designate its location.

cellobiose

More Practice: Try Problems 14.9, 14.10, 14.33a, 14.34a, 14.39a.

The hydrolysis of a disaccharide cleaves the C—O glycosidic linkage and forms two monosaccharides. For example, hydrolysis of maltose yields two molecules of glucose.

Cleave the glycosidic linkage.

maltose   +   H—OH   ⟶   glucose   +   glucose

Water is added in hydrolysis.

Concept Check 14.5

What monosaccharides are formed when cellobiose (Practice Problem 14.9) is hydrolyzed with water?

14.5A FOCUS ON HEALTH & MEDICINE
Lactose Intolerance

Lactose is the principal disaccharide found in milk from both humans and cows. Unlike many mono- and disaccharides, lactose is not appreciably sweet. Lactose consists of one galactose ring and one glucose ring, joined by a 1→4-β-glycoside linkage.

Individuals who are lactose intolerant can drink lactose-free milk. Tablets that contain the lactase enzyme can also be taken when ice cream or other milk products are ingested. *Jill Braaten/McGraw Hill*

D. Hurst/Alamy Stock Photo

Lactose is digested in the body by first cleaving the 1→4-β-glycoside bond using the enzyme *lactase.* **Individuals who are lactose intolerant no longer produce this enzyme,** and so lactose cannot be properly digested, causing abdominal cramps and diarrhea. Lactose intolerance is especially prevalent in Asian and African populations whose diets have not traditionally included milk beyond infancy.

Concept Check 14.6	What products are formed when lactose is hydrolyzed with water?

14.5B FOCUS ON HEALTH & MEDICINE
Sucrose and Artificial Sweeteners

Sucrose, the disaccharide found in sugarcane and the compound generally referred to as "sugar," is the most common disaccharide in nature. It contains one glucose ring and one fructose ring. Unlike maltose and lactose, which contain only six-membered rings, sucrose contains one six-membered and one five-membered ring.

sugarcane

Dinodia/Pixtal/age fotostock

two varieties of refined sugar

Elite Images/McGraw Hill

sucrose

Table 14.1 Relative Sweetness of Some Carbohydrates and Artificial Sweeteners

Compound	Relative Sweetness
Sorbitol	0.60
Glucose	0.75
Sucrose	1.00
Fructose	1.75
Aspartame	150
Saccharin	350
Sucralose	600

Sucrose's pleasant sweetness has made it a widely used ingredient in baked goods, cereals, bread, and many other products. It is estimated that the average American ingests 100 lb of sucrose annually. Like other carbohydrates, however, sucrose contains many calories. To reduce caloric intake while maintaining sweetness, a variety of artificial sweeteners have been developed. These include aspartame, saccharin, and sucralose (Figure 14.3). These compounds are much sweeter than sucrose, so only a small amount of each compound is needed to achieve the same level of perceived sweetness. A relative sweetness scale ranks the sweetness of carbohydrates and synthetic sweeteners, as shown in Table 14.1.

Figure 14.3 Artificial Sweeteners

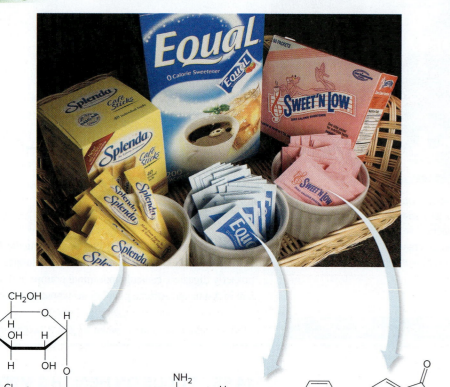

sucralose
(Trade name: Splenda)

aspartame
(Trade name: Equal)

saccharin
(Trade name: Sweet'n Low)

The sweetness of these artificial sweeteners was discovered accidentally. The sweetness of sucralose was discovered in 1976 when a chemist misunderstood his superior, and so he *tasted* rather than *tested* his compound. Aspartame was discovered in 1965 when a chemist licked his dirty fingers in the lab and tasted its sweetness. Saccharin, the oldest known artificial sweetener, was discovered in 1879 by a chemist who failed to wash his hands after working in the lab. Saccharin was not used extensively until sugar shortages occurred during World War I. Although there were concerns in the 1970s that saccharin causes cancer, there is no proven link between cancer occurrence and saccharin intake at normal levels. *Jill Braaten/McGraw Hill*

14.6 Polysaccharides

Polysaccharides contain three or more monosaccharides joined together. Three prevalent polysaccharides in nature are **cellulose, starch,** and **glycogen,** each of which consists of repeating glucose units joined by glycosidic bonds.

14.6A   Cellulose

Cellulose is found in the cell walls of nearly all plants, where it gives support and rigidity to wood, plant stems, and grass (Figure 14.4). Wood, cotton, and flax are composed largely of cellulose.

Figure 14.4   Cellulose

layers of microfibrils in the plant cell wall

cell wall

cellulose fiber

plant cell

cellulose molecules

glucose

long chains of cellulose molecules hydrogen bonded together

BioPhoto Associates/Science Source

Cellulose is an unbranched polymer composed of repeating glucose units joined in a 1→4-β-glycosidic linkage. The β glycosidic linkages create long linear chains of cellulose molecules that stack in sheets, making an extensive three-dimensional array.

β glycosidic linkage (shown in blue)

cellulose

In some cells, cellulose is hydrolyzed by an enzyme that cleaves all of the β glycoside bonds, forming glucose. **Humans do not possess this enzyme, and therefore *cannot* digest cellulose.** Ruminant animals, on the other hand, such as cattle, deer, and camels, have bacteria containing this enzyme in their digestive systems, so they can derive nutritional benefit from eating grass and leaves.

Much of the insoluble fiber in our diet is cellulose, which passes through the digestive system without being metabolized. Foods rich in cellulose include whole wheat bread, brown rice, and bran cereals. Fiber is an important component of the diet even though it gives us no nutrition; fiber adds bulk to solid waste, so that it is eliminated more readily.

14.6B Starch

Starch is the main carbohydrate found in the seeds and roots of plants. Corn, rice, wheat, and potatoes are common foods that contain a great deal of starch. **Starch is a polymer composed of repeating glucose units joined in α glycosidic linkages.** The two common forms of starch are **amylose** and **amylopectin.**

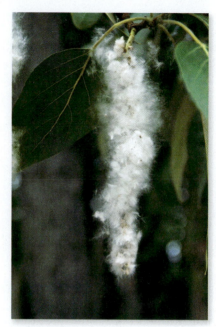

The outer bark of the cottonwood tree is largely the polysaccharide cellulose, whereas the inner bark also contains a naturally occurring analgesic related to aspirin. Native Americans of the Pacific Northwest used cottonwood bark and leaves to treat wounds. Cottonwood seeds are embedded in cotton fluff that allows the seeds to be blown a large distance for dispersal. *Daniel C. Smith*

α glycosidic linkage (shown in **red**)

amylose (the linear form of starch)

Two polysaccharide chains are connected at a branch point along one chain.

α glycosidic linkage (shown in **red**)

amylopectin (the branched form of starch)

Amylose, which comprises about 20% of starch molecules, has an unbranched skeleton of glucose molecules with **1→4-α-glycoside bonds.** Because of this linkage, an amylose chain adopts a helical arrangement, giving it a very different three-dimensional shape from the linear chains of cellulose (Figure 14.5).

Amylopectin, which comprises about 80% of starch molecules, consists of a backbone of glucose units joined in **α glycosidic bonds,** but it also contains considerable branching along the chain. The linear linkages of amylopectin are formed by **1→4-α-glycoside bonds,** similar to amylose.

Both forms of starch are water soluble. Because the OH groups in these starch molecules are not buried in a three-dimensional network, they are available for hydrogen bonding with water molecules, leading to greater water solubility than cellulose.

Both amylose and amylopectin are hydrolyzed to glucose with cleavage of the glycosidic bonds. The human digestive system has the necessary amylase enzymes needed to catalyze this process. Bread and pasta made from wheat flour, rice, and corn tortillas are all sources of starch that are readily digested.

Figure 14.5

Starch—Amylose and Amylopectin

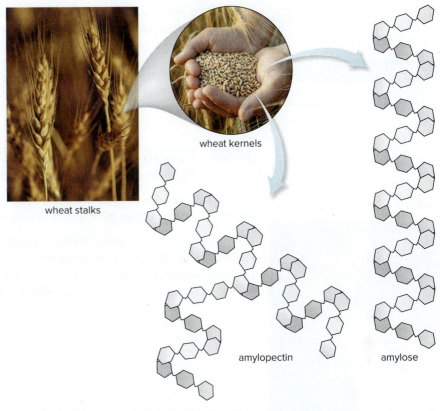

wheat stalks

wheat kernels

amylopectin

amylose

(left): PhotoLink/Photodisc/Getty Images; (right): Frolova_Elena/Shutterstock

14.6C   Glycogen

Glycogen is the major form in which polysaccharides are stored in animals. Glycogen, a polymer of glucose containing **α glycosidic bonds,** has a branched structure similar to amylopectin, but the branching is much more extensive (Figure 14.6).

Glycogen is stored principally in the liver and muscle. When glucose is needed for energy in the cell, glucose units are hydrolyzed from the ends of the glycogen polymer, and then further

Figure 14.6

Glycogen

glycogen in liver cells

branching

glycogen

Biophoto Associates/Science Source

metabolized with the release of energy. Because glycogen has a highly branched structure, there are many glucose units at the ends of the branches that can be cleaved whenever the body needs them.

Cellulose is water *in*soluble, despite its many OH groups. Based on its three-dimensional structure, why do you think this is so?

14.6D FOCUS ON HEALTH & MEDICINE
Human Milk Oligosaccharides

HEALTH NOTE

The World Health Organization recommends that children are exclusively breast fed until six months of age, and then nursed along with other forms of nutrition until a child is two years old.
Daniel C. Smith

Although the benefits of breast milk to a newborn have been known for a long time, how individual components impact the health of a newborn has only recently been discovered.

Human milk oligosaccharides (HMOs), which contain three or four monosaccharides joined together, constitute a group of carbohydrates found in breast milk. 2'-Fucosyllactose is the most prevalent component, making up about 30% of all HMOs. The two glycosidic linkages that join the three monosaccharides together are shown in blue.

2'-fucosyllactose

HMOs, often called the fiber of breast milk, are not hydrolyzed by the gastric juices of the stomach, nor are they absorbed in the intestine. Nevertheless, they play a key role in the health of a newborn, by helping to establish the presence of beneficial bacteria in the colon of an infant. Moreover, harmful pathogens attach to the surface of HMOs and are eliminated in the feces of the nursing infant.

Ongoing research continues to study the hundreds of unique components of human breast milk, in an effort to understand its benefits to both the mother and the child, even after infancy.

Label all glycosidic linkages in 3-fucosyllactose, another HMO found in breast milk.

3-fucosyllactose

The blood type of a blood donor and recipient must be compatible, so donated blood is clearly labeled with the donor's blood type. *DieterMeyrl/iStock/Getty Images*

14.7 FOCUS ON THE HUMAN BODY
Blood Type

Human blood is classified into one of four types—A, B, AB, and O. An individual's blood type is determined by three or four monosaccharides attached to a membrane protein of red blood cells. These monosaccharides include:

D-galactose   L-fucose   N-acetyl-D-glucosamine   N-acetyl-D-galactosamine

Each blood type is associated with a different carbohydrate structure, as shown in Figure 14.7. The short polysaccharide chains distinguish one type of red blood cell from another, and signal the cells about foreign viruses, bacteria, and other agents. When a foreign substance enters the blood, the body's immune system uses antibodies to attack and destroy the invading substance so that it does the host organism no harm.

Figure 14.7   Carbohydrates and Blood Types

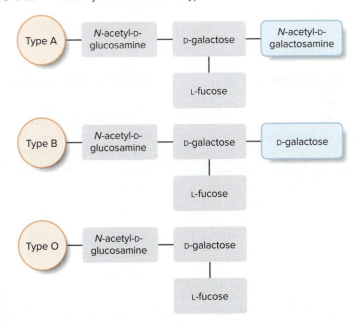

Each blood type is characterized by a different polysaccharide that is covalently bonded to a membrane protein of the red blood cell. There are three different carbohydrate sequences, one each for A, B, and O blood types. Blood type AB contains the sequences for both blood type A and blood type B.

Knowing an individual's blood type is necessary before receiving a blood transfusion. Because the blood of an individual may contain antibodies to another blood type, the types of blood that can be given to a patient are often limited. An individual with blood type A produces antibodies

Table 14.2 Compatibility Chart of Blood Types		
Blood Type	Can Receive Blood Type:	Can Donate to Blood Type:
A	A, O	A, AB
B	B, O	B, AB
AB	A, B, AB, O	AB
O	O	A, B, AB, O

to type B blood, and an individual with blood type B produces antibodies to type A blood. Type AB blood contains no antibodies to other blood types, whereas type O blood contains antibodies to both types A and B. As a result:

- Individuals with type O blood are called *universal donors* because type O blood can be given to individuals of any blood type.
- Individuals with type AB blood are called *universal recipients* because individuals with type AB blood can receive blood of any type.

Table 14.2 lists what blood types can be safely given to an individual. Blood must be carefully screened to make sure that the blood types of the donor and recipient are compatible.

CHAPTER REVIEW

KEY TERMS

Alditol (14.4)

Aldonic acid (14.4)

Aldose (14.2)

Benedict's reagent (14.4)

Carbohydrate (14.1)

Disaccharide (14.1)

Glycosidic linkage (14.5)

Haworth projection (14.3)

Hexose (14.2)

Ketose (14.2)

Monosaccharide (14.1)

Pentose (14.2)

Polysaccharide (14.1)

Reducing sugar (14.4)

Tetrose (14.2)

Triose (14.2)

KEY CONCEPTS

1 Characteristics of monosaccharides (14.2)

1 Classifying a monosaccharide	**2 Drawing a Fischer projection**	**3 Labeling a monosaccharide as D or L**
• **Aldoses** are monosaccharides with a carbonyl group at C1. • **Ketoses** are monosaccharides with a carbonyl at C2. • The terms triose, tetrose, and so forth are used to indicate the number of carbons in the chain.	• Replace each chirality center with a **cross** such that the horizontal bonds come forward (on wedges) and vertical bonds go back (on dashed wedges). • The carbonyl carbon is drawn at or near the top of the structure.	• A **D** sugar has the OH group of the chirality center farthest from the carbonyl on the *right* side. • An **L** sugar has the OH group of the chirality center farthest from the carbonyl on the *left* side.

② Cellulose, starch, and glycogen (14.6)

Cellulose, starch, and glycogen are all polymers of the monosaccharide glucose.

① Cellulose (14.6A)	② Starch (14.6B)	③ Glycogen (14.6C)
• Cellulose is an unbranched polymer composed of repeating glucose units joined in **1→4-β-glycosidic linkages.** • Cellulose forms long chains that stack in three-dimensional sheets. • The human digestive system does not contain the needed enzyme to metabolize cellulose.	• There are two forms of starch—**amylose,** which is an unbranched polymer, and **amylopectin,** which is a branched polysaccharide polymer. • Both forms contain **1→4-α-glycosidic linkages,** and the polymer winds in a helical arrangement. • Starch is digestible because the human digestive system has the needed enzyme to catalyze hydrolysis.	• Glycogen resembles amylopectin but is **more extensively branched.** • The linear portion of glycogen contains **1→4-α-glycosidic linkages.** • Glycogen is the major form in which polysaccharides are stored in animals. • When glucose is needed for energy in a cell, the glucose units at the ends of the chain are hydrolyzed and then metabolized with release of energy.

KEY REACTIONS

KEY SKILLS

① Determining the stereochemical features of a monosaccharide (14.2)

① Locating chirality centers	② Labeling a monosaccharide as D or L	③ Drawing the enantiomer
• **A chirality center has four different groups around carbon.** • The CHO and CH₂OH groups on the ends of a monosaccharide are *not* chirality centers. • The two middle C's (labeled in blue) that are bonded to H and OH are chirality centers.	• A **D sugar** has the OH group of the chirality center farthest from the carbonyl on the *right* side. • An **L sugar** has the OH group of the chirality center farthest from the carbonyl on the *left* side.	• The enantiomer is the mirror image, drawn so that each group is a reflection of the group in the original compound.

2 Drawing the cyclic forms of a D monosaccharide (14.3)

1 Draw a hexagon with an O atom in the upper right corner.	**2** Draw the new chirality center on the first carbon clockwise from the O atom.	**3** Add the OH groups and H atoms to the three remaining carbons (C2–C4).
• Add the CH₂OH to the first carbon to the left of the O atom. • For a D sugar, the CH₂OH is drawn *up*, above the plane of the ring.	• For a D sugar, the OH group is drawn *down* for the α isomer. • For a D sugar, the OH group is drawn *up* for the β isomer.	• The substituents on the *right* side of the carbon skeleton (shown in **red**) are drawn *down* in the Haworth projection. • The substituents on the *left* side of the carbon skeleton (shown in **blue**) are drawn *up* in the Haworth projection.

3 Characterizing the glycosidic linkage in a disaccharide (14.5)

1 Locate the glycosidic linkage and number the carbons in the rings.	**2** Classify the glycosidic linkage as α or β and use numbers to designate its location.
• The **glycosidic linkage** joins the two monosaccharides and contains a carbon bonded to two oxygens. • Each ring is numbered beginning at the carbon bonded to two oxygens.	• When the C—O bond is drawn *down*, the disaccharide has an α glycosidic linkage. The given monosaccharide has an α glycosidic linkage. • When the C—O bond is drawn *up*, the disaccharide has a β glycosidic linkage. • In the given monosaccharide, the glycosidic linkage joins C1 of one ring to C4 of the other ring, so it is called a **1→4-α-glycosidic linkage.**

glycosidic linkage shown in **red**

4 Drawing the hydrolysis products of a disaccharide (14.5)

1 Locate the glycosidic linkage.	**2** Cleave the C—O bond by adding the elements of H₂O across the bond.
• The glycosidic linkage contains the C—O that joins the two monosaccharides.	• Two monosaccharides are formed.

disaccharide
glycosidic linkage shown in **red**

CHAPTER 14 SELF-TEST

The Self-Test consists of short-answer questions that test an understanding of definitions, equations, and other material encountered within the chapter. Answers to each question are provided at the end of the chapter. For any questions you are unable to answer correctly, revisit the appropriate section(s) of the text; then proceed to working the end-of-chapter problems to further reinforce your understanding of the chapter material.

1. Which term from "Key Terms" answers each question?
   a. What term is used for a monosaccharide that contains an aldehyde and four carbons?
   b. What type of product is formed when an aldose is treated with Benedict's reagent?
   c. What are the flat rings used to represent the cyclic forms of monosaccharides called?
   d. What type of carbohydrate contains two monosaccharide units joined together?
   e. What type of monosaccharide contains a carbonyl group at C2?

2. Consider aldotetroses **A**, **B**, and **C**. (a) How many chirality centers does **A** contain? (b) Which compounds are D monosaccharides? (c) Which two compounds are enantiomers of each other?

A, B, C Fischer projection structures

3. Draw the structure of the product formed when **D** is treated with each reagent: (a) H_2, Pd; (b) Cu^{2+}, OH^-.

Structure **D** (Fischer projection)

4. Consider compounds **E–H**. (a) Classify **E** by the type of carbonyl group and the number of carbons in the chain. (b) Which structure represents the α isomer of **E**? (c) Which structure represents the β isomer of **E**? (d) Is **E** a reducing or nonreducing sugar?

Structures **E**, **F**, **G**, **H**

UNDERSTANDING KEY CONCEPTS

14.1 (a) Classify each monosaccharide by the type of carbonyl group and number of carbons in the chain. (b) Locate the chirality centers in each compound.

A   B

14.2 (a) Classify each monosaccharide by the type of carbonyl group and number of carbons in the chain. (b) Locate the chirality centers in each compound.

C   D

14.3 Draw the structure of each type of compound.
   a. an L-aldopentose   c. a five-carbon alditol
   b. a D-aldotetrose

14.4 Draw the structure of each type of compound.
   a. a D-aldotriose   c. a four-carbon aldonic acid
   b. an L-ketohexose

14.5 Consider the following monosaccharide.

Structure (Fischer projection)

   a. Is the monosaccharide a D or L sugar?
   b. Classify the monosaccharide by the type of carbonyl and the number of atoms in the chain.
   c. Label the chirality centers.
   d. Draw the enantiomer.
   e. Draw the α isomer of the cyclic form.
   f. What product is formed when the monosaccharide is treated with Benedict's reagent?
   g. What product is formed when the monosaccharide is treated with H_2 and Pd?

14.6 Answer the questions in Problem 14.5 using the following monosaccharide.

14.7 If a monosaccharide is represented by a single blue sphere, which of the following representations corresponds to (a) lactose; (b) cellulose; (c) amylopectin?

A   **B**   **C**

14.8 If a monosaccharide is represented by a single blue sphere, which of the following representations corresponds to (a) a disaccharide; (b) glycogen; (c) amylose?

A   **B**   **C**

14.9 (a) Locate the glycosidic linkage in the following disaccharide. (b) Number the carbon atoms in both rings. (c) Classify the glycosidic linkage as α or β, and use numbers to designate its location.

14.10 Answer the questions in Problem 14.9 using the following disaccharide.

ADDITIONAL PROBLEMS

Monosaccharides

14.11 Classify each monosaccharide by the type of carbonyl group and the number of carbons in the chain.

a. CHO / HO—C—H / HO—C—H / CH₂OH

b. CH₂OH / C=O / H—C—OH / HO—C—H / CH₂OH

14.12 Classify each monosaccharide by the type of carbonyl group and the number of carbons in the chain.

a. CHO / HO—C—H / CH₂OH

b. CHO / H—C—OH / HO—C—H / HO—C—H / CH₂OH

14.13 For each compound in Problem 14.11: [1] label all the chirality centers; [2] classify the compound as a D or L monosaccharide; [3] draw the enantiomer; [4] draw a Fischer projection.

14.14 For each compound in Problem 14.12: [1] label all the chirality centers; [2] classify the compound as a D or L monosaccharide; [3] draw the enantiomer; [4] draw a Fischer projection.

14.15 Are α-D-glucose and β-D-glucose enantiomers? Explain your choice.

14.16 Are D-fructose and L-fructose enantiomers? Explain your choice.

14.17 Consider monosaccharides **A, B,** and **C.**

A. CHO / H—C—OH / H—C—OH / CH₂OH

B. CHO / H—C—OH / HO—C—H / CH₂OH

C. CH₂OH / C=O / H—C—OH / CH₂OH

A   **B**   **C**

a. Which two monosaccharides are stereoisomers?
b. Identify two compounds that are constitutional isomers.
c. Draw the enantiomer of **B.**
d. Draw a Fischer projection for **A.**

14.18 Consider monosaccharides **D, E,** and **F.**

D. CHO / H—C—OH / H—C—OH / H—C—OH / CH₂OH

E. CH₂OH / C=O / HO—C—H / HO—C—H / CH₂OH

F. CHO / HO—C—H / HO—C—H / H—C—OH / CH₂OH

D   **E**   **F**

a. Which two monosaccharides are stereoisomers?
b. Identify two compounds that are constitutional isomers.
c. Draw the enantiomer of **F.**
d. Draw a Fischer projection for **D.**

14.19 Using Haworth projections, draw the α and β isomers of the cyclic form of the following D monosaccharide.

```
        CHO
   HO—C—H
    H—C—OH
   HO—C—H
    H—C—OH
        CH₂OH
```

14.20 Using Haworth projections, draw the α and β isomers of the cyclic form of the following D monosaccharide.

```
        CHO
    H—C—OH
    H—C—OH
   HO—C—H
    H—C—OH
        CH₂OH
```

14.21 Consider the following cyclic monosaccharide.

```
        CH₂OH
     OH     O    OH
         H
        OH   OH
      H            H
         H    H
```

a. Label the monosaccharide as an α or β isomer.
b. Draw the other cyclic form.

14.22 Consider the following cyclic monosaccharide.

```
          CH₂OH
      H       O    H
          H
         OH   OH
   HO            OH
          H    H
```

a. Label the monosaccharide as an α or β isomer.
b. Draw the other cyclic form.

14.23 Label each monosaccharide as an α or β isomer.

a.
```
       CH₂OH
    H      O   OH
      H
     H     H
   HO          H
     OH    OH
```

b.
```
   CH₂OH  O  CH₂OH
      OH       H
    H            OH
       H    OH
```

14.24 Label each monosaccharide as an α or β isomer.

a.
```
       CH₂OH
    H      O    H
       H
      OH   OH
   HO          OH
      H    H
```

b.
```
   CH₂OH  O    OH
      H       H
    H           CH₂OH
       OH    OH
```

Reactions of Carbohydrates

14.25 Label each compound as a monosaccharide, an aldonic acid, or an alditol.

a.

c.

b.

14.26 Label each compound as a monosaccharide, an aldonic acid, or an alditol.

a.

c.

b.

14.27 Draw the organic products formed when each monosaccharide is treated with each of the following reagents: [1] H_2, Pd; [2] Cu^{2+}, OH⁻.

a.
```
        CHO
   HO—C—H
    H—C—OH
        CH₂OH
```

b.
```
        CHO
    H—C—OH
    H—C—OH
   HO—C—H
   HO—C—H
        CH₂OH
```

14.28 Draw the organic products formed when each monosaccharide is treated with each of the following reagents: [1] H_2, Pd; [2] Cu^{2+}, OH⁻.

a.
```
        CHO
   HO—C—H
    H—C—OH
    H—C—OH
        CH₂OH
```

b.
```
        CHO
    H—C—OH
    H—C—OH
    H—C—OH
    H—C—OH
        CH₂OH
```

14.29 What product is formed when the given aldopentose is treated with Benedict's reagent?

```
        CHO
    H—C—OH
    H—C—OH
   HO—C—H
        CH₂OH
```

14.30 What product is formed when the given aldopentose is treated with Benedict's reagent?

Disaccharides and Polysaccharides

14.31 What monosaccharides are formed when the given disaccharide is hydrolyzed?

14.32 What monosaccharides are formed when the given disaccharide is hydrolyzed?

14.33 (a) Locate the glycosidic linkage in the following disaccharide, and label it as α or β. (b) Draw the structure of the monosaccharides formed on hydrolysis.

14.34 Answer the questions in Problem 14.33 for the disaccharide lactulose.

14.35 Draw the structure of a disaccharide that contains two six-membered rings and an α glycosidic linkage.

14.36 Draw the structure of a disaccharide that contains one six-membered ring and one five-membered ring, as well as a β glycosidic linkage.

14.37 What is the major disaccharide or polysaccharide in each of the following items?

a.   b.   c.

(a): Foodcollection; (b): UpperCut Images/SuperStock; (c): Rachel Weill/Getty Images

14.38 Describe the similarities and differences in the structures of lactose and sucrose.

14.39 Consider the disaccharide isomaltose.

isomaltose

a. Classify the glycosidic linkage as α or β.
b. What monosaccharides are formed when isomaltose is hydrolyzed?

14.40 Draw the structure of a disaccharide formed from two galactose units joined by a 1→4-β-glycosidic linkage. (The cyclic structure of galactose appears in Sample Problem 14.3.)

14.41 Identify polysaccharides **A** and **B** as either amylopectin or glycogen. Then decide if each of the following statements refers to one, both, or neither of the polysaccharides.

A   **B**

a. This polysaccharide contains 1→4-α-glycosidic bonds.
b. This polysaccharide is stored in the liver.
c. This polysaccharide is found in bread and pasta.
d. This polysaccharide is composed of galactose units.

14.42 Identify polysaccharides **C** and **D** as either amylose or cellulose. Then decide if each of the following statements refers to one, both, or neither of the polysaccharides.

C   **D**

a. This polysaccharide contains 1→4-α-glycosidic bonds.
b. This polysaccharide is not digestible by humans.
c. This polysaccharide is stored in the liver and muscle.
d. This polysaccharide is composed of glucose units.

Applications

14.43 Describe the difference between lactose intolerance and galactosemia.

14.44 Explain why cellulose is a necessary component of our diet even though we don't digest it.

14.45 Explain why fructose is called a reduced calorie sweetener whereas sucralose is called an artificial sweetener.

14.46 How do oxidation reactions help an individual with diabetes monitor blood glucose levels?

14.47 Fill in the following blood-compatibility chart. [Y = the blood types are compatible; N = the blood types are not compatible.]

Recipient	Donor			
	O	A	B	AB
O				
A				
B				
AB				

14.48 Why can an individual with type B blood receive only blood types B and O, but he or she can donate to individuals with either type B or AB blood?

CHALLENGE PROBLEMS

14.49 Draw a short segment of a polysaccharide that contains three galactose units joined together in 1→4-α-glycosidic linkages. (The cyclic structure of galactose appears in Sample Problem 14.3.)

14.50 Convert each cyclic form to an acyclic monosaccharide.

BEYOND THE CLASSROOM

14.51 In addition to the compounds shown in Figure 14.3, cyclamate, Truvia, acesulfame K, and alitame are artificial or low-calorie sweeteners. Research one or more of these compounds and draw its chemical structure. Is the compound isolated from a natural source or synthesized in the laboratory? How sweet is the compound compared to sucrose? Is the compound used in any commercial products? Does this sweetener offer any advantages or disadvantages over current widely used artificial sweeteners?

14.52 Pick a recipe for a favorite cookie, cake, or other baked good. Determine the number of grams of sugar used for one batch of the recipe, as well as the number of Calories that this amount contains. How much sucralose or saccharin would be used instead of sugar for an equivalent level of sweetness? Conversion factors you will need: density of granulated sugar = 0.70 g/cc; Calories from carbohydrates = 4 Cal/g.

CHAPTER 14 ANSWERS

Concept Check Answers

14.1 hexane < 1-decanol < glucose

14.2

```
      CHO
HO ——|—— H
      CH₂OH
```

14.3 a.
```
      CHO
 H ——|—— OH
HO ——|—— H
HO ——|—— H
 H ——|—— OH
      CH₂OH
```
b.
```
      CHO
HO ——|—— H
 H ——|—— OH
 H ——|—— OH
HO ——|—— H
      CH₂OH
```

14.4 a,b: α isomer   c. β isomer

14.5

14.6 glucose and galactose

14.7 Cellulose's many OH groups are already hydrogen bonded to other OH groups in the interior of the three-dimensional structure, so they are less available for hydrogen bonding to water.

14.8

3-fucosyllactose
glycosidic linkages shown in **red**

Practice Problem Answers

14.1 a. aldopentose   b. aldotetrose   c. ketotetrose

14.2 a., b., c.

14.3 a., c.

b.

* chirality center

14.4 a., b., c.

* chirality center

14.5 a., b., c.

14.6 a., b.

14.7 a., b.

14.8 a., b.

14.9 1⟶ 4-β-disaccharide

glycosidic linkage

Self-Test Answers

1. a. aldotetrose   c. Haworth projections   e. ketose
    b. aldonic acid   d. disaccharide

2. a. two   b. **B, C**   c. **A, C**

3. a., b.

4. a. aldohexose   b. **H**   c. **F**   d. reducing sugar

Odd-Numbered Problem Answers

14.1

A   **B**
aldotetrose   aldohexose
* chirality center

14.3 a., b., c.

14.5 a. D sugar

b. aldohexose

c., d.

* chirality center

enantiomer

e., f., g.

14.7 a. **A**   b. **C**   c. **B**

14.9 a,b:

glycosidic linkage

c. 1→4-α-glycosidic linkage

14.11 a. aldotetrose   b. ketopentose

14.13 a.

CHO — HO—C*—H, HO—C*—H, CH₂OH L-tetrose

CHO — H—C—OH, H—C—OH, CH₂OH enantiomer

CHO — HO——H, HO——H, CH₂OH Fischer projection

b.

CH₂OH, C=O, H—C*—OH, HO—C*—H, CH₂OH L-pentose

CH₂OH, C=O, HO—C—H, H—C—OH, CH₂OH enantiomer

CH₂OH, C=O, H——OH, HO——H, CH₂OH Fischer projection

* chirality center

14.15 α-D-Glucose and β-D-glucose are not enantiomers because they differ in the orientation of only one OH at C1.

14.17 a. **A** and **B**

b. **B** and **C** (or **A** and **C**)

c.

CHO, HO—C—H, H—C—OH, CH₂OH

d.

CHO, H——OH, H——OH, CH₂OH

14.19

β isomer

α isomer

14.21 a. β isomer   b.

14.23 a. β isomer   b. α isomer

14.25 a. aldonic acid   b. alditol   c. monosaccharide

14.27 a. [1] CH₂OH, HO—C—H, H—C—OH, CH₂OH

[2] COOH, HO—C—H, H—C—OH, CH₂OH

b. [1] CH₂OH, H—C—OH, H—C—OH, HO—C—H, HO—C—H, CH₂OH

[2] COOH, H—C—OH, H—C—OH, HO—C—H, HO—C—H, CH₂OH

14.29

COOH, H—C—OH, H—C—OH, HO—C—H, CH₂OH

14.31

+

14.33 a.

β glycosidic linkage

b.

+

14.35

14.37 a. starch   b. cellulose   c. lactose

14.39 a. α glycoside

b.

+

14.41 **A** = glycogen   **B** = amylopectin

a. both   b. **A**   c. **B**   d. neither

14.43 Lactose intolerance results from a lack of the enzyme lactase. It causes abdominal cramping and diarrhea. Galactosemia results from the inability to metabolize galactose, so it accumulates in the liver, causing cirrhosis, and in the brain, leading to mental retardation.

14.45 Fructose is a naturally occurring sugar with more perceived sweetness per gram than sucrose. Sucralose is a synthetic sweetener; that is, it is not naturally occurring.

14.47 O: YNNN; A: YYNN; B: YNYN; AB: YYYY

14.49

15

Lipids

The cocoa butter in chocolate is rich in **triacylglycerols,** the most abundant lipids. *Fcafotodigital/ iStock/Getty Images*

CHAPTER OUTLINE

CHAPTER GOALS

In this chapter you will learn how to:

1. Describe the general characteristics of lipids
2. Classify fatty acids and describe the relationship between melting point and the number of double bonds
3. Draw the structure of a wax and identify the carboxylic acid and alcohol components
4. Draw the structure of triacylglycerols and describe the difference between a fat and an oil
5. Draw the hydrolysis products of triacylglycerols
6. Identify the structural features of phospholipids
7. Describe the structure of a cell membrane, as well as different mechanisms of transport across the membrane
8. Recognize the main structural features of steroids like cholesterol
9. Define what a hormone is and list several examples of steroid hormones
10. Identify fat-soluble vitamins

I n Chapter 15, we turn our attention to lipids, biomolecules that are soluble in organic solvents and insoluble in water. Unlike the carbohydrates in Chapter 14, lipids contain few functional groups and are composed mainly of carbon–carbon and carbon–hydrogen bonds. Because lipids are the biomolecules that most closely resemble the hydrocarbons discussed in Chapters 10 and 11, we have already learned many facts that directly explain their properties. In Chapter 15 we learn more details about the chemistry of lipids.

Why Study ...

Lipids?

HEALTH NOTE

Common lipids include triacylglycerols in vegetable oils, cholesterol in egg yolk, and vitamin E in leafy greens.
Jill Braaten/McGraw Hill

Table 15.1	**Summary of Lipid Chemistry Prior to Chapter 15**
Topic	**Section**
Fatty acids	11.3B
Oral contraceptives	11.4
Margarine and butter	11.6
Aspirin and prostaglandins	13.4B
Soap	13.5B
Lipid hydrolysis	13.6C

15.1   Introduction to Lipids

- *Lipids* are biomolecules that are soluble in organic solvents and insoluble in water.

Lipids are unique among organic molecules because their identity is defined on the basis of a *physical property* and not by the presence of a particular functional group. Because of this, lipids come in a wide variety of structures and they have many different functions.

Lipids contain a large number of nonpolar carbon–carbon and carbon–hydrogen bonds. In addition, most lipids have a few polar bonds that may be found in a variety of functional groups. As a result, lipids are nonpolar or weakly polar molecules that are very soluble in organic solvents like hexane (C_6H_{14}) and carbon tetrachloride (CCl_4), and insoluble in a polar medium like water.

Because lipids share many properties with hydrocarbons, several features of lipid structure and properties have been discussed in previous chapters, as summarized in Table 15.1.

Lipids can be categorized as hydrolyzable or nonhydrolyzable.

1. *Hydrolyzable lipids* can be converted into smaller molecules by hydrolysis with water. We will examine three subgroups: waxes, triacylglycerols, and phospholipids.

Most hydrolyzable lipids contain an ester.

2. *Nonhydrolyzable lipids* cannot be cleaved into smaller units by aqueous hydrolysis. Nonhydrolyzable lipids tend to be more varied in structure. We will examine two different types: steroids and fat-soluble vitamins.

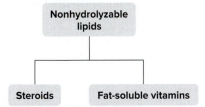

Lipids have many important roles in biological systems. Lipids release over twice the amount of energy per gram than carbohydrates or proteins (9 kcal/g for lipids compared to 4 kcal/g for carbohydrates and proteins, Section 4.1), so lipids are an excellent source of energy. Moreover, lipids are key components of the cell membrane, and they serve as chemical messengers in the body.

In which solvents or solutions will a lipid be soluble: (a) CH_2Cl_2; (b) 5% aqueous NaCl solution; (c) $CH_3CH_2CH_2CH_2CH_3$?

15.2 Fatty Acids

Hydrolyzable lipids are derived from **fatty acids,** carboxylic acids that were first discussed in Section 11.3B.

- *Fatty acids* are carboxylic acids (RCOOH) with long carbon chains of 12–20 carbon atoms.

Palmitic acid is a common 16-carbon fatty acid whose structure is given in condensed, skeletal, and three-dimensional representations.

$$CH_3CH_2CH_2CH_2CH_2CH_2CH_2CH_2CH_2CH_2CH_2CH_2CH_2CH_2CH_2COOH$$
palmitic acid
$$C_{16}H_{32}O_2$$

skeletal structure

polar C—O and O—H bonds
hydrophilic portion

nonpolar C—C and C—H bonds

hydrophobic portion

The nonpolar part of the molecule (comprised of C—C and C—H bonds) is not attracted to water, so it is said to be *hydrophobic* (water fearing). The polar part of the molecule is attracted to water, so it is said to be *hydrophilic* (water loving). In a lipid, the hydrophobic portion is always much larger than the hydrophilic portion.

Naturally occurring fatty acids have an even number of carbon atoms. There are two types of fatty acids.

- *Saturated* fatty acids have no double bonds in their long hydrocarbon chains.
- *Unsaturated* fatty acids have one or more double bonds in their long hydrocarbon chains. Generally, double bonds in naturally occurring fatty acids are cis.

Recall from Section 11.3 that a **cis** alkene has two alkyl groups on the *same* side of the double bond, whereas a **trans** alkene has two alkyl groups on *opposite* sides of the double bond.

Three-dimensional models for stearic acid, a saturated fatty acid, and oleic acid, an unsaturated fatty acid with one cis double bond, are shown. Table 15.2 lists the structures of the most common saturated and unsaturated fatty acids.

stearic acid

a saturated fatty acid
no double bonds in the long chain

cis double bond

oleic acid

an unsaturated fatty acid
one cis double bond in the long chain

Table 15.2 Common Fatty Acids

Number of C's	Number of C=C's	Structure	Name	Mp (°C)
		Saturated Fatty Acids		
12	0	$CH_3(CH_2)_{10}COOH$	Lauric acid	44
14	0	$CH_3(CH_2)_{12}COOH$	Myristic acid	58
16	0	$CH_3(CH_2)_{14}COOH$	Palmitic acid	63
18	0	$CH_3(CH_2)_{16}COOH$	Stearic acid	71
20	0	$CH_3(CH_2)_{18}COOH$	Arachidic acid	77
		Unsaturated Fatty Acids		
16	1	$CH_3(CH_2)_5CH=CH(CH_2)_7COOH$	Palmitoleic acid	1
18	1	$CH_3(CH_2)_7CH=CH(CH_2)_7COOH$	Oleic acid	16
18	2	$CH_3(CH_2)_4CH=CHCH_2CH=CH(CH_2)_7COOH$	Linoleic acid	−5
18	3	$CH_3CH_2CH=CHCH_2CH=CHCH_2CH=CH(CH_2)_7COOH$	Linolenic acid	−11
20	4	$CH_3(CH_2)_4(CH=CHCH_2)_4(CH_2)_2COOH$	Arachidonic acid	−49

The most common saturated fatty acids are palmitic and stearic acid. The most common unsaturated fatty acid is oleic acid. Linoleic and linolenic acids are called **essential fatty acids** because humans cannot synthesize them and must acquire them in our diets.

Oils formed from omega-3 fatty acids may provide health benefits to individuals with cardiovascular disease, as discussed in Section 15.4.

Unsaturated fatty acids are sometimes classified as **omega-n acids,** where n is the carbon at which the first double bond occurs in the carbon chain, beginning at the end of the chain that contains the CH_3 group. Thus, linoleic acid is an omega-6 acid and linolenic acid is an omega-3 acid.

first C=C at C**6** ⟶ an **omega-6 acid**
linoleic acid

first C=C at C**3** ⟶ an **omega-3 acid**
linolenic acid

As we learned in Section 11.3B, the presence of cis double bonds affects the melting point of these fatty acids greatly.

- As the number of double bonds in the fatty acid *increases,* the melting point *decreases.*

Sample Problem 15.1

Determining the Characteristics of a Fatty Acid

(a) Draw a skeletal structure of gadoleic acid, $CH_3(CH_2)_9CH=CH(CH_2)_7COOH$, a 20-carbon fatty acid obtained from fish oils. (b) Label the hydrophobic and hydrophilic portions. (c) Predict how its melting point compares to the melting points of arachidic acid and arachidonic acid (Table 15.2).

Analysis

- Skeletal structures have a carbon at the intersection of two lines and at the end of every line. The double bond must have the cis arrangement in an unsaturated fatty acid.
- The nonpolar C—C and C—H bonds comprise the hydrophobic portion of a molecule and the polar bonds comprise the hydrophilic portion.
- For the same number of carbons, *increasing* the number of double bonds *decreases* the melting point of a fatty acid.

Gadoleic acid (Sample Problem 15.1) and DHA (Practice Problem 15.2) are two fatty acids derived from tuna fish oil.
Tono Balaguer/123RF

Solution

a. and b. The skeletal structure for gadoleic acid is drawn below. The hydrophilic portion is the COOH group and the hydrophobic portion is the rest of the molecule.

c. Gadoleic acid has one cis double bond, giving it a lower melting point than arachidic acid (no double bonds), but a higher melting point than arachidonic acid (four double bonds).

Practice Problem 15.1

(a) Draw a skeletal structure for each fatty acid. (b) Label the hydrophobic and hydrophilic portions of each molecule. (c) Without referring to Table 15.2, which fatty acid has the higher melting point, **A** or **B**? Explain your choice.

$$CH_3(CH_2)_{16}COOH \qquad CH_3(CH_2)_4CH=CHCH_2CH=CH(CH_2)_7COOH$$
$$\textbf{A} \qquad\qquad\qquad\qquad \textbf{B}$$

Practice Problem 15.2

DHA (4,7,10,13,16,19-docosahexaenoic acid) is a common fatty acid present in tuna fish oil. (a) Draw a skeletal structure for DHA. (b) Give the omega-*n* designation for DHA.

$$CH_3CH_2CH=CHCH_2CH=CHCH_2CH=CHCH_2CH=CHCH_2CH=CHCH_2CH=CHCH_2CH_2COOH$$
$$DHA$$

Practice Problem 15.3

Give the omega-*n* designation for (a) oleic acid; (b) arachidonic acid (Table 15.2).

More Practice: Try Problems 15.15–15.18; 15.23a, b; 15.24a, b; 15.47a; 15.48a; 15.51; 15.52.

When commercial whaling was commonplace, spermaceti wax obtained from sperm whales was used extensively in cosmetics and candles.
Willyam Bradberry/Shutterstock

Water beads up on the surface of a leaf because of its waxy coating. *Daniel C. Smith*

15.3 Waxes

Waxes are the simplest hydrolyzable lipids.

- *Waxes* are esters (RCOOR') formed from a fatty acid (RCOOH) and a high-molecular-weight alcohol (R'OH).

For example, spermaceti wax, isolated from the heads of sperm whales, is largely cetyl palmitate, an ester with the structure $CH_3(CH_2)_{14}COO(CH_2)_{15}CH_3$. Cetyl palmitate is formed from a 16-carbon fatty acid [$CH_3(CH_2)_{14}COOH$] and a 16-carbon alcohol [$CH_3(CH_2)_{15}OH$].

Waxes form a protective coating on the feathers of birds to make them water repellent, and on leaves to prevent water evaporation. Beeswax, a complex mixture of over 200 different compounds, contains the wax myricyl palmitate as its major component.

CH₃(CH₂)₁₄ — C(=O) — O(CH₂)₂₉CH₃

myricyl palmitate

Jill Braaten/McGraw Hill   Ralf Hettler/E+/Getty Images

Sample Problem 15.2   Drawing the Structure of a Wax

Draw the structure of cetyl myristate, a minor component of spermaceti wax, formed from a 14-carbon fatty acid and a 16-carbon straight-chain alcohol.

Analysis

To draw the wax, arrange the carboxyl group of the fatty acid (RCOOH) next to the OH group of the alcohol (R'OH) with which it reacts. Then, **replace the OH group of the fatty acid with the OR' group of the alcohol,** forming an ester RCOOR' with a new C—O bond at the carbonyl carbon.

Solution

Draw the structures of the fatty acid and the alcohol, and replace the OH group of the 14-carbon acid with the O(CH₂)₁₅CH₃ group of the alcohol.

new C—O bond

CH₃(CH₂)₁₂ — C(=O) — OH   +   H—O(CH₂)₁₅CH₃   ⟶   CH₃(CH₂)₁₂ — C(=O) — O(CH₂)₁₅CH₃   +   H₂O

14-carbon fatty acid   16-carbon alcohol   cetyl myristate

Practice Problem 15.4

Draw the structure of a wax formed from stearic acid [CH₃(CH₂)₁₆COOH] and each alcohol.

a. CH₃(CH₂)₉OH   b. CH₃(CH₂)₁₁OH   c. CH₃(CH₂)₂₉OH

Practice Problem 15.5

One component of jojoba oil is a wax formed from eicosenoic acid [CH₃(CH₂)₇CH=CH(CH₂)₉CO₂H] and CH₃(CH₂)₇CH=CH(CH₂)₈OH. Draw the structure of the wax.

More Practice: Try Problems 15.3, 15.21, 15.22, 15.23d, 15.24d, 15.47b, 15.48b.

The seeds of the jojoba plant (Practice Problem 15.5) grown in the southwestern United States are rich in waxes used in cosmetics and personal care products.
Source: USDA, ARS, National Genetic Resources Program

Like other esters, **waxes (RCOOR') are hydrolyzed with water in the presence of acid or base to re-form the carboxylic acid (RCOOH) and alcohol (R'OH)** from which they are prepared (Section 13.6). Thus, hydrolysis of cetyl palmitate in the presence of H₂SO₄ forms a fatty acid and a long-chain alcohol by cleaving the carbon–oxygen single bond of the ester.

The C—O bond is broken.

CH₃(CH₂)₁₄ — C(=O) — O(CH₂)₁₅CH₃   +   H—OH   —H₂SO₄→   CH₃(CH₂)₁₄ — C(=O) — OH   +   H—O(CH₂)₁₅CH₃

cetyl palmitate   fatty acid   alcohol

What hydrolysis products are formed when cetyl laurate is treated with aqueous acid?

cetyl laurate

15.4 Triacylglycerols—Fats and Oils

Animal fats and vegetable oils, the most abundant lipids, are composed of **triacylglycerols.**

- *Triacylglycerols,* or triglycerides, are triesters formed from glycerol and three molecules of fatty acids.

glycerol   fatty acids   triacylglycerol   block diagram

R groups have 11–19 C's.

[Three ester groups are labeled in red.]

15.4A General Features

Triacylglycerols may be composed of three identical fatty acid side chains, or they may be derived from two or three different fatty acids. The fatty acids may be saturated or unsaturated. **Fats** and **oils** are triacylglycerols with different physical properties.

- Fats have higher melting points—they are *solids* at room temperature.
- Oils have lower melting points—they are *liquids* at room temperature.

The identity of the three fatty acids in the triacylglycerol determines whether it is a fat or an oil. *Increasing* **the number of double bonds in the fatty acid side chains** *decreases* **the melting point of the triacylglycerol.**

- Fats are derived from fatty acids having *few* double bonds.
- Oils are derived from fatty acids having a *larger* number of double bonds.

Figure 15.1 illustrates the fatty acid composition of some common fats and oils. Solid fats have a relatively high percentage of saturated fatty acids and are generally animal in origin. Thus, lard (hog fat), butter, and whale blubber contain a high percentage of saturated fats. With no double bonds, the three side chains of the saturated lipid lie parallel with each other, leading to a high melting point.

Figure 15.1

Fatty Acid Composition of Some Fats and Oils

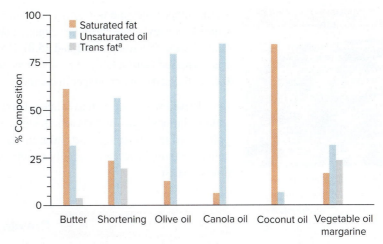

ªTrans double bonds are formed when triacylglycerols are partially hydrogenated, as described in Section 11.6.
[Source: Harvard School of Public Health, but similar data are available from many other sources.]

animal products
source of saturated fats

Jill Braaten/McGraw Hill

no double bonds in the long carbon chains

a saturated triacylglycerol

Liquid oils have a higher percentage of unsaturated fatty acids and are generally vegetable in origin. Thus, oils derived from corn, soybeans, and olives contain more unsaturated lipids. In the unsaturated lipid, a cis double bond places a kink in the side chain, making it more difficult to pack efficiently in the solid state, thus leading to a lower melting point.

vegetable oil
source of unsaturated oils

Image Source/Getty Images

one (or more) cis double bonds in the long carbon chains

an unsaturated triacylglycerol

Unlike other vegetable oils, oils from palm and coconut trees are very high in saturated fats. Considerable evidence currently suggests that diets high in saturated fats lead to a greater risk of heart disease (Sections 15.4B and 15.8). For this reason, the demand for coconut and palm oil has decreased considerably in recent years, and many coconut plantations previously farmed in the South Pacific are no longer in commercial operation.

Oils derived from fish such as salmon, herring, mackerel, and sardines are very rich in polyunsaturated triacylglycerols. These triacylglycerols pack so poorly that they have very low melting points, and they remain liquids even in very cold water. Fish oils derived from omega-3 fatty acids are thought to be especially beneficial for individuals at risk for developing coronary artery disease.

Sample Problem 15.3   Drawing the Structure of a Triacylglycerol

Draw the structure of a triacylglycerol formed from glycerol, one molecule of stearic acid, and two molecules of oleic acid. Bond the stearic acid to the 2° OH group (OH on the middle carbon atom) of glycerol.

Analysis

To draw the triacylglycerol, arrange each OH group of glycerol next to the carboxyl group of a fatty acid. Then **join each O atom of glycerol to a carbonyl carbon of a fatty acid,** to form three new C—O bonds.

Solution

Form three new ester bonds (RCOOR') from OH groups of glycerol and the three fatty acids (RCOOH).

oleic acid (R group in blue)
stearic acid (R group in red)

+ 3 H₂O

triacylglycerol

Practice Problem 15.6

Draw the structure of a triacylglycerol that contains three molecules of stearic acid.

In addition to triacylglycerols, the cocoa used to prepare chocolate contains antioxidants and caffeine. One ounce of unsweetened chocolate has as much caffeine as two ounces of brewed coffee. *Ingram Publishing/SuperStock*

Practice Problem 15.7

Unlike many fats and oils, the cocoa butter used to make chocolate is remarkably uniform in composition. All triacylglycerols contain oleic acid esterified to the 2° OH group of glycerol, and either palmitic acid or stearic acid esterified to the 1° OH groups. Draw the structure of a possible triacylglycerol that composes cocoa butter.

More Practice: Try Problems 15.27, 15.28, 15.29a–d, 15.30a–d.

15.4B FOCUS ON HEALTH & MEDICINE
Fats and Oils in the Diet

Fats and oils in our diet come from a variety of sources—meat, dairy products, seeds and nuts, salad dressing, fried foods, and any baked good or packaged food made with oil. Some fat is required in the diet. Fats are the building blocks of cell membranes, and stored body fat insulates an organism and serves as an energy source that can be used at a later time.

Currently, the United States Food and Drug Administration recommends that no more than 20–35% of an individual's calorie intake come from lipids. Moreover, a high intake of *saturated* triacylglycerols is linked to an increased incidence of heart disease. Saturated fats stimulate cholesterol synthesis in the liver and transport to the tissues, resulting in an increase in cholesterol concentration in the blood. An elevated cholesterol level in the blood can lead to cholesterol deposits or plaques on arteries, causing a narrowing of blood vessels, heart attack, and stroke (Section 15.8).

In contrast, *unsaturated* triacylglycerols lower the risk of heart disease by decreasing the amount of cholesterol in the blood. Unsaturated triacylglycerols from omega-3 fatty acids appear to reduce the risk of heart attack, especially in individuals who already have heart disease.

Dietary recommendations on fat intake must also take into account trans triacylglycerols, so-called *trans fats.* As we learned in Section 11.6, trans fats are formed when liquid oils are partially hydrogenated to form semi-solid triacylglycerols. The three-dimensional structure of a trans triacylglycerol shows its similarity to saturated triacylglycerols. Like saturated fats, trans fats also *increase* the amount of cholesterol in the bloodstream, thus increasing an individual's risk of developing coronary artery disease.

a trans double bond in the long carbon chain
a trans triacylglycerol

Thus, while lipids are a necessary part of anyone's diet, the amount of saturated fat and trans fat should be limited. Calories from lipids should instead be obtained from unsaturated oils. Figure 15.2 illustrates some foods high in saturated, unsaturated, and trans triacylglycerols.

Nutrition Facts
Serving Size 11 Crackers (31g)
Servings Per Container About 7

Amount Per Serving	
Calories 150 Calories from Fat 50	

	% Daily Value*
Total Fat 6g	**9%**
Saturated Fat 1g	**5%**
Trans Fat 0g	
Polyunsaturated Fat 3g	
Monounsaturated Fat 1g	
Cholesterol 0mg	**0%**
Sodium 270mg	**11%**
Total Carbohydrate 21g	**7%**
Dietary Fiber 1g	**4%**
Sugars 4g	

Jill Braaten/McGraw Hill

Figure 15.2 Saturated, Unsaturated, and Trans Triacylglycerols in the Diet

Food source

Triacylglycerol

Foods rich in saturated triacylglycerols:
fatty red meat, cheese, butter, fried foods, ice cream

- no double bonds in the carbon chains
- solid at room temperature
- increases blood cholesterol level

Foods rich in unsaturated triacylglycerols:
plant oils, nuts, soybeans, fish (salmon, herring, mackerel)

- one (or more) double bonds in the carbon chains
- liquid at room temperature
- decreases blood cholesterol level

Foods rich in trans triacylglycerols:
margarine, processed foods, fried foods, some baked goods

- one (or more) trans double bonds in the carbon chains
- semi-solid at room temperature
- increases blood cholesterol level

(top to bottom): Jill Braaten/McGraw Hill

Concept Check 15.3

Draw the structure of a triacylglycerol that fits each description:

a. a saturated triacylglycerol formed from three 12-carbon fatty acids

b. a trans triacylglycerol that contains a trans double bond in each hydrocarbon chain

Concept Check 15.4

Which food fits each description: [1] almonds; [2] whipped cream; [3] Cool Whip (imitation whipped topping made with hydrogenated vegetable oil)? More than one food may fit a description.

a. It is high in unsaturated triacylglycerols.

b. It is high in triacylglycerols derived from saturated fatty acids.

c. It likely contains little or no trans fats.

d. It can cause an increase in blood cholesterol level.

How does the triacylglycerol content of peanuts and peanut butter (made from partially hydrogenated peanut oil) differ with respect to the amount of unsaturated triacylglycerols and trans fats?

15.5 Hydrolysis of Triacylglycerols

Like other esters, **triacylglycerols are hydrolyzed with water in the presence of acid, base, or enzymes (in biological systems) to form glycerol and three molecules of fatty acids.** Thus, hydrolysis of tristearin with aqueous sulfuric acid forms glycerol and three molecules of stearic acid.

The three bonds drawn in red are broken in hydrolysis.

$+ \ 3\,H{-}OH$

tristearin

H_2SO_4

glycerol $+$ 3 HO

stearic acid

Hydrolysis cleaves the three single bonds between the carbonyl carbons and the oxygen atoms of the esters (Section 13.6). Because tristearin contains three identical R groups on the carbonyl carbons, three molecules of a single fatty acid, stearic acid, are formed. Triacylglycerols that contain different R groups bonded to the carbonyl carbons form mixtures of fatty acids, as shown in Sample Problem 15.4.

Drawing the Products of Triacylglycerol Hydrolysis

Draw the products formed when the given triacylglycerol is hydrolyzed with water in the presence of sulfuric acid.

Analysis

To draw the products of ester hydrolysis, **cleave the three C—O single bonds at the carbonyl carbons to form glycerol and three fatty acids (RCOOH).**

Solution

Hydrolysis forms glycerol and stearic, palmitic, and palmitoleic acids.

Bonds broken
during hydrolysis

3 **H—OH** | H_2SO_4

glycerol

stearic acid

palmitic acid

palmitoleic acid

Practice Problem 15.8   Draw the products formed from hydrolysis of the following triacylglycerol.

More Practice: Try Problems 15.7a, 15.8a, 15.29e, 15.30e, 15.33a, 15.34a.

15.5A   FOCUS ON THE HUMAN BODY
Metabolism of Triacylglycerols

Humans store energy in the form of triacylglycerols, kept in a layer of fat cells, called **adipose cells,** below the surface of the skin, in bone marrow, in the breast area (of women), around the kidneys, and in the pelvis (Figure 15.3). In adulthood, the number of adipose cells is constant. When weight is lost or gained the amount of stored lipid in each cell changes, but the number of adipose cells does not change.

Figure 15.3

The Storage and Metabolism of Triacylglycerols

triacylglycerol

metabolism

CO_2 + H_2O + **Energy** 9 kcal/g

adipose cells

Triacylglycerols are stored in adipose cells below the skin and concentrated in some regions of the body. The average fat content of men and women is ~20% and ~25%, respectively. This stored fat provides two to three months of the body's energy needs. *Dennis Strete/McGraw Hill*

A grizzly bear uses its stored body fat as its sole energy source during its many months of hibernation. *Daniel C. Smith*

Adipose tissue serves to insulate the organism, as well as provide energy for its metabolic needs for long periods of time. The first step in the metabolism of a triacylglycerol is hydrolysis of the ester bonds to form glycerol and three fatty acids. **This reaction is simply ester hydrolysis.** In cells, this reaction is carried out with enzymes called **lipases.**

triacylglycerol   glycerol   three fatty acids

ENVIRONMENTAL NOTE

When the price of crude oil is high, the use of **biofuels** such as biodiesel becomes economically attractive. Biofuels are prepared from renewable resources such as vegetable oils and animal fats. *AP Wide World Photos*

Complete metabolism of a triacylglycerol yields CO_2 and H_2O, and a great deal of energy. This overall reaction is reminiscent of the combustion of alkanes in fossil fuels, a process that also yields CO_2 and H_2O and provides energy to heat homes and power automobiles (Section 10.10). Fundamentally, both processes convert C—C and C—H bonds to C—O and O—H bonds, a highly exothermic reaction. Carbohydrates provide an energy boost, but only for the short term, such as during strenuous exercise. Our long-term energy needs are provided by triacylglycerols, because they store 9 kcal/g, whereas carbohydrates and proteins store only 4 kcal/g.

Because triacylglycerols release heat on combustion, they can in principle be used as fuels for vehicles. In fact, coconut oil was used as a fuel during both World Wars I and II, when gasoline and diesel supplies ran short. Because coconut oil is more viscous than petroleum products and freezes at 24 °C, engines must be modified to use it and it can't be used in cold climates. Nonetheless, a limited number of trucks and boats can now use vegetable oils, sometimes blended with diesel—*biodiesel*—as a fuel source.

15.5B Soap Synthesis

As we learned in Section 13.5, soaps are metal salts of carboxylic acids that contain many carbon atoms in a long hydrocarbon chain; that is, **soaps are metal salts of fatty acids.** For example, sodium stearate is the sodium salt of stearic acid, an 18-carbon saturated fatty acid.

sodium stearate

ionic end
polar head

long hydrocarbon chain
nonpolar tail

Soap is prepared by the basic hydrolysis (saponification) of a triacylglycerol. Heating an animal fat or vegetable oil with aqueous base hydrolyzes the three esters to form glycerol and sodium salts of three fatty acids.

triolein

+ 3 NaOH

H_2O

glycerol

sodium oleate

soap

(top): Jupiter Images/Image Source/Getty Images; (bottom): David Buffington/Getty Images

---

Saponification comes from the Latin *sapo* meaning *soap*.

These carboxylate salts are **soaps,** which clean away dirt, as shown in Figure 13.2. The nonpolar tail dissolves grease and oil and the polar head makes it soluble in water. Most triacylglycerols have two or three different R groups in their hydrocarbon chains, so soaps are usually mixtures of two or three different carboxylate salts.

Concept Check 15.6

What is the composition of the soap formed by basic hydrolysis of the following triacylglycerol?

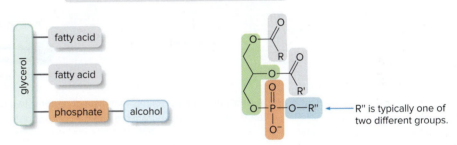

phosphodiester

phosphodiester in cells

15.6 Phospholipids

Phospholipids are lipids that contain a phosphorus atom. Phospholipids can be considered organic derivatives of phosphoric acid (H_3PO_4), formed by replacing two of the H atoms by R groups. This type of functional group is called a **phosphodiester**. In cells, the remaining OH group on phosphorus loses its proton, giving the phosphodiester a net negative charge.

Phosphoacylglycerols (or **phosphoglycerides**) are the most common type of phospholipid. They form the principal lipid component of most cell membranes. Their structure resembles the triacylglycerols of the preceding section with one important difference. Only *two* of the hydroxyl groups of glycerol are esterified with fatty acids. The third OH group is part of a phosphodiester, which is also bonded to an alkyl group (R") derived from a low-molecular-weight alcohol.

General structure of a phosphoacylglycerol

glycerol

— fatty acid

— fatty acid

— phosphate — alcohol

R" is typically one of two different groups.

There are two prominent types of phosphoacylglycerols. They differ in the identity of the R" group in the phosphodiester.

- When R" = $CH_2CH_2NH_3^+$, the compound is called a *cephalin*.
- When R" = $CH_2CH_2N(CH_3)_3^+$, the compound is called a *lecithin*.

cephalin

lecithin

Derived from the alcohol: ethanolamine

Derived from the alcohol: choline

The phosphorus side chain of a phosphoacylglycerol makes it different from a triacylglycerol. **The two fatty acid side chains form two nonpolar "tails" that lie parallel to each other, while the phosphodiester end of the molecule is a charged or polar "head."** A three-dimensional structure of a phosphoacylglycerol is shown in Figure 15.4.

Figure 15.4 Three-Dimensional Structure of a Phosphoacylglycerol

A phosphoacylglycerol has two distinct regions: two nonpolar tails due to the long-chain fatty acids, and a very polar head from the charged phosphodiester.

Sample Problem 15.5

Drawing the Structure of a Phosphoacylglycerol

Draw the structure of a cephalin formed from two molecules of stearic acid.

Analysis

Substitute the 18-carbon saturated fatty acid stearic acid for the R and R' groups in the general structure of a cephalin molecule. In a cephalin, $-CH_2CH_2NH_3^+$ forms part of the phosphodiester.

Solution

Practice Problem 15.9

Draw the structure of two different cephalins containing oleic acid and palmitic acid as fatty acid side chains.

More Practice: Try Problems 15.35, 15.36.

15.7 Cell Membranes

The cell membrane is a beautifully complex example of how chemistry comes into play in a biological system.

15.7A Structure of the Cell Membrane

The basic unit of living organisms is the **cell.** The cytoplasm is the aqueous medium inside the cell, separated from water outside the cell by the **cell membrane.** The cell membrane serves two apparently contradictory functions. It acts as a barrier to the passage of ions and molecules into and out of the cell, but it is also selectively permeable, allowing nutrients in and waste out.

Phospholipids are the major component of the cell membrane. **Phospholipids contain a hydrophilic polar head and two nonpolar tails composed of C—C and C—H bonds.** When phospholipids are mixed with water, they assemble in an arrangement called a **lipid bilayer,** with the ionic heads oriented on the outside and the nonpolar tails on the inside. The polar heads electrostatically interact with the polar solvent H_2O, whereas the nonpolar tails are held in close proximity by numerous London dispersion forces.

membrane phospholipid

polar head

nonpolar tails

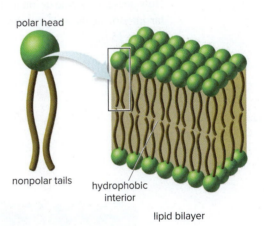

polar head

nonpolar tails

hydrophobic interior

lipid bilayer

Cell membranes are composed of these lipid bilayers (Figure 15.5). The charged heads of the phospholipids are oriented toward the aqueous interior and exterior of the cell. The nonpolar tails form the hydrophobic interior of the membrane, thus serving as an insoluble barrier that protects the cell from the outside.

While the phospholipid bilayer forms the main fabric of the cell membrane, proteins and cholesterol (Section 15.8) are embedded in the membrane as well. **Peripheral proteins** are embedded within the membrane and extend outward on one side only. **Integral proteins** extend through the entire bilayer.

Concept Check 15.7   Why are phospholipids rather than triacylglycerols present in cell membranes?

Figure 15.5

Composition of the Cell Membrane

Cell membranes are composed of a lipid bilayer having the hydrophilic polar heads of phospholipids arranged on the exterior of the bilayer, where they can interact with the polar aqueous environment inside and outside the cell. The hydrophobic tails of the phospholipid are arranged in the interior of the bilayer, forming a "greasy" layer that is only selectively permeable to the passage of species from one side to the other.

15.7B   Transport Across a Cell Membrane

How does a molecule or ion in the water on one side of a cell membrane pass through the nonpolar interior of the cell membrane to the other side? A variety of transport mechanisms occur (Figure 15.6).

Figure 15.6

How Substances Cross a Cell Membrane

Small molecules like O_2 and CO_2 can simply diffuse through the cell membrane, traveling from the side of higher concentration to the side of lower concentration. With larger polar molecules and some ions, simple diffusion is too slow or not possible, so a process of **facilitated transport** occurs. Ions such as Cl^- or HCO_3^- and glucose molecules travel through the channels created by integral proteins.

Some ions, notably Na^+, K^+, and Ca^{2+}, must move across the cell membrane against the concentration gradient—that is, from a region of lower concentration to a region of higher concentration. To move an ion across the membrane in this fashion requires energy input, and the process is called **active transport.** Active transport occurs whenever a nerve impulse causes a muscle to contract. In this process, energy is supplied to move K^+ ions from outside to inside a cell, against a concentration gradient.

Concept Check 15.8   Why don't ions readily diffuse through the interior of the cell membrane?

HEALTH NOTE

Plants do not synthesize cholesterol, so fresh fruits and vegetables, nuts, and whole grains are cholesterol free.
Purestock/SuperStock

Table 15.3 Cholesterol Content in Some Foods

Food	Serving Size	Cholesterol (mg)
Boiled egg	1	225
Cream cheese	1 oz	27
Cheddar cheese	1 oz	19
Butter	3.5 oz	250
Beefsteak	3.5 oz	70
Chicken	3.5 oz	60
Ice cream	3.5 oz	45
Sponge cake	3.5 oz	260

15.8 FOCUS ON HEALTH & MEDICINE
Cholesterol, the Most Prominent Steroid

The **steroids** are a group of lipids whose carbon skeletons contain three six-membered rings and one five-membered ring. This tetracyclic carbon skeleton is drawn below.

steroid skeleton

numbering the steroid skeleton

Many steroids also contain two methyl groups that are bonded to the rings. The steroid rings are lettered **A, B, C,** and **D,** and the 17 ring carbons are numbered as shown. The two methyl groups are numbered C18 and C19. Steroids differ in the identity and location of the substituents attached to the skeleton.

cholesterol

Cholesterol, the most prominent member of the steroid family, is synthesized in the liver and found in almost all body tissues. Cholesterol is obtained in the diet from a variety of sources, including meat, cheese, butter, and eggs. Table 15.3 lists the cholesterol content in some foods. While the American Heart Association currently recommends that the daily intake of cholesterol should be less than 300 mg, the average American diet includes 400–500 mg of cholesterol each day.

Concept Check 15.9   (a) Label the rings of the steroid nucleus in cholesterol. (b) Give the number of the carbon to which the OH group is bonded. (c) Between which two carbons is the double bond located? (d) Label the polar bonds in cholesterol and explain why it is insoluble in water.

While health experts agree that the amount of cholesterol in the diet should be limited, it is also now clear that elevated *blood* cholesterol (serum cholesterol) can lead to coronary artery disease. High blood cholesterol levels are associated with an increased risk of developing coronary artery disease, heart attack, and stroke. To understand the relationship between cholesterol and heart disease, we must learn about how cholesterol is transported through the bloodstream.

Like other lipids, cholesterol is insoluble in the aqueous medium of the blood, because it has only one polar OH group and many nonpolar C—C and C—H bonds. In order for it to be transported from the liver where it is synthesized, to the tissues, cholesterol combines with phospholipids and proteins to form small water-soluble spherical particles called **lipoproteins.**

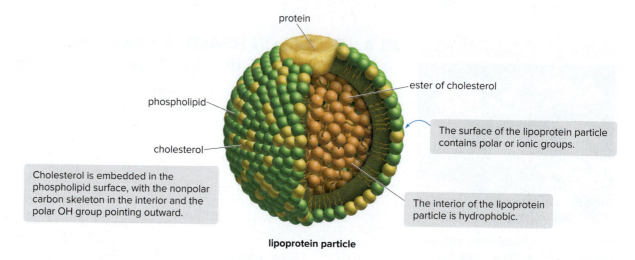

protein

ester of cholesterol

phospholipid

The surface of the lipoprotein particle contains polar or ionic groups.

cholesterol

Cholesterol is embedded in the phospholipid surface, with the nonpolar carbon skeleton in the interior and the polar OH group pointing outward.

The interior of the lipoprotein particle is hydrophobic.

lipoprotein particle

In a lipoprotein, the polar heads of phospholipids and the polar portions of protein molecules are arranged on the surface. The nonpolar molecules are buried in the interior of the particle. In this way, the nonpolar material is "dissolved" in an aqueous environment.

Lipoproteins are classified on the basis of their density, with two types being especially important in determining serum cholesterol levels.

- Low-density lipoproteins (LDLs) transport cholesterol from the liver to the tissues.
- High-density lipoproteins (HDLs) transport cholesterol from the tissues back to the liver.

Figure 15.7 shows a cross-section of both an HDL and LDL particle. HDL particles contain much more protein and much less cholesterol (and esters of cholesterol) than LDL particles.

Figure 15.7 Cross-Section of an HDL and LDL Particle

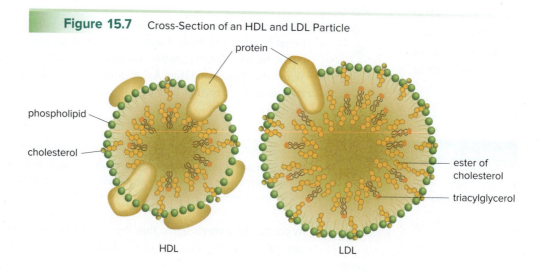

protein

phospholipid

cholesterol

ester of cholesterol

triacylglycerol

HDL

LDL

LDL particles transport cholesterol to tissues where it is incorporated in cell membranes. When LDLs supply more cholesterol than is needed, LDLs deposit cholesterol on the wall of arteries, forming plaque (Figure 15.8). Atherosclerosis is a disease that results from the buildup of these fatty deposits, restricting the flow of blood, increasing blood pressure, and increasing the likelihood of a heart attack or stroke. As a result, LDL cholesterol is often called "bad" cholesterol.

Figure 15.8   Plaque Formation in an Artery

a. Open artery

b. Blocked artery

Cross-section of a clear artery with no buildup of plaque

Artery almost completely blocked by the buildup of plaque

(a–b): Ed Reschke

HDL particles transport excess cholesterol from the tissues back to the liver, where it is converted to other substances or eliminated. Thus, HDLs reduce the level of serum cholesterol, so HDL cholesterol is often called "good" cholesterol.

Several drugs called **statins** are now available to reduce the level of cholesterol in the bloodstream. These compounds act by blocking the synthesis of cholesterol at its very early stages. Two examples include atorvastatin (Lipitor) and simvastatin (Zocor).

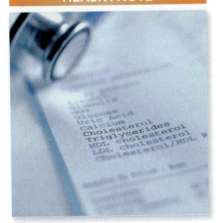
Generic name: atorvastatin
Trade name: Lipitor

Generic name: simvastatin
Trade name: Zocor

Concept Check 15.10

Would you expect triacylglycerols to be contained in the interior of a lipoprotein particle or on the surface with the phospholipids? Explain your choice.

15.9   Steroid Hormones

Many biologically active steroids are hormones secreted by the endocrine glands. **A *hormone* is a molecule that is synthesized in one part of an organism, which then elicits a response at a different site.** Two important classes of steroid hormones are the **sex hormones** and the **adrenal cortical steroids.**

There are two types of female sex hormones, **estrogens** and **progestins.**

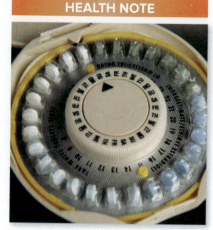

Oral contraceptives are lipids that artificially elevate hormone levels in a woman, thereby preventing pregnancy (Section 11.4). *McGraw Hill*

estradiol   estrone   progesterone

- Estradiol and estrone are estrogens synthesized in the ovaries. They control the development of secondary sex characteristics in females and regulate the menstrual cycle.
- Progesterone is a progestin often called the "pregnancy hormone." It is responsible for the preparation of the uterus for implantation of a fertilized egg.

The male sex hormones are called **androgens.**

testosterone   androsterone

- Testosterone and androsterone are androgens synthesized in the testes. They control the development of secondary sex characteristics in males—growth of facial hair, increase in muscle mass, and deepening of the voice.

Some body builders use anabolic steroids to increase muscle mass. Long-term or excessive use can cause many health problems, including high blood pressure, liver damage, and cardiovascular disease. *Comstock/JupiterImages*

Synthetic androgen analogues, called **anabolic steroids,** promote muscle growth. They were first developed to help individuals whose muscles had atrophied from lack of use following surgery. They have since come to be used by athletes and body builders, although their use is not permitted in competitive sports. Many physical and psychological problems result from their prolonged use.

Anabolic steroids, such as stanozolol, nandrolone, and tetrahydrogestrinone have the same effect on the body as testosterone, but they are more stable, so they are not metabolized as quickly. Tetrahydrogestrinone (also called THG or The Clear), the performance-enhancing drug used by track star Marion Jones during the 2000 Sydney Olympics, was considered a "designer steroid" because it was initially undetected in urine tests for doping. After its chemical structure and properties were determined, it was added to the list of banned anabolic steroids in 2004.

stanozolol   nandrolone   tetrahydrogestrinone

Both urine and blood tests are performed at the Olympics, coordinated by the World Anti-Doping Agency (WADA), to detect banned substances used by competitive athletes. *Andrew Brookes/Getty Images*

A second group of steroid hormones includes the **adrenal cortical steroids.** Three examples of these hormones are aldosterone, cortisone, and cortisol. All of these compounds are synthesized in the outer layer of the adrenal gland. Aldosterone regulates blood pressure and volume by controlling the concentration of Na^+ and K^+ in body fluids. Cortisone and cortisol serve as anti-inflammatory agents and they regulate carbohydrate metabolism.

Cortisone and related compounds are used to suppress organ rejection after transplant surgery and to treat many allergic and autoimmune disorders. Prolonged use of these steroids can have undesired side effects, including bone loss and high blood pressure. Prednisone, a widely used synthetic alternative, has similar anti-inflammatory properties but can be taken orally.

prednisone
(synthetic steroid)

A.T. Willett/Alamy Stock Photo

Concept Check 15.11

Compare the structures of estrone and progesterone. (a) Identify the differences in the A ring of these hormones. (b) How do these hormones differ in functionality at C17?

15.10   FOCUS ON HEALTH & MEDICINE
Fat-Soluble Vitamins

Vitamins are organic compounds required in small quantities for normal metabolism. Our cells cannot synthesize these compounds, so they must be obtained in the diet. Vitamins can be categorized as fat soluble or water soluble. **The fat-soluble vitamins are lipids.**

The four fat-soluble vitamins—**A, D, E,** and **K**—are found in fruits and vegetables, fish, liver, and dairy products. Although fat-soluble vitamins must be obtained from the diet, they do not have to be ingested every day. Excess vitamins are stored in adipose cells, and then used when needed. Table 15.4 summarizes the dietary sources and recommended daily intake of the fat-soluble vitamins.

Table 15.4 Fat-Soluble Vitamins

Vitamin	Food Source	Recommended Daily Intake
A	Liver, kidney, oily fish, dairy products, eggs, fortified breakfast cereals	900 µg (men) 700 µg (women)
D	Fortified milk and breakfast cereals	5 µg
E	Sunflower and safflower oils, nuts, beans, whole grains, leafy greens	15 mg
K	Cauliflower, soybeans, broccoli, leafy greens, green tea	120 µg (men) 90 µg (women)

Source: Data from Harvard School of Public Health.

Vitamin A is obtained from liver, oily fish, and dairy products, and is synthesized from β-carotene, the orange pigment in carrots. In the body, vitamin A is converted to 11-*cis*-retinal, the light-sensitive compound responsible for vision in all vertebrates. It is also needed for healthy mucous membranes. A deficiency of vitamin A causes night blindness, as well as dry eyes and skin.

vitamin A

Pixtal/age fotostock

vitamin D

Stockbyte/Getty Images

Vitamin D, strictly speaking, is not a vitamin because it can be synthesized in the body from cholesterol. Nevertheless, it is classified as such, and many foods (particularly milk) are fortified with vitamin D so that we get enough of this vital nutrient. Vitamin D helps regulate both calcium and phosphorus metabolism. A deficiency of vitamin D causes rickets, a bone disease characterized by knock-knees, spinal curvature, and other skeletal deformities.

Vitamin E is an antioxidant, and in this way it protects unsaturated side chains in fatty acids from unwanted oxidation. A deficiency of vitamin E causes numerous neurological problems, although it is rare for vitamin E deficiency to occur.

vitamin E

IT Stock/Getty Images

vitamin K

Comstock/Getty Images

Vitamin K regulates the synthesis of prothrombin and other proteins needed for blood to clot. A severe deficiency of vitamin K leads to excessive and sometimes fatal bleeding because of inadequate blood clotting.

Concept Check 15.12   Why is it much easier to overdose on a fat-soluble vitamin than a water-soluble vitamin?

Concept Check 15.13   Vitamin D is synthesized in the body from a steroid. Which of the steroid rings—A, B, C, or D—are intact in vitamin D and which ring has been cleaved?

CHAPTER REVIEW

KEY TERMS

Active transport (15.7)
Adrenal cortical steroid (15.9)
Anabolic steroid (15.9)
Androgen (15.9)
Cell membrane (15.7)
Cephalin (15.6)
Estrogen (15.9)
Facilitated transport (15.7)
Fat (15.4)

Fat-soluble vitamin (15.10)
Fatty acid (15.2)
High-density lipoprotein (15.8)
Hormone (15.9)
Hydrolyzable lipid (15.1)
Hydrophilic (15.2)
Hydrophobic (15.2)
Lecithin (15.6)
Lipase (15.5)

Lipid (15.1)
Lipid bilayer (15.7)
Lipoprotein (15.8)
Low-density lipoprotein (15.8)
Nonhydrolyzable lipid (15.1)
Oil (15.4)
Omega-n acid (15.2)
Phosphoacylglycerol (15.6)
Phosphodiester (15.6)

Phospholipid (15.6)
Progestin (15.9)
Saponification (15.5)
Saturated fatty acid (15.2)
Soap (15.5)
Steroid (15.8)
Triacylglycerol (15.4)
Unsaturated fatty acid (15.2)
Wax (15.3)

KEY CONCEPTS

① Characteristics of lipids (15.1)

① General features	**② Hydrolyzable lipids**	**③ Nonhydrolyzable lipids**
• Lipids are biomolecules that contain many nonpolar C—C and C—H bonds. • Lipids are *soluble* in organic solvents and *insoluble* in water.	• Hydrolyzable lipids can be converted to smaller molecules on reaction with water. • Hydrolyzable lipids include waxes (high-molecular-weight esters), triacylglycerols, and phospholipids (key components of cell membranes).	• Nonhydrolyzable lipids cannot be cleaved into smaller units by hydrolysis. • Nonhydrolyzable lipids include steroids and fat-soluble vitamins (A, D, E, and K).

② Comparing triacylglycerols and phospholipids

① Triacylglycerols (15.4)	**② Phosphoacylglycerols (15.6)**
 triacylglycerol	 R'' = H, CH$_3$ phosphoacylglycerol
• **Triacylglycerols,** or triglycerides, are **triesters** formed from **glycerol** and three molecules of fatty acids. • **Fats** are triacylglycerols derived from fatty acids having **few double bonds,** making them **solids** at room temperature. Fats are generally obtained from animal sources. • **Oils** are triacylglycerols derived from fatty acids having a **larger number of double bonds,** making them **liquids** at room temperature. Oils are generally obtained from plant sources.	• **A phosphoacylglycerol** has two nonpolar side chain tails and one ionic head. • **Glycerol** is the backbone, and there are two esters formed with two fatty acids. • The third OH group is part of a **phosphodiester,** which is also bonded to an alkyl group derived from a low-molecular-weight alcohol. • The low-molecular-weight alcohol is either ethanolamine (HOCH$_2$CH$_2$NH$_3$)$^+$ or choline [HOCH$_2$CH$_2$N(CH$_3$)$_3$]$^+$.

③ Examples of steroids

① General features (15.8)	**② Cholesterol (15.8)**	**③ Steroid hormones (15.9)**
• Steroids are **tetracyclic lipids** that contain three six-membered rings and one five-membered ring. • The rings are lettered **A, B, C,** and **D.** • Many steroids contain two methyl groups bonded to the rings.	• Cholesterol, the most prominent steroid, is synthesized in the liver and found in almost all body tissues. • Cholesterol is a component of healthy cell membranes, and it serves as the starting material for the synthesis of all other steroids and vitamin D.	• Many steroids are **hormones,** molecules synthesized in one part of an organism, which then elicit a response at another site. • Two types of female sex hormones are called **estrogens** and **progestins.** • Male sex hormones are called **androgens. Testosterone** is an androgen.

KEY REACTIONS

1

$$R-C\overset{O}{\underset{OR'}{\Vert}} + H-OH \xrightarrow[\text{hydrolysis}]{H_2SO_4} R-C\overset{O}{\underset{OH}{\Vert}} + R'OH$$

wax   (15.3)   fatty acid   alcohol

2

triacylglycerol + 3 H—OH $\xrightarrow[\substack{\text{or} \\ \text{lipase} \\ \text{hydrolysis}}]{H_2SO_4}$ glycerol + 3 fatty acid

[Bonds in **red** are broken.]   (15.5)

3

triacylglycerol + 3 Na⁺ OH⁻ $\xrightarrow[\substack{\text{hydrolysis} \\ \text{in base}}]{H_2O}$ glycerol + 3 soap

[Bonds in **red** are broken.]   (15.5B)

KEY SKILLS

1 Determining the characteristics of a fatty acid (15.2); example: eicosatrienoic acid $[CH_3CH_2(CH=CHCH_2)_3(CH_2)_8CO_2H]$

1 Drawing a skeletal structure	**2** Labeling hydrophobic and hydrophilic regions	**3** Determining relative melting points
• Eicosatrienoic acid is an **unsaturated fatty acid** because it contains C=C's in its hydrocarbon chain. • Double bonds in naturally occurring fatty acids are generally cis.	• The **COOH** group of a fatty acid is the *hydrophilic* portion (shown in red). • The nonpolar C—C and C—H bonds of the carbon chain constitute the *hydrophobic* portion. • Fatty acids are water insoluble because the nonpolar hydrophobic region is much larger than the hydrophilic region.	• As the number of double bonds in the fatty acid *increases,* the melting point *decreases.* • Eicosatrienoic acid has a *lower* melting point than arachidic acid $[CH_3(CH_2)_{18}CO_2H]$, a saturated fatty acid with the same number of carbon atoms. • Eicosatrienoic acid (3 C=C's) has a *higher* melting point than arachidonic acid (4 C=C's), an unsaturated fatty acid with more double bonds, but the same number of carbon atoms.

eicosatrienoic acid

2 Drawing the structure of a triacylglycerol (15.4); example: a triacylglycerol formed from three molecules of lauric acid $[CH_3(CH_2)_{10}CO_2H]$

1 Draw the starting materials.	**2** Draw the product by forming three new ester C—O bonds.
• Arrange the carboxyl groups of the fatty acids close to the OH groups of glycerol.	• Form three new esters using the **OH** groups of glycerol and the acyl groups (**RCO**) of the fatty acids.

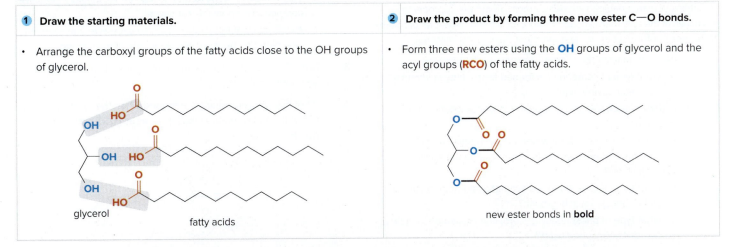

glycerol   fatty acids   new ester bonds in **bold**

3 Drawing the products formed from basic hydrolysis of a triacylglycerol (15.5B)

1 Locate the three ester C—O bonds.	**2** Cleave the three C—O single bonds at the carbonyl carbon.
three ester C—O bonds in **red**	glycerol sodium carboxylates • Basic hydrolysis forms glycerol and the sodium salts of three carboxylate anions (soaps). • In this example, hydrolysis forms two equivalents of one salt and one equivalent of another.

4 Drawing the structure of a phosphoacylglycerol (15.6); example: a lecithin formed form myristic acid [CH$_3$(CH$_2$)$_{12}$CO$_2$H]

1 Draw the general structure of a lecithin.	**2** Substitute R from the fatty acid to complete the structure.
from myristic acid lecithin • The phosphodiester of a lecithin is derived from choline [HOCH$_2$CH$_2$N(CH$_3$)$_3$$^+$]. • The R groups are derived from a fatty acid.	lecithin • In this example, both R groups are derived from the C$_{14}$ fatty acid myristic acid.

CHAPTER 15 SELF-TEST

The Self-Test consists of short-answer questions that test an understanding of definitions, equations, and other material encountered within the chapter. Answers to each question are provided at the end of the chapter. For any questions you are unable to answer correctly, revisit the appropriate section(s) of the text; then proceed to working the end-of-chapter problems to further reinforce your understanding of the chapter material.

1. Which term from "Key Terms" answers each question?
   a. What type of compound is formed from a fatty acid and a high-molecular-weight alcohol?
   b. Basic hydrolysis of a triacylglycerol yields what type of compound?
   c. What term is used for a water-soluble particle that contains proteins and phospholipids and is used to transport cholesterol?
   d. What type of compound is a synthetic androgen analogue that promotes muscle growth?
   e. What type of phosphoacylglycerol contains ethanolamine as part of its phosphodiester?

2. What vitamin fits each description?
   a. This vitamin is synthesized in the body from cholesterol.
   b. This vitamin is converted to 11-*cis*-retinal, the light-sensitive compound responsible for vision.
   c. A deficiency of this vitamin leads to excessive and sometimes fatal bleeding because of inadequate blood clotting.

3. Draw the structure of a lecithin formed from two molecules of pentanoic acid.

4. What hydrolysis products are formed when the following triacylglycerol is treated with (a) aqueous H$_2$SO$_4$; (b) aqueous NaOH?

5. Label each compound as a hydrolyzable or nonhydrolyzable lipid: (a) oleic acid; (b) phosphoacylglycerol; (c) vitamin K; (d) testosterone.

6. Label each compound as a fat or an oil: (a) a triacylglycerol derived from three molecules of palmitic acid; (b) a triacylglycerol derived from three molecules of linoleic acid.

UNDERSTANDING KEY CONCEPTS

15.1 (a) Which food is high in unsaturated triacylglycerols? (b) Which food is high in saturated fat? (c) Which food likely contains trans fats?

A

B

C

(a–c): Mark Dierker/McGraw Hill

15.2 Match the nutrition label to the type of food: (a) corn oil; (b) butter; (c) margarine.

A

B

C

(a–c): Mark Dierker/McGraw Hill

15.3 Draw the structure of a wax formed from palmitic acid $[CH_3(CH_2)_{14}COOH]$ and $CH_3(CH_2)_{21}OH$.

15.4 What hydrolysis products are formed when the wax $CH_3(CH_2)_{12}COO(CH_2)_{15}CH_3$ is treated with aqueous sulfuric acid?

15.5 Consider the following three types of compounds: [1] fatty acids; [2] soaps; [3] waxes. For each type of compound: (a) give the general structure; (b) draw the structure of a specific example; (c) label the compound as water soluble or water insoluble; (d) label the compound as soluble or insoluble in the organic solvent hexane $[CH_3(CH_2)_4CH_3]$.

15.6 How do fats and oils compare with respect to each of the following features?
   a. identity and number of functional groups present
   b. number of carbon–carbon double bonds present
   c. melting point
   d. natural source

15.7 Draw the products formed when the following triacylglycerol is hydrolyzed under each of the following conditions: (a) water and H_2SO_4; (b) water and NaOH.

15.8 Answer the questions from Problem 15.7 for the following triacylglycerol.

15.9 Block diagrams representing the general structures of two types of lipids are drawn. Which terms describe each diagram: (a) phospholipid; (b) triacylglycerol; (c) hydrolyzable lipid; (d) phosphoacylglycerol? More than one term may apply to a diagram.

A

B

15.10 Which "cartoon" represents a soap and which represents a phosphoacylglycerol? What structural features are present in the polar head and nonpolar tails of each compound?

A

B

ADDITIONAL PROBLEMS

General Characteristics of Lipids

15.11 Label each compound as a hydrolyzable or nonhydrolyzable lipid.

a. triacylglycerol

b. vitamin A

c. lecithin

d. cholesterol

e. phospholipid

f. cephalin

g. wax

h. estrogen

15.12 In which solvents or solutions might a steroid be soluble: (a) blood plasma; (b) CCl_4; (c) 5% NaCl solution?

Fatty Acids, Waxes, and Triacylglycerols

15.13 For the food product shown in the accompanying photo, label each statement as true or false.

TristanBM/Shutterstock

a. It contains trans triacylglycerols.

b. It contains saturated triacylglycerols.

c. It is likely high in triacylglycerols that are derived from stearic acid.

d. A diet high in this food is linked to an increased incidence of heart disease.

15.14 For the food product shown in the accompanying photo, label each statement as true or false.

Bewaru/Shutterstock

a. It contains trans triacylglycerols.

b. It contains unsaturated triacylglycerols.

c. It is likely high in triacylglycerols that are derived from palmitic acid.

d. A diet high in this food is linked to an increased incidence of heart disease.

15.15 Rank the fatty acids in order of increasing melting point: $CH_3(CH_2)_{16}COOH$, $CH_3(CH_2)_7CH=CH(CH_2)_7COOH$, and $CH_3(CH_2)_5CH=CH(CH_2)_7COOH$.

15.16 How does each of the following affect the melting point of a fatty acid: (a) increasing the number of carbon atoms; (b) increasing the number of double bonds?

15.17 Rank the fatty acids in order of increasing melting point.

15.18 How would you expect the melting points of the following unsaturated fatty acids to compare? Explain your choice.

15.19 Is a fatty acid a hydrolyzable lipid? Explain your choice.

15.20 Why are soaps water soluble, but the fatty acids from which they are derived, water insoluble?

15.21 Draw the structure of a wax formed from palmitic acid $[CH_3(CH_2)_{14}COOH]$ and each alcohol.

a. $CH_3(CH_2)_{11}OH$

b. $CH_3(CH_2)_9OH$

15.22 Draw the structure of a wax formed from a 30-carbon straight-chain alcohol and each carboxylic acid.

a. lauric acid

b. myristic acid

15.23 Consider fatty acid **A** depicted in the ball-and-stick model.

A

a. Draw a skeletal structure for **A** and clearly show the cis double bond.

b. What type of omega-*n* acid is this fatty acid?

c. Draw a stereoisomer of **A**.

d. Draw the structure of the wax formed from **A** and $CH_3(CH_2)_{10}OH$.

15.24 Consider fatty acid **B** depicted in the ball-and-stick model.

B

a. Draw a skeletal structure for **B** and clearly show the cis double bond.

b. What type of omega-*n* acid is this fatty acid?

c. Draw a stereoisomer of **B**.

d. Draw the structure of the wax formed from **B** and $CH_3(CH_2)_8OH$.

15.25 What hydrolysis products are formed when each wax is treated with aqueous sulfuric acid?

a. $CH_3(CH_2)_{16}COO(CH_2)_{17}CH_3$

b. $CH_3(CH_2)_{12}COO(CH_2)_{25}CH_3$

15.26 What hydrolysis products are formed when each wax is treated with aqueous sulfuric acid?

a. $CH_3(CH_2)_{18}COO(CH_2)_{29}CH_3$

b. $CH_3(CH_2)_{24}COO(CH_2)_{23}CH_3$

15.27 Draw a triacylglycerol that is formed from lauric, myristic, and linoleic acids.

15.28 Draw a polyunsaturated triacylglycerol formed from three molecules of linoleic acid.

15.29 Answer the following questions about the given triacylglycerol.

a. What fatty acids are used to form this triacylglycerol?

b. Would you expect this triacylglycerol to be a solid or a liquid at room temperature?

c. What regions are hydrophobic?

d. What regions are hydrophilic?

e. What hydrolysis products are formed when the triacylglycerol is treated with aqueous sulfuric acid?

15.30 Answer the questions from Problem 15.29 for the given triacylglycerol.

15.31 Draw the structure of a triacylglycerol that yields only the sodium carboxylate depicted in the ball-and-stick model when it is hydrolyzed in base.

Na^+

15.32 Draw the structure of a triacylglycerol that yields only the sodium carboxylate depicted in the ball-and-stick model when it is hydrolyzed in base.

Na^+

15.33 Draw the products formed when the given triacylglycerol is hydrolyzed under each of the following conditions: (a) water and H_2SO_4; (b) water and NaOH.

15.34 Draw the products formed when the given triacylglycerol is hydrolyzed under each of the following conditions: (a) water and H_2SO_4; (b) water and NaOH.

Phospholipids and Cell Membranes

15.35 Draw a cephalin formed from two molecules of palmitoleic acid.

15.36 Draw a lecithin formed from two molecules of oleic acid.

Steroids

15.37 Draw the structure of the anabolic steroid 4-androstene-3,17-dione, also called "andro," from the following description. Andro contains the tetracyclic steroid skeleton with carbonyl groups at C3 and C17, a double bond between C4 and C5, and methyl groups bonded to C10 and C13.

15.38 Draw the structure of the anabolic steroid methenolone from the following description. Methenolone contains the tetracyclic steroid skeleton with a carbonyl group at C3, a hydroxyl at C17, a double bond between C1 and C2, and methyl groups bonded to C1, C10, and C13.

15.39 Why must cholesterol be transported through the bloodstream in lipoprotein particles?

15.40 Why are LDLs soluble in the blood?

15.41 Describe the role of HDLs and LDLs in cholesterol transport in the blood. What is the relationship of HDL and LDL levels to cardiovascular disease?

15.42 What are anabolic steroids? Give an example. What adverse effects arise from using anabolic steroids?

15.43 (a) Draw the structure of an estrogen and an androgen. (b) What structural features are similar in the two steroids? (c) What structural features are different? (d) Describe the biological activity of each steroid.

15.44 (a) Draw the structure of an androgen and a progestin. (b) What structural features are similar in the two steroids? (c) What structural features are different? (d) Describe the biological activity of each steroid.

Vitamins

15.45 Answer each question with regards to vitamins A and D.
 a. How many tetrahedral carbons does the vitamin contain?
 b. How many trigonal planar carbons does the vitamin contain?
 c. Identify the functional groups.
 d. Label all polar bonds.
 e. What function does the vitamin serve in the body?
 f. What problems result when there is a deficiency of the vitamin?
 g. Give a dietary source.

15.46 Answer each question in Problem 15.45 for vitamins E and K.

General Questions

15.47 Give an example of each type of lipid.
 a. an unsaturated fatty acid with one C=C
 b. a wax that contains a total of 30 carbons

15.48 Give an example of each type of lipid.
 a. an unsaturated fatty acid with more than one C=C
 b. a wax derived from a 12-carbon fatty acid

15.49 Why are phosphoacylglycerols more water soluble than triacylglycerols?

15.50 How are soaps and phosphoacylglycerols similar in structure? How do they differ?

15.51 Some fish oils contain triacylglycerols formed from the polyunsaturated fatty acid, 7,10,13,16,19-docosapentaenoic acid.

7,10,13,16,19-docosapentaenoic acid

 a. Label the hydrophobic and hydrophilic portions of the fatty acid.
 b. How does the melting point of this fatty acid compare to its all trans isomer?
 c. Would you expect this fatty acid to be a solid or a liquid at room temperature?
 d. What type of omega-n acid is this fatty acid?

15.52 Some marine plankton contain triacylglycerols formed from the polyunsaturated fatty acid, 3,6,9,12,15-octadecapentaenoic acid.

3,6,9,12,15-octadecapentaenoic acid

 a. Label the hydrophobic and hydrophilic portions of the fatty acid.
 b. How does the melting point of this fatty acid compare to the melting point of oleic acid?
 c. Would you expect this fatty acid to be a solid or a liquid at room temperature?
 d. What type of omega-n acid is this fatty acid?

Applications

15.53 The main fatty acid component of the triacylglycerols in coconut oil is lauric acid, $CH_3(CH_2)_{10}COOH$. Explain why coconut oil is a liquid at room temperature despite the fact that it contains a large fraction of this saturated fatty acid.

15.54 Can an individual survive on a cholesterol-free diet?

15.55 Can an individual survive on a completely fat-free diet?

15.56 Why should saturated fats in the diet be avoided?

CHALLENGE PROBLEMS

15.57 If the serum cholesterol level in an adult is 167 mg/dL, how many grams of cholesterol are contained in 5.0 L of blood?

15.58 How many triacylglycerols can be prepared from three different fatty acids? Draw all possible structures, excluding stereoisomers.

BEYOND THE CLASSROOM

15.59 Pick a favorite snack food such as potato chips, corn chips, crackers, or peanut butter. Using the nutrition information on the package, determine how much "fat" (i.e., triacylglycerol) is contained in a serving. Calculate what percentage of saturated, unsaturated, and trans fat the product contains per serving. Calculate the number of grams of fat you would consume if you ate this snack food every day for a month.

15.60 Locate a biodiesel facility in your state. Where do the oils processed in the facility come from, and how are the oils converted to usable fuels? In what ways is the biodiesel used? What are the advantages and disadvantages of using biodiesel as an energy source? What fraction of the energy needs in your area is supplied by biodiesel?

CHAPTER 15 ANSWERS

Concept Check Answers

15.1 a, c.

15.2 $CH_3(CH_2)_{10}CO_2H + HO(CH_2)_{15}CH_3$

15.3 a.

b.

15.4 a. [1]   b. [2], [3]   c. [1], [2]   d. [2], [3]

15.5 peanuts: higher in unsaturated triacylglycerols; peanut butter: higher in trans triacylglycerols

15.6 $CH_3(CH_2)_{12}CO_2^-$ Na$^+$ + $CH_3(CH_2)_5CH=CH(CH_2)_7CO_2^-$ Na$^+$ + $CH_3(CH_2)_7CH=CH(CH_2)_7CO_2^-$ Na$^+$

15.7 Phospholipids have a polar head and two nonpolar tails to form the lipid bilayer. Triacylglycerols have no polar head to attract H_2O on the outside of the cell membrane.

15.8 Ions are insoluble in the hydrophobic interior of the cell membrane.

15.9 a,b,c:

OH at C3   double bond between C5 and C6

d. The C—O and O—H bonds are polar. The large number of C—C and C—H bonds makes cholesterol water insoluble.

15.10 Triacylglycerols would be found in the hydrophobic interior.

15.11 a. Estrone has a phenol and progesterone has a C=O and C=C in the A ring. Progesterone has a methyl group on C10.
   b. Estrone has a C=O at C17. C17 of progesterone is bonded to one C and one H.

15.12 Water-soluble vitamins are excreted in the urine, whereas fat-soluble vitamins are stored in the body.

15.13 A, C, D: intact, B cleaved

Practice Problem Answers

15.1 a,b:

c. **A** will have the higher melting point because **A** is saturated and the molecules can pack together better.

15.2 a.

b. omega-3

15.3 a. omega-9   b. omega-6

15.4 a. $CH_3(CH_2)_{16}CO_2(CH_2)_9CH_3$ c. $CH_3(CH_2)_{16}CO_2(CH_2)_{29}CH_3$
   b. $CH_3(CH_2)_{16}CO_2(CH_2)_{11}CH_3$

15.5 $CH_3(CH_2)_7CH=CH(CH_2)_9CO_2(CH_2)_8CH=CH(CH_2)_7CH_3$

15.6

15.7

15.8 $HOCH(CH_2OH)_2 + CH_3(CH_2)_{12}CO_2H + CH_3(CH_2)_5CH=CH(CH_2)_7CO_2H$
$+ CH_3(CH_2)_7CH=CH(CH_2)_7CO_2H$

15.9

[1] $R = (CH_2)_{14}CH_3$, $R' = (CH_2)_7CH=CH(CH_2)_7CH_3$
[2] $R = (CH_2)_7CH=CH(CH_2)_7CH_3$, $R' = (CH_2)_{14}CH_3$

Self-Test Answers

1. a. wax c. lipoprotein e. cephalin
    b. soap d. anabolic steroid

2. a. vitamin D b. vitamin A c. vitamin K

3.

4. a. $HOCH(CH_2OH)_2 + 3\ CH_3(CH_2)_6CO_2H$
    b. $HOCH(CH_2OH)_2 + 3\ CH_3(CH_2)_6CO_2^-\ Na^+$

5. a,c,d: nonhydrolyzable b. hydrolyzable

6. a. fat b. oil

Odd-Numbered Problem Answers

15.1 a. **B** b. **C** c. **A**

15.3 $CH_3(CH_2)_{14}CO_2(CH_2)_{21}CH_3$

15.5 [1] Fatty acid: (a) RCOOH; (b) $CH_3(CH_2)_{10}CO_2H$; (c) No; (d) Yes
    [2] Soap: (a) $RCOO^-Na^+$; (b) $CH_3(CH_2)_{10}CO_2^-Na^+$; (c) Yes; (d) No
    [3] Wax: (a) RCOOR'; (b) $CH_3(CH_2)_6CO_2(CH_2)_7CH_3$; (c) No; (d) Yes

15.7 a. $HOCH(CH_2OH)_2 + CH_3(CH_2)_{14}CO_2H +$
      $CH_3(CH_2)_7CH=CH(CH_2)_7CO_2H +$
      $CH_3(CH_2)_5CH=CH(CH_2)_7CO_2H$
    b. $HOCH(CH_2OH)_2 + CH_3(CH_2)_{14}CO_2^-\ Na^+ +$
      $CH_3(CH_2)_7CH=CH(CH_2)_7CO_2^-\ Na^+ +$
      $CH_3(CH_2)_5CH=CH(CH_2)_7CO_2^-\ Na^+$

15.9 **A:** a,c,d **B:** b,c

15.11 a,c,e,f,g: hydrolyzable b,d,h: nonhydrolyzable

15.13 a. false b,c,d: true

15.15 $CH_3(CH_2)_5CH=CH(CH_2)_7COOH,$
    $CH_3(CH_2)_7CH=CH(CH_2)_7COOH, CH_3(CH_2)_{16}COOH$

15.17 **B < A < C**

15.19 Fatty acids are not hydrolyzable because they contain a very
    long hydrocarbon chain attached to a carboxylic acid group.

15.21 a. $CH_3(CH_2)_{14}CO_2(CH_2)_{11}CH_3$
    b. $CH_3(CH_2)_{14}CO_2(CH_2)_9CH_3$

15.23 a.

    b. omega-3 acid

    c.

    d. $CH_3CH_2CH=CH(CH_2)_5CO_2(CH_2)_{10}CH_3$

15.25 a. $CH_3(CH_2)_{16}COOH + HO(CH_2)_{17}CH_3$
    b. $CH_3(CH_2)_{12}COOH + HO(CH_2)_{25}CH_3$

15.27

15.29 a. arachidic acid, stearic acid, and lauric acid
    b. solid
    c. The long hydrocarbon chains are hydrophobic.
    d. The ester linkages are hydrophilic.
    e. $HOCH(CH_2OH)_2 + HOOC(CH_2)_{10}CH_3 + HOOC(CH_2)_{16}CH_3 +$
      $HOOC(CH_2)_{18}CH_3$

15.31

    $R = (CH_2)_3CH=CH(CH_2)_3CH_3$

15.33 a. $HOCH(CH_2OH)_2 + CH_3(CH_2)_8CO_2H + CH_3(CH_2)_6CO_2H +$
      $CH_3(CH_2)_{10}CO_2H$

15.35

    $R = (CH_2)_7CH=CH(CH_2)_5CH_3$

15.37

15.39 Cholesterol is insoluble in the aqueous medium of the bloodstream. By being bound to a lipoprotein particle, it can be transported in the aqueous solution.

15.41 Low-density lipoproteins (LDLs) transport cholesterol from the liver to the tissues where it is incorporated in cell membranes. High-density lipoproteins (HDLs) transport cholesterol from the tissues back to the liver. When LDLs supply more cholesterol than is needed, LDLs deposit cholesterol on the wall of arteries, forming plaque, potentially resulting in atherosclerosis.

15.43 a.

estrone   testosterone

b. The estrogen (left) and androgen (right) both contain the four rings of the steroid skeleton and a methyl group bonded to C13.

c. The estrogen has an aromatic A ring and a hydroxyl group on this ring. The androgen has a carbonyl in the A ring, a C=C in the A ring, and an additional CH_3 group at C10. The D rings are also different. The estrogen contains a carbonyl at C17 and the androgen has an OH group.

d. Estrogens control the menstrual cycle and secondary sexual characteristics of females. Androgens control the development of male secondary sexual characteristics.

15.45

	Vitamin A	Vitamin D
a.	10	21
b.	10	6
c.	Five alkenes, one hydroxyl group	Three alkenes, one hydroxyl group
d.	Polar C–O and O–H bonds	Polar C–O and O–H bonds
e.	Required for normal vision	Regulates calcium and phosphorus metabolism
f.	Night blindness	Rickets and skeletal deformities
g.	Liver, kidney, oily fish, dairy	Milk and breakfast cereals

15.47 a. $CH_3(CH_2)_5CH=CH(CH_2)_7COOH$

b. $CH_3(CH_2)_{12}COO(CH_2)_{15}CH_3$

15.49 They contain an ionic head, making them more polar than triacylglycerols.

15.51 a. The hydrophilic portion is the COOH group and the hydrophobic portion is the carbon skeleton with nonpolar C–C and C–H bonds.

b. The melting point would be lower than the melting point of the trans isomer.

c. liquid

d. omega-3 fatty acid

15.53 The hydrocarbon chains have only 12 carbons in them, making them short enough so that the triacylglycerol remains a liquid at room temperature.

15.55 No, certain fatty acids and fat-soluble vitamins are required in the diet.

15.57 8.4 g

16

Amino Acids, Proteins, and Enzymes

The peptide oxytocin stimulates the flow of milk in nursing mothers. *Daniel C. Smith*

CHAPTER OUTLINE

CHAPTER GOALS

In this chapter you will learn how to:

1. Identify the general structural features of amino acids
2. Describe the acid–base properties of amino acids
3. Label the N- and C-terminal amino acids of simple peptides
4. Describe the characteristics of the primary, secondary, tertiary, and quaternary structure of proteins
5. Describe the features of fibrous proteins like α-keratin and collagen
6. Describe the features of globular proteins like hemoglobin and myoglobin
7. Draw the products of protein hydrolysis
8. Describe protein denaturation
9. Describe the main features of enzymes and classify an enzyme into one of six categories
10. Describe the use of enzymes to diagnose and treat disease

Of the four major groups of biomolecules—lipids, carbohydrates, proteins, and nucleic acids—proteins have the widest array of functions. **Keratin** and **collagen,** for example, form long insoluble fibers, giving strength and support to tissues. Hair, horns, hooves, and fingernails are all made up of keratin. **Collagen** is found in bone, connective tissue, tendons, and cartilage. **Membrane proteins** transport small organic molecules and ions across cell membranes. **Insulin,** the hormone that regulates blood glucose levels, and **hemoglobin,** which transports oxygen from the lungs to tissues, are proteins. **Enzymes** are proteins that catalyze and regulate all aspects of cellular function. In Chapter 16 we discuss proteins and their primary components, the amino acids.

16.1 Introduction

Proteins are biomolecules that contain many amide bonds, formed by joining amino acids together.

amino acid

protein
[Amide bonds are shown in red.]

Proteins occur widely in the human body, accounting for approximately 50% of its dry weight (Figure 16.1). Fibrous proteins, like keratin in hair, skin, and nails and collagen in connective tissue, give support and structure to tissues and cells. Protein hormones and enzymes regulate the body's metabolism. Transport proteins carry substances through the blood, and storage proteins store elements and ions in organs. Contractile proteins control muscle movements, and immunoglobulins are proteins that defend the body against foreign substances.

Figure 16.1 Proteins in the Human Body

keratin
- fibrous protein in hair, skin, and nails

hemoglobin
- transport protein that carries O_2 in the blood

myoglobin
- protein that stores O_2 in tissues

collagen
- fibrous protein in connective tissue
- found in tendons, bone, cartilage, and blood vessels

insulin
- protein hormone synthesized in the pancreas
- controls blood glucose levels

ferritin
- protein that stores iron in the liver

actin and myosin
- proteins that control muscle contractions

Table 16.1 Recommended Daily Protein Intake

Group	Daily Protein Intake (g protein/kg body weight)
Children (1–3 years)	1.1
Children (4–13 years)	0.95
Children (14–18 years)	0.85
Adult	0.8

Source: Data from U.S. Food and Drug Administration.

Unlike lipids and carbohydrates, which the body stores for use when needed, protein is not stored so it must be consumed on a daily basis. The current recommended daily intake for adults is 0.8 grams of protein per kilogram of body weight. Because children need protein for both growth and maintenance, the recommended daily intake is higher, as shown in Table 16.1.

16.2 Amino Acids

To understand protein properties and structure, we must first learn about the amino acids that compose them.

16.2A General Features of Amino Acids

Amino acids contain two functional groups—an amino group (NH_2) and a carboxyl group ($COOH$). In most naturally occurring amino acids, the amino group is bonded to the α carbon, the carbon adjacent to the carbonyl group, making them **α-amino acids.**

The 20 amino acids that occur naturally in proteins differ in the identity of the R group bonded to the α carbon. The R group is called the **side chain** of the amino acid. The simplest amino acid, called **glycine,** has R = H. Other side chains may be simple alkyl groups, or have additional functional groups such as OH, SH, COOH, or NH_2. Table 16.2 lists the structures of the 20 common amino acids that occur in proteins.

- Amino acids with an additional COOH group in the side chain are called acidic amino acids.
- Those with an additional basic N atom in the side chain are called basic amino acids.
- All others are neutral amino acids.

All amino acids have common names, which are abbreviated by a three-letter or one-letter designation. For example, glycine is often written as the three-letter abbreviation **Gly,** or the one-letter abbreviation **G.** These abbreviations are also given in Table 16.2.

Amino acids never exist in nature as neutral molecules with all uncharged atoms. Because amino acids contain a base (NH_2 group) and an acid ($COOH$), proton transfer from the acid to the base forms a salt called a **zwitterion,** which contains both a positive and a negative charge. These salts have high melting points and are water soluble.

The acid–base chemistry of amino acids is discussed in greater detail in Section 16.3. The structures in Table 16.2 show the charged form of the amino acids at the physiological pH of the blood.

Humans can synthesize only 10 of the 20 amino acids needed for proteins. The remaining 10, called **essential amino acids,** must be obtained from the diet and consumed on a regular, almost daily basis. Diets that include animal products readily supply all of the needed amino acids. Because plant sources generally do not have sufficient amounts of all of the essential amino

Table 16.2 The 20 Common Naturally Occurring Amino Acids

Neutral Amino Acids					
Name	**Structure**	**Abbreviations**	**Name**	**Structure**	**Abbreviations**
Alanine	$H_3\overset{+}{N}-\underset{CH_3}{\overset{H}{C}}-COO^-$	Ala A	Phenylalanine*	$H_3\overset{+}{N}-\underset{CH_2}{\overset{H}{C}}-COO^-$	Phe F
Asparagine	$H_3\overset{+}{N}-\underset{CH_2CONH_2}{\overset{H}{C}}-COO^-$	Asn N	Proline	$\overset{+}{N}H_2 \; COO^-$	Pro P
Cysteine	$H_3\overset{+}{N}-\underset{CH_2SH}{\overset{H}{C}}-COO^-$	Cys C	Serine	$H_3\overset{+}{N}-\underset{CH_2OH}{\overset{H}{C}}-COO^-$	Ser S
Glutamine	$H_3\overset{+}{N}-\underset{CH_2CH_2CONH_2}{\overset{H}{C}}-COO^-$	Gln Q	Threonine*	$H_3\overset{+}{N}-\underset{CH(OH)CH_3}{\overset{H}{C}}-COO^-$	Thr T
Glycine	$H_3\overset{+}{N}-\underset{H}{\overset{H}{C}}-COO^-$	Gly G	Tryptophan*	$H_3\overset{+}{N}-\underset{CH_2}{\overset{H}{C}}-COO^-$	Trp W
Isoleucine*	$H_3\overset{+}{N}-\underset{CH(CH_3)CH_2CH_3}{\overset{H}{C}}-COO^-$	Ile I	Tyrosine	$H_3\overset{+}{N}-\underset{CH_2}{\overset{H}{C}}-COO^- \;\; OH$	Tyr Y
Leucine*	$H_3\overset{+}{N}-\underset{CH_2CH(CH_3)_2}{\overset{H}{C}}-COO^-$	Leu L	Valine*	$H_3\overset{+}{N}-\underset{CH(CH_3)_2}{\overset{H}{C}}-COO^-$	Val V
Methionine*	$H_3\overset{+}{N}-\underset{CH_2CH_2SCH_3}{\overset{H}{C}}-COO^-$	Met M			

Essential amino acids are labeled with an asterisk (*).

(continued on next page)

Table 16.2 *(continued)*

Acidic Amino Acids			Basic Amino Acids		
Name	**Structure**	**Abbreviations**	**Name**	**Structure**	**Abbreviations**
Aspartic acid	$H_3\overset{+}{N}-\overset{H}{\underset{CH_2COO^-}{C}}-COO^-$	Asp D	Arginine*	$H_3\overset{+}{N}-\overset{H}{\underset{(CH_2)_3-N-C-NH_2}{C}}-COO^- \; \overset{+}{N}H_2$	Arg R
Glutamic acid	$H_3\overset{+}{N}-\overset{H}{\underset{CH_2CH_2COO^-}{C}}-COO^-$	Glu E	Histidine*	$H_3\overset{+}{N}-\overset{H}{\underset{CH_2}{C}}-COO^-$	His H
			Lysine*	$H_3\overset{+}{N}-\overset{H}{\underset{(CH_2)_4\overset{+}{N}H_3}{C}}-COO^-$	Lys K

Essential amino acids are labeled with an asterisk (*).

A diet of rice and tofu provides all essential amino acids. A peanut butter sandwich on wheat bread does the same. *Foodcollection*

acids, vegetarian diets must be carefully balanced. Grains—wheat, rice, and corn—are low in lysine, and legumes—beans, peas, and peanuts—are low in methionine, but a combination of these foods provides all the needed amino acids.

Concept Check 16.1

In addition to the amino and carboxyl groups, what other functional groups are present in each amino acid: (a) asparagine; (b) serine; (c) cysteine?

16.2B Stereochemistry of Amino Acids

Except for the simplest amino acid, glycine, **all other amino acids have a chirality center—a carbon bonded to four different groups—on the α carbon.** Thus, an amino acid like alanine ($R = CH_3$) has two possible enantiomers, drawn below in both three-dimensional representations with wedges and dashed wedges, and Fischer projections.

L isomer = L-alanine | D-alanine = D isomer

naturally occurring enantiomer

Like monosaccharides, the prefixes **D** and **L** are used to designate the arrangement of groups on the chirality center of amino acids. When drawn with a vertical carbon chain having the –COO⁻ group at the top and the R group at the bottom,

- **L** Amino acids have the –NH₃⁺ group on the *left* side in the Fischer projection. Common naturally occurring amino acids are **L** isomers.
- **D** Amino acids have the –NH₃⁺ group on the *right* side in the Fischer projection. **D** Amino acids occur infrequently in nature.

The essential amino acid leucine (Sample Problem 16.1) is sold as a dietary supplement that is used by body builders to help prevent muscle loss and heal muscle tissue after injury. *Jill Braaten*

MSG (monosodium glutamate), the sodium salt of glutamic acid (Table 16.2), is a common food additive used as a flavor enhancer in canned soups and other processed products. *Comstock/Getty Images*

Sample Problem 16.1	Drawing Fischer Projections

Draw the Fischer projection for each amino acid: (a) L-leucine; (b) D-cysteine.

Analysis

To draw an amino acid in a Fischer projection, **place the –COO⁻ group at the top and the R group at the bottom.** The **L isomer** has the $-NH_3^+$ on the *left* side and the **D isomer** has the $-NH_3^+$ on the *right* side.

Solution

a. For leucine, R = $CH_2CH(CH_3)_2$

$$COO^-$$
$$H_3\overset{+}{N}\text{—}\overset{|}{\text{C}}\text{—}H$$
$$CH_2CH(CH_3)_2$$

$\overset{+}{N}H_3$ on left
L isomer

b. For cysteine, R = CH_2SH

$$COO^-$$
$$H\text{—}\overset{|}{\text{C}}\text{—}\overset{+}{N}H_3$$
$$CH_2SH$$

$\overset{+}{N}H_3$ on right
D isomer

Practice Problem 16.1

Draw both enantiomers of each amino acid in Fischer projections and label them as D or L:
(a) phenylalanine; (b) methionine.

Practice Problem 16.2

Which of the following amino acids is naturally occurring? By referring to the structures in Table 16.2, name each amino acid and include its D or L designation in the name.

a.
$$COO^-$$
$$H_3\overset{+}{N}\text{—}H$$
$$CH_2OH$$

b.
$$COO^-$$
$$H\text{—}\overset{+}{N}H_3$$
$$CH_2CH_2COO^-$$

c.
$$COO^-$$
$$H_3\overset{+}{N}\text{—}H$$
$$CH_2COO^-$$

More Practice: Try Problems 16.1a, b; 16.2a, b; 16.3; 16.4; 16.13–16.18.

16.3 Acid–Base Behavior of Amino Acids

As mentioned in Section 16.2, an amino acid contains both a basic amino group (NH_2) and an acidic carboxyl group (COOH). As a result, **proton transfer from the acid to the base forms a zwitterion,** a salt that contains both a positive and a negative charge. The zwitterion is *neutral;* that is, the net charge on the salt is zero.

The lone pair can bond to a proton.

The carboxyl group can donate a proton.

zwitterion

In actuality, **an amino acid can exist in different forms, depending on the pH of the aqueous solution in which it is dissolved.** When the pH of a solution is around 6, alanine (R = CH_3) and other neutral amino acids exist in their zwitterionic form (**A**), having no net charge. In this form, the carboxyl group bears a net negative charge—it is a **carboxylate anion**—and the amino group bears a net positive charge (an **ammonium cation**).

alanine
no net charge
A
pH ≈ 6

When strong acid is added to lower the pH to 2 or less, the carboxylate anion gains a proton and the **amino acid has a net positive charge** (form **B**).

The carboxylate anion picks up a proton.

Adding acid:

H_3N^+-C-C CH$_3$ O$^-$ → H$^+$ → H_3N^+-C-C CH$_3$ OH   pH ≤ 2

A **B** overall **+1** charge

When strong base is added to **A** to raise the pH to 10 or higher, the ammonium cation loses a proton and the **amino acid has a net negative charge** (form **C**).

The ammonium cation loses a proton.

Adding base:

H_3N^+-C-C CH$_3$ O$^-$ → OH$^-$ → H_2N-C-C CH$_3$ O$^-$ + H$_2$O   pH ≥ 10

A **C** overall **−1** charge

Thus, **alanine exists in one of three different forms depending on the pH of the solution in which it is dissolved.** At the physiological pH of 7.4, neutral amino acids are primarily in their zwitterionic forms.

- The pH at which the amino acid exists primarily in its neutral form is called its *isoelectric point,* abbreviated as p*I*.

The isoelectric points of neutral amino acids are generally around 6. Acidic amino acids (Table 16.2), which have an additional carboxyl group that can lose a proton, have lower p*I* values (around 3). The three basic amino acids, which have an additional basic nitrogen atom that can accept a proton, have higher p*I* values (7.6–10.8).

Sample Problem 16.2   **Determining the Structure of an Amino Acid at a Given pH**

Draw the structure of the amino acid glycine at each pH: (a) 6; (b) 2; (c) 11.

Analysis

- **A neutral amino acid exists in its zwitterionic form (no net charge) at its isoelectric point,** which is pH = 6. The zwitterionic forms of neutral amino acids appear in Table 16.2.
- At **low pH (≤ 2),** the carboxylate anion is protonated and the amino acid has a **net positive (+1) charge.**
- At **high pH (≥ 10),** the ammonium cation loses a proton and the amino acid has a **net negative (−1) charge.**

Solution

a. At pH = 6, the neutral, zwitterionic form of glycine predominates.

b. At pH = 2, glycine has a net +1 charge.

c. At pH = 11, glycine has a net −1 charge.

| **neutral** pH = 6 | **+1 charge** pH = 2 | **−1 charge** pH = 11 |

Practice Problem 16.3

Draw the structure of the amino acid valine at each pH: (a) 6; (b) 2; (c) 11. Which form predominates at valine's isoelectric point?

Practice Problem 16.4

L-Thyroxine, a thyroid hormone and oral medication used to treat thyroid hormone deficiency, is an amino acid that does not occur in proteins. Draw the neutral, positively charged, and negatively charged forms of L-thyroxine.

L-thyroxine

More Practice: Try Problems 16.1c, 16.2c, 16.19–16.22.

16.4 Peptides

When amino acids are joined together by amide bonds, they form larger molecules called **peptides** and **proteins.**

- A *dipeptide* has two amino acids joined together by *one* amide bond.
- A *tripeptide* has three amino acids joined together by *two* amide bonds.

dipeptide   tripeptide

[Amide bonds are shown in red.]

Polypeptides and **proteins** both have many amino acids joined together in long linear chains, but the term **protein** is usually reserved for polymers of more than 40 amino acids.

- The amide bonds in peptides and proteins are called *peptide bonds.*
- The individual amino acids are called *amino acid residues.*

To form a dipeptide, **the −NH₃⁺ group of one amino acid forms an amide bond with the carboxylate (−COO⁻) of another amino acid, and the elements of H₂O are removed.** For example, reaction of the −COO⁻ group of alanine with the −NH₃⁺ group of serine forms a dipeptide with one new amide bond, as shown. The dipeptide has an ammonium cation ($-NH_3^+$) at one end of its chain and a carboxylate anion ($-COO^-$) at the other.

Ala   Ser   A

new amide bond in red

reacting functional groups

- The amino acid with the free $-NH_3^+$ group on the α carbon is called the N-terminal amino acid.
- The amino acid with the free $-COO^-$ group on the α carbon is called the C-terminal amino acid.

By convention, **the N-terminal amino acid is always written at the *left* end of the chain and the C-terminal amino acid at the *right*.**

N-terminal amino acid   C-terminal amino acid

alanylserine
Ala–Ser

Peptides are named as derivatives of the C-terminal amino acid. To name a peptide:

- Name the C-terminal amino acid using the names in Table 16.2.
- Name all other amino acids from left to right as substituents of the C-terminal amino acid. Change the *-ine* or *-ic acid* ending of the amino acid name to the suffix *-yl.*

Thus, the dipeptide, which has serine as its C-terminal amino acid, is named as *alanylserine.*

The peptide can be abbreviated by writing the one- or three-letter symbols for the amino acids in the chain from the N-terminal to the C-terminal end. Thus, Ala–Ser has alanine at the N-terminal end and serine at the C-terminal end.

Sample Problem 16.3   **Identifying the N-Terminal and C-Terminal Amino Acids in a Peptide**

Label the N-terminal and C-terminal amino acids in each peptide.

a.

b. Tyr–Pro–Glu–His   c. TSFMMQNA

Analysis

- The amino acid with the free $-NH_3^+$ group on the α carbon is the **N-terminal amino acid,** drawn at the *left* end of the chain.
- The amino acid with the free $-COO^-$ group on the α carbon is the **C-terminal amino acid,** drawn at the *right* end of the chain.

Solution

The N-terminal amino acids are labeled in blue, and the C-terminal amino acids are labeled in red.

a.

b. Tyr–Pro–Glu–His   c. TSFMMQNA

Practice Problem 16.5

Identify the N-terminal and C-terminal amino acid in each peptide.

a.
$$H_3\overset{+}{N}-CH-\underset{\underset{\displaystyle CH(CH_3)_2}{|}}{\underset{\displaystyle CH_2}{|}}\overset{\overset{\displaystyle O}{\|}}{C}-N-\underset{H}{|}CH-\underset{CH_3}{|}\overset{\overset{\displaystyle O}{\|}}{C}-O^-$$

b. Arg–His–Asn–Tyr

c.
$$H_3\overset{+}{N}-CH-\overset{\overset{\displaystyle O}{\|}}{C}-N-CH-\overset{\overset{\displaystyle O}{\|}}{C}-N-CH-\overset{\overset{\displaystyle O}{\|}}{C}-O^-$$
with side chains CH_2 (phenyl), H CH_3, H $CH_2CH_2CONH_2$

d. Val–Thr–Pro–Phe

More Practice: Try Problems 16.5a, b; 16.6a, b; 16.23b, c; 16.24b, c; 16.25[1]; 16.26[1]; 16.27a, b; 16.28a, b.

Sample Problem 16.4

Identifying Amino Acids in a Peptide

Identify the individual amino acids. What is the name of the tripeptide?

$$H_3\overset{+}{N}-CH-\overset{\overset{\displaystyle O}{\|}}{C}-N-CH-\overset{\overset{\displaystyle O}{\|}}{C}-N-CH-\overset{\overset{\displaystyle O}{\|}}{C}-O^-$$

Analysis

- To identify the individual amino acids, locate the amide bonds, and compare the side chain of each amino acid with the structures in Table 16.2.
- To name the peptide: [1] Name the C-terminal amino acid. [2] Name the other amino acids as substituents by changing the *-ine* (or *-ic acid*) ending to the suffix *-yl.* Place the names of the substituent amino acids in order from left to right.

Solution

The amide bonds that join the amino acids together are shown in red. The tripeptide contains leucine (N-terminal), alanine, and tyrosine (C-terminal).

leucine   alanine

C-terminal amino acid

tyrosine

The tripeptide is named as a derivative of the C-terminal amino acid, tyrosine, with leucine and alanine as substituents; thus, the tripeptide is named: **leucylalanyltyrosine.**

Practice Problem 16.6

(a) Identify the N-terminal amino acid in the tetrapeptide alanylglycylleucylmethionine. (b) What is the C-terminal amino acid? (c) Write the peptide using three-letter symbols for the amino acids.

Practice Problem 16.7

Identify the individual amino acids in each dipeptide, and then name the dipeptide using three-letter abbreviations.

Practice Problem 16.8

Locate the peptide bond in the dipeptide shown in the ball-and-stick model, and identify the amino acids that form the dipeptide. Name the dipeptide using the three-letter abbreviations for the amino acids.

More Practice: Try Problems 16.5c, 16.6c, 16.23a, 16.24a, 16.25[2], 16.26[2], 16.27c, 16.28c, 16.29, 16.30.

16.5 FOCUS ON THE HUMAN BODY
Biologically Active Peptides

Many relatively simple peptides have important biological functions.

16.5A Neuropeptides—Enkephalins and Pain Relief

Enkephalins, peptides synthesized in the brain, act as pain killers and sedatives by binding to pain receptors. Two enkephalins that differ in the identity of only one amino acid are known. Met-enkephalin contains a C-terminal methionine residue, whereas leu-enkephalin contains a C-terminal leucine.

Tyr–Gly–Gly–Phe–Met

met-enkephalin

Tyr–Gly–Gly–Phe–Leu

leu-enkephalin

The addictive narcotic analgesics morphine and heroin bind to the same receptors as the enkephalins, and thus produce a similar physiological response. Enkephalins are related to a group of larger polypeptides called **endorphins** that contain 16–31 amino acids. Endorphins also block pain and are thought to produce the feeling of well-being experienced by an athlete after excessive or strenuous exercise.

Concept Check 16.2	(a) Label the four amide bonds in met-enkephalin. (b) What N-terminal amino acid is present in both enkephalins?

16.5B Peptide Hormones—Oxytocin and Vasopressin

Oxytocin and **vasopressin** are cyclic peptide hormones secreted by the pituitary gland. Their sequences are identical except for two amino acids.

oxytocin

vasopressin

- The N-terminal amino acid is labeled in red.
- The amino acids that differ are labeled in blue.

Oxytocin, mentioned in the chapter opener, stimulates the contraction of uterine muscles, and it initiates the flow of milk in nursing mothers (Figure 16.2). Oxytocin, sold under the trade names Pitocin and Syntocinon, is used to induce labor.

Figure 16.2

Physiological Effects of Oxytocin

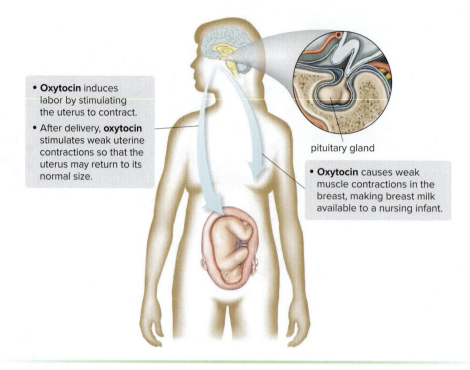

• **Oxytocin** induces labor by stimulating the uterus to contract.

• After delivery, **oxytocin** stimulates weak uterine contractions so that the uterus may return to its normal size.

pituitary gland

• **Oxytocin** causes weak muscle contractions in the breast, making breast milk available to a nursing infant.

Vasopressin, also called antidiuretic hormone (ADH), targets the kidneys and helps to keep the electrolytes in body fluids in the normal range. Vasopressin is secreted when the body is dehydrated and causes the kidneys to retain fluid, thus decreasing the volume of the urine (Figure 16.3).

The N-terminal amino acid in both hormones is a cysteine residue, and the C-terminal residue is glycine. Instead of a free carboxylate ($-COO^-$), both peptides have an amide ($-CONH_2$) at the C-terminal end, so this is indicated with the additional NH_2 group drawn at the end of the chain. The structure of both peptides includes a **disulfide bond,** a form of covalent bonding in which the $-SH$ groups from two cysteine residues are oxidized to form a sulfur–sulfur bond (Section 12.7). In oxytocin and vasopressin, the disulfide bonds make the peptides cyclic.

Figure 16.3

Vasopressin—An Antidiuretic Hormone

pituitary gland

• Secretion of **vasopressin** by the pituitary gland causes the kidneys to retain fluid, resulting in decreased urine output.

decreased urine output

Beefy meaty peptide is an octapeptide that gives a meaty flavor to foods. Its flavor-enhancing properties are similar to MSG, but high production costs have limited its widespread culinary use. Name the octapeptide using the three-letter abbreviations of its amino acids.

beefy meaty peptide

16.6 Proteins

To understand proteins, the large polymers of amino acids that are responsible for so much of the structure and function of all living cells, we must learn about four levels of structure, called the **primary, secondary, tertiary, and quaternary structure** of proteins.

16.6A Primary Structure

The *primary structure* of a protein is the particular sequence of amino acids that is joined together by peptide bonds. The most important element of this primary structure is the amide bond that joins the amino acids.

The carbonyl carbon of the amide has **trigonal planar** geometry. All six atoms involved in the peptide bond lie in the same plane. All bond angles are 120° and the C=O and N—H bonds are oriented 180° from each other. As a result, the backbone of the protein adopts a zigzag arrangement as shown in the three-dimensional structure of a portion of a protein molecule.

peptide bond

These six atoms lie in a plane.

Amide bonds are highlighted.

The primary structure of a protein—the exact sequence of amino acids—determines all properties and function of a protein. As we will see in Section 16.7, substitution of a single amino acid by a different amino acid can result in very different properties.

Can two *different* proteins be composed of the same number and type of amino acids?

16.6B Secondary Structure

The three-dimensional arrangement of localized regions of a protein is called its secondary structure. These regions arise due to hydrogen bonding between the N—H proton of one amide and the C=O oxygen of another. Two arrangements that are particularly stable are called the **α-helix** and the **β-pleated sheet.**

The **α-helix** forms when a peptide chain twists into a right-handed or clockwise spiral, as shown in Figure 16.4a. The C=O group of one amino acid is hydrogen bonded to an N—H group four amino acid residues farther along the chain. The R groups of the amino acids extend outward from the core of the helix.

Figure 16.4 Secondary Structure of Proteins

a. The right-handed α-helix

b. The β-pleated sheet

hydrogen bond

In the α-helix, all C=O bonds are pointing up and all N—H bonds are pointing down.

The β-pleated sheet consists of extended strands of the peptide chains held together by hydrogen bonding. The C=O and N—H bonds lie in the plane of the sheet, whereas the R groups (shown as yellow balls) alternate above and below the plane.

Both the myosin in muscle and the α-keratin in hair are proteins composed almost entirely of α-helices.

The **β-pleated sheet** forms when two or more peptide chains, called **strands,** line up side-by-side, as shown in Figure 16.4b. Hydrogen bonding often occurs between the N—H and C=O groups of nearby amino acid residues. The R groups are oriented above and below the plane of the sheet, and alternate from one side to the other along a given strand.

The β-pleated sheet arrangement is favored by amino acids with small R groups, like alanine and glycine. With larger R groups, steric interactions prevent the chains from getting close together, so the sheet cannot be stabilized by hydrogen bonding.

Most proteins have regions of α-helix and β-pleated sheet, in addition to other regions that cannot be characterized by either of these arrangements. Shorthand symbols are often used to indicate these regions of secondary structure. In particular, a flat helical ribbon is used for the α-helix, while a flat wide arrow is used for the β-pleated sheet. These representations are often used in **ribbon diagrams** to illustrate protein structure.

α-helix   β-sheet   random region

Proteins are drawn in a variety of ways to show different aspects of their structure. Figure 16.5 illustrates three different representations of the protein lysozyme, an enzyme found in both plants and animals. Lysozyme catalyzes the hydrolysis of bonds in bacterial cell walls, weakening them, often causing the bacteria to burst.

Figure 16.5 Lysozyme

a. Ball-and-stick model   b. Space-filling model   c. Ribbon diagram

(a) The ball-and-stick model of lysozyme shows the protein backbone with color-coded C, N, O, and S atoms. Individual amino acids are most clearly located using this representation. (b) The space-filling model uses color-coded balls for each atom in the backbone of the enzyme and illustrates how the atoms fill the space they occupy. (c) The ribbon diagram shows regions of α-helix and β-sheet that are not clearly in evidence in the other two representations.

16.6C Tertiary and Quaternary Structure

The three-dimensional shape adopted by the entire peptide chain is called its tertiary structure. A peptide generally folds into a shape that maximizes its stability. In the aqueous environment of the cell, proteins often fold in such a way as to place a large number of polar and charged groups on their outer surface, to maximize the dipole–dipole and hydrogen bonding interactions with water. This generally places most of the nonpolar side chains in the interior of the protein, where **London dispersion forces** between these hydrophobic groups help stabilize the molecule, too.

In addition, polar functional groups hydrogen bond with each other (not just water), and amino acids with charged side chains like –COO⁻ and –NH₃⁺ can stabilize tertiary structure by **electrostatic interactions.**

Finally, **disulfide bonds are the only covalent bonds that stabilize tertiary structure.** As mentioned in Section 16.5, these strong bonds form by the oxidation of two cysteine residues on either the same polypeptide chain or another polypeptide chain of the same protein.

Insulin, for example, consists of two separate polypeptide chains (labeled the **A** and **B** chains) that are covalently linked by two intermolecular disulfide bonds, as shown in Figure 16.6. The **A** chain, which also has an intramolecular disulfide bond, has 21 amino acid residues, whereas the **B** chain has 30.

Figure 16.7 schematically illustrates the many different kinds of intramolecular forces that stabilize the secondary and tertiary structures of polypeptide chains. Nearby amino acid residues that have only nonpolar carbon–carbon and carbon–hydrogen bonds are stabilized by London dispersion forces. Amino acids that contain hydroxyl (OH) and amino groups (NH₂) in their side chains can intermolecularly hydrogen bond to each other.

The many disulfide bonds in the proteins that compose fingernails make nails strong and hard. *Diffused Productions/Alamy Stock Photo*

Figure 16.6 Insulin

Insulin is a small protein consisting of two polypeptide chains (designated as the **A** and **B** chains), held together by two disulfide bonds. An additional disulfide bond joins two cysteine residues within the **A** chain.

Synthesized by groups of cells in the pancreas called the islets of Langerhans, insulin is the protein that regulates blood glucose levels. A relative or complete lack of insulin results in diabetes. Many of the abnormalities associated with this disease can be controlled by the injection of insulin.

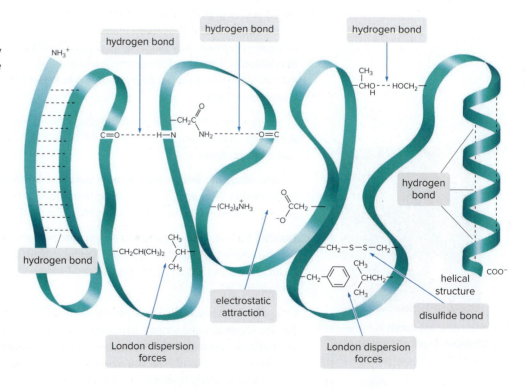

pancreas   islets of Langerhans   space-filling model of insulin

Figure 16.7

The Stabilizing Interactions in Secondary and Tertiary Protein Structure

The shape adopted when two or more folded polypeptide chains come together into one protein complex is called the **quaternary structure** of the protein. Each individual polypeptide chain is called a **subunit** of the overall protein. **Hemoglobin,** for example, consists of two α and two β subunits held together by intermolecular forces in a compact three-dimensional shape. The unique function of hemoglobin is possible only when all four subunits are together.

The four levels of protein structure are summarized in Figure 16.8.

Figure 16.8   The Primary, Secondary, Tertiary, and Quaternary Structure of Proteins

β-pleated sheet

α-helix

amino acid sequence

3-D shape of a polypeptide chain

protein complex of polypeptide chains

| Primary structure | Secondary structure | Tertiary structure | Quaternary structure |

Concept Check 16.5

What types of intermolecular forces are present between the side chains of each pair of amino acids? Identify the side chains of the amino acids from Table 16.2, and use Figure 16.7 as a guide.

   a. Ser and Tyr   b. Val and Leu   c. 2 Phe residues

Concept Check 16.6

The fibroin proteins found in silk fibers consist of large regions of β-pleated sheets stacked one on top of another. The polypeptide sequence in these regions has the amino acid glycine at every other residue. Explain how this allows the β-pleated sheets to stack on top of each other.

16.7   FOCUS ON THE HUMAN BODY
Common Proteins

Proteins are generally classified according to their three-dimensional shapes.

- *Fibrous proteins* are composed of long linear polypeptide chains that are bundled together to form rods or sheets. These proteins are insoluble in water and serve structural roles, giving strength and protection to tissues and cells.

- *Globular proteins* are coiled into compact shapes with hydrophilic outer surfaces that make them water soluble. Enzymes and transport proteins are globular to make them soluble in blood and other aqueous environments.

16.7A α-Keratins

α-Keratins are the proteins found in hair, hooves, nails, skin, and wool. They are composed almost exclusively of long sections of α-helix units, having large numbers of alanine and leucine residues. These nonpolar amino acids extend outward from the α-helix, so these proteins are very insoluble in water. Two α-keratin helices coil around each other, forming a structure called a **supercoil** or **superhelix.** These, in turn, form larger and larger bundles of fibers, ultimately forming a strand of hair, as shown schematically in Figure 16.9.

Figure 16.9 Anatomy of a Hair

strand of hair

supercoil

A supercoil is formed from two α-helices wound around each other.

Bundles of supercoils come together to form a hair.

The α-keratin in hair is made up largely of α-helix.

Photosindia/Getty Images

Figure 16.10

The Triple Helix of Collagen

In collagen, three polypeptide chains having an unusual left-handed helix wind around each other in a right-handed triple helix.

α-Keratins also have a number of cysteine residues, and because of this, disulfide bonds are formed between adjacent helices. The number of disulfide bridges determines the strength of the material. Claws, horns, and fingernails have extensive networks of disulfide bonds, making them extremely hard.

16.7B Collagen

Collagen, the most abundant protein in vertebrates, is found in connective tissues such as bone, cartilage, tendons, teeth, and blood vessels. Glycine and proline account for a large fraction of its amino acid residues. Collagen forms an elongated left-handed helix, and then three of these helices wind around each other to form a right-handed **superhelix** or **triple helix.** The side chain of glycine is only a hydrogen atom, so the high glycine content allows the collagen superhelices to lie compactly next to each other, thus stabilizing the superhelices via hydrogen bonding (Figure 16.10).

Vitamin C is required for a reaction that modifies the original amino acids incorporated into the collagen chain so that strong hydrogen bonds form between the helices. When there is a deficiency of vitamin C in the diet, the collagen fibers do not form properly and scurvy results. Weakened blood vessels and poorly formed cartilage lead to spongy and bloody gums and dark purple skin lesions.

16.7C   Hemoglobin and Myoglobin

heme

Hemoglobin and **myoglobin,** two globular proteins, are called **conjugated proteins** because they are composed of a protein unit and a nonprotein molecule. In hemoglobin and myoglobin, the nonprotein unit is called **heme,** a complex organic compound containing the Fe^{2+} ion complexed with a large nitrogen-containing ring system. The Fe^{2+} ion of hemoglobin and myoglobin binds oxygen. Hemoglobin, which is present in red blood cells, transports oxygen to wherever it is needed in the body, whereas myoglobin stores oxygen in tissues.

Myoglobin has 153 amino acid residues in a single polypeptide chain (Figure 16.11a). It has eight separate α-helical sections and a heme group held in a cavity inside the polypeptide. Myoglobin gives cardiac muscle its characteristic red color.

Hemoglobin consists of four polypeptide chains (two α subunits and two β subunits), each of which carries a heme unit (Figure 16.11b). Hemoglobin has more nonpolar amino acid residues than myoglobin. When each subunit is folded, some of these remain on the surface. The London dispersion forces between these hydrophobic groups are what stabilize the quaternary structure of the four subunits.

Carbon monoxide is poisonous because it binds to the Fe^{2+} of hemoglobin 200 times more strongly than does oxygen. Hemoglobin complexed with CO cannot carry O_2 from the lungs to the tissues. Without O_2 available to the tissues for metabolism, cells cannot function, and they die.

The properties of all proteins depend on their three-dimensional shape, and their shape depends on their primary structure—that is, their amino acid sequence. This is particularly well exemplified by comparing normal hemoglobin with **sickle cell hemoglobin,** a mutant variation in which a single amino acid of both β subunits is changed from glutamic acid to valine. The replacement of one acidic amino acid (Glu) with one nonpolar amino acid (Val) changes the shape of hemoglobin, which has profound effects on its function. Red blood cells with sickle cell hemoglobin become elongated and crescent shaped, and they are unusually fragile. As a result, they rupture and occlude capillaries, causing pain and inflammation, leading to severe anemia and organ damage. The end result is often a painful and premature death.

The high concentration of myoglobin in a whale's muscles allows it to remain underwater for long periods of time.
Daniel C. Smith

ENVIRONMENTAL NOTE

Leghemoglobin, an oxygen-carrying protein (like hemoglobin) found in soybean roots, is used in plant-derived burgers. These meat alternatives are marketed to provide consumers with the nutritional benefits of meat without the negative environmental impact that occurs in large-scale livestock farming.
John D. Ivanko/Alamy Stock Photo

Figure 16.11   Protein Ribbon Diagrams for Myoglobin and Hemoglobin

a. Myoglobin

heme

Myoglobin consists of a single polypeptide chain with a heme unit shown in a ball-and-stick model.

b. Hemoglobin

heme

Hemoglobin consists of two α and two β chains shown in red and blue, respectively, and four heme units shown in ball-and-stick models.

When red blood cells take on a "sickled" shape in persons with sickle cell disease, they block capillaries causing organ injury and they break easily leading to profound anemia. This devastating illness results from the change of a single amino acid in hemoglobin. Note the single sickled cell surrounded by three red cells with normal shape.
Janice Haney Carr/CDC

This disease, called **sickle cell anemia,** is found almost exclusively among people originating from central and western Africa, where malaria is an enormous health problem. Sickle cell hemoglobin results from a genetic mutation in the DNA sequence that is responsible for the synthesis of hemoglobin. Individuals who inherit this mutation from both parents develop sickle cell anemia, whereas those who inherit it from only one parent are said to have the sickle cell trait. They do not develop sickle cell anemia and they are more resistant to malaria than individuals without the mutation. The relative benefit of this mutation apparently accounts for this detrimental gene being passed on from generation to generation.

Concept Check 16.7

Why is hemoglobin more water soluble than α-keratin?

Concept Check 16.8

Why is it possible to discuss the quaternary structure of hemoglobin but not the quaternary structure of myoglobin?

16.8 Protein Hydrolysis and Denaturation

The properties of a protein are greatly altered and often entirely destroyed when any level of protein structure is disturbed.

16.8A Protein Hydrolysis

Like other amide bonds, the peptide bonds in proteins are hydrolyzed by treatment with aqueous acid, base, or certain enzymes.

- The hydrolysis of the amide bonds in a protein forms the individual amino acids that make up the primary structure.

For example, hydrolysis of the amide bonds in the tripeptide Ile–Gly–Phe forms the amino acids isoleucine, glycine, and phenylalanine. When each amide bond is broken, the elements of H_2O are added, forming a carboxylate anion (–COO⁻) in one amino acid and an ammonium cation (–NH₃⁺) in the other.

The first step in the digestion of dietary protein is hydrolysis of the amide bonds of the protein backbone. The enzyme **pepsin** in the acidic gastric juices of the stomach cleaves some of the amide bonds to form smaller peptides, which pass into the small intestine and are further broken down into individual amino acids by the enzymes **trypsin** and **chymotrypsin.**

Proteins in the diet serve a variety of nutritional needs. Like carbohydrates and lipids, proteins can be metabolized for energy. Moreover, the individual amino acids formed by hydrolysis are used as starting materials to make new proteins that the body needs. Likewise the N atoms in the amino acids are incorporated into other biomolecules that contain nitrogen.

| Sample Problem 16.5 | Drawing the Products of Peptide Hydrolysis |

Draw the structures of the amino acids formed by hydrolysis of the neuropeptide leu-enkephalin (Section 16.5).

Analysis

Locate each amide bond in the protein or peptide backbone. To draw the hydrolysis products, **break each amide bond** by adding the elements of H_2O to form a **carboxylate anion (–COO⁻)** in one amino acid and an **ammonium cation (–NH₃⁺)** in the other.

Solution

[1] Locate the amide bonds in the peptide backbone.
[2] Break each bond by adding H_2O.

leu-enkephalin

Tyr   Gly   Gly   Phe   Leu

| Practice Problem 16.9 | Draw the structure of the products formed by hydrolysis of each tripeptide: (a) Ala–Leu–Gly; (b) Ser–Thr–Phe. |

More Practice: Try Problems 16.31–16.34.

16.8B   Protein Denaturation

When the secondary, tertiary, or quaternary structure of a protein is disturbed, the properties of a protein are also altered and the biological activity is often lost.

- *Denaturation* is the process of altering the shape of a protein without breaking the amide bonds that form the primary structure.

High temperature, acid, base, and even agitation can disrupt the noncovalent interactions that hold a protein in a specific shape. Heat breaks up weak London forces between nonpolar amino acids. Heat, acid, and base disrupt hydrogen-bonding interactions between polar amino acids, which account for much of the secondary and tertiary structure. As a result, denaturation causes a globular protein to uncoil into an undefined randomly looped structure.

coiled globular protein

heat

loose coils and loops

Cooking or whipping egg whites denatures the globular proteins they contain, forming insoluble protein. *Jill Braaten/McGraw Hill*

Denaturation often makes globular proteins less water soluble. Globular proteins are typically folded with hydrophobic regions in the interior to maximize the interaction of polar residues on the outside surface with water. This makes them water soluble. When the protein is denatured, more hydrophobic regions are exposed and the protein often loses water solubility.

We witness many examples of protein denaturation in the kitchen. As milk ages it becomes sour from enzymes that produce lactic acid. The acid also denatures milk proteins, which precipitate as an insoluble curd. Ovalbumin, the major protein in egg white, is denatured when an egg is boiled or fried, forming a solid. Even vigorously whipping egg whites denatures its protein, forming the stiff meringue used to top a lemon meringue pie.

Concept Check 16.9

Heating collagen, a water-insoluble fibrous protein, forms the jelly-like substance called gelatin. Explain how this process may occur.

16.9 Enzymes

We conclude the discussion of proteins with *enzymes,* **proteins that serve as biological catalysts for reactions in all living organisms.** Like all catalysts (Section 5.10B), enzymes increase the rate of reactions, but they themselves are not permanently changed in the process. Enzymes are crucial to the biological reactions that occur in the body, which would otherwise often proceed too slowly to be of any use. In humans, enzymes must catalyze reactions under very specific physiological conditions, usually a pH around 7.4 and a temperature of 37 °C.

16.9A Characteristics of Enzymes

Enzymes are generally water-soluble, globular proteins that exhibit two characteristic features.

- Enzymes greatly enhance reaction rates.
- Enzymes are very specific.

The specificity of an enzyme varies. Some enzymes, such as catalase, catalyze a single reaction. Catalase catalyzes the conversion of hydrogen peroxide (H_2O_2) to O_2 and H_2O (Section 5.1).

$$2\ H_2O_2(aq) \xrightarrow{\text{catalase}} 2\ H_2O(l)\ +\ O_2(g)$$

Other enzymes, such as carboxypeptidase A, catalyze a particular type of reaction with a variety of substrates. **Carboxypeptidase A,** a digestive enzyme that breaks down proteins, catalyzes the

When a bloody wound is cleaned with hydrogen peroxide, the enzyme catalase in the blood converts the H_2O_2 to H_2O and O_2, forming a white foam of oxygen bubbles. *Daniel C. Smith*

hydrolysis of a specific type of peptide bond—the amide bond closest to the C-terminal end of the protein.

Only this amide bond is broken.

The conversion of lactate to pyruvate illustrates other important features of enzyme-catalyzed reactions. The enzyme for this process is *lactate dehydrogenase,* a name derived from the substrate (lactate) and the type of reaction, a **dehydrogenation**—that is, the removal of two hydrogen atoms.

To carry out this reaction a **cofactor** is needed.

- A *cofactor* is a metal ion or a nonprotein organic molecule needed for an enzyme-catalyzed reaction to occur.

NAD^+ (nicotinamide adenine dinucleotide) is the cofactor that oxidizes lactate to pyruvate. **An organic compound that serves as an enzyme cofactor is called a *coenzyme*.** While lactate dehydrogenase greatly speeds up this oxidation reaction, NAD^+ is the coenzyme that actually oxidizes the substrate, lactate. **NAD^+ is a common biological oxidizing agent** used as a coenzyme (Section 18.3).

The need for metal ions as enzyme cofactors explains why trace amounts of certain metals must be present in our diet. For example, enzymes that catalyze oxidation reactions require either Fe^{2+} or Cu^{2+} as a cofactor needed for transferring electrons.

16.9B Classification of Enzymes

Enzymes are classified into six categories by the type of reaction they catalyze, as summarized in Table 16.3.

- Oxidoreductases catalyze oxidation–reduction reactions.

When a substrate is oxidized or reduced, a coenzyme is required. The coenzyme serves as the oxidizing or reducing agent. Oxidoreductases are subclassified as **oxidases** when they catalyze oxidations, **reductases** when they catalyze reduction, or **dehydrogenases** when two hydrogen atoms are removed or added during the redox reaction.

- Transferases catalyze the transfer of a group from one molecule to another.

For example, a **transaminase** catalyzes the transfer of an amino group (NH_2, Section 18.11A), and a **kinase** catalyzes the transfer of a phosphate (Section 18.4).

- Hydrolases catalyze hydrolysis reactions—the cleavage of bonds with water.

For example, the digestion of a triacylglycerol, which begins with the hydrolysis of its three esters to glycerol and fatty acids (Section 15.5A), occurs with hydrolases called **lipases.** In addition to lipases, other subclasses of hydrolases include **proteases,** which catalyze protein hydrolysis,

Table 16.3 Classification of Enzymes

Enzyme Class or Subclass	Reaction Catalyzed
Oxidoreductases	**Oxidation–reduction**
• Oxidases	• Oxidation
• Reductases	• Reduction
• Dehydrogenases	• Addition or removal of 2 H's
Transferases	**Transfer of a group**
• Transaminases	• Transfer of an NH_2 group
• Kinases	• Transfer of a phosphate
Hydrolases	**Hydrolysis**
• Lipases	• Hydrolysis of lipid esters
• Proteases	• Hydrolysis of amide bonds in proteins
• Nucleases	• Hydrolysis of nucleic acids
Isomerases	**Isomerization**
Lyases	**Addition to a double bond or elimination to give a double bond**
• Dehydrases	• Removal of H_2O
• Decarboxylases	• Removal of CO_2
• Synthases	• Addition of a small molecule to a double bond
Ligases	**Bond formation accompanied by ATP hydrolysis**
• Carboxylases	• Bond formation between a substrate and CO_2

Meat tenderizers contain **papain**, an enzyme that hydrolyzes peptide bonds, making meat more tender by breaking down the protein in tough fibers. Papain is obtained from the fruit of the papaya tree. *David Moyer/McGraw Hill*

and **nucleases,** which catalyze the hydrolysis of phosphate bonds in nucleic acids (Chapter 17). Carboxypeptidase A (Section 16.9A) is a protease.

• Isomerases catalyze the conversion of one isomer to another.

• Lyases catalyze the addition of a molecule to a double bond or the elimination of a molecule to give a double bond.

Lyases are subclassified as **dehydrases** when they catalyze the removal of H_2O, **decarboxylases** when they catalyze the removal of CO_2, and **synthases** when they catalyze the addition of a small molecule such as H_2O to a double bond.

• Ligases catalyze bond formation accompanied by energy release from a hydrolysis reaction.

A ligase enzyme catalyzes the joining together of two molecules. Because this process is often energetically unfavorable, the energy released in a hydrolysis reaction is used to drive the reaction. We will examine reactions of this sort in Section 18.3.

Sample Problem 16.6   **Classifying Enzymes**

Classify the enzyme used in each of the following reactions.

a. (pyruvate) ⟶ (acetaldehyde) + CO_2

b. $H_3\overset{+}{N}$—CH—C—N—CH$_2$—C + H_2O ⟶ $H_3\overset{+}{N}$—C—C + $H_3\overset{+}{N}$—CH$_2$—C

Analysis

Concentrate on the functional groups that change in the reaction, and use Table 16.3 to classify the enzyme.

Solution

a. The reaction involves the loss of a molecule (CO_2) from a double bond, so the enzyme is a **lyase.** Because CO_2 is removed, the lyase is subclassified as a **decarboxylase.**

b. The reaction involves hydrolysis—that is, cleavage of a bond with water—so the enzyme is a **hydrolase.** Because hydrolysis cleaves the amide bond in a dipeptide, the hydrolase is subclassified as a **protease.**

Practice Problem 16.10

What class of enzyme catalyzes each of the following reactions?

a.

succinate   fumarate

b.

malate   oxaloacetate

More Practice: Try Problems 16.45, 16.46.

16.9C Naming Enzymes

The names of most enzymes end in the suffix -*ase*. The name of an enzyme generally has two parts.

- The first part identifies the substrate on which the enzyme acts.
- The second part identifies the class of reaction catalyzed (Table 16.3).

Thus, *succinate dehydrogenase* catalyzes the removal of two hydrogen atoms from the substrate succinate, and *pyruvate decarboxylase* catalyzes the loss of CO_2 from the substrate pyruvate.

Some older names of enzymes are still in common use. The names are derived from the name of the substrate by adding the suffix **-*ase*.** Thus, the addition of water to the substrate fumarate is catalyzed by the lyase fumarase.

fumarate   fumarase   enzyme

In addition, some enzymes are still identified by names that are not related to the substrate or type of reaction they catalyze. Examples include the digestive enzymes trypsin, chymotrypsin, and pepsin, which catalyze the hydrolysis of proteins.

Concept Check 16.10

From the name alone, decide which of the following might be enzymes: (a) sucrose; (b) sucrase; (c) lactose; (d) lactase; (e) phosphofructokinase.

16.10 How Enzymes Work

An enzyme contains a region called the **active site** that binds the substrate, forming an **enzyme–substrate complex.** The active site is often a small cavity that contains amino acids that are attracted to the substrate with various types of intermolecular forces. Sometimes polar amino acids of the enzyme hydrogen bond to the substrate or nonpolar amino acids have stabilizing hydrophobic interactions.

16.10A Enzyme Specificity

Two models have been proposed to explain the specificity of a substrate for an enzyme's active site: the **lock-and-key model** and the **induced-fit model.**

In the lock-and-key model, the shape of the active site is rigid. The three-dimensional geometry of the substrate must exactly match the shape of the active site for catalysis to occur. The lock-and-key model explains the high specificity observed in many enzyme reactions.

In the induced-fit model, the shape of the active site is more flexible. It is thought that when the substrate and the enzyme interact, the shape of the active site can adjust to fit the shape of the substrate.

This model is often used to explain why some enzymes catalyze reactions with a wider variety of substrates. The shape of the active site and the substrate must still be reasonably similar, but once bound, the shape of the active site is different from the active site in the unbound enzyme. A well-characterized example of the induced-fit model is seen in the binding of glucose to the enzyme hexokinase, shown in Figure 16.12.

Figure 16.12 Hexokinase—An Example of the Induced-Fit Model

a. Shape of the active site before binding

b. Shape of the active site after binding

active site   glucose

glucose in the active site

The open cavity of the active site of the enzyme closes around the substrate for a tighter fit.

(a) The free enzyme has a small open cavity that fits a glucose molecule. (b) Once bound, the cavity encloses the glucose molecule more tightly in the enzyme–substrate complex.

16.10B Factors That Affect Enzyme Activity

The activity of an enzyme is affected by both temperature and pH.

As we learned in Section 5.10, an increase in temperature increases the rate of most chemical reactions. **Enzyme-catalyzed reactions also exhibit an increase in rate with increasing temperature until a maximum activity occurs,** often around 37 °C, and then activity falls off rapidly. At higher temperatures the enzyme begins to denature, and at 50–60 °C its activity is usually destroyed. Figure 16.13a illustrates an example of how the rate of an enzyme-catalyzed reaction changes with temperature.

Enzyme activity is also affected by pH. Each enzyme has an optimum pH at which catalysis occurs at a maximum rate. Although enzymes in cells have an optimum pH around 7.4, a digestive enzyme like pepsin, which is secreted in the gastric juices of the stomach, has an optimum pH of 2. Figure 16.13b illustrates how the activity of an enzyme with an optimum pH of 5 varies with pH.

Figure 16.13 Factors Affecting Enzyme Activity

a. Temperature

b. pH

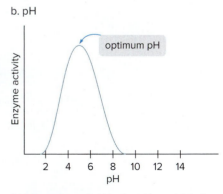

(a) An enzyme-catalyzed reaction increases in rate with increasing temperature until the enzyme begins to denature. Then, the reaction rate decreases rapidly. The maximum rate for the enzyme depicted in this graph occurs at approximately 37 °C.

(b) The optimum pH is the pH at which the catalytic activity of the enzyme is at its maximum. The enzyme depicted in this graph has maximum activity at pH 5.

Concept Check 16.11

Estimate the optimum pH for the digestive enzyme trypsin, which hydrolyzes proteins in the small intestine, from the following graph.

Concept Check 16.12

Using the graph in Figure 16.13a, determine which temperature in each pair gives the higher rate of reaction: (a) 25 °C or 35 °C; (b) 35 °C or 45 °C.

16.10C Enzyme Inhibitors

Some substances bind to enzymes and in the process, greatly alter or destroy the enzyme's activity.

- An *inhibitor* is a molecule that causes an enzyme to lose activity.

An inhibitor can bind to an enzyme reversibly or irreversibly.

- A *reversible* inhibitor binds to an enzyme but then enzyme activity is restored when the inhibitor is released.
- An *irreversible* inhibitor covalently binds to an enzyme, permanently destroying its activity.

Penicillin is an antibiotic that kills bacteria because it *irreversibly* binds to glycopeptide transpeptidase, an enzyme required for the synthesis of a bacterial cell wall (Section 13.10B). Penicillin binds to a hydroxyl group (OH) of the enzyme, thus inactivating it, halting cell wall construction, and killing the bacterium.

penicillin G

Penicillin is covalently bound to the enzyme, inactivating it.

enzyme

inactive enzyme

Reversible inhibition can be **competitive** or **noncompetitive.**

A *noncompetitive* inhibitor binds to the enzyme but does not bind at the active site. The inhibitor causes the enzyme to change shape so that the active site can no longer bind the substrate. When the noncompetitive inhibitor is no longer bound to the enzyme, normal enzyme activity resumes.

Without the inhibitor, the substrate fits into the active site.

When the inhibitor is bound, the active site changes shape and the substrate no longer fits.

A *competitive* inhibitor has a shape and structure similar to the substrate, so it competes with the substrate for binding to the active site. The antibiotic **sulfanilamide** is a competitive inhibitor of the enzyme needed to synthesize the vitamin folic acid from *p*-aminobenzoic acid in bacteria. When sulfanilamide binds to the active site, *p*-aminobenzoic acid cannot be converted to folic acid, so a bacterium cannot grow and reproduce. Sulfanilamide does not affect human cells because humans do not synthesize folic acid, and must obtain it in the diet.

sulfanilamide
a competitive inhibitor

p-aminobenzoic acid
PABA

folic acid

Both compounds can bind to the active site.

Concept Check 16.13

Use the given representations for an enzyme, substrate, and inhibitor to illustrate the process of competitive inhibition.

active site

enzyme   substrate   inhibitor

Concept Check 16.14

The nerve gas sarin acts as a poison by covalently bonding to a hydroxyl group in the active site of the enzyme acetylcholinesterase. This binding results in a higher-than-normal amount of acetylcholine, resulting in muscle spasms. From this description, would you expect sarin to be a competitive, noncompetitive, or irreversible inhibitor?

16.11　FOCUS ON HEALTH & MEDICINE
Using Enzymes to Diagnose and Treat Diseases

Measuring enzyme levels in the blood and understanding the key role of enzymes in biological reactions have aided greatly in both diagnosing and treating diseases.

16.11A　Enzyme Levels as Diagnostic Tools

Certain enzymes are present in a higher concentration in particular cells. When the cells are damaged by disease or injury, the cells rupture and die, releasing the enzymes into the bloodstream. Measuring the activity of the enzymes in the blood then becomes a powerful tool to diagnose the presence of disease or injury in some organs.

Measuring enzyme levels is now routine for many different conditions. For example, when a patient comes into an emergency room with chest pain, measuring the level of creatine phosphokinase (CPK) and other enzymes indicates whether a heart attack, which results in damage to some portion of the heart, has occurred. Common enzymes used for diagnosis are listed in Table 16.4.

Table 16.4　Common Enzymes Used for Diagnosis

Enzyme	Condition
Creatine phosphokinase	Heart attack
Alkaline phosphatase	Liver or bone disease
Acid phosphatase	Prostate cancer
Amylase, lipase	Diseases of the pancreas

16.11B　Treating Disease with Drugs That Interact with Enzymes

Molecules that inhibit an enzyme can be useful drugs. The antibiotics penicillin and sulfanilamide are two examples discussed in Section 16.10. Drugs used to treat high blood pressure and acquired immune deficiency syndrome (AIDS) are also enzyme inhibitors.

ACE inhibitors are a group of drugs used to treat individuals with high blood pressure. Angiotensin is a peptide that narrows blood vessels, thus increasing blood pressure. Angiotensin is formed from angiotensinogen by action of **ACE, the angiotensin-converting enzyme,** which cleaves two amino acids from the inactive peptide.

ACE cleaves here.

Arg–Arg–Val–Tyr–Ile–His–Pro–Phe—His–Leu　⟶　Arg–Arg–Val–Tyr–Ile–His–Pro–Phe　+　His–Leu

angiotensinogen　　　　　　　　　　　　　　　　　angiotensin

increases blood pressure

By blocking the conversion of angiotensinogen to angiotensin, blood pressure is decreased. Several effective ACE inhibitors are currently available, including captopril and enalapril.

Generic name: captopril
Trade name: Capoten

Generic name: enalapril
Trade name: Vasotec

Several enzyme inhibitors are also available to treat human immunodeficiency virus (HIV), the virus that causes AIDS. The most effective treatments are **HIV protease inhibitors.** These drugs inhibit the action of the HIV protease enzyme, an essential enzyme needed by HIV to make copies of itself that go on to infect other cells. For example, the anti-HIV drug fosamprenavir (trade name: Lexiva) is metabolized to amprenavir, the active drug that inhibits the HIV protease

HEALTH NOTE

ACE inhibitors such as lisinopril (trade name: Zestril) decrease the concentration of angiotensin, thus decreasing blood pressure.
Sally Anderson/Alamy Stock Photo

enzyme, so the virus cannot replicate. A ribbon diagram of the HIV-1 protease enzyme with amprenavir at the active site is shown in Figure 16.14.

fosamprenavir

Generic name: amprenavir
Trade name: Agenerase

Figure 16.14   The HIV Protease Enzyme

a. Ball-and-stick model of the HIV protease enzyme

b. Ribbon diagram with the protease inhibitor amprenavir in the active site

CHAPTER REVIEW

KEY TERMS

Active site (16.9)

Amino acid (16.2)

Coenzyme (16.9)

Cofactor (16.9)

Competitive inhibitor (16.10)

Conjugated protein (16.7)

C-Terminal amino acid (16.4)

Decarboxylase (16.9)

Dehydrase (16.9)

Dehydrogenase (16.9)

Denaturation (16.8)

Dipeptide (16.4)

Enzyme (16.9)

Enzyme–substrate complex (16.10)

Fibrous protein (16.7)

Globular protein (16.7)

α-Helix (16.6)

Heme (16.7)

Hydrolase (16.9)

Induced-fit model (16.10)

Inhibitor (16.10)

Irreversible inhibitor (16.10)

Isoelectric point (16.3)

Isomerase (16.9)

Kinase (16.9)

Ligase (16.9)

Lipase (16.9)

Lock-and-key model (16.10)

Lyase (16.9)

Noncompetitive inhibitor (16.10)

N-Terminal amino acid (16.4)

Nuclease (16.9)

Oxidase (16.9)

Oxidoreductase (16.9)

Peptide (16.4)

Peptide bond (16.4)

β-Pleated sheet (16.6)

Primary structure (16.6)

Protease (16.9)

Protein (16.1)

Quaternary structure (16.6)

Reductase (16.9)

Reversible inhibitor (16.10)

Secondary structure (16.6)

Synthase (16.9)

Tertiary structure (16.6)

Transaminase (16.9)

Transferase (16.9)

Tripeptide (16.4)

Zwitterion (16.2)

KEY CONCEPTS

① Characteristics of amino acids (16.2)

① General features	② Stereochemistry
• Amino acids contain an amino group (NH_2) on the α carbon to the carboxyl group (COOH). • Because they contain both an acid and a base, amino acids exist in their neutral form as **zwitterions.** • The R group of an amino acid is called the **side chain.** The 20 amino acids that occur in proteins are classified as acidic, basic, or neutral depending on the R group, and are identified by a three-letter or one-letter abbreviation (Table 16.2). • Because they are salts, amino acids are water soluble and have high melting points.	• All amino acids except glycine (R = H) have a chirality center on the α carbon, so there are two possible enantiomers. • Fischer projections can be used to show the stereochemistry on the α carbon, and the prefixes D and L are used to designate the arrangement of groups. • When the amino acid is drawn as a vertical chain with the COO^- at the top and R group at the bottom, **L amino acids** (which are naturally occurring) have the NH_3^+ group on the *left* in the Fischer projection and D amino acids have the NH_3^+ group on the *right*.

α-amino acid → proton transfer → zwitterion salt

L-amino acid   Fischer projection

② Characteristics of dipeptides (16.4)

① Peptide structure	② N- and C-terminal amino acids	③ Naming a dipeptide
• To form a dipeptide, the carboxyl group of one amino acid bonds to the amino group of another amino acid in an **amide (peptide) bond.**	• The amino acid that contains the free NH_3^+ group on the α carbon is called the **N-terminal amino acid.** • The amino acid that contains the free COO^- group on the α carbon is called the **C-terminal amino acid.** • Peptide **A** contains alanine as the N-terminal amino acid and glycine as the C-terminal amino acid.	• Peptides are written from left to right, from the N-terminal to the C-terminal amino acid. • Using three-letter abbreviations, peptide **A** is Ala–Gly. • A peptide is named as a derivative of the C-terminal amino acid, with other amino acids named as substituents. Peptide **A** is named alanylglycine.

Ala   Gly   peptide bond in gray
                A

③ Levels of protein structure (16.6)

① Primary	② Secondary	③ Tertiary	④ Quaternary
• The primary structure of a protein is the particular **sequence of amino acids** joined together by amide bonds. • The exact sequence of amino acids determines the properties of the protein.	• The two most common types of secondary structure are the **α-helix** and the **β-pleated sheet.** Both structures are stabilized by hydrogen bonds between the N—H and C=O groups.	• The tertiary structure is the **three-dimensional shape** adopted by the entire peptide chain. London forces, hydrogen bonding, ionic interactions, and disulfide bonds stabilize the three-dimensional structure.	• When a protein contains more than one polypeptide chain, the quaternary structure describes the shape of the **protein complex** formed by two or more chains.

α-helix

β-pleated sheet

4 **Common proteins (16.7)**

1 **Fibrous proteins** are composed of long linear polypeptide chains that serve structural roles and are water insoluble.	**2** **Globular proteins** have compact shapes with polar amino acids on the outside to make them water soluble.

α-Keratin	Collagen	Hemoglobin	Myoglobin
• α-Keratin in hair is a fibrous protein composed almost exclusively of α-helix units that wind together to form a superhelix. Disulfide bonds between chains make the resulting bundles of protein chains strong.	• Collagen, found in connective tissue and the most abundant protein in vertebrates, is composed of a superhelix formed from three elongated left-handed helices.	• Hemoglobin is a conjugated protein consisting of a protein unit and a nonprotein molecule called a heme. • Hemoglobin, which is present in red blood cells, transports oxygen.	• Myoglobin is a conjugated protein that contains a heme unit that complexes oxygen. • Myoglobin stores oxygen in tissues and gives cardiac muscle its characteristic red color.

5 **Characteristics of enzymes (16.9)**

1 **General features**	**2** **Classes of enzymes**
• Enzymes are **biological catalysts** that greatly increase the rate of biological reactions and are highly specific for a substrate or a type of substrate. • An enzyme binds a substrate at its active site, forming an enzyme–substrate complex by either the **lock-and-key model** or the **induced-fit model**.	• Enzymes are classified into six categories by the type of reaction they catalyze: **oxidoreductases, transferases, hydrolases, isomerases, lyases,** and **ligases** (Table 16.3). • Some enzyme types are subclassified. For example, a **kinase** is a transferase that catalyzes the transfer of a phosphate.

6 **Enzyme activity (16.10)**

1 **Factors affecting enzyme activity**	**2** **Enzyme inhibitors**
• **Temperature:** The rate of an enzyme-catalyzed reaction *increases* with *increasing* temperature, until a maximum activity occurs, often around 37 °C. • **pH:** Each enzyme has an optimum pH at which maximum activity occurs. Pepsin in the stomach has an optimum pH of 2, whereas the catalytic activity of enzymes in cells is often a maximum at pH = 7.4.	• An **irreversible enzyme inhibitor** covalently bonds to an enzyme and permanently destroys its activity. • A **noncompetitive reversible inhibitor** binds to the enzyme, but not at the active site. The inhibitor causes the enzyme to change shape so that the active site can no longer bind to the substrate. • A **competitive reversible inhibitor** competes with the substrate for binding to the active site.

KEY SKILLS

1 **Drawing an amino acid at different pH's (16.3); example: leucine at low pH, its isoelectric point, and high pH**

1 **An amino acid at pH $\leq$ 2**	**2** **An amino acid at its isoelectric point**	**3** **An amino acid at pH $\geq$ 10**
• At low pH ($\leq$ 2), the carboxylate anion is protonated and the amino acid has a net positive (+1) charge. CO_2H $H_3\overset{+}{N}-\underset{\underset{CH_2CH(CH_3)_2}{\mid}}{\overset{\mid}{C}}-H$ (+1) charge	• A neutral amino acid exists in its zwitterionic form (no net charge) at its isoelectric point. CO_2^- $H_3\overset{+}{N}-\underset{\underset{CH_2CH(CH_3)_2}{\mid}}{\overset{\mid}{C}}-H$ 0 charge	• At high pH ($\geq$ 10), the ammonium cation loses a proton and the amino acid has a net negative (−1) charge. CO_2^- $H_2N-\underset{\underset{CH_2CH(CH_3)_2}{\mid}}{\overset{\mid}{C}}-H$ (−1) charge

2 **Drawing the products of peptide hydrolysis (16.8); example: hydrolysis products of Gly–Ala–Ser**

1 **Locate each amide bond in the peptide.**	**2** **Break each amide bond by adding H_2O.**
amide bond amide bond Gly–Ala–Ser	• Break the amide C—N bonds by adding H_2O across the bond, forming a carboxylate anion (COO^-) in one amino acid and an ammonium cation (NH_3^+) in the other. Gly Ala Ser

CHAPTER 16 SELF-TEST

The Self-Test consists of short-answer questions that test an understanding of definitions, equations, and other material encountered within the chapter. Answers to each question are provided at the end of the chapter. For any questions you are unable to answer correctly, revisit the appropriate section(s) of the text; then proceed to working the end-of-chapter problems to further reinforce your understanding of the chapter material.

1.   Which term from "Key Terms" answers each question?

   a. The particular sequence of amino acids that is joined together by peptide bonds constitutes what level of protein structure?

   b. What term is used for the pH at which an amino acid exists primarily in its neutral form?

   c. What type of secondary structure forms when a peptide chain twists into a right-handed or clockwise spiral?

   d. What type of enzyme catalyzes a hydrolysis reaction?

   e. What type of inhibitor has a shape and structure similar to the substrate and binds at the active site?

2.   For the given tripeptide: (a) give the three-letter abbreviation for the N-terminal amino acid; (b) give the three-letter abbreviation for the C-terminal amino acid; (c) name the tripeptide as a derivative of the C-terminal amino acid.

$$H_3\overset{+}{N}-\underset{\underset{+}{(CH_2)_4NH_3}}{\overset{\overset{H}{|}}{C}}-\overset{\overset{O}{\|}}{C}-\underset{H}{N}-\underset{\underset{}{CH(CH_3)_2}}{\overset{\overset{H}{|}}{C}}-\overset{\overset{O}{\|}}{C}-\underset{H}{N}-\underset{\underset{}{CH_3}}{\overset{\overset{H}{|}}{C}}-\overset{\overset{O}{\|}}{C}\underset{O^-}{}$$

3.   For the amino acid serine: (a) draw the neutral form of the L enantiomer in a Fischer projection; (b) give the three-letter abbreviation; (c) label the amino acid as acidic, basic, or neutral.

4.   What class of enzyme catalyzes each of the following reactions?

   a. addition of water to a double bond

   b. hydrolysis of the ester bonds in a triacylglycerol

   c. conversion of $CH_3COCO_2^-$ to CH_3CHO and CO_2

5.   Label each statement as True or False.

   a. A tetrapeptide contains four peptide bonds.

   b. The net charge of a neutral amino acid at pH = 2 is +1.

   c. If a digestive enzyme has an optimum pH of 2, enzyme activity decreases when the pH is increased to 7.4.

   d. Fibrous proteins are generally water soluble because they contain a hydrophilic outer surface that hydrogen bonds with water.

   e. When a protein is denatured by heating, its peptide bonds are broken.

UNDERSTANDING KEY CONCEPTS

16.1   (a) Identify the amino acid shown with all uncharged atoms in the ball-and-stick model. (b) Give the three-letter and one-letter abbreviations for the amino acid. (c) Draw the form present at the isoelectric point of the amino acid.

16.2   (a) Identify the amino acid shown with all uncharged atoms in the ball-and-stick model. (b) Give the three-letter and one-letter abbreviations for the amino acid. (c) Draw the form of the amino acid present at pH 11.

16.3   For each amino acid: [1] draw the L enantiomer in a Fischer projection; [2] classify the amino acid as neutral, acidic, or basic; [3] give the three-letter symbol; [4] give the one-letter symbol.

   a. leucine   b. tryptophan

16.4   For each amino acid: [1] give the name; [2] give the three-letter abbreviation; [3] give the one-letter abbreviation; [4] classify the amino acid as neutral, acidic, or basic.

a. $H_3\overset{+}{N}-\underset{\underset{}{CH_2CH_2SCH_3}}{\overset{\overset{COO^-}{|}}{\underset{|}{C}}}-H$ b. $H_3\overset{+}{N}-\underset{\underset{}{CH_2OH}}{\overset{\overset{COO^-}{|}}{\underset{|}{C}}}-H$

16.5   For the given tripeptide: (a) identify the amino acids that form the peptide; (b) label the N- and C-terminal amino acids; (c) name the tripeptide using three-letter symbols.

$$H_3\overset{+}{N}-\underset{\underset{}{CH(CH_3)_2}}{CH}-\overset{\overset{O}{\|}}{C}-\underset{H}{N}-\underset{H}{CH}-\overset{\overset{O}{\|}}{C}-\underset{H}{N}-\underset{\underset{}{CH_2}}{CH}-\overset{\overset{O}{\|}}{C}-O^-$$

16.6   For the given tripeptide: (a) identify the amino acids that form the peptide; (b) label the N- and C-terminal amino acids; (c) name the tripeptide using three-letter symbols.

$$H_3\overset{+}{N}-\underset{\underset{\underset{}{COO^-}}{CH_2}}{CH}-\overset{\overset{O}{\|}}{C}-\underset{H}{N}-\underset{\underset{}{CH_3}}{CH}-\overset{\overset{O}{\|}}{C}-\underset{H}{N}-\underset{\underset{}{CH(CH_3)CH_2CH_3}}{CH}-\overset{\overset{O}{\|}}{C}-O^-$$

16.7 Label the regions of secondary structure in the following protein ribbon diagram.

16.8 Label the regions of secondary structure in the following protein ribbon diagram.

16.9 What type of interactions occur at each of the labeled sites in the portion of the protein ribbon shown?

16.10 Using the given representations for an enzyme and substrate, which diagram represents (a) the enzyme–substrate complex; (b) competitive inhibition; (c) noncompetitive inhibition?

ADDITIONAL PROBLEMS

Amino Acids

16.11 The amino acid alanine is a solid at room temperature and has a melting point of 315 °C, whereas pyruvic acid (CH_3COCO_2H) has a similar molecular weight but is a liquid at room temperature with a boiling point of 165 °C. Account for the difference.

16.12 Why is phenylalanine water soluble but 4-phenylbutanoic acid ($C_6H_5CH_2CH_2CH_2COOH$), a compound of similar molecular weight, water insoluble?

16.13 Draw the structure of a naturally occurring amino acid that:
   a. contains a 1° alcohol
   b. contains an amide
   c. is an essential amino acid with an aromatic ring

16.14 Draw both enantiomers of the amino acid asparagine and label them as D or L.

16.15 For each amino acid: [1] draw the L enantiomer in a Fischer projection; [2] classify the amino acid as neutral, acidic, or basic; [3] give the three-letter symbol; [4] give the one-letter symbol.
   a. lysine   b. aspartic acid

16.16 For each amino acid: [1] draw the L enantiomer in a Fischer projection; [2] classify the amino acid as neutral, acidic, or basic; [3] give the three-letter symbol; [4] give the one-letter symbol.
   a. arginine   b. tyrosine

16.17 For each amino acid: [1] give the name; [2] give the three-letter abbreviation; [3] give the one-letter abbreviation; [4] classify the amino acid as neutral, acidic, or basic.

a. $H_3\overset{+}{N}$—|—H   b. $H_3\overset{+}{N}$—|—H
          | |
     $CH_2CH_2CONH_2$ $CH(OH)CH_3$
     (COO⁻)   (COO⁻)

16.18 Which of the following Fischer projections represent naturally occurring amino acids? Name each amino acid and designate it as a D or L isomer.

a. $H_3\overset{+}{N}$—|—H   b. H—|—$\overset{+}{N}H_3$
          | |
     $CH_2CH(CH_3)_2$ CH_2COO^-
     (COO⁻)   (COO⁻)

Acid–Base Properties of Amino Acids

16.19 Draw the amino acid leucine at each pH: (a) 6; (b) 10; (c) 2. Which form predominates at leucine's isoelectric point?

16.20 Draw the amino acid isoleucine at each pH: (a) 6; (b) 10; (c) 2. Which form predominates at isoleucine's isoelectric point?

16.21 Draw the structure of the neutral, positively charged, and negatively charged forms of the amino acid tyrosine. Which form predominates at pH 1? Which form predominates at pH 11? Which form predominates at the isoelectric point?

16.22 Draw the structure of the neutral, positively charged, and negatively charged forms of the amino acid proline. Which form predominates at pH 1? Which form predominates at pH 11? Which form predominates at the isoelectric point?

Peptides

16.23 Melittin, the principal toxin of bee venom, consists of a peptide with the sequence GIGAVLKVLTTGLPALISWIKRKRQQ (using one-letter abbreviations for the amino acids).

Bigemrg/Getty Images

    a. How many amide bonds join the amino acids in the peptide chain?

    b. Give the name of the N-terminal amino acid.

    c. Give the name of the C-terminal amino acid.

16.24 α-Cobratoxin is a neurotoxin found in the venom of the Chinese cobra, *Naja atra,* a prevalent venomous snake that inhabits China and Taiwan. A partial sequence of its 71 amino acid chain is shown.

IRCFITPDIT— **51 amino acids** —CNPFPTRKRP

Mitja Derenda/Getty Images

    a. How many amide bonds join the amino acids in the peptide chain?

    b. Give the name of the N-terminal amino acid.

    c. Give the name of the C-terminal amino acid.

16.25 For each tripeptide: [1] identify the N-terminal and C-terminal amino acids; [2] name the peptide using three-letter symbols for the amino acids.

    a. leucylvalyltryptophan   b. alanylglycylvaline

16.26 For each tripeptide: [1] identify the N-terminal and C-terminal amino acids; [2] name the peptide using three-letter symbols for the amino acids.

    a. tyrosylleucylisoleucine   b. methionylisoleucylcysteine

16.27 For the given tripeptide: (a) identify the amino acids that form the peptide; (b) label the N- and C-terminal amino acids; (c) name the tripeptide using three-letter symbols.

$$\underset{\underset{\underset{\underset{OH}{|}}{CH(CH_3)_2}}{\underset{|}{CH_2}}}{\overset{+}{H_3N}-CH}-\overset{\overset{O}{\|}}{C}-N-\underset{H}{CH}-\overset{\overset{O}{\|}}{C}-N-\underset{CH_2CH_2SCH_3}{CH}-\overset{\overset{O}{\|}}{C}-O^-$$

16.28 For the given tripeptide: (a) identify the amino acids that form the peptide; (b) label the N- and C-terminal amino acids; (c) name the tripeptide using three-letter symbols.

$$\overset{+}{H_3N}-\underset{CH_2OH}{CH}-\overset{\overset{O}{\|}}{C}-N-\underset{\underset{C_6H_5}{CH_2}}{\underset{H}{CH}}-\overset{\overset{O}{\|}}{C}-N-\underset{\underset{CH_2CH_2CONH_2}{}}{\underset{H}{CH}}-\overset{\overset{O}{\|}}{C}-O^-$$

16.29 (a) Identify the amino acids that form the dipeptide shown in the ball-and-stick model. (b) Name the dipeptide using the three-letter abbreviations for the amino acids.

16.30 Identify the N-terminal and C-terminal amino acids in the dipeptide shown in the ball-and-stick model. Name the dipeptide using the one-letter abbreviations for the amino acids.

16.31 Draw the structures of the amino acids formed when the tripeptide in Problem 16.27 is hydrolyzed.

16.32 Draw the structures of the amino acids formed when the tripeptide in Problem 16.28 is hydrolyzed.

16.33 Give the three-letter abbreviations for the amino acids formed by hydrolysis of the tripeptide depicted in the ball-and-stick model.

16.34 Give the three-letter abbreviations for the amino acids formed by hydrolysis of the tripeptide depicted in the ball-and-stick model.

Proteins

16.35 What type of intermolecular forces exist between the side chains of each of the following pairs of amino acids?

    a. isoleucine and valine   b. Lys and Glu

16.36 Which of the following pairs of amino acids can have intermolecular hydrogen bonding between the functional groups in their side chains?

    a. two tyrosine residues   b. alanine and threonine

16.37 Which peptide in each pair contains amino acids with side chains that exhibit predominantly London dispersion forces?

    a. Met–Gly–Leu–Phe–Gln–Ala or Lys–Gly–Arg–Tyr–Trp–Glu

    b. Tyr–Asp–Leu–Lys–His or Phe–Asn–Leu–Leu–Met

16.38 Decide if the side chains of the amino acid residues in the following peptides are nonpolar or polar, and label the hydrophobic and hydrophilic end of each peptide.

    a. VLLFGEDEK   b. RKYSFLGAA

16.39 Compare α-keratin and hemoglobin with regards to each of the following: (a) secondary structure; (b) water solubility; (c) function; (d) location in the body.

16.40 Compare collagen and myoglobin with regards to each of the following: (a) secondary structure; (b) water solubility; (c) function; (d) location in the body.

16.41 When a protein is denatured, how is its primary, secondary, tertiary, and quaternary structure affected?

16.42 Hydrogen bonding stabilizes both the secondary and tertiary structures of a protein. (a) What functional groups hydrogen bond to stabilize secondary structure? (b) What functional groups hydrogen bond to stabilize tertiary structure?

16.43 Describe the function or biological activity of each protein or peptide: (a) insulin; (b) myoglobin; (c) α-keratin; (d) chymotrypsin; (e) oxytocin.

16.44 Describe the function or biological activity of each protein or peptide: (a) collagen; (b) hemoglobin; (c) vasopressin; (d) pepsin; (e) met-enkephalin.

Enzymes

16.45 What class of enzyme catalyzes each of the following reactions?

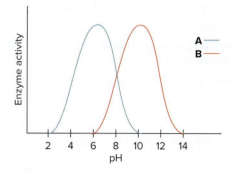

16.46 What class of enzyme catalyzes each of the following reactions?

    a. hydrolysis of Val–Gly to Val and Gly

    b. transfer of a phosphate from ATP to fructose 6-phosphate

    c. conversion of CH_3CH_2OH to CH_3CHO with NAD^+

16.47 Use the graph of enzyme activity versus pH for enzymes **A** and **B** to answer the questions.

    a. What is the optimum pH for enzyme **A**?

    b. What is the optimum pH for enzyme **B**?

    c. Which enzyme has the higher activity at pH 4?

    d. Which enzyme has the higher activity at pH 9?

16.48 How will each of the following changes affect the rate of an enzyme-catalyzed reaction whose optimum pH is 7.4?

    a. decreasing the temperature from 37 °C to 5 °C

    b. increasing the temperature from 37 °C to 90 °C

    c. increasing the pH from 7.4 to 10

    d. increasing the pH from 5 to 7.4

16.49 Use the given representations for an enzyme, substrate, and inhibitor to illustrate the process of noncompetitive inhibition.

16.50 Use the given representations for an enzyme and substrate to illustrate the difference between the lock-and-key model and the induced-fit model of enzyme specificity.

16.51 How are enzyme inhibitors used to treat high blood pressure? Give a specific example of a drug used and an enzyme inhibited.

16.52 How are enzyme inhibitors used to treat HIV? Give a specific example of a drug used and an enzyme inhibited.

Applications

16.53 What structural feature in α-keratin makes fingernails harder than skin?

16.54 Why does the α-keratin in hair contain many cysteine residues?

16.55 Why must vegetarian diets be carefully balanced?

16.56 Why does cooking meat make it easier to digest?

16.57 Sometimes an incision is cauterized (burned) to close the wound and prevent bleeding. What does cauterization do to protein structure?

16.58 Why is insulin administered by injection instead of taken in tablet form?

16.59 How is sickle cell disease related to hemoglobin structure?

16.60 The silk produced by a silkworm is a protein with a high glycine and alanine content. With reference to the structure, how does this make the silk fiber strong?

16.61 Explain the difference in the mechanism of action of penicillin and sulfanilamide. How is enzyme inhibition involved in both mechanisms?

16.62 How are blood enzyme levels used to diagnose certain diseases? Give an example of a specific condition and enzyme used for diagnosis.

CHALLENGE PROBLEMS

16.63 Explain why two amino acids—aspartic acid and glutamic acid—have a +1 net charge at low pH, but a −2 net charge at high pH.

16.64 Trandolapril is an ACE inhibitor that contains several functional groups.
  a. Identify all the functional groups in trandolapril.
  b. Hydrolysis of the ester in trandolapril forms trandolaprilat, which is the active ACE inhibitor. What is the structure of trandolaprilat?

  c. What products are formed when both the ester and amide are hydrolyzed?

trandolapril

BEYOND THE CLASSROOM

16.65 Pick an ethnic diet and give two or three examples of vegetarian dishes that would provide all of the essential amino acids.

16.66 Greek yogurt is considered especially nutritious, because it has more protein and fewer carbohydrates than most regular yogurts. Compare the protein and carbohydrate content of a popular Greek yogurt with that of a regular yogurt. How does the fat and sodium content of the two varieties compare? If you ingest one 8-oz serving of yogurt each day for a month, then what is the difference in protein intake in the two varieties? Research how Greek yogurt is made, and suggest reasons why Greek yogurt is generally more expensive than regular yogurt. Does your research indicate that there are considerable health benefits derived from regularly eating Greek yogurt?

CHAPTER 16 ANSWERS

Concept Check Answers

16.1 a. amide   b. alcohol   c. thiol

16.2 a. Each amide has a N atom bonded to a carbonyl C.
        b. tyrosine

16.3 Lys–Gly–Asp–Glu–Glu–Ser–Leu–Ala

16.4 Yes, the amino acids may be ordered differently.

16.5 a. hydrogen bonding between 2 OH groups
        b,c: London dispersion forces between the nonpolar side chains

16.6 Glycine has no large side chain, so the β-pleated sheets can pack together well.

16.7 Hemoglobin is a globular protein with polar groups on its exterior. Keratin is a fibrous protein with nonpolar groups on its exterior.

16.8 Hemoglobin has four chains that combine in a quaternary structure. Myoglobin has no quaternary structure because it is a single chain.

16.9 Heating disrupts hydrogen bonding and other intermolecular forces. The superhelix unwinds, making it less ordered, turning it into a jelly-like substance.

16.10 b,d,e

16.11 8

16.12 a. 35 °C   b. 35 °C

16.13 The inhibitor and substrate are similar in shape, so the inhibitor fits into the active site, blocking the substrate from binding there.

16.14 irreversible inhibitor

Practice Problem Answers

16.1

L structure: $H_3\overset{+}{N}$—C—H with COO^- top and R, R bottom (labeled L)
D structure: H—C—$\overset{+}{N}H_3$ with COO^- top and R bottom (labeled D)

a. R = CH₂—(phenyl)   b. R = CH₂CH₂SCH₃

16.2
a. L-serine, naturally occurring
b. D-glutamic acid
c. L-aspartic acid, naturally occurring

16.3
a. $H_3\overset{+}{N}$—C—COO^- with H, CH(CH₃)₂
predominant at p*I*

b. $H_3\overset{+}{N}$—C—COOH with H, CH(CH₃)₂

c. H_2N—C—COO^- with H, CH(CH₃)₂

16.4

(thyroxine structure: iodinated diphenyl ether, HO— ring with I's)

= R

R—CO_2^- with $\overset{+}{N}H_3$: neutral
R—CO_2H with $\overset{+}{N}H_3$: net (+1) charge
R—CO_2^- with NH_2 : net (−1) charge

16.5
a. N-terminal: leucine   C-terminal: alanine
b. N-terminal: arginine   C-terminal: tyrosine
c. N-terminal: phenylalanine   C-terminal: glutamine
d. N-terminal: valine   C-terminal: phenylalanine

16.6
a. alanine   b. methionine   c. Ala–Gly–Leu–Met

16.7
a. alanine and isoleucine, Ala–Ile
b. tyrosine and valine, Tyr–Val

16.8 The peptide bond joins the C=O of one amino acid to the N atom of the second amino acid. The dipeptide (Ser–Gly) is composed of serine and glycine.

16.9
a. $H_3\overset{+}{N}$—C—COO^- with CH₃; $H_3\overset{+}{N}$—C—COO^- with CH₂CH(CH₃)₂; $H_3\overset{+}{N}$—C—COO^- with H

b. $H_3\overset{+}{N}$—C—COO^- with CH₂OH; $H_3\overset{+}{N}$—C—COO^- with CH(OH)CH₃; $H_3\overset{+}{N}$—C—COO^- with CH₂—(phenyl)

16.10 a,b: dehydrogenase

Self-Test Answers

1.
a. primary
b. isoelectric point
c. α-helix
d. hydrolase
e. competitive inhibitor

2.
a. Lys   b. Ala   c. lysylvalylalanine

3.
a. $H_3\overset{+}{N}$—C—COO^- with CH₂OH
b. Ser
c. neutral

4.
a. lyase   b. lipase   c. decarboxylase

5.
a,d,e: False   b,c: True

Odd-Numbered Problem Answers

16.1 a,b,c: $H_3\overset{+}{N}$—C—H with COO^- top and CH₂CH₂SCH₃
methionine
Met, M

16.3
a. [1] $H_3\overset{+}{N}$—C—H with COO^- and CH₂CH(CH₃)₂
[2] neutral [3] Leu [4] L

b. [1] $H_3\overset{+}{N}$—C—H with COO^- and CH₂—(indole ring)
[2] neutral [3] Trp [4] W

16.5
a,b: valine: N-terminal, glycine, phenylalanine: C-terminal
c. Val–Gly–Phe

16.7

α-helix, random, β-sheet, random, β-sheet, random, α-helix

16.9 **A** hydrogen bonding, **B** London dispersion forces, **C** electrostatic attraction

16.11 Alanine is an ionic salt with strong electrostatic forces, leading to its high melting point, and making it a solid at room temperature. Pyruvic acid has hydrogen bonding, a much weaker intermolecular force, so it is a liquid at room temperature.

16.13
a. $H_3\overset{+}{N}$—C—COO^- with CH₂OH
b. $H_3\overset{+}{N}$—C—COO^- with CH₂CONH₂
c. $H_3\overset{+}{N}$—C—COO^- with CH₂—(phenyl)

16.15
a. [1] $H_3\overset{+}{N}$—C—H with COO^- and $(CH_2)_4\overset{+}{N}H_3$
[2] basic [3] Lys [4] K

b. [1] $H_3\overset{+}{N}$—C—H with COO^- and CH₂COO⁻
[2] acidic [3] Asp [4] D

16.17
a. [1] glutamine   [2] Gln   [3] Q   [4] neutral
b. [1] threonine   [2] Thr   [3] T   [4] neutral

16.19
a. $H_3\overset{+}{N}$—C—COO^- with CH₂CH(CH₃)₂
predominant form at p*I*

b. H_2N—C—COO^- with CH₂CH(CH₃)₂

c. $H_3\overset{+}{N}$—C—COOH with CH₂CH(CH₃)₂

16.21

neutral
predominant form at p*I*

positive charge
pH = 1

negative charge
pH = 11

16.23 a. 25   b. glycine   c. glutamine

16.25 a. [1] N-terminal: leucine
C-terminal: tryptophan

b. [1] N-terminal: alanine
C-terminal: valine

[2] Leu–Val–Trp

[2] Ala–Gly–Val

16.27 a,b: leucine: N-terminal, tyrosine, methionine: C-terminal
c. Leu–Tyr–Met

16.29 a. phenylalanine, alanine   b. Phe–Ala

16.31

16.33 alanine, cysteine, glycine

16.35 a. London dispersion forces   b. electrostatic

16.37 a. Met–Gly–Leu–Phe–Gln–Ala   b. Phe–Asn–Leu–Leu–Met

16.39

	a. Secondary Structure	b. H$_2$O Solubility	c. Function	d. Location
Hemo-globin	Globular with much α-helix	Soluble	Carries oxygen to tissues	Blood
Keratin	α-Helix	Insoluble	Firm tissues	Nail, hair

16.41 When heated, a protein's primary structure is unaffected. The 2°, 3°, and 4° structures may be altered.

16.43 a. Insulin is a hormone that controls glucose levels.
b. Myoglobin stores oxygen in muscle.
c. α-Keratin forms hard tissues such as hair and nails.
d. Chymotrypsin is a protease that hydrolyzes peptide bonds.
e. Oxytocin is a hormone that stimulates uterine contractions and induces the release of breast milk.

16.45 a. dehydrase   b. isomerase

16.47 a. 6   b. 10   c. **A**   d. **B**

16.49 substrate

The inhibitor alters active site configuration and the substrate cannot enter.

inhibitor

16.51 Captopril inhibits the angiotensin-converting enzyme, blocking the conversion of angiotensinogen to angiotensin. This reduces the concentration of angiotensin, which in turn lowers blood pressure.

16.53 The α-keratin in nails has more cysteine residues to form disulfide bonds. The larger the number of disulfide bonds, the harder the substance.

16.55 Humans cannot synthesize the amino acids methionine and lysine. Plant sources generally do not have sufficient amounts of all the essential amino acids. Grains are low in lysine, and legumes are low in methionine, but a combination of these foods provides all the needed amino acids.

16.57 Cauterization denatures the proteins in a wound.

16.59 In sickle hemoglobin there is a substitution of a single amino acid—namely, valine for glutamic acid.

16.61 Penicillin inhibits the formation of the bacterial cell wall by irreversibly binding to an enzyme needed for its construction. Sulfanilamide inhibits the production of folic acid and therefore reproduction in bacteria.

16.63 Both aspartic acid and glutamic acid have two carboxylic acid groups. At low pH they have a +1 charge (from –NH$_3^+$ and 2 –COOH), but at a high pH both acid groups are ionized, leading to a net charge of −2 (from –NH$_2$ and 2 –COO$^-$).

Design Credits: DNA molecule structure background for Chapter Opener: Watchara/Shutterstock

Nucleic Acids and Protein Synthesis

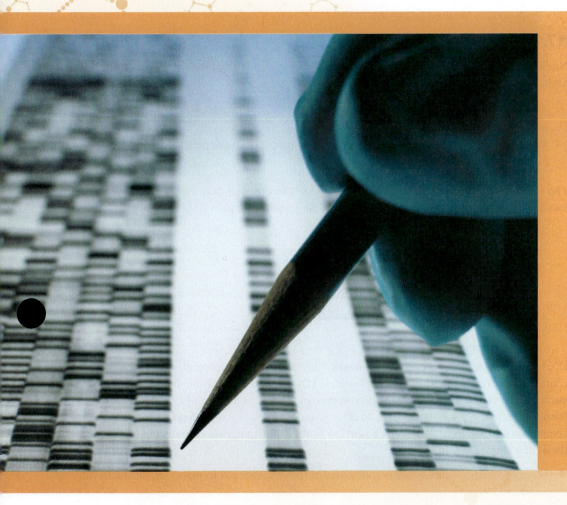

Techniques to analyze DNA samples are commonly utilized in criminal investigations and to screen for genetic diseases.
Rafe Swan/Cultura Creative/Alamy

17

CHAPTER GOALS

In this chapter you will learn how to:

1. Draw the structure of nucleosides and nucleotides
2. Draw short segments of the nucleic acids DNA and RNA
3. Describe the basic features of the DNA double helix
4. Outline the main steps of replication
5. List the three types and functions of RNA molecules
6. Explain the process of transcription
7. Describe the basic elements of the genetic code
8. Explain the process of translation
9. Define the terms "mutation" and "genetic disease"
10. Describe the basic features of DNA fingerprinting
11. Describe the main characteristics of viruses

Whether you are tall or short, fair-skinned or dark-complexioned, blue-eyed or brown-eyed, your unique characteristics are determined by the nucleic acid polymers that reside in the chromosomes of your cells. The nucleic acid **DNA** stores the genetic information of a particular organism, whereas the nucleic acid **RNA** translates this genetic information into the synthesis of proteins needed by cells for proper function and development. Even minor alterations in the nucleic acid sequence can have significant effects on an organism, sometimes resulting in devastating diseases like sickle cell anemia and cystic fibrosis. In Chapter 17, we study nucleic acids and learn how the genetic information stored in DNA is translated into protein synthesis.

17.1 Nucleosides and Nucleotides

Nucleic acids are unbranched polymers composed of repeating monomers called *nucleotides*. There are two types of nucleic acids.

- DNA, deoxyribonucleic acid, stores the genetic information of an organism and transmits that information from one generation to another.
- RNA, ribonucleic acid, translates the genetic information contained in DNA into proteins needed for all cellular functions.

The nucleotide monomers that compose DNA and RNA consist of three components—a **monosaccharide, a nitrogen-containing base, and a phosphate group.**

DNA molecules contain several million nucleotides whereas RNA molecules are much smaller, containing perhaps a few thousand nucleotides. DNA is contained in the chromosomes of the nucleus, each chromosome having a different type of DNA. The number of chromosomes differs from species to species. Humans have 46 chromosomes (23 pairs). An individual chromosome is composed of many genes. **A *gene* is a portion of the DNA molecule responsible for the synthesis of a single protein.**

We begin our study of nucleic acids with a look at the structure and formation of the nucleotide monomers.

17.1A Nucleosides—Joining a Monosaccharide and a Base

The nucleotides of both DNA and RNA contain a five-membered ring monosaccharide, often called simply the *sugar* component.

- In RNA, the monosaccharide is the aldopentose D-ribose.
- In DNA the monosaccharide is D-2-deoxyribose, an aldopentose that lacks a hydroxyl group at C2.

D-ribose
(present in **RNA**)

D-2-deoxyribose
(present in **DNA**)

The prefix *deoxy* means *without oxygen.*

Only five common nitrogen-containing bases are present in nucleic acids. Three bases with one ring (**cytosine, uracil,** and **thymine**) are derived from the parent compound **pyrimidine.** Two bases with two rings (**adenine** and **guanine**) are derived from the parent compound **purine.** Each base is designated by a one-letter abbreviation as shown.

| pyrimidine (parent compound) | cytosine **C** | uracil **U** (RNA only) | thymine **T** (DNA only) |

| purine (parent compound) | adenine **A** | guanine **G** |

Uracil (U) occurs only in RNA, whereas thymine (T) occurs only in DNA. As a result:

- **DNA contains the bases A, G, C, and T.**
- **RNA contains the bases A, G, C, and U.**

| **Concept Check 17.1** | 5-Fluorouracil is an anticancer drug that interferes with DNA synthesis because its structure resembles a DNA base that is needed for normal DNA replication. 5-Fluorouracil is sold as a cream to treat basal cell carcinoma, the most common form of skin cancer. What DNA base does 5-fluorouracil resemble? |

Peter Jordan/Alamy Stock Photo

A *nucleoside* **is formed by joining a carbon of the monosaccharide with a nitrogen atom of the base.** A nucleoside is called an *N*-glycoside. Primes (') are used to number the carbons of the monosaccharide in a nucleoside.

| D-ribose | + | cytosine **C** | → | cytidine **a ribonucleoside** |

| D-2-deoxyribose | + | adenine **A** | → | deoxyadenosine **a deoxyribonucleoside** |

For example, joining cytosine with ribose forms the ribonucleoside **cytidine.** Joining adenine with 2-deoxyribose forms the deoxyribonucleoside **deoxyadenosine.**

Nucleosides are named as derivatives of the bases from which they are formed.

> • To name a nucleoside derived from a pyrimidine base, use the suffix *-idine* (cytosine → cyt*idine*).
> • To name a nucleoside derived from a purine base, use the suffix *-osine* (adenine → aden*osine*).
> • For deoxyribonucleosides, add the prefix *deoxy-,* as in *deoxy*adenosine.

Sample Problem 17.1 | **Naming a Nucleoside**

Identify the base and monosaccharide used to form the following nucleoside, and then name it.

Analysis

• The sugar portion of a nucleoside contains the five-membered ring. If there is an OH group at C2′, the sugar is **ribose,** and if there is no OH group at C2′, the sugar is **deoxyribose.**
• The base is bonded to the five-membered ring. A pyrimidine base has one ring, and is derived from either cytosine, uracil, or thymine. A purine base has two rings, and is derived from either adenine or guanine.
• Nucleosides derived from pyrimidines end in the suffix **-idine.** Nucleosides derived from purines end in the suffix **-osine.**

Solution

• The sugar contains no OH at C2′, so it is derived from deoxyribose.
• The base is thymine.
• To name the deoxyribonucleoside, change the suffix of the base to *-idine* and add the prefix *deoxy;* thus, thymine → deoxythymidine.

Practice Problem 17.1

Identify the base and monosaccharide used to form the following nucleosides, and then assign names.

Practice Problem 17.2

Draw the structure of guanosine. Classify the compound as a ribonucleoside or a deoxyribonucleoside.

More Practice: Try Problems 17.15; 17.16; 17.17a; 17.18a, b.

17.1B Nucleotides—Joining a Nucleoside with a Phosphate

Nucleotides are formed by adding a phosphate group to the 5′-OH of a nucleoside. Nucleotides are named by adding the term *5′-monophosphate* to the name of the nucleoside from which they are derived. **Ribonucleotides** are derived from ribose, whereas **deoxyribonucleotides** are derived from 2-deoxyribose.

cytidine
a ribonucleoside

cytidine 5′-monophosphate
a ribonucleotide

CMP

deoxyadenosine
a deoxyribonucleoside

deoxyadenosine 5′-monophosphate
a deoxyribonucleotide

dAMP

Because of the lengthy names of nucleotides, three- or four-letter abbreviations are commonly used instead. Thus, **c**yti**d**ine 5′-**m**ono**p**hosphate is **CMP** and **d**eoxy**a**denosine 5′-**m**ono**p**hosphate is **dAMP.**

Table 17.1 summarizes the information about nucleic acids and their components learned thus far. Table 17.2 summarizes the names and abbreviations used for the bases, nucleosides, and nucleotides needed in nucleic acid chemistry.

Table 17.1 Summary of the Components of Nucleosides, Nucleotides, and Nucleic Acids

Type of Compound	Components
Nucleoside	• A monosaccharide + a base • A **ribo**nucleoside contains the monosaccharide **ribose.** • A **deoxy**ribonucleoside contains the monosaccharide 2-**deoxy**ribose.
Nucleotide	• A nucleoside + phosphate = a monosaccharide + a base + phosphate • A **ribo**nucleotide contains the monosaccharide **ribose.** • A **deoxy**ribonucleotide contains the monosaccharide 2-**deoxy**ribose.
DNA	• A polymer of **deoxy**ribonucleotides • The monosaccharide is 2-**deoxy**ribose. • The bases are **A, G, C,** and **T.**
RNA	• A polymer of **ribo**nucleotides • The monosaccharide is **ribose.** • The bases are **A, G, C,** and **U.**

Table 17.2 Names of Bases, Nucleosides, and Nucleotides in Nucleic Acids

Base	Abbreviation	Nucleoside	Nucleotide	Abbreviation
DNA				
Adenine	A	Deoxyadenosine	Deoxyadenosine 5'-monophosphate	dAMP
Guanine	G	Deoxyguanosine	Deoxyguanosine 5'-monophosphate	dGMP
Cytosine	C	Deoxycytidine	Deoxycytidine 5'-monophosphate	dCMP
Thymine	T	Deoxythymidine	Deoxythymidine 5'-monophosphate	dTMP
RNA				
Adenine	A	Adenosine	Adenosine 5'-monophosphate	AMP
Guanine	G	Guanosine	Guanosine 5'-monophosphate	GMP
Cytosine	C	Cytidine	Cytidine 5'-monophosphate	CMP
Uracil	U	Uridine	Uridine 5'-monophosphate	UMP

Concept Check 17.2

Which nucleic acid (DNA or RNA) contains each of the following components?

a. the sugar ribose   c. the base T   e. the nucleotide GMP

b. the sugar deoxyribose   d. the base U   f. the nucleotide dCMP

Concept Check 17.3

Identify each component as a base, nucleoside, or nucleotide.

a. adenine   b. cytidine   c. uridine 5'-monophosphate   d. deoxythymidine

Di- and triphosphates can also be prepared from nucleosides by adding two and three phosphate groups, respectively, to the 5'-OH. For example, adenosine can be converted to adenosine 5'-diphosphate and adenosine 5'-triphosphate, abbreviated as ADP and ATP, respectively. We will learn about the central role of these phosphates, especially ATP, in energy production in Chapter 18.

adenosine 5'-diphosphate
ADP

adenosine 5'-triphosphate
ATP

Sample Problem 17.2

Naming a Nucleotide

Identify the base and monosaccharide in the following nucleotide. Name the nucleotide and give its three- or four-letter abbreviation.

Analysis

- Identify the sugar and base as in Sample Problem 17.1.
- Name the nucleotide by naming the nucleoside and adding the term *5'-monophosphate,* because the molecule contains only one phosphorus bonded to the 5'-OH.
- Use the abbreviations from Table 17.2. When the sugar portion has an OH group at C2', a three-letter abbreviation is used. When the sugar portion has no OH group at C2', the four-letter abbreviation that begins with "d" is used.

Solution

The sugar portion is derived from ribose and the base is guanine.

guanosine 5'-monophosphate
GMP

Practice Problem 17.3

Identify the base and monosaccharide in each nucleotide. Name the nucleotide and give its three- or four-letter abbreviation.

Practice Problem 17.4

Give the name that corresponds to each abbreviation: (a) GTP; (b) dCDP; (c) dTTP; (d) UDP.

Practice Problem 17.5

Label each statement about the compound deoxycytidine as true or false.

a. Deoxycytidine is a nucleotide.

b. Deoxycytidine is a nucleoside.

c. Deoxycytidine contains a phosphate at its 5'-OH group.

d. Deoxycytidine contains a pyrimidine base.

More Practice: Try Problems 17.3, 17.4, 17.18c, 17.19, 17.20.

---

phosphodiester

17.2   Nucleic Acids

Nucleic acids—both DNA and RNA—are polymers of nucleotides, formed by joining the 3'-OH group of one nucleotide with the 5'-phosphate of a second nucleotide in a **phosphodiester** linkage (Section 15.6).

For example, joining the 3'-OH group of dCMP (deoxycytidine 5'-monophosphate) and the 5'-phosphate of dAMP (deoxyadenosine 5'-monophosphate) forms a dinucleotide that contains a 5'-phosphate on one end (called the **5' end**) and a 3'-OH group on the other end (called the **3' end**).

As additional nucleotides are added, the nucleic acid grows, each time forming a new phosphodiester linkage that holds the nucleotides together. Figure 17.1 illustrates the structure of a polynucleotide formed from four different nucleotides. Several features are noteworthy.

- A polynucleotide contains a backbone consisting of alternating sugar and phosphate groups. All polynucleotides contain the same sugar–phosphate backbone.
- A polynucleotide has one free phosphate group at the 5' end.
- A polynucleotide has a free OH group at the 3' end.

Figure 17.1   Primary Structure of a Polynucleotide

The name of a polynucleotide is read from the 5' end to the 3' end, using the one-letter abbreviations for the bases it contains. Drawn is the structure of the polynucleotide CATG.

The **primary structure** of a polynucleotide is the sequence of nucleotides that it contains. This sequence, which is determined by the identity of the bases, is unique to a nucleic acid. **In DNA, the sequence of bases carries the genetic information of the organism.**

Polynucleotides are named by the sequence of the bases they contain, beginning at the 5' end and using the one-letter abbreviation for the bases. Thus, the polynucleotide in Figure 17.1 contains the bases cytosine, adenine, thymine, and guanine, in order from the 5' end; thus, it is named CATG.

Sample Problem 17.3	Drawing and Naming a Dinucleotide

(a) Draw the structure of a dinucleotide formed by joining the 3'-OH group of AMP to the 5'-phosphate in GMP. (b) Label the 5' and 3' ends. (c) Name the dinucleotide.

Analysis

Draw the structure of each nucleotide. In this case the sugar is ribose because the names of the mononucleotides do not contain the prefix *deoxy*. **Bond the 3'-OH group to the 5'-phosphate to form the phosphodiester bond.** The name of the dinucleotide begins with the nucleotide that contains the free phosphate at the 5' end.

Solution

a. and b.

c. Polynucleotides are named beginning at the 5' end, so this dinucleotide is named AG.

Practice Problem 17.6	Draw the structure of a dinucleotide formed by joining the 3'-OH group of dTMP to the 5'-phosphate in dGMP.

Practice Problem 17.7	Label the 5' end and the 3' end in each polynucleotide: (a) ATTTG; (b) CGCGUU; (c) GGACTT.

Practice Problem 17.8	Label each statement about the polynucleotide ATGGCG as true or false.

   a. The polynucleotide has six nucleotides.

   b. The polynucleotide contains six phosphodiester linkages.

   c. The nucleotide at the 5' end contains the base guanine.

   d. The nucleotide at the 3' end contains the base guanine.

   e. The polynucleotide could be part of a DNA molecule.

   f. The polynucleotide could be part of an RNA molecule.

More Practice: Try Problems 17.5, 17.6, 17.21, 17.22.

17.3   The DNA Double Helix

Our current understanding of the structure of DNA is based on the model proposed initially by James Watson and Francis Crick in 1953 (Figure 17.2).

> • DNA consists of two polynucleotide strands that wind into a right-handed double helix.

The sugar–phosphate backbone lies on the outside of the helix and the bases lie on the inside, perpendicular to the axis of the helix. The two strands of DNA run in *opposite* directions; that is, one strand runs from the 5' end to the 3' end, while the other runs from the 3' end to the 5' end.

Figure 17.2   The Three-Dimensional Structure of DNA—A Double Helix

DNA consists of a double helix of polynucleotide chains. In view (a), the three-dimensional molecular model shows the sugar–phosphate backbone with the red (O), black (C), and white (H) atoms visible on the outside of the helix. In view (b), the bases on the interior of the helix are labeled. *(a): Molekuul/ SPL/age fotostock*

The double helix is stabilized by hydrogen bonding between the bases of the two DNA strands as shown in Figure 17.3. A purine base on one strand always hydrogen bonds with a pyrimidine base on the other strand. Two bases hydrogen bond together in a predictable manner, forming **complementary base pairs.**

> • Adenine pairs with thymine using two hydrogen bonds, forming an A–T base pair.
> • Cytosine pairs with guanine using three hydrogen bonds, forming a C–G base pair.

Figure 17.3 Hydrogen Bonding in the DNA Double Helix

hydrogen bonding between base pairs

Hydrogen bonding of base pairs (**A–T** and **C–G**) holds the two strands of DNA together.

Because of this consistent pairing of bases, knowing the sequence of one strand of DNA allows us to write the sequence of the other strand, as shown in Sample Problem 17.4.

Sample Problem 17.4	Writing the Sequence of the Complementary Strand of a DNA Molecule

Write the sequence of the complementary strand of the following portion of a DNA molecule: 5'–TAGGCTA–3'.

Analysis

The complementary strand runs in the opposite direction, from the 3' to the 5' end. Use base pairing to determine the corresponding sequence on the complementary strand: **A pairs with T** and **C pairs with G.**

Solution

Original strand: 5'–T A G G C T A–3'

Complementary strand: 3'–A T C C G A T–5'

Practice Problem 17.9

Write the complementary strand for each of the following strands of DNA.

a. 5'–AAACGTCC–3'   c. 5'–ATTGCACCCGC–3'

b. 5'–TATACGCC–3'   d. 5'–CACTTGATCGG–3'

More Practice: Try Problems 17.25, 17.26.

Identical twins have the same genetic makeup, so that characteristics determined by DNA—such as hair color, eye color, or complexion—are also identical. *Daniel C. Smith*

The enormously large DNA molecules that compose the **human genome**—the total DNA content of an individual—pack tightly into the nucleus of the cell. The double-stranded DNA helices wind around a core of protein molecules called histones to form a chain of nucleosomes, as shown in Figure 17.4. The chain of nucleosomes winds into a supercoiled fiber called chromatin, which composes each of the 23 pairs of chromosomes in humans.

Figure 17.4 The Structure of a Chromosome

The **genetic information of an organism is stored in the sequence of bases of its DNA molecules.** How is this information transferred from one generation to another? How, too, is the information stored in DNA molecules used to direct the synthesis of proteins?

To answer these questions we must understand three key processes.

- *Replication* is the process by which DNA makes a copy of itself when a cell divides.
- *Transcription* is the ordered synthesis of RNA from DNA. In this process, the genetic information stored in DNA is passed onto RNA.
- *Translation* is the synthesis of proteins from RNA. In this process, the genetic message contained in RNA determines the specific amino acid sequence of a protein.

Each chromosome contains many **genes,** those portions of the DNA molecules that result in the synthesis of specific proteins. Only a small fraction (1–2%) of the DNA in a chromosome contains genetic messages or genes that result in protein synthesis.

17.4 Replication

How is the genetic information in the DNA of a parent cell passed onto new daughter cells during replication?

During replication, the strands of DNA separate and each serves as a template for a new strand. Thus, **the original DNA molecule forms two DNA molecules, each of which contains one strand from the parent DNA and one new strand.** The sequence of both strands of the daughter DNA molecules exactly matches the sequence in the parent DNA.

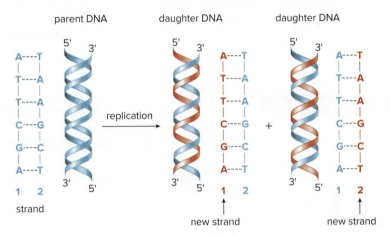

The first step in replication is the unwinding of the DNA helix to expose the bases on each strand. Unwinding breaks the hydrogen bonds that hold the two strands of the double helix together. Once bases have been exposed on the unwound strands of DNA, the enzyme DNA polymerase catalyzes the replication process using the four nucleoside triphosphates (derived from the bases A, T, G, and C) that are available in the nucleus (Figure 17.5).

- The identity of the bases on the template strand determines the order of the bases on the new strand: A must pair with T, and G must pair with C.
- Replication occurs in only one direction on the template strand, from the 3' end to the 5' end.

Because replication proceeds in only one direction—that is, from the 3' end to the 5' end of the template—the two new strands of DNA must be synthesized by somewhat different techniques. One strand, called the **leading strand,** grows continuously. The other strand, called the **lagging strand,** is synthesized in small fragments, which are then joined together by a ligase enzyme. The end result is two new strands of DNA, one in each of the daughter DNA molecules, both with complementary base pairs joining the two DNA strands together.

Figure 17.5

DNA Replication

Replication proceeds along both strands of unwound DNA. The leading strand grows continuously, while the lagging strand must be synthesized in fragments that are joined together by a DNA ligase enzyme.

Sample Problem 17.5	**Determining the Sequence of a Newly Synthesized Strand of DNA**

What is the sequence of a newly synthesized DNA segment if the template strand has the sequence 3'–TGCACC–5'?

Analysis

The newly synthesized strand runs in the opposite direction, from the 5' end to the 3' end in this example. Use base pairing to determine the corresponding sequence on the new strand: **A pairs with T** and **C pairs with G**.

Solution

Template strand: 3'–T G C A C C–5'

New strand: 5'–A C G T G G–3'

Practice Problem 17.10	

What is the sequence of a newly synthesized DNA segment if the template strand has each of the following sequences?

   a. 3'–AGAGTCTC–5'   b. 5'–ATTGCTC–3'   c. 3'–ATCCTGTAC–5'   d. 5'–GGCCATACTC–3'

More Practice: Try Problems 17.31, 17.32.

17.5   RNA

While RNA is also composed of nucleotides, there are important differences between DNA and RNA. In RNA,

- The sugar is ribose.
- U (uracil) replaces T (thymine) as one of the bases.
- RNA is single stranded.

RNA molecules are much smaller than DNA molecules. Although RNA contains a single strand, the chain can fold back on itself, forming loops, and intermolecular hydrogen bonding between paired bases on a single strand can form helical regions. When base pairing occurs within an RNA molecule (or between RNA and DNA), **C and G form base pairs,** and **A and U form base pairs.**

There are three different types of RNA molecules.

- Ribosomal RNA (rRNA)
- Messenger RNA (mRNA)
- Transfer RNA (tRNA)

Ribosomal RNA, the most abundant type of RNA, is found in the ribosomes in the cytoplasm of the cell. Each ribosome is composed of one large subunit and one small subunit that contain both RNA and protein. rRNA provides the site where polypeptides are assembled during protein synthesis.

Messenger RNA is the carrier of information from DNA (in the cell nucleus) to the ribosomes (in the cytoplasm). Each gene of a DNA molecule corresponds to a specific mRNA molecule. The sequence of nucleotides in the mRNA molecule determines the amino acid sequence in a particular protein.

Transfer RNA, the smallest type of RNA, interprets the genetic information in mRNA and brings specific amino acids to the site of protein synthesis in the ribosome. Each amino acid is recognized by one or more tRNA molecules, which contain 70–90 nucleotides. tRNAs have two important sites. The 3' end, called the **acceptor stem,** always contains the nucleotides ACC

Figure 17.6   Transfer RNA

Each tRNA binds a specific amino acid to its 3' end and contains an anticodon that identifies that amino acid for protein synthesis.

and has a free OH group that binds a specific amino acid. Each tRNA also contains a sequence of three nucleotides called an **anticodon,** which is complementary to three bases in an mRNA molecule, and identifies what amino acid must be added to a growing polypeptide chain.

tRNA molecules are often drawn in the cloverleaf fashion shown in Figure 17.6. The acceptor stem and anticodon region are labeled. Folding creates regions of the tRNA in which nearby complementary bases hydrogen bond to each other.

Table 17.3 summarizes the characteristics of the three types of RNAs.

Table 17.3 Three Types of RNA Molecules

Type of RNA	Abbreviation	Function
Ribosomal RNA	rRNA	The site of protein synthesis, found in the ribosomes
Messenger RNA	mRNA	Carries the information from DNA to the ribosomes
Transfer RNA	tRNA	Brings specific amino acids to the ribosomes for protein synthesis

17.6   Transcription

The conversion of the information in DNA to the synthesis of proteins begins with *transcription—* **that is, the synthesis of messenger RNA from DNA.**

RNA synthesis begins in the same manner as DNA replication: the double helix of DNA unwinds (Figure 17.7). Because RNA is single stranded, however, only one strand of DNA is needed for RNA synthesis. The **template** strand is the strand of DNA used for RNA synthesis.

Figure 17.7

Transcription

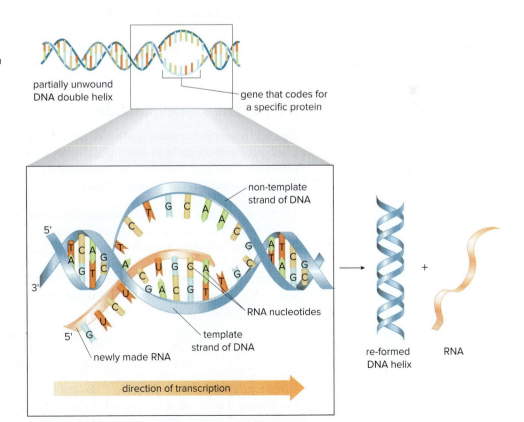

In transcription, the DNA helix unwinds and RNA polymerase catalyzes the formation of mRNA along the DNA template strand. Transcription forms an mRNA molecule with base pairs complementary to the DNA template strand.

Each mRNA molecule corresponds to a small segment of a DNA molecule. Transcription proceeds from the 3' end to the 5' end of the template strand. Complementary base pairing determines what RNA nucleotides are added to the growing RNA chain: **C pairs with G, T pairs with A,** and **A pairs with U.** Thus, the RNA chain grows from the 5' to 3' direction. When transcription is completed, the new mRNA molecule is released and the double helix of the DNA molecule re-forms.

- Transcription forms a messenger RNA molecule with a sequence that is *complementary* to the DNA template from which it is prepared.

Sample Problem 17.6	Determining the Sequence of an mRNA Molecule

If a portion of the template strand of a DNA molecule has the sequence 3'–CTAGGATAC–5', what is the sequence of the mRNA molecule produced from this template?

Analysis

mRNA has a base sequence that is complementary to the template from which it is prepared.

Solution

Template strand of DNA: 3'–C T A G G A T A C–5'
� complementary
mRNA sequence: 5'–G A U C C U A U G–3'

Practice Problem 17.11	What is the sequence of the mRNA molecule synthesized from each DNA template?

   a. 3'–TGCCTAACG–5'   b. 3'–GACTCC–5'   c. 3'–TTAACGCGA–5'   d. 3'–CAGTGACCGTAC–5'

Practice Problem 17.12	What is the sequence of the DNA template strand from which each of the following mRNA strands was synthesized?

   a. 5'–UGGGGCAUU–3'   b. 5'–GUACCU–3'   c. 5'–CCGACGAUG–3'   d. 5'–GUAGUCACG–3'

More Practice: Try Problems 17.35, 17.36.

17.7 The Genetic Code

Once the genetic information of DNA has been transcribed in a messenger RNA molecule, RNA can direct the synthesis of an individual protein. How can RNA, which is composed of only four different nucleotides, direct the synthesis of polypeptides that are formed from 20 different amino acids? The answer lies in the **genetic code.**

- The genetic code is the set of three-nucleotide units (a triplet) in mRNA called codons that correspond to particular amino acids. As a result, a series of codons in mRNA determines the amino acid sequence in a protein.

For example, the codon UAC in an mRNA molecule codes for the amino acid serine, and the codon UGC codes for the amino acid cysteine. The same genetic code occurs in almost all organisms, from bacteria to whales to humans.

Given four different nucleotides (A, C, G, and U), there are 64 different ways to combine them into groups of three, so there are 64 different codons. Sixty-one codons code for specific amino acids, so many amino acids correspond to more than one codon, as shown in Table 17.4. For example, the codons GGU, GGC, GGA, and GGG all code for the amino acid glycine. Three codons—UAA, UAG, and UGA—do not correspond to any amino acids; they are called **stop codons** because they signal the termination of protein synthesis.

Table 17.4 The Genetic Code—Triplets in Messenger RNA

First Base (5' end)	Second Base								Third Base (3' end)
	U		C		A		G		
U	UUU	Phe	UCU	Ser	UAU	Tyr	UGU	Cys	U
	UUC	Phe	UCC	Ser	UAC	Tyr	UGC	Cys	C
	UUA	Leu	UCA	Ser	UAA	Stop	UGA	Stop	A
	UUG	Leu	UCG	Ser	UAG	Stop	UGG	Trp	G
C	CUU	Leu	CCU	Pro	CAU	His	CGU	Arg	U
	CUC	Leu	CCC	Pro	CAC	His	CGC	Arg	C
	CUA	Leu	CCA	Pro	CAA	Gln	CGA	Arg	A
	CUG	Leu	CCG	Pro	CAG	Gln	CGG	Arg	G
A	AUU	Ile	ACU	Thr	AAU	Asn	AGU	Ser	U
	AUC	Ile	ACC	Thr	AAC	Asn	AGC	Ser	C
	AUA	Ile	ACA	Thr	AAA	Lys	AGA	Arg	A
	AUG	Met	ACG	Thr	AAG	Lys	AGG	Arg	G
G	GUU	Val	GCU	Ala	GAU	Asp	GGU	Gly	U
	GUC	Val	GCC	Ala	GAC	Asp	GGC	Gly	C
	GUA	Val	GCA	Ala	GAA	Glu	GGA	Gly	A
	GUG	Val	GCG	Ala	GAG	Glu	GGG	Gly	G

Concept Check 17.4

What amino acid is coded for by each codon?

a. GCC   b. AAU   c. CUA   d. AGC   e. CAA   f. AAA

Concept Check 17.5

What codons code for each amino acid?

a. glycine   b. isoleucine   c. lysine   d. glutamic acid

Codons are written so that reading from left to right, the first triplet codes for the N-terminal amino acid in a protein, and the last triplet codes for the C-terminal amino acid. Sample Problem 17.7 illustrates the conversion of a sequence of bases in mRNA to a sequence of amino acids in a peptide.

Sample Problem 17.7

Deriving an Amino Acid Sequence from mRNA

Derive the amino acid sequence that is coded for by the following mRNA sequence.

CAU AAA ACG GUG UUA AUA

Analysis

Use Table 17.4 to identify the codons that correspond to each amino acid. Codons correspond to a peptide written from the N-terminal to C-terminal end.

Solution

Practice Problem 17.13

Derive the amino acid sequence that is coded for by each mRNA sequence.

   a. CAA GAG GUA UCC UAC AGA

   b. GUC AUC UGG AGG GGC AUU

   c. CUA UGC AGU AGG ACA CCC

Practice Problem 17.14

Write a possible mRNA sequence that codes for each of the following peptides.

   a. Met–Arg–His–Phe   b. Gly–Ala–Glu–Gln   c. Gln–Asn–Gly–Ile–Val   d. Thr–His–Asp–Cys–Trp

Practice Problem 17.15

Considering the given sequence of nucleotides in an mRNA molecule, (a) what is the sequence of the DNA template strand from which the RNA was synthesized? (b) What peptide is synthesized by this mRNA sequence?

GAG CCC GUA UAC GCC ACG

More Practice: Try Problems 17.39–17.42, 17.45a, 17.46a.

17.8 Translation and Protein Synthesis

The translation of the information in messenger RNA to protein synthesis occurs in the ribosomes. Each type of RNA plays a role in protein synthesis.

- mRNA contains the sequence of codons that determines the order of amino acids in the protein.
- Individual tRNAs bring specific amino acids to add to the peptide chain.
- rRNA contains binding sites that provide the platform on which protein synthesis occurs.

Each individual tRNA contains an **anticodon** of three nucleotides that is complementary to the codon in mRNA and identifies individual amino acids (Section 17.5). For example, a codon of UCA in mRNA corresponds to an anticodon of AGU in a tRNA molecule, which identifies serine as the amino acid. Other examples are shown in Table 17.5.

Concept Check 17.6

For each codon: [1] Write the anticodon. [2] What amino acid does each codon represent?

   a. CGG   b. GGG   c. UCC   d. AUA   e. CCU   f. GCC

Table 17.5 Relating Codons, Anticodons, and Amino Acids

mRNA Codon		tRNA Anticodon		Amino Acid
ACA	⟶	UGU	⟶	threonine
GCG	⟶	CGC	⟶	alanine
AGA	⟶	UCU	⟶	arginine
UCC	⟶	AGG	⟶	serine

There are three stages in translation: initiation, elongation, and termination. Figure 17.8 depicts the main features of translation.

Figure 17.8

Translation—The Synthesis of Proteins from RNA

- **Initiation** consists of the binding of the ribosomal subunits to mRNA and the arrival of the first tRNA carrying its amino acid.
- The protein is synthesized during **elongation.** One by one a tRNA with its designated amino acid binds to a site on the ribosome adjacent to the first tRNA. A peptide bond forms and a tRNA is released. The ribosome shifts to the next codon and the process continues.
- **Termination** occurs when a stop codon is reached. The synthesis is complete and the protein is released from the complex.

[1] Initiation

Translation begins when an mRNA molecule binds to the smaller subunit of the ribosome and a tRNA molecule carries the first amino acid of the peptide chain to the binding site. Translation always begins at the codon AUG, which codes for the amino acid methionine. The arriving tRNA contains an anticodon with the complementary base sequence UAC.

[2] Elongation

The next tRNA molecule containing an anticodon for the second codon binds to mRNA, delivering its amino acid, and a peptide bond forms between the two amino acids. The first tRNA molecule, which has delivered its amino acid and is no longer needed, dissociates from the complex. The ribosome shifts to the next codon along the mRNA strand and the process continues when a new tRNA molecule binds to the mRNA.

[3] Termination

Translation continues until a stop codon is reached. There is no tRNA that contains an anticodon complementary to any of the three stop codons (UAA, UAG, and UGA), so protein synthesis ends and the protein is released from the ribosome. Often the first amino acid in the chain, methionine, is not needed in the final protein and so it is removed after protein synthesis is complete.

Figure 17.9 shows a representative segment of DNA, and the mRNA, tRNA, and amino acid sequences that correspond to it.

Figure 17.9 Comparing the Sequence of DNA, mRNA, tRNA, and a Polypeptide

DNA template strand ⟶ 3' end	TAC	AAC	CCT	CGG	CCT	AGT	5' end
mRNA ⟶ 5' end	AUG	UUG	GGA	GCC	GGA	UCA	3' end
tRNA anticodons ⟶	UAC	AAC	CCU	CGG	CCU	AGU	
Polypeptide ⟶	Met	Leu	Gly	Ala	Gly	Ser	

Sample Problem 17.8

Using mRNA to Determine an Amino Acid Sequence and the Identity of tRNA Anticodons

What sequence of amino acids would be formed from the following mRNA sequence: CAA AAG ACG UAC CGA? List the anticodons contained in each of the needed tRNA molecules.

Analysis

Use Table 17.4 to determine the amino acid that is coded for by each codon. The anticodons contain complementary bases to the codons: **A pairs with U, and C pairs with G.**

Solution

Amino acid sequence: Gln Lys Thr Tyr Arg

mRNA codons: 5' CAA AAG ACG UAC CGA 3'

tRNA anticodons: GUU UUC UGC AUG GCU

Practice Problem 17.16

What sequence of amino acids would be formed from each mRNA sequence? List the anticodons contained in each of the needed tRNA molecules.

a. CCA CCG GCA AAC GAA GCA   b. GCA CCA CUA AGA GAC

More Practice: Try Problems 17.7, 17.8, 17.37–17.40.

Sample Problem 17.9

Deriving an Amino Acid Sequence from DNA

What polypeptide would be synthesized from the following template strand of DNA: CGG TGT CTT TTA?

Analysis

To determine what polypeptide is synthesized from a DNA template, two steps are needed.

- Use the DNA sequence to determine the transcribed mRNA sequence: **C pairs with G, T pairs with A, and A (on DNA) pairs with U (on mRNA).**
- Use the codons in Table 17.4 to determine what amino acids are coded for by a given codon in mRNA.

Solution

DNA template strand:	⟶	3'	CGG TGT CTT TTA	5'	
mRNA:	⟶	5'	GCC ACA GAA AAU	3'	
Polypeptide:	⟶		Ala — Thr — Glu — Asn		

Practice Problem 17.17

What polypeptide would be synthesized from each of the following template strands of DNA?

a. TCT CAT CGT AAT GAT TCG   b. GCT CCT AAA TAA CAC TTA

More Practice: Try Problems 17.9, 17.10, 17.43, 17.44, 17.55, 17.56, 17.59, 17.60.

17.9 Mutations and Genetic Diseases

Although replication provides a highly reliable mechanism for making an exact copy of DNA, occasionally an error occurs, thus producing a DNA molecule with a slightly different nucleotide sequence.

- A *mutation* is a change in the nucleotide sequence in a molecule of DNA.

If the mutation occurs in a nonreproductive cell, the mutation is passed on to daughter cells within the organism, but is not transmitted to the next generation. If the mutation occurs in an egg or sperm cell, it is passed on to the next generation of an organism. Some mutations are random events, whereas others are caused by **mutagens,** chemical substances that alter the structure of DNA. Exposure to high-energy radiation such as X-rays or ultraviolet light can also produce mutations.

Mutations can be classified according to the change that results in a DNA molecule.

- A *point* mutation is the *substitution* of one nucleotide for another.

Original DNA:	〜〜〜 G A **G** T T C 〜〜〜
	replacement of **G** by **C**
Point mutation:	〜〜〜 G A **C** T T C 〜〜〜

- A *deletion* mutation occurs when one or more nucleotides is *lost* from a DNA molecule.

Original DNA:	〜〜〜 G A **G** T T C 〜〜〜
	loss of **G**
Deletion mutation:	〜〜〜 G A T T C 〜〜〜

- An *insertion* mutation occurs when one or more nucleotides is *added* to a DNA molecule.

Original DNA:	〜〜〜 G A G T T C 〜〜〜
	addition of **C**
Insertion mutation:	〜〜〜 G A G **C** T T C 〜〜〜

A mutation can have a negligible, minimal, or catastrophic effect on an organism. To understand the effect of a mutation, we must determine the mRNA sequence that is transcribed from the DNA sequence as well as the resulting amino acid for which it codes. A point mutation in the three-base sequence CTT in a gene that codes for a particular protein illustrates some possible outcomes of a mutation.

The sequence CTT in DNA is transcribed to the codon GAA in mRNA, and using Table 17.4, this triplet codes for the amino acid glutamic acid. If a point mutation replaces CTT by CTC in DNA, CTC is transcribed to the codon GAG in mRNA. Because GAG codes for the *same* amino acid—glutamic acid—this mutation does not affect the protein synthesized by this segment of DNA. Such a mutation is said to be **silent.**

Alternatively, suppose a point mutation replaces CTT by CAT in DNA. CAT is transcribed to the codon GUA in mRNA, and GUA codes for the amino acid valine. Now, the mutation produces a protein with one *different* amino acid—namely, valine instead of glutamic acid. In some proteins this alteration of the primary sequence may have little effect on the protein's secondary and tertiary structure. In other proteins, such as hemoglobin, the substitution of valine for glutamic acid produces a protein with vastly different properties, resulting in the fatal disease sickle cell anemia (Section 16.7).

Finally, suppose a point mutation replaces CTT by ATT in DNA. ATT is transcribed to the codon UAA in mRNA, and UAA is a stop codon. This terminates protein synthesis and no more amino acids are added to the protein chain. In this case, a needed protein is not synthesized and depending on the protein's role, the organism may die.

Sample Problem 17.10	**Determining the Result of a Point Mutation**

(a) What dipeptide is produced from the following segment of DNA: AGAGAT? (b) What happens to the dipeptide when a point mutation occurs and the DNA segment contains the sequence ATAGAT instead?

Analysis

Transcribe the DNA sequence to an mRNA sequence with complementary base pairs. Then use Table 17.4 to determine what amino acids are coded for by each codon.

Solution

a. Because UCU codes for serine and CUA codes for leucine, the dipeptide Ser–Leu results.

AGA GAT ⟶ UCU CUA ⟶ Ser — Leu

DNA   mRNA   dipeptide

b. Because UAU codes for tyrosine, the point mutation results in the synthesis of the dipeptide Tyr–Leu.

ATA GAT ⟶ UAU CUA ⟶ Tyr — Leu

DNA   mRNA   dipeptide

Practice Problem 17.18	Consider the following sequence of DNA: AACTGA. (a) What dipeptide is formed from this DNA after transcription and translation? (b) How is the amino acid sequence affected when point mutations produce each of the following DNA sequences: [1] AACGGA; [2] ATCTGA; [3] AATTGA?

More Practice: Try Problems 17.45–17.48.

A patient with cystic fibrosis must regularly receive chest physiotherapy, a procedure that involves pounding on the chest to dislodge thick mucus clogging the lungs. *BSIP/Newscom*

When a mutation causes a protein deficiency or results in a defective protein and the condition is inherited from one generation to another, a **genetic disease** results. For example, cystic fibrosis, the most common genetic disease in Caucasians, is caused by a mutation resulting in the synthesis of a defective protein—cystic fibrosis transmembrane conductance regulator (CFTR)—needed for proper passage of ions across cell membranes. Individuals with cystic fibrosis have decreased secretions of pancreatic enzymes, resulting in a failure to thrive (poor growth), and they produce thick sticky mucus in the lungs that leads to devastating lung infections and a shortened life span.

17.10 Modern DNA Technology

Modern advances in DNA technology have yielded impressive results in the last 20 years.

17.10A FOCUS ON THE HUMAN BODY
DNA Fingerprinting

Because the DNA of each individual is unique, **DNA fingerprinting** (Figure 17.10) is now routinely used as a method of identification.

Figure 17.10

DNA Fingerprinting in Forensic Analysis

Each vertical lane (numbered 1–8) corresponds to a DNA sample.

- Lanes 1 and 7 are called DNA ladders. They correspond to DNA fragments of known size and are used to show the approximate size of the DNA fragments of the fingerprints.
- Lane 2 is the DNA obtained from a female assault victim and lane 5 is the female DNA from the crime scene. The two lanes match, indicating that they are from the same individual.
- Two individuals (**A** and **B**), whose DNA appears in lanes 3 and 4, were considered suspects.
- The male DNA obtained at the crime scene is shown in lane 6. The horizontal bands correspond to those of suspect **A**, incriminating him and eliminating individual **B** as a suspect.

Each band corresponds to a fragment of DNA.

Decreasing size of DNA fragments

Almost any type of cell—skin, saliva, semen, blood, and so forth—can be used to obtain a DNA fingerprint. The DNA is cut into fragments with various enzymes and the fragments are separated by size using a technique called gel electrophoresis. DNA fragments can be visualized after they react with a radioactive probe. The result is an image consisting of a set of horizontal bands, each band corresponding to a segment of DNA, sorted from low to high molecular weight.

To compare the DNA of different individuals, samples are placed next to each other on the same gel and the position of the horizontal bands compared. DNA fingerprinting is now routinely used in criminal cases to establish the guilt or innocence of a suspect. Only identical twins have identical DNA, but related individuals have several similar DNA fragments. Thus, DNA fingerprinting can be used to establish paternity by comparing the DNA of a child with that of each parent. DNA is also used to identify a body when no other means of identification is possible. DNA analysis was instrumental in identifying human remains found in the rubble of the World Trade Center after the towers collapsed on September 11, 2001.

HEALTH NOTE

17.10B The Human Genome Project

Modern advances in DNA technology have resulted in the sequencing of the entire human genome.

The **Human Genome Project,** launched in 1990 and completed in 2003, was an international undertaking to identify and sequence all three billion nucleotide base pairs in the human genome. This project has yielded much vital information. We now know, for example, that only 2% of the base pairs constitute the 20,000 genes of the human genome, a much smaller fraction than originally assumed. A multinational research effort has been in place for over a decade to determine the role of the "junk" DNA—DNA that does not encode for a protein. Results have shown that this DNA is responsible for a variety of necessary functions in the cell.

Many medical advances have also resulted from this project. The genetic basis of thousands of genetic diseases is now known, and in the United States, all newborns are screened for numerous genetic disorders by a simple blood test shortly after birth. When genetic disorders are discovered within a family, parents can undergo genetic testing and learn how likely it is that their offspring will inherit an abnormality.

Newborn screening can detect life-threatening disorders that are treatable if detected early. *Ted Horowitz/Getty Images*

An enormous amount of DNA sequence information is now available, and as technology has improved, thousands of genomes in the plant and animal world have been sequenced. A genome that might have taken years to sequence in 1990 can now be determined in a matter of days. The cost of a partial genetic screen is no longer prohibitive, so individuals can learn more about their ancestry or their risk of developing certain genetic diseases by a blood test or saliva swab. Using the sophisticated technology available today, as well as the competitive pricing offered by several commercial enterprises, the National Human Genome Research Institute estimated that in 2016, the DNA sequence of an organism could be obtained for under $1000.

17.11 FOCUS ON HEALTH & MEDICINE
Viruses

A *virus* **is an infectious agent consisting of a DNA or RNA molecule that is contained within a protein coating.** A virus contains no enzymes or free nucleotides of its own, so it is incapable of replicating. When it invades a host organism, however, it takes over the biochemical machinery of the host.

A virus that contains DNA uses the materials in the host organism to replicate its DNA, transcribe DNA to RNA, and synthesize a protein coating, thus forming new virus particles. These new virus particles leave the host cell and infect new cells and the process continues. Many prevalent diseases, including the common cold, influenza, and herpes are viral in origin.

Widely used vaccines contain an inactive form of a virus that causes an individual's immune system to produce antibodies to the virus to ward off infection. Many childhood diseases that were once very common, including mumps, measles, and chickenpox, are now prevented by vaccination. Polio has been almost completely eradicated, even in remote areas worldwide, by vaccination.

17.11A Coronaviruses

Coronaviruses are a group of viruses that cause a variety of diseases that include the common cold, SARS (severe acute respiratory syndrome), and MERS (Middle East respiratory syndrome). At the end of 2019, a novel coronavirus, **SARS-CoV-2,** was first reported to cause an unusual viral pneumonia known as **COVID-19,** coronavirus disease 2019. Because the disease is highly transmissible and can be spread by an individual before symptoms are evident, SARS-CoV-2 has led to a global pandemic, resulting in the infection and death of millions worldwide.

SARS-CoV-2 is a **retrovirus,** a virus that contains a core of RNA along with proteins that enable it to replicate. Surrounding the inner core is a viral envelope composed of a lipid bilayer, and projecting from the viral core are club-like spike proteins that allow the virus to attach to and thus infect cells of the host (Figure 17.11).

Once a retrovirus invades a host organism, it must first make DNA by the process of **reverse transcription** (Figure 17.12). After viral DNA has been synthesized, the DNA can transcribe RNA, which can direct protein synthesis. New retrovirus particles are thus prepared and released to infect other cells.

The first vaccines approved to fight COVID-19 are unlike vaccines used to fight many childhood diseases, which use an inactive form of a virus. COVID-19 vaccines contain synthetic mRNA encapsulated in lipid nanoparticles. These mRNA vaccines instruct cells to make unusual proteins that the immune system does not recognize, resulting in an immune response that protects the bodies against infection.

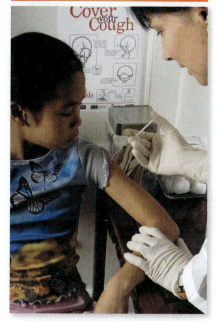
Figure 17.11   Schematic of SARS-CoV-2, the Retrovirus That Causes COVID-19

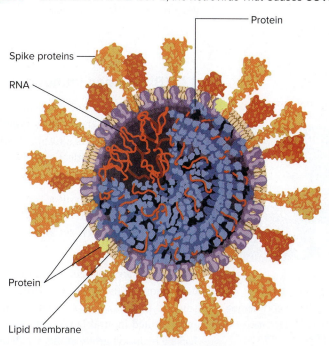

• SARS-CoV-2 consists of an RNA and protein core, surrounded by a lipid bilayer. The spike proteins that project from the surface latch onto receptors in human lung cells and move inside the cell.

Figure 17.12

How a Retrovirus Infects an Organism

A retrovirus (containing viral RNA and a reverse transcriptase) binds and infects a host cell. Reverse transcription forms viral DNA from RNA. One strand of viral DNA becomes a template to transcribe viral RNA, which is then translated into viral proteins. A new virus is assembled and the virus leaves the host cell to infect other cells.

17.11B Human Immunodeficiency Virus (HIV)

HEALTH NOTE

AZT, also called zidovudine and originally sold under the trade name Retrovir, has been available since the 1990s for the treatment of HIV. *Source: Bill Branson/Jim Minor, NIH Pharmacy/National Cancer Institute (NCI)*

AIDS (acquired immune deficiency syndrome) is caused by HIV (human immunodeficiency virus), a retrovirus that attacks lymphocytes central to the body's immune response against invading organisms. As a result, an individual infected with HIV becomes susceptible to life-threatening bacterial infections. HIV is spread by direct contact with the blood or other body fluids of an infected individual.

HIV is often treated with a "cocktail" of drugs designed to destroy the virus at different stages of its reproductive cycle. One group of drugs, the protease inhibitors such as amprenavir (Section 16.11), act as enzyme inhibitors that prevent viral RNA from synthesizing needed proteins.

Other drugs are designed to interfere with reverse transcription, an essential biochemical process unique to the virus. Two drugs in this category are **AZT** (azidodeoxythymidine) and **ddI** (dideoxyinosine). The structure of each drug closely resembles the nucleotides that must be incorporated in viral DNA, and thus they are inserted into a growing DNA strand. Each drug lacks a hydroxyl group at the 3' position, however, so no additional nucleotide can be added to the DNA chain, thus halting DNA synthesis.

azidodeoxythymidine
AZT

no OH group

dideoxyinosine
ddI

no OH group

Concept Check 17.7	Explain why antibiotics such as sulfanilamide and penicillin (Section 16.10) are effective in treating bacterial infections but are completely ineffective in treating viral infections.

CHAPTER REVIEW

KEY TERMS

Anticodon (17.5, 17.8)

Codon (17.7)

Complementary base pairs (17.3)

Deletion mutation (17.9)

Deoxyribonucleic acid
  (DNA, 17.1, 17.3)

Deoxyribonucleoside (17.1)

Deoxyribonucleotide (17.1)

DNA fingerprinting (17.10)

Gene (17.1)

Genetic code (17.7)

Genetic disease (17.9)

Insertion mutation (17.9)

Lagging strand (17.4)

Leading strand (17.4)

Messenger RNA (mRNA, 17.5)

Mutation (17.9)

Nucleic acid (17.1, 17.2)

Nucleoside (17.1)

Nucleotide (17.1)

Point mutation (17.9)

Polynucleotide (17.2)

Replication (17.4)

Retrovirus (17.11)

Reverse transcription (17.11)

Ribonucleic acid (RNA, 17.1, 17.5)

Ribonucleoside (17.1)

Ribonucleotide (17.1)

Ribosomal RNA (rRNA, 17.5)

Transcription (17.6)

Transfer RNA (tRNA, 17.5)

Translation (17.8)

Virus (17.11)

KEY CONCEPTS

1 **Nucleotides and polynucleotides (17.1, 17.2)**

1 **Nucleotides**

- A nucleotide contains a monosaccharide joined to a nitrogen-containing base, and a phosphate bonded to the 5'-OH group of the monosaccharide.
- The monosaccharide is either ribose or 2-deoxyribose, and the bases are abbreviated as A, G, C, T, and U.

uridine 5'-monophosphate
UMP

2 **Polynucleotides**

- A polynucleotide contains a backbone consisting of alternating sugar and phosphate groups.
- The identity and order of the bases distinguish one polynucleotide from another.
- A polynucleotide has one free phosphate group at the 5' end.
- A polynucleotide has a free OH group at the 3' end.

dinucleotide **TC**

2 **A comparison of DNA and RNA (17.2, 17.3, 17.5)**

Nucleic acid	DNA	RNA
1 Structure	• DNA is composed of a right-handed **double helix with two strands** of deoxyribonucleotides winding in the opposite direction. • The double helix is stabilized by complementary base pairing: **A pairs with T and G pairs with C.** • The genetic information of an organism is stored in the sequence of the DNA bases.	• RNA contains a **single strand** of ribonucleotides. • There are three types of RNA. **Ribosomal RNA (rRNA)** is located in the ribosomes and is the site where amino acids are assembled into proteins. **Messenger RNA (mRNA)** is the carrier of genetic information from DNA in the nucleus to the ribosomes in the cytoplasm. **Transfer RNA (tRNA)** brings specific amino acids to the site of protein synthesis.
2 Monosaccharide	• The monosaccharide component of DNA is **deoxyribose.**	• The monosaccharide component of RNA is **ribose.**
3 Bases	• DNA contains the bases A, G, C, and **T.**	• RNA contains the bases A, G, C, and **U.**

3 **The main features of replication, transcription, and translation**

1 Replication (17.4)	**2** Transcription (17.6)	**3** Translation (17.8)
• **Replication is the process by which DNA makes a copy of itself.** • Each strand of DNA serves as a template for a new strand. The original DNA molecule forms two DNA molecules, each of which contains one strand from the parent DNA and one new strand. • The identity of the bases on the template strand determines the order of the bases on the new strand: **A must pair with T, and G must pair with C.**	• **Transcription is the synthesis of mRNA from DNA.** RNA synthesis occurs on the template strand of DNA. • Complementary base pairing determines what RNA nucleotides are added to the growing RNA chain: **C pairs with G, T pairs with A, and A pairs with U.**	• **Translation is the synthesis of proteins from RNA.** Each type of RNA plays a role in this process. • mRNA contains the sequence of **codons** that determines the order of amino acids in the protein. • Individual tRNAs contain **anticodons** that bind to specific amino acids to add to the peptide chain. • rRNA contains binding sites that provide the platform on which protein synthesis occurs.

KEY SKILLS

1 Drawing the newly synthesized strand of DNA formed during replication (17.4); example: using 5'–GCGATTCCGT–3' as the original segment of DNA

1 Write the original segment of DNA.	**2** Use base pairing to write the sequence of the segment of DNA formed during replication.
Original strand: 5'–G C G A T T C C G T–3'	• Write the complementary strand from the 3' end to the 5' end. • **A** pairs with **T** and **G** pairs with **C.** **Original strand:** 5'–G C G A T T C C G T–3' ↓ ↓ ↓ ↓ ↓ ↓ ↓ ↓ ↓ ↓ **Complementary strand:** 3'–C G C T A A G G C A–5'

2 Determining the sequence of an mRNA molecule from a DNA template (17.6); example: using 3'–TACACATTC–5' as the DNA template

1 Write the original segment of DNA.	**2** Use base pairing to write the mRNA sequence after transcription.
• mRNA is transcribed from the template strand of DNA. **Template strand:** 3'–T A C A C A T T C–5'	• mRNA has a base sequence that is complementary to the template from which it is prepared. • In transcription, **C pairs with G, T (on DNA) pairs with A (on RNA), and A (on DNA) pairs with U (on RNA).** **Template strand:** 3'–T A C A C A T T C–5' ↓ ↓ ↓ ↓ ↓ ↓ ↓ ↓ ↓ **mRNA:** 5'–A U G U G U A A G–3'

3 Deriving an amino acid sequence from DNA (17.8); example: using 3'–CCGTATCTT–5' as the DNA template

1 Use the DNA sequence to determine the transcribed mRNA sequence.	**2** Use the codons in Table 17.4 to determine what amino acids are coded for by a given codon in mRNA.
• **G** pairs with **C, T** pairs with **A,** and **A** (on DNA) pairs with **U** (on RNA). **DNA template strand:** 3'– CCG TAT CTT –5' **mRNA:** 5'– GGC AUA GAA –3' Triplets correspond to RNA codons.	**mRNA:** 5'– GGC AUA GAA –3' ↓ ↓ ↓ **Polypeptide:** Gly − Ile − Glu

4 Determining the effect of a point mutation (17.9)

Mutations are changes in the nucleotide sequence in a DNA molecule. A mutation that causes an inherited condition may result in a genetic disease.

1 Determine the dipeptide produced from an original segment of DNA—CAGATG.	**2** Determine the dipeptide formed when a point mutation occurs to form the DNA segment CTGATG.
• Transcribe the DNA sequence to an mRNA sequence with complementary base pairs. • Use the codons in Table 17.4 to determine what amino acids are coded for by each codon. In this example, the dipeptide Val–Tyr is formed.	• When a point mutation converts C**A**G in DNA to C**T**G, transcription forms the codon GAC, which codes for the amino acid Asp (aspartic acid). • Thus, the point mutation leads to the formation of a different dipeptide, Asp–Tyr.
CAG ATG → **GUC UAC** → Val Tyr DNA mRNA dipeptide	**CTG ATG** → **GAC UAC** → Asp Tyr DNA mRNA dipeptide

CHAPTER 17 SELF-TEST

The Self-Test consists of short-answer questions that test an understanding of definitions, equations, and other material encountered within the chapter. Answers to each question are provided at the end of the chapter. For any questions you are unable to answer correctly, revisit the appropriate section(s) of the text; then proceed to working the end-of-chapter problems to further reinforce your understanding of chapter material.

1. Which term from "Key Terms" answers each question?
   a. What compound serves as the carrier of genetic information from DNA in the nucleus to the ribosomes in the cytoplasm?
   b. What process converts the genetic message contained in RNA to the specific amino acid sequence in a protein?
   c. What term is used for an infectious agent that consists of an RNA molecule that is contained within a protein coating?
   d. What type of compound consists of the aldopentose D-ribose and a nitrogen-containing base?
   e. What process results in a change in the nucleotide sequence of DNA that occurs when a nucleotide is lost from the DNA molecule?

2. Give the name and three- or four-letter abbreviation for the given molecule.

3. Write the sequence of the complementary strand of the following portion of a DNA molecule: 5'–AGGTATC–3'.

4. If a portion of the template strand of a DNA molecule has the sequence 3'–TACACATTC–5', then what is the sequence of the mRNA molecule produced from this template?

5. Derive the amino acid sequence that is coded for by the following mRNA sequence: UUGCAGAGAGCACCGGAG.

6. What sequence of amino acids would be formed from the following mRNA sequence: ACAUAUCAGACCUGG? List the anticodons contained in each of the needed tRNA molecules.

UNDERSTANDING KEY CONCEPTS

17.1 Label each statement as pertaining to DNA, RNA, or both.

a. The polynucleotide is double stranded.

b. The polynucleotide may contain adenine.

c. The polynucleotide may contain dGMP.

d. The polynucleotide is a polymer of ribonucleotides.

17.2 Label each statement as pertaining to DNA, RNA, or both.

a. The polynucleotide is single stranded.

b. The polynucleotide may contain guanine.

c. The polynucleotide may contain UMP.

d. The polynucleotide is a polymer of deoxyribonucleotides.

17.3 (a) Identify the base and monosaccharide in the following nucleotide. (b) Give the name and three- or four-letter abbreviation for the compound.

17.4 (a) Identify the base and monosaccharide in the following nucleotide. (b) Give the name and three- or four-letter abbreviation for the compound.

17.5 Consider the given dinucleotide.

a. Identify the bases present in the dinucleotide.

b. Label the 5' and 3' ends.

c. Give the three- or four-letter abbreviations for the two nucleotides that form the dinucleotide.

d. Is this dinucleotide a ribonucleotide or a deoxyribonucleotide? Explain your choice.

e. Name the dinucleotide.

17.6 Answer Problem 17.5 for the following dinucleotide.

17.7 Fill in the missing information in the schematic of a tRNA during the elongation phase of translation.

17.8 Fill in the missing information in the schematic of a tRNA during the elongation phase of translation.

17.9 Fill in the codon, anticodon, or amino acid needed to complete the following table that relates the sequences of DNA, mRNA, tRNA, and the resulting polypeptide.

DNA template strand:	3' end	TTG	ATA	GGT	TGC	TTC	TAC	5' end
mRNA codons:	5' end							3' end
tRNA anticodons:								
Polypeptide:								

17.10 Fill in the codon, anticodon, or amino acid needed to complete the following table that relates the sequences of DNA, mRNA, tRNA, and the resulting polypeptide.

DNA template strand:	3' end	TCC	GAC	TTG	TGC	CAT	CAC	5' end
mRNA codons:	5' end							3' end
tRNA anticodons:								
Polypeptide:								

17.11 The lanes in the given gel show the DNA of a father and three children. Which lane represents the DNA of the father?

17.12 Which lane in the given gel represents an identical twin of the individual whose DNA is shown in Lane 1?

ADDITIONAL PROBLEMS

Nucleosides, Nucleotides, and Nucleic Acids

17.13 What is the difference between a gene and a chromosome?

17.14 What is the difference between uracil and uridine?

17.15 Identify the base and monosaccharide in the following compound and then name it.

17.16 Identify the base and monosaccharide in the following compound and then name it.

17.17 (a) Give the name of each compound shown as a ball-and-stick model. (b) Classify the compound as a base, nucleoside, or nucleotide. (c) Would the compound be a component of DNA, RNA, or both?

A   **B**

17.18 Give the name, abbreviation, or structure of each of the following:
a. a purine base
b. a nucleoside that contains 2-deoxyribose and a pyrimidine base
c. a nucleotide that contains ribose and a purine base

17.19 Classify each molecule as a nucleoside or nucleotide.
a. adenosine   c. GDP
b. deoxyguanosine   d. dTDP

17.20 Classify each molecule as a nucleoside or nucleotide.
a. uridine   c. dGMP
b. deoxycytidine   d. UTP

17.21 Draw the structure of the dinucleotide formed by joining the 3'-OH group of dTMP with the 5'-phosphate of dAMP.

17.22 Draw the structure of the dinucleotide formed by joining the 5'-phosphate of UMP with the 3'-OH of AMP.

17.23 Describe in detail the DNA double helix with reference to each of the following features: (a) the sugar–phosphate backbone; (b) the functional groups at the end of each strand; (c) the hydrogen bonding between strands.

17.24 Describe in detail the DNA double helix with reference to each of the following features: (a) the location of the bases; (b) the complementary base pairing; (c) the phosphodiester linkages.

17.25 Write the sequence of the complementary strand of each segment of a DNA molecule.
a. 5'–AAATAAC–3'   c. 5'–CGATATCCCG–3'
b. 5'–ACTGGACT–3'   d. 5'–TTCCCGGGATA–3'

17.26 Fill in the complementary bases at the labeled sites of the portion of DNA shown.

17.27 If 27% of the nucleotides in a sample of DNA contain the base adenine (A), what are the percentages of bases T, G, and C?

17.28 If 19% of the nucleotides in a sample of DNA contain the base cytosine (C), what are the percentages of bases G, A, and T?

Replication, Transcription, Translation, and Protein Synthesis

17.29 Consider the following simple graphic that illustrates the main features of DNA replication. Label each of the letters (**A–G**) with one of the following terms: leading strand, lagging strand, template strand, 5' end, or 3' end. A term may be used more than once.

17.30 Consider the following graphic that illustrates the transcription of DNA to RNA. Label each of the letters (**A–F**) with one of the following terms: nucleotide, RNA, template strand, RNA polymerase, 5' end, or 3' end.

17.31 What is the sequence of a newly synthesized DNA segment if the template strand has the sequence 3'–ATGGCCTATGCGAT–5'?

17.32 What is the sequence of a newly synthesized DNA segment if the template strand has the sequence 3'–CGCGATTAGATATTGCCGC–5'?

17.33 Explain the roles of messenger RNA and transfer RNA in converting the genetic information coded in DNA into protein synthesis.

17.34 What are the two main structural features of transfer RNA molecules?

17.35 What is the sequence of the mRNA molecule synthesized from each DNA template?
 a. 3'–ATGGCTTA–5'   b. 3'–CGGCGCTTA–5'

17.36 What is the sequence of the mRNA molecule synthesized from each DNA template?
 a. 3'–GGCCTATA–5'   b. 3'–GCCGAT–5'

17.37 For each codon, give its anticodon and the amino acid for which it codes.
 a. CUG   b. UUU   c. AAG   d. GCA

17.38 For each codon, give its anticodon and the amino acid for which it codes.
 a. GUU   b. AUA   c. CCC   d. GCG

17.39 Derive the amino acid sequence that is coded for by each mRNA sequence.
 a. CCA ACC UGG GUA GAA   c. GUC GAC GAA CCG CAA
 b. AUG UUU UUA UGG UGG

17.40 Derive the amino acid sequence that is coded for by each mRNA sequence.
 a. AAA CCC UUU UGU
 b. CCU UUG GAA GUA CUU
 c. GGG UGU AUG CAC CGA UUG

17.41 Write a possible mRNA sequence that codes for each peptide.
 a. Ile–Met–Lys–Ser–Tyr   b. Pro–Gln–Glu–Asp–Phe

17.42 Write a possible mRNA sequence that codes for each peptide.
 a. Phe–Phe–Leu–Lys   b. Val–Gly–Gln–Asp–Asn

17.43 Using the given DNA template strand in the following representation, fill in the transcribed mRNA sequence and the polypeptide that would be synthesized from this template.

17.44 Using the given mRNA sequence in the following representation, (a) fill in the DNA template strand from which this mRNA was synthesized. (b) What polypeptide would be synthesized from this mRNA sequence?

Mutations

17.45 Consider the following mRNA sequence: CUU CAG CAC.
 a. What amino acid sequence is coded for by this mRNA?
 b. What is the amino acid sequence if a mutation converts CAC to AAC?
 c. What occurs when a mutation converts CAG to UAG?

17.46 Consider the following mRNA sequence: ACC UUA CGA.
 a. What amino acid sequence is coded for by this mRNA?
 b. What is the amino acid sequence if a mutation converts UUA to UCA?
 c. What is the amino acid sequence if a mutation converts CGA to AGA?

17.47 Consider the following sequence of DNA: 3'–TTA CGG–5'.
 a. What dipeptide is formed from this DNA after transcription and translation?
 b. If a mutation converts CGG to AGG in DNA, what dipeptide is formed?

17.48 Consider the following sequence of DNA: 3'–ATA GGG–5'.
 a. What dipeptide is formed from this DNA after transcription and translation?
 b. What occurs when a mutation converts ATA to ATT in DNA?

DNA Fingerprinting

17.49 The given gel contains the DNA fingerprint of a mother (lane 1), father (lane 2), and four children (lanes 3–6). One child is adopted, two children are the offspring of both parents, and one child is the offspring of one parent only. Explain your answer in each part.

a. Which lanes, if any, represent a biological child of both parents?
b. Which lanes, if any, represent an adopted child?
c. Which lanes (if any) represent a biological child of the mother only?
d. Which lanes (if any) represent a biological child of the father only?

17.50 With references to the gel in Problem 17.49, which lanes (if any) represent twins? Explain your choice.

Viruses

17.51 What is a retrovirus?

17.52 What is reverse transcription?

17.53 Lamivudine, also called 3TC, is a nucleoside analogue used in the treatment of HIV and the virus that causes hepatitis B. (a) What nucleoside does lamivudine resemble? (b) How does lamivudine inhibit reverse transcription?

lamivudine
3TC

17.54 How does a vaccine protect an individual against certain viral infections?

General Questions

17.55 Fill in the bases, codon, anticodon, or amino acid needed to complete the following table that relates the sequences of DNA, mRNA, tRNA, and the resulting polypeptide.

DNA template strand:	3' end	CAT				5' end
mRNA codons:	5' end		UCA		AUG	3' end
tRNA anticodons:				GUG		
Polypeptide:			Thr			

17.56 Fill in the bases, codon, anticodon, or amino acid needed to complete the following table that relates the sequences of DNA, mRNA, tRNA, and the resulting polypeptide.

DNA template strand:	3' end	CAA			GTC	5' end
mRNA codons:	5' end		UAC			3' end
tRNA anticodons:				ACA		
Polypeptide:				Lys		

17.57 If there are 325 amino acids in a polypeptide, how many bases are present in a single strand of the gene that codes for it, assuming that every base is transcribed and then translated to polypeptide?

17.58 If a single strand of a gene contains 678 bases, how many amino acids result in the polypeptide prepared from it, assuming every base of the gene is transcribed and then translated?

17.59 Met-enkephalin (Tyr–Gly–Gly–Phe–Met) is a pain killer and sedative. What is a possible nucleotide sequence in the template strand of the gene that codes for met-enkephalin, assuming that every base of the gene is transcribed and then translated?

17.60 Leu-enkephalin (Tyr–Gly–Gly–Phe–Leu) is a pain killer and sedative. What is a possible nucleotide sequence in the template strand of the gene that codes for leu-enkephalin, assuming that every base of the gene is transcribed and then translated?

CHALLENGE PROBLEMS

17.61 Give a possible nucleotide sequence in the template strand of the gene that codes for the following tripeptide.

17.62 Like proteins, DNA becomes denatured when it is heated. From what you learned about complementary base pairing and hydrogen bonding in Section 17.3, propose a reason that explains why the temperature required for denaturing increases as the G–C content of the double helix increases.

BEYOND THE CLASSROOM

17.63 Although James Watson and Francis Crick receive much of the credit for unraveling the structure of the double helix of DNA, much early research in this area was carried out by Rosalind Franklin, a woman who was ineligible to receive the Nobel Prize because she had passed away before it was awarded for this work in 1962. Few women have earned the Nobel Prize since it was first awarded in 1901. Locate a list of Nobel Prize winners on the web or in the library, and pick one woman who has received this honor in Chemistry or in Physiology or Medicine. Describe the nature of the work for which she received this achievement.

17.64 What were the goals and scope of the Human Genome Project (Section 17.10B), and how was this project carried out? List at least three scientific conclusions that were made from the results. What legal, ethical, and social issues surrounded this undertaking?

CHAPTER 17 ANSWERS

Concept Check Answers

17.1 thymine

17.2 a,d,e: RNA   b,c,f: DNA

17.3 a. base   b,d: nucleoside   c. nucleotide

17.4 a. Ala   c. Leu   e. Gln
b. Asn   d. Ser   f. Lys

17.5 a. GGU, GGC, GGA, GGG   c. AAA, AAG
b. AUU, AUC, AUA   d. GAA, GAG

17.6 a. [1] GCC   [2] Arg
b. [1] CCC   [2] Gly
c. [1] AGG   [2] Ser
d. [1] UAU   [2] Ile
e. [1] GGA   [2] Pro
f. [1] CGG   [2] Ala

17.7 Antibiotics act upon enzymes found in bacteria. Viruses use materials in the host to reproduce, so antibiotics are ineffective.

Practice Problem Answers

17.1 a. base = uracil, sugar = ribose, nucleoside = uridine
b. base = guanine, sugar = deoxyribose,
nucleoside = deoxyguanosine

17.2

ribonucleoside

17.3 a. base = uracil, sugar = ribose,
uridine 5'-monophosphate, UMP
b. base = thymine, sugar = deoxyribose,
deoxythymidine 5'-monophosphate, dTMP

17.4 a. guanosine 5'-triphosphate
b. deoxycytidine 5'-diphosphate
c. deoxythymidine 5'-triphosphate
d. uridine 5'-diphosphate

17.5 a,c: false   b,d: true

17.6

17.7 a. 5'–ATTTG–3'   c. 5'–GGACTT–3'
b. 5'–CGCGUU–3'

17.8 a,d,e: true   b,c,f: false

17.9 a. 3'–TTTGCAGG–5'   c. 3'–TAACGTGGGCG–5'
b. 3'–ATATGCGG–5'   d. 3'–GTGAACTAGCC–5'

17.10 a. 5'–TCTCAGAG–3'   c. 5'–TAGGACATG–3'
b. 3'–TAACGAG–5'   d. 3'–CCGGTATGAG–5'

17.11 a. 5'–ACGGAUUGC–3'   c. 5'–AAUUGCGCU–3'
b. 5'–CUGAGG–3'   d. 5'–GUCACUGGCAUG–3'

17.12 a. 3'–ACCCCGTAA–5'   c. 3'–GGCTGCTAC–5'
b. 3'–CATGGA–5'   d. 3'–CATCAGTGC–5'

17.13 a. Gln–Glu–Val–Ser–Tyr–Arg
b. Val–Ile–Trp–Arg–Gly–Ile
c. Leu–Cys–Ser–Arg–Thr–Pro

17.14 a. 5'–AUGCGCCAUUUU–3'   c. 5'–CAAAAUGGAAUUGUG–3'
b. 5'–GGCGCCGAACAA–3'   d. 5'–ACUCAUGAUUGCUGG–3'

17.15 a. CTC GGG CAT ATG CGG TGC
b. Glu–Pro–Val–Tyr–Ala–Thr

17.16 a. Pro–Pro–Ala–Asn–Glu–Ala
GGU GGC CGU UUG CUU CGU
b. Ala–Pro–Leu–Arg–Asp
CGU GGU GAU UCU CUG

17.17 a. Arg–Val–Ala–Leu–Leu–Ser
b. Arg–Gly–Phe–Ile–Val–Asn

17.18 a. Leu–Thr   b. [1] Leu–Pro
[2] Stop codon, so no dipeptide is formed.
[3] Leu–Thr

Self-Test Answers

1. a. messenger RNA   c. retrovirus   e. deletion mutation
    b. translation   d. ribonucleoside

2. deoxythymidine 5'-monophosphate, dTMP

3. 3'–TCCATAG–5'

4. 5'–AUGUGUAAG–3'

5. Leu–Gln–Arg–Ala–Pro–Glu

6. amino acid sequence: Thr–Tyr–Gln–Thr–Trp; tRNA anticodons: UGU, AUA, GUC, UGG, ACC

Odd-Numbered Problem Answers

17.1 a,c: DNA   b. both   d. RNA

17.3 a. thymine, deoxyribose
b. deoxythymidine 5'-monophosphate, dTMP

17.5 a. uracil, adenine
b. The 5' end has the free phosphate and the 3' end has two OH groups on the five-membered sugar ring.
c. UMP, AMP
d. This dinucleotide is a ribonucleotide because the sugar rings contain an OH group on C2'.
e. UA

17.7 anticodon: CGC; amino acid: alanine

17.9

DNA template strand:	3' end	TTG	ATA	GGT	TGC	TTC	TAC	5' end
mRNA codons:	5' end	AAC	UAU	CCA	ACG	AAG	AUG	3' end
tRNA anticodons:		UUG	AUA	GGU	UGC	UUC	UAC	
Polypeptide:		Asn	Tyr	Pro	Thr	Lys	Met	

17.11 lane 3

17.13 A gene is a portion of the DNA molecule responsible for the synthesis of a single protein. Many genes form each chromosome.

17.15 base: cytosine
monosaccharide: deoxyribose
deoxycytidine

17.17 **A:** a. deoxyadenosine   b. nucleoside   c. DNA
B: a. cytosine   b. base   c. both DNA and RNA

17.19 a,b: nucleoside   c,d: nucleotide

17.21 a.

17.23 a. deoxyribose as the sugar with the sugar–phosphate groups on the outside
b. 5' end: phosphate; 3' end: OH group
c. hydrogen bonding in the interior of the helix: A pairs with T and G pairs with C

17.25 a. 3'–TTTATTG–5'   c. 3'–GCTATAGGGC–5'
b. 3'–TGACCTGA–5'   d. 3'–AAGGGCCCTAT–5'

17.27 27% T, 23% G, 23% C

17.29 **A:** 3' end   **E:** lagging strand
B: template strand   **F:** template strand
C: 5' end   **G:** 5' end
D: leading strand

17.31 5'–TACCGGATACGCTA–3'

17.33 Messenger RNA carries the specific sequence of the DNA code from the nucleus to the ribosomes to make a protein. Each transfer RNA brings a specific amino acid to the growing protein chain on the ribosome determined by the sequence of the mRNA.

17.35 a. 5'–UACCGAAU–3'   b. 5'–GCCGCGAAU–3'

17.37 a. GAC:Leu   b. AAA:Phe   c. UUC:Lys   d. CGU:Ala

17.39 a. Pro–Thr–Trp–Val–Glu
b. Met–Phe–Leu–Trp–Trp
c. Val–Asp–Glu–Pro–Gln

17.41 a. AUU AUG AAA AGU UAU   b. CCU CAA GAA GAU UUU

17.43 mRNA: UUGCAGGAGUGCUAA
peptide: Leu–Gln–Glu–Cys

17.45 a. Leu–Gln–His
b. Leu–Gln–Asn
c. This is a stop codon so the chain is terminated.

17.47 a. Asn–Ala   b. Asn–Ser

17.49 a. lanes 3 and 5   c. lane 6
b. lane 4   d. none

17.51 A retrovirus is a virus that has RNA rather than DNA in its core. Once it invades a cell it uses reverse transcriptase to synthesize DNA for replication.

17.53 a. cytidine
b. Lamivudine is a nucleoside analogue that gets incorporated into a DNA chain, but because it does not contain a 3' hydroxyl group, synthesis is terminated.

17.55

DNA template strand:	3' end	CAT	AGT	TGA	GTG	TAC	5' end
mRNA codons:	5' end	GUA	UCA	ACU	CAC	AUG	3' end
tRNA anticodons:		CAU	AGU	UGA	GUG	UAC	
Polypeptide:		Val	Ser	Thr	His	Met	

17.57 975

17.59 ATA CCA CCA AAA TAC

17.61 TAA CCT AAA

Design Credits: DNA molecule structure background for Chapter Opener: Watchara/Shutterstock

18 Energy and Metabolism

The stored fat in penguins and other marine animals insulates them in icy waters and serves as an energy source when they are unable to regularly search for food. *Daniel C. Smith*

CHAPTER GOALS

In this chapter you will learn how to:

1. Define metabolism and describe the four stages of catabolism
2. Explain the role of ATP in energy production
3. Describe the roles of the main coenzymes used in metabolism
4. Describe the main aspects of glycolysis
5. List the pathways for pyruvate metabolism
6. List the main features of the citric acid cycle
7. Describe the main components of the electron transport chain and oxidative phosphorylation
8. Calculate the energy yield from glucose metabolism
9. Summarize the process of the β-oxidation of fatty acids
10. Identify the structures of ketone bodies and describe their role in metabolism
11. Describe the main components of amino acid catabolism

Why Study ...

Energy and
Metabolism?

Despite the wide diversity among life forms, virtually all organisms contain the same types of biomolecules—lipids, carbohydrates, proteins, and nucleic acids—and use the same biochemical reactions. These reactions provide both the raw materials and the energy for growth and maintenance. The metabolism of ingested food begins with the hydrolysis of large biomolecules into small compounds that can be absorbed through the intestinal wall. Then specific pathways that often involve many steps convert these compounds into lower-molecular-weight molecules and generate energy for movement, thought, and a myriad of other processes. In Chapter 18, we learn about the metabolism of carbohydrates, lipids, and proteins.

18.1   An Overview of Metabolism

Each moment thousands of reactions occur in a living cell: large molecules are broken down into smaller components, small molecules are converted into larger molecules, and energy changes occur.

- *Metabolism* is the sum of all of the chemical reactions that take place in an organism.

There are two types of metabolic processes called **catabolism** and **anabolism.**

- *Catabolism* is the breakdown of large molecules into smaller ones. Energy is generally released during catabolism.
- *Anabolism* is the synthesis of large molecules from smaller ones. Energy is generally absorbed during anabolism.

The oxidation of glucose ($C_6H_{12}O_6$) to carbon dioxide (CO_2) and water (H_2O) is an example of catabolism, whereas the synthesis of a protein from component amino acids is an example of anabolism. An organized series of consecutive reactions that converts a starting material to a final product is called a **metabolic pathway.**

18.1A   Energy Production in the Cell

Catabolism breaks down the carbohydrates, lipids, and proteins in food into smaller molecules, releasing energy to supply the body's needs. Where does energy production occur in cells?

A typical animal cell, such as the one shown in Figure 18.1, is surrounded by a cell membrane (Section 15.7) and has a nucleus that contains DNA in chromosomes (Section 17.3).

A complex set of biochemical pathways converts ingested carbohydrates, lipids, and proteins to usable materials and energy to meet the body's needs.
Dennis Gray/Cole Group/Getty Images

Figure 18.1   The Mitochondrion in a Typical Animal Cell

A mitochondrion is a small organelle located in the cytoplasm of the cell. Cellular energy production occurs in the mitochondria.

The **cytoplasm,** the region of the cell between the cell membrane and the nucleus, contains various specialized structures called **organelles,** each of which has a specific function. **Mitochondria** are small sausage-shaped organelles in which energy production takes place. Mitochondria contain an outer membrane and an inner membrane with many folds. The area between these two membranes is called the **intermembrane space.** Energy production occurs within the **matrix,** the area surrounded by the inner membrane of the mitochondrion. The number of mitochondria in a cell varies depending on its energy needs. Cells in the heart, brain, and muscles of the human body typically contain many mitochondria.

18.1B   The Four Stages of Catabolism

The catabolic pathways can be organized into four stages, as shown in Figure 18.2.

<table>
<tr><td>

Figure 18.2

The Four Stages in Biochemical Catabolism and Energy Production

</td><td>

</td></tr>
</table>

- **Carbohydrates: Glycolysis** converts glucose to pyruvate, which is then metabolized to acetyl CoA (Sections 18.4–18.5).
- **Lipids:** Fatty acids are oxidized by a stepwise procedure to form acetyl CoA (Section 18.9).
- **Amino acids:** The amino acids formed from protein hydrolysis are often assembled into new proteins without any other modification. Excess amino acids are catabolized for energy as discussed in Section 18.11. The amino groups (NH_2) are converted to urea [$(NH_2)_2C{=}O$], which is excreted in urine.

(left): Fuse/Corbis/Getty Images; (middle): Nancy R. Cohen/Getty Images; (right): Lew Robertson/Brand X Pictures/Getty Images

Stage [1]—Digestion

The catabolism of food begins with **digestion,** which is catalyzed by enzymes in the saliva, stomach, and small intestine (Figure 18.3). The hydrolysis of carbohydrates to monosaccharides begins with amylase enzymes in the saliva, and continues in the small intestine. Protein digestion begins in the stomach, where acid denatures the protein, and the protease pepsin begins to cleave the protein backbone into smaller polypeptides and amino acids. Digestion continues in the small intestine, where trypsin and chymotrypsin further cleave the protein backbone to form amino acids. Triacylglycerols, the most common lipids, are first emulsified by bile secreted by the liver, and then hydrolyzed to glycerol and fatty acids by lipases in the small intestine.

These small molecules are each absorbed through the intestinal wall into the bloodstream and transported to other cells in the body. Some substances such as cellulose are not metabolized as they travel through the digestive tract, so they pass into the large intestine and are excreted.

Figure 18.3 Digestion of Carbohydrates, Proteins, and Triacylglycerols

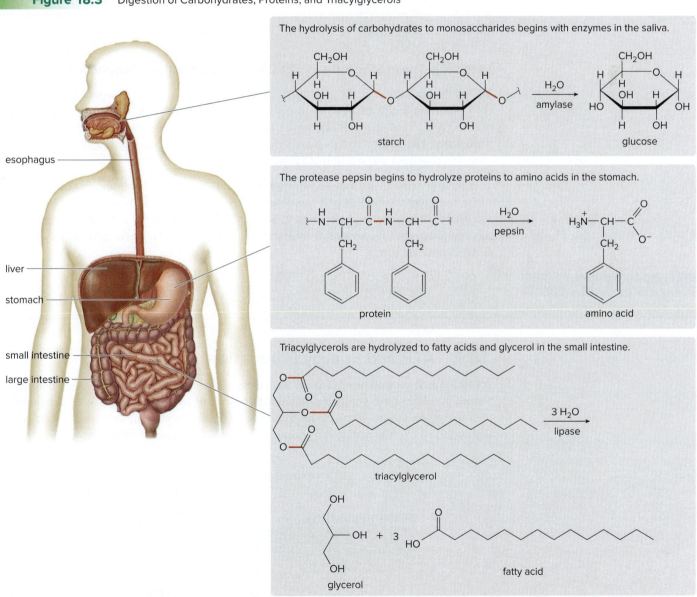

The first stage of catabolism, **digestion,** is the hydrolysis of large molecules to small molecules: polysaccharides such as starch are hydrolyzed to monosaccharides (Section 14.6), proteins are hydrolyzed to their component amino acids (Section 16.8), and triacylglycerols are hydrolyzed to glycerol and fatty acids (Section 15.5). Each of these molecules enters its own metabolic pathway to be further broken down into smaller components, releasing energy.

Once small molecules are formed, catabolism continues to break down each type of molecule to smaller units releasing energy in the process.

Stage [2]—Formation of Acetyl CoA

Monosaccharides, amino acids, and fatty acids are degraded into **acetyl groups (CH_3CO-),** two-carbon units that are bonded to coenzyme A (a coenzyme), forming **acetyl CoA.** Details of the structure of acetyl CoA are given in Section 18.3C.

The product of the catabolic pathways is the *same* for all three types of molecules. As a result, a common catabolic pathway, the **citric acid cycle,** continues the processing of all types of bio-molecules to generate energy.

Stage [3]—The Citric Acid Cycle

The citric acid cycle is based in the mitochondria. In this biochemical cycle, the acetyl groups of acetyl CoA are oxidized to carbon dioxide. Some energy produced by this process is stored in the bonds of a nucleoside triphosphate (Section 17.1) and reduced coenzymes, whose structures are shown in Section 18.3.

Stage [4]—The Electron Transport Chain and Oxidative Phosphorylation

Within the mitochondria, the electron transport chain and oxidative phosphorylation produce **ATP**—adenosine 5'-triphosphate—the primary energy-carrying molecule in metabolic pathways. Oxygen combines with hydrogen ions and electrons from the reduced coenzymes to form water. The result of catabolism is that biomolecules are converted to CO_2 and H_2O and energy is produced and stored in ATP molecules.

In order to understand metabolism, we must first learn about the structure and properties of some of the molecules involved. In particular, Section 18.2 is devoted to a discussion of ATP and how it is used to supply energy in reactions. In Section 18.3 we examine the structure and reactions of the key coenzymes nicotinamide adenine dinucleotide (NAD^+), flavin adenine dinucleotide (FAD), and coenzyme A.

Concept Check 18.1	What advantage might there be to funneling all catabolic pathways into a single common pathway, the citric acid cycle?

18.2 ATP and Energy Production

As we learned in Section 17.1, **ATP,** adenosine 5'-triphosphate, is a nucleoside triphosphate formed by adding three phosphates to the 5'-OH group of adenosine, a nucleoside composed of the sugar ribose and the base adenine. Similarly, **ADP,** adenosine 5'-diphosphate, is a nucleoside diphosphate formed by adding two phosphates to the 5'-OH group of adenosine.

In metabolic pathways, the interconversion of ATP and ADP is the most important process for the storage and release of energy.

> • Hydrolysis of ATP cleaves one phosphate group, forming ADP and hydrogen phosphate, HPO_4^{2-}, often abbreviated as P_i (inorganic phosphate). This reaction *releases* 7.3 kcal/mol of energy.

This bond is broken.

ATP + H_2O → (hydrolysis) → ADP + P_i

Any process, such as walking, running, swallowing, or breathing, is fueled by the release of energy from the hydrolysis of ATP to ADP. ATP is the most prominent member of a group of "high-energy" molecules, reactive molecules that release energy by cleaving a bond.

> • The reverse reaction, phosphorylation, adds a phosphate group to ADP, forming ATP. Phosphorylation *requires* 7.3 kcal/mol of energy.

$$ADP + P_i \longrightarrow ATP + H_2O$$

Figure 18.4 summarizes the reactions and energy changes that occur when ATP is synthesized and hydrolyzed. As we learned in Section 5.9, when energy is *released* in a reaction, the energy change is reported as a *negative* (−) value, so the energy change for ATP hydrolysis is −7.3 kcal/mol. When energy is *absorbed* in a reaction, the energy change is reported as a *positive* (+) value, so the energy change for the phosphorylation of ADP is +7.3 kcal/mol. The synthesis of ATP with P_i is the reverse of ATP hydrolysis, and the energy changes for such reactions are *equal* in value but *opposite* in sign.

ATP is constantly synthesized and hydrolyzed. It is estimated that each ATP molecule exists for about a minute before it is hydrolyzed and its energy released.

Concept Check 18.2

GTP, guanosine 5′-triphosphate, is another high-energy molecule that releases 7.3 kcal of energy when it is hydrolyzed to GDP. Write the equation for the hydrolysis of GTP to GDP. Is the energy change reported as a positive (+) or negative (−) value?

The hydrolysis of ATP to ADP is a favorable reaction because energy is released and lower energy products are formed. In metabolism, this reaction provides the energy to drive reactions that require energy.

Figure 18.4 ATP Hydrolysis and Synthesis

						Energy change
ATP Hydrolysis	ATP + H_2O	⟶	ADP + P_i			−7.3 kcal/mol
ATP Synthesis	ADP + P_i	⟶	ATP + H_2O			+7.3 kcal/mol

Energy is **released**.   Energy is **absorbed**.

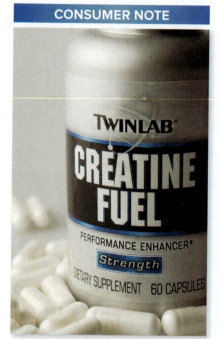

Since the 1990s, creatine supplements have been used by some athletes who wish to increase muscle mass and strength. Increased creatine intake increases the amount of creatine phosphate (Concept Check 18.4) in the muscle. *Jill Braaten*

Reactions that use ATP or coenzymes (Section 18.3) are often drawn using a combination of horizontal and curved arrows. The principal organic reactants and products are drawn from left to right with a reaction arrow as usual, but additional compounds like ATP and ADP are drawn on a curved arrow. This technique is meant to emphasize the organic substrates of the reaction, while making it clear that other materials are needed for the reaction to occur.

For example, the phosphorylation of glucose with ATP forms glucose 6-phosphate and ADP. In this reaction, glucose and glucose 6-phosphate are separated by a horizontal reaction arrow, and ATP and ADP are drawn on a curved arrow.

$$\text{glucose} \xrightarrow{\quad \overset{\text{ATP} \quad \text{ADP}}{\curvearrowright} \quad} \text{glucose 6-phosphate}$$

Concept Check 18.3

Use curved arrow symbolism to write the reaction of fructose with ATP to form fructose 6-phosphate and ADP.

Concept Check 18.4

Creatine phosphate is stored in muscles. When existing supplies of ATP are depleted during strenuous exercise, creatine phosphate reacts with ADP to form creatine and a new supply of ATP as a source of more energy. (a) Write this reaction using curved arrow symbolism. (b) If the energy change in this reaction is −3.0 kcal/mol, is energy absorbed or released?

18.3 Coenzymes in Metabolism

Many reactions in metabolic pathways involve coenzymes. As we learned in Section 16.9, a *coenzyme* **is an organic compound needed for an enzyme-catalyzed reaction to occur.** Some coenzymes serve as important oxidizing and reducing agents (Sections 18.3A and 18.3B), whereas coenzyme A activates acetyl groups (CH_3CO-), resulting in the transfer of a two-carbon unit to other substrates (Section 18.3C).

18.3A Coenzymes NAD^+ and NADH

Many coenzymes are involved in oxidation and reduction reactions. A coenzyme may serve as an oxidizing agent or a reducing agent in a biochemical pathway.

- An oxidizing agent causes an oxidation reaction to occur, so the oxidizing agent is reduced.
- A reducing agent causes a reduction reaction to occur, so the reducing agent is oxidized.

In examining what happens to a coenzyme during oxidation and reduction, it is convenient to think in terms of hydrogen atoms being composed of protons (H^+) and electrons (e^-).

- When a coenzyme gains hydrogen atoms—that is, H^+ and e^-—the coenzyme is reduced; thus, the coenzyme is an *oxidizing* agent.
- When a coenzyme loses hydrogen atoms—that is, H^+ and e^-—the coenzyme is oxidized; thus, the coenzyme is a *reducing* agent.

The coenzyme **nicotinamide adenine dinucleotide, NAD^+,** is a common biological oxidizing agent. When NAD^+ reacts with two hydrogen atoms, it gains one proton and two electrons and one proton is left over. Thus, NAD^+ is reduced and a new C—H bond is formed in the product, written as **NADH,** and referred to as the *reduced form* of nicotinamide adenine dinucleotide.

Add 2 H⁺ and 2 e⁻.

new C—H bond

NAD⁺
nicotinamide adenine dinucleotide

NADH
(reduced form of NAD⁺)

Curved arrow symbolism is often used to depict reactions with coenzymes.

1 C—O bond

ethanol

NAD⁺   NADH + H⁺

2 C—O bonds

acetaldehyde

The conversion of ethanol to acetaldehyde is an **oxidation** because the number of C—O bonds in the substrate *increases.* **NAD⁺ serves as the oxidizing agent, and in the process, is reduced to NADH.** The reduced form of the coenzyme, **NADH, is a biological reducing agent.** When NADH reacts, it forms NAD⁺ as a product. Thus, **NAD⁺ and NADH are interconverted by oxidation and reduction reactions.**

Sample Problem 18.1

Determining the Coenzyme Needed for an Oxidation or Reduction

Label the reaction as an oxidation or reduction, and give the reagent, NAD⁺ or NADH, that would be used to carry it out.

pyruvate   →   lactate

Analysis

Count the number of C—O bonds in the starting material and product. Oxidation *increases* the number of C—O bonds and reduction *decreases* the number of C—O bonds. **NAD⁺ is the coenzyme needed for an oxidation, and NADH is the coenzyme needed for a reduction.**

Solution

The conversion of pyruvate to lactate is a reduction, because the product has one fewer C—O bond than the reactant. To carry out the reduction, the reducing agent NADH could be used.

NADH + H⁺   NAD⁺

pyruvate   →   lactate

Practice Problem 18.1

Label each reaction as an oxidation or reduction, and give the reagent, NAD⁺ or NADH, that would be used to carry out the reaction.

a.   C=O   →   CH₃OH

b.   →

More Practice: Try Problems 18.3, 18.19c, 18.20–18.22.

18.3B   Coenzymes FAD and FADH₂

Flavin adenine dinucleotide, FAD, is another common biological oxidizing agent. When it acts as an oxidizing agent, FAD is reduced by adding two hydrogen atoms, forming **FADH₂**, the *reduced form* of flavin adenine dinucleotide.

FAD
flavin adenine dinucleotide

FADH₂
(reduced form of FAD)

Leafy green vegetables, soybeans, and almonds are good sources of riboflavin, vitamin B₂. Because this vitamin is light sensitive, riboflavin-fortified milk contained in glass or clear plastic bottles should be stored in the dark. *Jill Braaten/McGraw Hill*

FAD is synthesized in cells from **vitamin B₂, riboflavin.** Riboflavin is a yellow, water-soluble vitamin obtained in the diet from leafy green vegetables, soybeans, almonds, and liver. When large quantities of riboflavin are ingested, excess amounts are excreted in the urine, giving it a bright yellow appearance.

riboflavin
vitamin B₂

Table 18.1 summarizes the common coenzymes used in oxidation and reduction reactions.

Table 18.1 Coenzymes in Oxidation and Reduction

Coenzyme	Role
NAD⁺	Oxidizing agent
NADH	Reducing agent
FAD	Oxidizing agent
FADH₂	Reducing agent

Concept Check 18.5

What makes riboflavin a water-soluble vitamin?

Concept Check 18.6

Which reagent, FAD or FADH₂, would be used to carry out the following reaction?

18.3C   Coenzyme A

Coenzyme A differs from other coenzymes in this section because it is not an oxidizing or a reducing agent. In addition to many other functional groups, coenzyme A contains a **sulfhydryl** group (SH group), making it a **thiol (RSH).** To emphasize this functional group, we sometimes abbreviate the structure as **HS–CoA.**

coenzyme A

= **HS —CoA**

thiol

The sulfhydryl group of coenzyme A reacts with acetyl groups (CH₃CO–) or other acyl groups (RCO—) to form **thioesters, RCOSR'.** When an acetyl group is bonded to coenzyme A, the product is called **acetyl coenzyme A,** or simply **acetyl CoA.**

$$CH_3-\overset{O}{\overset{\|}{C}}- \;+\; HS-CoA \longrightarrow CH_3-\overset{O}{\overset{\|}{C}}-S-CoA \qquad R-\overset{O}{\overset{\|}{C}}-S-R'$$

acetyl CoA

General structure of a thioester

Thioesters such as acetyl CoA are another group of high-energy compounds that release energy on reaction with water. In addition, acetyl CoA reacts with other substrates in metabolic pathways to deliver its two-carbon acetyl group, as in the citric acid cycle in Section 18.6. Coenzyme A is synthesized in cells from **pantothenic acid, vitamin B₅.**

pantothenic acid
vitamin B₅

Avocados are an excellent dietary source of pantothenic acid, vitamin B₅. *Digital Vision/Getty Images*

Concept Check 18.7   Predict the water solubility of vitamin B₅.

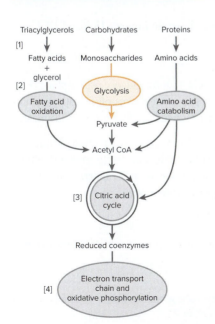

18.4 Glycolysis

The metabolism of monosaccharides centers around glucose. Whether it is obtained by the hydrolysis of ingested polysaccharides or stored glycogen (Section 14.6), glucose is the principal monosaccharide used for energy in the human body.

- *Glycolysis* is a linear, 10-step pathway that converts glucose, a six-carbon monosaccharide, to two molecules of pyruvate ($CH_3COCO_2^-$).

Glycolysis is an anaerobic (without air) pathway that takes place in the cytoplasm and can be conceptually divided into two parts (Figure 18.5).

- Steps [1]–[5] make up the *energy-investment phase*. The addition of two phosphate groups requires the energy stored in two ATP molecules. Cleavage of a carbon–carbon bond forms two three-carbon products.
- Steps [6]–[10] make up the *energy-generating phase*. Each of the three-carbon products is ultimately oxidized, forming NADH, and two high-energy phosphate bonds are broken to form two ATP molecules.

Figure 18.5

An Overview of Glycolysis

In the energy-investment phase of glycolysis, ATP supplies energy needed for steps [1] and [3]. In the energy-generating phase, each three-carbon product from step [5] forms one NADH and two ATP molecules. Because two glyceraldehyde 3-phosphate molecules are formed from each glucose molecule, a total of two NADH and four ATP molecules are formed in the energy-generating phase.

18.4A   The Steps in Glycolysis

The specific steps and all needed enzymes in glycolysis are shown in Figures 18.6 and 18.7.

Glycolysis: Steps [1]–[5]

Glycolysis begins with the phosphorylation of glucose with ATP to form glucose 6-phosphate (Figure 18.6). Isomerization of glucose 6-phosphate to fructose 6-phosphate takes place with an isomerase enzyme in step [2]. Phosphorylation with ATP in step [3] yields fructose 1,6-bisphosphate.

Cleavage of the six-carbon chain of fructose 1,6-bisphosphate forms two three-carbon products—dihydroxyacetone phosphate and glyceraldehyde 3-phosphate. Because only glyceraldehyde 3-phosphate continues on in glycolysis, dihydroxyacetone phosphate is isomerized to glyceraldehyde 3-phosphate in step [5], completing the energy-investment phase of glycolysis.

In summary:

- The first phase of glycolysis converts glucose to *two* molecules of glyceraldehyde 3-phosphate.
- The energy from two ATP molecules is utilized.

Concept Check 18.8   Identify the type of carbonyl groups present in dihydroxyacetone phosphate and glyceraldehyde 3-phosphate. Classify the OH groups in each compound as 1°, 2°, or 3°.

Figure 18.6   Glycolysis: Steps [1]–[5]

- All –PO$_3^{2-}$ groups in glycolysis are abbreviated as —(P).
- The energy from two ATP molecules is used for phosphorylation in steps [1] and [3].
- Cleavage of a carbon–carbon bond and isomerization form two molecules of glyceraldehyde 3-phosphate from glucose, completing the energy-investment phase of glycolysis.
- The atoms of a compound that are changed in a reaction are highlighted in gray.

Glycolysis: Steps [6]–[10]

Each three-carbon aldehyde (glyceraldehyde 3-phosphate) produced in step [5] of glycolysis is carried through a series of five reactions that ultimately form pyruvate (Figure 18.7).

In step [6], oxidation of the –CHO group of glyceraldehyde 3-phosphate and phosphorylation form 1,3-bisphosphoglycerate. In this process, the oxidizing agent NAD$^+$ is reduced to NADH.

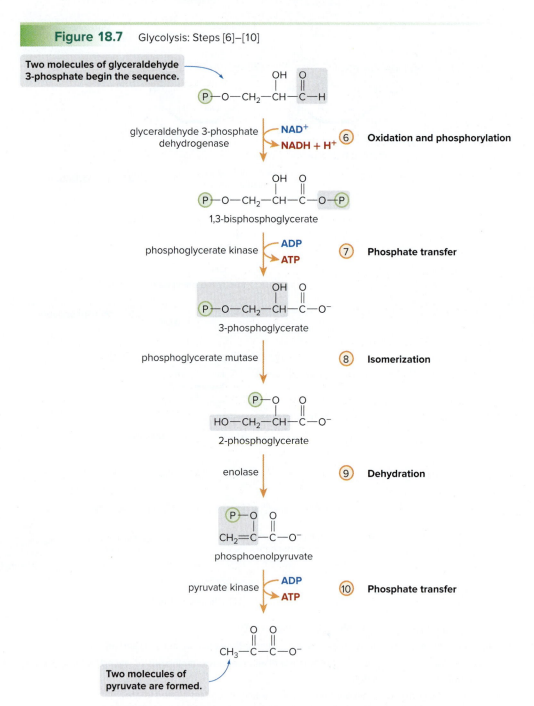

Figure 18.7 Glycolysis: Steps [6]–[10]

Two molecules of glyceraldehyde 3-phosphate begin the sequence.

glyceraldehyde 3-phosphate dehydrogenase — NAD$^+$ → NADH + H$^+$ — (6) **Oxidation and phosphorylation**

1,3-bisphosphoglycerate

phosphoglycerate kinase — ADP → ATP — (7) **Phosphate transfer**

3-phosphoglycerate

phosphoglycerate mutase — (8) **Isomerization**

2-phosphoglycerate

enolase — (9) **Dehydration**

phosphoenolpyruvate

pyruvate kinase — ADP → ATP — (10) **Phosphate transfer**

Two molecules of pyruvate are formed.

- In the energy-generating phase of glycolysis, NADH is formed in step [6] and ATP is formed in steps [7] and [10].
- Because one glucose molecule yields two molecules of glyceraldehyde that begin the sequence, **2 CH$_3$COCO$_2^-$, 2 NADH**, and **4 ATP** are ultimately formed in steps [6]–[10].
- The atoms of a compound that are changed in a reaction are highlighted in gray.

Transfer of a phosphate group from 1,3-bisphosphoglycerate to ADP forms 3-phosphoglycerate and generates ATP in step [7]. Isomerization of the phosphate group in step [8] and loss of water in step [9] form phosphoenolpyruvate. Finally, transfer of a phosphate to ADP forms ATP and pyruvate in step [10]. Thus, **one NADH molecule is produced in step [6] and two ATPs are formed in steps [7] and [10] for each glyceraldehyde 3-phosphate.**

- Because each glucose molecule yielded *two* glyceraldehyde 3-phosphate molecules in step [5], overall *two* NADH molecules and *four* ATP molecules are formed in the energy-generating phase of glycolysis.

Concept Check 18.9

A kinase is an enzyme that catalyzes the transfer of a phosphate group from one substrate to another. Identify all of the reactions of glycolysis that utilize kinases. What species is phosphorylated in each reaction?

Concept Check 18.10

(a) Convert the ball-and-stick model of **A,** one of the intermediates in glycolysis, to a structural formula, and name the compound. (b) What is the immediate precursor of **A** in glycolysis; that is, what compound forms **A** as a reaction product? (c) What compound is formed from **A** in glycolysis?

A

18.4B The Net Result of Glycolysis

Three major products are formed in glycolysis—**ATP, NADH,** and **pyruvate.**

- Two ATP molecules are used in the energy-investment phase (steps [1] and [3]), and four molecules of ATP are formed in the energy-generating phase (steps [7] and [10]). The net result is the *synthesis of two molecules of ATP from glycolysis.*

- Two molecules of NADH are formed from two glyceraldehyde 3-phosphate molecules in step [6]. The NADH formed in glycolysis must be transported from the cytoplasm to the mitochondria for use in the electron transport chain to generate more ATP.

- Two three-carbon molecules of pyruvate ($CH_3COCO_2^-$) are formed from the six carbon atoms of glucose. The fate of pyruvate depends on oxygen availability, as discussed in Section 18.5.

The overall process of glycolysis can be summarized in the following equation.

$$C_6H_{12}O_6 \ + \ 2\,NAD^+ \quad \xrightarrow[\text{2 ADP}]{\text{2 ATP}} \quad 2\ CH_3{-}\overset{\overset{O}{\|}}{C}{-}\overset{\overset{O}{\|}}{C}{-}O^- \ + \ 2\,NADH \ + \ 2\,H^+$$

glucose   pyruvate

Although glycolysis is an ongoing pathway in cells, the rate of glycolysis depends on the body's need for the products it forms—that is, pyruvate, ATP, and NADH. When ATP levels are high, glycolysis is inhibited at various stages. When ATP levels are depleted, such as during strenuous exercise, glycolysis is activated so that more ATP is synthesized.

18.5 The Fate of Pyruvate

Although pyruvate is the end product of glycolysis, it is not the final product of glucose metabolism. What happens to pyruvate depends on the existing conditions and the organism. In particular, there are three possible products:

$$CH_3-\overset{\overset{O}{\|}}{C}-\overset{\overset{O}{\|}}{C}-O^- \longrightarrow CH_3-\overset{\overset{O}{\|}}{C}-SCoA \quad \text{or} \quad CH_3-\overset{\overset{OH}{|}}{\underset{|}{\overset{|}{C}}}-\overset{\overset{O}{\|}}{C}-O^- \quad \text{or} \quad CH_3CH_2OH$$

pyruvate   acetyl CoA   lactate   ethanol

- **Acetyl CoA**, $CH_3COSCoA$, is formed under aerobic conditions.
- **Lactate**, $CH_3CH(OH)CO_2^-$, is formed under anaerobic conditions.
- **Ethanol**, CH_3CH_2OH, is formed in fermentation.

18.5A Conversion to Acetyl CoA

When oxygen is plentiful, oxidation of pyruvate by NAD^+ in the presence of coenzyme A forms acetyl CoA and carbon dioxide. Although oxygen is not needed for this specific reaction, oxygen is needed to oxidize NADH back to NAD^+. Without an adequate supply of NAD^+, this pathway cannot occur. The acetyl CoA formed in this process then enters the common metabolic pathways—the citric acid cycle, the electron transport chain, and oxidative phosphorylation—to generate a great deal of ATP.

$$CH_3-\overset{\overset{O}{\|}}{C}-\overset{\overset{O}{\|}}{C}-O^- + HS-CoA \xrightarrow[\quad\quad]{NAD^+ \quad NADH + H^+} CH_3-\overset{\overset{O}{\|}}{C}-SCoA + CO_2$$

pyruvate   acetyl CoA

18.5B FOCUS ON HEALTH & MEDICINE
Conversion to Lactate

When oxygen levels are low, the metabolism of pyruvate must follow a different course. Oxygen is needed to oxidize the NADH formed in step [6] of glycolysis back to NAD^+. If there is not enough O_2 to re-oxidize NADH, cells must get NAD^+ in a different way. The conversion of pyruvate to lactate provides the solution.

Reduction of pyruvate with NADH forms lactate and NAD^+, which can now re-enter glycolysis and oxidize glyceraldehyde 3-phosphate at step [6].

NAD⁺ can re-enter glycolysis at step [6].

$$CH_3-\overset{\overset{O}{\|}}{C}-\overset{\overset{O}{\|}}{C}-O^- \xrightarrow[\quad\quad]{NADH + H^+ \quad NAD^+} CH_3-\overset{\overset{OH}{|}}{\underset{\underset{H}{|}}{C}}-\overset{\overset{O}{\|}}{C}-O^-$$

pyruvate   lactate

During periods of strenuous exercise, when ATP needs are high, there is an inadequate level of oxygen to re-oxidize NADH. At these times, pyruvate is reduced to lactate for the sole purpose of re-oxidizing NADH to NAD^+ to maintain glycolysis. Under anaerobic conditions, therefore, lactate becomes the main product of glucose metabolism and only two ATP molecules are formed per glucose.

$$\text{Anaerobic conditions} \quad C_6H_{12}O_6 \xrightarrow[\quad \text{2 ADP} \quad \text{2 ATP} \quad]{} 2\ CH_3-\overset{OH}{\underset{H}{\overset{|}{\underset{|}{C}}}}-CO_2^-$$

glucose   lactate

Anaerobic metabolism leads to an increase in lactate levels in muscles, which in turn is associated with soreness and cramping. During these periods an "oxygen debt" is created. When vigorous activity ceases, an individual inhales deep breaths of air to repay the oxygen debt caused by heavy exercise. Lactate is then gradually re-oxidized to pyruvate, which can once again be converted to acetyl CoA, and muscle soreness, fatigue, and shortness of breath resolve.

In any tissue deprived of oxygen, pyruvate is converted to lactate rather than acetyl CoA. The pain produced during a heart attack, for example, is caused by an increase in lactate concentration that results when the blood supply to part of the heart muscle is blocked (Figure 18.8). The lack of oxygen delivery to heart tissue results in the anaerobic metabolism of glucose to lactate rather than acetyl CoA.

Measuring lactate levels in the blood is a common diagnostic tool used by physicians to assess how severely ill an individual is. **A higher-than-normal lactate concentration generally indicates inadequate oxygen delivery to some tissues.** Lactate levels increase transiently during exercise, but can remain elevated because of lung disease, congestive heart failure, or the presence of a serious infection.

Figure 18.8   Lactate Production During a Heart Attack

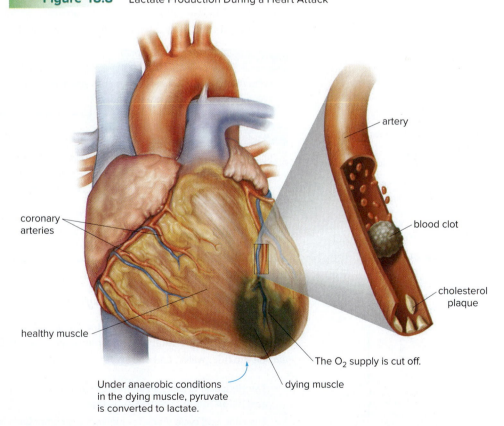

artery

coronary arteries

blood clot

cholesterol plaque

healthy muscle

The O_2 supply is cut off.

Under anaerobic conditions in the dying muscle, pyruvate is converted to lactate.

dying muscle

Fermentation plays a key role in the production of bread, wine, and cheese.
John Thoeming/McGraw Hill

18.5C   Conversion to Ethanol

In yeast and other microorganisms, pyruvate is converted to ethanol (CH_3CH_2OH) and carbon dioxide (CO_2) by a two-step process: **decarboxylation** (loss of CO_2) to acetaldehyde followed by reduction to ethanol.

The NAD^+ generated during reduction can enter glycolysis as an oxidizing agent in step [6]. As a result, glucose can be metabolized by yeast under anaerobic conditions: glycolysis forms pyruvate, which is further metabolized to ethanol and carbon dioxide. Two molecules of ATP are generated during glycolysis.

• *Fermentation* is the anaerobic conversion of glucose to ethanol and CO_2.

The ethanol in beer, wine, and other alcoholic beverages is obtained by the fermentation of sugar, quite possibly the oldest example of chemical synthesis. The carbohydrate source determines the type of alcoholic beverage formed.

Fermentation plays a role in forming other food products. Cheese is produced by fermenting curdled milk, whereas yogurt is prepared by fermenting fresh milk. When yeast is mixed with flour, water, and sugar, the enzymes in yeast carry out fermentation to produce the CO_2 that causes bread to rise. Some of the characteristic and "intoxicating" odor associated with freshly baked bread is due to the ethanol that has evaporated during the baking process.

Concept Check 18.11

What role does NADH play in the conversion of pyruvate to lactate? What role does NADH play in the conversion of pyruvate to ethanol?

Concept Check 18.12

(a) In what way(s) is the conversion of pyruvate to acetyl CoA similar to the conversion of pyruvate to ethanol? (b) In what way(s) are the two processes different?

Concept Check 18.13

(a) In what way(s) is the conversion of pyruvate to lactate similar to the conversion of pyruvate to ethanol? (b) In what way(s) are the two processes different?

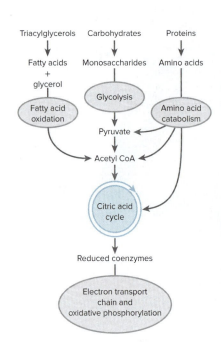

18.6   The Citric Acid Cycle

The **citric acid cycle,** a series of enzyme-catalyzed reactions that occur in mitochondria, makes up the third stage of the catabolism of biomolecules—carbohydrates, lipids, and amino acids—to carbon dioxide, water, and energy.

• The citric acid cycle is an eight-step cyclic metabolic pathway that begins with the addition of acetyl CoA to a four-carbon substrate.
• The citric acid cycle produces high-energy compounds for ATP synthesis in stage [4] of catabolism.

18.6A Overview of the Citric Acid Cycle

The citric acid cycle is also called the tricarboxylic acid cycle (TCA cycle) or the Krebs cycle. The general scheme (Figure 18.9) illustrates the key features.

- The citric acid cycle begins when two carbons of acetyl CoA ($CH_3COSCoA$) react with a four-carbon organic substrate to form a six-carbon product (step [1]).
- Two carbon atoms are removed to form two molecules of CO_2 (steps [3] and [4]).
- Four molecules of reduced coenzymes (NADH and $FADH_2$) are formed in steps [3], [4], [6], and [8]. These molecules serve as carriers of electrons to the electron transport chain in stage [4] of catabolism, which ultimately results in the synthesis of a great deal of ATP.
- One mole of GTP is synthesized in step [5]. GTP is a high-energy nucleoside triphosphate similar to ATP.

Figure 18.9   General Features of the Citric Acid Cycle

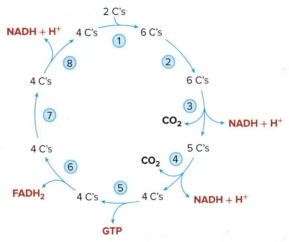

The citric acid cycle begins with the addition of two carbons from acetyl CoA in step [1] to a four-carbon organic substrate, drawn at the top of the cyclic pathway. Each turn of the citric acid cycle forms two molecules of CO_2, four molecules of reduced coenzymes (3 NADH and 1 $FADH_2$), and one high-energy GTP molecule.

18.6B Specific Steps of the Citric Acid Cycle

The eight reactions of the citric acid cycle, which can be conceptually divided into two parts, are shown in Figure 18.10. The first part of the cycle includes the addition of acetyl CoA to oxaloacetate to form the six-carbon product citrate, which undergoes two separate **decarboxylations**—reactions that give off CO_2. In part [2], functional groups are added and oxidized to re-form oxaloacetate, the substrate needed to begin the cycle again. Each step is enzyme catalyzed.

Part [1] of the Citric Acid Cycle

As shown in Figure 18.10, acetyl CoA enters the cycle by reaction with oxaloacetate at step [1] of the pathway. This reaction adds two carbons to oxaloacetate, forming citrate. In step [2], the 3° alcohol of citrate is isomerized to the 2° alcohol isocitrate. These first two steps add carbon atoms and rearrange functional groups.

Loss of two carbon atoms begins in step [3], by decarboxylation of isocitrate. The oxidizing agent NAD^+ also converts the 2° alcohol to a ketone to form α-ketoglutarate, which now contains one fewer carbon atom. This reaction forms NADH and H^+, which will carry electrons and protons gained in this reaction to the electron transport chain. In step [4], decarboxylation releases a

Figure 18.10

Steps in the Citric Acid Cycle

Each step of the citric acid cycle is enzyme catalyzed. The net result of the eight-step cycle is the conversion of the two carbons added to oxaloacetate to two molecules of CO_2. Reduced coenzymes (NADH and $FADH_2$) are also formed, which carry electrons to the electron transport chain to synthesize ATP. One molecule of high-energy GTP is synthesized in step [5]. Gray boxes highlight the atoms that are changed at each step of the cycle.

second molecule of CO_2. Also, oxidation with NAD^+ in the presence of coenzyme A forms the thioester succinyl CoA. **By the end of step [4], two carbons are lost as CO_2 and two molecules of NADH are formed.**

The overall result of steps [1]–[4] of the citric acid cycle:

- The C's from acetyl CoA in step [1] are shown in blue.
- Two C's of oxaloacetate (shown in red) are lost as CO_2.

Part [2] of the Citric Acid Cycle

Part [2] consists of four reactions that manipulate the functional groups of succinyl CoA to re-form oxaloacetate. In step [5], succinyl CoA is hydrolyzed to form succinate, releasing energy to convert GDP to GTP. GTP, guanosine 5'-triphosphate, is similar to ATP: GTP is a high-energy molecule that releases energy during hydrolysis. This is the only step of the citric acid cycle that directly generates a triphosphate.

In step [6], succinate is converted to fumarate with FAD. This reaction forms the reduced coenzyme $FADH_2$, which will carry electrons and protons to the electron transport chain. Addition of water in step [7] forms malate and oxidation of the 2° alcohol in malate with NAD^+ forms oxaloacetate in step [8]. Another molecule of NADH is also formed in step [8]. By the end of step [8], two more molecules of reduced coenzymes ($FADH_2$ and NADH) are formed. **Because the product of step [8] is the starting material of step [1], the cycle can continue as long as additional acetyl CoA is available for step [1].**

The overall result of steps [5]–[8] of the citric acid cycle:

Steps 5 – 8

succinyl CoA → oxaloacetate

- The C's from acetyl CoA in step [1] are shown in blue.
- The number of C's stays the same.
- A CH_2 group is oxidized to a C=O.
- The thioester is hydrolyzed to CO_2^-.

Overall, the citric acid cycle results in formation of

- two molecules of CO_2
- four molecules of reduced coenzymes (3 NADH and 1 $FADH_2$)
- one molecule of GTP

The net equation for the citric acid cycle can be written as shown. The ultimate fate of each product is also indicated.

$$CH_3-C(=O)-SCoA \ + \ 2 H_2O \ + \ GDP \ + \ 3 NAD^+ \ + \ P_i \ + \ FAD$$

acetyl CoA

citric acid cycle

$$2 CO_2 \ + \ HSCoA \ + \ GTP \ + \ 3 NADH + 3 H^+ \ + \ FADH_2$$

exhaled gas | coenzyme that re-enters the cycle | energy source | reduced coenzymes

- The main function of the citric acid cycle is to produce reduced coenzymes that enter the electron transport chain and ultimately produce ATP.

The rate of the citric acid cycle depends on the body's need for energy. When energy demands are high and the amount of available ATP is low, the cycle is activated. When energy demands are low and NADH concentration is high, the cycle is inhibited.

Although the citric acid cycle is complex, many individual reactions can be understood by applying the basic principles of organic chemistry learned in previous chapters.

Sample Problem 18.2 — Analyzing an Individual Step of the Citric Acid Cycle

(a) Write out the reaction that converts succinate to fumarate with FAD using curved arrow symbolism.
(b) Classify the reaction as an oxidation, reduction, or decarboxylation.

Analysis

Use Figure 18.10 to draw the structures for succinate and fumarate. Draw the organic reactant and product on the horizontal arrow and the oxidizing reagent FAD, which is converted to $FADH_2$, on the curved arrow. Oxidation reactions result in a loss of electrons, a loss of hydrogen, or a gain of oxygen. Reduction reactions result in a gain of electrons, a gain of hydrogen, or a loss of oxygen. A decarboxylation results in the loss of CO_2.

Solution

a. Equation:

$$
\begin{array}{ccc}
\overset{\displaystyle CO_2^-}{\underset{\displaystyle |}{}} & & \overset{\displaystyle CO_2^-}{\underset{\displaystyle |}{}} \\
CH_2 & \quad FAD \quad FADH_2 \quad & CH \\
| & \longrightarrow & \| \\
CH_2 & & HC \\
| & & | \\
CO_2^- & & CO_2^- \\
\text{succinate} & & \text{fumarate}
\end{array}
$$

b. Because succinate contains four C—H bonds and fumarate contains only two C—H bonds, hydrogen atoms have been lost, making this reaction an oxidation. In the process FAD is reduced to $FADH_2$.

Practice Problem 18.2

(a) Draw out the structures for the reaction that converts malate to oxaloacetate with NAD^+ using curved arrow symbolism. (b) Classify the reaction as an oxidation, reduction, or decarboxylation.

Practice Problem 18.3

(a) Convert the ball-and-stick model of **A,** one of the eight synthetic intermediates in the citric acid cycle, to a structural formula, and name the compound. (b) What is the immediate precursor of **A** in the citric acid cycle; that is, what compound forms **A** as a reaction product? (c) What compound is formed from **A** in the citric acid cycle?

A

More Practice: Try Problems 18.40, 18.60.

18.7 The Electron Transport Chain and Oxidative Phosphorylation

Most of the energy generated during the breakdown of biomolecules is formed during stage [4] of catabolism. Because oxygen is required, this process is called **aerobic respiration.** There are two facets to this stage:

- **the electron transport chain, or the respiratory chain**
- **oxidative phosphorylation**

The reduced coenzymes formed in the citric acid cycle enter the **electron transport chain** and the electrons they carry are transferred from one molecule to another by a series of oxidation–reduction reactions. Each reaction releases energy until electrons and protons react with oxygen to form water. Electron transfer also causes H+ ions to be pumped across the inner mitochondrial cell membrane, creating an energy reservoir that is used to synthesize ATP by the **phosphorylation** of ADP.

18.7A The Electron Transport Chain

The electron transport chain is a multistep process that relies on four enzyme systems, called **complexes I, II, III, and IV,** as well as mobile electron carriers. Each complex is composed of enzymes, additional protein molecules, and metal ions that can gain and lose electrons in oxidation and reduction reactions. The complexes are situated in the inner membrane of the mitochondria, arranged so that electrons can be passed to progressively stronger oxidizing agents (Figure 18.11).

The electron transport chain begins with the reduced coenzymes—3 NADH and 1 FADH₂—formed during the citric acid cycle. These reduced coenzymes are electron rich and as such, they are capable of donating electrons to other species. Thus, **NADH and FADH₂ are *reducing agents* and when they donate electrons, they are oxidized.** When NADH donates two electrons, it is oxidized to NAD+, which can re-enter the citric acid cycle. Likewise, when FADH₂

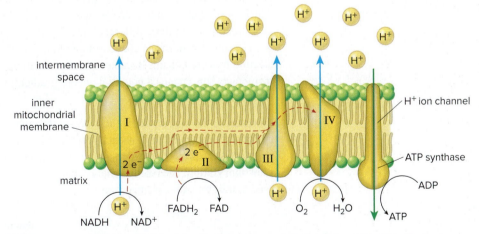

The four enzyme complexes (I–IV) of the electron transport chain are located within the inner membrane of a mitochondrion, between the matrix and the intermembrane space. Electrons enter the chain when NADH and FADH₂ are oxidized and then transported through a series of complexes along the pathway shown in red. The electrons ultimately combine with O₂ to form H₂O. Protons (H+) are pumped across the inner membrane into the intermembrane space at three locations shown by blue arrows. The energy released when protons return to the matrix by traveling through a channel (in green) in the ATP synthase enzyme is used to convert ADP to ATP.

Hydrogen cyanide (HCN) is a poison because it disrupts the electron transport chain. The pits of apricots and peaches contain amygdalin, which forms HCN in the presence of certain enzymes.
Jill Braaten/McGraw Hill

donates two electrons, it is oxidized to FAD, which can be used as an oxidant in step [6] of the citric acid cycle once again.

Once in the electron transport chain, the electrons are passed down from complex to complex in a series of redox reactions, and small packets of energy are released along the way. At the end of the chain, **the electrons and protons (obtained from the reduced coenzymes or the matrix of the mitochondrion) react with inhaled oxygen to form water** and this facet of the process is complete.

Because oxygen is needed for the final stage of electron transport, this process is **aerobic.**

Concept Check 18.14	If NADH and FADH$_2$ were not oxidized in the electron transport chain, what would happen to the citric acid cycle?
Concept Check 18.15	At several points in the electron transport chain, iron cations gain or lose electrons by reactions that interconvert Fe^{2+} and Fe^{3+} cations. (a) When Fe^{3+} is converted to Fe^{2+}, is the reaction an oxidation or a reduction? (b) Is Fe^{3+} an oxidizing agent or a reducing agent?

18.7B   ATP Synthesis by Oxidative Phosphorylation

Although the electron transport chain illustrates how electrons carried by reduced coenzymes ultimately react with oxygen to form water, we have still not learned how ATP is synthesized. The answer lies in what happens to the H$^+$ ions in the mitochondrion.

H$^+$ ions generated by reactions in the electron transport chain, as well as H$^+$ ions present in the matrix of the mitochondria, are pumped across the inner mitochondrial membrane into the inter-membrane space at three different sites (Figure 18.11). This process requires energy, because it moves protons against the concentration gradient. The energy comes from redox reactions in the electron transport chain. Because the concentration of H$^+$ ions is then higher on one side of the membrane, this creates a potential energy gradient, much like the potential energy of water that is stored behind a dam.

To return to the matrix, the H$^+$ ions travel through a channel in the ATP synthase enzyme. ATP synthase is the enzyme that catalyzes the phosphorylation of ADP to form ATP. The energy released as the protons return to the matrix converts ADP to ATP. This process is called **oxidative phosphorylation,** because the energy that results from the oxidation of the reduced coenzymes is used to transfer a phosphate group.

> **Energy released from H$^+$ movement fuels phosphorylation.**

$$ADP \; + \; P_i \; \xrightarrow[\text{ATP synthase}]{} \; ATP \; + \; H_2O$$

Concept Check 18.16	In which region of the mitochondrion—the matrix or the intermembrane space—would the pH be lower? Explain your choice.

18.7C   ATP Yield from Oxidative Phosphorylation

How much ATP is generated during stage [4] of catabolism?

> • Each NADH enters the electron transport chain at complex I in the inner mitochondrial membrane and the resulting cascade of reactions produces enough energy to synthesize 2.5 ATPs.
> • $FADH_2$ enters the electron transport chain at complex II, producing energy for the synthesis of 1.5 ATPs.

How much ATP is generated for each acetyl CoA fragment that enters the entire common catabolic pathway—that is, stages [3] and [4] of catabolism?

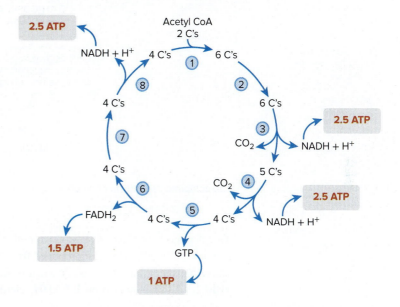

For each turn of the citric acid cycle, three NADH molecules and one $FADH_2$ molecule are formed. In addition, one GTP molecule is produced directly during the citric acid cycle (step [5]); one GTP molecule is equivalent in energy to one ATP molecule. These facts allow us to calculate the total number of ATP molecules formed for each acetyl CoA.

$$
\begin{array}{rcl}
3\ \text{NADH} \times 2.5\ \text{ATP/NADH} &=& 7.5\ \text{ATP} \\
1\ FADH_2 \times 1.5\ \text{ATP/}FADH_2 &=& 1.5\ \text{ATP} \\
1\ \text{GTP} &=& 1\ \ \text{ATP} \\
\hline
&& \mathbf{10}\ \ \ \mathbf{ATP}
\end{array}
$$

> • Complete catabolism of each acetyl CoA molecule that enters the citric acid cycle forms **10 ATP** molecules.

Each ATP molecule can now provide energy for other reactions.

18.8   The ATP Yield from Glucose

How much ATP is generated from the complete catabolism of glucose ($C_6H_{12}O_6$) to carbon dioxide (CO_2)? To answer this question we must take into account the number of ATP molecules formed in the following sequential pathways:

> • the glycolysis of glucose to two pyruvate molecules
> • the oxidation of two pyruvate molecules to two molecules of acetyl CoA
> • the citric acid cycle
> • the electron transport chain and oxidative phosphorylation

Figure 18.12

The ATP Yield from the Aerobic Metabolism of Glucose to CO_2

The complete catabolism of glucose forms six CO_2 molecules and 32 ATP molecules.

To calculate how much ATP is generated, we must consider both the ATP formed directly in reactions, as well as the ATP produced from reduced coenzymes (NADH and $FADH_2$) after oxidative phosphorylation. As we learned in Section 18.7, **each NADH provides the energy to yield 2.5 ATPs, whereas each $FADH_2$ yields 1.5 ATPs.**

This calculation must also take into account that glucose is split into *two* three-carbon molecules at step [4] of glycolysis, so the **ATP yield in each reaction must be *doubled* after this step.** With this information and Figure 18.12 in hand, we can now determine the total yield of ATP from the complete catabolism of glucose.

- Glycolysis yields a net of two ATP molecules. The two molecules of NADH formed during step [6] of glycolysis yield five additional ATPs.

$$C_6H_{12}O_6 \longrightarrow 2\ CH_3COCO_2^- + \textbf{2 ATP} + \textbf{2 NADH}$$
glucose   pyruvate

$2 \times (2.5\ \text{ATP/NADH}) = \textbf{5 ATP}$

- Oxidation of two molecules of pyruvate to acetyl CoA in the mitochondria forms two NADH molecules that yield five ATP molecules.

$$2\ CH_3COCO_2^- \longrightarrow 2\ CH_3COSCoA + 2\ CO_2 + \textbf{2 NADH}$$
pyruvate   acetyl CoA

$2 \times (2.5\ \text{ATP/NADH}) = \textbf{5 ATP}$

- Beginning with two acetyl CoA molecules, the citric acid cycle (Figure 18.10) forms two GTP molecules, the energy equivalent of two ATPs, in step [5]. The six NADH molecules and two $FADH_2$ molecules also formed yield an additional 18 ATPs from the electron transport chain and oxidative phosphorylation. Thus, 20 ATPs are formed from two acetyl CoA molecules.

$$2 \times (1.5\ \text{ATP/FADH}_2) = \textbf{3 ATP}$$

$$\underset{\text{acetyl CoA}}{2\ \text{CH}_3\text{COSCoA}} \longrightarrow 4\ \text{CO}_2 + \textbf{2 GTP} + \textbf{6 NADH} + \textbf{2 FADH}_2$$

2 ATP

$$6 \times (2.5\ \text{ATP/NADH}) = \textbf{15 ATP}$$

Adding up the ATP formed in each pathway gives a **total of 32 molecules of ATP for the complete catabolism of each glucose molecule.** Most of the ATP generated from glucose metabolism results from the citric acid cycle, electron transport chain, and oxidative phosphorylation.

Glucose is the main source of energy for cells and the only source of energy used by the brain. When energy demands are low, glucose is stored as the polymer glycogen in the liver and muscles. When blood levels of glucose are low, glycogen is hydrolyzed to keep adequate blood glucose levels to satisfy the body's energy needs.

Blood glucose levels are carefully regulated by two hormones. When blood glucose concentration rises after a meal, **insulin** stimulates the passage of glucose into cells for metabolism. When blood glucose levels are low, the hormone **glucagon** stimulates the conversion of stored glycogen to glucose.

Concept Check 18.17

How much ATP results from each transformation?

    a. glucose → 2 pyruvate   c. glucose → 2 acetyl CoA

    b. pyruvate → acetyl CoA   d. 2 acetyl CoA → 4 CO_2

Concept Check 18.18

What three reactions form CO_2 when glucose is completely catabolized?

Concept Check 18.19

What three reactions form a nucleoside triphosphate (GTP or ATP) directly when glucose is completely catabolized?

Concept Check 18.20

What is the difference in ATP generation between the aerobic oxidation of glucose to CO_2 and the anaerobic conversion of glucose to lactate?

18.9 The Catabolism of Triacylglycerols

The first step in the catabolism of triacylglycerols, the most common lipids, is the hydrolysis of the three ester bonds to form glycerol and fatty acids (Figure 18.3).

18.9A Fatty Acid Catabolism by β-Oxidation

Fatty acids are catabolized by **β-oxidation, a process in which two-carbon acetyl CoA units are sequentially cleaved from the fatty acid.** Key to this process is the oxidation of the β carbon to the carbonyl group of a thioester (RCOSR'), which then undergoes cleavage between the α and β carbons.

Triacylglycerols   Carbohydrates   Proteins

Fatty acids + glycerol   Monosaccharides   Amino acids

Fatty acid oxidation

Glycolysis

Amino acid catabolism

Pyruvate

Acetyl CoA

Citric acid cycle

Reduced coenzymes

Electron transport chain and oxidative phosphorylation

This carbon is oxidized.   These C's come off as acetyl CoA.

$$-\text{CH}_2-\text{CH}_2-\text{CH}_2-\underset{\beta}{\text{CH}_2}-\underset{\alpha}{\text{CH}_2}-\overset{\overset{\text{O}}{\|}}{\text{C}}-$$

This C—C bond is cleaved.

Fatty acid oxidation begins with conversion of the fatty acid to a thioester with coenzyme A. This process requires energy, which comes from the hydrolysis of two phosphate bonds in ATP to form AMP, adenosine *mono*phosphate. Much like the beginning of glycolysis requires an energy investment, so, too, the initial step of fatty acid oxidation requires energy input.

$$CH_3(CH_2)_{14}-CH_2-CH_2-\overset{\overset{\displaystyle O}{\|}}{C}-OH \quad + \quad HS-CoA \xrightarrow[\text{acyl CoA synthetase}]{\text{ATP} \quad \text{AMP} + 2\,P_i} CH_3(CH_2)_{14}-CH_2-CH_2-\overset{\overset{\displaystyle O}{\|}}{C}-SCoA$$

stearic acid   an acyl CoA

Once the product, an **acyl CoA,** is inside the mitochondrion, the process of β-oxidation is set to begin. β-Oxidation requires four steps to cleave a two-carbon acetyl CoA unit from the acyl CoA, as shown with the 18-carbon fatty acid stearic acid in Figure 18.13.

In step [1], FAD removes two hydrogen atoms to form $FADH_2$ and a double bond between the α and β carbons of the thioester. Water is added to the double bond in step [2] to place an OH group on the β carbon to the carbonyl group, which is then oxidized in step [3] to form a carbonyl group. The NAD^+ oxidizing agent is reduced to NADH in step [3] as well. Finally, cleavage of the bond between the α and β carbons forms acetyl CoA and a 16-carbon acyl CoA in step [4].

- As a result, a new acyl CoA having two carbons fewer than the original acyl CoA is formed.

Figure 18.13 β-Oxidation of a Fatty Acid

- Atoms involved in each reaction are labeled in gray.

The following equation summarizes the important components of β-oxidation for a general acyl CoA, $RCH_2CH_2COSCoA$. **Each four-step sequence forms one molecule each of acetyl CoA, NADH, and $FADH_2$.**

Once the 16-carbon acyl CoA is formed in step [4], it becomes the substrate for a new four-step β-oxidation sequence. The process continues until a four-carbon acyl CoA is cleaved to form *two* acetyl CoA molecules. As a result:

- An **18-carbon** acyl CoA is cleaved to *nine* two-carbon acetyl CoA molecules.
- A total of *eight* cycles of β-oxidation are needed to cleave the eight carbon–carbon bonds.

Thus, complete β-oxidation of the acyl CoA derived from stearic acid forms:

- **9 $CH_3COSCoA$ molecules** (from the 18-carbon fatty acid)
- **8 NADH** (from eight cycles of β-oxidation)
- **8 $FADH_2$** (from eight cycles of β-oxidation)

β-Oxidation of unsaturated fatty acids proceeds in a similar fashion, although an additional step is required. **Ultimately every carbon in the original fatty acid ends up as a carbon atom of acetyl CoA.**

| Sample Problem 18.3 | Determining the Outcome of β-Oxidation of a Fatty Acid |

Consider lauric acid, $CH_3(CH_2)_{10}CO_2H$. (a) How many molecules of acetyl CoA are formed from complete β-oxidation? (b) How many cycles of β-oxidation are needed for complete catabolism?

Analysis

The number of carbons in the fatty acid determines the number of molecules of acetyl CoA formed and the number of times β-oxidation occurs.

- **The number of molecules of acetyl CoA equals one-half the number of carbons in the original fatty acid.**
- Because the final turn of the cycle forms *two* molecules of acetyl CoA, **the number of cycles is one fewer than the number of acetyl CoA molecules formed.**

Solution

Because lauric acid has 12 carbons, it forms six molecules of acetyl CoA from five cycles of β-oxidation.

Practice Problem 18.4

For each fatty acid: [1] How many molecules of acetyl CoA are formed from complete catabolism? [2] How many cycles of β-oxidation are needed for complete oxidation?

  a. arachidic acid ($C_{20}H_{40}O_2$)   b. palmitoleic acid ($C_{16}H_{30}O_2$)

More Practice: Try Problems 18.9; 18.10; 18.43; 18.44; 18.47c, d; 18.48.

18.9B   The Energy Yield from Fatty Acid Oxidation

How much energy—in terms of the number of molecules of ATP formed—results from the complete catabolism of a fatty acid? To determine this quantity, we must take into account the ATP cost for the conversion of the fatty acid to the acyl CoA, as well as the ATP production from the coenzymes (NADH and FADH$_2$) and acetyl CoA formed during β-oxidation. The steps are shown in the accompanying *How To* procedure.

How To **Determine the Number of Molecules of ATP Formed from a Fatty Acid**

Example How much ATP is formed by the complete catabolism of stearic acid, $C_{18}H_{36}O_2$?

Step [1] Determine the amount of ATP required to synthesize the acyl CoA from the fatty acid.

- Because the conversion of stearic acid ($C_{17}H_{35}COOH$) to an acyl CoA ($C_{17}H_{35}COSCoA$) requires the hydrolysis of two phosphate bonds, this is equivalent to the energy released when 2 ATPs are converted to 2 ADPs.
- Thus, the **first step in catabolism costs the equivalent of 2 ATPs—that is, −2 ATPs.**

Step [2] Add up the ATP generated from the coenzymes produced during β-oxidation.

- As we learned in Section 18.9A, each cycle of β-oxidation produces one molecule each of NADH and FADH$_2$. To cleave eight carbon–carbon bonds in stearic acid requires eight cycles of β-oxidation, so 8 NADH and 8 FADH$_2$ are produced.

$$8 \text{ NADH} \times 2.5 \text{ ATP/NADH} = 20 \text{ ATP}$$
$$8 \text{ FADH}_2 \times 1.5 \text{ ATP/FADH}_2 = \underline{12 \text{ ATP}}$$
$$\text{From reduced coenzymes: } \quad \textbf{32 ATP}$$

- Thus, 32 ATPs would be produced from oxidative phosphorylation after the reduced coenzymes enter the electron transport chain.

Step [3] Determine the amount of ATP that results from each acetyl CoA, and add the results for steps [1]–[3].

- From Section 18.9A, stearic acid generates nine molecules of acetyl CoA, which then enter the citric acid cycle and go on to produce ATP by the electron transport chain and oxidative phosphorylation. As we learned in Section 18.7C, each acetyl CoA results in 10 ATPs.

$$9 \text{ acetyl CoA} \times 10 \text{ ATP/acetyl CoA} = \textbf{90 ATP}$$

- Totaling the values obtained in steps [1]–[3]:

$$(-2) + 32 + 90 = \textbf{120 ATP} \text{ molecules from stearic acid}$$

Answer

Palmitic acid (Concept Check 18.21) is the main constituent of palm oil, an edible oil obtained from the fruit of oil palms native to west and southwest Africa. *KhunJompol/iStock/Getty Images*

18.10   Ketone Bodies

When carbohydrates do not meet energy needs, the body turns to catabolizing stored triacylglycerols, which generate acetyl CoA by β-oxidation of fatty acids. Normally the acetyl CoA is metabolized in the citric acid cycle. When acetyl CoA levels exceed the capacity of the citric acid cycle, however, acetyl CoA is converted to three compounds that are collectively called **ketone bodies**—acetoacetate, β-hydroxybutyrate, and acetone.

HEALTH NOTE

Ketostix are a brand of test strips that detect ketone bodies in urine. *Jill Braaten/ McGraw Hill*

- *Ketogenesis* is the synthesis of ketone bodies from acetyl CoA.

Ketone bodies are produced in the liver, and because they are small molecules that can hydrogen bond with water, they are readily soluble in blood and urine. Once they reach tissues, β-hydroxybutyrate and acetoacetate can be re-converted to acetyl CoA and metabolized for energy.

Under some circumstances—notably starvation, vigorous dieting, and uncontrolled diabetes—when glucose is unavailable or cannot pass into a cell for use as fuel, ketone bodies accumulate, a condition called **ketosis.** As a result, ketone bodies are eliminated in urine and the sweet odor of acetone can be detected in exhaled breath. Sometimes the first indication of diabetes in a patient is the detection of excess ketone bodies in a urine test.

An abnormally high concentration of ketone bodies can lead to **ketoacidosis**—that is, a lowering of the blood pH caused by the increased level of β-hydroxybutyrate and acetoacetate.

Low carbohydrate diets, popularized by Dr. Robert Atkins in a series of diet books published in the 1990s, restrict carbohydrate intake to induce the utilization of the body's stored fat as its main energy source. This increased level of fatty acid metabolism leads to an increased concentration of ketone bodies in the blood and urine.

HEALTH NOTE

Low carbohydrate diets such as the Atkins plan induce the use of stored fat for energy production to assist weight loss. *Jill Braaten/McGraw Hill*

18.11 Amino Acid Metabolism

After proteins are hydrolyzed in the stomach and intestine, the individual amino acids are reassembled to new proteins or converted to intermediates in other metabolic pathways. The catabolism of amino acids provides energy when the supply of carbohydrates and lipids is exhausted.

The catabolism of amino acids can be conceptually divided into two parts: the fate of the amino group and the fate of the carbon atoms. As shown in Figure 18.14, amino acid carbon skeletons are converted to pyruvate, acetyl CoA, or various carbon compounds that are part of the citric acid cycle.

Figure 18.14 An Overview of the Catabolism of Amino Acids

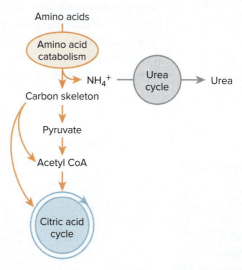

The breakdown of amino acids forms NH_4^+, which enters the urea cycle to form urea, and a carbon skeleton that is metabolized to either pyruvate, acetyl CoA, or an intermediate in the citric acid cycle.

18.11A Degradation of Amino Acids—The Fate of the Amino Group

The catabolism of carbohydrates and triacylglycerols deals with the oxidation of carbon atoms only. With amino acids, an **amino group (–NH$_2$)** must be metabolized, as well. The catabolism of amino acids begins with the removal of the amino group from the carbon skeleton to form an α-keto acid and an ammonium ion (NH_4^+) by a stepwise procedure.

$$\underset{\text{amino acid}}{R-\overset{\overset{+}{N}H_3}{\underset{H}{C}}-CO_2^-} \longrightarrow \underset{\text{α-keto acid}}{R-\overset{O}{C}-CO_2^-} \;+\; NH_4^+$$

For example, removal of the amino group from alanine forms pyruvate and NH_4^+.

$$\underset{\text{alanine}}{CH_3-\overset{\overset{+}{N}H_3}{\underset{H}{C}}-CO_2^-} \longrightarrow \underset{\text{pyruvate}}{CH_3-\overset{O}{C}-CO_2^-} \;+\; NH_4^+$$

The amino group is removed from the amino acid.

α-Keto acids like pyruvate are then degraded along the catabolic pathways described in Section 18.11B. The ammonium ion (NH_4^+) enters the **urea cycle,** where it is converted to urea, $(NH_2)_2C{=}O$, in the liver. Urea is then transported to the kidneys and eliminated in urine.

Sample Problem 18.4	Drawing the Products Formed from Amino Acid Metabolism

What products are formed when the amino group of leucine is removed during the early stages of amino acid catabolism?

$$(CH_3)_2CHCH_2-\underset{\underset{H}{|}}{\overset{\overset{+NH_3}{|}}{C}}-CO_2^-$$

leucine

Analysis

To draw the organic product, replace the C—H and C—NH_3^+ on the α carbon of the amino acid by C=O. NH_4^+ is formed from the amino group.

Solution

$$\underset{\text{α carbon}}{(CH_3)_2CHCH_2-\underset{\underset{H}{|}}{\overset{\overset{+NH_3}{|}}{C}}-CO_2^-} \longrightarrow (CH_3)_2CHCH_2-\overset{\overset{O}{\|}}{C}-CO_2^- \ + \ NH_4^+$$

Practice Problem 18.5	What products are formed when the amino group of each amino acid is removed during the early stages of amino acid catabolism: (a) threonine; (b) glycine? Use the structures in Table 16.2.

More Practice: Try Problems 18.11, 18.12, 18.53, 18.54.

18.11B Degradation of Amino Acids—The Fate of the Carbon Skeleton

Once the nitrogen has been removed from an amino acid, the carbon skeletons of individual amino acids are catabolized in a variety of ways. There are three common fates of the carbon skeletons of amino acids, shown in Figure 18.15:

- conversion to pyruvate, $CH_3COCO_2^-$
- conversion to acetyl CoA, $CH_3COSCoA$
- conversion to an intermediate in the citric acid cycle

Some amino acids such as alanine (Section 18.11A) are catabolized to pyruvate. Pyruvate can be broken down for energy or used to synthesize glucose. In considering catabolism, amino acids are often divided into two groups.

- *Glucogenic* amino acids are catabolized to pyruvate or an intermediate in the citric acid cycle. Glucogenic amino acids can be used to synthesize glucose.

- *Ketogenic* amino acids are converted to acetyl CoA, or the related thioester acetoacetyl CoA, $CH_3COCH_2COSCoA$. These catabolic products cannot be used to synthesize glucose, but they can be converted to ketone bodies and yield energy by this path.

We will not examine the specific pathways that convert the carbon skeletons of individual amino acids into other products. Figure 18.15 illustrates where each amino acid feeds into the metabolic pathways we have already discussed.

Figure 18.15

Amino Acid Catabolism

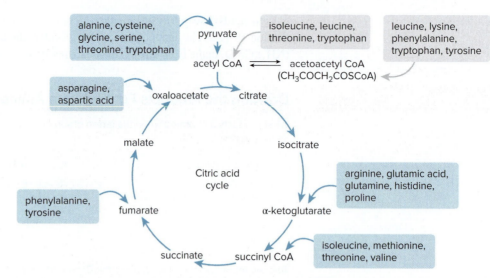

Glucogenic amino acids are highlighted in blue, whereas ketogenic amino acids are highlighted in gray. Amino acids that appear more than once in the scheme can be degraded by multiple routes.

Concept Check 18.25	What metabolic intermediate is produced from the carbon atoms of each amino acid?
	a. cysteine b. aspartic acid c. valine d. threonine

Concept Check 18.26	Which amino acid(s) in Concept Check 18.25 are ketogenic?

CHAPTER REVIEW

KEY TERMS

Acetyl CoA (18.3)
Acyl CoA (18.9)
Adenosine 5'-diphosphate (ADP, 18.2)
Adenosine 5'-triphosphate (ATP, 18.2)
Anabolism (18.1)
Catabolism (18.1)

Citric acid cycle (18.6)
Coenzyme A (18.3)
Electron transport chain (18.7)
Fermentation (18.5)
Flavin adenine dinucleotide (FAD, 18.3)
Glucogenic amino acid (18.11)
Glycolysis (18.4)

Ketogenesis (18.10)
Ketogenic amino acid (18.11)
Ketone bodies (18.10)
Ketosis (18.10)
Metabolism (18.1)
Mitochondrion (18.1)
NADH (18.3)

Nicotinamide adenine dinucleotide (NAD+, 18.3)
β-Oxidation (18.9)
Oxidative phosphorylation (18.7)
Phosphorylation (18.2)
Urea cycle (18.11)

KEY CONCEPTS

① **Stages in metabolism (18.1)**

Metabolism is the sum of all the chemical reactions that take place in an organism. Catabolic reactions break down large molecules and anabolic reactions synthesize larger molecules.

① **Stage [1]**	**②** **Stage [2]**	**③** **Stage [3]**	**④** **Stage [4]**
• Metabolism begins with **digestion.** • Large molecules—polysaccharides, proteins, and triacylglycerols—are hydrolyzed to smaller molecules—monosaccharides, amino acids, fatty acids, and glycerol.	• Biomolecules are degraded into two-carbon **acetyl units** (CH₃CO–).	• The **citric acid cycle** converts two carbon atoms to two molecules of CO_2, and forms reduced coenzymes NADH and FADH₂ that carry electrons to the electron transport chain.	• The **electron transport chain** and **oxidative phosphorylation** produce ATP, and oxygen is converted to water.

2 **Key organic compounds in metabolism**

1 ATP and ADP (18.2)	**2** NAD$^+$/NADH and FAD/FADH$_2$ (18.3A, 18.3B)	**3** Coenzyme A and acetyl CoA (18.3C)
• **The interconversion of ATP and ADP is the central method of energy transfer in cells.** • Hydrolysis of ATP to ADP *releases* energy that can be used to drive reactions that require energy input. • ATP is synthesized from ADP by a phosphorylation reaction that *requires* energy input.	• NAD$^+$ and FAD are biological **oxidizing** agents. • NADH and FADH$_2$ are biological **reducing** agents. • Each NADH that enters Stage [4] of catabolism provides the energy to synthesize **2.5 ATPs.** • Each FADH$_2$ that enters Stage [4] provides the energy to synthesize **1.5 ATPs.**	• Coenzyme A (**HSCoA**) is a thiol that combines with an acetyl group to form acetyl CoA (**CH$_3$COSCoA**). • Acetyl CoA is a **thioester** that is the product of Stage [2] of catabolism for lipids, carbohydrates, and amino acids. • Acetyl CoA is a high-energy compound that releases energy on reaction with water.

3 **Glycolysis (18.4)**

1 Key features	**2** Overall reaction
• Glycolysis converts glucose to two molecules of pyruvate by a 10-step pathway. • In the energy-investment phase, steps [1]–[5], the energy from two ATP molecules is used for phosphorylation and two three-carbon products are formed. • In the energy-generating phase, steps [6]–[10], two pyruvate molecules (CH$_3$COCO$_2^-$), 2 NADHs, and 4 ATPs are generated.	

4 **Pathways for pyruvate metabolism (18.5)**

1 Aerobic conditions (18.5A)	**2** Anaerobic conditions (18.5B)	**3** Fermentation (18.5C)
• When oxygen is plentiful, pyruvate is converted to **acetyl CoA,** which can enter the citric acid cycle.	• When the oxygen level is low, the anaerobic metabolism of pyruvate forms **lactate and NAD$^+$.**	• In yeast and other microorganisms, pyruvate is converted to **ethanol and CO$_2$** by fermentation.

5 **The citric acid cycle (18.6)**

1 Key features	**2** Overall reaction
• The citric acid cycle is an eight-step cyclic pathway that begins with the addition of acetyl CoA to a four-carbon substrate. • Two carbons are converted to CO$_2$, and four molecules of reduced coenzymes (3 NADH + FADH$_2$) are formed. One molecule of GTP, a high-energy nucleoside triphosphate, is also formed. • The main function of the citric acid cycle is to generate reduced coenzymes that enter the electron transport chain and ultimately produce ATP.	

6 **Key features of the electron transport chain and oxidative phosphorylation (18.7)**

1 The electron transport chain (18.7A)	**2** ATP synthesis and oxidative phosphorylation (18.7B)
• In the electron transport chain, electrons from reduced coenzymes are passed from one molecule to another in a series of redox reactions, releasing energy along the way. • At the end of the chain, electrons and protons react with inhaled oxygen to form water. Because oxygen is needed for the final stages of electron transport, this process is **aerobic**.	• H+ ions are pumped across the inner membrane of the mitochondrion, forming a high concentration of H+ ions in the intermembrane space, creating a potential energy gradient. • When the H+ ions travel through the channel in the **ATP synthase enzyme,** this energy is used to convert ADP to ATP—a process called oxidative phosphorylation.

7 **ATP yield from the aerobic metabolism of glucose (18.8)**

To determine the amount of ATP generated, we must consider the ATP formed directly in reactions, as well as the ATP formed from reduced coenzymes after oxidative phosphorylation; **total ATPs formed = 32.**

1 From glycolysis	**2** From the conversion of pyruvate to acetyl CoA	**3** From the citric acid cycle
• Glycolysis converts glucose to pyruvate and forms **2 ATPs** directly. The two molecules of NADH yield an additional **5 ATPs** after Stage [4] of catabolism.	• When two molecules of pyruvate are oxidized and decarboxylated to two molecules of acetyl CoA, two molecules of NADH are formed, which yield an additional **5 ATPs** after Stage [4].	• The citric acid cycle converts two molecules of acetyl CoA to two GTPs, the energy equivalent of **2 ATPs.** Six NADH molecules and two FADH$_2$ molecules yield an additional **18 ATPs** from Stage [4]. Thus, **20 ATPs** are formed from two acetyl CoA molecules.

8 **Fatty acid catabolism (18.9)**

1 Key features	**2** The overall reaction for each four-step β-oxidation of a fatty acid
• Fatty acids are metabolized by **β-oxidation,** which sequentially cleaves two-carbon acetyl CoA units from an acyl CoA derived from a fatty acid. • Each cycle of β-oxidation consists of a four-step sequence that forms one molecule each of acetyl CoA, NADH, and FADH$_2$.	

9 **Ketone bodies (18.10)**

1 Ketogenesis	**2** Ketosis	**3** Ketoacidosis
• **Ketogenesis is the synthesis of ketone bodies from acetyl CoA.** • The term ketone body refers to three compounds—**acetoacetate, β-hydroxybutyrate,** and **acetone.**	• **Ketosis is the accumulation of ketone bodies,** a condition that results from starvation, vigorous dieting, or uncontrolled diabetes.	• **Ketoacidosis** is a condition that results when there is an abnormally high level of ketone bodies. • Increased levels of acetoacetate and β-hydroxybutyrate result in a **lowering of the pH of the blood.**

KEY SKILLS

1 Writing equations with curved arrow symbolism (18.2); example: oxidation of isocitrate with NAD^+

1 Draw the organic reactants and products with a horizontal reaction arrow as usual.	**2** Draw additional compounds with a curved arrow.
• Oxidation of isocitrate forms oxalosuccinate.	• The coenzyme NAD^+ is converted to NADH, which are added on a curved arrow.

2 Determining the products of β-oxidation of a fatty acid (18.9); example: decanoic acid ($C_9H_{19}CO_2H$)

1 Determine the number of molecules of acetyl CoA formed.	**2** Determine the number of molecules of NADH and $FADH_2$ formed.
• The number of molecules of acetyl CoA equals **one-half** the number of C's in the fatty acid.	• The number of cycles of β-oxidation is **one fewer** than the number of acetyl CoA molecules. • The number of cycles **equals** the number of molecules of NADH and $FADH_2$ formed.

3 Determining the amount of ATP formed during fatty acid catabolism (18.9); example: decanoic acid ($C_9H_{19}CO_2H$)

1 Determine the ATP used to form the fatty acyl CoA.	**2** Determine the ATP formed from reduced coenzymes.	**3** Determine the amount of ATP from each acetyl CoA and add up the results.
• Forming an acyl CoA requires the energy equivalent of 2 ATPs. $C_{10}H_{20}O_2 \xrightarrow{-2\,ATP} C_9H_{19}COSCoA$	• β-Oxidation forms 4 NADH and 4 $FADH_2$. $4\ NADH \times 2.5\ ATP/NADH = \mathbf{10}\ ATPs$ $4\ FADH_2 \times 1.5\ ATP/FADH_2 = \mathbf{6}\ ATPs$ From reduced coenzymes: **16** ATPs	• Each cycle of the citric acid cycle forms 10 ATPs. • Five acetyl CoA molecules are formed from a 10-C fatty acid. 5 acetyl CoA × 10 ATP/acetyl CoA = **50** ATPs • Total ATP production: $(-2) + 16 + 50 = \mathbf{64}$ ATPs

4 Drawing the products formed from the early stages of metabolism of an amino acid (18.11)

The catabolism of amino acids involves two parts. First, the amino group is removed and the NH_4^+ ion formed enters the urea cycle, where it is converted to urea and eliminated in urine. The carbon skeletons of the amino acids are then catabolized by a variety of pathways to yield pyruvate, acetyl CoA, or an intermediate in the citric acid cycle.

1 Locate the α carbon of the amino acid.	**2** Remove the amino group and form a C=O on the α carbon.
	• Replace the C—H and C—NH_3^+ on the α carbon of the amino acid by C=O. • NH_4^+ is formed from the amino group.

CHAPTER 18 SELF-TEST

The Self-Test consists of short-answer questions that test an understanding of definitions, equations, and other material encountered within the chapter. Answers to each question are provided at the end of the chapter. For any questions you are unable to answer correctly, revisit the appropriate section(s) of the text; then proceed to working the end-of-chapter problems to further reinforce your understanding of the chapter material.

1. Which term from "Key Terms" answers each question?
   a. What process involves the anaerobic conversion of glucose to ethanol and CO_2?
   b. What cyclic metabolic pathway consists of eight steps and begins with the addition of acetyl CoA to a four-carbon compound?
   c. What catabolic process converts glucose to pyruvate?
   d. By what process are two-carbon acetyl units cleaved from a fatty acid?
   e. What process results in the synthesis of ketone bodies from acetyl CoA?

2. Classify each compound as an oxidizing agent, a reducing agent, or neither: (a) NADH; (b) ADP; (c) FAD; (d) a coenzyme that gains hydrogen atoms.

UNDERSTANDING KEY CONCEPTS

18.1 In what stage of catabolism does each of the following processes occur?
   a. cleavage of a protein with chymotrypsin
   b. oxidation of a fatty acid to acetyl CoA
   c. oxidation of malate to oxaloacetate with NAD^+
   d. conversion of ADP to ATP with ATP synthase
   e. hydrolysis of starch to glucose with amylase

18.2 In what stage of catabolism does each of the following processes occur?
   a. conversion of a monosaccharide to acetyl CoA
   b. hydrolysis of a triacylglycerol with lipase
   c. reaction of oxygen with protons and electrons to form water
   d. conversion of succinate to fumarate with FAD
   e. degradation of a fatty acid to acetyl CoA

18.3 When a substrate is oxidized, is NAD^+ oxidized or reduced? Is NAD^+ an oxidizing agent or a reducing agent?

18.4 When a substrate is reduced, is $FADH_2$ oxidized or reduced? Is $FADH_2$ an oxidizing agent or a reducing agent?

18.5 What steps in the citric acid cycle have each of the following characteristics?
   a. The reaction generates NADH.
   b. CO_2 is removed.
   c. The reaction utilizes FAD.
   d. The reaction forms a new carbon–carbon single bond.

18.6 What steps in the citric acid cycle have each of the following characteristics?
   a. The reaction generates $FADH_2$.
   b. The organic substrate is oxidized.
   c. The reaction utilizes NAD^+.
   d. The reaction breaks a carbon–carbon bond.

3. (a) How many molecules of acetyl CoA are formed from the β-oxidation of a fatty acid with molecular formula $C_{22}H_{44}O_2$? (b) How many cycles of β-oxidation are needed for complete oxidation? (c) How many molecules of ATP are formed?

4. What products are formed from each of the following compounds during Stage [4] of catabolism: (a) NADH; (b) $FADH_2$; (c) ADP; (d) O_2?

5. Use curved arrow symbolism to write out the reaction of fructose 6-phosphate with ATP to form fructose 1,6-bisphosphate and ADP.

6. What products are formed when the amino group of isoleucine is removed during the early stages of amino acid metabolism?

$$H_3\overset{+}{N}-\overset{\overset{\displaystyle H}{|}}{\underset{\underset{\displaystyle CH(CH_3)CH_2CH_3}{|}}{C}}-CO_2^-$$
isoleucine

7. What organic product and coenzymes are formed from glucose as a result of glycolysis?

18.7 Compare the energy-investment phase and the energy-generating phase of glycolysis with regards to each of the following: (a) the reactant that begins the phase and the final product formed; (b) the amount of ATP used or formed; (c) the number of reduced coenzymes used or formed.

18.8 Write the overall equation with key coenzymes for each process.
   a. glucose → pyruvate
   b. glucose → ethanol
   c. pyruvate → lactate

18.9 (a) How many molecules of acetyl CoA are formed from complete β-oxidation of the fatty acid depicted in the ball-and-stick model? (b) How many cycles of β-oxidation are needed for complete oxidation?

18.10 (a) How many molecules of acetyl CoA are formed from complete β-oxidation of the fatty acid depicted in the ball-and-stick model? (b) How many cycles of β-oxidation are needed for complete oxidation?

18.11 What products are formed when the amino group is removed from methionine in the early stages of amino acid catabolism?

18.12 What products are formed when the amino group is removed from tyrosine in the early stages of amino acid catabolism?

18.13 How might pyruvate be metabolized in the cornea, tissue that has little blood supply?

18.14 How is pyruvate metabolized in red blood cells, which contain no mitochondria?

ADDITIONAL PROBLEMS

Metabolism

18.15 Place the following steps in the catabolism of carbohydrates in order: (a) the electron transport chain; (b) the conversion of glucose to acetyl CoA; (c) the hydrolysis of starch; (d) oxidative phosphorylation; (e) the citric acid cycle.

18.16 What is the difference between metabolism and digestion?

ATP and Coenzymes

18.17 (a) Using curved arrow symbolism, write the equation for the reaction of fructose 1,6-bisphosphate with ADP to form fructose 6-phosphate and ATP. (b) If the energy change in this reaction is +3.4 kcal/mol, does this reaction absorb or release energy? (c) Is this reaction energetically favorable?

18.18 (a) Using curved arrow symbolism, write the equation for the reaction of 1,3-bisphosphoglycerate with ADP to form 3-phosphoglycerate and ATP. (b) If this reaction releases 4.5 kcal/mol of energy, is the energy change reported as a positive (+) or negative (–) value?

18.19 Classify each substance as an oxidizing agent, a reducing agent, or neither: (a) $FADH_2$; (b) ATP; (c) NAD^+.

18.20 Draw a product that could form when 6-hydroxy-2-heptanone is treated with each coenzyme: (a) NAD^+; (b) NADH.

6-hydroxy-2-heptanone

18.21 Label the reaction as an oxidation or reduction and give the coenzyme, NAD^+ or NADH, which might be used to carry out the transformation. Write the reaction using skeletal structures with curved arrow symbolism.

18.22 Label the reaction as an oxidation or reduction and give the coenzyme, NAD^+ or NADH, which might be used to carry out the transformation. Write the reaction using skeletal structures with curved arrow symbolism.

Glucose Metabolism

18.23 Considering the individual steps in glycolysis:
a. Which steps form ATP?
b. Which steps use ATP?
c. Which steps form a reduced coenzyme?
d. Which step breaks a C—C bond?

18.24 Explain the role of the coenzymes NAD^+ and NADH in each reaction.
a. pyruvate → acetyl CoA
b. pyruvate → lactate
c. pyruvate → ethanol

18.25 Glucose is completely metabolized to six molecules of CO_2. What specific reactions generate each molecule of CO_2?

18.26 Why is glycolysis described as an anaerobic process?

18.27 Consider the aerobic and anaerobic avenues of pyruvate metabolism in the human body.
a. Where do the carbon atoms of pyruvate end up in each pathway?
b. What coenzymes are used and formed?

18.28 What is the main purpose for the conversion of pyruvate to lactate under anaerobic conditions?

18.29 What metabolic products are formed from pyruvate in each case: (a) anaerobic conditions in the body; (b) anaerobic conditions in yeast; (c) aerobic conditions?

18.30 Why must the NADH produced in glycolysis be re-oxidized to NAD^+? How is this accomplished aerobically? How is this accomplished anaerobically?

18.31 Explain in detail how 32 ATP molecules are generated during the complete catabolism of glucose to CO_2.

18.32 Which compound(s) are formed during the fermentation of glucose to ethanol?

A   B   C

Citric Acid Cycle, Electron Transport Chain, and Oxidative Phosphorylation

18.33 (a) Which intermediate(s) in the citric acid cycle contain two chirality centers? (b) Which intermediate(s) contain a 2° alcohol?

18.34 What products of the citric acid cycle are funneled into the electron transport chain?

18.35 Why are the reactions that occur in stage [4] of catabolism sometimes called aerobic respiration?

18.36 What is the role of each of the following in the electron transport chain: (a) NADH; (b) O_2; (c) complexes I–IV; (d) H^+ ion channel?

18.37 What is the role of each of the following in the electron transport chain: (a) $FADH_2$; (b) ADP; (c) ATP synthase; (d) the inner mitochondrial membrane?

18.38 Explain the importance of the movement of H^+ ions across the inner mitochondrial membrane and then their return passage through the H^+ ion channel in ATP synthase.

18.39 What are the final products of the electron transport chain?

18.40 The conversion of isocitrate to α-ketoglutarate in step [3] of the citric acid cycle actually occurs by a two-step process: isocitrate is converted first to oxalosuccinate, which then goes on to form α-ketoglutarate.

| isocitrate | oxalosuccinate | α-ketoglutarate |

a. Classify reaction [3a] as an oxidation, reduction, or decarboxylation.

b. Classify reaction [3b] as an oxidation, reduction, or decarboxylation.

c. For which step is NAD^+ necessary?

d. Why is the enzyme that catalyzes this reaction called isocitrate dehydrogenase?

Triacylglycerol Metabolism

18.41 How much ATP is used or formed when a fatty acid is converted to an acyl CoA? Explain your reasoning.

18.42 How much ATP is ultimately generated from each cycle of β-oxidation of a fatty acid?

18.43 How many molecules of acetyl CoA are formed from complete β-oxidation of myristic acid, $C_{13}H_{27}CO_2H$, obtained from coconut oil? How many cycles of β-oxidation are needed for complete oxidation?

18.44 How many molecules of acetyl CoA are formed from complete β-oxidation of oleic acid, $C_{17}H_{33}CO_2H$, obtained from olive oil? How many cycles of β-oxidation are needed for complete oxidation?

18.45 How much ATP is generated by the complete catabolism of myristic acid in Problem 18.43?

18.46 How much ATP is generated by the complete catabolism of oleic acid in Problem 18.44?

18.47 Consider decanoic acid, $C_9H_{19}CO_2H$.
   a. Label the α and β carbons.
   b. Draw the acyl CoA derived from this fatty acid.
   c. How many acetyl CoA molecules are formed by complete β-oxidation?
   d. How many cycles of β-oxidation are needed for complete oxidation?
   e. How many molecules of ATP are formed from the complete catabolism of this fatty acid?

18.48 Fill in the boxes with the number of moles of each substance formed during the metabolism of $CH_3(CH_2)_{20}CO_2H$.

$$CH_3(CH_2)_{20}CO_2H \xrightarrow{\beta\text{-oxidation}} \boxed{a.}\ CH_3COSCoA + \boxed{b.}\ NADH + \boxed{c.}\ FADH_2$$

Ketone Bodies

18.49 What is the difference between ketosis and ketogenesis?

18.50 How is the production of ketone bodies related to ketoacidosis?

18.51 Why are more ketone bodies produced in an individual whose diabetes is poorly managed?

18.52 Why do some individuals use test strips to measure the presence of ketone bodies in their urine?

Amino Acid Metabolism

18.53 Draw the structure of the α-keto acid formed by removal of the amino group during the catabolism of each amino acid: (a) glycine; (b) phenylalanine.

18.54 Draw the structure of the α-keto acid formed by removal of the amino group during the catabolism of each amino acid: (a) tyrosine; (b) asparagine.

18.55 What metabolic intermediate is formed from the carbon skeleton of each amino acid?

    a. phenylalanine   c. asparagine

    b. glutamic acid   d. glycine

18.56 What metabolic intermediate is formed from the carbon skeleton of each amino acid?

    a. lysine   c. methionine

    b. tryptophan   d. serine

18.57 What is the difference between ketogenic and glucogenic amino acids?

18.58 Can an amino acid be considered both glucogenic and ketogenic? Explain your choice.

General Questions and Applications

18.59 (a) Convert the ball-and-stick model of **A,** one of the intermediates in glycolysis, to a structural formula, and name the compound. (b) What is the immediate precursor of **A** in glycolysis; that is, what compound forms **A** as a reaction product? (c) What compound is formed from **A** in glycolysis?

A

18.60 (a) Convert the ball-and-stick model of **B,** one of the eight synthetic intermediates in the citric acid cycle, to a structural formula, and name the compound. (b) What is the immediate precursor of **B** in the citric acid cycle; that is, what compound forms **B** as a reaction product? (c) What compound is formed from **B** in the citric acid cycle?

B

18.61 What is the cause of the pain and cramping in a runner's muscles?

18.62 Explain the reaction that occurs during the baking of bread that causes the bread to rise.

18.63 Why is the Atkins low carbohydrate diet called a ketogenic diet?

18.64 What metabolic conditions induce ketogenesis?

CHALLENGE PROBLEMS

18.65 How many moles of ATP can be synthesized from ADP using the 500. Calories ingested during a fast-food lunch? How many molecules of ATP does this correspond to?

18.66 Determine the amount of ATP generated per gram of glucose (molar mass 180.16 g/mol) compared to the amount of ATP generated per gram of stearic acid (284 g/mol) during catabolism. Does your result support or refute the fact that lipids are more effective energy-storing molecules than carbohydrates?

BEYOND THE CLASSROOM

18.67 Examine the nutrition label of a favorite breakfast cereal. Determine the number of grams of digestible carbohydrates in a serving by subtracting the number of grams of dietary fiber (due to indigestible cellulose) from the total carbohydrate content. Assume this value gives the number of grams of glucose per serving. How many moles of ATP are formed from the glucose in one serving of cereal? (Molar mass of glucose = 180.16 g/mol.)

18.68 Many examples of interactions between foods, herbal supplements, and various medications are now well known. Because chemicals in food or herbal remedies can alter the activity of liver enzymes or drug delivery systems in the body, patients are sometimes advised to avoid a certain item in their diet while taking a medication. As an example, individuals who take the cholesterol-lowering drug atorvastatin (trade name: Lipitor) are told to avoid grapefruit juice. Research what is known on the molecular level about the grapefruit juice effect. Is a similar effect observed with other cholesterol-lowering medications such as simvastatin (Zocor), lovastatin (Mevacor), and ezetimibe (Zetia)? Can you find other examples of how foods and herbal supplements affect the properties of a prescription medication?

CHAPTER 18 ANSWERS

Concept Check Answers

18.1 The same reactions and enzymes can be used to metabolize all types of biomolecules.

18.2 The energy change is a negative value;
$$GTP + H_2O \longrightarrow GDP + P_i$$

18.3
$$\text{fructose} \xrightarrow[\quad]{\text{ATP} \quad \text{ADP}} \text{fructose 6-phosphate}$$

18.4 a. $\text{creatine phosphate} \xrightarrow[\quad]{\text{ADP} \quad \text{ATP}} \text{creatine}$
b. Energy is released.

18.5 The many NH and OH groups can hydrogen bond to water, making riboflavin water soluble.

18.6 FAD

18.7 water soluble

18.8 dihydroxyacetone phosphate: ketone, 1° alcohol
glyceraldehyde 3-phosphate: aldehyde, 2° alcohol

18.9 [1] glucose; [3] fructose 6-phosphate; [7] ADP; [10] ADP

18.10 a.

$$^{2-}O_3PO \!-\! \overset{}{\underset{\underset{O}{\|}}{C}} \!-\! O^-$$

phosphoenolpyruvate

b. 2-phosphoglycerate
c. pyruvate

18.11 NADH is the reducing agent in both processes.

18.12 a. In both conversions, CO_2 is lost.
b. The production of acetyl CoA forms NADH, whereas the production of ethanol uses NADH.

18.13 a. Both conversions use NADH.
b. One process involves decarboxylation and one process involves reduction.

18.14 The citric acid cycle would cease.

18.15 a. reduction   b. oxidizing agent

18.16 The intermembrane space has a lower pH because $[H^+]$ is higher.

18.17 a. 7   b. 2.5 for each pyruvate   c. 12   d. 20

18.18 conversion of pyruvate to acetyl CoA and steps [3] and [4] of the citric acid cycle

18.19 steps [7] and [10] of glycolysis and step [5] of the citric acid cycle

18.20 aerobic oxidation: 32 ATPs; anaerobic oxidation: 2 ATPs

18.21 106 ATPs

18.22 134 ATPs

18.23 All ketone bodies contain a carbonyl group.

18.24 β-Hydroxybutyrate does not contain a ketone.

18.25 a. pyruvate   c. succinyl CoA
b. oxaloacetate   d. acetyl CoA, pyruvate, succinyl CoA

18.26 threonine

Practice Problem Answers

18.1 a. reduction using NADH   b. oxidation using NAD^+

18.2 a.

$$\text{NAD}^+ \quad \text{NADH} + \text{H}^+$$

$$\begin{array}{c} CO_2^- \\ | \\ HO-C-H \\ | \\ CH_2 \\ | \\ CO_2^- \end{array} \longrightarrow \begin{array}{c} CO_2^- \\ | \\ C=O \\ | \\ CH_2 \\ | \\ CO_2^- \end{array}$$

b. oxidation

18.3 a.

$$\begin{array}{c} H \qquad CO_2^- \\ \diagdown \quad / \\ C=C \\ / \quad \diagdown \\ {}^-O_2C \qquad H \end{array}$$

A
fumarate

b. succinate   c. malate

18.4 a. [1] 10; [2] 9 cycles   b. [1] 8; [2] 7 cycles

18.5 a. $CH_3CH(OH)\!-\!\overset{\overset{O}{\|}}{C}\!-\!CO_2^- \ + \ NH_4^+$

b. $H\!-\!\overset{\overset{O}{\|}}{C}\!-\!CO_2^- \ + \ NH_4^+$

Self-Test Answers

1. a. fermentation   c. glycolysis   e. ketogenesis
b. citric acid cycle   d. β-oxidation

2. a. reducing agent   b. neither   c,d: oxidizing agent

3. a. 11   b. 10   c. 148

4. a. NAD^+ b. FAD   c. ATP   d. H_2O

5. $\text{fructose 6-phosphate} \xrightarrow[\quad]{\text{ATP} \quad \text{ADP}} \text{fructose 1,6-bisphosphate}$

6. $CH_3CH_2CH(CH_3)COCO_2^- \ + \ NH_4^+$

7. $2\ CH_3COCO_2^- \ + \ 2\ NADH \ + \ 2\ ATP$

Odd-Numbered Problem Answers

18.1 a,e: stage [1]   c. stage [3]
b. stage [2]   d. stage [4]

18.3 When a substrate is oxidized, NAD^+ is reduced. NAD^+ is an oxidizing agent.

18.5 a. steps [3], [4], and [8]   c. step [6]
b. steps [3] and [4]   d. step [1]

18.7 Energy-investment phase: (a) glucose, glyceraldehyde 3-phosphate, (b) 2 ATP utilized, (c) no coenzymes used or formed
Energy-generating phase: (a) glyceraldehyde 3-phosphate, pyruvate, (b) 4 ATP/glucose produced, (c) 2 NAD^+ used and 2 NADH produced per glucose

18.9 a. 3   b. 2

18.11 $CH_3SCH_2CH_2 \overset{\overset{O}{\|}}{-}C-CO_2^- + NH_4^+$

18.13 Pyruvate is metabolized anaerobically to lactate.

18.15 c, b, e, a, d

18.17 a. fructose 1,6-bisphosphate $\xrightarrow[\text{ADP} \quad \text{ATP}]{}$ fructose 6-phosphate

b. The reaction absorbs energy.

c. This reaction is energetically unfavorable.

18.19 a. reducing agent   b. neither   c. oxidizing agent

18.21

18.23 a. 7,10   b. 1,3   c. 6   d. 4

18.25 The conversion of pyruvate to acetyl CoA forms one CO_2 for each pyruvate. The conversion of isocitrate to α-ketoglutarate and α-ketoglutarate to succinyl CoA in the citric acid cycle forms one CO_2 for each acetyl CoA.

18.27 a. aerobic: CO_2; anaerobic: lactate

b. aerobic: NAD^+ and FAD are used and NADH and $FADH_2$ are formed; anaerobic: NADH is used to convert pyruvate to lactate and NAD^+ is formed.

18.29 a. lactate   b. ethanol   c. CO_2

18.31 During glycolysis: 2 ATP and 2 NADH generated.
Each NADH yields 2.5 ATP, so glycolysis yields 7 ATP total.
Conversion of 2 pyruvate to 2 acetyl CoA yields 2 NADH, which gives 5 ATP.
In the citric acid cycle, 20 additional ATP are formed for
2 $CH_3COSCoA$: 2 GTP (ATP), 6 NADH $\longrightarrow$ 15 ATP,
2 $FADH_2$ $\longrightarrow$ 3 ATP.
Total: 32 ATP.

18.33 a. isocitrate   b. malate, isocitrate

18.35 They require oxygen.

18.37 a. $FADH_2$ donates electrons to the electron transport chain.

b. ADP is a substrate for the formation of ATP.

c. ATP synthase catalyzes the formation of ATP from ADP.

d. The inner mitochondrial membrane contains the four complexes for the electron transport chain and the ATP synthase enzyme.

18.39 NAD^+, FAD, and H_2O

18.41 Two phosphate bonds of ATP are hydrolyzed, forming AMP; therefore, 2 ATP equivalents are used.

18.43 7 molecules of acetyl CoA   6 cycles

18.45 92

18.47 a. $CH_3(CH_2)_6 \underset{\beta}{-}CH_2 \underset{\alpha}{-}CH_2-CO_2H$

b. $CH_3(CH_2)_6 \underset{\beta}{-}CH_2 \underset{\alpha}{-}CH_2-COSCoA$

c. 5   d. 4   e. 64

18.49 Ketosis is the condition under which ketone bodies accumulate. Ketogenesis is the synthesis of ketone bodies from acetyl CoA.

18.51 In poorly controlled diabetes, glucose cannot be metabolized. Fatty acids are used for metabolism and ketone bodies are formed to a greater extent.

18.53 a. $H\overset{\overset{O}{\|}}{-}C-CO_2^-$ b.

18.55 a. acetoacetyl CoA, fumarate   c. oxaloacetate

b. α-ketoglutarate   d. pyruvate

18.57 Ketogenic amino acids are converted to acetyl CoA or a related thioester and can be converted to ketone bodies. Glucogenic amino acids are catabolized to pyruvate or another intermediate in the citric acid cycle.

18.59 a.

A
fructose 6-phosphate

b. glucose 6-phosphate

c. fructose 1,6-bisphosphate

18.61 Cramping is due to lactic acid buildup formed from anaerobic metabolism of glucose.

18.63 The diet calls for ingestion of protein and fat, so in the absence of carbohydrates to be metabolized, ketone bodies are formed.

18.65 68 moles, 4.1×10^{25} molecules

The Elements

Element	Symbol	Atomic Number	Relative Atomic Mass*	Element	Symbol	Atomic Number	Relative Atomic Mass*
Actinium	Ac	89	(227)	Mendelevium	Md	101	(258)
Aluminum	Al	13	26.98	Mercury	Hg	80	200.59
Americium	Am	95	(243)	Molybdenum	Mo	42	95.95
Antimony	Sb	51	121.76	Moscovium	Mc	115	(289)
Argon	Ar	18	39.79	Neodymium	Nd	60	144.24
Arsenic	As	33	74.92	Neon	Ne	10	20.18
Astatine	At	85	(210)	Neptunium	Np	93	(237)
Barium	Ba	56	137.33	Nickel	Ni	28	58.69
Berkelium	Bk	97	(249)	Nihonium	Nh	113	(285)
Beryllium	Be	4	9.012	Niobium	Nb	41	92.91
Bismuth	Bi	83	208.98	Nitrogen	N	7	14.01
Bohrium	Bh	107	(272)	Nobelium	No	102	(259)
Boron	B	5	10.81	Oganesson	Og	118	(294)
Bromine	Br	35	79.90	Osmium	Os	76	190.23
Cadmium	Cd	48	112.41	Oxygen	O	8	16.00
Calcium	Ca	20	40.08	Palladium	Pd	46	106.42
Californium	Cf	98	(252)	Phosphorus	P	15	30.97
Carbon	C	6	12.01	Platinum	Pt	78	195.08
Cerium	Ce	58	140.12	Plutonium	Pu	94	(239)
Cesium	Cs	55	132.91	Polonium	Po	84	(210)
Chlorine	Cl	17	35.45	Potassium	K	19	39.10
Chromium	Cr	24	52.00	Praseodymium	Pr	59	140.91
Cobalt	Co	27	58.93	Promethium	Pm	61	(145)
Copernicium	Cn	112	(285)	Protactinium	Pa	91	231.04
Copper	Cu	29	63.55	Radium	Ra	88	(226)
Curium	Cm	96	(247)	Radon	Rn	86	(222)
Darmstadtium	Ds	110	(281)	Rhenium	Re	75	186.21
Dubnium	Db	105	(262)	Rhodium	Rh	45	102.91
Dysprosium	Dy	66	162.5	Roentgenium	Rg	111	(280)
Einsteinium	Es	99	(252)	Rubidium	Rb	37	85.47
Erbium	Er	68	167.26	Ruthenium	Ru	44	101.07
Europium	Eu	63	151.96	Rutherfordium	Rf	104	(261)
Fermium	Fm	100	(257)	Samarium	Sm	62	150.36
Flerovium	Fl	114	(287)	Scandium	Sc	21	44.96
Fluorine	F	9	19.00	Seaborgium	Sg	106	(266)
Francium	Fr	87	(223)	Selenium	Se	34	78.97
Gadolinium	Gd	64	157.25	Silicon	Si	14	28.08
Gallium	Ga	31	69.72	Silver	Ag	47	107.87
Germanium	Ge	32	72.63	Sodium	Na	11	22.99
Gold	Au	79	196.97	Strontium	Sr	38	87.62
Hafnium	Hf	72	178.49	Sulfur	S	16	32.06
Hassium	Hs	108	(277)	Tantalum	Ta	73	180.95
Helium	He	2	4.003	Technetium	Tc	43	(98)
Holmium	Ho	67	164.93	Tellurium	Te	52	127.6
Hydrogen	H	1	1.008	Tennessine	Ts	117	(293)
Indium	In	49	114.82	Terbium	Tb	65	158.93
Iodine	I	53	126.90	Thallium	Tl	81	204.38
Iridium	Ir	77	192.22	Thorium	Th	90	232.04
Iron	Fe	26	55.85	Thulium	Tm	69	168.93
Krypton	Kr	36	83.80	Tin	Sn	50	118.71
Lanthanum	La	57	138.91	Titanium	Ti	22	47.88
Lawrencium	Lr	103	(262)	Tungsten	W	74	183.84
Lead	Pb	82	207.2	Uranium	U	92	238.03
Lithium	Li	3	6.938	Vanadium	V	23	50.94
Livermorium	Lv	116	(291)	Xenon	Xe	54	131.29
Lutetium	Lu	71	174.97	Ytterbium	Yb	70	173.05
Magnesium	Mg	12	24.30	Yttrium	Y	39	88.91
Manganese	Mn	25	54.94	Zinc	Zn	30	65.38
Meitnerium	Mt	109	(276)	Zirconium	Zr	40	91.22

*Values in parentheses represent the mass number of the most stable isotope.

Appendix B

Periodic Table of the Elements

1A (1)	2A (2)	3B (3)	4B (4)	5B (5)	6B (6)	7B (7)	8B (8)	8B (9)	8B (10)	1B (11)	2B (12)	3A (13)	4A (14)	5A (15)	6A (16)	7A (17)	8A (18)
1 **H** 1.008																	2 **He** 4.003
3 **Li** 6.938	4 **Be** 9.012											5 **B** 10.81	6 **C** 12.01	7 **N** 14.01	8 **O** 16.00	9 **F** 19.00	10 **Ne** 20.18
11 **Na** 22.99	12 **Mg** 24.30											13 **Al** 26.98	14 **Si** 28.08	15 **P** 30.97	16 **S** 32.06	17 **Cl** 35.45	18 **Ar** 39.79
19 **K** 39.10	20 **Ca** 40.08	21 **Sc** 44.96	22 **Ti** 47.87	23 **V** 50.94	24 **Cr** 52.00	25 **Mn** 54.94	26 **Fe** 55.85	27 **Co** 58.93	28 **Ni** 58.69	29 **Cu** 63.55	30 **Zn** 65.38	31 **Ga** 69.72	32 **Ge** 72.63	33 **As** 74.92	34 **Se** 78.97	35 **Br** 79.90	36 **Kr** 83.80
37 **Rb** 85.47	38 **Sr** 87.62	39 **Y** 88.91	40 **Zr** 91.22	41 **Nb** 92.91	42 **Mo** 95.95	43 **Tc** (98)	44 **Ru** 101.07	45 **Rh** 102.91	46 **Pd** 106.42	47 **Ag** 107.87	48 **Cd** 112.41	49 **In** 114.82	50 **Sn** 118.71	51 **Sb** 121.76	52 **Te** 127.6	53 **I** 126.90	54 **Xe** 131.29
55 **Cs** 132.91	56 **Ba** 137.33	57–71 Lanthanides	72 **Hf** 178.49	73 **Ta** 180.95	74 **W** 183.84	75 **Re** 186.21	76 **Os** 190.23	77 **Ir** 192.22	78 **Pt** 195.08	79 **Au** 196.97	80 **Hg** 200.59	81 **Tl** 204.38	82 **Pb** 207.2	83 **Bi** 208.98	84 **Po** (210)	85 **At** (210)	86 **Rn** (222)
87 **Fr** (223)	88 **Ra** (226)	89–103 Actinides	104 **Rf** (261)	105 **Db** (262)	106 **Sg** (266)	107 **Bh** (272)	108 **Hs** (277)	109 **Mt** (276)	110 **Ds** (281)	111 **Rg** (280)	112 **Cn** (285)	113 **Nh** (285)	114 **Fl** (287)	115 **Mc** (289)	116 **Lv** (291)	117 **Ts** (293)	118 **Og** (294)

Lanthanides (6)

57 **La** 138.91	58 **Ce** 140.12	59 **Pr** 140.91	60 **Nd** 144.24	61 **Pm** (145)	62 **Sm** 150.36	63 **Eu** 151.96	64 **Gd** 157.25	65 **Tb** 158.93	66 **Dy** 162.5	67 **Ho** 164.93	68 **Er** 167.26	69 **Tm** 168.93	70 **Yb** 173.05	71 **Lu** 174.97

Actinides (7)

89 **Ac** (227)	90 **Th** 232.04	91 **Pa** 231.04	92 **U** 238.03	93 **Np** (237)	94 **Pu** (239)	95 **Am** (243)	96 **Cm** (247)	97 **Bk** (249)	98 **Cf** (252)	99 **Es** (252)	100 **Fm** (257)	101 **Md** (258)	102 **No** (259)	103 **Lr** (262)

Legend: nonmetal · metalloid · metal

Useful Prefixes, Conversion Factors, Physical Constants, and Common Element Colors

Common Metric Prefixes

Prefix	Symbol	Numerical Value
giga-	G	10^9
mega-	M	10^6
kilo-	k	10^3
deci-	d	10^{-1}
centi-	c	10^{-2}
milli-	m	10^{-3}
micro-	μ	10^{-6}
nano-	n	10^{-9}

Useful Conversion Factors

Length	Mass	Volume
1 m = 100 cm = 1,000 mm	1 kg = 1,000 g	1 L = 1,000 mL
1 cm = 10 mm	1 g = 1,000 mg	1 mL = 1 cm^3 = 1 cc
2.54 cm = 1 in.	1 kg = 2.205 lb	946.4 mL = 1 qt
1 m = 39.37 in.	453.6 g = 1 lb	1 L = 1.057 qt
1 km = 0.6214 mi	28.35 g = 1 oz	29.57 mL = 1 fl oz

Temperature	Pressure	Energy
$T_F = \dfrac{9}{5}(T_C) + 32$	1 atm = 14.70 psi	1 kcal = 1,000 cal
$T_C = \dfrac{5}{9}(T_F - 32)$	1 atm = 760. mm Hg	1 cal = 4.184 J
$T_K = T_C + 273.15$	1 mm Hg = 1 torr	

Physical Constants

Avogadro's number	6.022×10^{23} particles/mol
Universal gas constant (R)	0.08206 L · atm/(mol · K) or 62.4 L · mm Hg/(mol · K)
Volume of one mole of gas at STP	22.4 L

Common Element Colors Used in Molecular Art

C · · · H · · · O · · · N · · · F · · · Cl · · · Br · · · I · · · S · · · P

Appendix D

Useful Mathematical Concepts

Three common mathematical concepts are needed to solve many problems in chemistry:

- Using scientific notation
- Determining the number of significant figures
- Using a scientific calculator

Scientific Notation

To write numbers that contain many leading zeros (at the beginning) or trailing zeros (at the end), scientists use **scientific notation.**

- In scientific notation, a number is written as $y \times 10^x$, where y (the coefficient) is a number between 1 and 10, and x is an exponent, which can be any positive or negative whole number.

To convert a standard number to scientific notation:

1. Move the decimal point to give a number between 1 and 10.
2. Multiply the result by 10^x, where x is the number of places the decimal point was moved.
   - If the decimal point is moved to the **left,** x is **positive.**
   - If the decimal point is moved to the **right,** x is **negative.**

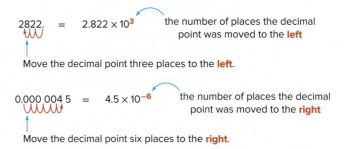

To convert a number in scientific notation to a standard number, use the value of x in 10^x to indicate the number of places to move the decimal point in the coefficient.

- **Move the decimal point to the right when x is positive.**
- **Move the decimal point to the left when x is negative.**

Table D.1 Numbers in Standard Form and Scientific Notation

Number	Scientific Notation
26,200	2.62×10^4
0.006 40	6.40×10^{-3}
3,000,000	3×10^6
0.000 000 139	1.39×10^{-7}
2,000.20	2.00020×10^3

Table D.1 shows how several numbers are written in scientific notation.

Often, numbers written in scientific notation must be multiplied or divided.

- To multiply two numbers in scientific notation, *multiply* the coefficients together and *add* the exponents in the powers of 10.

- To divide two numbers in scientific notation, *divide* the coefficients and *subtract* the exponents in the powers of 10.

$$\text{Divide coefficients.} \quad \frac{6.0 \times 10^6}{2.0 \times 10^{10}} \quad \text{Subtract exponents.} \quad = \quad 3.0 \times 10^{-4}$$
$$(6.0 \div 2.0) \qquad\qquad (6 - 10)$$

Table D.2 Calculations Using Numbers Written in Scientific Notation

Calculation	Answer
$(3.5 \times 10^3) \times (2.2 \times 10^{22}) =$	7.7×10^{25}
$(3.5 \times 10^3) \div (2.2 \times 10^{22}) =$	1.6×10^{-19}
$(3.5 \times 10^3) \times (2.2 \times 10^{-10}) =$	7.7×10^{-7}
$(3.5 \times 10^3) \div (2.2 \times 10^{-10}) =$	1.6×10^{13}

Table D.2 shows the result of multiplying or dividing several numbers written in scientific notation.

Significant Figures

Whenever we measure a number, there is a degree of uncertainty associated with the result. The last number (farthest to the right) is an estimate. **Significant figures** are all of the digits in a measured number including one estimated digit. How many significant figures are contained in a number?

- All nonzero digits are always significant.
- A zero *counts* as a significant figure when it occurs between two nonzero digits, or at the end of a number with a decimal point.
- A zero does *not* count as a significant figure when it occurs at the beginning of a number, or at the end of a number that does not have a decimal point.

Table D.3 lists the number of significant figures in several quantities.

Table D.3 Examples Illustrating Significant Figures

Quantity	Number of Significant Figures	Quantity	Number of Significant Figures
1,267 g	Four	203 L	Three
24,345 km	Five	6.10 atm	Three
1.200 mg	Four	0.3040 g	Four
0.000 001 mL	One	1,200 m	Two

The number of significant figures must also be taken into account in calculations. To avoid reporting a value with too many digits, we must often **round off the number** to give the correct number of significant figures. Two rules are used in rounding off numbers.

- If the first number that must be dropped is 4 or fewer, drop it and all remaining numbers.
- If the first number that must be dropped is 5 or greater, *round the number up* by adding one to the last digit that will be retained.

To round 63.854 to two significant figures:

These digits must be retained.

$$63.\underline{854}$$ first digit to be dropped

These digits must be dropped.

- Because the first digit to be dropped is **8** (5 or greater), add 1 to the first digit to its left.
- The number 63.854 rounded to two digits is **64.**

Table D.4 gives other examples of rounding off numbers.

Table D.4 Rounding Off Numbers

Original Number	Rounded to	Rounded Number
15.2538	Two places	15
15.2538	Three places	15.3
15.2538	Four places	15.25
15.2538	Five places	15.254

The first number to be dropped is indicated in **red** in each original number.

The number of significant figures in the answer of a problem depends on the type of mathematical calculation—multiplication (and division) or addition (and subtraction).

- **In multiplication and division, the answer has the same number of significant figures as the original number with the *fewest* significant figures.**

five significant figures   first digit to be dropped

$$5.5067 \quad \times \quad 2.6 \quad = \quad 14.\underline{3}1742 \text{ rounded to } \textbf{14}$$

two significant figures

The answer must contain only **two** significant figures.

- **In addition and subtraction, the answer has the same number of decimal places as the original number with the *fewest* decimal places.**

two digits after the decimal point

$$10.17 \quad + \quad 3.5 \quad = \quad 13.\underline{6}7 \text{ rounded to } \textbf{13.7}$$

one digit after the decimal point

last significant digit   The answer can have only **one** digit after the decimal point.

Table D.5 lists other examples of calculations that take into account the number of significant figures.

Table D.5 Calculations Using Significant Figures

Calculation	Answer
3.2 × 699 =	2,236.8 rounded to **2,200**
4.66892 ÷ 2.13 =	2.191981221 rounded to **2.19**
25.3 + 3.668 + 29.1004 =	58.0684 rounded to **58.1**
95.1 − 26.335 =	68.765 rounded to **68.8**

Using a Scientific Calculator

A scientific calculator is capable of carrying out more complicated mathematical functions than simple addition, subtraction, multiplication, and division. For example, these calculators allow the user to convert a standard number to scientific notation, as well as readily determine the logarithm (log) or antilogarithm (antilog) of a value. Carrying out these operations is especially important in determining pH or hydronium ion concentration in Chapter 8.

Described in this section are the steps that can be followed in calculations with some types of calculators. Consult your manual if these steps do not produce the stated result.

Converting a Number to Scientific Notation

To convert a number from its standard form to a number in scientific notation, follow the steps in columns [1]–[3].

	① Number to enter	② Buttons to press	③ Display	Meaning
A number ≥ 1:	1,200	2nd SCI	1.2^{03}	1.2×10^3
A **decimal** < 1:	0.052	2nd SCI	5.2^{-02}	5.2×10^{-2}

Entering a Number Written in Scientific Notation

To enter a number in scientific notation, follow the steps in columns [1]–[3].

	① Number to enter	② Buttons to press	③ Display	Meaning
A number with a **positive** exponent, as in 1.5×10^8	1.5	EE 8	1.5^{08}	150,000,000
A number with a **negative** exponent, as in 3.5×10^{-4}	3.5	EE 4 Change sign (+ to –)	3.5^{-04}	0.000 35

Taking the Logarithm of a Number: Calculating pH from a Known [H₃O⁺]

Because $pH = -\log [H_3O^+]$, a calculator is used to determine pH values. To calculate pH from a known hydronium ion concentration, follow the steps in columns [1]–[3].

[H₃O⁺]	① Enter [H₃O⁺] in scientific notation	② Display	③ Buttons to press	pH
2.0×10^{-3} M	Press: 2.0 EE 3 Change sign (+ to –)	2.0^{-03}	LOG Change sign (+ to –)	2.70
7.3×10^{-8} M	Press: 7.3 EE 8 Change sign (+ to –)	7.3^{-08}	LOG Change sign (+ to –)	7.14

Taking the Antilogarithm of a Number: Calculating $[H_3O^+]$ from a Known pH

Because $[H_3O^+]$ = antilog($-$pH), a calculator is used to calculate an antilogarithm—that is, the number that has a given logarithm value—to determine the hydronium ion concentration from a given pH. To calculate $[H_3O^+]$ from a known pH, follow the steps in columns [1]–[3].

pH	① Number to enter	② Buttons to press	③ Display	$[H_3O^+]$
8.50	8.50	Change sign (+ to –) 2ⁿᵈ LOG	3.16227766017^{-09}	3.2×10^{-9} M
3.91	3.91	Change sign (+ to –) 2ⁿᵈ LOG	1.23026877081^{-04}	1.2×10^{-4} M

Table D.6 lists pH values that correspond to given $[H_3O^+]$ values. You can practice using a calculator to determine pH or $[H_3O^+]$ by entering a value in one column, following the listed steps, and then checking to see if you obtain the corresponding value in the other column.

Table D.6 The pH of a Solution from a Given Hydronium Ion Concentration $[H_3O^+]$

$[H_3O^+]$	pH	$[H_3O^+]$	pH
1.8×10^{-10}	9.74	4.0×10^{-13}	12.40
3.8×10^{-2}	1.42	6.6×10^{-4}	3.18
5.0×10^{-12}	11.30	2.6×10^{-9}	8.59
4.2×10^{-7}	6.38	7.3×10^{-2}	1.14

Glossary

A

Acetyl CoA (18.3) A compound formed when an acetyl group (CH_3CO-) is bonded to coenzyme A (HS–CoA); $CH_3COSCoA$.

Achiral (12.11) Being superimposable on a mirror image.

Acid (8.1) In the Arrhenius definition, a substance that contains a hydrogen atom and dissolves in water to form a hydrogen ion, H^+.

Acidic solution (8.4) A solution in which $[H_3O^+] > [OH^-]$; thus, $[H_3O^+] > 10^{-7}$ M.

Actinides (2.4) A group of 15 elements in the periodic table beginning with the element actinium (Ac, $Z = 89$).

Active site (16.10) The region in an enzyme that binds a substrate, which then undergoes a very specific reaction with an enhanced rate.

Active transport (15.7) The process of moving an ion across a cell membrane that requires energy input.

Acyclic alkane (10.5) An alkane with molecular formula C_nH_{2n+2}, which contains a chain of carbon atoms but no rings. An acyclic alkane is also called a saturated hydrocarbon.

Acyl CoA (18.9) The thioester formed from a fatty acid and coenzyme A that undergoes β-oxidation in mitochondria; general structure RCOSCoA.

Addition reaction (11.5) A reaction in which elements are added to a compound.

Adenosine 5'-diphosphate (ADP, 18.2) A nucleoside diphosphate formed by adding two phosphates to the 5'-OH group of adenosine.

Adenosine 5'-triphosphate (ATP, 18.2) A nucleoside triphosphate formed by adding three phosphates to the 5'-OH group of adenosine. ATP is the most prominent member of a group of "high-energy" molecules that release energy during hydrolysis.

Adrenal cortical steroid (15.9) A steroid hormone synthesized in the outer layer of the adrenal gland.

Alcohol (10.4, 12.1) A compound containing a hydroxyl group (OH) bonded to a tetrahedral carbon atom; general formula ROH.

Aldehyde (10.4, 12.1) A compound that has a hydrogen atom bonded directly to a carbonyl carbon; general formula RCHO.

Alditol (14.4) A compound produced when the carbonyl group of an aldose is reduced to a 1° alcohol.

Aldonic acid (14.4) A compound produced when the aldehyde carbonyl of an aldose is oxidized to a carboxyl group.

Aldose (14.2) A monosaccharide with an aldehyde carbonyl group at C1.

Alkali metal (2.4) An element located in group 1A (group 1) of the periodic table. Alkali metals include lithium (Li), sodium (Na), potassium (K), rubidium (Rb), cesium (Cs), and francium (Fr).

Alkaline earth element (2.4) An element located in group 2A (group 2) of the periodic table. Alkaline earth elements include beryllium (Be), magnesium (Mg), calcium (Ca), strontium (Sr), barium (Ba), and radium (Ra).

Alkaloid (13.7) A naturally occurring amine derived from a plant source.

Alkane (10.4) A compound having only C—C and C—H single bonds.

Alkene (10.4, 11.1) A compound having a carbon–carbon double bond.

Alkyl group (10.6) A group formed by removing one hydrogen from an alkane.

Alkyl halide (10.4, 12.6) A compound with the general structure R–X that contains a halogen atom (X = F, Cl, Br, or I) bonded to a tetrahedral carbon.

Alkyne (10.4, 11.1) A compound with a carbon–carbon triple bond.

Alpha (α) particle (9.1) A high-energy particle that is emitted from a radioactive nucleus and contains two protons and two neutrons.

Amide (10.4, 13.1) A compound that contains a nitrogen atom bonded directly to a carbonyl carbon; general structure $RCONR'_2$, where R' = H or alkyl.

Amine (10.4, 13.1) An organic compound that contains a nitrogen atom bonded to one, two, or three alkyl groups; general structure RNH_2, R_2NH, or R_3N.

Amino acid (16.2) A compound that contains two functional groups—an amino group (NH_2) and a carboxyl group (COOH) bonded to the same carbon.

Amino group (10.4) An $-NH_2$ group.

Ammonium ion (3.6) An NH_4^+ ion.

Ammonium salt (13.8) An ionic compound that contains a positively charged ammonium ion and an anion. $(CH_3CH_2CH_2NH_3)^+Cl^-$ is an ammonium salt.

Amphoteric compound (8.2) A compound that contains both a hydrogen atom and a lone pair of electrons, so that it can be either an acid or a base.

Anabolic steroid (15.9) A synthetic androgen analogue that promotes muscle growth.

Anabolism (18.1) The synthesis of large molecules from smaller ones in a metabolic pathway.

Androgen (15.9) A hormone that controls the development of secondary sex characteristics in males.

Anion (3.2) A negatively charged ion with more electrons than protons.

Anticodon (17.8) Three nucleotides in a tRNA molecule that are complementary to the codon in mRNA and identify an individual amino acid.

Antioxidant (11.10) A compound that prevents an unwanted oxidation reaction from occurring.

Aqueous solution (7.1) A solution with water as the solvent.

Aromatic compound (10.4, 11.8) A compound that contains a benzene ring, a six-membered ring with three double bonds.

Atmosphere (6.1) A unit used to measure pressure; 1 atm = 760. mm Hg.

Atom (2.2) The basic building block of matter composed of a nucleus and an electron cloud.

Atomic mass unit (2.2) A unit abbreviated as amu, which equals one-twelfth the mass of a carbon atom that has six protons and six neutrons; 1 amu = 1.661×10^{-24} g.

Atomic number (2.2) The number of protons in the nucleus of an atom; symbolized as Z.

Atomic weight (2.3) The weighted average of the mass of all naturally occurring isotopes of a particular element, reported in atomic mass units.

Avogadro's law (6.6) A gas law that states that the volume of a gas is proportional to the number of moles present when the pressure and temperature are held constant.

Avogadro's number (5.5) A quantity that contains 6.022×10^{23} items—usually atoms, molecules, or ions.

B

Balanced chemical equation (5.2) An equation written so that an equal number of atoms of each element is present on both sides.

Barometer (6.1) A device for measuring atmospheric pressure.

Base (8.1) In the Arrhenius definition, a substance that contains hydroxide and dissolves in water to form OH^-.

Basic solution (8.4) A solution in which $[OH^-] > [H_3O^+]$; thus, $[OH^-] > 10^{-7}$ M.

Becquerel (9.4) An SI unit used to measure radioactivity, abbreviated as Bq; 1 Bq = 1 disintegration/s.

Benedict's reagent (14.4) A Cu^{2+} reagent that oxidizes the aldehyde carbonyl of an aldose to a carboxyl group, yielding an aldonic acid.

Beta (β) particle (9.1) A high-energy electron emitted from a radioactive nucleus.

Boiling point (4.4) The temperature at which a liquid is converted to the gas phase.

Bonding (3.1) The joining of two atoms in a stable arrangement.

Boyle's law (6.2) A gas law that relates pressure and volume. Boyle's law states that for a fixed amount of gas at constant temperature, the pressure and volume of the gas are inversely related.

Branched-chain alkane (10.5) An alkane that contains one or more carbon branches bonded to a carbon chain.

Brønsted–Lowry acid (8.1) A proton donor.

Brønsted–Lowry base (8.1) A proton acceptor.

Buffer (8.8) A solution whose pH changes very little when acid or base is added. Most buffers are solutions composed of approximately equal amounts of a weak acid and the salt of its conjugate base.

Building-block element (2.1) One of the four nonmetals—oxygen, carbon, hydrogen, and nitrogen—that comprise 96% of the mass of the human body.

C

Calorie (4.1) A unit of energy that equals the amount of energy needed to raise the temperature of 1 g of water by 1 °C; abbreviated as cal, where 1 cal = 4.184 J.

Carbohydrate (14.1) A polyhydroxy aldehyde or ketone, or a compound that can be hydrolyzed to a polyhydroxy aldehyde or ketone.

Carbonate (3.6) A polyatomic anion with the structure CO_3^{2-}.

Carbonyl group (10.4, 12.1) A carbon–oxygen double bond (C=O).

Carboxylase (16.9) An enzyme that catalyzes bond formation between a substrate and CO_2.

Carboxylate anion (13.5) The conjugate base of a carboxylic acid; general structure $RCOO^-$.

Carboxyl group (10.4, 13.1) A COOH group.

Carboxylic acid (10.4, 13.1) A compound that contains an OH group bonded directly to the carbonyl carbon; general structure RCOOH or RCO_2H.

Catabolism (18.1) The breakdown of large molecules into smaller ones during metabolism.

Catalyst (5.10) A substance that increases the rate of a reaction but is recovered unchanged at the end of the reaction.

Cation (2.8) A positively charged particle with fewer electrons than protons.

Cell membrane (15.7) The semipermeable membrane that surrounds the cell, composed of a lipid bilayer.

Celsius scale (1.9) One of three temperature scales in which water freezes at 0 °C and boils at 100 °C.

Cephalin (15.6) A phosphoacylglycerol in which the identity of the R group esterified to the phosphodiester is $-CH_2CH_2NH_3^+$.

Chain reaction (9.6) The process by which each neutron produced during fission can go on to bombard three other nuclei to produce more nuclei and more neutrons.

Charles's law (6.3) A gas law that states that for a fixed amount of gas at constant pressure, the volume of the gas is proportional to its Kelvin temperature.

Chemical equation (5.1) An expression that uses chemical formulas and other symbols to illustrate what reactants constitute the starting materials in a reaction and what products are formed.

Chemical formula (2.1) A representation that uses element symbols to show the identity of elements in a compound, and subscripts to show the number of atoms of each element contained in the compound.

Chemical properties (1.2) Those properties that determine how a substance can be converted to another substance by a chemical reaction.

Chemistry (1.1) The study of matter—its composition, properties, and transformations.

Chiral (12.11) Not superimposable on a mirror image.

Chirality center (12.11) A carbon atom bonded to four different groups.

Chlorofluorocarbons (CFCs, 2.4, 6.9) Compounds containing the elements of carbon, chlorine, and fluorine that are implicated in the destruction of ozone in the upper atmosphere.

Cis isomer (11.3) An alkene with two R groups on the same side of the double bond.

Citric acid cycle (18.6) A cyclic metabolic pathway that begins with the addition of acetyl CoA to a four-carbon substrate and ends when the same four-carbon compound is formed as a product eight steps later.

Codon (17.7) A sequence of three nucleotides (triplet) in mRNA that codes for a specific amino acid.

Coenzyme (16.9) An organic molecule needed for an enzyme-catalyzed reaction to occur.

Coenzyme A (18.3) A coenzyme that contains a sulfhydryl group (SH group) making it a thiol (RSH), and abbreviated as HS–CoA.

Cofactor (16.9) A metal ion or a nonprotein organic molecule needed for an enzyme-catalyzed reaction to occur.

Colloid (7.1) A homogeneous mixture with large particles, often having an opaque appearance.

Combination reaction (5.3) A reaction in which two or more reactants are joined to form a single product.

Combined gas law (6.5) A gas law that relates pressure, volume, and temperature. For a constant number of moles, the product of pressure and volume divided by temperature is a constant.

Combustion (10.10) An oxidation reaction in which carbon-containing compounds react with oxygen to form carbon dioxide (CO_2) and water.

Competitive inhibitor (16.10) An enzyme inhibitor that has a shape and structure similar to the substrate, and competes with the substrate for binding to the active site.

Complementary base pairs (17.3) The predictable pairing of bases between two strands of DNA. Adenine pairs with thymine using two hydrogen bonds, forming an A–T base pair, and cytosine pairs with guanine using three hydrogen bonds, forming a C–G base pair.

Compound (1.3) A pure substance formed by chemically combining two or more elements.

Concentration (7.5) The amount of solute dissolved in a given amount of solution.

Condensation (4.6) The conversion of a gas to a liquid.

Condensed structure (10.3) A representation used for a compound having a chain of atoms bonded together. The atoms are drawn in, but the two-electron bond lines and lone pairs on heteroatoms are generally omitted.

Conjugate acid (8.2) The product formed by the gain of a proton by a base.

Conjugate acid–base pair (8.2) Two species that differ by the presence of a proton.

Conjugate base (8.2) The product formed by loss of a proton from an acid.

Conjugated protein (16.7) A compound composed of a protein unit and a nonprotein molecule.

Constitutional isomers (10.5) Isomers that differ in the way the atoms are connected to each other.

Conversion factor (1.7) A term that converts a quantity in one unit to a quantity in another unit.

Cooling curve (4.7) A graph that shows how the temperature of a substance changes as heat is removed.

Coronaviruses (17.11) A group of retroviruses that cause diseases such as the common cold, SARS, MERS, and COVID-19.

Covalent bond (3.7) A chemical bond that results from the sharing of electrons between two atoms.

Critical mass (9.6) The amount of a radioactive element required to sustain a chain reaction.

Cross formula (12.11) A Fischer projection formula that replaces a chirality center with a cross. The horizontal lines represent wedged bonds and the vertical lines represent dashed wedges.

C-Terminal amino acid (16.4) In a peptide, the amino acid with the free –COO⁻ group on the α carbon.

Cubic centimeter (1.4) A unit of volume equal to one milliliter; one cubic centimeter = 1 cm^3 = 1 cc.

Curie (9.4) A unit used to measure radioactivity and equal to 3.7×10^{10} disintegrations/s. A curie corresponds to the decay rate of one gram of the element radium.

Cycloalkane (10.5) A compound with the general formula C_nH_{2n} that contains carbons joined in one or more rings.

D

Dalton's law (6.8) A law that states that the total pressure (P_{total}) of a gas mixture is equal to the sum of the partial pressures of its component gases.

Decarboxylase (16.9) An enzyme that catalyzes the removal of CO_2.

Decomposition reaction (5.3) A reaction in which a single reactant is converted to two or more products.

Dehydrase (16.9) An enzyme that catalyzes the removal of water.

Dehydration (12.5) The loss of water (H_2O) from a starting material.

Dehydrogenase (16.9) An enzyme that catalyzes the addition or removal of two hydrogen atoms.

Deletion mutation (17.9) The loss of one or more nucleotides from a DNA molecule.

Denaturation (16.8) The process of altering the shape of a protein without breaking the amide bonds that form the primary structure.

Density (1.10) A physical property that relates the mass of a substance to its volume; density = g/(mL or cc).

Deoxyribonucleic acid (DNA, 17.1, 17.3) A polymer of deoxyribonucleotides that stores the genetic information of an organism and transmits that information from one generation to another.

Deoxyribonucleoside (17.1) A compound that contains the monosaccharide 2-deoxyribose and a purine or pyrimidine base.

Deoxyribonucleotide (17.1) A compound that contains the monosaccharide 2-deoxyribose bonded to a purine or pyrimidine base, as well as a phosphate at the 5'-OH group.

Deposition (4.6) The conversion of a gas directly to a solid.

Deuterium (2.3) A hydrogen atom having one proton and one neutron, giving it a mass number of two.

Dialysis (7.8) A process that involves the selective passage of substances across a semipermeable membrane, called a dialyzing membrane.

Diatomic molecule (3.7) A molecule that contains two atoms. Hydrogen (H_2) is a diatomic molecule.

Dilution (7.7) The addition of solvent to a solution to decrease the concentration of solute.

Dipeptide (16.4) A peptide formed from two amino acids joined together by one amide bond.

Dipole (3.11) The partial separation of charge in a bond or molecule.

Dipole–dipole interactions (4.3) The attractive intermolecular forces between the permanent dipoles of two polar molecules.

Disaccharide (14.1) A carbohydrate composed of two monosaccharides joined together.

Dissociation (8.3) The process that occurs when an acid or base dissolves in water to form ions.

Disulfide (12.7) A compound that contains a sulfur–sulfur bond.

DNA fingerprinting (17.10) A technique in which DNA is cut into fragments that are separated by size using gel electrophoresis. This forms a set of horizontal bands, each band corresponding to a segment of DNA, sorted from low to high molecular weight.

Double bond (3.8) A multiple bond that contains four electrons—that is, two two-electron bonds.

Double replacement reaction (5.3) A reaction in which two compounds exchange "parts"—atoms or ions—to form two new compounds.

E

Electrolyte (7.2) A substance that conducts an electric current in water.

Electron (2.2) A negatively charged subatomic particle.

Electron cloud (2.2) The space surrounding the nucleus of an atom, which contains electrons and comprises most of the volume of an atom.

Electron-dot symbol (2.7) A symbol that shows the number of valence electrons around an atom.

Electronegativity (3.11) A measure of an atom's attraction for electrons in a bond.

Electronic configuration (2.6) The arrangement of electrons in an atom's orbitals.

Electron transport chain (18.7) A series of reactions that transfers electrons from reduced coenzymes to progressively stronger oxidizing agents, ultimately converting oxygen to water.

Element (1.3) A pure substance that cannot be broken down into simpler substances by a chemical reaction.

Elimination (12.5) A reaction in which elements of the starting material are "lost" and a new multiple bond is formed.

Enantiomers (12.11) Mirror images that are not superimposable.

Endothermic (4.6, 5.9) Absorbing energy. In an endothermic chemical reaction, ΔH is positive (+) and energy is absorbed.

Energy (4.1) The capacity to do work.

Energy diagram (5.9) A schematic representation of the energy changes in a reaction, which plots energy on the vertical axis and the progress of the reaction—the reaction coordinate—on the horizontal axis.

Energy of activation (5.9) The difference in energy between the reactants and the transition state; symbolized by E_a.

English system of measurement (1.4) A system of measurement used primarily in the United States in which units are not systematically related to each other and require memorization.

Enthalpy change (5.9) The energy absorbed or released in any reaction— also called the heat of reaction and symbolized by ΔH.

Enzyme (16.9) A biological catalyst composed of one or more chains of amino acids in a very specific three-dimensional shape.

Enzyme–substrate complex (16.10) A structure composed of a substrate bonded to the active site of an enzyme.

Equilibrium (5.11) A reaction that consists of forward and reverse reactions that have equal reaction rates, so the concentration of each species does not change.

Equivalent (7.2) The number of moles of charge that a mole of ions contributes to a solution.

Ester (10.4, 13.1) A compound that contains an OR group bonded directly to the carbonyl carbon; general structure RCOOR.

Estrogen (15.9) A hormone that controls the development of secondary sex characteristics in females and regulates the menstrual cycle.

Ether (10.4, 12.1) A compound that has two alkyl groups bonded to an oxygen atom; general structure ROR.

Exact number (1.5) A number that results from counting objects or is part of a definition.

Exothermic (4.6, 5.9) Releasing energy. In an exothermic reaction, energy is released and ΔH is negative (–).

F

Facilitated transport (15.7) The process by which some ions and molecules travel through the channels in a cell membrane created by integral proteins.

FAD (flavin adenine dinucleotide, 18.3) A biological oxidizing agent synthesized in cells from vitamin B_2, riboflavin. FAD is reduced by adding two hydrogen atoms, forming $FADH_2$.

Fahrenheit scale (1.9) One of three temperature scales in which water freezes at 32 °F and boils at 212 °F.

Fat (11.3, 15.4) A triacylglycerol with few double bonds, making it a solid at room temperature.

Fat-soluble vitamin (15.10) A vitamin that dissolves in an organic solvent but is insoluble in water. Vitamins A, D, E, and K are fat soluble.

Fatty acid (11.3, 15.2) A carboxylic acid (RCOOH) with a long carbon chain, usually containing 12–20 carbon atoms.

Fermentation (18.5) The anaerobic conversion of glucose to ethanol and CO_2.

Fibrous protein (16.7) A water-insoluble protein composed of long linear polypeptide chains that are bundled together to form rods or sheets.

Fischer esterification (13.6) Treatment of a carboxylic acid (RCOOH) with an alcohol (R'OH) and an acid catalyst to form an ester (RCOOR').

Fischer projection formula (12.11) A method of drawing chiral compounds with the chirality center at the intersection of a cross. The horizontal bonds are assumed to be wedges and the vertical bonds are assumed to be dashed wedges.

Formula weight (5.6) The sum of the atomic weights of all the atoms in a compound, reported in atomic mass units (amu).

Forward reaction (5.11) In equilibrium, a reaction that proceeds from left to right as drawn.

Freezing (4.6) The conversion of a liquid to a solid.

Functional group (10.4) An atom or a group of atoms with characteristic chemical and physical properties.

G

Gamma (γ) ray (9.1) High-energy radiation released from a radioactive nucleus.

Gas (1.2) A state of matter that has no definite shape or volume. The particles of a gas move randomly and are separated by a distance much larger than their size.

Gas laws (6.2) A series of laws that relate the pressure, volume, and temperature of a gas.

Gay–Lussac's law (6.4) A gas law that states for a fixed amount of gas at constant volume, the pressure of the gas is proportional to its Kelvin temperature.

Geiger counter (9.4) A small portable device used for measuring radioactivity.

Gene (17.1) A portion of a DNA molecule responsible for the synthesis of a single protein.

Genetic code (17.7) The sequence of nucleotides in mRNA (coded in triplets) that specifies the amino acid sequence of a protein. Each triplet is called a codon.

Genetic disease (17.9) A disease resulting from a mutation that causes a condition to be inherited from one generation to another.

Globular protein (16.7) A protein that is coiled into a compact shape with a hydrophilic outer surface to make it water soluble.

Glucogenic amino acid (18.11) An amino acid that can be used to synthesize glucose.

Glycolysis (18.4) A linear, 10-step pathway that converts glucose, a six-carbon monosaccharide, to two three-carbon pyruvate molecules.

Glycosidic linkage (14.5) The C—O bond that joins two monosaccharides together. The carbon in a glycosidic linkage is bonded to two oxygens.

Gram (1.4) The basic unit of mass in the metric system; abbreviated as g.

Gray (9.4) A unit that measures absorbed radiation; abbreviated as Gy.

Ground state (2.6) The lowest-energy arrangement of electrons.

Group (2.4) A column in the periodic table.

Group number (2.4) A number that identifies a particular column in the periodic table.

H

Half-life (9.3) The time it takes for one-half of a sample to decay.

Half reaction (5.4) An equation written for an individual oxidation or reduction that shows how many electrons are gained or lost.

Halogen (2.4) An element located in group 7A (group 17) of the periodic table. Halogens include fluorine (F), chlorine (Cl), bromine (Br), iodine (I), and astatine (At).

Haworth projection (14.3) A planar, six-membered ring used to represent the cyclic form of glucose and other sugars.

Heating curve (4.7) A graph that shows how the temperature of a substance changes as heat is added.

Heat of fusion (4.6) The amount of energy needed to melt one gram of a substance.

Heat of reaction (5.9) The energy absorbed or released in any reaction and symbolized by ΔH—also called the enthalpy change.

Heat of vaporization (4.6) The amount of energy needed to vaporize one gram of a substance.

α-Helix (16.6) A secondary structure of a protein formed when a peptide chain twists into a right-handed or clockwise spiral.

Heme (16.7) A complex organic compound containing an Fe^{2+} ion complexed with a large nitrogen-containing ring system.

Henry's law (7.4) A law that states that the solubility of a gas in a liquid is proportional to the partial pressure of the gas above the liquid.

Heteroatom (10.2) Any atom in an organic compound that is not carbon or hydrogen.

Heterogeneous mixture (7.1) A mixture that does not have a uniform composition throughout a sample.

Hexose (14.2) A monosaccharide with six carbons.

High-density lipoprotein (15.8) A spherical particle that transports cholesterol from the tissues to the liver.

Homogeneous mixture (7.1) A mixture that has a uniform composition throughout a sample.

Hormone (15.9) A compound synthesized in one part of an organism, which then travels through the bloodstream to elicit a response at a target tissue or organ.

Human Genome Project (17.10) An international project completed in 2003 that sequenced the entire human genome.

Hydration (11.5) The addition of water to a molecule.

Hydrocarbon (10.4) A compound that contains only the elements of carbon and hydrogen.

Hydrogenation (11.5) The addition of hydrogen (H_2) to an alkene.

Hydrogen bonding (4.3) An attractive intermolecular force that occurs when a hydrogen atom bonded to O, N, or F is electrostatically attracted to an O, N, or F atom in another molecule.

Hydrolase (16.9) An enzyme that catalyzes a hydrolysis reaction.

Hydrolysis (13.6) A cleavage reaction that uses water.

Hydrolyzable lipid (15.1) A lipid that can be converted to smaller molecules by hydrolysis with water.

Hydronium ion (3.6) The H_3O^+ ion.

Hydrophilic (15.2) The polar part of a molecule that is attracted to water.

Hydrophobic (15.2) The nonpolar part of a molecule (C—C and C—H bonds) that is not attracted to water.

Hydroxide (3.6) The OH⁻ ion.

α-Hydroxy acid (13.4) A compound that contains a hydroxyl group on the α carbon to a carboxyl group.

Hydroxyl group (10.4, 12.1) An OH group.

Hypertonic solution (7.8) A solution that has a higher osmotic pressure than body fluids.

Hypotonic solution (7.8) A solution that has a lower osmotic pressure than body fluids.

I

Ideal gas law (6.7) A gas law that relates the pressure (P), volume (V), temperature (T), and number of moles (n) of a gas in a single equation; $PV = nRT$, where R is a constant.

Incomplete combustion (10.10) An oxidation reaction that forms carbon monoxide (CO) instead of carbon dioxide (CO_2) because insufficient oxygen is available.

Induced-fit model (16.10) The binding of a substrate to an enzyme such that the shape of the active site adjusts to fit the shape of the substrate.

Inexact number (1.5) A number that results from a measurement or observation and contains some uncertainty.

Inhibitor (16.10) A molecule that causes an enzyme to lose activity.

Inner transition metal elements (2.4) A group of elements consisting of the lanthanides and actinides.

Insertion mutation (17.9) The addition of one or more nucleotides to a DNA molecule.

Intermolecular forces (4.3) The attractive forces that exist between molecules.

Ion (3.1) A charged species in which the number of protons and electrons in an atom is not equal.

Ion–dipole interaction (7.3) The attraction of an ion to a dipole in another molecule.

Ionic bond (3.1) A bond that results from the transfer of electrons from one element to another.

Ionization energy (2.8) The energy needed to remove an electron from a neutral atom.

Ion–product constant (8.4) The product of the concentrations of H_3O^+ and OH^- in water or an aqueous solution—symbolized by K_w and equal to 1×10^{-14}.

Irreversible inhibitor (16.10) An inhibitor that covalently binds to an enzyme, permanently destroying its activity.

Isoelectric point (16.3) The pH at which an amino acid exists primarily in its neutral form; abbreviated as pI.

Isomers (10.5) Two different compounds with the same molecular formula.

Isotonic solution (7.8) Two solutions with the same osmotic pressure.

Isotopes (2.3) Atoms of the same element having a different number of neutrons.

IUPAC nomenclature (10.6) A systematic method of naming compounds developed by the International Union of Pure and Applied Chemistry.

J

Joule (4.1) A unit of measurement for energy; abbreviated as J, where 1 cal = 4.184 J.

K

Kelvin scale (1.9) A temperature scale commonly used by scientists. The Kelvin scale is divided into kelvins (K); $T_K = T_C + 273.15$.

Ketogenesis (18.10) The synthesis of ketone bodies from acetyl CoA.

Ketogenic amino acid (18.11) An amino acid that cannot be used to synthesize glucose, but can be converted to ketone bodies.

Ketone (10.4, 12.1) A compound that has two alkyl groups bonded to the carbonyl group; general structure RCOR.

Ketone bodies (18.10) Three compounds—acetoacetate, β-hydroxybutyrate, and acetone—formed when acetyl CoA levels exceed the capacity of the citric acid cycle.

Ketose (14.2) A monosaccharide with a carbonyl group at C2.

Ketosis (18.10) The accumulation of ketone bodies during starvation and uncontrolled diabetes.

Kinase (16.9) An enzyme that catalyzes the transfer of a phosphate from one substrate to another.

Kinetic energy (4.1) The energy of motion.

Kinetic–molecular theory (6.1) A theory that describes the fundamental characteristics of gas particles.

L

β-Lactam (13.10) An amide contained in a four-membered ring.

Lagging strand (17.4) The strand of DNA synthesized in small fragments during replication, which are then joined together by an enzyme.

Lanthanides (2.4) A group of 15 elements in the periodic table beginning with the element lanthanum (La, $Z = 57$).

Law of conservation of energy (4.1) A law that states that the total energy in a system does not change. Energy cannot be created or destroyed.

Law of conservation of matter (5.1) A law that states that atoms cannot be created or destroyed in a chemical reaction.

LD$_{50}$ (9.4) The lethal dose of radiation (or a poison) that kills 50% of a population.

Leading strand (17.4) The strand of DNA that grows continuously during replication.

Le Châtelier's principle (5.11) A principle that states that if a chemical system at equilibrium is disturbed or stressed, the system will react in the direction that counteracts the disturbance or relieves the stress.

Lecithin (15.6) A phosphoacylglycerol in which the identity of the R group esterified to the phosphodiester is $-CH_2CH_2N(CH_3)_3^+$.

Lewis structure (3.7) An electron-dot structure for a molecule that shows the location of all valence electrons in the molecule, both the shared electrons in bonds and the nonbonded electron pairs.

Ligase (16.9) An enzyme that catalyzes bond formation accompanied by energy release from a hydrolysis reaction.

"Like dissolves like" (7.3) The principle in solubility that compounds dissolve in solvents having similar types of intermolecular forces.

Lipase (15.5, 16.9) An enzyme that catalyzes the hydrolysis of the esters in a triacylglycerol.

Lipid (13.6, 15.1) A biomolecule that is soluble in organic solvents and insoluble in water.

Lipid bilayer (15.7) The basic structure of the cell membrane formed from two layers of phospholipids having their ionic heads oriented on the outside and their nonpolar tails on the inside.

Lipoprotein (15.8) A small water-soluble spherical particle composed of proteins and lipids.

Liquid (1.2) A state of matter that has a definite volume, but takes on the shape of the container it occupies. The particles of a liquid are close together but they can randomly move past each other.

Liter (1.4) The basic unit of volume in the metric system; abbreviated as L.

Lock-and-key model (16.10) The binding of a substrate to a rigid active site, such that the three-dimensional geometry of the substrate exactly matches the shape of the active site.

London dispersion forces (4.3) Very weak intermolecular interactions due to the momentary changes in electron density in a molecule.

Lone pair (3.7) An unshared electron pair.

Low-density lipoprotein (15.8) A spherical particle containing proteins and lipids, which transports cholesterol from the liver to the tissues.

Lyase (16.9) An enzyme that catalyzes the addition of a molecule to a double bond or the elimination of a molecule from a double bond.

M

Main group element (2.4) An element in groups 1A–8A of the periodic table.

Major mineral (macronutrient, 2.1) One of the seven elements present in the body in small amounts (0.1–2% by mass) and needed in the daily diet.

Markovnikov's rule (11.5) The rule that states that in the addition of H_2O to an unsymmetrical alkene, the H atom bonds to the less substituted carbon atom.

Mass (1.4) A measure of the amount of matter in an object.

Mass number (2.2) The total number of protons and neutrons in a nucleus; symbolized as *A*.

Matter (1.1) Anything that has mass and takes up volume.

Melting (4.6) The conversion of a solid to a liquid.

Melting point (4.4) The temperature at which a solid is converted to the liquid phase.

Messenger RNA (mRNA, 17.5) The carrier of information from DNA (in the cell nucleus) to the ribosomes (in the cell cytoplasm). Each gene of a DNA molecule corresponds to a specific mRNA molecule.

Metabolism (18.1) The sum of all of the chemical reactions that take place in an organism.

Meta isomer (11.9) A 1,3-disubstituted benzene.

Metal (2.1) A shiny element that is a good conductor of heat and electricity.

Metalloid (2.1) An element with properties intermediate between a metal and a nonmetal. Metalloids include boron (B), silicon (Si), germanium (Ge), arsenic (As), antimony (Sb), and tellurium (Te).

Meter (1.4) A unit used to measure length; abbreviated as m.

Metric system (1.4) A measurement system in which each type of measurement has a base unit and all other units are related to the base unit by a prefix that indicates if the unit is larger or smaller than the base unit.

Micelle (13.5) A spherical droplet formed when soap is dissolved in water. The ionic heads of the soap molecules are oriented on the surface and the nonpolar tails are packed in the interior.

Millimeters mercury (6.1) A unit used to measure pressure; abbreviated as mm Hg and also called "torr."

Mitochondrion (18.1) A small sausage-shaped organelle within a cell in which energy production takes place.

Mixture (1.3) Matter composed of more than one component.

Molarity (7.6) The number of moles of solute per liter of solution; abbreviated as M.

Molar mass (5.6) The mass of one mole of any substance, reported in grams per mole.

Mole (5.5) A quantity that contains 6.022×10^{23} items—usually atoms, molecules, or ions.

Molecular formula (3.8) A formula that shows the number and identity of all of the atoms in a compound, but it does not indicate what atoms are bonded to each other.

Molecular weight (5.6) The formula weight of a covalent compound.

Molecule (3.1) A compound or element containing two or more atoms joined together with covalent bonds.

Monomers (11.7) Small molecules that covalently bond together to form polymers.

Monosaccharide (14.1) A carbohydrate that cannot be hydrolyzed to simpler compounds.

Mutation (17.9) A change in the nucleotide sequence in a molecule of DNA.

N

NAD^+ (nicotinamide adenine dinucleotide, 18.3) A biological oxidizing agent and coenzyme synthesized from the vitamin niacin. NAD^+ and NADH are interconverted by oxidation and reduction reactions.

NADH (18.3) A biological reducing agent and coenzyme formed when NAD^+ is reduced.

Net ionic equation (8.6) An equation that contains only the species involved in a reaction.

Neutralization reaction (8.6) An acid–base reaction that produces a salt and water as products.

Neutral solution (8.4) Any solution that has an equal concentration of H_3O^+ and OH^- ions and a pH = 7.

Neutron (2.2) A neutral subatomic particle in the nucleus.

Noble gases (2.4) Elements located in group 8A (group 18) of the periodic table. The noble gases are helium (He), neon (Ne), argon (Ar), krypton (Kr), xenon (Xe), and radon (Rn).

Nomenclature (3.4) The system of assigning an unambiguous name to a compound.

Nonbonded electron pair (3.7) An unshared electron pair or lone pair.

Noncompetitive inhibitor (16.10) An inhibitor that binds to an enzyme but does not bind at the active site.

Nonelectrolyte (7.2) A substance that does not conduct an electric current when dissolved in water.

Nonhydrolyzable lipid (15.1) A lipid that cannot be cleaved into smaller units by aqueous hydrolysis.

Nonmetal (2.1) An element that does not have a shiny appearance and poorly conducts heat and electricity.

Nonpolar bond (3.11) A bond in which electrons are equally shared.

N-Terminal amino acid (16.4) In a peptide, the amino acid with the free $-NH_3^+$ group on the α carbon.

Nuclear fission (9.6) The splitting apart of a nucleus into lighter nuclei and neutrons.

Nuclear fusion (9.6) The joining together of two nuclei to form a larger nucleus.

Nuclear reaction (9.2) A reaction that involves the subatomic particles of the nucleus.

Nuclease (16.9) An enzyme that catalyzes the hydrolysis of nucleic acids.

Nucleic acid (17.1, 17.2) An unbranched polymer composed of nucleotides. DNA and RNA are nucleic acids.

Nucleoside (17.1) A compound formed by joining a monosaccharide with a nitrogen atom of a purine or pyrimidine base.

Nucleotide (17.1) A compound formed by adding a phosphate group to the 5'-OH of a nucleoside.

Nucleus (2.2) The dense core of the atom that contains protons and neutrons.

O

Octet rule (3.2) The rule in bonding that states that main group elements are especially stable when they possess eight electrons (an octet) in the outer shell.

Oil (11.3, 15.4) A triacylglycerol that is liquid at room temperature.

Omega-*n* acid (15.2) An unsaturated fatty acid where *n* is the carbon at which the first double bond occurs in the carbon chain. The numbering begins at the end of the chain with the CH_3 group.

Orbital (2.5) A region of space where the probability of finding an electron is high.

Organic chemistry (10.1) The study of compounds that contain the element carbon.

Ortho isomer (11.9) A 1,2-disubstituted benzene.

Osmosis (7.8) The selective diffusion of solvent (usually water) across a semipermeable membrane from a less concentrated solution to a more concentrated solution.

Osmotic pressure (7.8) The pressure that prevents the flow of additional solvent into a solution on one side of a semipermeable membrane.

Oxidase (16.9) An enzyme that catalyzes an oxidation.

Oxidation (5.7, 12.5) The loss of electrons from an atom. Oxidation may result in a gain of oxygen atoms or a loss of hydrogen atoms.

β-Oxidation (18.9) A process in which two-carbon acetyl CoA units are sequentially cleaved from a fatty acid.

Oxidative phosphorylation (18.7) The process by which the energy released from the oxidation of reduced coenzymes is used to convert ADP to ATP using the enzyme ATP synthase.

Oxidizing agent (5.4) A compound that gains electrons (i.e., is reduced), causing another compound to be oxidized.

Oxidoreductase (16.9) An enzyme that catalyzes an oxidation–reduction reaction.

P

Para isomer (11.9) A 1,4-disubstituted benzene.

Parent name (10.6) The root that indicates the number of carbons in the longest continuous carbon chain in a molecule.

Partial hydrogenation (11.6) The hydrogenation of some, but not all, of the double bonds in a molecule.

Partial pressure (6.8) The pressure exerted by one component of a mixture of gases.

Parts per million (7.5) A concentration term (abbreviated ppm)—the number of "parts" in 1,000,000 parts of solution.

Penicillin (13.10) An antibiotic that contains a β-lactam and interferes with the synthesis of the bacterial cell wall.

Pentose (14.2) A monosaccharide with five carbons.

Peptide (16.4) A compound that contains many amino acids joined together by amide bonds.

Peptide bond (16.4) An amide bond in peptides and proteins.

Period (2.4) A row in the periodic table.

Periodic table (2.1) A schematic arrangement of all known elements that groups elements with similar properties.

Phenol (11.10) A compound that contains an OH group bonded to a benzene ring.

Pheromone (10.5) A chemical substance used for communication in a specific animal species, most commonly an insect population.

Phosphate (3.6) A PO_4^{3-} anion.

Phosphoacylglycerol (15.6) A lipid with a glycerol backbone that contains two of the hydroxyls esterified with fatty acids and the third hydroxyl as part of a phosphodiester.

Phosphodiester (15.6) A derivative of phosphoric acid (H_3PO_4) that is formed by replacing two of the H atoms by R groups.

Phospholipid (15.6) A lipid that contains a phosphorus atom.

Phosphorylation (18.2) A reaction that adds a phosphate group to a molecule.

pH scale (8.5) The scale used to report the H_3O^+ concentration; pH = –log [H_3O^+].

Physical properties (1.2) Those properties of a substance that can be observed or measured without changing the composition of the material.

β-Pleated sheet (16.6) A secondary structure formed when two or more peptide chains, called strands, line up side-by-side.

Point mutation (17.9) The substitution of one nucleotide for another.

Polar bond (3.11) A bond in which electrons are unequally shared and pulled toward the more electronegative element.

Polyatomic ion (3.6) A cation or anion that contains more than one atom.

Polymer (11.7) A large molecule made up of repeating units of smaller molecules—called monomers—covalently bonded together.

Polymerization (11.7) The joining together of monomers to make polymers.

Polynucleotide (17.2) A polymer of nucleotides that contains a sugar–phosphate backbone.

Polysaccharide (14.1) Three or more monosaccharides joined together.

_p_ Orbital (2.5) A dumbbell-shaped orbital higher in energy than an _s_ orbital in the same shell.

Positron (9.1) A radioactive particle that has a negligible mass and a +1 charge.

Potential energy (4.1) Energy that is stored.

Pressure (6.1) The force (*F*) exerted per unit area (*A*); symbolized by *P*.

Primary (1°) alcohol (12.2) An alcohol having the general structure RCH_2OH.

Primary (1°) alkyl halide (12.6) An alkyl halide having the general structure RCH_2X.

Primary (1°) amide (13.9) A compound having the general structure $RCONH_2$.

Primary (1°) amine (13.7) A compound having the general structure RNH_2.

Primary structure (16.6) The particular sequence of amino acids that is joined together by peptide bonds in a protein.

Product (5.1) A substance formed in a chemical reaction.

Progestin (15.9) A hormone responsible for the preparation of the uterus for implantation of a fertilized egg.

Prostaglandins (13.4) A group of carboxylic acids that contain a five-membered ring, are synthesized from arachidonic acid, and have a wide range of biological activities.

Protease (16.9) An enzyme that catalyzes the hydrolysis of the amide bonds in proteins.

Proteins (16.1) Biomolecules that contain many amide bonds, formed by joining amino acids together.

Proton (2.2) A positively (+) charged subatomic particle that resides in the nucleus of the atom.

Proton transfer reaction (8.2) A Brønsted–Lowry acid–base reaction in which a proton is transferred from an acid to a base.

Pure substance (1.3) A substance that contains a single component, and has a constant composition regardless of the sample size.

Q

Quaternary structure (16.6) The shape adopted when two or more folded polypeptide chains come together into one protein complex.

R

Rad (9.4) The radiation absorbed dose; the amount of radiation absorbed by one gram of a substance.

Radioactive decay (9.2) The process by which an unstable radioactive nucleus emits radiation, forming a nucleus of new composition.

Radioactive isotope (9.1) An isotope that is unstable and spontaneously emits energy to form a more stable nucleus.

Radioactivity (9.1) The energy emitted by a radioactive isotope.

Radiocarbon dating (9.3) A method to date artifacts that is based on the ratio of the radioactive carbon-14 isotope to the stable carbon-12 isotope.

Reactant (5.1) The starting material in a reaction.

Reaction rate (5.10) A measure of how fast a chemical reaction occurs.

Redox reaction (5.4) A reaction that involves the transfer of electrons from one element to another.

Reducing agent (5.4) A compound that loses electrons (i.e., is oxidized), causing another compound to be reduced.

Reducing sugar (14.4) A carbohydrate that is oxidized with Benedict's reagent.

Reductase (16.9) An enzyme that catalyzes a reduction.

Reduction (5.4) The gain of electrons by an atom. Reduction may result in the loss of oxygen atoms or the gain of hydrogen atoms.

Rem (9.4) The radiation equivalent for man; the amount of radiation absorbed by a substance that also factors in its energy and potential to damage tissue.

Replication (17.4) The process by which DNA makes a copy of itself when a cell divides.

Retrovirus (17.11) A virus that contains a core of RNA.

Reverse reaction (5.11) In equilibrium, a reaction that proceeds from right to left as drawn.

Reverse transcription (17.11) A process by which a retrovirus produces DNA from RNA.

Reversible inhibitor (16.10) An inhibitor that binds to an enzyme, but enzyme activity is restored when the inhibitor is released.

Reversible reaction (5.11) A reaction that can occur in either direction, from reactants to products or from products to reactants.

Ribonucleic acid (RNA, 17.1, 17.5) A polymer of ribonucleotides that translates genetic information to protein synthesis.

Ribonucleoside (17.1) A compound that contains the monosaccharide ribose and a purine or pyrimidine base.

Ribonucleotide (17.1) A compound that contains the monosaccharide ribose bonded to either a purine or pyrimidine base as well as a phosphate at the 5'-OH group.

Ribosomal RNA (rRNA, 17.5) The most abundant type of RNA. rRNA is found in the ribosomes of the cell and provides the site where polypeptides are assembled during protein synthesis.

S

Saponification (13.6, 15.5) The basic hydrolysis of an ester.

Saturated fatty acids (15.2) Fatty acids that have no double bonds in their long hydrocarbon chains.

Saturated hydrocarbon (10.5) An alkane with molecular formula C_nH_{2n+2} that contains a chain of carbon atoms but no rings.

Saturated solution (7.3) A solution that has the maximum number of grams of solute that can be dissolved.

Scientific notation (1.6) A system in which numbers are written as $y \times 10^x$, where y is a number between 1 and 10 and x can be either positive or negative.

Secondary (2°) alcohol (12.2) An alcohol having the general structure R_2CHOH.

Secondary (2°) alkyl halide (12.6) An alkyl halide having the general structure R_2CHX.

Secondary (2°) amide (13.9) A compound that has the general structure RCONHR'.

Secondary (2°) amine (13.7) A compound that has the general structure R_2NH.

Secondary structure (16.6) The three-dimensional arrangement of localized regions of a protein. The α-helix and β-pleated sheet are two kinds of secondary structure.

Semipermeable membrane (7.8) A membrane that allows only certain molecules to pass through.

Shell (2.5) A region where an electron that surrounds a nucleus is confined. A shell is also called a principal energy level.

Sievert (9.4) A unit that measures absorbed radiation; abbreviated as Sv.

Significant figures (1.5) All of the digits in a measured number, including one estimated digit.

Single replacement reaction (5.3) A reaction in which one element replaces another element in a compound to form a different compound and element as products.

SI units (1.4) The International System of Units formally adopted as the uniform system of units for the sciences.

Skeletal structure (10.3) A shorthand method used to draw organic compounds in which carbon atoms are assumed to be at the junction of any two lines or at the end of a line, and all H's on C's are omitted.

Soap (13.5, 15.5) A salt of a long-chain carboxylic acid.

Solid (1.2) A state of matter that has a definite shape and volume. The particles of a solid lie close together, and are arranged in a regular, three-dimensional array.

Solubility (7.3) The amount of solute that dissolves in a given amount of solvent.

Solute (7.1) The substance present in the lesser amount in a solution.

Solution (7.1) A homogeneous mixture that contains small particles. Liquid solutions are transparent.

Solvation (7.3) The process of surrounding particles of a solute with solvent molecules.

Solvent (7.1) The substance present in the larger amount in a solution.

s Orbital (2.5) A spherical orbital that is lower in energy than other orbitals in the same shell.

Specific gravity (1.10) A unitless quantity that compares the density of a substance with the density of water at 4 °C.

Specific heat (4.5) The amount of heat energy needed to raise the temperature of 1 g of a substance 1 °C.

Spectator ion (8.6) An ion that appears on both sides of an equation but undergoes no change in a reaction.

Standard molar volume (6.6) The volume of one mole of any gas at STP—22.4 L.

States of matter (1.2) The forms in which most matter exists—that is, gas, liquid, and solid.

Stereochemistry (12.11) The three-dimensional structure of molecules.

Stereoisomers (11.3, 12.11) Isomers that differ only in their three-dimensional arrangement of atoms.

Steroid (15.8) A lipid whose carbon skeleton contains three six-membered rings and one five-membered ring.

STP (6.6) Standard conditions of temperature and pressure—1 atm (760 mm Hg) for pressure and 273.15 K (0 °C) for temperature.

Straight-chain alkane (10.5) An alkane that has all of its carbons in one continuous chain.

Sublimation (4.6) A phase change in which the solid phase enters the gas phase without passing through the liquid state.

Sulfate (3.6) An SO_4^{2-} ion.

Sulfhydryl group (12.1) An SH group.

Supersaturated solution (7.4) A solution that contains more than the predicted maximum amount of solute at a given temperature.

Suspension (7.1) A heterogeneous mixture that contains large particles (> 1 μm in diameter) suspended in a liquid.

Synthase (16.9) An enzyme that catalyzes the addition of a small molecule to a double bond.

T

Temperature (1.9) A measure of how hot or cold an object is.

Tertiary (3°) alcohol (12.2) An alcohol that has the general structure R_3COH.

Tertiary (3°) alkyl halide (12.6) An alkyl halide having the general structure R_3CX.

Tertiary (3°) amide (13.9) A compound that has the general structure $RCONR'_2$.

Tertiary (3°) amine (13.7) A compound that has the general structure R_3N.

Tertiary structure (16.6) The three-dimensional shape adopted by an entire peptide chain.

Tetrose (14.2) A monosaccharide with four carbons.

Thiol (12.1) A compound that contains a sulfhydryl group (SH group) bonded to a tetrahedral carbon atom; general structure RSH.

Titration (8.7) A technique for determining an unknown molarity of an acid by adding a base of known molarity to a known volume of acid.

Tollens reagent (12.10) An oxidizing agent that contains silver(I) oxide (Ag_2O) in aqueous ammonium hydroxide (NH_4OH).

Trace element (micronutrient, 2.1) An element required in the daily diet in small quantities—usually less than 15 mg.

Transaminase (16.9) An enzyme that catalyzes the transfer of an amino group.

Transcription (17.6) The process that synthesizes RNA from DNA.

Transferase (16.9) An enzyme that catalyzes the transfer of a group from one molecule to another.

Transfer RNA (tRNA, 17.5) The smallest type of RNA, which brings a specific amino acid to the site of protein synthesis on a ribosome.

Trans isomer (11.3) An alkene with two R groups on opposite sides of a double bond.

Transition metal element (2.4) An element contained in one of the 10 columns in the periodic table numbered 1B–8B.

Transition state (5.9) The unstable energy maximum located at the top of the energy hill in an energy diagram.

Translation (17.8) The synthesis of proteins from RNA.

Triacylglycerol (13.6, 15.4) A triester formed from glycerol and three molecules of fatty acids.

Triose (14.2) A monosaccharide with three carbons.

Tripeptide (16.4) A peptide that contains three amino acids joined together by two amide bonds.

Triple bond (3.8) A multiple bond that contains six electrons—that is, three two-electron bonds.

Tritium (2.3) A hydrogen atom that has one proton and two neutrons, giving it a mass number of three.

U

Universal gas constant (6.7) The constant, symbolized by R, that equals the product of the pressure and volume of a gas, divided by the product of the number of moles and Kelvin temperature; $R = PV/nT$.

Unsaturated fatty acid (15.2) A fatty acid that has one or more double bonds in its long hydrocarbon chain.

Unsaturated hydrocarbon (11.1) A compound that contains fewer than the maximum number of hydrogen atoms per carbon.

Unsaturated solution (7.3) A solution that has less than the maximum number of grams of solute.

Urea cycle (18.11) The process by which an ammonium ion is converted to urea, $(NH_2)_2C{=}O$.

V

Valence electron (2.7) An electron in the outermost shell that takes part in bonding and chemical reactions.

Valence shell electron pair repulsion (VSEPR) theory (3.10) A theory that predicts molecular geometry based on the fact that electron pairs repel each other; thus, the most stable arrangement keeps these groups as far away from each other as possible.

Vaporization (4.6) The conversion of a liquid to a gas.

Virus (17.11) An infectious agent consisting of a DNA or RNA molecule that is contained within a protein coating.

Volume/volume percent concentration (7.5) The number of milliliters of solute dissolved in 100 mL of solution.

W

Wax (15.3) An ester (RCOOR') formed from a fatty acid (RCOOH) and a high-molecular-weight alcohol (R'OH).

Weight (1.4) The force that matter feels due to gravity.

Weight/volume percent concentration (7.5) The number of grams of solute dissolved in 100 mL of solution.

X

X-ray (9.7) A high-energy form of radiation.

Z

Zwitterion (16.2) A neutral compound that contains both a positive and a negative charge.

Index

Page numbers followed by *f* indicate figures; *mn,* marginal notes; and *t,* tables. Page numbers preceded by the letters A through D indicate pages in the Appendix.